1st International Annual Meeting on STEM Education (IAMSTEM 2018)

Journal of Physics: Conference Series Volume 1340

Khon Kaen, Thailand
13 - 15 August 2018

ISBN: 978-1-7138-0008-8
ISSN: 1742-6588

Printed from e-media with permission by:

Curran Associates, Inc.
57 Morehouse Lane
Red Hook, NY 12571

CURRAN ASSOCIATES INC. proceedings.com

Some format issues inherent in the e-media version may also appear in this print version.

Additional copies of this publication are available from:

Curran Associates, Inc.
57 Morehouse Lane
Red Hook, NY 12571 USA
Phone: 845-758-0400
Fax: 845-758-2633
Email: curran@proceedings.com
Web: www.proceedings.com

TABLE OF CONTENTS

It gives us great pleasure to announce that the 1st International Annual Meeting on STEM education (IAMSTEM 2018) held 13-15 August 2018 at AVANI hotel, Khon Kaen, Thailand was a resounding success, with over 200 participants from throughout Thailand and from other countries. On behalf of the organizers, ASEAN Comparative Educational Research Network (ACER-N), Science Education Association of Thailand (SEAT) co-host from Malaysia - Universiti Kebangsaan Malaysia; co-host from Indonesia - Universitas Negeri Padang, and Sekolah Tinggi Agama Islam Negeri (STAIN) Sorong; co-host from Vietnam - Thai Nguyen University of Education, Hanoi National University of Education, and Can Tho University; co-host from Thailand - Office of Basic Education Commission (OBEC), Rajabhat Mahasarakham University, and Khon Kaen University. as well as other corporate sponsors whose generous contributions made this a successful meeting. IAMSTEM 2018 provided a unique opportunity for experts, academics, and educators to exchange knowledge and build friendships. The meeting consisted of four keynote speakers, and a number of contributed oral and poster presentations. Resulting from the success of IAMSTEM 2018 are qualified full research articles from the presentations, which have been reviewed and compiled in this IOP Journal of Physics: Conference Series. It is our sincere hope that these proceedings will contribute to advancing our understanding of STEM education and applications and serve as initiators for even greater achievements in the field.

IAMSTEM 2018 Editorial Team

IOP Conf. Series: Journal of Physics: Conf. Series **1340** (2019) 011002 doi:10.1088/1742-6596/1340/1/011002

Peer review statement

All papers published in this volume of Journal of Physics: Conference Series have been peer reviewed through processes administered by the proceedings Editors. Reviews were conducted by expert referees to the professional and scientific standards expected of a proceedings journal published by IOP Publishing.

IOP Conf. Series: Journal of Physics: Conf. Series **1340** (2019) 012001 doi:10.1088/1742-6596/1340/1/012001

Supporting STEM Curriculum Implementation with Professional Learning: The University of Sydney STEM Teacher Enrichment Academy

Judy Anderson
Director STEM Teacher Enrichment Academy
Sydney School of Education and Social Work
The University of Sydney

the corresponding author's e-mail address: judy.anderson@sydney.edu.au

Abstract. Based on the need for schools to improve students' capabilities in science, technology, engineering and mathematics (STEM) and to develop students' 21st Century skills, school systems are implementing a range of STEM education initiatives. As there is no STEM curriculum in many countries, teachers are faced with designing new lessons and new subjects as well as developing new learning approaches. While secondary school teachers may have expertise in one of the STEM subjects, few have expertise in more than one so the need for professional learning support is critical to the success of such initiatives. A group of experts at the University of Sydney has developed one program that has produced positive outcomes for teachers and students. Members of the group from the STEM Teacher Enrichment Academy work with school teams to help them identify a clear purpose for their STEM initiative, to design a program to address the purpose and to support them in implementing and evaluating the program with their students.

1. Introduction

Unlike other countries, it is not compulsory to study mathematics or science in the senior secondary school (grades 11 and 12) in most Australian states and territories. While enrolment patterns in mathematics and science subjects revealed small declines in the 1990's, there have been larger declines since 2000[15] and in the state of NSW, there has been a 13% decline since 2001 of students studying the more challenging calculus-based mathematics courses – necessary pre-requisites for many tertiary STEM degrees[17]. There are also declining enrolments in some STEM-related degrees at university level with declines in the number of mathematicians and scientists in the workforce and predictions that Australia will need many more to meet workplace demands into the future [19].Coupled with these concerns, Australia's performance on the international assessments of TIMSS and PISA, has declined with fewer students meeting the highest benchmarks [21].

There are many factors influencing student engagement in secondary schooling. An investigation was conducted intothe lower participation of students in senior mathematics in Australia, by surveying teachers and careers advisors, and conducting focus group meetings with students [18]. Of the four main factors identified in the report, poor pedagogical practices, perceived level of difficulty, and perceived relevance of mathematicsarethree key issues worthy of investigation in the STEM context. Many students also perceive science subjects to be difficult and uninteresting and frequently make decisions early in their schooling to discontinue their study as soon as possible [14,8].One strategy to counteract these issues suggests mathematics and science should be taught using inquiry-based learning through real-world problems, allowing students to see the relevance of the content they are learning, particularly if the subjects are connected or integrated [8]. While subject integration is not a new concept, connecting mathematics and science curriculum with the other STEM subjects of technology and engineering has the potential to enhance student engagement and develop students' 21st Century skills [6,23].

Integrated learning can be implemented in classrooms in multiple ways [8] which could involve connecting content, connecting processes or using themes to link key ideas. Further, integrating curriculumcould involve a multidisciplinary approach, where teachers from two or more of the STEM subjects co-design integrated tasks, lessons or units of work so that students have a synthesised, integrated approach to learning STEM content [6]. To date, there has been little research conducted into the efficacy of STEM integration and application in secondary classrooms [4,12] but there is some evidence to suggest that STEM integration is successful in increasing student engagement within mathematics classrooms [24]. Based on the assumption that students benefit from opportunities to connect knowledge across the curriculum, a professional learning approach was developed to support teachers in planning and implementing connected approaches in primary and secondary schools.

Since its inception in 2014, the STEM Teacher Enrichment Academy has held 11 programs reaching more than 650 teachers in 120 secondary and primary schools in NSW, Australia. This paper describes the approach for secondary school teachers, presents the findings and subsequent changes to the program from the first two trialswith 130 teachers from 25 secondary schools, describes the retrospective evaluationstrategy for the secondary school programs delivered between 2014 and 2017, presents the overall large scale data evaluation strategy for 2018 onwards, and outlines some early evidence of the success of the STEM Academy for teachers and their students.

2. The STEM Teacher Enrichment Academy: Setting the Context

Since 2014, the Faculty of Education and Social Work has been collaborating with the Faculties of Science, and Engineering and Information Technology, to build the nation's STEM capacity through teacher enrichment and professional development with the establishment of the STEM Teacher Enrichment Academy – see http://sydney.edu.au/stem/academy/. The academy's flagship is a multi-day residential program for up to 70 secondary teachers of mathematics, science and technology designed to be foundational in enhancing teachers' knowledge of content and pedagogy, and inspiring them to reinvigorate their classroom practice. The overall Academy aims were to:

1. support teachers' knowledge and understanding of, and abilities to implement, pedagogical strategies promoting student engagement in STEM subjects;
2. enhance teachers' knowledge of content and approaches to teaching mathematics, science and technology, demonstrated in their design and implementation of tasks and units of work;
3. encourage the development of multidisciplinary units of work and/or projects;
4. develop a community of practice between cohorts of STEM teachers by offering ongoing support and engagement through an evolving website, newsletters, and STEM events; and
5. develop teachers' knowledge of STEM-related projects in the broader community as well as knowledge of STEM programs at University and STEM careers.

With a focus on examining content and processes from the STEM subjects, Academy sessions were facilitated by the University's academic specialists and STEM leaders. The program involved a three-day residential session at the University followed by up to two full school terms developing, planning and implementing STEM strategies in school-based teams. Teachers then returned for a further two-day session at the University to share their experiences, present evidence of teacher and student learning, discuss issues and challenges, and consider future initiatives. Each cross-disciplinary school team of two mathematics, two science and two technology teachers worked together to develop inquiry-based learning approaches to implement both within their subject discipline as well as across the subject disciplines [16]. Additional support was provided to school teams by professional mentors who visited participating teachers prior to, during and in between the two on-campus sessions.

Modelled on commonly agreed core features, the Academy professional learning approach incorporated a content focus, active learning, coherence, duration and collective participation [10]. Each session within the program focused on content from either one of the separate curriculum STEM subject areas, or sought ways to demonstrate how the content connected, overlapped or complemented curriculum topics. Participants were actively engaged in discussing practice, sharing ideas, and critiquing examples of STEM activities, projects, lessons or units of work. They also worked in their teams to review scope and sequence charts and school programs, to redesign their approach to implementing STEM on their return to school. With common shared goals, there was a clear purpose to our work providing coherence and clarity for individual teachers as well as school teams. The program was delivered over at least six months of a school year allowing time for design and trialling of programs, sharing of outcomes, and redesign for future implementation. Finally, we sought to strengthen individual teacher's connections to their school teams, to encourage schools to share ideas and collaborate with other schools who identified similar challenges, and to develop a community of practice facilitated by an online Edmodo site where plans, resources and ideas were shared.

To evaluate the efficacy of the Academy approach, the first two trial programs were externally evaluated by academics from other Australian universities [2]. Both external teams used surveys of teacher participants and interviews with a sample of teachers but the second team also conducted two case studies of exemplary schools to seek students' perspectives and to identify key drivers of school change supporting the implementation of STEM programs. A summary of the results of the two external evaluations is presented in the next section of this paper.

3. Outcomes from the First Two Trial Programs in Secondary Schools

For the first Academy, 60 teachers from 13 schools visited the University in November 2014 and returned in March 2015 (see Table 1 for sector representation). While most schools were Sydney based, four were clustered near a country town in the central West of NSW. This small country hub of schools enabled greater opportunity for collegiality, an essential ingredient given the small size of these schools with some teachers reporting feeling isolated and with limited access to quality professional learning. Like the first Academy, the second involved 70 teachers from 12 schools with a country hub of two larger schools from a regional town and took place in November 2015 with a subsequent return to the University in May 2016. When selecting each group of schools, we sought diversity in socio-economic status, gender composition, and size to provide a range of knowledge and experiences.

Table 1. School sector representation and gender composition for the first two STEM Academies

	Department of Education	Catholic Systemic	Independent	Total schools
2014/15	8 (1 female)	1	4 (2 male, 2 female)	13
2015/16	7 (1 male)	2 (1 female)	3 (1 male, 1 female)	12

Analyses of teacher data revealed the features most supportive of school team design and delivery of STEM approaches included the provision of planning time, mentor input, and the structure and content of the program which began with a focus on the separate STEM subjects, allowing teachers to develop new skills and pedagogical strategies, before exploring multidisciplinary approaches. Initially focusing on the individual STEM subjects was adopted because mathematics and science teachers make limited use of inquiry-based learning in lessons than is recommended in research and in curriculum documents [1,3].

IOP Conf. Series: Journal of Physics: Conf. Series **1340** (2019) 012001 doi:10.1088/1742-6596/1340/1/012001

Feedback from the external evaluations indicated teachers wanted more examples of STEM integration including sample tasks, projects, and lessons – interestingly, when we did provide such examples, it was not always evident to teachers how they might use them and how the tasks connected with curriculum requirements. Indeed, there appears to be a need to make the connections between the STEM subjects more transparent for teachers [12], particularly when they are presented with already-prepared multidisciplinary tasks. These observations further highlighted the siloed nature of secondary school teaching with teachers being most comfortable with their subject specialisation – to adopt a STEM curriculum perspective, teachers require horizontal expertise and they need to "boundary cross – stepping into unfamiliar domains" [7]. Teachers also need to come to terms with the issue of inconsistency in language in curriculum documents [12] although some have addressed this by focusing on the engineering design process or systems thinking [6]. Regardless of these concerns, all school teams developed and implemented a STEM approach and were willing to share their experiences (successes and issues) when they returned for the second on-campus session.

Because the schools were diverse, particularly in relation to whether the teachers from different subjects had worked together before coming to the Academy, the approaches they initially adopted were equally diverse. Some of the approaches used by Academy schools included:

1. embedding more cross-curriculum applications within regular subject-specific lessons (egs, exploring half-life in mathematics lessons; using virtual worlds in science to collect data to model and investigate real world ecological problems);
2. conducting multidisciplinary investigations in several STEM subject lessons to design solutions to problems (egs, improving the recycling system at the school; designing a new grandstand for the school football field);
3. undertaking an extended investigation over several weeks or school terms to design an artifact (eg, designing an energy efficient home on a nearby plot of land);
4. redesigning the STEM curriculum program for a whole grade around themes or big ideas such as mission to another planet; human diseases and prosthetics; and better parks and gardens;
5. creating a STEM elective for grade 9 and 10 students; and
6. inviting STEM speakers to the school to share their career pathways and experiences within a range of workplaces.

While this list may appear to be a rather eclectic set of approaches without any real cohesion, it recognises and accepts that schools were at different places in designing integrated curriculum and in embracing substantial change to curriculum design and delivery. Our acceptance of such diversity acknowledges that schools need to consider the needs of their students, the competence and interest of teachers, the overwhelming influence of siloed assessment in schools, and that real change takes time.

Our experiences from the first two Academy programs revealed some schools moved more quickly to developing multidisciplinary STEM approaches because of earlier experiences of writing integrated units of work, and working together as a team. Some teams were cohesive and had already worked on projects together,some teams were dominated by one or two teachers who already had a plan which they wanted to implement regardless of other teachers' views, while other teams had never worked together on creative programming and curriculum design and needed to spend time getting to know each other.Team building and effective whole school planning have now become critical components of the Academy and these begin with each school before they attend the first session at the University.Preliminary planning meetings include the school principal and other school leaders who need to play a key role in supporting the development of STEM initiatives which frequently have implications for timetabling, teacher allocation to classes, alignment of STEM subjects on timetable lines, and resourcing. During the preliminary meeting, schools are encouraged to use school-based data to begin their planning.

A critical first component of designing a STEM strategy in any school is defining a clear purpose or vision for the school. We encourage schools to use data to help develop a STEM plan and to identify any issues they may want to address – the types of data they might use include:

- enrolment patterns in the STEM subjects in grades 11 and 12;

- post-school university degrees and other programs their students pursue;
- National Assessment Program for Literacy and Numeracy (NAPLAN)data and other school-based assessment data;
- student attitude, engagement and aspirations data;
- parent and community interests and expectations; and
- teacher knowledge, interests and experiences.

School team members may also wish to use research to help develop their plan including information about:

- improving attitudes to mathematics and science [22]
- breaking down silos to develop 21st century skills [9];
- promoting 'real world' problem solving in teams [20];
- improving the uptake of higher level mathematics, science and technology in senior grades [17]
- increasing female student enrolments in STEM subjects in the senior grades [17];
- increasing Indigenous student enrolments in STEM subjects [11];
- developing STEM literate citizens [6];
- preparing students for the workplace [6].

A vision statement could be developed from a purpose such as [6]:

> If we want students to learn how to apply knowledge, their education experiences must involve them in both learning the knowledge of STEM disciplines and reacting to situations that require them to apply that knowledge in contexts appropriate to their age and stage of development (p. x).

Having developed a vision for STEM, school teams then need to develop policies, programs and practices for implementation [6]. These should all be accompanied by appropriate resourcing including team leadership, staffing, materials and support from other stakeholders such as parents and community members.

One of our greatest challenges in the STEM Academy has been building a community of practice between schools and participating teachers. While on campus, teachers willingly discussed ideas with teachers from other schools, and engaged in worthwhile sharing of resources and teaching ideas but after returning to school, teaching and other priorities quickly take over and appear to limit teachers' capacity for ongoing sharing in the online community. In some schools, finding time to meet as a school team was enough of a challenge and proved to be an inhibiting factor in moving plans forward. For schools to become STEM Academy participants, we had requested principals provide time for teachers to work on their projects but this was not always achieved and remains another challenge to be addressed.

The two case studies of exemplary schools from the second STEM Academy program demonstrated how change in curriculum structures can be achieved with an enthusiastic and committed team leader who has school executive support. Both schools held regular STEM team meetings, developed strategies to share planning documents and files, were flexible enough to accommodate last minute changes to school routines which frequently took valuable time away from classrooms, informed parents and community members about their STEM projects and invited them to be involved, and took a broader perspective than just one class aiming to embed STEM project work across a whole grade with as many teachers involved as possible. Students were more engaged as they could showcase their work to the rest of the school, better understood how STEM can help to solve real-world problems, with at least some students reconsidering their aspirations and post-school STEM pathways. While the data presented in this section are limited to the first two programs, they did provide useful feedback and led to several changes in the Academy program – this approach to using teacher feedback for improvement has been ongoing with each program evolving to best meet the needs of teachers. However, our fundamental principles have remained the same with a focus on inquiry-based learning, embedded in curriculum requirements, involving cross-disciplinary teams of teachers from each school, collaboratively working together with Academy personnel to design and deliver quality multidisciplinary programs for their students.

4. A Retrospective Evaluation for STEM Academy Programs from 2014-2017

Since the first two Academy programs, we have been working with a post-doctoral research fellow (Dr Debbie Tully) to develop an overall evaluation strategywhich was designed in two phases. Phase 1took a retrospective approach to evaluation through interviews with past participating secondary schools from 2014-2017. Within this period 306 teachers from 57 schools participated and from this pool, semi-structured interviews were completed with participants from 20 schools: five principals, 18 STEM team leaders/teachers and nine STEM teachers. Interviews were conducted between six months to three years after their school's involvement with the Academy to probe the STEM initiatives in their school, their perceptions of the impact of the Academy, adjustments made to their school's curriculum to accommodate STEM teaching and learning, their involvement with communities of practice, partnerships with industry, and efforts to sustain STEM initiatives in their schools. All interviews were transcribed so that content analysis could be undertaken, as data were coded and labelled with the primary themes outlined in the STEM Academy's aims and goals. Secondary themes were determined through an open coding process. The focus of Phase 1 was to evaluate how, and in what ways, the aims and goals of the STEM Academy were accomplished in schools after completion of the Academy program. Key findings are listed below under the two themes of sustainability and growth of STEM programs and the key drivers leading to sustaining and expanding STEM programs since participation.

4.1. Sustainability and growth of schools' STEM programs post Academy

Interviews with teachers from the 20 schools indicated:

- 95% of schools continue to offer their original STEM Academy project in their curriculum;
- 90% of schools have expanded STEM curriculum and projects beyond their initial Academy project;
- 65% of schools have added a grade 9 and 10 STEM elective subject;
- 85% of schools have an ongoing STEM professional learning team that regularly meets to plan, design and discuss the implementation of STEM curriculum; and
- 70% of schools have established links in forming a partnership with industry and/or local community organisations.

In addition, most schools are involved in STEM related student outreach programs and/or STEM related competitions with their local universities. While timetable issues were present in most schools, teachers and school leaders were creative in ways to work flexibly with their team to ensure program delivery.For most schools, their targeted STEM students (typically grades 7 and 8) have yet to reach their final year of school, but a few schools from the earlier Academy programs have indicated an increase in student interest to select STEM related subjects in the senior school.

4.2. Key Drivers in Partner Schools in sustaining and expanding STEM programs post Academy

As was indicated through the evaluation of the first two trial Academy programs, program success hinges on the support of school leadership and the executive team in providing time and resources to further STEM teaching and learning. This includes a team of dedicated teachers who meet regularly in a professional learning community to plan and implement school-based STEM initiatives/curriculum, and a designated STEM leader who is given a time allocation in this role.Programs are likely to be sustained and expand if a collegial hub or community of practice of STEM teachers is created which includes other teachers in the school who did not attend the Academy led by a passionate and dedicated STEM leader. Typically, there was growth in STEM curriculum with expanded opportunities for students to experience authentic, "real-world" STEM learning throughout grades 7 to 12 andfor students to experience STEM learning outside the classroom through STEM clubs, STEM competitions and STEM based excursions.There was also evidence of continuing involvement with university STEM outreach programs that connects students to STEM learning as well as growth in partnerships with businesses and community.

5. The Large-scale Evaluation Strategy for Programs from 2018

Phase 2 of the evaluation strategy involved large-scale data collection from teachers and their students in all programs from 2018 onwards [2].The design was informed by research into the factors influencing curriculum innovation in schools including supportive school leadership; quality, targeted professional learning; changes in teachers' values and beliefs; collaboration between teachers; student-centred learning environments; and instructional guidance that advances learning [5]. These key factors and the Academy program objectives informed the overall evaluation plan for the STEM program, which is designed as a mixed-methods protocol, with both questionnaire and interview components, measuring outcomes for principals/school leaders, teachers and students. Parent and industry partners are also considered (Figure 1). Based on a large STEM project conducted by the Friday Institute at the North Carolina University, we adapted questionnaires designed and validated by that research group, for teachers, students and parents – the T-STEM survey, the S-STEM survey and the P-STEM survey [24]. We are currently in the process of collecting and analyzing these data.

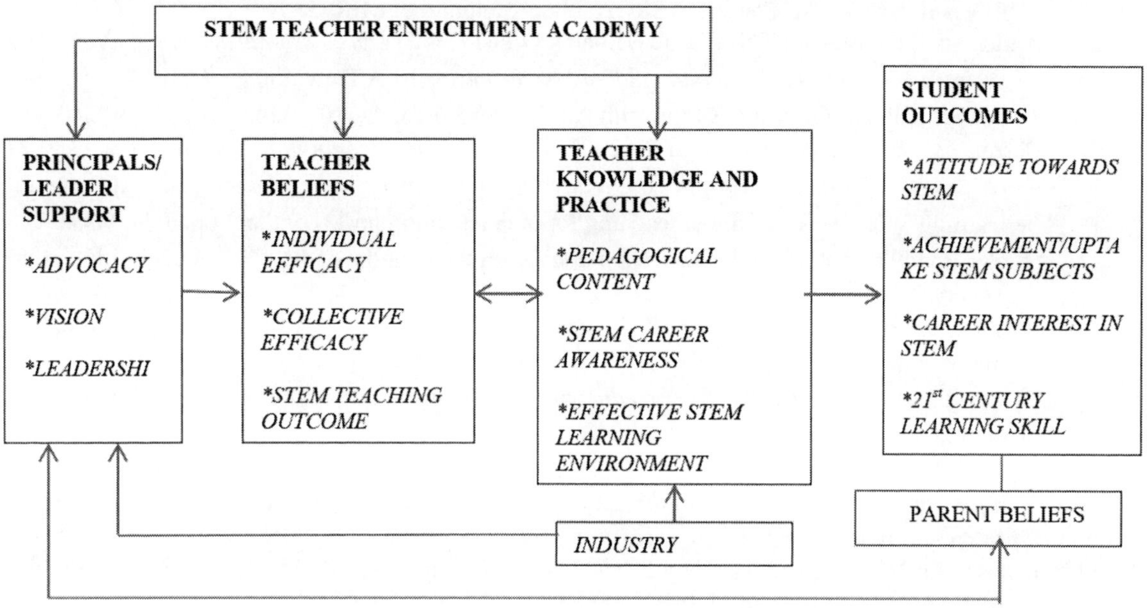

Figure 1. STEM Academy program evaluation model

6. Summary and Conclusions

Our approach to professional learning of secondary STEM teachers is a multi-day residential program delivered on the University campus or in large regional towns in NSW. To date we have worked with about 650 primary and secondary school teachers to develop their capacity to design and deliver multi-disciplinary STEM programs to their students. Aimed at improving student engagement, by enhancing teachers' knowledge and understanding, early data suggest we are having impact on student uptake of STEM subjects via the increased selection of STEM electives in grades 9 and 10 and intentions to enrol in higher level mathematics and science in grades 11 and 12. Our large-scale data collection strategy in 2018 will provide additional information about student attitudes and engagement in the STEM subjects before and after schools implement their STEM strategy.

Schools have adopted a wide variety of approaches to implementing STEM education – frequently these decisions have been based on available personnel, teacher interest and resources but school structures can act as impediments to innovative practices. In summary, our evaluations to date indicate successful schools have

IOP Conf. Series: Journal of Physics: Conf. Series **1340** (2019) 012001 doi:10.1088/1742-6596/1340/1/012001

- a clear vision and purpose;
- support from school leadership;
- an energetic and resourceful STEM team leader;
- space and time to meet and plan together;
- a STEM plan with a timeline and tasks allocated to team members;
- a commitment to move slowly and purposefully;
- intention to bring others with them (teachers, parents, community); and
- capacity to find solutions to challenges.

7. References

[1] Anderson J 2005 *Implementing problem solving in mathematics classrooms: What support do teachers want?* (In P Clarkson et al EdsBuilding connections: Theory, research and practice Procs of the 28th An Conf of MERGA 89-96 Melbourne: MERGA)

[2] Anderson J, Holmes K, Tully DandWilliams G 2017 *STEM professional learning: Evaluating secondary school teachers' and students' experiences* (In A Downton et al Eds 40 years on: We are still learning!Procs of the 40th An Conf of MERGA 586-603 Melbourne: MERGA)

[3] Barron B and Darling-Hammond L 2008 *Teaching for meaningful learning: A review of research on inquiry-based and cooperative learning* (In L Darling-Hammond et alPowerful learning: What we know about teaching for understanding pp 11-70 San Fran: John Wiley)

[4] Bruder Rand Prescott A 2013 Research evidence on the benefits of IBLZDM (Math Edn vol 45) chapter 8 pp 11-22

[5] Bryk A, Sebring P, Allensworth E, Easton Jand Luppescu S 2010 Organizing schools for improvement: Lessons from Chicago (Chicago: Uni of Chicago Press)

[6] Bybee R 2013 *The case for STEM education: Challenges and opportunities* (Arlingon, VA: National Science Teachers Association)

[7] Clarke D 2014 *Disciplinary inclusivity in educational research design: Permeability and affordances in STEM education* (Keynote address at STEM Education and our Planet: Making Connections Across Contexts)

[8] Davison D, Miller K and Metheny D 1995 *What does integration in science and mathematics really mean?* (Scl Sc and Math vol 95) pp 226-230

[9] de Bruin L and Harris A 2017 *Fostering creative ecologies in Australasian secondary schools* (Aust J of Ter Edn vol 42) pp 23-43

[10] Desimone L 2009 *Improving impact studies of teachers' professional development: Towards better conceptualisations and measures* (Edal Rer vol 38) pp 181-99

[11] DreiseTand Thomson S 2014 *Unfinished business: PISA shows Indigenous youth are being left behind* (ACER Oc Essays)

[12] English L 2016 *STEM education K-12: Perspectives on integration* (Int J of STEM Edn vol 3)

[13] Friday Inst for Edal Inn 2012 Teacher efficacy and attitudes toward STEM survey (NC: Au)

[14] Jenkins Eand Neslson N 2005 *Important but not for me: Students' attitudes towards secondary school science in England* (Res in Sc and Tech Edn vol 23) pp 41-57

[15] Kennedy J, Lyons Tand Quinn F 2014 *The continuing decline of science and mathematics enrolments in Australian high schools* (Teaching Sc vol 60) 34-46

[16] Maaß Kand Artigue M 2013 *Implementation of inquiry-based learning in day-to-day teaching: A synthesis* (ZDM Math Edn vol 45) pp 779-95

[17] Mack Jand Walsh B 2014 *Mathematics and science combinations NSW HSC 2001-2011 by gender. Technical paper*

[18] McPhan G, Morony W, Pegg J, Cooksey R and Lynch T 2008 *Maths? Why not* (Canberra: Department of Education, Employment and Workplace Relations)
[19] Office of the Chief Scientist 2016 *Australia's STEM workforce: Science, technology, engineering and mathematics* (Canberra: Commonwealth of Australia)
[20] Sanders M 2008 *STEM, STEM education, STEMania* (The Tech Ter vol 68) pp 20-26
[21] Thomson S, Wernert N, O'Grady Eand Rodrigues S 2016 *TIMSS 2015: A first look at Australia's results* (Melbourne, Vic: ACER)
[22] Tytler R, Osborne J, Williams G, Tytler, Kand Cripps Clark J 2008 Opening up pathways: Engagement in STEM across the primary-secondary school transition (Canb: AustrDEEWP)
[23] Vasquez J, Sneider C and Comer, M 2013 *STEM lesson essentials, Grades 3-8: integrating science, technology, engineering, and mathematics* (New York: Heinemann)
[24] Venville G, Wallace J, Rennie Land Malone J 1998 *The integration of science, mathematics, and technology in a discipline-based culture* (Scl Sc and Math vol 98) pp 294-302

IOP Conf. Series: Journal of Physics: Conf. Series **1340** (2019) 012002 doi:10.1088/1742-6596/1340/1/012002

STEM Education Landscape: The Case of Singapore

Tang Wee Teo[1*]

[1]Multi-Centric Education Research and Industry STEM Centre at NIE (meriSTEM@NIE), Natural Sciences and Science Education (Academic Group), National Institute of Education, Nanyang Technological University, Singapore

* Corresponding auther's email address: tangwee.teo@nie.edu.sg

Abstract. This conference proceeding is based upon the keynote delivered at the inaugural International Annual Meeting on STEM Education 2018 in Khon Kaen, Thailand. The title of the keynote was STEM Education and Research Landscape. During the talk, I provided an overview of STEM education in the United States, Singapore and a few other countries. STEM education in the United States has been extensively reported elsewhere (e.g., see National Academies Press). To scope this proceeding, I focus on STEM education in Singapore. Here, I offer a broad overview of the current state of STEM education in Singapore. Additionally, I provide some new information on this topic, which was not presented at the time of the talk. New STEM education-related events and efforts are emerging as I write. After all, the STEM education landscape is fast evolving in Singapore and the nearby regions.

1. The Role of the Ministry of Education

Many people have asked me: How does STEM education look like in Singapore? My summative response is that STEM education efforts in Singapore is wholistic. By wholistic, I mean that multiple stakeholders and agencies participate in the offer of diverse STEM programmes for different groups of clienteles. Hence, the isn't one, but diverse and collective efforts to promote and support STEM education. A key agent driving STEM education is the Ministry of Education (MOE), Singapore.

The Applied Learning Programme (ALP) was started by the MOE in 2013 to promote more authentic and practice-oriented learning experiences for students in Primary (Grades 1-6, aged 7-12) and Secondary schools (Grades 7-10, aged 13-16). Schools that embarked on the ALP design their own programmes emphasizing on the relevance of what is being taught to the current and future social and industry needs and trends. This is achieved through providing hands-on or experiential learning opportunities to students to enact the real-life situations so that in the process, they become equipped with the skills to apply the knowledge. Schools may design their own ALP curriculum or work with the industry, community-based entities, institutions of higher learning, and/or professional training bodies to offer the curriculum. Due to the popularity of the ALP, the former Minister of Education (Schools), Mr Ng Chee Meng, announced in Parliament that all primary schools will have an ALP in 2023 [1]. Among the ALP, which could cover STEM, languages, humanities, business, entrepreneurship, aesthetics, and so on, STEM ALP is one of the most popular programme in schools.

To support STEM education in schools, STEM Inc., the abbreviation for Science, Technology, Engineering, Mathematics innovation and creativity, or Incorporation, was established in January 2014. STEM Inc. was charged with the mandate to ignite students' passion for STEM and they play a strong supporting role in the STEM ALP in secondary schools. They have STEM lesson packages that can be used in the classrooms during the STEM ALP lessons. Schools can engage STEM Inc. to conduct STEM ALP lessons in their schools or at the Science Centre Singapore.

While the STEM ALP is an informal curriculum, there are formal secondary school applied subjects that are examinable at the GCE O-levels and GCE N-level national examinations (Tan, Teo, & Poon, 2017). The MOE Applied Subjects taken at O-levels include computing, electronics, exercise and sports science, and drama. There are also MOE-ITE (Institute of Technical Education) Applied Subjects taken at Normal (Technical)-Level for students in the Normal Technical tracks and these include mobile robotics, smart electrical technology and retail operations.

2. Selective STEM Schools and STEM Education in non-STEM-focused Schools

According to a U.S. report by The Committee on Highly Successful Schools or Programs for K-12 STEM Education [2], which was supported by the National Science Foundation, there are four types of STEM schools or programmes in the United States. These are STEM selective schools, inclusive STEM-focused schools, STEM-focussed Career and Technical Education, and STEM education in non-STEM-focused schools. Based on my scan of the STEM landscape in Singapore, the first and fourth types of school or programme exist locally.

In 2005, the National University of Singapore High School of Mathematics and Science (NUS High) was established to offer advanced mathematics and science programmes for academically talented students in these disciplines. NUS High is thus, similar to the elite or selective STEM-focused schools or specialized STEM school in the United States (e.g., Illinois Mathematics and Science Academy and North Carolina School of Science and Mathematics), Australia (e.g., Australian Science and Mathematics School), and the Philippines (e.g., Philippines Science High School). Other than its advanced curriculum, NUS High shares similar characteristics as other elite STEM-focused schools in terms of having highly qualified teachers with Master's and doctorate degrees, like-minded peers interested in STEM disciplines, and students having the opportunity to do independent research projects under the supervision of their teachers, scientists, or university faculty.

Following NUS High, a second STEM-focused school, the School of Science and Technology (SST), that offers a four-year GCE O-level programme [3] was established in 2010. SST adopts an integrated approach to applied learning and a cornerstone of SST pedagogy is the use of ICT (information communication technology). SST works closely with the local polytechnics to offer applied subjects such as Biotechnology, Computing, Design Studies and Electronics to students.

While NUS High and SST are similar to the elite or selective STEM schools, some schools in Singapore also offer specialized STEM-related programmes as a niche area for a selected group of academically talented students. For example, Hwa Chong Institution has a School Scientist Programme where students embark on independent research projects with an in-house School Scientist. Raffles Institution has established the Raffles Science Institute intended to nurture students to become future leaders and pioneers of the science community. The school has an OpenLab, which is staffed by researchers with doctorate. As teacher-researchers, they supervise the brightest science students in projects such as water technology, viral biology, marine ecology, and gene-chip technologies.

3. Informal STEM Programmes

Other informal STEM programmes include STEM-based competitions and research attachments. Examples of STEM-based competitions include the Shell Singapore: The Bright Ideas Challenge held in 2017 and 2018 for secondary school students (Grades 7-10, aged 13-16) to pitch creative ideas for future cities that are vibrant, healthy, and clean to live in. The winner team could also win the chance to visit attend a STEM-related event in London. The Science Centre Singapore have also organised the Drone Odyssey Challenge which is a game-based competition to inculcate technical skillsets, critical thinking and develop appreciation of new and disruptive technologies relevant to the modern world. The competition is designed for primary and secondary school students and there are workshops and demonstrations to lead up to the competition so that students are prepared for the tasks. There are also

Maker Faire Singapore organized to bring practitioners (e.g., tech enthusiasts, crafters, educators, tinkerers, hobbyists, engineers, science clubs, commercial exhibitors, students, artists) of the maker movement to show what innovation products they have made and what they have learned.

4. STEM at Higher Education

The STEM wave has also caught on the tertiary education sector. In 2009, the fourth autonomous university, Singapore University of Technology and Design, was established. SUTD offers a design-centric and interdisciplinary curriculum through cohort-based learning approach. The Institute of Technical Education, which is a post-secondary institution for school leavers and working adults, also offers short-term STEM courses such as three-dimensional printing, creating computer games, programming mobile Applications (Apps), and mobile robotics.

5. meriSTEM@NIE

In 2018, the Multi-centric Education Research and Industry STEM Centre at the National Institute of Education (meriSTEM@NIE) was established. NIE is Singapore's national teacher education institute and plays a key role in teacher preparation and the provision of professional and school leadership development programmes. meriSTEM@NIE is a research centre within NIE. The mission of the centre is:

- To enhance the quality of STEM literacy in Singapore through cross-disciplinary partnerships in research, teaching, and outreach so that future generations of educators, learners, and citizens are able to harness relevant STEM knowledge and skills in addressing current and emerging challenges for self and others.

meriSTEM@NIE is the first STEM centre in Singapore that engages in research, research-informed STEM education and evidence-based STEM education outreach. The Centre's work is guided by the Integrated STEM Instructional and Evaluation Framework, which is used to inform the design of integrated lesson packages for teachers and students. The same Framework can be used to evaluate the fidelity of implementation of integrated STEM.

6. Summary

In essence, the STEM education landscape in Singapore is enriched by supportive efforts from the Ministry of Education and ground-up efforts from private and public organizations. Even as you read this piece, the STEM landscape is evolving.

References

[1] Chia L 2018 *All primary schools to set up Applied Learning Programmes by 2023:* Ng Chee Meng Retrieved on March 9, 2019 from https://www.channelnewsasia.com/news/singapore/all-primary-schools-to-set-up-applied-learning-programmes-by-10014282

[2] National Research Council 2011 *Successful K-12 STEM Education: Identifying Effective Approaches in Science, Technology, Engineering, and Mathematics* National Academies Press

[3] Tan K C D, Teo T W and Poon C L 2016 Singapore science education. In M.-H. Chiu (Ed.), *Science Education Research and Practice in Asia: Challenges and Opportunities* (Singapore: Springer) pp 155–174

IOP Conf. Series: Journal of Physics: Conf. Series **1340** (2019) 012003 doi:10.1088/1742-6596/1340/1/012003

STEM Education Teaching approach: Inquiry from the Context Based

Sukanya Sutaphan and Chokchai Yuenyong*
Faculty of Education, Khon Kaen University, Khon Kaen, Thailand

*Corresponding author's e-mail address: ychok@kku.ac.th

Abstract. This study aimed to develop theoretical framework for STEM teaching strategies in school setting. The paper clarified what the worldwide leaders and educators' beliefs, focuses, and actions of STEM education; what the conceptions of STEM education should be provided for teachers; Suggested STEM education teaching in aspect of practicing knowledge in real world through STS approach, concept of technology and engineering reserved for STEM education; the implementation of Yuenyong (2006)' s STS approach for enhancing students' applying scientific and other knowledge for designing the solutions in real-world problems. Based on these literatures, this paper will propose the practical idea of STEM education teaching approach which integration key concepts among problem based, project based, inquiry, design, and a real world problem solving as STS. The paper, therefore, propose the theoretical framework of STEM education teaching approach as inquiry from the context based. The teaching approach consists of 7 stages. These include (1) Identification of social issues, (2) Identification of potential solution, (3) Need for knowledge, (4) Decision-making, (5) Development of prototype or product, (6) Test and evaluate the Solution, and (7) Socialization and completion decision stage. The paper discussed more details of each stage.

Keywords: STEM education, pedagogy, context based

1. Introduction

STEM education has been pushed to improved education in the early twenty-first century. Obviously, international concerns for advancing STEM education have increased in recent years. Not only educators but also policy developers, non-government organization, and business and industry organization are stressing the urgency for improving STEM skills to meet current and future social and economic challenges. There are many concerning with the STEM education in shaping innovation and development. Innovation and production of agriculture, industrial, military, and consumer products are the strength of the nation economics. This push the policy of many countries continues under the orientation of STEM education. In the USA for example, the 2013 reported from the Committee on STEM Education stressed that "The jobs of the future are STEM jobs," with STEM competencies increasingly required not only within but also outside of specific STEM occupations [41]. This showed the needs of developing competencies in the STEM disciplines. To do so, the goal of education systems have driven to prepare the future STEM workforce. Another reason of pursuing STEM education is being emphasized is because of the lack of student achievement in STEM subjects [22], as determined by their low rankings on TIMSS (Trends in International Mathematics and Science Study) assessments (IAE 2011) [20]; [46].

However, the leaders and educators in each country held the varied beliefs, focuses, and actions of STEM education. These included [7]; [8]; [10]; [23]; [26]; [31]; [46]; [48]; [54]:

- STEM educational reform,
- preparing for STEM workforce and STEM literacy,
- structuring STEM as a track of school curriculum,
- providing a broad range of content and practices such as:
    - specific STEM subjects – integrative STEM,
    - integrate STEM into single courses,
    - merging the content of science and technology education,
    - the development of engineering programs into schools,
    - integrating engineering content into science, technology, and mathematics education,
    - additional applications of computer science into the K-12 curriculum
    - Adding art to form STEAM
    - a STEM project-based,
- exploration of different teaching strategies for STEM such as:
    - STEM team teaching
    - Teaching STEM through a real-world problem solving,
    - proposed a national STEM education strategic approach,
    - exploration of new curriculum and methods for delivering STEM concepts,
    - communications about careers needing different levels of STEM knowledge and abilities
- teacher preparation for STEM education teaching approach such as:
    - STEM education teacher training
    - providing materials for STEM teachers
    - Partner organization for STEM in primary and secondary school levels

Internationally, there are some movements for STEM educational reform. There were found that many foundations have been established by government, schools, and enterprises to foster STEM education and communications. The new ways of integrating content and practices among technology, engineering, science, and mathematics were organized. This generates the principles to enhance student learning of complex concepts. The hands-on nature of this school subject has allowed for learning to become conceptualize knowledge for students and brings it into real world uses. It has much to offer to the STEM educational reform efforts. In France, for example, the La Main a\` la Pa\^te program (asking for collaborative hands-on work) is an inquiry-based project that aims to teach young students scientific concepts and procedures using innovative STEM teaching methods. The Sciences a\` l'E´ cole (Science at School) project is another supported by the French government. It has worked to encourage project-based learning programs and scientific subjects at the post-secondary levels (colleges, vocational schools, and preparatory classes). It developed multidisciplinary programs to promote educational innovation to increase linkages with scientific vocations. These initiatives have become core activities to promote STEM educational reform in France. Its programs have focused on the cultural dimensions of science in the curriculum, connecting knowledge within societal contexts, and establishing linkages between real-life problems and industry [46]. In England, STEM education reform has been aligned to two main objectives: (a) the flow of qualified people into the STEM workforce and (b) STEM literacy for the general population [18]. In Scotland, the Science and Engineering Education Advisory Group (2012) offered recommendations for STEM educational reform. The advisory group has focused attention on the separate STEM subjects. A report from Australia revealed establish national STEM education policies and curriculum programs in Australia and some strategies of Australia's STEM educational reform [31]. Like Thailand movement, many government organizations (e.g. IPST and OBEC), schools, universities, and private sectors provided some teacher training and materials for STEM education. STEM education have developed strategic national STEM policy framework which provide positive conditions for variety of activities. These included centrally driven and funded STEM programs of MOE, IPST, and others; and initiatives and partnerships and engagement (Chevron, EGAT, and others) that link STEM activities in schools, vocational and higher education with industry, business and the professions [24]; [33]; [14]. This could conceptualize also STEM educational reform [61].

The preparations of STEM workforce and STEM literacy were developed. STEM education should not focus only on students who are going to choose STEM careers but also all needs to interact natural and material setting with at least a set of knowledge, skills, and attitudes involving STEM. The preparation for STEM workforce involves multidimensions of students, STEM concepts, skills, ways of thinking, the 21st skills, and attitude towards STEM careers. [12]; [40]. For example, in England, the curriculum provided clear vision of STEM as two main objectives (1) the flow of qualified people into the STEM workforce and (2) STEM literacy for the general population. Some countries hope to increase the number of students who elected to study STEM majors while in college and then transition into the future STEM workforce, while others hope to improve the methods teachers use to present STEM subjects to their learners and increase students' understanding and use of this knowledge. In Taiwan, science and technology was combined as a learning area for junior high schools. This was done to help students experience the relationships between science and technology and how this knowledge is integrated in business and industrial practices [46].

The view of studying science and technology research have emphasized on the integration among disciplines for decades. The key driver of the future economy and concomitant creation of jobs will be innovation, largely derived from advances in science, technology, mathematics and engineering (STEM) [40]. The STEM workforce is a powerful component of this innovation pipeline. The jobs of the future are integrally STEM driven and the foundation of STEM knowledge student receive has been directly linked to the prosperity of the country [50]. Structuring STEM as a track of school curriculum is another issue that some countries tried to prepare STEM workforce and action on the STEM education. For example, in the Philippines, the department of education approved the K-12 track series in year 2013. There are four tracks including Academic, Sports, Arts and Design, and Technical-Vocational-Livelihood tracks. Four different strands fall under the Academic track: Accounting, Business and Management (ABM) strand, Science, Technology, Engineering and Mathematics (STEM) strand, Humanities and Social Sciences (HUMSS) strand and the General Academics strand (GAS). Through the STEM strand, students will get to learn the basics of STEM-related disciplines such as Calculus, Biology, Physics and Chemistry. Students will get an overview of the foundation of most STEM fields which basically deal with several branches of physical, Earth and life sciences, and pure and applied mathematics. And, a future career of students in STEM stand is about bio researcher, engineer, microbiologist, statistician, programmer, software developer, database administrator [57].

It seems that the concept of STEM education covers a broad range of contents and practices. Literatures suggested that STEM can become a multidisciplinary approach to learning that can be used to integrate knowledge from the separate STEM subjects into existing science, technology and engineering, or mathematics instructional programs using project-based learning, engineering design-based studies, or problem-based learning strategies. Many policy discussions of STEM have focused on the improved teaching of separate subjects, especially those of science and mathematics [10].

It is believed that developing engineering programs or integrating engineering content into science, technology, and mathematics education will enhance students' learning and interest in STEM subjects. Some countries have experimented with the development of engineering programs for their schools in an effort to bring context to science, mathematics, and technology content. Other nations are integrating engineering content into technology education curriculum (e.g., Canada, Sweden, United States). For examples, the Next Generation Science Standards (NGSS) has integrated engineering design concepts and activities into its structure for future K-12 science education programs [42]. Project Lead the Way (PLTW 2014) courses were designed to promote engineering careers and STEM learning at the K-12 levels. Engineering by Design, which was developed by the International Technology and Engineering Educators Association's STEM Center for Teaching and Learning, is a standard-based curriculum for technology and engineering education, which integrates mathematics and science standards content into technology and engineering courses for the K-12 levels [46]. Another example in Thailand, IPST and OBEC enhanced teachers to provide the engineering design process into science, technology, and mathematics education in order to support students to learn STEM education [61].

In some cases, STEM has incorporated technology and engineering into its framework as a means to show how scientific applications can be incorporated within science and mathematics lessons [54]. Some countries have merged the content of science and technology education (e.g., France, Israel, The Netherlands; [35]). And, Thailand also just merged the content of science and technology education in the year of 2018 [25]. In Taiwan's most recent planning for K-12 compulsory education, technology education will be separated from science in the curriculum. In the new plan, technology education will be combined with information technology, thus creating a new learning area [39]. The Israel government is encouraging additional applications of computer science into the K-12 curriculum. It established a 260-h course for all students and an in-depth study (450-h course) for those who express interest in further study of computer sciences. The Israeli curriculum includes both theoretical material and hands-on applications within their computer science courses [2]. The Korean education system combined the subjects of science, technology, and home economics as one learning area in 2011. The arts were added to STEM to restructure the Korean curriculum and it now encompasses these subjects to form STEAM [26].

And, some see it taught as a new integrative subject labeled as STEM. In this interpretation, educational practices would use multiple STEM subjects and integrate these into single courses. This is known as integrated or integrative STEM [38]; [47].

Internationally, there are explorations of strategies for STEM teaching such as communications about careers needing different levels of STEM knowledge and abilities, teaching STEM through a real-world problem solving, proposed a national STEM education strategic approach, exploration of new curriculum and methods for delivering STEM concepts, and STEM team teaching. In UK, for example, the ASE website provided some information and activities to communicate STEM careers. These include providing information of rotes into STEM careers, career profiles and interviews with STEM professionals, STEM learning careers posters, STEM ambassadors, STEM insight and so on [5].

There are many proposed STEM education strategic approach. In Australia, the Australian Academy of Technological Sciences and Engineering developed an innovative STEM program (The Science and Technology Education Leveraging Relevance) that addresses the problems within secondary school level science and mathematics education. Meanwhile, the Office of the Chief Scientist (2013) also proposed a national STEM education strategic approach. This approach emphasizes four essential elements (Education, Knowledge, Innovation, and Influence) in STEM education [31]. In UK, the National STEM Centre in the United Kingdom was established to provide a wide range of high-quality support materials for STEM teachers and to explore new curriculum and methods for delivering STEM concepts to students [7]. In France, proposed the STEM education programs which focused on the cultural dimensions of science in the curriculum, connecting knowledge within societal contexts, and establishing linkages between real-life problems and industry [46]. In Thailand, some government and non-government agency proposed some STEM strategic approach. They provided some strategies of inquiry or solving problem with a little bit different steps of teaching. For example, IPST proposed the engineering design process including Identification, Related Information Source, Design, Planning and Development, Testing, Evaluation and Design Improvement, and Presentation. And, the EDP of OBEC STEM consists of 9 steps including (1) Identify the problem, (2) analyzing five capitals, (3) explore information, (4) knowledge sharing, (5) model solution, (6) planning and development, (7) testing and evaluating the solution, (8) present the solution, and (9) reflection and revise [61].

Some cases organized the integrated learning activities as the ideas of STEM team teaching. For example, in China Beijing's Yew Chung International School teachers provided team teaching. The team teaching includes classes in science, mathematics, geography, and other subjects. Strategies used for teaching mathematics have one teacher explaining step-by-step procedures in English, while the second teacher shows students shortcuts for solving the same problems through the Mandarin language. These strategies have resulted in better understandings by students and higher student test scores [2].

Teacher preparation for STEM education has been also considered. For example, in UK schools, Bell (2012) found that STEM has not had the impact that was planned. Many teachers are unaware of the existing STEM endeavors and are unsure of the contributions they can make for students to better understand and learn STEM subjects. Some organizations and ministries of education have developed educational materials to assist teachers in instructing from a STEM perspective. The ASE (2010) provided the STEM learning website which suggested good resources and ideas of STEM education. The ASE suggested resources of science learning and practicing through the SATIS revisited version. This version was has updated and revitalized the groundbreaking SATIS (Science and Technology in Society) units from the 1980's and 90's. The units provide a wide range of strategies to cover the program of study for How Science Works and the wide spectrum of science. Materials consist of bright stimulus visuals, student activities including research, discussion, role plays, quizzes and board games with background information sheets to suit a wide range of student interests and abilities. They also contain guidance for teachers and an overview of the science considered. Another example, in Thailand, the government organizations have pushed the projects to prepare STEM teachers across the nation. The STEM professional development projects have been provided for in-service teachers by many organizations such as the institute for the promotion of teaching science and technology (IPST), the office of basic education commission (OBEC), the office of higher education commission (OHEC), the ministry of energy, the Electricity Generating Authority of Thailand (EGAT), the chevron enjoy science project, and the schools and universities STEM education projects. These projects could provide approximately 5,000 teachers (in the year of 2017) who could generate the learning activities through integrating knowledge, concepts and skills systematically as STEM education [61]. There are some suggestions of partner organization for STEM in primary and secondary school levels. The Partners in Research (PIR) has also broadened its programs to implement science, technology, engineering, and mathematics education at primary and secondary school levels [27].

As above mentioned about a boarded view of perceptions, beliefs and actions on STEM, this paper may share some ideas based on those STEM education review. This paper will share an approach for STEM teaching and learning. But, the paper will focused on how teachers may provide some kinds of STEM learning activities in the single subjects as the integrated STEM. The paper will propose the framework of STEM pedagogy. The ideas of proposing STEM teaching approach are about enhancing students' to apply science, mathematics, technology, and other knowledge for finding solutions or products through inquiry from the context based. In order to provide some theoretical framework, the concept of STEM teaching and learning and suggested ideas for literatures about related to STEM will be clarified.

2. Suggested ideas of STEM teaching and learning

What should the conceptions of STEM education be provided for teachers? It found that descriptions of what comprises STEM content and practices and what STEM conceptions look like range in the literature, especially conceptions of STEM education [9]. However, there are three consensus issues of discussion about STEM education including (a) instructionally (b) as a set of integrated or interconnected disciplines, or (c) as more dependent on the stakeholders or context in which it is viewed or conceptualized. For this study, we will hold Moore et.al. [36] definition of STEM education for enhancing teachers. They define it as ''the teaching and learning of the content and practices of disciplinary knowledge which include science and/or mathematics through the integration of the practices of engineering and engineering design of relevant technologies.'' To them, five characteristics distinguish integrated STEM instruction from other teacher pedagogy: (a) the content and practices of one or more anchor science and mathematics disciplines define some of the primary learning goals; (b) the integrator is the engineering practices and engineering design of technologies as the context; (c) the engineering design or engineering practices related to relevant technologies requires the use of scientific and mathematical concepts through design justification; (d) the development of 21st century skills is emphasized; and (e) the context of instruction requires solving a real-world problem or task through teamwork. This conceptualization of STEM is grounded in learning research [61].

It seems that the engineer design process was widely recognized for STEM education. Engineering can act as a connector for meaningfully learning the content of mathematics and science [37]. Engineering design has been treated as a pedagogical strategy to bridge science and mathematics concepts in use of solving ill-defined (open-ended) problems, developing creative thinking, formulating solutions and making decisions, and considering alternative solutions to meet a variety of constrains [53]; [56]. Engineering requires the use of scientific and mathematical concepts to address the types of ill-structured and open-ended problems that occur in the real world [52]. Using engineering design as a context for these problems is a natural way for students to learn through STEM integration. Using engineering contexts as spaces for students to develop real-world representations can be the catalyst for developing related scientific and mathematical concepts through using multiple representations (concrete models, pictures, language, and symbols) and facilitating translations between and among them [37]. The engineer design process was also introduced in Thailand by government agencies. For example, the EDP of OBEC STEM consists of 9 steps including (1) Identify the problem, (2) analyzing five capitals, (3) explore information, (4) knowledge sharing, (5) model solution, (6) planning and development, (7) testing and evaluating the solution, (8) present the solution, and (9) reflection and revise [61]. In Malaysia, the application of engineering design process in the module is based on the five cycles as; (1) ask, (2) imagine, (3) create, (4) test, and (5) improve. The application of STEM content knowledge during the design processes will be the key component of students' learning in solving engineering-based problem. The context of instruction requires solving the real-world problem or task through teamwork [50].

Another conceptualized idea of teaching and learning in STEM education was suggested by Williams (2019). He provided some practical ideas of STEM teaching and learning which focuses on processes and engagement of students in collaborative activity. He discussed that the principles of teaching and learning in STEM Education: (1) involves the integration of science, technology and mathematics, (2) is student centred, (3) engages students in collaborative activity, (4) focusses on processes, (5) occurs within the curriculum (is not extra-curricular), and is project and/or problem based.

STEM education teaching and learning should focus on the processes that could be provided as a way of structuring student activity. These processes included problem based, project based, inquiry, and design. The processes will engage students to generate ideas, research and investigate, evaluate, model their ideas, identify needs, solve problems, document what they do, and communicate. It is through engagement with these activities that students will learn not only crystalized knowledge but also get to practice the important fluid skills they need for their future [55]. Therefore, teachers may foster students not only strongly hold crystalized knowledge which mentioned already in curriculums, books, or facts; but also support them to develop the fluid skills which need to think flexibly. In order to be success on finding solutions or developing prototypes or products, students may develop also some fluid skills such as creative thinking, cooperation in group working, time management, communicate and convince to people, and so on.

Williams (2019) also reminded some implications when students move on the processes of doing STEM activities. STEM education should focus more on process rather than content. The STEM education important learning occurs through the activities of the process. However, content could be applied when the learning of content is necessary through an activity to a situation. A focus on the activities of a process will also enable students to learn from failure. Therefore, teachers should not expect too much too soon from students. It will take time for them to get used to this complexity, and the progressive development from a simple approach to something more complex (and therefore more realistic) takes many years. There are skills involved in what we require of students as they progress through these stages and these skills need to be taught. Therefore, it is not only teaching about content based curriculum but also procedural knowledge. So each STEM project may have a focus on the development of a particular procedural skill, and over time students develop this repertoire of skills. STEM activities are diverse; even within one particular focus such as design processes, there is much diversity. The process can begin and end at different places. An inquiry approach may begin with an environmental problem, a design approach may begin with a product, a problem based approach may begin with a problem. The conclusion or goal of the process may also vary: a project based activity will usually conclude with a prototype, but an inquiry approach may end with a presentation.

In view of teaching, STEM education is teaching for learning. So, STEM education should not be some teaching is to get through the curriculum or to ensure students achieve well in an examination. Therefore, student center need to be considered in STEM learning activities. Students need more scaffolding as they progress through various activities of a process. This requires smart teaching, by ensuring the more naturally creative students have the flexibility to be creative in their pursuit of solutions, while other students, who are not sure what to do next, are provided with the support and directions they need [55].

To enhance students to learn in STEM education, the activities are engaged in a STEM process. The importance thing is the moving students back and forth between their thinking and doing. Students will begin by thinking about a problem or issue or need, and then in order to progress their thinking, they do something with that idea. As they test their idea by talking to others, sketching, modelling, etc., the idea is refined, enhanced, clarified or validated. This back and forth between thinking and doing is the way that procedural progress is made, and the extent to which it is facilitated, increases the potential of more sophisticated solutions and higher order thinking. The role of the teacher then becomes the facilitation of the transition between thinking and doing, allowing students time to think, and allowing them time (and structuring activities) to act on their thoughts [55].

3. Suggested Ideas of Context based inquiry as STS approach

Regarding on the conceptions of STEM education, the literatures indicated that STEM teaching and learning involved the integrator is the engineering practices and engineering design of technologies as the context; the engineering design or engineering practices related to relevant technologies requires the use of scientific and mathematical concepts through design justification; and the context of instruction requires solving a real-world problem or task through teamwork [35]; [61]. And, it should focuses on the design processes that are much diversity. The STEM teaching and learning should provide an inquiry approach that may begin with an environmental problem, a design approach may begin with a product, a problem based approach may begin with a problem. Then, the related contents (e.g. science and mathematics) content could be applied when the learning of content is necessary through an activity to a situation [55].

Another view of Science technology and society (STS) is a teaching approach to enhancing students' practicing knowledge in real world. The STS allows students to learn science through solving problem and scientific inquiry from related to societal issues or technological issues or products [62]; [59]. In UK, an example of introduce STS for STEM education, the ASE (2010) suggested the STS unit called SATIS (Science and Technology in Society) units in the UK STEM learning website to assist teachers in instructing from a STEM perspective. The SATIS revisited version was has updated and revitalized the groundbreaking SATIS units from the 1980's and 90's. The units provide a wide range of strategies to cover the program of study for How Science Works and the wide spectrum of science. Materials consist of bright stimulus visuals, student activities including research, discussion, role plays, quizzes and board games with background information sheets to suit a wide range of student interests and abilities. They also contain guidance for teachers and an overview of the science considered.

The STS unit always engages students through the social issues or technological issues. Then, students could generate some problems or questions based on those issues. The unit, then, move students to learn science and other related knowledge with aiming to find some knowledge for developing the possible solutions and select the appropriated solution for their society. It seems that STS teachers required holding strongly understanding of concept of technology in order to balance science teaching by its nature and science teaching related to society. Regarding to the philosophy of STS, the T-technology provided strongly linking up between science and society (Yuenyong, 2018). To establish well STS teaching, the philosophy of technology need to be discussed a little bit more.

Ankiewicz and de Swardt (2006) clarified that regarding on Mitcham (1994)' tension between the two uses of technology provided a narrow and a wide meaning in technology present discourse, which roughly coincides with the manner in which two professional groups, namely engineers and social scientists, respectively. The concept ''technology'' has mainly been reserved by engineers to refer to a direct involvement with material construction and the manipulation of artefacts. In the social sciences the concept ''technology'' also includes making material artefacts, the completed objects, their usages and to a limited extent also the intellectual and social contexts. Sometimes technology is defined in such a way that it includes making non-material things like laws and language.

Mitcham's framework is now accepted to be a 'point of reference' in the field of philosophy of technology, consists of the four modes of the manifestation of technology: technology as object, technology as knowledge, technology as activity, and technology as volition [17].

Technology as object could be seen through the nature of technological artefacts which attribute to a dual nature – physical and functional nature. Artifacts have a physical nature such as length, weight, colour, structure, geometry, and so on. And, at the same time, artifacts have a functional nature; for example, some objects allow us to use them for a specific purpose. In fact the purpose of design can be described as finding a right match between these two. In other words, designers seek for realising objects that have a physical nature that fits with the intended functional nature. But users in fact also deal with the relationship between the physical and the functional nature when they seek ways to use the artifact [17].

Technology as knowledge can be differentiated according to various types of knowledge, for example maxims, rules, theories etc. [32]. There are two types of knowledge in technology including conceptual knowledge (''knowing that'') and procedural knowledge (''knowing how''). However, these two types of knowledge cannot be separated. Conceptual or descriptive knowledge relates to the links between knowledge items, implying that when learners can identify these links, we can say that they have conceptual understanding. For example, in the area of 'gearing' we hope that students will see the relationship among the 'direction of rotation', 'change of speed', and 'torque''' [34]. Conceptual knowledge relevant to technology therefore includes justification from other subjects such as science [17]. Procedural knowledge referred to as tacit, personal or implicit knowledge. ''Design, modelling, problem solving, systems approaches, project planning, quality assurance and optimisation are all candidates for technological procedural knowledge ...'' [34]. Simon (1969, in [3]) who linked classic design methodology with problem solving theories from computer science and psychology, summarised the essential design knowledge base for the training of designers. These included evaluation of designs (e.g. statistics), computational methods (e.g. algorithms – linear programming, control theory), formal logic of design, search for alternatives, theory of structure and design organization, and representation of design problems. However, procedural knowledge cannot be taught. It can be gained by thorough practice only [17].

Technology as activity (methodology), in particular, provides insights in the procedural knowledge in technology [17]. The activity, for example, includes designing, making, and using or evaluating [29].

Volition is another view of technology. Technological aspects, therefore, are skills to support people with their decision making. The understanding of what, why, and how ideas, design, systems, volition of application scientific knowledge and other knowledge (e.g., philosophical, sociological, political, economic, humanistic aspects and so on) work becomes key aspects of surviving in society

[17]. For example, citizens have to make decision about energy issues (e.g. energy use, saving energy, and pollution) that requires technological ideas and understanding. This aspect public rhetoric about science and technology becomes dominated by dichotomies like 'chances and problem', 'advantages and disadvantages', or 'uses and abuses' [60]. This view generated the relationship between technology and our human will. These include (1) technology support extension of the human body and mind; (2) technological impact on culture or society; and (3) technology depends on values (e.g. ethical and aesthetical values) [17].

Thinking about technology according to its impact on entities stretching outside its essential nature is frequently found: as the impact of technology on the environment and society, but also the impact of human values and needs on technology [15]. Technology as knowledge, the second mode in which technology is manifested, has most frequently been the subject of analytical investigations in the epistemology or theory of knowledge, according to Mitcham (1994, p. 192). An epistemological aspect frequently coming to the fore is the nature of the relationship between the technological knowledge and knowledge of the natural sciences. By taking particular social implications of technology into account, the Science–Technology relationship can be extended to the field of Science, Technology and Society (STS) studies. STS studies are grounded in socio-technological understanding, that is, systematic knowledge of the mutual relationship between technical objects, the natural environment and social practice. Because technology is a key element of STS, it is expected that the philosophy of technology (in particular the 'volition' domain; see the previous paragraph) will have implications for STS studies. The dynamic nature of technology as such leaves its own philosophy in a tentative or flexible state. However, the implications of the philosophy of technology, which is in a development phase at the moment and certain changes in emphasis occur, for STS studies ought to be determined continuously. By taking particular social implications of technology into account, the Science-Technology relationship can be extended to the field of Science, Technology and Society (STS) studies. Based on the philosophy of technology, it could be mentioned that the STS teaching provided students to apply science and other knowledge to solve problems of human needs. This way of thinking is related to engineers [3].

Based on the second author research experiences on STS, he suggested school science teachers provided STS teaching approach in 2006 that consisted of 5 stages including (1) identification of social issues; (2) identification of potential solutions; (3) need for knowledge; (4) decision-making; and (5) socialization stage. Those science teachers, who followed Yuenyong (2006)'s STS approach, had showed some results of enhancing students' practicing knowledge in real world. The using of Yuenyong (2006)'s STS approach for science teaching allowed students using applying scientific and other knowledge for designing the solutions, and provided the context of instruction requires solving a real-world problem or task through teamwork. It also found that those of practicing of Yuenyong (2006)'s STS approach in school enhanced students to design some products, systems, rules, models, or volitions as possible solutions. These studies indicated that students could design methodology in technology which usually used by engineers (Klahan and Yuenyong, 2012). The examples of those studies could be highlighted as following.

Chantaranima and Yuenyong (2014) revealed how they enhance students' analytical thinking to develop some prototypes or products for social issues. On the way of developing product, it revealed that students required knowledge of sound and others for designing. They reported the results of enhancing Grade 11 students' analytical thinking ability in learning about sound unit through STS approach. The sound unit was developed based the five stages of Yuenyong (2006)'s STS approach. The STS sound unit enhanced students' analytical thinking through learning about nature of sound and property from the social context of applying scientific and other knowledge for space mission. Identification of social issues in this unit provided students to identify issues or problems from the situation of "Mission Cronossa". This event was assumed that it is about "a spaceship had fallen down to a planet named 'Leonpha' during a mission to survey the 'Kronessa' planet. There were several astronauts injured and died because they could not contact principal spaceship." Supposing that they were aerospace engineers assigned to design some equipment to help people in the space by NASA, students were challenged. The invented equipment must be able to send some noise in space. This situation was used to stimulate students to identify the problem aspects. Those students must know what the problems in creating the equipment to voice in space are. It was found that most of students could identify the cause of problem. They suggested that sound cannot be transmitted naturally in space because it is a mechanical wave (the wave depends on medium for travelling). In identification of potential solutions, students were working in group to find possible equipment based on their thinking criteria. And, students' planning showed that they will arrange equipment by using external attributes and functions as criteria regardless the principle of inventing and functioning new equipment. Their ideas of inventing could be categorized into three categories including decorations, sound receiver or transmitter, musical instrument, clothes, animal-imitated equipment. The ideas about decorations included rings, watches, belts, dental braces pendant, bangle, and necklace. The ideas about sound receiver or transmitter included walky-talky and Bluetooth. The musical instruments included guitar and space flute. The ideas related to clothes are space helmet and space clothes. The ideas related to animal-imitated equipment is Gumbie 1 (dragonfly). Based on these ideas, students had to find out the appropriate reasons to support their design and invention and to convince friends to accept their own ideas. In need for knowledge stage, students learned about scientific knowledge – sound and property in order to find good reason of choosing the best equipment to develop a model. In decision-making stage, students can designate what they think it is an importance, and among them in the group must make the decision together and concern the followings (1) How can we build the equipment to send some noise in space? (2) What should the equipment look like? And (3) What can the equipment do? And, in socialization stage, students organized an exhibition displaying their model of helping equipment equipped with sound transmitter in space. In exhibition, students from each group had to explain to and discuss with audients attending the exhibition to understand what those students designed and invented, how they did it, and what benefits it can provide?

Another example, Klahan and Yuenyong (2012) reported the implementation of Yuenyong (2006) STS approach. The study enhanced Grade 12 students' technological capabilities in teaching and learning about electromagnetics through science technology and society (STS) approach. Findings revealed that teaching and learning about electromagnetic wave through Yuenyong (2006)'s STS approach supported students' model forming on technological process including designing, acting, and applying. The process of model forming generated students' performances that could be indicated how they listed possible solutions for issues related to electromagnetic wave. Regarding Jones (1997), this process could explain students' capabilities under three ways including (1) student response to the requirements of the technological activity; (2) the total process of students carrying out the technological activity; and (3) influences on that process.

Student response, the student response to the requirements of the technological activity was analysed using significant features including (1) proposing model or task, (2) reasons for choice of model or task, and (3) selected model or task. Data analysis found that students proposed various kinds of model or tasks depending on their interest. These models could be classified into three categories. These include (1) safety such as thief warning signal, weapon scanner, electromagnetic wave absorber; (2) gaining knowledge such as short film, radio program, and village radio; and (3) luxury devices such as wave losing weight set, lightening zip, control car toy for pet. Total process, analysis of total process of 6 student models reflected what their process of developing those models. Total process of students carried out the technological activity that could be categorized into 4 pathways. Influence on that process could be interpreted from applying scientific and other knowledge and skills while students were constructing models. These influences included students' concept of electromagnetic, special knowledge and skills, culture of learning, goal of students' learning, learning resources, and scientific equipment. The influences on that process usually affected the students' pathways and achieving of constructing models [29].

And, some implementations of Yuenyong (2006) STS approach gave students chance to apply their scientific and other knowledge for problem solving in context of entrepreneurship. It revealed that students not only apply science and mathematics knowledge for solving problem of social issues but also values, culture, designing/procedure knowledge, marketing, financial thing, commence, economic, law, and so on. It could be mentioned that the STS unit enhance students to get ideas of linking between their innovative ideas to entrepreneurship that required applying the innovation to bring the ideas to life. Entrepreneurship refers to the process of developing new business ventures, or growing existing ones. Central to this activity are entrepreneurs: innovative and risk-taking individuals who seek to bring about change and new opportunities, both for themselves and for the business communities in which they operate [6].

Ahmad and Seymour (2008) analyzed key definitions of entrepreneurship activity. They categorized three key definitions of entrepreneurship activity. These are: (1) the creation of innovative products and markets through transformation of resources [19]; [49]; (2) the emergence and development of new firms [30]; and (3) the pursuit of opportunity through risk taking and alertness to asymmetric information [28]; [51]. An entrepreneur can work in their environment to try new things, recognize opportunities, and serve new markets. This type of innovative activity, within an established organization, is known as "intrapreneurship" [44]; [48]. Intrapreneurship may result in the formation of a new organization; however, it can also generate new administrative techniques or strategies, newly competitive attitudes, and new products, services or technologies [4]. de Jong and Wennekers (2008) provide a comprehensive listing of intrepreneurial activities and behavioral attributes. The activities include "opportunity perception, idea generation, designing a new product or another recombination of resources, internal coalition building, persuading the management, resource acquisition, planning and organizing" (p. 2).

Chachiyo et.al. (2019) reported their study about developing fermented bamboo shoots STEM education learning unit through Yuenyong (2006) STS approach. It revealed that students provided knowledge about chemical reaction in the process products of fermented bamboo shoots (FBSs). In the FBSs process product, students also took in to account local values and culture, marketing, financial thing, and commence. The identification stage, the unit provided the issues of the OTOP product in Khon Kaen and situation of market need for FBSs. This social issue allowed students to search and think more about how to make FBSs in your context, family or community. Identification of potential solution stage, students worked in team to design/solve the issue how to ferment bamboo shoots and how to design the FBSs packaging. On the way of developing potential solutions, students have analyzed 5 capitals (Natural, Physical, Social/Technology, Human and Financial) to develop lists of possible solutions. Need for knowledge stage, students learned about chemical reaction in mystery box, experiment about chemical reaction, doing simulation through PhET, and doing exercise about chemical formula. Based these activities, students may get some information if the change of the Fermented substances cause a chemical reaction. In decision making stage, it's time to apply students' knowledge to make FBSs by themself, based on knowledge of Chemical Reaction, rate of reaction, and factors of reaction. To design some products and packages, students require considering not only chemical reaction but also culture, market, bamboo supply and species. ISAN culture is another factor of designing because ISAN people accepted some taste of FBSs and some ways of cooking FBSs. Bamboo shoot species and supply should be concerned based on ISAN culture. The market and logistics also may influence the product process. Regarding on many ISAN people stay overseas, the product process may consider how much temperature of freezing for long time of traveling. Students, therefore, came up with various ideas of the FBSs process and products. Socialization stage, students share their ideas in school exhibition where students could redesign for the complete FBSs products.

4. Conceptual framework of context based STEM education teaching approach

It was suggested from the literatures to provide teaching and learning on STEM education. Moore et.al. (2015) clarified five characteristics distinguish integrated STEM instruction from other teacher pedagogy: (a) the content and practices of one or more anchor science and mathematics disciplines define some of the primary learning goals; (b) the integrator is the engineering practices and engineering design of technologies as the context; (c) the engineering design or engineering practices related to relevant technologies requires the use of scientific and mathematical concepts through design justification; (d) the development of 21st century skills is emphasized; and (e) the context of instruction requires solving a real-world problem or task through teamwork. This conceptualization of STEM is grounded in learning research [61]. However, there are number of processes, which have a long history of research and implementation in education, which are particularly appropriate for STEM Education. They include problem based, project based, inquiry, and design [55].

The STEM educations should provide not only the ways of investigation and solving problem but also a real world problem solving. Based on these literatures, this paper will propose the practical idea of STEM education teaching approach which integration key concepts among problem based, project based, inquiry, design, and a real world problem solving as STS. The paper, therefore, propose the theoretical framework of STEM education teaching approach as inquiry from the context based. This paper proposed a STEM education teaching approach that developed ideas through learning as inquiry from the context based. The framework of stages of teaching was adapted from Yuenyong (2006) the 5 stages of STS approach. These include (1) Identification of social issues, (2) Identification of potential solution, (3) Need for knowledge, (4) Decision-making, (5) Socialization stage. The EDP could be integrated in order to develop a framework of STEM education teaching strategy. We, therefore, come up to the 7 stages of STEM (Science, Technology, Engineering, and Mathematics) strategies. In this study focused on teaching approach as inquiry in context. This approach including (1) Identification of social issues, (2) Identification of potential solution, (3) Need for knowledge, (4) Decision-making, (5) Development of prototype or product, (6) Test and evaluation of the solution, and (7) Socialization and completion decision stage.

(1) Identification of social issues stage, in order to enhance students to begin with problem and/or issue from the context. The characteristics of social issues may relate to context local issues or designing social, cultural, and technological products. The issues may include – disasters (e.g. flooding, earthquake, tsunami etc.), pollutions, environmental issues, biotechnology, health, cosmetics, ecosystem, energy or nuclear power plant. Or, the issues related designing may be introduced; for examples, toy designing, playground, precise agriculture farming, medicines, local food products, packaging, or other technological product or system issues. And, the important thing, it makes sure that these issues may allow students to chronologically develop various solutions or designing. To starting social issues, teacher may engage their students by present the pictures and VDO clip-showed about specific subject matter. The pictures, VDO clip and/or advance organizer expected to motivate students to engage students identify what the problems of these social issues. Teachers may provide some questions to probe students to perceive the key ideas of problems. Not only teacher ask students (Q&A activities) to discuss and/or brainstorming but also high expectation engage in inquiry, students can learn about problem/social issue and the setting of the questions by themselves.

(2) Identification of potential solution, this stage enhance students to develop possible solutions based on existing their knowledge, skills, experiences, creativity and so on. Learning activities need to be provided to motivate students to explore/ search, study and research prior-knowledge link to new information to support their ideas about the ways to solve this problem (chances/opportunities to learn more). After identify the problem, students should have chance to plan how to solve the problems or raise the alternatives for the solutions. Scientific and other knowledge may be listed to support the issues. What the need of knowledge may also be clarified by students. The stage requires students to think of what, why, and how ideas, design, system, volition of applying scientific knowledge and others should be worked for social problems. These reflect that students need to be enhanced to construct meaning of technological concept in this stage. Learning activities may motivate students to clarify what they focus on combine mathematics and science knowledge to design prototype and/or product (innovation) for facilitating our life or solving (social issues)/problems) and also another disciplines such as socio-cultural issues link to social-society, culture, economics, art, environment, and so on. Then, students may engage the need of further studying about those related knowledge, particularly science and mathematics. Teachers need also to provide learning activities in order to develop students' ideas about the plausibility of designing. Students, for example, discuss and/or brainstorming about what their 5 capitals (physical capital, financial capital, social capital, human

capital, and natural capital) to support their list of possible solutions. Students need to think of 5W+1H questions (What, Where, When, Who, Why, and How-how to) that depends on each group created by their own designing prototypes and/or products (innovations).

(3) Need for knowledge stage, learning activities should provide related to what kinds of knowledge were raised by students in the identification of potential solution stage. Particularly, students' scientific and mathematical knowledge should be provided as need for decision making on how to do alternatives and the best solutions. Learning activities should be organized regarding on nature of science and mathematics. Scientific and mathematical concepts have been formulated in many ways to help students to understand technology and social issues. The majority of them perceived that STEM education need to be provided as integration knowledge among interdisciplinary STEM in context. It's expected, STEM concepts was formulated through activities of demonstration, centripetal, experiments, and inside and out-side inquiry STEM classroom to search-study, research and prepare themselves to present their all possible solutions, and try to make decision which one is the appropriated solution.

(4) Decision-making, the stage is designed students to apply scientific and mathematical knowledge and others for an appropriated solution. Students should have chance to argue about science and mathematics related to technological and societal issues. Students will list the possible solutions each group of students presented their solutions and then select an appropriated solution regarding their justification. This stage guideline will help students to use all the knowledge and make decision that they got to solve real-life problems. And also sharing their result of searching, studying, researching and decision-making-what and why they choose this one as an appropriated solution. Students involves in making a decision on how to use and/or apply the knowledge and technology. This aspect public rhetoric about their problems/issues related technological and social issue. Students' argumentation becomes dominated by dichotomies like 'chances and problem', 'advantages and disadvantages', or uses and abuses'. Students have to select between alternatives of solution by systematically comparing as many relevant pro's and con's as possible. Then, students' ideas, designing, proposing system, or volition will be drawn or constructed. STEM education, teaching strategies as inquiry in context may be used discussion among students' group, and brainstorming to allow students designing the prototypes and/or products (innovations). On the process of students' designing, they need to be enhanced to develop their possible solutions, select the best solution, and construct a prototypes and/or products together to learn-inquires what and how to be appropriated problem solver in this situation/context.

(5) Development of prototype or product, students' on tasks of designing, will provide models, simulations or others for solving this problem lead to making their innovation regarding on their planning and designing, and presenting their prototypes and/or products. Teachers need to provide activities regarding on engineering design as design-based learning. STEM activity requires students who could integrate and support creativity within activity. Teacher guided, supported, and suggested each group of students to provide learning activities regarding on engineering process design (EPD). The conceptual framework of the EPD allowed them to develop ideas to enhance students solving problem and applying STEM disciplines and relate to other knowledge through problem based learning, socio-scientific issues, and STS approach. The learning activities of enhancing students to develop prototypes or products need to support students to linking between two types of knowledge in technology including conceptual knowledge (''knowing that'') and procedural knowledge (''knowing how''). Conceptual knowledge relevant to technology therefore includes justification from other subjects such as science and mathematics. Procedural knowledge referred to as tacit, personal or implicit knowledge. Design, modelling, problem solving, systems approaches, project planning, quality assurance and optimisation are all candidates for technological procedural knowledge.

(6) Test and evaluation of the solution, students have to provide the methods, materials/equipment appropriate for proofing their prototypes and/or products. And also students will be asked to develop their ideas what and how to test and evaluation of the solution. Concept of the test and evaluation should be provided concerning on science and mathematics knowledge, human needs, business and

industrial practices. The learning activities may remind students to provide the test and evaluation to make sure about "Is it working?" planning and designing, and prepare good presenting about what and how your prototypes and/or products its' effective? Students may apply science and mathematics for doing experiments, providing mathematical models and simulations. The hands-on activities and/or experiment to convince teacher and their friends should not only provide inside but also out-side-classroom/school as inquiry in context. They may provide some story such as short films or storytelling to test or proof for human needs, business and industrial practices.

(7) Socialization and completion decision stage, as inquiry the end with a presentation, students presented their process of their planning, designing and making innovations, they have to share their prototypes and/or products (innovations). The stage is designed to allow students to validate their values and STEM knowledge for solutions during their sharing in society. Students reflected what aspects of STEM knowledge were applied on their designing and from your best, the possible ways to make it happen and can used in real-life. Teacher assessed students' creative thinking on students' planning, designing and making innovation. Present ideas to design and build, present how to test and evaluate of the solution. And, also present how the prototypes and/or products (innovations) could be explained as the real one. Students need to do something as people who are a part of society by reporting their proposal for solving problem. Students might exhibit their solution in public. Or, students may communicate their products, science projects, ideas, designing, proposing system, or volition to the head of community, or share to social media and newspaper. Then, students need to report what they learned from the validation or sharing their proposal or product through socialization process. Then, students share what they learn from the comment and/or feedback. Then, students need to be asked to write what they learn from those reflections which they may get some more ideas to re-design and completion decision.

5. Conclusion

According to literature review about STEM education, aspect of practicing knowledge in real world through STS approach and concept of technology and engineering, the paper have proposed STEM education pedagogy. The paper, then, proposed the theoretical framework of STEM education teaching approach as inquiry from the context based. The teaching approach consists of 7 stages. These include (1) Identification of social issues, (2) Identification of potential solution, (3) Need for knowledge, (4) Decision-making, (5) Development of prototype or product, (6) Test and evaluate the Solution, and (7) Socialization and completion decision stage.

Regarding on the 7 stages of context based STEM education teaching approach; it may provide a referent of STEM education teaching in school setting and STEM project teaching. Teachers may perceive concrete ways of getting start the STEM education learning activities when they are thinking about the identification of social issues. Then, the activities could be started from social issues and/or human needs such as disasters, pollutions, environmental issues, biotechnology, health, cosmetics, ecosystem, energy, market, commerce, and designing some technological products. The 7 stages will allow students using applying scientific and other knowledge for designing the solutions, and provided the context of instruction requires solving a real-world problem or task through teamwork. And, students will have also chance to apply their scientific and other knowledge for problem solving in context of entrepreneurship.

And, because the stages of teaching developed regarding on framework of STS and EPD, the learning activities could balance problem solving based on concept of technology and engineering. The stages of teaching also keep four modes of the manifestation of technology to design learning activities. Taking technology as object, therefore, reminds students to provide artifacts which attribute to a dual nature – physical and functional nature. Considering conceptual and procedural knowledge of technology give chance students to learn related knowledge and thinking and practicing about knowing how. As provided activities regarding on technological procedural knowledge (tacit, personal or implicit knowledge); students will perceive about design, modelling, problem solving, systems approaches, project planning, quality assurance and optimisation during developing prototypes or products. Taking technology as activity supports students to learn about activity of designing, making, and using or evaluating. View of technology as volition will enhance students to generate the relationship between technology and our human will. These include (1) technology support extension of the human body and mind; (2) technological impact on culture or society; and (3) technology depends on values. As mentioned above, the 7 stages of context based STEM education teaching approach may suggest teachers to design STEM education learning activities in school setting.

References

[1] Ahmad, N., and Seymour, R. (2008). Defining entrepreneurial activity: Definitions supporting frameworks for data collection. *OECD statistics working papers*, 2008/01. Paris: OECD Publishing. https://doi. org/10.1787/243164686763.

[2] American Society for Engineering Education (ASEE). (2011). *Preparing future engineers around the world*. Prism, 21(5), 26–34.

[3] Ankiewicz, P. and de Swardt, E. (2006). Some Implications of the Philosophy of Technology for Science, Technology and Society (STS) Studies. *International Journal of Technology and Design Education*, 16, 117–141

[4] Antoncic, B., & Hisrich, R. D. (2001). Intrapreneurship: Construct refinement and cross-cultural validation. *Journal of Business Venturing*, 16(5), 495–527.

[5] Association for Science Education (ASE) (2010). STEM resources – SATIS Revisited. Retrieved from: https://www.stem.org.uk/resources/collection/2885/satis-revisited

[6] Aziz, A.A.A. and Rowland, S. (2018). The entrepreneurship skills that biotechnology graduates need: findings from entrepreneurial employees in a developing economy. *Entrepreneurship Education*, 1: 61–83.

[7] Barlex, D. (2009, August). The STEM Programme in England—Help or hindrance for design and technology education? Paper presented at the PATT22 conference, Delft, Netherlands. Retrieved from http://www.iteaconnect.org/Conference/PATT/PATT22/Barlex.pdf.

[8] Bell, D. (2012). Perceptions of STEM, within the context of teaching D&T in secondary schools: A phenomenographically inspired study. In T. Ginner, J. Hallstro¨m, & M. Hulte´n (Eds.), *Technology education in the 21st century* (pp. 74–79). Linko¨ping: Linko¨ping Electronic Conference Proceedings.

[9] Breiner J.M., Harkness, S.S., Johnson, C.C., Koehler, C.M., 2012. What is STEM? A discussion about conceptions of STEM in education and partnerships. *Sch Sci Math*, 112(1):3–11.

[10] Bybee, R. W. (2013). *The case for STEM education: Challenges and opportunities*. Arlington: NSTA Press.

[11] Chachiyo, T. Sutaphan, S. and Yuenyong, C. (2019, in press). Developing Fermented Bamboo Shoots STEM Education Learning Unit through STS approach. *Journal for the Education of Gifted Young Scientists*,

[12] Chamrat, S., Manokarn, M., and Thammaprateep, J. (2019). STEM Literacy Questionnaire as an Instrument for STEM Education Research Field: Development, Implementation and Utility. *AIP Proceeding*, 2081: 030013-1 - 030013-11

[13] Chantaranima, T. and Yuenyong, C. (2014). The Outcomes of Teaching and Learning About Sound Based on Science Technology and Society (STS) Approach. *Procedia - Social and Behavioral Sciences*, 116 (2014): 2286 – 2292

[14] Chevron Enjoy Science project (2016). The 21st century skill. Retrieved from: https://www.enjoy-science.org/en

[15] Custer, R. L.: 1995, 'Examining the Dimensions of Technology', *International Journal of Technology and Design Education*, 5(3), 219–244.

[16] de Jong, J., and Wennekers, S. (2008). Intrapreneurship. Conceptualising entrepreneurial employee behavior. Zoetermeer: SCALES (Scientific Analysis of Entrepreneurship and SMEs).

[17] de Vries, M.C. (2007) Relations between science education and technology education: what can we learn from the philosophy of technology? Document for lecturing at Science Education program, Khon Kaen University, during 14 - 15 September 2007.
[18] Department of Education and Skills. (2006). The science, technology, engineering and mathematics (STEM) programme report. Retrieved from http://www.nationalstemcentre.org.uk/res/documents/page/050110 114146stem_programme_report_2006.pdf.
[19] Drucker, P. F. (1985). The discipline of innovation. *Harvard Business Review*, 63(3), 67–72
[20] English, L.D. (2016). STEM education K-12: perspectives on integration. *International Journal of STEM Education*, 3:3, 1-8
[21] Fraenkel, J. R., Wallen, N. E., and Hyun, H. H. (2012). *How to design and evaluate research in education* (8th ed.). New York: McGraw-Hill. France: Troubled economic outlook heightens the need for multi-stakeholder collaboration in STEM education.
[22] Fraser, B. J., Tobin, K., & McRobbie, C. J. (Eds.). (2012). *Second international handbook on science education*. New York: Springer.
[23] Herschbach, D. R. (2011). The STEM initiative: Constraints and challenges. *Journal of STEM Teacher Education*, 48(1), 96–122.
[24] Institute for promoting science and technology teaching (IPST). (2012). Technological Process. Retrieved from www.designtechnology.ipst.ac.th/index.php/?option=com _content&view= article&id=69&Itemid=108.
[25] Institute for promoting science and technology teaching (IPST). (2018). Science Content Standard. Bangkok: Thailand
[26] Kim, J. S. (2011, November). New national curriculum of technology education in Korea. Paper presented at the 2011 international conference on technology education in the Pacific-Rim Countries, Nagoya, Japan.
[27] Krug, D. H. (2012). STEM education and sustainability in Canada and the United States. Paper presented at the 2nd international STEM in education conference, Beijing, China. Retrieved from http://stem2012.bnu.edu.cn/data/long%20paper/stem2012_87.pdf.
[28] Kirzner, I. M. (1973). *Competition and entrepreneurship*. Chicago: University of Chicago Press.
[29] Klahan, T. and Yuenyong, C. (2012). An analysis of grade 12 students' technological capability in learning about electromagnetics through science technology and society approach (STS approach). *Procedia - Social and Behavioral Sciences*. 46: 5085 – 509
[30] Lumpkin, G. T., & Dess, G. G. (1996). Clarifying the entrepreneurial orientation construct and linking it to performance. *Academy of Management Review*, 21(11), 135–172.
[31] Marginson, S., Tytler, R., Freeman, B. and Roberts, K. (2013), STEM: country comparisons: international comparisons of science, technology, engineering and mathematics (STEM) education. Final report. Australian Council of Learned Academies, Melbourne, Vic.
[32] Mitcham, C.: 1994, *Thinking Through Technology*, The University of Chicago Press, Chicago.
[33] The Ministry of Education Thailand (2015). Handbook of STEM education smart trainer team. Bangkok:
[34] McCormick, R.: 1997, 'Conceptual and Procedural Knowledge', *International Journal of Technology and Design Education*, 7(1–2), 141–159.
[35] Moon, B., Brown, S., & Ben-Peretz, M. (Eds.). (2000). *Routledge international companion to education*. London: Taylor and Francis Books.
[36] Moore T.J., Johnson C.C., Peters-Burton E.E., Guzey, S.S., (2015). The need for a STEM road map. In: Johnson CC, Peters-Burton EE, Moore TJ (eds) *STEM road map: a framework for integrated STEM education*. Routledge, p 1.
[37] Moore, T. J., Tank, K. M., Glancy, A. W., & Kersten, J. A. (2013, January). A framework for implementing engineering standards in K-12. Paper to be presented at the international conference of Association for Science Teacher Education (ASTE), Charleston, SC.

[38] Morrison, J., & Bartlett, R. (2009). STEM as curriculum: An experiential approach. *Education Week*, 23, 28– 31.
[39] National Academy for Educational Research. (2014). Grade 1– 12 curriculum guidelines master Plan (draft). Retrieved from http://www.naer.edu.tw/ezfiles/ 0/1000/attach/19/pta_2670_4235131_77332.pdf.
[40] National Research Council (NRC) (2011). *Successful K-12 STEM education: Identifying effective approaches in science, technology, engineering, and mathematics*. Washington, DC: National Academies Press. doi:10.17226/13158
[41] National Science and Technology Council. (2013). *A report from the committee on STEM education*. Washington, D.C: National Science and Technology Council.
[42] NGSS Lead States. (2013). Next generation science standards: For states, by states. Washington, DC: National Academies Press.
[43] Office of the Chief Scientist. (2013). Science, technology, engineering and mathematics in the national interest: A strategic approach. Retrieved from http://www.chiefscientist.gov.au/wp-content/uploads/ STEMstrategy290713FINALweb.pdf.
[44] Pinchot, G. I. (1985). *Intrapreneuring: Why you don't have to leave the corporation to become an entrepreneur*. New York: Harper and Row.
[45] Ritz, J. M. (2011, November). Technology education in U.S.: Report of progress. Paper presented at the 2011 international conference on technology education in the Pacific-Rim Countries, Nagoya, Japan.
[46] Ritz, J. M. and Fan, S. C. (2015). STEM and technology education: international state-of-the-art. *Int J Technol Des Educ* (2015) 25:429–451
[47] Roberts, A., & Cantu, D. (2012, June). Applying STEM instructional strategies to design and technology curriculum. In T. Ginner, J. Hallstro¨m, & M. Hulte´n (Eds.), Technology education in the 21st century (pp. 111–118). Linko¨ping: Linko¨ping Electronic Conference Proceedings.
[48] Sanders, M. (2009). STEM, STEM education STEMmania. *The Technology Teacher*, 68(4), 20–26.
[49] Schumpeter, J. (1911/1934). The theory of economic development: An inquiry into profits, capital, credit, interest, and the business cycle. New Brunswick (USA) & London (UK): Transaction Publishers.
[50] Shahali, E.H.M, Halim, L., Rasul, M.S., Osman, K., and Zulkifeli, M. A. (2017). STEM Learning through Engineering Design: Impact on Middle Secondary Students' Interest towards STEM. EURASIA *Journal of Mathematics Science and Technology Education*. 13(5):1189-1211
[51] Shane, S., & Venkataraman, S. (2000). The promise of entrepreneurship as a feld of research. *Academy of Management Review*, 25(1), 217–226.
[52] Sheppard, S. D, Macantangay, K., Colby, A. & Sullivan, W. M. (2009). *Educating engineers: Designing for the future of the field*. San Francisco, CA: Jossey-Bass.
[53] Wang, H., Moore, T., Roehrig, G., & Park, M.S. (2011). STEM Integration: Teacher perceptions and practice. *Journal of Pre-College Engineering Education Research*, 1(2), 1-13.
[54] Williams, P. J. (2011). STEM education: Proceed with caution. *Design and Technology Education*, 16(1), 26–35.
[55] Williams, P.J. (2019). The principles of teaching and learning in STEM education. *AIP Proceeding*, 2081: 020001-1 - 020001-7.
[56] Yasin, R. M., Halim, L., and Ishar, A. (2012). Effects of problem-solving strategies in the teaching and learning of engineering drawing subject. *Asian Social Science*, 8(16), 65.
[57] Yu, N. (2019). The K-12 Academic Track Series: STEM. Edukasyon.ph BLOG. Retrieved from: https://blog.edukasyon.ph/senior-high/the-k-12-academic-track-series-stem/
[58] Yuenyong, C. (2006). Teaching and Learning about Energy: Using STS Approach. Bangkok, Thailand: Thesis of Doctoral Degree in Science Education, Kasetsat University.
[59] Yuenyong, C. (2013). Enhancing Scientific Literacy in Thailand. *Global Studies of Childhood*, 3(1): 86 – 98

[60] Yuenyong, C. (2017). Enhancing Thai Students' Thinking Skills about Energy issues: Influence of Local Values. Chemistry: *Bulgarian Journal of Science Education*, 26 (3): 363 – 376
[61] Yuenyong, C. (2019). Lesson learned of building up community of practice for STEM education in Thailand. *AIP Proceeding*, 2081: 020002-1- 020001-6.
[62] Yuenyong, C. and Narjaikaew, P. (2009) Scientific Literacy and Thailand Science Education, *International Journal of Environmental and Science Education*, 4(3), 335-349.

IOP Conf. Series: Journal of Physics: Conf. Series **1340** (2019) 012004 doi:10.1088/1742-6596/1340/1/012004

STEM - Play, Learn and Work: STEM Education in Academic Club Based on Community and Local Products to Improve a Positive Attitude towards STEM

Punyaphat Chaiwongsa[1], Nittayaporn Kinboon[2] and Nessara Yanasarn[3]

[1,2] Science and Technology Department, Phuwiengwittayakom School, Khon Kaen, Thailand

[3] Mathematics Department, Phuwiengwittayakom School, Khon Kaen, Thailand

E-mail : semistar7@hotmail.com, kroonitta@gmail.com, sara11104@hotmail.com

Abstract. This paper developed a practical learning activity named "STEM - Play, Learn and Work" to increases students' interest and engagement in the learning of STEM (Science, Technology, Engineering, and Mathematics). Because we need to show that STEM knowledge and skills can solve everyday problems and promote community and local products, one popular product in our local community is selected as a representative to be the main theme of learning. We designed the project-based learning activity combined various STEM subjects through steps of production for the representative from the beginning to the end. We used an experimental "STEM - Play, Learn and Work" to 19 students in grade ninth – twelfth in class of academic club. The learning outcomes including student performance and attitude towards STEM are discussed. The assessment results (one assessment taken before the activity, one after) show that the activity improved students' knowledge in STEM subjects significantly. It also increased students' positive attitude towards STEM and interest in STEM careers. Students have the opportunity to apply their meaningful STEM knowledge and skills to solve problems in everyday life through this practical activity.

1. Introduction

STEM Education has been in the mainstream of the global learning in the 21st century [3]. It emphasizes on combining different subject areas of science, technology, engineering, and mathematics in a way that integrates them together. Such subjects especially science and mathematics, students cannot link them to apply in real-life problems. This brings to declined Thailand national students' achievement and negative attitude on the subjects in STEM [6]. Today as never before, in addition to academic subject areas, 21st century interdisciplinary themes are equally important in promoting understanding of academic content. These themes include entrepreneurial literacy, world education systems are failing to adequately prepare all student to succeed life because they are lack of essential skills. One important skill in the 21st century is entrepreneurial literacy [3]. In this paper, we present an example of activity that increases our students' interest and engagement in the learning of STEM through improving their familiar community and local products that can be accessible easily in their everyday life. As you know, most of schools in Khon Kaen province are located near sources of water for irrigation such as rivers, reservoirs, and groundwater. Ubolratana Dam is the most important one so that most of students' families are agricultural or inland fishery families. Fish is one of the main products for this area. By traditional experiences for many years, their family made the popular pickled fish to sell around their community. The process of pickled fish production is complex but interesting. This process can be applied in learning STEM and be explained in STEM knowledge so students get practice themselves in the subject of STEM through this attractive activity and it lets them recognized about their homeland.

2. Related work

Wong and Huen [7] presents a conceptual model to enhance the connection between the knowledge that students learn in school and solving real life problems. They suggests implementing STEM-related activities into the existing education curriculum instead of rewriting the whole courses. Knipprath et al. [4] and Hayden et al. [2] proposed STEM education which aims to foster STEM literacy and a positive attitude towards STEM. The project findings show that students experienced increased interest and attitudes toward STEM. Hauze and French [1] examined the effect on applied STEM education through guitar building.

Mahoney [5] studied students' attitude toward STEM. The questionnaire was developed for collected students' attitude. We will adapt it to use in our research.

3. Methodology

3.1. Purpose

The main goal of the activity was to improve STEM knowledge and a positive attitude towards STEM using integrating STEM subjects to real world problems by STEM – Play, Learn and Work model. The activity was designed based on 21st century skills to students who need challenge themselves in academic club class.

3.2. Learning theme and STEM concept design

According to our research objectives, we will develop STEM activity that help to furnish students with STEM related skills and a positive attitude towards STEM so that learning theme was considered by our STEM teaching team using the basis of promoting local community products so that one of community products was selected as theme's representative. Then the team will provide the fundamental STEM subject concepts related to the selected product and the concepts will be assigned to be responsible by each domain expert in our team. We plan to teach each concept followed by lesson plans consequently by domain experts.

Table 1. Related STEM concepts for learning theme of pickled fish.

	Concepts	Contents
1.Science	Food science	Acid-base, Food preservation
2.Technology	Computer graphics	Label and logo's brand design using Adobe Photoshop
3.Engineering	Packaging design	Packaging design
4.Mathematics	Proportion, linear function, volume	Ingredients proportion, Volume calculator for packaging, proportion prediction

3.3. Activity design and implementation

After we selected learning theme and the concept of STEM subjects, we move to focus on activity design. We need to know how to link STEM subject concepts and our selected local community product in class. We divide this activity into 3 main steps (Play, Learn and Work). That we will explain in detail as followed.

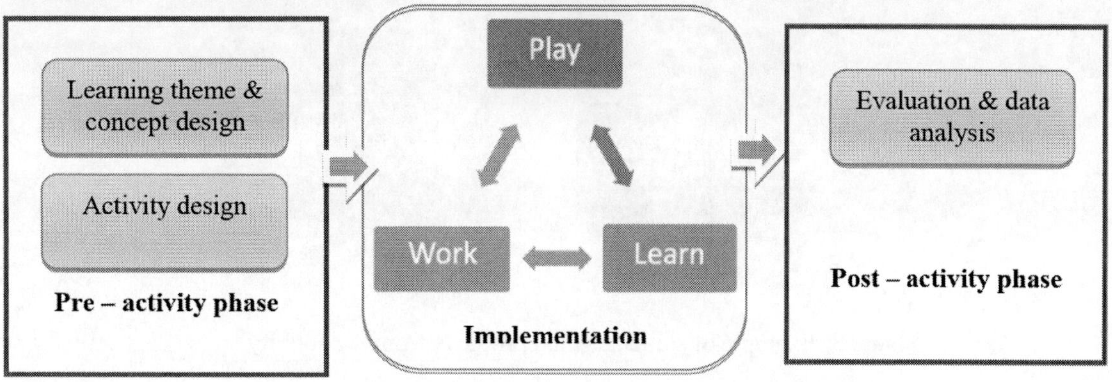

Figure 1. Proposed STEM – Play, Learn and Work model

3.3.1. Play stage. In this stage, teacher create interest and generate curiosity in the topic of study. For this study selected product (picked fish) is the key so that students should be asking questions about STEM subject concepts that are relevant to the product and production process of the product. This activity help students to make connections with the previous knowledge.

3.3.2. Learn stage. In this study, Learn stage was divided into 2 parts. During the beginning part of "Learn" stage students should be given opportunities to work together without direct instruction from the teacher. In the end part, teachers help students encourage them to explain concepts of STEM subjects in their own word. Teachers use prepared materials to improve students' understanding.

Figure 2. Some activity in class based on STEM concepts of technology and mathematics

3.3.3. Work stage. They had the chance to apply that knowledge on this stage. Our project-based activity, we inspire them from planning to finishing their output. In this case they need to prepare ingredients to do pickled fish following step by step that they learned from Learn stage. Work stage apply here as well because students should be using the previous information to ask questions, propose solutions, and make decisions and experiment.

Figure 3. Example of students' final products from the activity

3.4. Evaluation method

Student achievement of the 4 core STEM concepts was measured through the deployment of pre- and post-assessments evaluating student knowledge across the 4 core concepts. The data was analysed using descriptive statistics such as mean score, percentage and standard derivation. The overall achievement value of STEM concepts is computed by (1):

$$S_{STEM} = \alpha S_C + \beta T_C + \gamma E_C + \delta M_C \qquad (1)$$

where S_{STEM} is the percentage of overall achievement for STEM activity as well as S_C, T_C, E_C and M_C are the percentages of achievement for concepts in science, technology, engineering and mathematics respectively. This α, β, γ and δ are also the coefficients of S_C, T_C, E_C and M_C respectively. The STEM teaching team can also define weighted scores α, β, γ and δ to denote relative importance of the subjects; α, β, γ and δ are in [0, 100] and $\alpha + \beta + \gamma + \delta = 100$.

In addition, students' attitude towards STEM was gathered by giving all of participant students answered the questionnaire which questions expressed in a Likert scale from 1 to 5. We collected the information before activity and after activity in three aspects of attitudes i.e. interesting, ability and values. The data was also analysed using descriptive statistics such as mean score, percentage and standard derivation.

4. Results

At this point achievements and attitude toward STEM were observed for the four content areas of STEM. Student achievement percentages of mean on STEM all subject concept that taken before the activity is less than the one after as shown in Table 2. This show that student achievement of STEM concepts was improved slightly along the activity. The results of students' achievement taken before the activity for the content areas of science, technology, and engineering are 22.63%, 27.16%, 24.21% and 19.74% respectively and the after one are 70.18%, 73.68%, 68.68% and 56.58% respectively. Among the four content areas, student mean score of technology was consistently the highest in both taken before and after activity. The differences between taken before and after activity were 47.55%, 46.52%, 44.74% and 36.84. The most change of achievement was on the content areas of science. It reached 47.55%. This can be concluded that the activity made understanding of the content of science the most. However, the final overall achievement was calculated by equation [3]. We weighted the coefficients α, β, γ and δ are 25 because the contents of STEM subject are equally important, so that the overall achievements of STEM before and after activity were 23.44% and 67.35%, then the difference was 43.91%.

Table 2. Mean scores and standard deviations for the achievement of STEM.

Aspects	Contents area							
	Science		Technology		Engineering		Mathematics	
	M (%)	SD	M (%)	SD	M (%)	SD	M (%)	SD
Pre-assessment	22.63	3.39	27.16	3.35	24.21	3.39	19.74	3.39
Post-assessment	70.18	3.81	73.68	2.24	68.95	2.04	56.58	3.33
Differences	+47.55	0.42	+46.52	-1.11	+44.74	-1.35	+36.84	-0.06

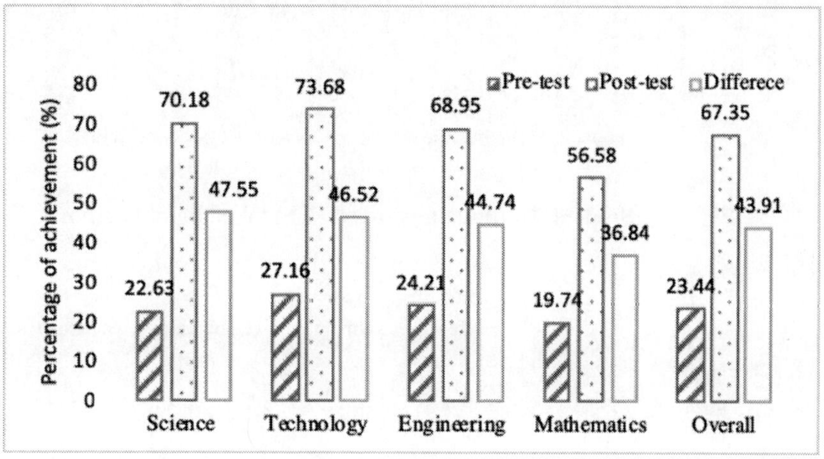

Figure 4. Comparison of Pre-Post-test student achievement on STEM subjects

We also studied the attitude towards STEM using the questionnaire which questions expressed in a Likert scale from 1 to 5. Students' answers to the questionnaire, shown in Table 3, attest that they appreciated the activity. The attitude result taken before activity for interest, ability and value aspect displayed the highest score in technology subject. The post-assessment for attitude shown the highest in technology subject for interest and ability, but the attitude on value aspect the highest score was in mathematics subject. In particular, the students' attitude on value in STEM has been appreciated the most.

Table 3. Mean scores and standard deviations for the attitude towards STEM.

Aspects	Contents area							
	Science		Technology		Engineering		Mathematics	
	M (%)	SD	M (%)	SD	M (%)	SD	M (%)	SD
Pre-assessment:								
interest	45.61	1.01	49.65	0.56	39.82	0.72	38.25	1.03
ability	32.11	0.41	36.49	0.67	32.28	0.76	29.47	0.99
value	57.37	0.51	59.47	0.99	57.19	0.89	55.96	0.40
Post-assessment:								
interest	78.60	0.33	83.33	0.99	62.11	0.85	67.54	0.94
ability	72.11	0.73	74.91	0.76	64.91	0.75	64.04	0.71
value	84.21	0.62	80.53	0.80	77.19	0.91	85.79	0.63

IOP Conf. Series: Journal of Physics: Conf. Series **1340** (2019) 012004 doi:10.1088/1742-6596/1340/1/012004

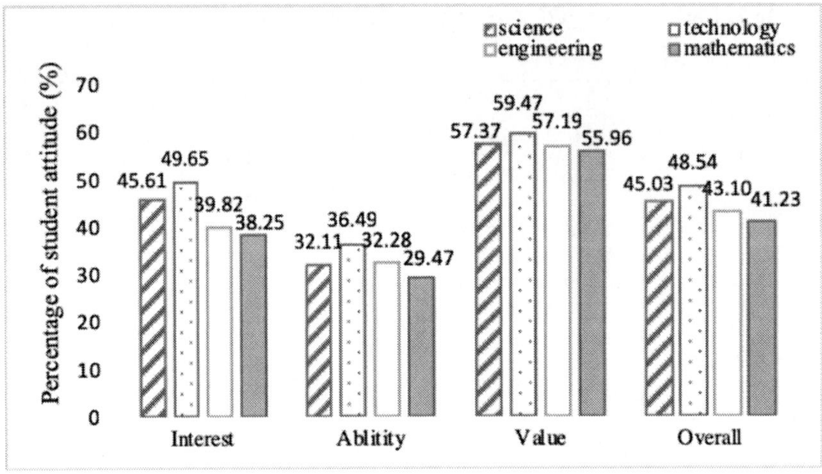

Figure 5. Students' attitude towards STEM taken before activity

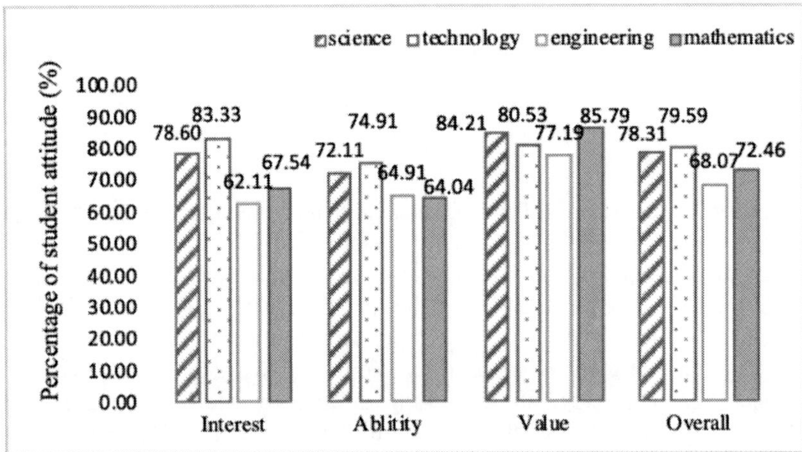

Figure 6. Students' attitude towards STEM taken after activity

The difference before and after activity's attitude on ability aspects for science was the most greatest. The second difference was 38.42% for ability aspect in technology. The lowest difference of the attitude on value aspect was 20.00% in engineering subject. This information of attitude scores told us about the how activity change their attitude towards each STEM subjects so that we can argue that the proposed activity increased students' attitude on every aspects.

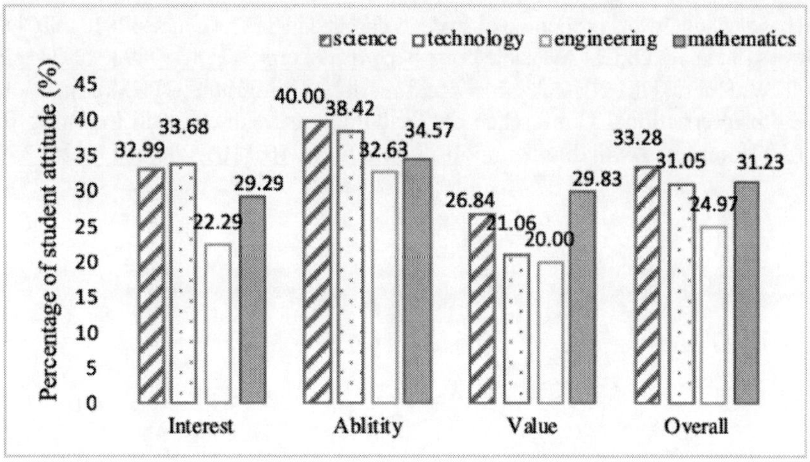

Figure 7. Differences of students' attitude towards STEM between taken before- after activity

5. Conclusions and future work

This paper presents a developed activity based on STEM subjects. To improve both STEM subjects and local community knowledge, we selected an impact product that be the theme of learning. To analyze the performance of our activities, we will conduct an empirical study among 19 students grade ninth – twelfth. The proposed activity increased students' positive attitude towards STEM and interest in STEM careers according to the research in [ ], [ ] and [ ]. Students have the opportunity to apply their meaningful STEM knowledge and skills to solve problems in everyday life through this practical activity.

We also expect to continue the activity for other products to expand knowledge on STEM education and local community. In addition, we plan to try this activity on new group of students in other nearby schools, and also improve evaluation method by using multiple experts. Finally we prefer to form STEM professional learning community in our community.

6. References

[1] Hauze S and French D 2018 Predictors of success in applied STEM education through Guitar building. 2018 IEEE Integrated STEM Education Conference (ISEC) [Internet] Mar Available from: http://dx.doi.org/10.1109/isecon.2018.8340506

[2] Hayden K, Ouyang Y, Scinski L, Olszewski B and Bielefeldt T 2011 Increasing student interest and attitudes in STEM: Professional development and activities to engage and inspire learners **Contemporary Issues in Technology and Teacher Education 11** Retrieved from http://www.citejournal.org/volume-11/issue-1-11/science/increasing-student-interest-and-attitudes-in-stem-professional-development-and-activities-to-engage-and-inspire-learners

[3] Kay K and Greenhill V 2010 Twenty-First Century Students Need 21st Century Skills. Bringing Schools into the 21st Century [Internet] **Springer Netherlands 27** pp 41–65 Available from: http://dx.doi.org/10.1007/978-94-007-0268-4_3

[4] Knipprath H, Thibaut L, Buyse M-P, Ceuppens S, Loof HD and Meester JD 2018 STEM Education in Flanders: How STEM@school Aims to Foster STEM Literacy and a Positive Attitude towards **STEM IEEE Instrumentation & Measurement Magazine [Internet] 21** 36–40 Available from: http://dx.doi.org/10.1109/mim.2018.8360917

[5] Mahoney MP 2010 Students' Attitudes Toward STEM: Development of an Instrument for High School STEM-Based Programs. **The Journal of Technology Studies [Internet] Virginia Tech Libraries 36** Available from: http://dx.doi.org/10.21061/jots.v36i1.a.4

[6] Thailand has a relatively low proportion of students enrolled in STEM courses 2018 OECD Development Pathways [Internet]. OECD Available from: http://dx.doi.org/10.1787/9789264293311-graph30-en
[7] Wong GKW, Huen JHM 2017 A conceptual model of integrated STEM education in K-12 2017 IEEE 6th International Conference on Teaching, Assessment, and Learning for Engineering (TALE) [Internet] Available from: http://dx.doi.org/10.1109/tale.2017.8252350

IOP Conf. Series: Journal of Physics: Conf. Series **1340** (2019) 012005 doi:10.1088/1742-6596/1340/1/012005

The design of STEM Education lessons on statistical mechanics for undergraduate students

Udom Tipparach[1]*, Makabodee Ruaysap[3], Jutamanee Tipparach[3], and Krittidej Chanthawara[4]

[1]Department of Physics, Faculty of Science, Ubon Ratchathani University, Ubon Ratchathani 34190, Thailand

[2]Chemistry Program, Faculty of Science, Ubon Ratchathani Rajabhat University, Ubon Ratchathani 34000, Thailand

[3]Business English Program, Faculty of Humanities and Social Sciences, Ubon Ratchathani Rajabhat University, Ubon Ratchathani 34000, Thailand

[4]Mathematics Program, Faculty of Science, Ubon Ratchathani Rajabhat University, Ubon Ratchathani 34000, Thailand

*Corresponding author's e-mail address: udom.t@ubu.ac.th

Abstract. Lessons combining of the integration of Science, Technology, Engineering and Mathematics or STEM Education were designed and used to enhance learning activities in a Statistical Mechanics classroom for undergraduate students at Ubon Ratchathani University, Thailand. Three STEM Education activities were introduced into the classroom including, solar drying project, magnetic induction, and heat radiation. The STEM Education activities will be described in details. The students were introduced to STEM Education and problem-based project learning. The students were trained to solve real problems by concentrating on engineering problems and then thoughtfully guided by instructor's questions to apply all subject matter of Statistical Mechanics (Science) and techniques and student's skills of Technology, Engineering and Mathematics. The students were able to gain and build up their own knowledge and skills necessary for 21st century. The students were impressed by the STEM Education activities. The students' satisfactory toward the STEM Education activities was in the highest level.

1. Introduction

Science is driven by human curiosity while engineering is driven by human needs and problem solving criteria. Children have natural curiosity about the world around them especially science. They enjoy doing experiments in science and exploring new things. They may want to observe or measure things around them. Such practical activities are motivating and engaging, especially if there are strong collaborative elements. Working with a classmate ensures that there is someone to share ideas, acquire knowledge, and also promote social development as well as emotional growth. Students' curiosity is fueled when they find out enough to know that there are still questions to ask and things to investigate.

However, many teachers and professors in schools and universities have experienced with a large number of students in introductory classes. The courses are rich in contents. Then, these courses often do not engage students in active, authentic scientific investigation, nor do they adequately concentrate on problem-solving process and inquiry skills required to further development. It has long been misunderstood that Physics has an image of being both difficult and boring [1, 2].

Decreasing motivation and competence in Science especially Physics studies among students at different levels has also been an issue broadly discussed by educators, researchers, and politicians. Some researchers found that one of the big problems was laboratory scarcity [3]. Our research shows we can use simple tools to do simple experiments and integrate subject matter in Physics and Mathematics so that the students may learn more contents [4]. We also show that these experiments may be used for the integration of science, technology, engineering, and mathematics (STEM) education. Then, instructors spend most of their time to lecture. As a result, these courses often do not engage students in active, authentic scientific investigation, nor do they adequately concentrate on problem-solving process and inquiry skills required to further development. Students may get concepts what they have learned, but they lack of skills necessary to survive in 21th century. Decreasing motivation and competence in academic proficiency and skills necessary to survive in 21th century among students at different levels especially college students are issues broadly discussed by educators, researchers, and politicians. The survival skills necessary to in 21th century include 6 areas: problem solving, creativity skills, self-help, rational thinking, technology literacy, and social skills.

Data analysis showed that the graduates of Ubon Ratchathani Rajabhat University (UBUR) and Ubon Ratchathani University (UBU) could not get jobs in their majors. The students could not link their knowledge they have learned to new knowledge what they are learning and could not build new knowledge based on background they have learned [5]. Therefore, in this work, we first intended to improve the teachers to develop teaching and coaching skills and then to improve teaching and learning methods by using the integration of science, technology, engineering, and mathematics (STEM) education into classrooms.

The purposes of this study were to develop learning management forms for developing STEM based instruction for higher education instructors on Statistical Mechanics for undergraduate students.

2. Theoretical Background

Faraday's law states that [6] the induced electro motive force, emf (ε), in a circuit is proportional to the rate of change of magnetic flux, ϕ_B, through any surface bounded by that circuit. emf (ε) can be expressed in the following equation,

$$\varepsilon = -N\frac{d\phi_B}{dt},$$

where N is the number of turn of the coil. Faraday's law experiment is set up as shown in Fig. 1.

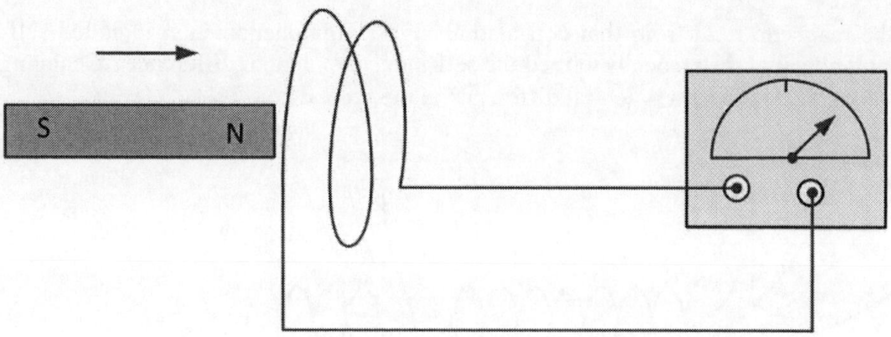

(a) When a magnet is moved near a closed circuit, current flows in the circuit.

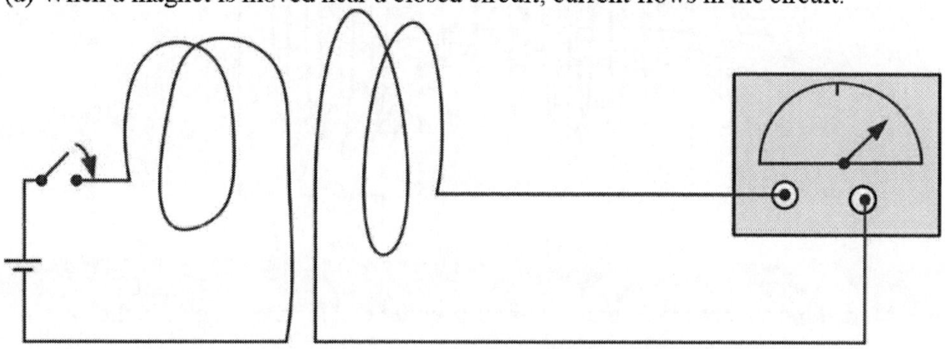

(b) When a current in an adjacent circuit changes, current flows (induced) in the circuit.

Figure 1. The illustration diagram of Faraday's law and an induced current.

Lenz's law states that [7] the induced current in a loop is in the direction that creates a magnetic field that opposes the change in magnetic flux through the area enclosed by the loop. Lenz's law experiment can be done as follows:

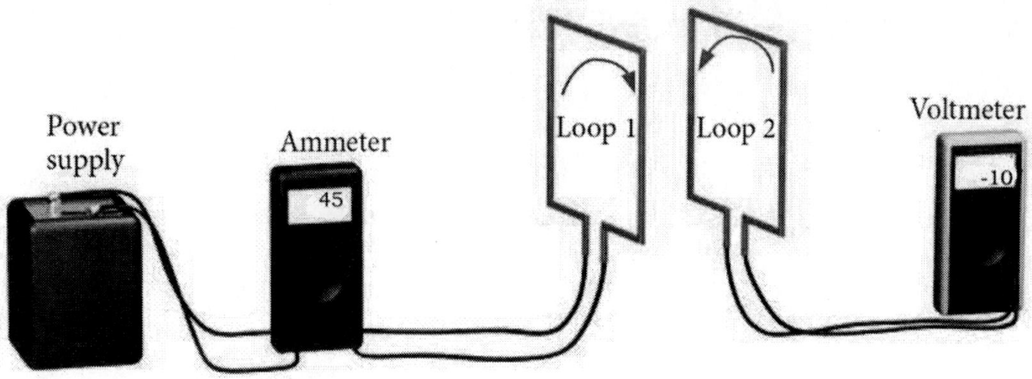

Figure 2. The illustration of Lenz's law experiment. The current flowing in loop 1 induces the current in opposite direction [7].

Mutual induction can be explained as follows: Consider two coils as shown below, or inductors are close to each other, and a changing current in the first coil produces a magnetic flux in the second coil. However, the changing current in the first coil also induces a potential difference in that coil, and thus the magnetic field from that coil also changes. This phenomenon is called self-induction. The resulting potential difference is termed the self-induced potential difference. Changing the current in the first coil also induces a potential difference in the second coil.

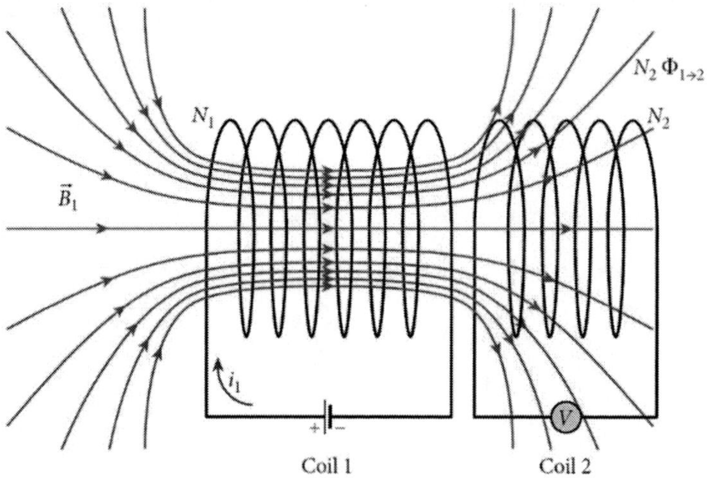

Figure 3. Mutual inductance. A changing current in either coil induces an emf in the other coil.

The magnetic property of materials can be studied by using metals with different magnetic susceptibility. The magnetic field, **B** , inside the material then depends on the external magnetic field, $\mathbf{B}_0$, and the magnetization, **M**

$$\boldsymbol{B} = \boldsymbol{B_0} + \mu_0\boldsymbol{M}$$

where μ_0 is again the magnetic permeability of free space. Instead of including the external magnetic field, $\mathbf{B}_0$, it is customary to use the magnetic field strength, **H:**

$$\boldsymbol{H} = \frac{\boldsymbol{B_0}}{\mu_0}$$

$$\boldsymbol{B} = \mu_0(\boldsymbol{H} + \boldsymbol{M})$$

Since the unit of magnetic field is [B] = T and the unit of the magnetic permeability is [μ_0] = T m/A, the units of magnetization and magnetic field strength are [M] = [H] = A/m.
For most materials this relationship is linear: $\mathbf{M} = \chi_m \mathbf{H}$, where the proportionality constant χ_m is called the magnetic susceptibility of the material.

$$\boldsymbol{B} = \mu_0(\boldsymbol{H} + \boldsymbol{M}) = \mu_0(\boldsymbol{H} + \chi_m\boldsymbol{H}) = \mu_0(1 + \chi_m)\boldsymbol{H} = \mu_0 K_m\boldsymbol{H}$$

where K_m is relative magnetic permeability.

$$K_m = 1 + \chi_m$$

IOP Conf. Series: Journal of Physics: Conf. Series **1340** (2019) 012005 doi:10.1088/1742-6596/1340/1/012005

Table 1 Values of magnetic susceptibility for some common diamagnetic, paramagnetic, and ferromagnetic materials.

materials		magnetic susceptibility (χ_m)
paramagnetic	Aluminum	2.3×10^{-5}
	Oxygen	0.0002×10^{-5}
diamagnetic	Copper	0.98×10^{-5}
	Nitrogen	-0.0005×10^{-5}
ferromagnetic	Iron	5,500
	Mu-metal (77%Ni-16%Fe-5%Cu-2%Cr)	100,000

In Statistical Mechanics class, student have learned probability of a dipole of magnetic materials pointing amarelle to an external magnetic field [8],

$$P(r) = \frac{e^{-\beta\epsilon_r}}{Z},$$

where Z is a partition function. The average energy of magnetic energy is

$$\bar{E} = \sum \epsilon_r \cdot P(r) = \sum_r \frac{\epsilon_r e^{-\beta\cdot\epsilon_r}}{\sum_r e^{-\beta\cdot\epsilon_r}} = -\frac{\partial \ln Z}{\partial\beta}$$

$$\bar{E} = \frac{\mu H\left(e^{\frac{\mu H}{k_B T}} - e^{-\mu H/k_B T}\right)}{e^{\mu H/k_B T} + e^{-\mu H/k_B T}}$$

These equations will be applied to increase energy of a magnetic field.

3. Methods

Statistical Mechanics deals with microscopic points of view to describe macroscopic properties of matter. For example, magnetic field or potential energy of magnets (macrostate) depends on each tiny magnetic dipole (microsite). Most students do not realize until they experience real world problems or experiments. In our lesson design, we present the design of simple sets of apparatus for teaching on electromagnetic induction in an undergraduate level, science shows, and in-service teachers' training courses. We show that useful experiments do not necessary to require expensive apparatus. The experiments include Faraday's law, Lenz's law, inductance, and magnetism of magnetic materials. We use inquiry method to challenge ability of new generation teachers and to use the integration of contents and pedagogical knowledge for teaching Physics. We will report the used of the experiment sets to demonstrate the concepts of the relationship between electricity and magnets through an inquiry method. This work can be used as a part of classroom innovation and the teaching-learning process reforms.

The engineering problems are as follows:

1. How can we induce magnetic field?
2. How can we increase the current?
3. How can we increase the magnetic field?
4. How can we observe the magnetic strength?
5. What materials should we use?
6. What temperature do we need?

These are STEM Education problems that students have to solve. The students were guided through the project. They have learned STEM skills as shown in table 2.

It is difficult to have magnetic flux in a coil changed by switching on and off manually. To do this, we apply an alternative current (AC) to the first coil and the change in magnetic flux in the first coil leads to generate an induced current in the other coil. To increase the magnetic flux in the coil, we use materials with high magnetic susceptibility. Students can perform an experiment by inserting metals such as aluminum, copper and iron as a core between two set of coils. The students can observe the effect of magnetic material cores by viewing the bright of the light bulb.

The student carried out the STEM project. The instructor provided materials, liquid nitrogen, light bulbs, copper wires, rings and others.

Table 2 STEM Education that the students have learned from the STEM project

STEM	Contents and skills
Science (S)	Concepts of statistical Mechanics (Physics), partition function, average of energy, effect of temperature on average of energy (macrostate)
Technology (T)	Methods, techniques, materials
Engineering (E)	Design, control, and process analysis, the use of engineering process
Mathematics (M)	Calculation, approximation

4. Results and Discussion

The students worked in group coached by the instructor. They all found that best materials to generate strong magnetic field by induction was ferromagnetic materials such as iron and nickel because they have high value of magnetic susceptibility. It could serve as a core for magnetic induction. They can design experiments to prove the idea. They also learned how to increase magnetic field by lower temperature as they were guided as shown in Statistical Mechanics texts. The pictures below show the real work done by the students.

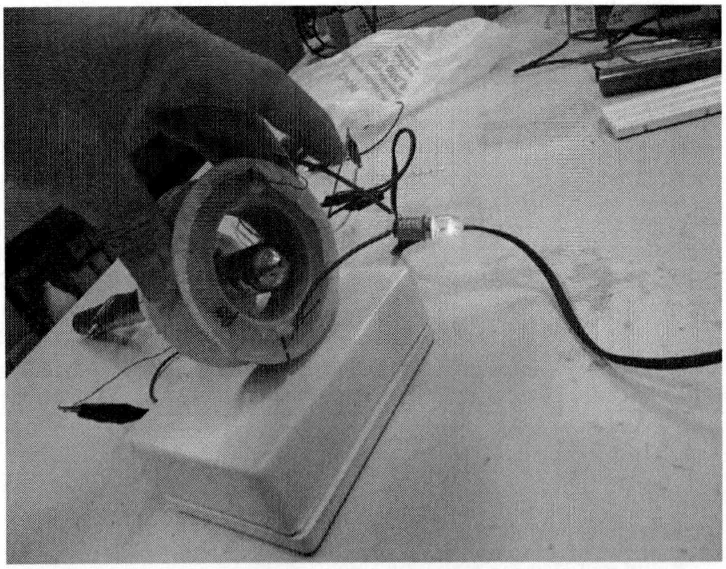

Figure 4. Simple experiment for testing different magnetic materials core for mutual inductance

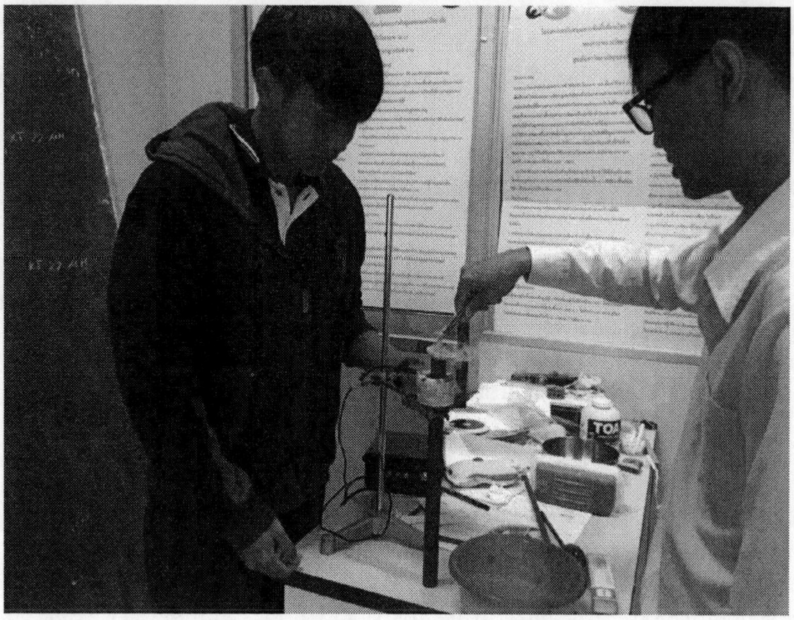

Figure 5. Lenz's law experiment with liquid nitrogen cooling down to increase magnetic field

The student used engineering process. The engineering process start with 1) identify the need or problem, 2) research the need or problem, 3) develop possible solutions, 4) select the best solution, 5) construct a prototype, 6) test and evaluate the solution(s), 7) communicate the solution(s), and 8) redesign to fit the need. They used engineering methods or process. For example, they used engineering processes in designing and planning to keep the lessons going. During teaching and coaching their students to learn so that their students could build new knowledge and skills, the lecturers found that they have learned new teaching methods and new experience. From the observation, the students appreciated the lessons. They enjoined working in group, presentation, and discussion. The students appreciated the lesson very much so that they have gone to teacher' trainer program at Faculty of Education, Khon Khean University after completing BS degree in Physics at UBU.

5. Introduction

We have developed STEM Education based instruction for undergraduate students at UBU which are located in Northwestern part of Thailand. The students were eager to acquire new method of teaching and learning, STEM Education for their classrooms. We have also applied new developed learning management lessons and plans on Statistical Mechanics to our classrooms. The students learned greatly and positively responded to the developed lessons. The students are able to solve real problems using STEM Education. The students' necessary skills for 21th century improved with significant satisfaction.

Acknowledgements

The authors would like to the National Research Council of Thailand for financial supports. We greatly appreciate Ubon Ratchathani Rajabhat University (UBRU) and Ubon Ratchathani University (UBU) for providing research space, classroom, and meeting rooms.

References

[1] R. Sillitto, L.M. MacKinnon, "Building a multimedia educational resource for physics learners", Physics Education, 35(3), 325-331.
[2] Rujira Karisook, Udom Tipparach, "The attitudes and expectations of students from Benchama Maharat School on learning Physics", Proceedings in 36th Congress on Science and Technology of Thailand, 26-28 October 2010.
[3] Sunee Clainil, Science Education in Thailand: Decay and regression, Institute for the promotion of teaching science and technology , 2555.
[4] Chalita Taksikanon and Udom Tipparach, Integration Science and Mathematics in Schools", Proceedings in Siam physics congress 2012 May 9-12, 2012, 266-268.
[5] Jutamanee Tipparach, Makabodee Ruaysap, Krittidej Chanthawara, Supath Kookiattikoon, and Wanchana Boonchom, Udom Tipparach, A need analysis in developing STEM based instruction for higher Education, Humanities and Social Sciences Journal, Ubon Ratchathani Ratjabhat University, Vol.9, No.1, 2018, p167-180.
[6] R. A. Serway, John W. Jewett, Physics for Scientists and Engineers 9th Ed., Cengage Learning, New York, 2014.
[7] W. Bauer, Gary D. Westfall, University physics with modern physics, McGraw-Hill Companies, Inc., New York, 2011.
[8] F. Reif, Fundamentals of statistical and thermal physics, McGraw-Hill, New York NY, 1965

IOP Conf. Series: Journal of Physics: Conf. Series **1340** (2019) 012006 doi:10.1088/1742-6596/1340/1/012006

The effect of SSI overlap STEM Education on Secondary Students' Socio-scientific decision making

Siripun Siribunnam
Secondary school science teacher at Phadungnaree school, Mahasarakham, Thailand
Email : crutongtang@gmail.com
Radka Bednarova
Secondary school english teacher at Phadungnaree school, Mahasarakham, Thailand
Email : bednarova.r@gmail.com
Prasart Nuangchalerm
Mahasarakham University, Thailand
Prasart.n@msu.ac.th

Abstract. The purpose of this study is to investigate the secondary students' decision making process when global socio-scientific issue (SSI) is applied to STEM-based learning. SSI overlap STEM framework adapted from the literature was designed to help students to take a responsible informed decision when considering global environmental issue as a part of their classroom activity. Thirty one students participated in the pre- and post- socio-scientific decision-making test. Data was also collected through class observation. Both quantitative and qualitative data were gathered and analysed; content analysis of qualitative data collected from student's reasoning helped to explain the initial quantitative findings. Based on the qualitative analysis, the results revealed that the students' abilities benefited from the SSI overlap STEM framework implementation. The results showed that there were considerable changes in the students' decision making before and after the activity. It also indicated that the SSI overlap STEM framework designed for this study was effective and could be used in science classes.

1.Introduction

The advancement of science, technology, and the proposition concerning the 21st century skills have great influence on educational paradigm changes. Education on all levels nowadays puts emphases on so-called *higher thinking skills*, such as critical thinking, problem solving, decision making and creativity, as well as developing students' social skills and communication, and use of technology as a tool for inquiry [12]. Integration of all these as well as the ability to learn, both in the classroom and in a real life, became the primary trend in education in order to make learning meaningful to students. Only that way, students will recognize the value of their studies and will be able to apply their knowledge in their everyday lives.

The study of science is essential to our lives. Furthermore, the advancement of science generates rapid changes in society. Learners these days need to be equipped with the ability to think rationally and to make decisions based on data obtained by information technology. They also need to be aware of global affairs and a *global* citizenship based upon the 21st century skills. The reformation of study of science is, therefore, necessary for improvement of scientific literacy [14] which includes such topics as, understanding of Nature of Science, appropriate application of scientific concepts, and implementation of scientific process in solving problems and making decisions.

Our understanding of science, as well as its practical application, could be divided into two parts, non-normative (e.g. gathering data, observation, prediction, scientific method and process) and normative (e.g. describing courses of action, choosing to create selected products, decision about what ought to be done). Both elements are closely connected. Recognising the importance of their mutual contribution to the understanding of the Nature of Science is, therefore, vital for formulating and implementing policies when teaching science. However, there seem to be trends in promoting non-normative to normative elements when practising science [19].

A prime example was illustrated in the examination of the Next Generation Science Standards (NRC 2012) by [19] who points out disintegrated views that allow those who practice science, or make decisions about the use of science in everyday life, to abandon their *sense of responsibility* by making science awkwardly independent on what is happening in the world [19].

There have been efforts to create guidelines (NRC 2011, BOSE at NAS 2012) that would improve current situation; however, they somehow fail to provide complex answers. The document compiled by the National Research Council (NRC 2011), for example, offers variety of generalised suggestions on improving the STEM learning experience while, unfortunately, the school settings remain separated from the cultural context thus failing to connect to the life environment of the students [19]. Both STS (science-technology-society) and STEM education have thus so far failed to pay enough attention to socio-scientific and sociocultural perspectives that play central role in forming one's sense of scientific identity which is necessary for developing moral responsibility [19],[13].

The issue thus remains. The students still depend on the teacher in regard to what to know and how to think, and / or the studied problems lack the connection to the students' interests and beliefs.

1.1 STEM Education

The educational acronym "STEM" represents a multidisciplinary educational perspective that combines Science, Technology, Engineering and Mathematics. It is widely used and is directed at those involved in education focusing on improving how STEM education is developed and delivered. Even though STEM education should incorporate all four disciplines, it often focuses on science and mathematics only. In these, students engage in group activities, laboratory and field studies, and projects, as they are expected to develop cognitive skills, such as adaptability, complex communication, non-routine problem solving, self-management, and systems thinking, in order to enhance their competitiveness in the modern world. However, there has been considerable lack of promoting technical knowledge and engineering. Using technologies and understanding of how things work is the real concept of STEM education and it has great economic importance [2].

To improve this situation, teachers involved in STEM education need to learn more about the STEM areas, and need to start showing the students how these are connected. To begin this transformation, teachers must become STEM thinkers who can demonstrate how STEM is involved in most areas of the students' lives [5].

Content knowledge of the subjects is essentially important for students to be able to express their intrinsic opinion on any STEM-related topic in a meaningful way which brings them to further inquiries, discussions, and evidence-based reasoning. The ability to put scientific issues in context, as well as real understanding of the context, demands the exercise of prudence, morality and character. It is the lack of these that elucidates the deficit thinking in STEM-based initiatives [2].

1.2 SSI and decision making

Socio-scientific issues (SSI) are a product of knowledge and technology discussed and criticised by public in regard of its relevance to the society. These include problems concerning health, general living conditions, limited energy supplies, economic growth etc. in contrast to environmental conservation [13]. Socio-scientific issues are applied to stimulate the interest in argument-based learning and in discussion based on scientific knowledge and data [11]. At present, the management of knowledge derived from issues concerning science and society is regarded important when studying science making learners realize that their learning is connected to their lives. Taking into account the context of socially responsible science, teaching should encourage the students to base their decisions on moral reasoning and on a wide scope of scientific literacy.

A current reformation of the study of science tends to support the scientific advancement towards establishing morality and ethics in a scientific dimension, instead of teaching solely science-based content. The use of socio-scientific issues responds to these scientific needs [16].

The use of scientific knowledge before applying it onto situations within social and moral context stimulates learners to find the truth while opening their minds to the opinions of others. They learn to negotiate carefully and systematically while using reasoning [17]. As a result of that, a meaningful way of learning is created for students.

Even though great variety of work has been done on socio-scientific issues, further investigation at a classroom level is still required. There is a need for identifying evidence-based pedagogic practices which could improve students' decision-making abilities. Using issue-based approach, the students would be guided towards making an informed decision. The focus here would be on enhancing their understanding of the concept and scientific inquiry, attitudes and values, and their ability to use rational arguments [8]. The SSI movement also seeks to engage students in decision-making regarding current social issues with moral implications embedded in scientific contexts [18].

To help students to look at the studied issue from different perspectives, taking into account relevant scientific knowledge, reasoning, and values concerning the possible outcome, a decision-making framework has been created [10]. However, even though most of the students, when making decision, do look at the issue from various perspectives, these are in the end considered unequally, relying more on ethical values. This is particularly true for controversial SSI, such as stem-cell research [4]. In addition to that, results by Walpuski et al. suggest that decision-making and argumentation are highly dependent on how frequently students are given the chance to argue in their science classes.

SSI is becoming part of the science curricula as an important way of achieving scientific literacy among secondary school students (e.g.,Hong Kong Curriculum Development Council, 2007; National Research Council [NRC], 1996, [9]. For example, Lee and Grace. Zeidler (2016) considers biodiversity conservation an important socio-scientific issue that is often thought of as a prerequisite to sustainable living and development. The foundation for citizens' understanding of conservation issues could be laid down in formal school education.

1.3 SSI overlap STEM framework

It has been mentioned above that STEM-related research and programs restrict their views only on science, technology, engineering and mathematics, mostly paying attention to *crosscutting connections* among those four areas. That however, leaves gab where particular scientific issue should be put into context of cultural and life experiences of ordinary students. Zeidler (2016) in his study uses a term *"STEM silos"* to describe this situation.

He offers a model ("STEAM") which comprises of STEM disciplines *supported* by the Arts. The Arts here are represented by disciplines, such as sociology, psychology, history, fine arts, philosophy and education, and serve as a means of sociocultural contextualization.

He supports the idea that only when bringing down the rigorous boundaries of discipline-specific silos of STEM education, proper integration of science and socio-cultural context is possible using sociology, psychology, fine arts, etc. to contextualize the knowledge in a way personally meaningful to students.

Zeidler (2016) also points out that to take responsible informed decision we need to have both, the knowledge as well as the ethical responsibility and the ability to use evidence-based reasoning. The concept of dividing science into normative and non-normative components is thus unrealistic if not dangerous. Only accepting moral responsibility makes us *scientifically literate* and ready to participate in a larger global society.

"*While literacy may not require a moral compass, scientific literacy, in the sense [he advocates], does.*" [19].

Teaching science in Thailand is not yet satisfactory. What is taught in schools remains virtually irrelevant to the reality failing to stimulate learners' interests and to connect the studied subjects with the students' daily lives. The little emphasis on knowledge and testing leaves the students poorly equipped with effective thinking skills, such as problem solving and decision making, when entering the labour market. This is especially true when dealing with socio-scientific issues.

According to the latest testing under the Programme for International Student Assessment (PISA), it was found out that the mean of scores earned by Thai students was below the international average. The results are even more alarming as the scores dropped notably in comparison to the previous years' evaluations. The reason for this is that the students are not familiar with tests consisting of analytical features which use reasoning as part of the decision-making process.

The aim of this study is to investigate decision making process in a class of secondary students when global socio-scientific issue is applied to STEM-based learning. The following questions were asked:

1. Does implementing the SSI overlap STEM framework help 13- to 14-year-old secondary students to make an informed socio-scientific decision?

2. What are the characteristics of the SSI overlap STEM framework that they use to make an informed decision?

2. Methodology

The study was conducted in an intact class of 31 secondary students aged between 13 and 14 years and it examined the decision making process that the students use to make decisions about a socio-scientific issue under the guidance of the SSI overlap STEM framework. The decision making process here includes perspectives that the students took into consideration as well as the way how they reached the various options and evaluated the significance of contextual concepts and values.

2.1. Presentation of the socio-scientific issue.

At the beginning of a lesson, the students are introduced with some basic facts about carbon dioxide and greenhouse effect, a real life *socio-scientific issue* which was presented to the students in a form of a short video clip. In the video atmospheric scientists demonstrate their experiences and evidence of the greenhouse effect and climate change, including global warming, obtained through their research.

Global warming is considered to be a consequence of expending greenhouse effect which is caused by gases such as carbon dioxide, methane and nitrous oxide that block heat from escaping from the atmosphere. Carbon dioxide (CO_2), as a minor but very important component of the atmosphere, is released through natural processes such as respiration and volcano eruptions. However, over the last century, human activities, that include deforestation and fossil fuels burning, have increased the concentration of CO_2 in the atmosphere by more than a third.

We decided to use *global warming* as a chosen topic presented to students, because it is of particular importance in the contemporary world due to its great impact on the global environment.

2.2. Data Collection Techniques and Analysis Methods

The development of the students' socio-scientific decision making abilities was tracked by analyzing data collected through class observation. For the analysis, both quantitative and qualitative data was used; content analysis of qualitative data collected during students' discussions helped to explain the initial quantitative results. The results of the socio-scientific decision making pre-test and post-test were compared.

2.3. SSI overlap STEM framework

At the beginning of a lesson the students were divided into six groups and were asked to suggest socio-scientific issue and their solution to it. This was treated as a pre-test with socio-scientific decision making reasoning test. Then the chosen topic, *global warming,* was presented to the class.

After being presented with the topic, the students then engaged in a decision-making exercise through group discussion guided by the SSI overlap STEM framework. This was designed drawing on works by Lee (2014) and Billiar et al (2014) who developed six-key-stage practice which is depicted in Figure 1.

Figure 1 : SSI overlap STEM framework used in this study

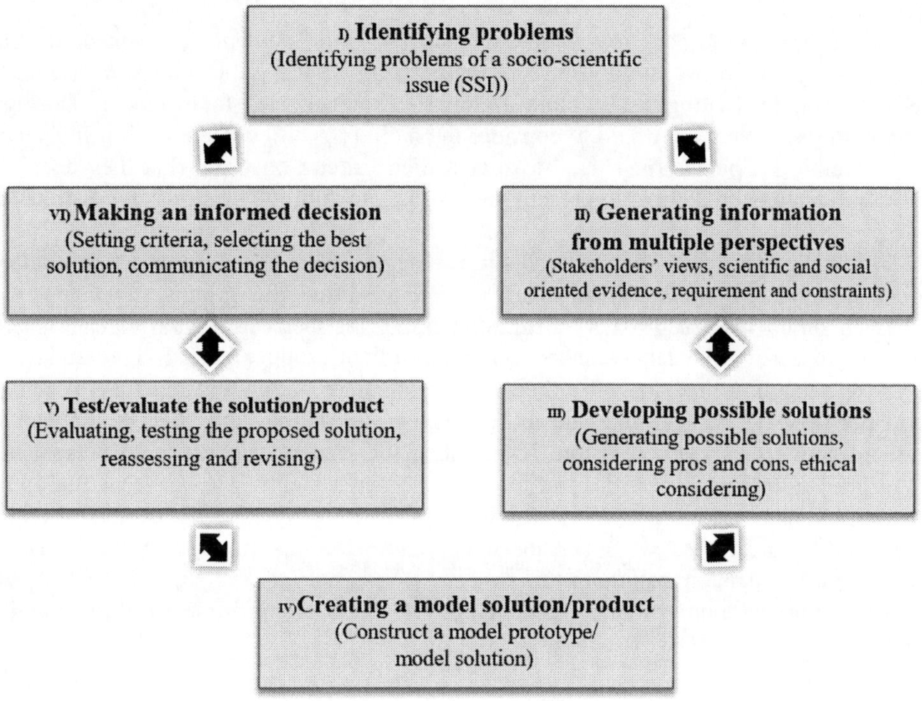

Stage one - *Identifying problems*- serves to encourage students to identify questions and problems within broader socio-cultural context of a larger global society. Students practice proposing questions such as: "How can we reduce CO_2 in the atmosphere?" or "How can we stop or at least slow down rising sea level?" These are the type of questions that can be given to secondary students to enhance their ability to make decisions about environmental conservation.

Since reasoning and argument are essential to finding the best solution to a problem, in stage two – *Generating information from multiple perspectives* – the students are encouraged to explore as many perspectives as possible through brainstorming. Practicing this should prevent them from focusing excessively on preconceived beliefs and not exploring other options. It also helps the students to understand the scientific evidence and to compile and evaluate all the evidence they have at hand [7].

In stage three – *Developing possible solution* – students are asked to present at least two options of a possible solution; that way they keep their minds open to a plausible solution. Later they engage in a discourse about advantages and disadvantages of each of their options. Among others, they need to consider the ethical aspects of their solutions and these must be discussed in a way relevant to the issue. The aim of this exercise is for students to enhance their critical thinking.

In stage four – *Creating a model solution / product* - students design a model product or a solution to the particular issue. During the process they are developing their problem-solving skills when applying scientific knowledge to a question or a technological problem, identifying mathematical components of the issue, and using reasoning to link evidence to an explanation. Students are required to work individually along with taking part in group activities.

In stage five – *Testing and evaluating solution / product* - students test and evaluate the model prototype or model solution they created. Students need to consider its effectivity, and balance competing criteria of required functions, technical viability, safety, costs and aesthetics, and keeping with legal requirements. There is not a simple answer but rather a range of more or less satisfactory solutions of which the optimal choice depends on how well the proposed solution meets the set criteria and restrictions.

In stage six – *Making an informed decision* – the students are asked to debate about specifications they would use to choose from their options. Each group needs to present their ideas both verbally and in writing using tables, graphs, drawings and their model. During extended discussion with peers, they are asked to consider the main pros and cons of their principal arguments, including the ethical aspects. They need to revisit their criteria ensuring that they considered every option. The groups then take turns to present their results, including decisions and justifications, to the whole class.

Following the presentation, students in each group were asked to discuss and reflect on their group's decision, drawing on the decisions and arguments of the other groups.

Each group was required to record the results of their discussion at each stage on their worksheet and to analyze the data collected in the test of their proposed solution or a model.

After the groups had arrived at their mutual decisions, individual students in each group were asked whether they agreed with the decision of their group. If they did, then the collective decision was recorded as the member's decision. If not, then the student was asked to present his/her own decision as well as to supply justification for his/her point of view. This gave us the opportunity to compare various solutions and determine which one of them met the specific design criteria and which design solved the problem best when considering given restrictions. In stage six this was treated as the post-socio-scientific decision making test.

During the lessons, students' feedback on the socio-scientific decision making process was collected.

3. Results

3.1 Parties affected by the issue as suggested by students

Table 1 shows the range of groups affected by the issue as identified by the students; these are listed in descending order of frequency. It also shows a number of groups of students that listed each particular party when considering the impact of global warming. The most frequently cited group was the 'environmentalists'. The fact the students mentioned 'humans/people' implied that the students are concerned with the future of humanity. Some group also mentioned the global ecosystem. Even more surprisingly one of the groups identified an 'atmosphere' as a party affected by the global warming reflecting on the role that the atmosphere plays in the analyzed issue. Taking these perspectives together, the students seemed to be thinking quite extensively about different views of a sufficiently wide range of affected parties.

Table 1: Parties affected as perceived by students

Parties affected	No of groups of students who listed particular party (N= 6)	Point of observation (as predicted by the students)
Environmentalists	5	Concerned with deforestation, biomass burning, using fossil fuels and cement production. People should become more environmentally aware.
Ecosystem	4	The ecosystems are changing dramatically as global temperature rises.
Plants	4	Plants use CO_2 and water to produce sugars which are then used to build more complex compounds. During this process oxygen is produced as a by-product.
Coal and fossil fuels	3	Most of the CO_2 that comes from human activities is released while burning coal and other fossil fuels. People should increase use of alternative sources of energy.
Public/government	3	Set policies for economic development and environmental conservation. They also design laws to make sure that people and businesses comply with the legislation.
Scientist	3	Human activities have caused CO_2 to increase above levels not seen in hundreds of thousands of years.
Cell	2	CO_2 is an end product of cellular respiration in organisms that metabolise sugars, fats and amino acids while obtaining energy.
Food	2	CO_2 is added to drinking water and carbonated beverages including beer and sparkling wine to add effervescence. The frozen form of CO_2, known as dry ice, is used as a refrigerant
Humans / people	2	Humans live more comfortably since the beginning of the industrial revolution. However, the impact of their activities on the environment is massive.
Atmosphere / oceans	1	CO_2 released from the burning of fossil fuels remains in the air and is not absorbed by the vegetation and the ocean

3.2. Knowledge and information researched by the students

Table 2 shows the types of information and knowledge that the students researched in order to provide more in depth arguments during a discussion with their peers that eventually led to resolving the problem. The type of information such as chemical and physical properties, carbon footprint and CO_2 emissions, sought out by the students, reflects their views and opinions on the issue in question and it could serve as background and contextual information for further discussion of possible strategies. Another type of information, such as that on photosynthesis and cellular respiration, reflects the will of some of the students to find solution to the issue.

Table 2: Types scientific and social oriented evidence, requirement and constraints thought to be useful by the students

Information / knowledge	Chemical and physical properties	Greenhouse gas	Photosynthesis	Carbon fixation	Fossil fuels	Human activities	Earth's atmosphere	CO_2 emissions	Carbon footprint	Cellular respiration	Electricity
No. of responses	4	3	5	1	4	5	5	5	4	3	5

3.3 Models designed by the students

Table 3 shows examples of models designed by the students according to the set criteria using respective arguments, counterarguments, and ethical considerations. These options can be categorized into four main types of action as shown in the following table. Various ways how to reduce CO_2 level in the atmosphere were suggested by the students, among others using less-electricity, RRR – reduce, reuse, recycle, growing more plants and reduce the use of fossil fuels thus reducing CO_2 emissions; these are described in the following table.

Table 3: Examples of the possible solutions with arguments, counterarguments and any ethical considering as put forward by the students

Option /Solution	Argument	Counterargument	Ethical considering	Evaluating model/revise
Model 1: Action: Using as little electricity as possible				
<u>Human-powered amplifier</u> (3)	-Humans should use less energy -Campaign for reducing the use of electric power.	-Humans live comfortably while using electric-powered machines	-Future benefits of reducing use of electricity -Cleaner environment	-Original model worked, however, it was not very efficient -Modifying materials -Change in size
Model 2: Action: Implementing RRR – reduce, reuse and recycle				
<u>Reduce, reuse and recycle things on daily basis</u> (4)	-Reduce the consumption of non-recyclable materials (plastic) -Educate about separating waste	-Not always possible –some materials in too small pieces. -Recycling certain kinds of plastic may spread toxins	-Save usage of natural resources -Preserving the environment -Help to lower greenhouse gas emissions	- Make the multi-purpose shelf model stronger. -Adapt future board shoes can be use in actual

Option /Solution	Argument	Counterargument	Ethical considering	Evaluating model/revise
Model 3: Action: Breaking down the CO_2 or turning it into another substance				
Growing plants (5)	Grow plants and trees to transform CO_2 into food and air	The problem is not solved completely people will continue with "unhealthy" habits / activities.	-People can live comfortably on Earth. -Taking care of the atmosphere should come first.	-Follow up plant Growth -Prove pot plant can be decomposed.
Model 4: Action: Reducing the use of fossil fuels and CO_2 emissions				
Eco car model (3)	-The public would get to know more about using of fossil fuels solution. -Save energy and money.	-Governments focus on economic development -Biodiesel gives about 10 percent less energy than diesel.	-Preserving the environment -People should focus on using alternative energy sources	-To make it efficient optimal use of lightweight materials is necessary. -Reducing air resistance

3.4 Criteria used to choose between the options

Table 4 states the main criteria chosen by the six groups *of students*. From the list presented in Table 4, the students chose explicit criteria that they used to make their decision. Groups 1 to 5 emphasized that their decision should be environmentally sound. On the contrary, group 6 was mostly concerned with saving money. This approach was consistent with the ethical arguments the students used to justify their final decision.

Table 4: Criteria adopted by the students that they used to make their final decision.

Group	Criteria for choosing between the option/solution
1	-Practical and effective way of reducing CO_2
2	-Taking care of the atmosphere
3	-General environmental awareness
4	-Use alternative sources of energy / use less fossil fuels
5	-CO_2 decomposition
6	-Saving money

Table 5 shows a more detailed comparison of the results. The action most often cited in the pre-test, at 38.7%, was 'reducing use of fossil fuels' (the Eco-car model); the second most often cited action was 'using less electricity', at 22.6%.

However, after implementing the SSI overlap STEM framework the students' most favorite action became 'braking down the CO_2' action which is represented by the growing plants model (41.9%). The second most cited action was again the 'using less electricity' action (25.8%).

IOP Conf. Series: Journal of Physics: Conf. Series **1340** (2019) 012006 doi:10.1088/1742-6596/1340/1/012006

Table 5: Frequencies of each type of action used in decision making process, comparing pre-test and post-test results

Type of action	Pre-test (N=31) No. of Students	Pre-test (N=31) percentage	Post-test (N=31) No. of Students	Post-test (N=31) percentage
Model type1 : The less-electric machine	5	22.6%	8	25.8%
Model type 2 : The reduce reuse and recycle	6	19.4%	6	19.4%
Model type3 : Growing plant	5	16.1%	13	41.9%
Model type 4 : The Eco cars model	12	38.7%	4	12.9%

3.5 Students' progress in reasoning

The audiotaped record of the group discussion captured nuances of the students' reasoning throughout the decision-making process. The following exchange is an example of the students engaging in argumentation and reasoning, which led them to become aware of the fact that any option is bound to have advantages and disadvantages as well as ethical aspects that need to be taken into consideration. They learned to debate about specifications they would use to choose from their suggested options, and eventually managed to evaluate the model prototype or model solution they had created.

Transcript 1 shows a sample exchange that the students use reasoning before arriving at a conclusion that is supported by their interaction with their peers.

Transcript 1:

S1: Using too much of fossil fuels is a big problem that leads to increasing CO_2 level.

S2: Although we use fossil fuels much, it makes it easier and more comfortable for us to go around.

S3: Everybody should be environmentally aware.

S4: Now that I know more about CO_2, I feel that I can create a possible solution to solve the problem.

S5: Finding more information about CO_2 makes it easier for me to design a solution.

S6: There are various options for us.

S2: Even though we proved that the model solution can decrease CO_2, I think it will work better if we adjust it.

S3: Solving the problem with a group of peers makes it easier than when working on my own.

S4: Global warming wouldn't be such a big problem if everybody "chipped in".

S3: Our group used criteria that helped us to decide with environmental awareness on our minds.

S4: Growing plants is the best solution if we want to reduce CO_2 effectively.

At the end of the lesson, the students communicated their decisions to the rest of the class. Their decision was to growing plants and their reason for it was that trees and other plants transform carbon dioxide into sugars (food) and air. Considering the specific criteria before making their final decision, this seemed to be practical and effective way of reducing CO_2.

4. Discussion

As we know from previous studies, the students are better motivated to learn new information from problems presented to them within broader socio-cultural context rather than from textbooks. These too often explain concepts without the contextual background information. This is especially true when students are given the chance to apply their newly acquired knowledge to solve problems based on the SSI overlap STEM framework, as occurred in this study. The discussion of socio-scientific issues in case studies like ours provides the students with social context for learning scientific concepts as well as developing their decision making skills to handle environmental issues. Although the chosen topic, global warming, is not covered in the secondary earth-system science curriculum, it helps to illustrate a number of key concepts, such as the importance of the environmental relationships in keeping the ecosystems in balance, that the students need to learn at this level.

In this study, the students were provided with the information about CO_2, and how to reduce its level. It could be argued that this information might have influenced the students' attitude towards the way how to resolve the issue. The analysis, however, show that while using the prescribed framework as a guideline for their decision making process, the students were able to better identify and understand problems within broader socio-scientific context. They were encouraged to explore number of perspectives through brainstorming suggesting number of views on parties affected by the issue in question. The knowledge acquired through studying various resources was also an important aspect that helped the students to develop conceptual understanding of the effect of CO_2 and global warming on the environment, the studied issue, and to detect any misconceptions about the said issue. Both the perspectives and the knowledge acquired by the students contributed to their ability to better formulate broader range of suggestions and possible solutions while considering their advantages and disadvantages as well as their ethical aspects. The most important part of this exercise was peer discussion. At the beginning of the exercise it was not easy for the students to specify their thoughts, especially regarding the ethical aspects, but their argumentation and reasoning improved significantly after applying the SSI overlap STEM framework. The arguments which were recorded during the post-test and which provided justification for the students' decisions also proved that the students managed to acquire the scientific knowledge necessary for making better informed decision. It may be useful to turn this classroom activity into a project-based task in which students search for information relevant to the issue by themselves. This independent kind of learning has the added value of developing students' inquiry skills, mainly searching for and evaluating information, and capacity to clarify their understanding of the impact of CO_2 thus assuming greater ownership of their learning.

The findings of this study have several implications for the teachers too. Firstly, it is the need of developing their pedagogical content knowledge which will necessarily lead to improving students' decision making skills in the context of environmental education. The inclusion of the SSI overlap STEM framework in the earth system science curriculum means that students would be expected to consider wider perspectives in tackling the earth-system science issues in addition to applying scientific knowledge. Teachers would thus need to readjust their roles and lead the students to study environmental issues, with the aim of developing both the scientific knowledge of students as well as a wide range of other skills, including reasoning, argumentation, and decision making. Lee and Grace (2010) in their work proposed students' reasoning processes in making decisions about an authentic, local socio-scientific issue: bat conservation. After evaluating their results they found out that there were substantial changes in students' decisions taken before and after the activity. That indicates changes in students' values moving from anthropocentric to wide-ranging perspectives which emphasize both pragmatic as well as biocentric values. That however poses conservational dilemma on students in which they need to choose between human needs and their impact on the world. On one hand, this encourages students to reflect on the main impacts of human actions in regard to global warming and how this influence, if left unnoticed, could affect other organisms, which might, in turn, affect humans in the long run. On the other hand, students need to consider the price that humans might need to pay for preserving wildlife and natural habitats. There are numerous studies on local as well as global level illustrating this conservation dilemma which could be used for classroom discussion activities. This could help the students to develop their abilities to make informed decisions and to further their understanding of the earth-system science concepts and values behind earth-system science.

5. Acknowledgement

We would like to thank for Director of Phadungnaree School and the Promotion of Teaching Science and Technology (IPST) Thailand

6. References

[1] Billiar, K., Hubelbank, J., Oliva, T., and Camesano, T. Teaching STEM by design. *Advance in Engineering Education*," 2014, (1), pp. 1-21.
[2] Bybee, R., M., "*The case for STEM Education Challenges and Opportunities*," National Science Teachers Association (NSTA), 2013, Virginia.
[3] Grace, M. "Developing high quality decision-making discussions about biological conservation in a normal classroom setting," *International Journal of Science Education*, 2009, vol.31, no.4, pp. 551-570.
[4] Halverson, K. L. and Siegel, M. A. and Freyermuth S. K. "Lenses for Framing Decisions: Undergraduates'decisionmaking about stem cell research,"*International Journal of Science Education*. 2009, vol.31, no.9,pp. 1249–1268.
[5] John, S., "*An Introduction to STEM Education Best Practices and lesson design#3, /ASTEP Training Suite,*" Academy for science Mathematics and Technology Education Professionals(ASTEP), IPST, 2018, Bangkok. pp. 1-21.
[6] Kortland, K. "An STS case study about students' decision making on the waste issue,"*Science Education*, 1996, vol 80, pp. 673-689.
[7] Kuhn, D. *Education for thinking*. Cambridge, Massachusetts : Harvard University Press, 2005.
[8] Lee, Y. C., "Developing decision-making skills for socio-scientific issues,"*Journal of Biological Education*, 2007, vol.41, no.4, pp. 170-177.
[9] Lee, Y. C., Students' Reasoning and Decision Making About a Socioscientific Issue : A Cross-Context Comparison : Published online 9 August 2012 in Wiley Online Library <http://wileyonlinelibrary.com> 2012.
[10] Lee, Y. C., and Grace, M. "Students' reasoning processes in making decision about an authentic, local socio-scientific issue: bat conservation," *EducationResearch*, 2010, vol.44, no.4, pp. 156-165.
[11] Nuangchalerm, P."Engaging students to Perceive Nature of Science through socio scientific issues-based Instruction,"*European Journal of Social Science*, 2010, vol.13, no.1, pp.34-37.
[12] Siriphutachai., P., "STEM Education and 21st Century Skills Development," *Executive Journal*, 2013, vol.33, no.2, pp. 49-55.
[13] Sadler, T. D., and Zeidler, D. L. "Patterns of informal reasoning in the context of socioscientific decision making,"*Journal of Research in Science Teaching*, 2005, vol.42, no.1, pp.112-138.
[14] Sadler,T. D., and Zeidler, D. L. 2004. "Student conceptualizations of the nature of science in response to socioscientific issue,"*International Journal of Science Education*, 2004, vol.26, no.4, pp. 387-409.

[15] Walpuski.M,Tepner.O, Sumfleth.E, Dollny.S, Hostenbachs.J, Pollende.T. "Multiple perspectives on students' scientific communication and reasoningin chemistry education,"*ActaDidacticaNorgeConferrence Vision Teaching,* 2012. vol. 6, no.1, University Duisburg-Essen.
[16] Yu, Y. *Adults' Decision-making about the Electronic Waste Issue: The Role of the Nature of Science Conceptualizations and Moral Concerns in Socio-scientific Decision-making.* Dissertation, Ph.D. (Art and Science). Columbia : Graduate School, 2010, Columbia University.
[17] Zeidler, D. and B. Nichols. "Socio scientific Issues Theory and Practice," *Journal of ElementaryScience Education*, 2009, vol.21, no.2, pp. 49-58.
[18] Zeidler, D. L., Sadler, T. D., Applebaum, S. and Callahan, B. E. "Advancing reflective judgment through socio scientific issues," *Journal of Research in Science Teaching*, 2009, vol.46, no.1, pp. 74-101.
[19] Zeidler, D. L.,"STEM education: A deficit framework for the twenty first century? A sociocultural socio scientific response," *Culture Study of Science Education,* 2016, vol. 11, pp. 11–26.

How to Apply Technology in STEM Education Activities

Artnarong Manosuttirit

Department of Innovation and Educational Technology, Faculty of Education, Burapha University, Chonburi, Thailand, 20131

E-mail : m_artnarong@hotmail.com

Abstract. STEM has been playing a part in Thailand for quite some time. STEM teaching and learning is originated from the US. Later, many countries have applied it to their own education systems. Since the arrival of STEM, ministry of education, under the supervision of teaching science and technology by Institute for the Promotion of Teaching Science and Technology, has used this teaching model. National Legislative Assembly of Thailand has also proposed STEM policy in order to gain tangible results. Until now, there is still no clarity and there are confusions with the organization of STEM teaching and learning. This article aims to propose the idea of the application technology in project-based STEM Activities. The content of this article contains concept of STEM, meaning of STEM, policy and principle of STEM for development of the nation, environment of STEM teaching and learning, and case study of applying technology with project-based STEM activities.

Keywords: How to Apply Technology, STEM Education

1. Introduction

Concept of STEM

Apisith Thongchai and et al (2012) concluded that STEM is an acronym for Science, Technology, Engineering and Mathematics. The concept of STEM is originated in the US from the gathering of the representatives from every prominent sectors in the country with the aim to improve the quality of the population by enhancing their skills in order to be able to compete with other nations. The government had the education policy to support learning and teaching by integrate these four subjects together. Up to now the word STEM is popularized as the government wished to promote this kind of teaching and learning. Nevertheless, there was still quite many misunderstandings toward STEM. Many people were more familiar with Science and Mathematics. General Education also concentrated on these two subjects. Moreover, many also misunderstood STEM as an invention or development of a way of learning that promote each individual subject. In fact, STEM is an integration of four subjects, namely Science, Technology, Engineering and Mathematics, which each of the subject holds the same importance. STEM engages learners to use all subjects to solve the problems, research and create or develop things in the present. The teaching is formed by teachers which come from many areas of expertises. In the real working world, knowledge from many areas needs to be combined and not separate into parts. This supports the essential skills of development in globalization. However, from the past to present, each subject is still being taught independently. [1]

IOP Conf. Series: Journal of Physics: Conf. Series **1340** (2019) 012007 doi:10.1088/1742-6596/1340/1/012007

Rakpol Thananuwong (2013) wrote that STEM Education is learning of content and the skills of Science, Mathematics, Engineering and Technology. These four cores are the subjects which promote learners to gain knowledge and has a quality living in the 21st century, a fast-pace, globalized and full of technology era. The four subjects are essential to enhance the capabilities of economic competition, the development of the life quality and the nation's security.

In the past, the learning of these four subjects in STEM are separated and independent, for example STEM teaching and learning in Thailand or in the US had courses of Science, Mathematics, Engineering and Technology separately with the lesser focus on the engineering. Educators and academics in the US have compared this separation of the four subjects as "silo", which lies independently in the field.[2]

It was found that at present the US was not the first when measuring nation's capacity like it used to be in the past. Many countries around the world have made progresses. The PISA results also showed that the US was worse than many countries. Moreover, the population of science, technology and engineering also dwindled. Thus, the government has implemented a policy which supports the development of STEM. They hoped to raise the PISA results. Furthermore, the swift changes of technology, especially in the communication, logistics, trading, etc, makes the world globalized so it is utmost importance to bring STEM teaching and learning to support the necessary skills in the 21st century. (Apisith Thongchai, et al 2012) [1]

Pornthip Siripatrachai (2013) said the necessary skills for the 21st century began from the assembly of academics from many expertises in the US. The government wanted to improve to quality of the people in order to elevate the capacity of the nation so it can compete with others. They also wanted their people to has quality and ability in the society so that they can live in the swiftly changing world. Thus, equipping learners with the 21st Century Student Outcomes, which contains knowledge, skills and expertise is essential. (The Partnership for 21st Century Skills, 2009). The outcomes are as follows, [3]

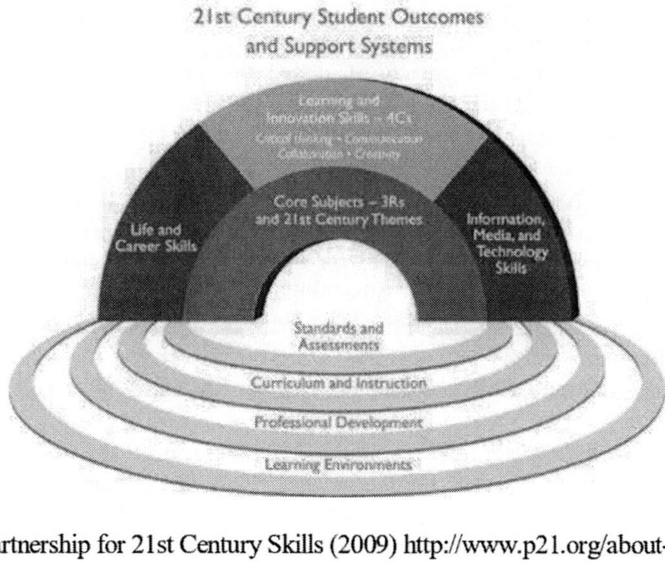

Figure 1. The Partnership for 21st Century Skills (2009) http://www.p21.org/about-us/p21-framework

1. Core Subjects and 21st Century Themes are English, Reading, Art of language using, foreign languages, mathematics, economics, science. Art, Geography, History, Citizenship, Political Science. These subjects also need to cover the new areas of content which will influence work and community. Education institutes also have not given enough importance toward subjects such as awareness toward the earth, fundamental course of finance, economics, administration, entrepreneurship, civil and awareness for health and welfare.

2. Learning and Innovation Skills, as follows

- Creativity and Innovation which covers creative design, collaborate creatively with others and application of creativity.
- Critical Thinking and Problem Solving which includes rational thinking, systematical thinking, decision making and solving problems.
- Communication and Collaboration which concentrates on communication by using various types of media and collaboration with others effectively and clearly.

3. Information, Media and Technology Skills. In the 21st century, which many technological advancements, learners need to know the these following skills.

- Information Literacy
- Media Literacy
- ICT (Information, Communications & Technology) Literacy

4. Life and Career Skills. In our daily and working lives, we need not only the people who have knowledge and thinking skills, but we also need the people who are able to work in a complex context. These necessary skills are as follows.

- Flexibility and Adaptability
- Initiative and Self Direction
- Social and Cross-cultural Skills
- Productivity and Accountability
- Leadership and Responsibility

2. STEM: Meaning and its importance

Originated from the US, the concept and principle of STEM was announced to be its national policy. After that many countries received and adapted STEM to their own nations. Especially in Thailand, STEM has been applied to teaching and learning of Science and Technology. There are many definitions to STEM such as

STEM is a way of organizing education that integrate four subjects, which are Science, Engineering, Technology and Mathematics. STEM concentrates on practicality of the knowledge to be used in real life, including the development of processes or products, which benefit the working and living. [4]

Wasinee Itsarasena Na Ayutthaya(2017) gave definition to STEM as a way of organizing education that is influencing the changes of human in the 21st century. STEM Education or STEM is derived from the integration of four subjects as S stands for Science, T for Technology, E for Engineering and M for Mathematics. STEM also means root, which can imply to the foundation that causes development and changes to human. These four subjects are necessary skills to human of the present and the future. The teaching of STEM is different from the past that required learners to recite. STEM Education is an integration of Science, Technology, Engineering and Mathematics which allows learners to experience, practice, and think creatively of how to solve the problems. STEM Education should be taught from kindergarten level as it trains learners to be curious for new knowledge and know to solve problems. Consequently, they grow up having a systematic thinking and they are aware of the changes and the advancement of technology in the world. Teaching STEM to learners from their young age will also create good attitude toward Science and Mathematics, which often seen from the learners as difficult subjects and being dislike. In the long term, this could influence the young generation to be interested in Sciences, Technology, Engineering and Mathematics. Consequently, more people will be working in the areas of Science-Mathematics which allows the number to be met with the needs in economic and technological markets at present and in the future. STEM Education is therefore very important for human development. Teaching and learning STEM with other subjects such as language, arts, social science and morality will create holistic knowledge and skills for people in order to improve the quality of our society.[5]

3. Policy of STEM education

The policy proposal of STEM education which aims at integrating knowledge of science, technology, engineering and mathematics emphasizing on problem solving in real life to promote additional experiences, life skills and creativities as well as to prepare the students to be ready for any practices that require scientific, technological and mathematical knowledge and process that lead to future innovation. It is an enjoyable learning and teaching process for students and benefits their future careers. STEM education helps establish manpower who has skills on problem solving, creativity and new creative innovation. Hence, it is the important foundation of innovative skill development and the significant mechanism in improving economy of Thailand to higher income level in the future and being able to integrate knowledge from real life and working.

Social stimulation of STEM educational significance would cause developmental mechanism for scientific, technological and innovative manpower for society. Public hearing of organizations, institutes and associations, especially those who use the products of STEM education, plays a role in developing manpower and pushing forward national policy to support Thailand's innovative systems. [6]

4. Organizing STEM teaching and learning

Apisit Thongchai et el (2012) elaborated the above mentioned of the necessary skills for the 21st century and the nature of the its science as follows,

- Science is a study of natural phenomena by using the process of Scientific Inquiry. Science is tools that make human understand more about nature. In May 2012, the US promoted the new Science K-12 Framework and published for online public hearing. The concept was to combine Technology and Engineering together with raising the importance of engineering design to be equal to scientific inquiry.

- Technology is a study of working process to solve, adjust and develop in order to serve the needs of human. The processes of solving the problems or work, in the technological term is called Engineering design or Design process, which has similar process to scientific enquiry. Organizing this learning pattern is a problem-based or project-based learning. Nevertheless, most of the people usually misunderstood that Technology is computers and ICT equipment. In fact, Technology means the process of solving the problems or work in order to create things that serve our needs. In the US, there is International Technology and Engineering Educators Association or ITEEA, who sets the standard of Technology so teachers can teach in the same direction.

- Engineering is a study of creation or innovation or other things that facilitates human's needs by applying the knowledge of Science, Mathematics and Technological process. However, even in the US, the subject of Engineering is not clearly observable in mandatory education but rather is being covered in the same umbrella as Technology. [1]

-Mathematic is an important subject. It has a clear nature of theory. Mathematics is a subject which can be well-combine with the other three subjects. For organizing teaching and learning, Rakpol Thananuwong (2013) proposed that the integration of STEM Education lies in the learning that link Science and Mathematics explicitly in Engineering Connection. Or it can be said that STEM learning is a design-based learning, which explicitly shown in Engineering knowledge. Integration of the four

subjects in STEM teaching and learning requires group practice, discussion and communication skills in order to present the work and result. It is similar to project-based learning, which has been studied that help learners to acquire "deeper learning". Deeper Learning contains five aspects as follows;

- Mastering Core content
- Think critically and Solving complex problems
- working collaboratively
- communicating effectively
- Self-directed and Incorporate feedback Apart from that, integration of STEM Education is also combine two aspects of teaching and learning.

IOP Conf. Series: Journal of Physics: Conf. Series **1340** (2019) 012007 doi:10.1088/1742-6596/1340/1/012007

- Context integration which integrates four subjects together with learner's daily life in one context.
- Content integration which integrates four subjects together with one "Big Idea". [2]

5. STEM teaching and learning condition

Artnarong Manosuttirit(2016) researched STEM teaching and learning condition. It was found that there are many factors that require supports and modification. The details are as follows.

1. Educational policy and curriculum: Each school cannot determine the direction and format of STEM instruction because it lack enough information and understanding in STEM education management and there is no clear guidelines provided. This makes each school worried that STEM instruction may affect the teaching and learning of other subjects, especially in terms of limited studying time which is not conducive to any other activities.
2. Integration and collaboration: As there is no clear curriculum and policy, it is not possible to determine details of STEM teaching and activities in term of time, credit, place, and budget. Many schools still have not received financial support or have to use their own budget. This is an ongoing problem that needs to be solved.
3. Responsible teachers: STEM education is a new concept that needs time to understand and adopt. Normally, the teachers already have a lot of responsibilities to take care of. If they are assigned to handle STEM teaching, they will have more burdens and it may have an effect on their daily jobs. Therefore, nobody are willing to take responsibility in this respect. This point should be clearly understood and personnel tasks should be certainly defined.
4. Support from responsible organizations: Many schools carry out their own STEM instruction and activities without correct guidance and support so they are not certain whether what they did are right or wrong. This situation leads to some mistakes in STEM teaching and has an impact on related factors, including place, budget, teaching personnel, teacher training and development, which are all vital to STEM education advancement.

STEM instructional guidelines: The data obtained from the experts and schools in STEM education network can be concluded as follows.

1. Open approach and horizontal learning should be applied. The teachers should control the situation, listen to the students' opinions, and allow the students to speak, think, practice, and research on their own.
2. Any surrounding topics can be applied to STEM activities, especially in application field as everything is already integrated. STEM activities should be opened without any limitation and should focus on problem-solving methods, concepts, and processes.
3. The teachers of each subject should design the activity together in holistic manner. All involved aspects should not be split and inserted with irrelevant ones. The teachers should emphasize understanding on integration, skills, and ways of thinking.
4. Activities need to be challenging. They should not be too difficult or too easy. The challenge should be moderate and appropriate with environment and restriction.
5. STEM should be used with other instructional activities and teaching methods in order to respond to skill and ability requirements such as using STEM together with project-based teaching to practice 21st century skills, critical thinking, and project executing.
6. Activities should be flexible and suitable for educational environment. They should neither have rigid format nor add burden to the teachers and students. All of the activities should be blended in routine work and inserted in regular class content concerning problem-solving in daily life and practical situation in the community.

From the circumstances and the problems occurred, there were still difficulties that make STEM teaching and learning and activities possible. With many factors that need to be integrated within the four subjects, it takes a huge amount of time in preparation, set up the right time and organize activities for STEM teaching and learning. A case study presents here is in a Computer class in Lower Secondary Level, called "Arduino STEM" which is a project-based activity.

6. Example of Technology application to STEM teaching and learning

Subject : Embedded System Arduino UNO R3 Grade: 8 Times: 40 hrs.(2 hrs/week)

Course Description:

Getting Started with Arduino, Arduino has become much more than just a tool for making, tinkering, and exploring physical and digital interaction, but an entire ecosystem supporting students across all disciplines. Arduino Education is committed to empowering educators with the necessary hardware and software tools to create a more hands-on, innovative learning experience.

Arduino's one-of-a-kind STEAM (Science, Technology, Engineering, Arts, and Mathematics)

Learning Objectives:

1. Solve problems with process
2. Programming language and can be analyzed by using the computer.
3. Could understand the working principles of Arduino design and function.
4. Responsible. Creative and local resources to apply them appropriately.
5. Designs created by making fantasy computer language properly.

Learning Contents:

- Lecture by media E-learning.
- Reporting group.
- Write a program and control the Arduino board Working with Functions on Arduino IDE Using LEDs.
- Write a program (Automatic Fan,Traffic Light)
- Create jobs Project

Learning Activities:

(Introduction)

1. The students meditate before class and learn about morals in daily life for 2 minutes. Brain Gym
2. The teacher states the learning objectives to the students.

(Instruction)

1. Teacher did the analysis of individual student results to improve their learning and performance.
2. The student did login to Moodle learning platform
3. Teachers make a plan to prepare students to analyze and solve problems with application programs.
4. Teachers teaching about Recommended program "Moodle system and Installing Arduino Using LEDs and App Inventor"

(Conclusion)

1. Ask random students to describe and demonstrate teaching content information from the lecture
2. The teacher separates the students into a team for group activity projects and show Conception Teachers and students to share a summary discussion.

Learning evaluation:

1. Presentation.
2. Quality work project.
3. Quiz

Example of Projects:

IOP Conf. Series: Journal of Physics: Conf. Series **1340** (2019) 012007 doi:10.1088/1742-6596/1340/1/012007

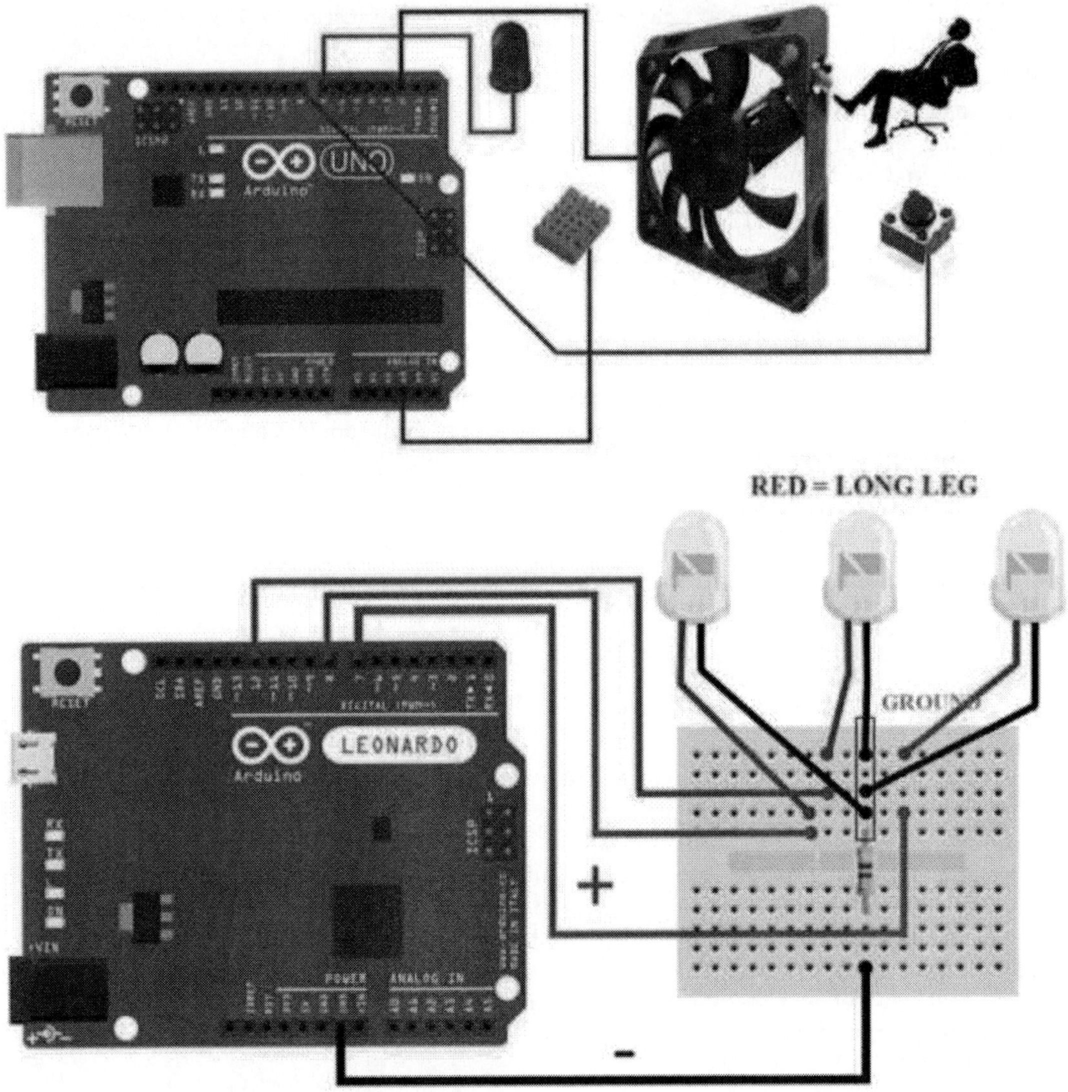

Figure 2. Arduino STEM (Automatic Fan,Traffic Light) www.acsp-cs.com

7. Conclusion

Although STEM Education has reached Thailand for quite some time and the relating government organizations have tried to push STEM Education's teaching and learning, in term of policy, personnel and other issues, it is still quite an abstract concept. Thus, it is necessary to study STEM and plan how STEM in Thailand should be. As Sutheera Prasertsil (2016, p.10) suggested evidences in the US shown that in 100 working people, 96 is hired by 4 people and those 4 people are educated in STEM, which is Science-Technology-Engineering-Mathematics. So, the US has tried to propel the education system by using STEM in its fundamental education and encourage STEM to be used in university level. The problems, when compare to the US, that stops STEM Education from succeed in Thailand, is Thai context, teachers' development and the ability of the teachers to deliver knowledge to learners. Sutheera added that "STEM teachers" need to equip themselves with these necessary skills. [8]

1. Awareness of the patent
2. Knowing various kinds of technology

3. Ability to see Mathematics as representative of Science and Simulation
4. Ability to see assumption
5. Understanding the subject thoroughly
6. Seeking opportunities
7. Understanding systematically
8. Creating a reasoning charts which can be explained by each subject's content
9. Application of knowledge
10. Five mantras of inventions (Sutheera Prasertsri ,2015)
    10.1. What are the previous inventions?
    10.2. What are the disadvantages of the previous inventions? What needs to be solved? What solution did we choose and why?
    10.3. What principle do we applied to the disadvantages? Why do we choose these principles?
    10.4. How do we apply with our invention?
    10.5. How can we prove that our inventions are better than the previous ones? [9]
11. Understanding liberal arts
12. ability to sketch
13. ability to write flow chart of the process with feedback
14. Computational thinking

These skills might only be some parts of the suggestion that STEM teachers should possess, further study and application to each and individual context also need to be done. This, however, needs assistance from many organizations in order to concretely improve Thailand STEM

8. References

[1] Apisith Thongchai and et al (2012) Composer. Special Lecture Handout in Science, Technology, Engineering, and Mathematics Education: Preparing students for the 21st Century. by Prof. Dr. Edward R. Reeve.15 May 2012, at Sanan Sumit Hall The Institute for the Promotion of Teaching Science and Technology

[2] Rakpol Thananuwong (2013) Composer . Training Report in STEM Education by Prof. Mitchell Nathan.10-11 Jan 2013

[3] Pornthip Siripatrachai (2013) STEM Education and Development in 21 Century : Bangkok University, Journal of Administration. Yrs. 33 ep.2 April-June 2013 ,p 49-56

[4] The Institute for the Promotion of Teaching Science and Technology. (2018). Knowing STEM(http://www.stemedthailand.org)online [24/6/2018]

[5] Wasinee Itsarasena Na Ayutthaya(2017) Knowledge of STEM Education. Chulalongkorn University Press, Bangkok

[6] National Legislative Assembly.2015. Report of Policy Recommendation STEM Education Proactive policy Youth development and manpower in science, technology, engineering and mathematics. Online, http://www.senate.go.th/w3c/senate/pictures/comm/1547 *ข้อเสนอเชิงนโยบาย*STEM.pdf [19/6/2018]

[7] Artnarong Manosuttirit(2016). A Study of Teaching STEM Education in Thai High School. Thammasart International Journal of Science and Technology. Vol 21 No.5, 1-6 page

[8] Sutheera Prasertsil (2016). Decode STEM Teaching. Poh Pan Punya Project. Online https://kruyaivijai.wordpress.com/ [9.7.2018]

[9] Sutheera Prasertsil (2015). STEM in Invention : Development of Vision and Imagination. Poh Pan Punya Project Songkhla 2015. Online https://kruyaivijai.wordpress.com/ [9.7.2018]

Using Facebook in Blended Learning in Vietnamese Undergraduate Students

Dang Thi Thu Huong[1*], Bui Thi Oanh[2], Pham Thi Kieu Oanh[2] and Luong Khanh Luong[1]

[1]National Academy of Education Management, Dhaka, Bangladesh
[2]Thai Nguyen University of Education, Vietnam

*Corresponding author's e-mail address: huongdangnn@gmail.com

Abstract. The study investigated the use of Facebook as an educational tool as well as the influence of Facebook integration into blended learning environment on English pronunciation teaching and learning activities. The subjects for the study consisted of 30 students who were the first-year English majors at Thai Nguyen University of Education. They were divided into two groups: the experimental group and the traditional group. Besides, there was a student teacher from the same university to be the subject of the experimental group. A combination of qualitative, quantitative and experimental approaches was employed to collect data and students' views on this model. This allowed the researcher to get hold of a transformation in students' pronunciation competence and their attitudes towards a blended format with supporting a social networking site (Facebook). The findings of the study indicate that Facebook retains outstanding features to offer a professional online course for English pronunciation learning, such as live lectures, group discussions, schedule events and etc. Furthermore, Facebook benefits the teachers in managing students' learning autonomy so that the teacher is likely to grasp the ability of individuals in a short time. However, the use of Facebook still remains obstacles in the path of carrying out English pronunciation activities, and when putting it into blended-learning, the model handles these obstacles and limitations in the current learning and teaching of English pronunciation as well.

1. Introduction

Teaching and learning English in Vietnam are burning questions and given much attention to by the Government and the Ministry of Education and Training (MOET). Therefore, the Ministry of Education and Training launched a project called "National Foreign Languages 2020" and set a fundamental goal that "By 2020 most Vietnamese students graduating from secondary, vocational schools, colleges and universities will be able to use a foreign language confidently in their daily communication, their study and work in an integrated, multi-cultural and multi-lingual environment, making foreign languages a competitive advantage of development for Vietnamese people in the cause of industrialization and modernization for the country". However, this project has been considered to be on the brink of failing to reach its targets by Minister Phung Xuan Nha for several reasons. One of the reasons is that teaching and learning English in Vietnam do not seem to follow the international standards. It seems that language communication or English language speaking, in general, is a challenge to most students in Vietnam and the fact shows that most of the graduates cannot communicate fluently in English. Many employers complained that they were really disappointed about the low English language proficiency and poor confidence of university graduates. It has derived from the fact that students have little opportunities to speak English in class. This is demonstrated that English teachers mainly focus on reading comprehension, vocabulary and structural patterns with the aims of passing the final exams although the MOET highlights the significance of communicative competence [8]. Moreover, it is argued that English pronunciation competence of numerous English major postgraduates and even those who are currently teaching English at schools is relatively low. This leads to students' bad English pronunciation that is taught by them while pronunciation is regarded as a crucial factor which decides the success in communication. Therefore, teaching and learning English pronunciation of English major students at Universities should be on the list of priorities when setting out the curriculum to help learners make progress in their communication as well as in their professional future. However, teaching and learning English also have some limitations. According to [7] students are not satisfied with learning English at university for several reasons: the limited time for teaching and learning in the curriculum, the unbalance in English competence of students in a class, poorly organized and qualified English classes, out-of-date textbooks and boring teaching methods. As above-mentioned problems, there is a strong need to have a solution to tackle these problems. Industry 4.0 brings both challenges and opportunities for the Vietnamese education system. Many educational institutions are applying new technologies to their system. Professor Nguyen Xuan Thu, former vice chancellor of RMIT University in Melbourne, co-founder of the RMIT International University in Vietnam and the Australia- Vietnam Education Promotion Association, pointed out that traditional universities can collaborate with online teaching establishments to meet the needs of Industry 4.0. The "e-learning" or "online learning" term has emerged as a phenomenon of the integration of new technologies in education. Beside positive consequences, e-learning, or online learning faces up to negative consequences such as lack of face-to-face communication, one of the main disadvantages of this learning model. These days, there are a lot of solutions proposed; however, blended-learning is above all others due to its striking features. Blended-learning is also in accordance with Vietnam's context which is the combination of traditional learning (face to face) and e-learning. Therefore, it is an ideal measure to overcome obstacles and difficulties in teaching and learning English in Viet Nam. According to [5], blended-learning is a formal educational program in which a student learns at least in part through online delivery of content and instruction with some elements of student control over time, place, path, and/or pace and at least in part at a supervised brick and mortar location away from home. To carry out "blended-learning" model, there is a requirement of using a social media. In this research, the researcher chooses Facebook as an e-learning tool to implement blended-learning for several reasons as below:

In Vietnam, Facebook makes up for by far the most interests and is becoming an indispensable part of university students. They use Facebook for making friends, discussing and updating status or pictures. However, some people spend a considerable amount of time on surfing Facebook without intention. What is more, Facebook is overwhelming the other activities including study.

Second, Facebook actually is not designed for education, but it is a potential tool for teachers and learners as well to make use of it owing to useful and valuable functions. [2] argued that the use of Facebook may be helpful in improving low self-efficacy and self-regulated learning. [6] discovered that by using Facebook students have more chances to find experts or fellow students to discuss and investigate a subject. It is a golden opportunity to approach and widen knowledge from professors and qualified people. High teacher self-disclosure leads students to higher levels of motivation and effective learning and leads to a more comfortable classroom climate [3]. Therefore, it is highly recommended to use it for academic purposes instead of wasting time on Facebook.

Third, as a replacement for the traditional teaching method, a new approach of using e-learning in teaching and learning, in particular, Facebook will fascinate and motivate them to study.

Last but not least, Facebook is free for everyone, so any student can have their own Facebook account to serve their learning without worrying about the financial situations.

The study was undertaken to give out the approach and the experience in using Facebook as an educational tool in the blended-learning environment to facilitate English pronunciation teaching and learning activities with the aims of evaluating the influences of using Facebook for English pronunciation learning (advantages and disadvantages). The other purpose was to explore the students' attitudes/feedback/comments towards using Facebook in blended-learning for learning English pronunciation. Three questions were employed in the study:

1. What are activities on Facebook practical for English pronunciation learning?
2. What are the benefits and drawbacks when using Facebook for English pronunciation learning activities?
3. What are students' attitudes/feedback/comments towards using Facebook in blended-learning for learning English pronunciation?

2. Methodology

The study consisted of 30 students from English 51 at Thai Nguyen University of Education chosen randomly to be the subjects. In other words, there were 15 students in the experimental group and the rest came to the traditional group. Besides, there is a volunteer who played a role as a teacher joining the experiment. An experiment was implemented in three weeks to examine the effects of Facebook in blended-learning to support English pronunciation teaching and learning activities on learners' English pronunciation competence. The instructional context was a series of theoretic lectures and home assignments posted on Facebook. Then, the products of practicing pronunciation were uploaded for an after-class program in which teaching and learning activities via Facebook had been designed. After that, pronunciation activities were carried out in the classroom to reinforce their competence after the assignments on Facebook. Coupled with the experimental method, the qualitative method was implemented by observation during the process of the experimental method and semi-structured interview after finishing the experiment to perceive a change in learners' pronunciation and students' attitudes toward this model. Finally, the quantitative method was undertaken to collect data and learners' feedback toward using Facebook integrated blended-learning course after finishing the experimental method because quantitative method contributes the overview of the research problem to data collection from a large population by using some numeric tools [1].

2.1. *Data collection procedure*

Some text.

- Pre-test and post-test

  The researcher used an app named "Elsa speak" to test the subjects' English pronunciation competence before and after the experiment.

  The process of carrying pre-test and post-test:

The researcher contacted the monitor of English K51 (the first-year students) class to make an appointment. Then, the researcher came to class and introduced the study to ask for their help. Because there were 15 students to join the experimental group, choosing the participants based on their willingness. Those selected students were informed about the test schedule from Elsa speak. Coupled with the subjects of the experimental group, the subjects from the traditional classroom group also needed to take the test in this app. To do this, the researcher launched the app and how to use the app on Facebook. After the students had completed the test, all the outcomes were collected.

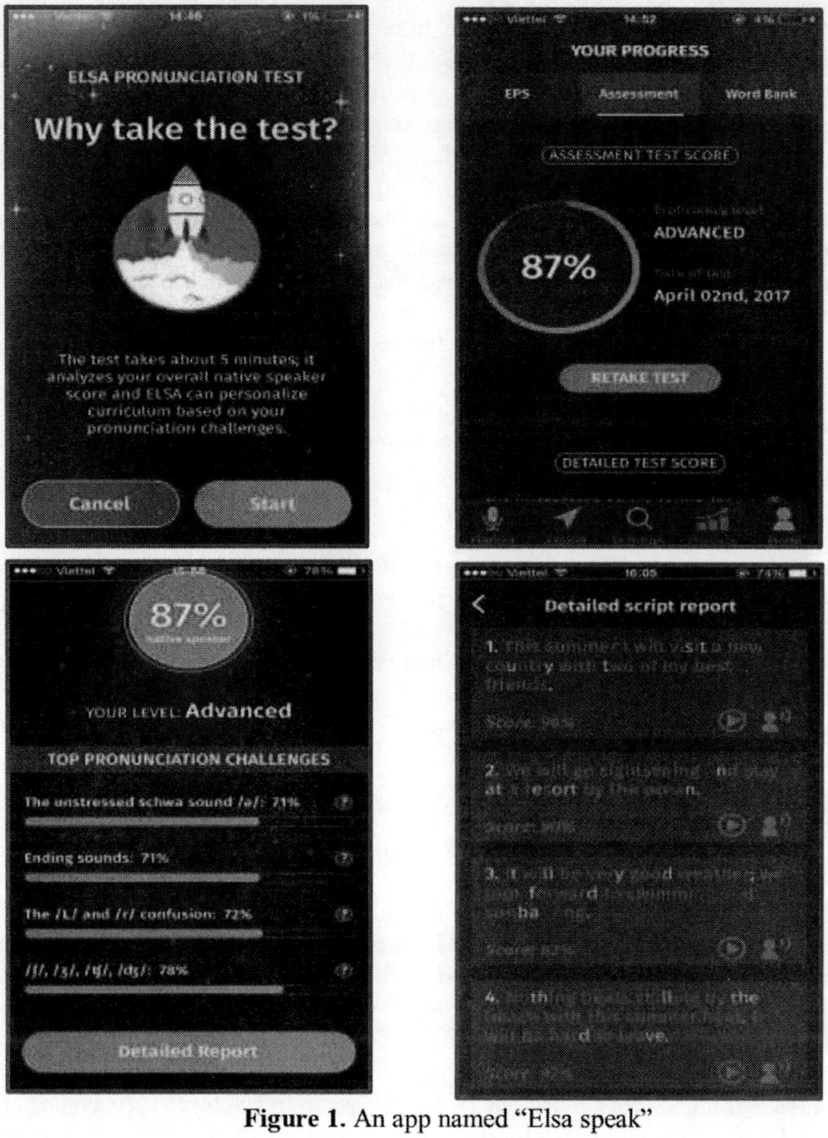

Figure 1. An app named "Elsa speak"

After finishing the experiment, 30 students belonging to the subjects of the study were required to retake the test to gather data so that the researcher made a comparison about the differences in English pronunciation competence between two groups.

Apart from the pre-test and post-test, the questionnaires and interviews were employed with the aims at providing further feedback/attitude/comments from the students and teachers.

- Questionnaires
  The questionnaire was conducted after finishing the experiment. To be more specific, the participants of the experimental group did the questionnaire in the last period of the offline class in blended-learning. Besides, the researcher was available in class to explain and instructed the participants to do the questionnaire. Finally, all the questionnaires were gathered and the researcher expressed the gratitude to their excellent participation during the research experiment.

- The semi-structured interview
  The semi-structured interview was also implemented after finishing the experiment. The researcher contacted the teacher to arrange the interview. The questions were open-ended to get further information about her ideas towards the model and the given issues from the answers of the participants based on the questionnaires for students.

2.2. *Data analysis*
After having the answers to the three research questions of: 1) practical activities on Facebook for English pronunciation learning; 2) their benefits and drawbacks; and 3) students' attitudes/ feedback/comments of using Facebook in blended learning for English pronunciation. The authors started processing the data.

For analyzing data, the researcher calculated the results in terms of percentage as well as Mean, Mode and SD (Standard Deviation) [4].

According to BusinessDictionary, the three values can be understood as follows:

Mean: is a simple average of a range of values or quantities, computed by dividing the total of all values by the number of values.

Mode: is a type of average that refers to the most common or most- frequently occurring value in a series of data.

SD (standard deviation): a measure of the unpredictability of a random variable expressed as the average deviation of a set of data from its arithmetic mean and computed as the positive square root of the variance.

Data were presented in a pie chart, bar charts and tables to give readers the overview of each part and make them easy to observe. Then, every perception/idea was analyzed independently to explain further on the phenomenon. Based on the data analysis, conclusions were given and several possible suggestions were recommended to tackle the obstacles of the model.

3. Findings and discussion
The study consisted of 30 students from English 51 at Thai Nguyen University of Education chosen randomly to be the subjects. In other words, there were 15 students in the experimental group and the rest came to the traditional group. Besides, there is a volunteer who played a role as a teacher joining the experiment. An experiment was implemented in three weeks to examine the effects of Facebook in blended-learning to support English pronunciation teaching and learning activities on learners' English pronunciation competence. The instructional context was a series of theoretic lectures and home assignments posted on Facebook. Then, the products of practicing pronunciation were uploaded for an after-class program in which teaching and learning activities via Facebook had been designed. After that, pronunciation activities were carried out in the classroom to reinforce their competence after the assignments on Facebook. Coupled with the experimental method, the qualitative method was implemented by observation during the process of the experimental method and semi-structured interview after finishing the experiment to perceive a change in learners' pronunciation and students' attitudes toward this model. Finally, the quantitative method was undertaken to collect data and learners' feedback toward using Facebook integrated blended-learning course after finishing the experimental method because quantitative method contributes the overview of the research problem to data collection from a large population by using some numeric tools [1].

3.1. *Background of the subjects*
In the number of 15 students from the experimental group, 93.3 percent of them confirmed that they have not experienced "blended-learning" before while there was only one person among 15 students, who has ever experienced ahead of the experiment. He gave his situation as follows:

"I learned online through Skype with my aunt and when she came back to Vietnam, I learned face-to-face with her"

This statistic reveals that Blended-learning is not widespread in Vietnam, especially at schools. It seems that the traditional class still dominates every learning and teaching method. Besides, 100 percent of the subjects committed themselves to use Facebook; however, years of having a Facebook account fluctuated from two to six years.

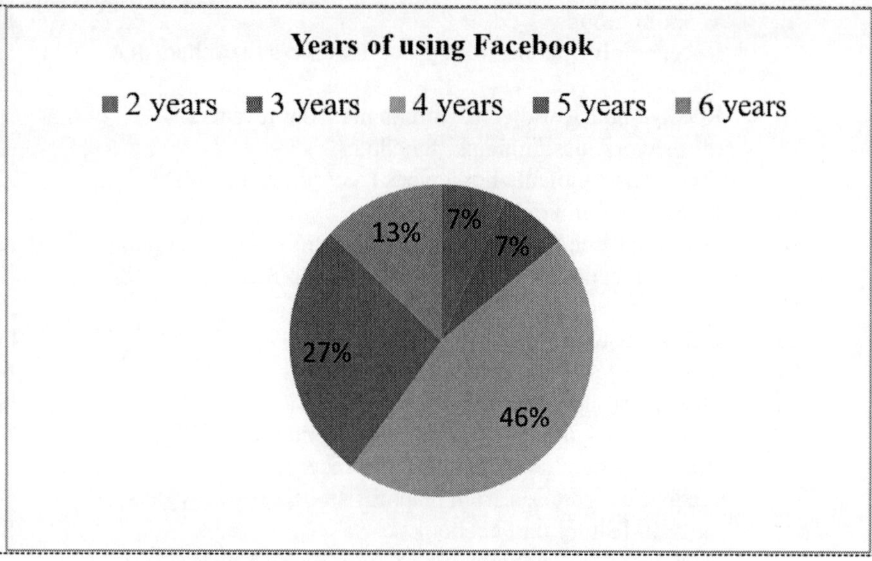

Figure 2. Years of using Facebook by the subjects.

More interestingly, when being asked the time to access to Facebook in a day, 100 percent of them selected the same period of time in a day, from one to five hours a day which they often spend on Facebook.

3.2. *Research question 1: What activities on Facebook are practical for English pronunciation learning?*

After the experiment finished, the researcher wanted to investigate the roles of Facebook for English pronunciation learning in blended-learning through activities that the participants often do on Facebook when learning.

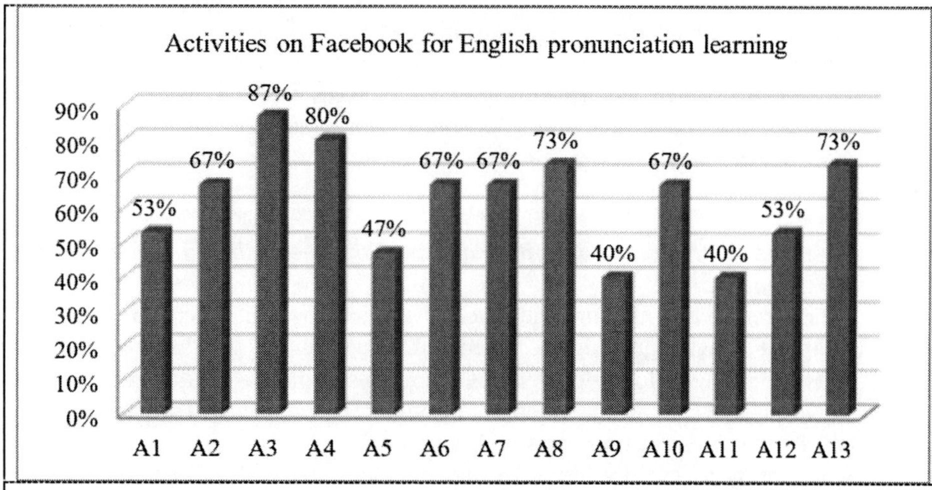

Figure 3. Facebook activities for English pronunciation learning.

Table 1. Facebook activities for English pronunciation learning.

No.	Code	Meaning	Number of participants
1	A1	Work in groups	8
2	A2	Receive information from group members (Teachers and classmates)	10
3	A3	Receive notifications/ requirements from teachers (homework, assignments, schedules,..)	13
4	A4	Upload assignments/homework (recordings or videos..)	12
5	A5	Share ideas and opinions	7
6	A6	Give feedback/ comments to other members' assignments	10
7	A7	Receive feedback/comments from group members on the content which you have posted	10
8	A8	Contact teachers and group members	11
9	A9	Watch lectures and interact with teachers	6
10	A10	Discuss group assignments	10
11	A11	Share knowledge and problems in pronouncing English	6
12	A12	Practice pronouncing English and receive feedback/correction from teachers and others via video call/call feature on Facebook	8
13	A13	Receive reference materials relating English pronunciation from teachers	11

As can be seen from the bar chart, the top four common activities were Receiving notifications/ requirements from teachers (homework, assignments, schedules,..), Uploading assignments/ homework(recordings or videos..), Contacting teachers and group members and Receiving reference materials relating English pronunciation from teachers respectively. By far the most common activity on Facebook for English pronunciation learning, Receiving notification/ requirements from teachers (homework, assignments, schedules,..) was performed at 87%. 80% of the participants reckoned that they often upload assignments/homework (recordings or videos..). On the other hand, the smallest proportion of participants' activities came to watching lectures and interacting with teachers and sharing knowledge and problems in pronouncing English at 40%. As I observed during the experiment, there were several reasons for each activity. First, when teaching English pronunciation, the teacher would encourage student-centered learning to boost students' cognition and competence. Therefore, lectures designed by the teacher were not conducted frequently. Moreover, some students shared that they could not watch lectures and join the lectures through live streaming sometimes because of the poor internet connection. Secondly, some students hesitated to share their thoughts/problems from the very first time of the experiment; however, they gradually changed after every lesson.

3.3. *Research question 2: What are the benefits and drawbacks when using Facebook for English pronunciation learning activities?*

The questionnaire is designed to measure the level of opinion to each statement. Each statement is attached to five scales (Strongly agree, agree, undecided, disagree, strongly disagree) in a likert-scale, and the participants are required to choose one of them. Based on the results from 15 participants' choices, the researcher would analyze the outcomes through Mean/Mode/SD (standard deviation).

In the study, the three values can be understood as follows:

Mean: is simple average assessed value of 15 students to a statement in five scales (Strongly agree, agree, undecided, disagree, and strongly disagree)

Mode: is a value which receives the most choices from 15 students to a statement and the mode value in the study ranges from one to five.

For example: If the mode of statement B1 is two, the scale "Agree" receives the most choices from 15 students.

SD (standard deviation): is a measure which shows the amount of variation of 15 students' choices from scale "Strongly agree" to scale "Strongly disagree" in a statement.

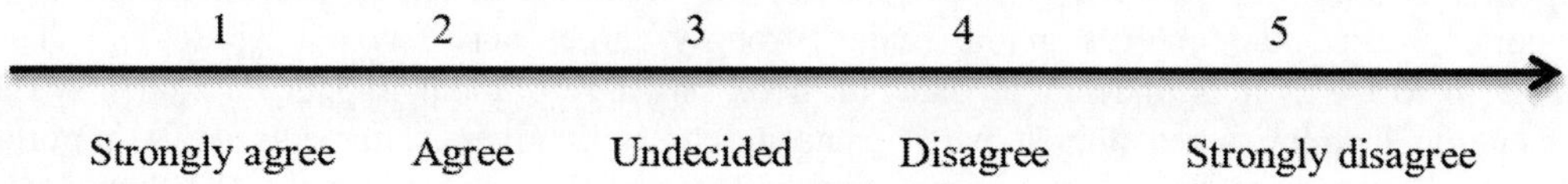

Table 2. The benefits of using Facebook for English pronunciation learning activities

No	Code	Meaning
1	B1	I would learn and discuss at any time and place
2	B2	I would not miss any lesson if I had been absent
3	B3	I would interact more with teachers and classmates
4	B4	I would feel more comfortable when discussing on Facebook
5	B5	Typing would give me time to think of more ideas
6	B6	I would feel interested in learning through live streaming
7	B7	I would update the schedules easily
8	B8	I would approach various learning resource
9	B9	Facebook messenger group allows members to chat directly and indirectly
10	B10	I would be corrected my pronunciation by calling directly to teachers on Facebook

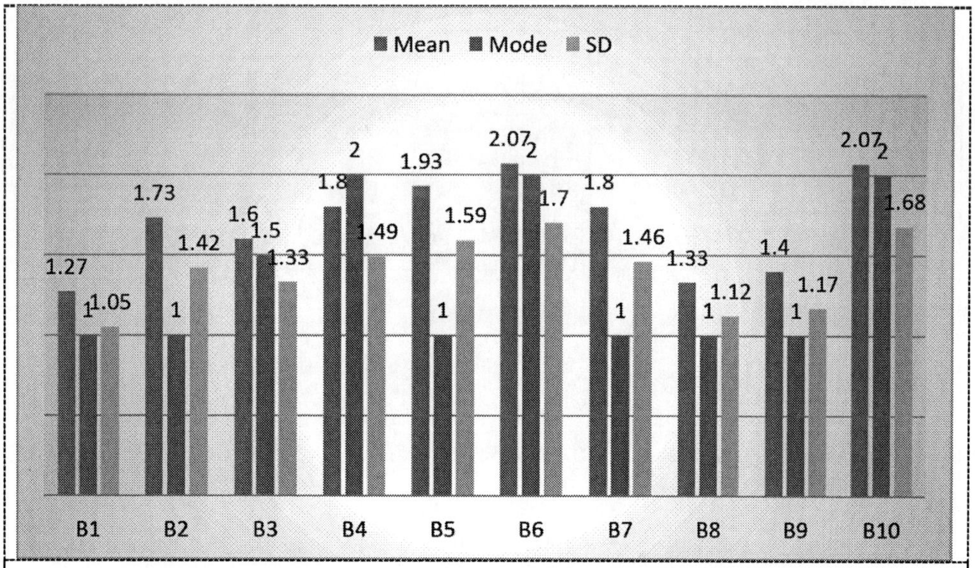

Figure 4. Benefits of using Facebook for English learning activities.

As can be seen from the chart, the mean ranged from 1.27 to 2.07. Overall, it is evident that most students showed their level of agreement with the scale "Strongly agree" and "Agree". Of the ten statements, students agreed on these ideas I would feel interested in learning through live streaming and I would be corrected my pronunciation by calling directly to teachers on Facebook the most with 2.07. In comparison, the lowest mean came to the perception that I would learn and discuss at any time and place (1.27). It means that a large number of students strongly agreed with this idea which would be convenient for the whole team, just ahead of the idea I would approach various learning resource with 1.33.

In respect of Mode, the scale "strongly agree" was received many students' choices in the six perceptions. Besides, the statement I would interact more with teachers and classmates had the same most common students 'choices in the scale "strongly agree" and "agree" (Mode 1.5). The rest belonged to Mode 2. It is undeniable that Facebook has a substantial impact on students' English pronunciation learning when almost participants agreed with given statements. It is important to analyze the SD value relating to the Mean value. The chart figures out that the SD fluctuated from 1.05 to 1.7. Generally, there was a big gap in each perception to five scales. The top two ranking were I would feel interested in learning through live streaming and I would be corrected my pronunciation by calling directly to teachers on Facebook with the similar SD score of 1.7 and 1.68 respectively. In contrast, the lowest SD value was for I would learn at any time and place (1.05).

Table 3. The drawbacks of using Facebook for English pronunciation learning activities.

No	Code	Meaning
1	C1	I would not follow teachers during live lectures because of internet connection
2	C2	I would be easily distracted from lectures by chatting or receiving notifications on Facebook
3	C3	I would not stay online as scheduled once
4	C4	Typing would be a waste of time
5	C5	The concentration would not be constant because we could be at any place
6	C6	I could not upload files with heavy properties

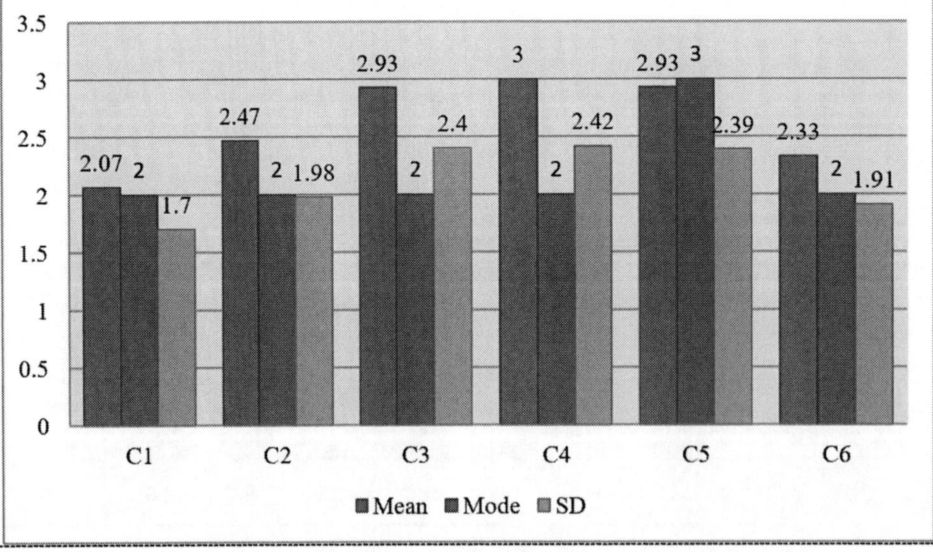

Figure 5. Drawbacks of using Facebook for English pronunciation learning activities.

As can be seen from the chart, the mean fluctuated from 2.07 to 3. It reveals that there was the large gap among students' opinions. The top three disadvantages that participants agreed were I would not follow teachers during live lectures because of internet connection, I cannot upload files with heavy properties and Typing could be a waste of time respectively. On the other hand, the largest number of students showing their disagreement belonged to I would not stay online as scheduled once. The scale "undecided" receiving the largest number of participants was the concentration would not be constant because we could be at any place. In terms of mode, participants 'choices scattered in five scales in general. However, the mode 2 was considerably chosen by students. This reveals that most of the students agreed with the given perceptions. In contrast, there was only one mode 3, which came to the concentration would not be constant because we could be at any place.

It is interesting to note that the SD value fluctuated from 1.7 to 2.42. In particular, the connection between the mean and the SD value when the higher the mean is, the higher the SD ranks. The highest SD value came to the waste of time when typing.

Table 4. Research question 3: What are students' attitudes/feedback/comments towards using Facebook in blended-learning for learning English pronunciation?

No	Code	Meaning
1	D1	I would reinforce knowledge continually
2	D2	This model would contribute to enjoyable and stimulating learning
3	D3	I would practice English pronunciation regularly
4	D4	The model boosted my confidence in learning English pronunciation
5	D5	It would save time on going to the class
6	D6	It would complement environmental communication
7	D7	The model built strong relationships between a member and the others in a group
8	D8	I would like to study English pronunciation through this model
9	D9	I would improve my pronunciation through blended-learning than traditional learning

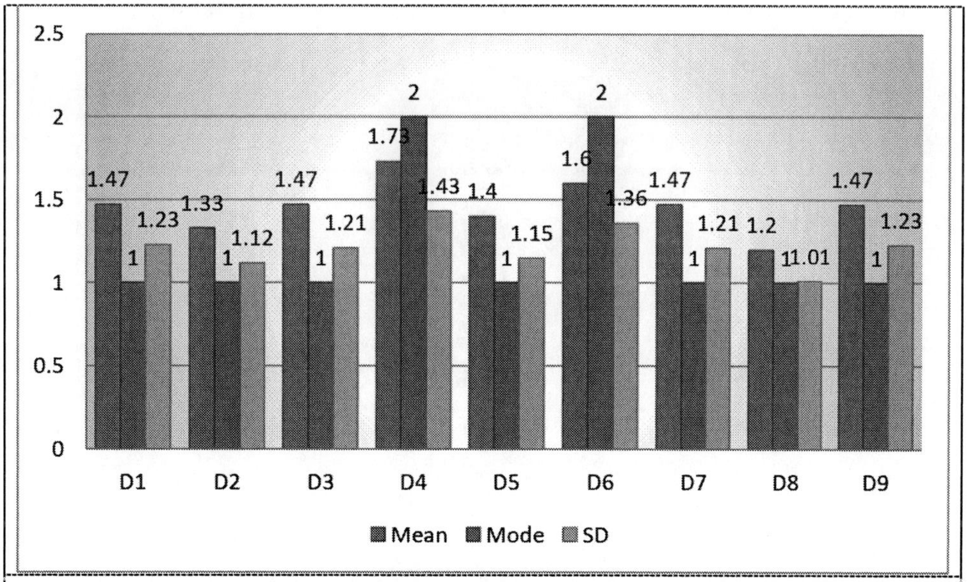

Figure 6. Students' attitudes/feedback/comments towards using Facebook in blended-learning for learning English pronunciation.

The provided chart shows the number of students expressing their opinions on 9 perceptions in the likert-scale. It is obvious that their choices for 9 ideas mainly focused on the scale "Strongly agree" and "Agree" and there were no choices for 9 perceptions with the scale "Strongly disagree" and "Disagree". These figures were illustrated by the mode of 1 for 7 perceptions and the mode of 2 for 2 perceptions. Specifically, students agreed on the ideas the model boosted my confidence in learning English pronunciation and It would complement environmental communication. The rest perceptions were strongly agreed by the large number of students.

What is more, to demonstrate perception D9 (I would improve my pronunciation through blended learning than traditional learning) to be true, the researcher piloted pre-test and post-test on 2 subjects (the experimental group and the traditional group) before and after the experiment to evaluate the effectiveness of using Facebook in blended-learning based on the test on app "Elsa speak" which calculates users' pronunciation level compared to a native speaker by percentage.

Table 5. The learning outcomes of the experimental group. (a)

Participants	Pre-test score	Post-test score
1	66%	75%
2	70%	77%
3	46%	52%
4	70%	77%
5	79%	82%
6	74%	80%
7	57%	61%
8	18%	64%
9	78%	84%
10	79%	81%
11	70%	78%
12	60%	72%
13	64%	75%
14	75%	79%
15	61%	61%
16	74%	78%
17	69%	70%
18	78%	82%
19	62%	68%
20	73%	76%

Table 5. The learning outcomes of the traditional group. (b)

Participants	Pre-test score	Post-test score
21	73%	73%
22	79%	80%
23	62%	66%
24	53%	56%
25	76%	76%
26	72%	74%
27	50%	48%
28	57%	62%
29	77%	79%
30	66%	66%

A glance at the tables provided above, it is clear that there was a big difference between pre-test scores and post-test scores of the participants in the first table while there was an insignificant gap between pre-test scores and post-test scores of 15 students in the second table. This demonstrated that although the experiment implemented in three weeks, participants' pronunciation competence improved when learning in blended-learning. To the traditional group, the improvement in pronunciation ability of 15 students seemed to be unremarkable. Therefore, the perception I would improve my pronunciation through blended-learning than traditional learning receiving the majority of students was completely right.

In conclusion, by analyzing the questionnaire responses gathered, the researcher realized that the new method stimulated students' interest and desire to study English pronunciation. To put it another way, they still kept a positive attitude, so it could come to the conclusion that the model put them under no pressure at all.

4. Findings and discussion

The research presents that students' English pronunciation competence was dramatically improved with the use of Facebook in the teaching and learning of English pronunciation through blended learning. Their pronunciation skills were proven to have bettered not only their responses to the questionnaires but also their scores in the pretest and posttest design. The results show that this new method had positive impacts on learning and teaching English pronunciation. As the research observed, students' interest in learning was maintained during the process of the experiment. Furthermore, their confidence and the closeness in the relationship between the teacher and students were increased. More interestingly, after the treatment period both the teacher and students supposed that they preferred the new method than the traditional method. However, there are still several problems in the use of Facebook for learning English pronunciation. Therefore, the teacher suggested some effective solutions to deal with these problems. Above all, despite possible problems, the use of Facebook in blended-learning in the teaching and learning English proves to be effective and might be a recommendation for teachers and learners at universities.

IOP Conf. Series: Journal of Physics: Conf. Series **1340** (2019) 012008 doi:10.1088/1742-6596/1340/1/012008

References

[1] Babbie E R 2010 *The* Practice of social research. 12th ed. Belmont, CA:Wadsworth Cengage, 2010; Muijs, Daniel. Doing Quantitative Research in Education with SPSS. 2nd edition. (London:SAGE) Retrieved from http://libguides.usc.edu/writingguide/quantitative.

[2] Browers-Campbell J 2008 *Cyber "Pokes": motivational antidote for developmental college readers.* Journal of college reading and learning.

[3] Mazer J, Murphy R and Simonds C 2007 *I'll see you on Facebook. The effects if computer-mediated teacher self-disclosure on student motivation, affecting learning, and classroom climate.*

[4] Mean, mode, standard deviation (SD). Retrieved from: http://www.businessdictionary.com/

[5] Michael B H 2014 Blended: *Using Disruptive innovation to improve schools.* San Francisco: Jossey-Bass

[6] Senges M 2008 *Facebook for teaching and learning*

[7] Tuyet T T 2013 *Factor affecting teaching and learning Englishin Vietnamese Universities*

[8] Van H V 2010 *The current situations and Issues of the Teaching of English in Viet Nam*

Assessment of In-service Science Teachers' Teaching Competencies in Integrated STEM Program

Tepporn Lomarak[1*] , Bancha Nuansai[2] , Worrawat Promden[3] and Arunrassame Sangsila[4]

[1]Head of General Science Program, Faculty of Education, Buriram Rajabhat University, TH
[2]Associate director of Research and Development Institute, Buriram Rajabhat University, TH
[3,4]Lecturrer at General Science Program, Faculty of Education, Buriram Rajabhat University, TH

*corresponding author's e-mail address: tlomarak@gmail.com

Abstract. The purpose of the study was to investigate the effects of a professional development instructional program Integrated STEM for elementary schools in a rural context of Thailand which carried out at the Faculty of Education, Buriram Rajabhat University during February 2018. Research instruments were (1) assessment observational record consisted of 5 elements of competencies with 4 levels of rubric criteria including beginning, approaching proficient, proficient, and distinguished levels, (2) achievement test related STEM designed to assess specific science content knowledge. The research findings revealed that the integrated STEM program was recommended to use for developing in-service science teachers' teaching competencies. From evaluation of experts, the results indicated that competencies' levels of teaching of in-service science teachers was achieved as approaching proficient level. Moreover, the results of one sample t-test indicated that teachers' achievement score on science content knowledge after implementing the integrated STEM program was significantly higher than before at the 0.01 level. It can be suggested that in-service training programs should be developed and used for science teachers to raise their awareness of the necessity of STEM education and to enhance their competencies in teaching preparation program.

1. Introduction

Ministry of Education launched project to engage in-service teachers with innovative classroom. The Institute for the Promotion of Teaching Science and Technology (IPST) has initiated a new approach by emphasizing knowledge and skills which are suitable to professional life in a highly competitive society. STEM education is an integrated approach which stands for science, technology, engineering, and mathematics. These disciplines are distinction in our daily lives, helps us to solve the problem by creative thinking based on contents clarification. STEM education is now distributing around the world, and also in Thailand education. IPST engages STEM education to science and mathematics curriculum, initially introduced with learning activities, and propagating STEM to school practices at national level. To prepare students, the IPST promotes the science process skills, mathematical skills and other skills for the 21st century among learners by way of the integrated science, technology, engineering, and mathematics curriculum. IPST settled STEM education into curriculum and educational policy in 2013. After some seminars, workshops and experimental activities with many countries, the IPST executive committee decided to initiate a campaign to promote integrated STEM teaching and learning in Thai K-12 schools under "the Development and Promotion of STEM Education Project" [17]. Consequently, there could be many unforeseen challenges facing researchers

IOP Conf. Series: Journal of Physics: Conf. Series 1340 (2019) 012009 doi:10.1088/1742-6596/1340/1/012009

and teachers who are involved in implementing this campaign. It was thought that the investigation of the opinions and suggestions from the teachers and educators who have used STEM strategies might reveal some implications that could lead to effective approach in adoption of STEM instructional practices nationwide and promoting STEM education experiences to increase student engagement and achievement.

Many different instructional practices and STEM approaches have been developed over the years, the lecture-based classroom is still a main delivery method that Thai teachers use to disseminate knowledge. However, teaching competencies such as critical thinking, interpersonal skills and intrapersonal skills are developed not only through obtaining knowledge but also through experience. In this regard, in-service science teachers need to provide an environment in which students take active roles in learning. Regional study on transversal competencies in education policy and practice conducted in Thailand revealed that teachers employed mainly lecture-based approaches, and rarely incorporated student-student interaction. This was the case across all of the subjects observed (mathematics, Thai, English, art and Thai culture) despite policy-level documents in Thailand (both national and school-based) emphasizing the importance of 'life skills' and student interaction [18].

There are many benefits that could improve the current situation of science, mathematics and technology education. For example, there are studies showing that learning science through STEM education could result in significant learning gains in science content [16-11]. Also, some studies have shown that students who learn scientific concepts through authentic engineering design process attained significantly higher achievement, more engaging experience and higher knowledge retention than those who learn through scripted inquiry-based teaching methods, especially for students who are underrepresented groups such as Black and Hispanic [5-20]. Moreover, it has shown that exposure to STEM activities positively impacts elementary students' perceptions and dispositions [2]. So that, programs focusing on integrated STEM approaches should a growing priority in elementary schools with aims to provide early exposure for students, especially in rural areas of Thailand.

Professional development has a significant role in preparing in-service science teachers with adequate instructional practices and promoting STEM education experiences. However, the current teaching practices in Thailand do not often respond to the needs of STEM learning environments, such as inquiry and student-centered approaches and the utilizing of integrated STEM curriculum in rural context of Thailand. These, in turn, require in-service science teacher to have skills for collaboration and ability of using STEM strategies for organizing and delivering instruction through integrating technology and engineering into science and mathematics. Although, IPST promotes concepts and some activities through national workshop, online-based training, and face to face by STEM experts and STEM core trainers nationwide. STEM education is less monitoring the need for supporting and assessing competencies' levels of teaching and evaluating specific content knowledge related STEM instruction to enhance in-service science teachers' competencies in rural context of Thailand. In addition, there is a lack of research on the effects of a professional development program integrated STEM for elementary schools in a rural context of Thailand. Based on the current situation, we have identified the main challenges in assessing need for supporting and building competencies' levels of teaching and evaluating specific content knowledge related STEM instruction to enhance in-service science teachers' competencies in rural context of Thailand.

2.Objectives

The aims of this research were:

(1) To assess teachers' competencies of teaching after participating integrated STEM workshop.

(2) To examine the effect of the integrated STEM program by assessing teachers' achievement score on science content knowledge after implementing the integrated STEM workshop.

3. Method

3.1 Research Design

In this study, we used qualitative research to examine competencies' levels of teaching STEM activity by using assessment observational rubric consisted of 5 elements of competencies with 4 levels of rubric criteria including beginning, approaching proficient, proficient, and distinguished levels. We developed and applied the STEM achievement test items which comprises multiple choice and short answer response achievement test items designed to assess specific science content knowledge regarding STEM activity. The One Sample t-test was employed to examine the effect of using the integrated STEM program.

3.2 Participants

The participants in this study included 30 teachers from STEM workshops of Research and Academic Services organized by Faculty of Education, Buriram Rajabhat University. Participant teachers came from public schools located near the Thai-Cambodian borders, and included 24 primary and 6 lower secondary teachers aged between 26 and 52 years old. In-service elementary science teachers (N = 30) were enrolled in the STEM workshops during 17-18 February 2018. These teachers are having 3 to 20 years of experience in elementary school level but they were usually not taught STEM as part of a normal curriculum before. All of these teachers had a bachelor degree and were assigned to teach more than one subject and more than one grade level.

3.3 Design of the STEM Workshop

The STEM workshop consisted of four activities lasting about two hours and 30 minutes each. Three areas were identified for the development of workshops. The areas were: (1) exploring STEM concepts, which was principally concepts, methodologies, tools, and applications; (2) engineering design process, which was a mindset that emphasizes open-ended problem solving and encourages learners to learn from failure; (3) good design, materials, and construction, which was principally preparation and testing of structures; (4) simple machines, as incorporated into the "Chain Reaction in Marble Run Activity (Figure 1)."

Figure 1. Simple machines, as incorporated into the "Chain Reaction in Marble Run Activity

Figure 2. Utilizing the engineering design process to design, build, and test their creations of the prototype

Figure 3. Facilitator asking the higher-order thinking questions during STEM workshop

The integrated STEM workshop lasting for 2 days (organized during 17-18 February 2018) was designed with a focus on encouraging rural in-service science teachers to solve an ill-defined problem utilizing the engineering design process to design, build, and test their creations of a chain reaction in marble run (see Figure 2). In the challenge situation, teachers would be asked to consider the constraints of the materials and time, identify the problem, think about what they already know, design, plan, construct, test and evaluate a physical prototype of their track design. Teachers were encouraged to construct a convincing argument for a particular solution as opposed to all other possible solutions and start out by drawing plan of their prototype with dimensions and details on a

IOP Conf. Series: Journal of Physics: Conf. Series 1340 (2019) 012009 doi:10.1088/1742-6596/1340/1/012009

chart paper provided. Each team consisted of 4-5 science teachers and received only 2 plastic tubes, 2 paper boxes, 30 pieces of dominoes, 1 plastic board, 1 pair of scissors, 1 roll clear sticky tape, 1 roll cotton rope and at least 1 marble. Researchers gave teams 20 minutes to plan their marble run track and give teams 30 minutes to build their marble run. They ask to create a track that keeps a marble rolling as long as possible (longer than 3 minutes). The 2 days STEM workshop consisted of integrated STEM activities lasting about 12 hours. Teachers also needed to answer the higher-order thinking questions during participating STEM activities (see Figure 3). In particularly, answering the questions inspired teachers to acquire new knowledge, competences and 21st century skills in group working. Some samples of questions used were: "In your opinion, if your tracks were constructed identical to this prototype, how to keep children safe while they are on? If yes/no, please explain why?" (Evaluation of cause-effect relationship based on science concepts); and "How can your prototype be modified in order to improve its results in the future?" (Analysis of experimental design). The questions were specially designed to evaluate teachers' analysis, critical thinking, evaluation and communication skills in connecting STEM activities with their daily life. Finally, in-service science teachers took the achievement test on science content knowledge after implementing the integrated STEM workshop.

3.4 Instruments
Research instruments were: (1) Assessment observational record used to assess teaching competencies adapted from Charlotte Danielson' framework for teaching evaluation instrument, 2013 edition [6]; (2) Multiple choice and short answer response test items designed to assess specific science content knowledge about STEM activity. The achievement test on science content knowledge related STEM (ATSCK-STEM) consisted of 43 multiple choice test items and 31 short answer response test items with 80 total score. The ATSCK-STEM was verified by three experts to find the Index of Item Objective Congruence between test items and learning objectives. The IOC was found to be 0.67-1.00. The reliability of the ATSCK measured by Cronbach alpha was 0.895. Discrimination indices ranged from 0.33-0.46. Difficulty indices ranged from 0.31 to 0.80 providing a wide range of difficulty in the items.

3.5 Data Collection
Items for the assessment observational record were developed and adapted from Charlotte Danielson' framework for teaching evaluation instrument, 2013 Edition. All criteria were later reviewed and validated by a group of 12 experienced science teachers and 3 university professors to give comment on the readability and coherent of the criteria in the assessment observational record before being adopted as research instruments. These experts agreed that all items were relevant and should remain in the study. After participating the STEM workshop, teachers have a responsibility to plan and manage their own STEM activities and practices in their classrooms. Researchers and experts followed up teachers to assess teaching competencies focused on the role of the teacher in the classroom, directly linked with the craft of teaching – with professional knowledge and skills for action in their classroom. The assessment observational record consisted of 5 elements of competencies with 4 levels of rubric criteria including beginning, approaching proficient, proficient, and distinguished levels.

The assessment observational record utilizing for data collection after implementing the STEM workshop, included 4 parts:

Part 1: asked about the background information of the participants (e.g., age, education background, teaching experience).

Part 2: asked about teachers' prior experiences, understanding of content knowledge regarding STEM activities and approaches. Researchers employed a multiple choice and short answer response test items designed to assess specific science content knowledge about STEM activity.

Part 3: in-service teachers selected their preference from seven steps of engineering design process (see Figure 4) represented by a diagram with brief description as following to create STEM activity for teaching in their classroom after participating the STEM workshop.

International Annual Meeting on STEM Education (I AM STEM) 2018 IOP Publishing
IOP Conf. Series: Journal of Physics: Conf. Series **1340** (2019) 012009 doi:10.1088/1742-6596/1340/1/012009

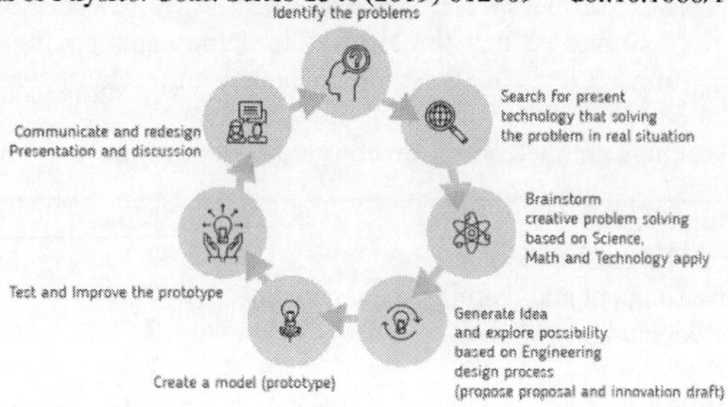

Figure 4. The seven steps of engineering design process

Part 4: the assessment observational record form consisted of 5 domains of competencies with 4 levels of rubric criteria including beginning, approaching proficient, proficient, and distinguished levels.

The use of coaching and mentoring approaches for deliberated school improvement involve both conversations and classroom observations. These approaches provide opportunities for in-school professional learning and support classroom walk through, by focusing on specific theme or topics. Then, this process is particularly to support the achievement of teaching practice and competencies of STEM approach. For assessing teaching competencies' level and performance, these processes were employed: (1) pre-observation meeting was to typically focus on the type of specific lesson being planned and aspects or criteria the teacher would like the observer to focus on; (2) classroom observations were intended to provide direct feedback on the specific pedagogies used by teachers in situ. The criteria and indicators an observer might use were outlined as followings: planning and preparation of STEM approach, classroom environment and management, instruction strategies, assessment in instruction and reflecting on teaching and feedback.

3.6 Data Analysis

In this research study, the effect of implementing the integrated STEM program was examined by using the ATSCK-STEM consisted of 43 multiple choice test items and 31 short answer response test items. Data was analyzed to examine the effect of the integrated STEM program by assessing teachers' achievement score on science content knowledge after implementing the integrated STEM workshop. The finding was used to justify whether in-service science teachers had a satisfactory achievement with respect to an established criterion at 75 % of total score (total score = 80).

Comparison of the in-service science teachers' achievement after implementing the integrated STEM program was performed statistically by using the one sample t-test to examine whether difference in achievement score exists.

Responses from the interviews, document, lesson plans and information on classroom observations were explored to assessing teachers' competencies after participating integrated STEM workshop. Coaching and mentoring approaches were use as the process of following up for deliberated teacher's improvement involve both conversations with teachers and classroom observations. After pre-observation meeting with school director and science teachers, classroom observations were organized to provide feedback on the specific pedagogies used by teachers. Assessment observational record were used as instrument to evaluate teaching competencies of in-service science teachers.

4. Results and Discussion

4.1 Teachers' competencies after participating STEM workshop followed by meetings during the academic year

Researchers followed up in-service science teachers lasting 8 weeks began on May 1st, 2018 and continues through June 22nd, 2018 by using coaching and mentoring approach for involving meeting, group discussion with conversations and classroom observations. By focusing on assessing teaching competencies' level and performance of in-service science teachers, classroom observations were

IOP Conf. Series: Journal of Physics: Conf. Series **1340** (2019) 012009 doi:10.1088/1742-6596/1340/1/012009
intended to provide feedback on the specific pedagogies used by teachers on-the-job training. As shown in table 1, the results revealed that the level of teaching skills among the in-service science teachers is approaching proficient level in all essential elements of teaching competencies.

Table 1. Teaching competencies' level of in-service science teachers (N=30)

Elements of competencies	Mean	S.D.
1. planning for STEM instruction	2.10	0.33
2. classroom environment and learning management	2.67	0.48
3. strategies for encouraging student to learn effectively	2.63	0.24
4. feedback and assessment	2.45	0.34
5. teaching reflection	2.49	0.40
Average	2.48	0.41

From experts' evaluation, the results indicated that competencies' levels of teaching and pedagogical content knowledge (PCK) of in-service science teachers was achieved as approaching proficient level (Mean = 2.48, S.D.=0.41) in five elements of competencies.

The results revealed, as shown in table 1, the level of teaching skills among the in-service science teachers was not achieved proficient level. It can be noted that there are common characteristics and content that unite the STEM activities, especially in science and technology, which establishes the need for collaboration and ways for these teachers to work together within schools. In-service science teachers who participating in long professional development program on STEM integration and felt that they also needed to work together with other teachers in other STEM disciplines because they only familiar to teach mathematics not STEM activity [21]. Given severe time constraints for evaluating teaching competencies in schools due to teacher workload [14], management lesson plan in schools will need to consider the creation of sufficient time to allow the collaboration to undertake their role during informal meeting, organizing lesson plan and classroom observing. This may be particularly problematic in some elementary where teachers were already engaged in full-time in-service duty. It is reasonable to assume that teachers in some schools, in this study, may be reluctant to take on the additional responsibilities inherent in teaching STEM activities according to follow-up process after implementing STEM workshop. Our observational data, classroom observation and assessment of teaching competencies are also not complete as the in-service science teachers may employ only a small part of their accumulation of PCK for teaching STEM and they may need more collaboration undertake the role in teaching STEM activity. In this study, the level of teaching competencies of in-service science teachers is approaching proficient level because PCK might develop along with teaching experience in science [3, 10, 19]. It is pointed out that teaching experience is an important factor that affect development of teaching competencies' level. Providing experience for in-service science teachers to practices with artificial class or micro-teaching will create an opportunity for them to use their theoretical science content knowledge and PCK in classroom environment. In this way, in-service science teachers will catch the chance of gaining experience through teaching STEM activity. Micro-teaching practices for in-service teachers also requires physical environment and technical equipment for preparation STEM lesson. At this point, schools should be provided with essential support.

4.2 Teachers' achievement score on science content knowledge related STEM

Teachers' achievement score was assessed by the ATSCK-STEM consisting of 43 multiple choice test items and 31 short answer response test items with 80 total score. The score of ATSCK-STEM were collected and analyzed after STEM workshop. The results of comparison of the post-test score and the criterion score of 75% (Table 2).

IOP Conf. Series: Journal of Physics: Conf. Series 1340 (2019) 012009 doi:10.1088/1742-6596/1340/1/012009

Table 2. The comparison of the mean score of in-service science teachers' achievement score and the criterion score

Score	N	Total score	$\bar{x}$	S.D.	% of Mean	t
Posttest	30	80	63.23	4.569	79.04	3.88[a]

[a] $t_{(0.01;\ df=29)}=2.756$

The results of t-test one sample indicated that teachers' achievement score on content knowledge related STEM after implementing the integrated STEM program was higher than the criterion score of 75% at the 0.01 level of significance. In this study during implementing STEM workshop, constructivist approaches, problem-based learning, and making connections to the real world often characterized effective STEM activity. In the classroom, constructivist approaches, problem-based learning, and making connections to the real world often characterize effective STEM education when implemented using inquiry-based strategies. Techniques such as active learning and forming cooperative learning groups are central to achieving the most important outcomes of STEM [15]. STEM activities and instructional strategies that were found to be effective include questioning techniques, guided independent research studies, and discussion groups. The results consistent with other research studies in high school level, in adoption of a STEM-focused curriculum, reform instructional strategies, project-based learning, and integrated and innovative technology are in use by independent STEM [4-9]. Elementary schools with successful STEM emphasis demonstrated that improved student learning in science, mathematics and reading shared 5 common elements including professional capacity of colleague and staff, learning communities, a student-centered learning environment, and instructional guidance [12]. In this study, in-service science teachers were provided opportunities to explore STEM concepts, procedures, tools, and applications and experience engineering design process, which was principally preparation and testing of structures, materials, and construction. Workshop planning emphasized open-ended problem solving and encourages in-service science teachers to learn simple and complex machines, as incorporated into the "Marble Run Activity". The time for doing integrated STEM workshop was designed appropriately for encouraging these teachers to solve an ill-defined problem utilizing the engineering design process to design, build, and test their creations during workshop. These findings suggest that engineering design activities that integrate the STEM components can help students contextualize the principles of science to improve learning achievement [7]. So that, teachers' achievement score on content knowledge related STEM after implementing the integrated STEM program was higher than the criterion score of 75% at the 0.01 level of significance.

5. Recommendation

5.1 Professional development program integrated STEM activities as partnership in school improvement

In-service science teachers suggested that a direct dialogue between teachers and researchers on STEM activities during informal meeting before classroom observation would be essential for promoting STEM initiative program in elementary schools. In-service science teachers suggested that this informal meeting, collaboration after STEM workshops, on-the jobs training at school and classroom observation can occur through a partnership between teachers and researchers. In addition, in-service science teachers recommended that good partnerships can replace the traditional professional development programs and that help teachers and researchers can work together to design and develop lesson plans, curriculum materials and deliver instruction to promoting STEM for elementary students in context of rural areas.

One of the important findings of this study showed that the school culture plays a key role concerning the implementation of STEM at elementary school. The study showed that implementing STEM education in non-STEM schools required a different preparation than that in STEM schools. The non-STEM school culture required collaboration among in-service science teachers and need more building a collaborative and supportive STEM community in school. In this study suggests that exchange of experience and constant dialogue between teachers and the administrators during observing teaching in the school were highly emphasized.

IOP Conf. Series: Journal of Physics: Conf. Series **1340** (2019) 012009 doi:10.1088/1742-6596/1340/1/012009

5.2 Recommendations for policy makers

Therefore, researchers can provide recommendations at the policy level to introduce STEM programs for in-service science teachers in rural context. This could lead to developing a STEM partnership-based professional development model of what teachers need in terms of content knowledge related STEM and how to assess PCK to enact STEM education in different school culture and local context.

6. Limitations of the study

This study only focuses on in-service science teachers engaged in STEM teaching in rural areas of Thailand, and therefore it is difficult to generalize. Besides, because of the nature of PCK, it is difficult to identify its components with clear boundaries. Our observational data, document analysis and classroom observation are also not complete as the in-service science teachers may employ only a small part of his/her accumulation of PCK for teaching STEM. Another difficulty is that analyzing teaching competencies is time-consuming to develop, administer, and analyze.

7. Acknowledgement

This research was conducted as part of the professional development program for in-service science teachers organized by General Science Program, Faculty of Education, with financial support from Research and Development Institute, Buriram Rajabhat University and National Research Council of Thailand (Grant number 85/2017).

References

[1] Apedoe X S, Reynolds B, Ellefson M R and Schunn C D 2008 Bringing engineering design into high school science classrooms :The heating/cooling unit *J .of Sci.educ.and Sci . Technol* .**17**)5 (pp 454-465

[2] Bagiati Aikaterini, Yoon So Yoon, Evangelou Demetra and Ngambeki Ida 2010 Engineering Curricula in Early Education :Describing the Landscape of Open Resources *Early .Child . Res.& Pract.***12**) 2(

[3] Baxter J A and Lederman N G 1999 Assessment and measurement of pedagogical content knowledge .In Gess-Newsome J and Lederman N G)Eds(., *"Examining pedagogical content knowledge)* "Dordrecht, The Netherlands :Kluwer Academic Publisher (pp 147-161

[4] Bruce-Davis M N, Gubbins E J, Gilson C M, Villanueva M, Foreman J L and Rubenstein L D 2014 STEM high school administrators', teachers', and students 'perceptions of curricular and instructional strategies and practices *J .of Advance .Academics* **25**)3 (pp 272-306

[5] Cantrell P, Pekcan G, Itani A and Velasquez-Bryant, N 2006 The effect of engineering modules on student learning in middle school science classroom *J .of Engineer .Educ .95)*4 (301

[6] Charlotte Danielson 2013 *Framework for Teaching Evaluation Instrument, 2013*)Princeton, NJ: The Danielson Group(

[7] English L 2016 STEM education K-12 :perspectives on integration *Int .Journal of STEM Education* **3**)1(

[8] Fortus D, Dershimer C, Krajcik J, Marx R and Mamlok-Naaman R 2004 Design-based science and student learning *J .of Res.in Sci .Teach* .**41**)10 (pp 1081-1110

[9] Lynch S J, Behrend T, Peters Burton E and Means B 2013 Inclusive STEM-focused high schools :STEM education policy and opportunity structures *.Paper Prepared for NARST 2013 Annual Conference.*

[10] Magnusson S, Krajacik J and Borko H 1999 Nature, sources, and development of PCK for science teaching In Gess-Newsome J and Lederman N G)eds" *(.Examining PCK :The construct and its implications for science education)* "Boston :Kluwer Academic Press (pp 95-120

[11] Mehalik M .M, Doppelt Y and Schuun C D 2008 Middle-school science through design-based learning versus scripted inquiry :Better overall science concept learning and equity gap reduction *J .of Engineer .Educ .97)*1 (pp 71-85

[12] National Research Council 2011 *Successful K-12 STEM Education :Identifying Effective*

IOP Conf. Series: Journal of Physics: Conf. Series **1340** (2019) 012009 doi:10.1088/1742-6596/1340/1/012009
Approaches in Science, Technology, Engineering, and Mathematics) Washington, DC : The National Academies Press

[13] Riskowski J L, Todd C V, Wee B, Dark M and Harbor J 2009 Exploring the effectiveness of an interdisciplinary water resources engineering module in an eighth-grade science course *International .J .of Engineer .Educ .25)*1 (pp 181-195

[14] Rhodes C P and Houghton-Hill S 2000 The Linkage of Continuing Professional Development and the Classroom Experience of Pupils :barriers perceived by senior managers in some secondary schools *J.of In-Service.Educ.* **26** pp 423-35

[15] Smith K, Douglas T C and Cox M F 2009 Supportive Teaching and Learning Strategies in STEM Education .Book chapter in *New Directions in Teaching and Learning :Creating Culture/Climate that Supports Undergraduate Teaching and Learning in Science, Technology, Engineering, and Mathematics* **117** pp 19-32

[16] Svarovsky G N and Shaffer D W 2007 Soda constructing knowledge through Exploratoids *J .of Res.in Sci .Teach* **.44)**1 (pp 133–153.

[17] The Institute for the Promotion of Teaching Science and Technology 2013 *The Development and Promotion of STEM Education Project)* Bangkok :IPST(

[18] UNESCO 2016 *School and Teaching Practices for Twenty-first Century Challenges :Lessons from the Asia-Pacific Region United Nations Educational, Scientific and Cultural Organization)* UNESCO :Bangkok Office(

[19] van Driel J H, de Jong O and Verloop N 2002 The development of preservice chemistry teachers 'pedagogical content knowledge *Sci .Educ* **.86** pp 572-590

[20] Vasquez J, Sneider C, and Comer M, 2013 *STEM Lesson Essentials :Integrating Science, Technology, Engineering, and Mathematics)* Portsmouth, NH :Heinemann(

[21] Wang H, Moore T J, Roehrig G H, Park 2011 STEM Integration :Teacher Perceptions and Practice *J.of Pre-College .Engineer .Educ.Res* **.1)**2 (pp 1-13

[22] Wendell K B and Lee H S 2010 Elementary students' learning of materials science practices through instruction based on engineering design tasks *J .of Sci.educ.and Sci .Technol* **.19)**6 (pp 580-601

Developing the Astronomy and Architecture Unit for Providing Students' Perception of the Relationship between Science, Technology, Engineering, and Mathematics

Russana Anantasook[1], Sakanan Anantasook[2], and chokchai Yuenyong[3]
[1] Rattanaburi Secondary School, Surin, Thailand
[2] Tabaowittaya Secondary School, Surin, Thailand
[3] Faculty of Education, Khon Kaen University, Thailand

the corresponding author's e-mail : [1] yoyasong@gmail.com, [2]anantasook@gmail.com, [3]ychok@kku.ac.th

Abstract. STEM education suggested that students should be enhanced to learn astronomy with integration between Science, Technology, Engineering and Mathematics. To help Thai students make sense of STEM education in astronomy learning, this paper presents learning activities of the Socio-cultural astronomy and architecture Unit. The developing of the Socio-cultural astronomy and architecture Unit is a part of research that aimed to enhance students' perception of the relationship between Science, Technology, Engineering, and Mathematics. This paper will discuss how to develop the astronomy and architecture Unit through Socio-cultural approach that regarding sociocultural perspective into 5 stages. These included (1) Engagement, (2) Investigation, (3) Discussion and conclusion, (3) Create production understanding, and (5) Learning community reflection. The learning activities could be highlighted as following. First stage, students focus on various religious buildings in the ancient time that built like a solar calendar structure. Then, they give opinions in this issue; (1) How to build a solar calendar for identify true season in your district? (2) Why do not build a house perpendicular the Sun's pathway? Second stage, students used Stellarium astronomical program to find the position of sunrise or sunset in each month. Then, study about the details of buildings in various global civilizations which aims to astronomical observations and create activities in relation to true seasons. Third stage, each student group conclude these ideas; the principles for building place like a solar calendar, building house and set the room activity that concern with the Sun's pathway and saving energy in their house. Fourth stage, students was assigned to create a model of the solar calendar in their district. The model should relationship with the sunrise or sunset position in important astronomy dates, support astronomical observations, and become landmark for tourism. Finally, each group presented their model to the committee and then, wrote committee's reflection, and their opinions. The paper may have implications for developing astronomy teaching with STEM education.

1. Introduction

In the 21st century, scientific and technological innovations have become increasingly important as we face the benefits and challenges of both globalization and a knowledge-based economy. To succeed in this new information-based and highly technological society, students need to develop their capabilities in STEM to levels much beyond what was considered acceptable in the past [15].

The goal and vision of the Thai science education suggests that science teaching and learning should lay emphasis on the relationship between science, technology and society; and lifelong learning [8]. And, the goal also aims to provide students with the intrigued study in Science, Technology,

Engineering and Mathematics. Thus, students should be use the knowledge in many subjects to solve real problems, research and improve many things in their real lives. All those issues need all knowledge they have, not individual one [1][17]. Therefore, integrated concept of science, technology, engineering and mathematics or STEM Education was recognized. Science is the subject that study from natural phenomena by scientific inquiry. Technology is the subject that applied all subjects to help solve the problems and also improving and developing to human need. Engineering is the subject that creative innovation and build many things to accommodate human by use the knowledge of Science, Mathematics and Technology to inventive. Mathematics is the subject that about calculation at this subject is very important foundation of education. It can be further in engineering [4][3].

Astronomy is one of area and now become the most effective catalysts that inspires young children to look upon the skies and marvel at the bounty of this celestial sphere has on offer. At the very young age, especially during their primary and secondary schools, children are more easily exposed to science and technology than other later stages in life and that is when their innovative mindsets and scientific capacities should be incubated. In addition, astronomy is an impetus to pave the way for the children to pursue their careers not only in the field of astronomy but also in other related sciences from engineering to biology or even space technology, adding more manpower into STEM workforce in the future [21]. Astronomy has been aptly called a "gateway drug" for STEM (Science, Technology, Engineering and Math) learning [6].

Astronomy is a natural science that deals with the study of celestial objects (such as the Sun, the Moon, stars, planets, comets, and galaxies) and phenomena that originate outside the Earth's atmosphere. Astronomy is one of the oldest sciences in the world [16][10][23]. Historical astronomers attempted to explain the actual motions of celestial objects from their observations [20]. So, many structural symbolic patterns were built to perceive and understand the sky. Archaeoastronomy is multidisciplinary science of architecture, archaeology, and astronomy [13]. It studies about pre-technological humankind's awareness of celestial phenomena and its influence on their societies such as how ancient architecture were built, early forms of calendars, and the development of mathematical concepts [19]. It was the earliest concept in astronomical knowledge and a decisive factor in the emergence of agriculture and navigation in early civilizations [7]. It is not only deeply rooted in almost every culture [18], but it is still the basis of our modern system of time measurement such as the seasons and the calendar and of navigation and surveying [12].

Astronomy provides tremendous opportunity for interdisciplinary learning so astronomy activities should be emphasis on the relationship between multidisciplinary such as science, technology, engineering, mathematics, and also archaeoastronomy. Some educators suggested the 7 STEM guidelines activities for enhance students to learn science content on the relationship between science, technology, engineering, and mathematics (STEM). These include; Integrate Science, Technology, Engineering and Mathematics Content; Link all the Science, Mathematics and Technology to real world; Engage in inquiry; Project base; Apply Technology Strategically; Focus on the skills of the Twenty first Century; Building an awareness and participation in the community [23]. It indicates that Sociocultural approach could support all these 7 STEM guidelines activities to provide astronomy learning activities on the relation between STEM and also archaeoastronomy.

2. Socio-cultural Approach of Astronomy Learning

In this study, the author developed astronomy and architecture unit through Sociocultural approach that consisted of five stages including engagement, investigation, discussion and conclusion, create production understanding, and learning community reflection stage. The detail in each stage will show in the following:

(1) Engagement stage. The Sociocultural astronomy and architecture Units has to begin in scientific fact. Teachers lead the discussion that the axis of the Earth is tilted by 23.5 degrees to the plane in which it travels around the Sun. Different parts of the Earth point toward or away from the Sun at different times of the year, the cause of seasons. The observer on the Earth saw the position of sunrise and sunset change in cycle during a year. Then, set them to realm of societal and cultural context. The teacher designed the situation which concern cultural context or mentioning everyday events. Students look at the image of religious places in various historical territories. Particularly, Prasat Phanomrung,

Hindu temple in Buriram province that built like a solar calendar structure. Students were encouraged to share their prior knowledge and understanding which shown the knowledge gaps. Finally, they give opinions in these issue; (1) How to build a solar calendar for identify true season in Nonnarai district? (2) Why do not build a house perpendicular the Sun's pathway?

(2) Investigation stage. The teacher assigned cooperative tasks. The students designed procedures to answer the question, solve the problem or complete the assignment with their group members. During learning activities, students were required for knowledge to research, share, and discuss with fellow group members. The teacher gives guidance and supports in manner to help students to finish tasks. Many strategies were helped students to understand some scientific content. For examples, cooperative learning, using the archaeoastronomy instrument unit, outdoor learning, and lecture. In this unit, students used Stellarium astronomical program in their computer notebook study the position of sunrise or sunset in each month. Then, study details of buildings in various global civilizations and in Thailand such as Prasat Phanomrung, Prasat Phu Phek which aims to astronomical observations and activities in relation to true seasons. In addition, know the reason why Thais believe that do not build a house perpendicular the Sun's pathway?

(3) Discussion and conclusion stage. The students in each group were discuss about results from previous stage, develop a group's conclusion, and shared its findings with the class. They use mind map to conclude these ideas; the principles for building place like a solar calendar, building house and set the room activity that concern with the Sun's pathway and saving energy in their house. During learning, teacher aided students to communicate their thinking by prodding with questions that attempted to get students to clarify their thinking and further evaluate the value of their thoughts. Also, the teacher might assist students by means of modeling, giving reinforcement, instructing, explaining, or guiding. In addition, to demonstrate their conclusions, students were encouraged to use drawing, acting, writing, and any other method or tool that would aid them.

(4) Create production understanding stage. Teacher gives the situation for student to solve, or created final product, or do mini project that for demonstrated their understanding. This was the time for students to demonstrate their full grasp of the meanings (concepts) by using signs and tools. Not everyone's understanding was identical, but they shared similar and important key ideas. In this stage, students can create production understanding both in and out of the classroom. It probably illustrated students' reasoning ability, and technological process. In the class, teachers give the Horizon table and assign students to create a model of the solar calendar in Nonnarai district. The model should relationship with the sunrise or sunset position in important astronomy dates such as Solstice or Equinox. Also, support astronomical observations and landmark for tourism. Student can use materials that can found in their community.

(5) Learning community reflection stage. Students need to act as people who are a part of society by presenting their production understanding with their learning communities. They might exhibit their product in public and receive feedback from expertise in that community. This process will help students rethinking on their ideas that it is effective for stakeholders or need to improve something on their products. Therefore, students probably exchanged their ideas or improved the product to the best. In the unit, student presented their model in front of the committee that includes; school director, astronomy teacher, social and culture teacher, art teacher. Then, write committee's reflection, and their opinions (run on the leisure time around 1 hour for all group).

Summary; an important learning activities based on sociocultural perspective is help to improve students learning behavior through interaction and cooperation with one another [5][11][24] in social contexts [25]. Students constructed knowledge as a form of "situated cognition" [2]. The concept of astronomy and architecture are related to students' contexts. In addition, the astronomy learning about astronomy and architecture would allow students to learn astronomy on the relation between concepts of science, technology, engineering, and mathematics. Because students will learning about the apparent motion of the Sun (science; the season concept), using the stellarium astronomical program to find the location of sunrise and sunset (technology), measurement the angle of the position of Equinox and Solstice on the horizon plate (Mathematics), inventing model of the solar calendar related to their location (Engineering).

3. Participants

The astronomy and architecture unit was trialed by the author in Naraikhumphongwittaya School. The school is located in the small Nonnarai district, Surin province, Thailand. This district is far from Surin province approximately 70 kilometres. The total number of Grade 7-12 students is approximately 700. Their parents work as laborers in Bangkok and as farmers. The Grade 9 participants in the second semester of 2014 academic year that include 32 students, 21 girls and 11 boys, who live in the district and other villages nearby.

4. Finding and Discussion

4.1 Student Understanding about astronomy and architecture concepts

The students have learned that ancient human looking the position of sunrise or sunset that changed every day and it is rising on the eastern horizon and setting on the western horizon. In the summer season, the Sun appear moving from true east to the north, and in the winter, it was veering to the south. Thus, human creating buildings or something such as places of worship that relating with the position of sunrise or sunset in someday to identify times of the season that is called Solar Calendar. Nowadays, the concept of housing in relation to the movement of the Sun during the year, it contributes to reduce the budget, and saving the energy which used in the house. Because of the rising and setting of the Sun in each season, it is affect the direction and the amount of light entering the building, and it also relationship with the temperature in the house living space. These knowledge could be encourage students interested in two ways; (1) build the house to save using energy, and (2) build each construction look like as solar calendar that to tell the time and sued for astronomical observations, furthermore make the house is sacred and more magical, as mentioned in the student's writing journal in the following:

- *" If the modern house was built based on the rising and setting of the Sun, the house will become special place like ancient castle in the Khmer Empire." (Chanat)*
- *"The main structure of the house that is not match the movement of the Sun during the year, the house will high temperature, the owner house will pay much money for the electricity." (Pornpimai)*

Teachers design activities to support students to collect the angle of the rising from true east and setting of the Sun from true west during the year (12 months) that using Stellarium astronomical program to stimulate. This activity allows students to know by themselves that the rising and setting of the Sun appears change the position every day in one year cycle. In addition, each area on the Earth receive difference angle of sunlight that cause the seasons on Earth. The astronomer identify that one astronomy year have four season, the point of four special day that include: Vernal Equinox (Vasantravisuwat; 21 March), Summer Solstice (Krisamayan; 21 June), Autumnal Equinox (Sashavisuwat; 23 September), Winter Solstice (Hemayan; 22 December) as shown in student's writing journal in the following:

- *"I never know before about name of four seasons in Thai language and English, I always know that Thailand has three seasons: summer, winter and rainy season." (Rungsiri)*
- *"I have seen with my own eyes that each month, sunrise and sunset does not in the same point location and it always not the same time. In summer, the Sun will appear close to the north celestial pole that from the true east. In this time, the Sun will quickly rise and slowly set, so long time this area have sunlight and so hot." (Adchariya)*
- *"I just know that, the Sun appears to move close the north celestial pole about 23.5 degrees from the true east and also moving on to close the south celestial pole about 23.5 degrees, I use Stellarium software to simulation that." (Tida)*
- *"Previous time, I only know that the Sun does not rise at the same location every day, but this is pattern of a different season that could be told the season, months or zodiac." (Chanat)*

After students was clear understanding that changing the position of the sunrise and sunset in a year could be identify the season, special holy days, and also help designing residential buildings to save energy. Teachers assign students to create a model of the solar calendar in Nonnarai district. The model should relationship with the sunrise or sunset position in important dates such as Solstice or Equinox. The model structure must support astronomical observations and landmark for tourism. Then, each group presented their model in front of the committee that includes; school director, astronomy teacher, social and culture teacher, art teacher. The researcher found that five models present about temples and ancient castles which some point relative with sunrise and sunset position on the day of Equinox, Summer Solstice, and Winter Solstice. However, only one male students group has designed a house structure that the main structure laying related to the Sun pathway (from east to west). The main objective is reducing much sunlight because if the main structure laying from north to south, the house will receive much directly sunlight on both the morning and evening. This activity helps students understand the concept of designing the ancient monastery as solar calendar that for tell the seasons, and including some ideas such as build housing related to the rising and setting of the Sun for saving the energy, as mentioned in the student's writing journal in the following:

- *"I understand that why the Angkor temple is facing to the east that because the sunlight in the Equinox day will shine to the centre of the castle. If it happens, that mean that day was the beginning of the year, or a day to do some special activities." (Peerakorn)*
- *"The positioning of the pagodas and temples in the ancient city that maybe they mark the positioning day of Summer Solstice, or Winter Solstice from the observation point as the centre of the main place." (Phuwanat)*
- *"When I become working man, I will design and build a custom home as the special and attractive places both in beliefs about the sunlight, and for saving the energy." (Sarawut)*

Considering in the learning objectives of astronomy and architecture unit, after learning, the students were able to (1) explain the meaning of the celestial sphere, equator, the north celestial pole, the south celestial pole, and Earth's axis. (2) Search and explain the cause that the rising and setting of the Sun is not exactly the same position, and the cause of seasons. The students could surely mention that the Earth's rotation axis makes an angle of about 66.5 degrees with the plane of its orbit around the Sun, or about 23.5 degrees from the perpendicular to the ecliptic plane. Therefore, the position of sunrise and sunset were changes every day in one year cycle. It effect to the people's lifestyle on the Earth that depend on their location on the Earth. Many people build the accommodations or landmarks related to the Sun's pathway for different reasons. In addition, students can use the astronomical program to simulate the motion of the Sun throughout the years in around 30 minutes. It could be helps students quickly understand the concept of season.

4.2 Students' reasoning and value clarification for astronomy and architecture

Teacher used pictures of religious places in several countries such as Prasat Phanomrung, Burirum province, city wall castle, sanctuary, pagoda, and religious place to motivate students to interest about the solar calendar. Then, suggested them to understand that ancient human observed the position of sunrise and sunset which change every day. In summer, sunrise appearance shift from true east to the north and in the shift to the south in winter. They took the construction such as the stones, or temple are aligned to the Sun's position for used as astronomical observatory and also used it is a solar calendar as a seasonal clock. These concepts also inherited from generation to generation and have the influence to the present building. Because of the position of sunrise and sunset in each season affects the direction and amount of sunlight entering the building, it can make high or low temperature in the house. Therefore, housing construction which correlated with the Sun's pathway during the year it could help the owner for saving the energy, do not use more air conditioner and save the electricity charge. After students understand the concept about season and solar calendar, teachers give the Horizon table and assign them to create a model of the solar calendar in Nonnarai district. The model should relationship with the sunrise or sunset position in important dates such as Solstice or Equinox. The model structure must support astronomical observations and landmark for tourism.

According to learning activities, it could be mention that students have scientific reasoning with other aspects, it combine season concept and others condition for buildings the landmark or the living place. For example, some students give the reason base on their belief such as if the house is consistent with the Sun's pathway, it will make the house auspicious. Another reason is for saving energy and electricity costs as mentioned in the student's writing journal in the following:

- *"I understand that many Angkor temples turned the main structure to the east because the sunlight in the Equinox day will shine inside to the center of the temple. It shows that the important day as the new year or a day to do some special activities." (Group 1)*
- *"The location of pagodas in temples and ancient palace really are position marker which related to the position of sunrise or sunset in summer solstice, or winter solstice that can observe from one location as the center in that place" (Group 2)*
- *"When I grow up and build my own house, I will design it to be charming in both auspicious depend on ancient beliefs and also saving energy." (Group 3)*
- *"If we build a religious place we will concern with the position of sunrise and sunset in important dates such as Equinox's day. It should be an important tourist attraction to visit" (Group 6)*

These assume that many groups of students held multidimensional value clarification for created the production understanding. Their reasoning depended on scientific knowledge, economy aspects, and the societal aspects such as believe and attractions.

4.3 Development of students' technological process and learning in STEM Education

According to the Sociocultural approach for teaching and learning astronomy and architecture that consists of five stages: (1) Engagement stage, (2) Investigation stage, (3) Discussion and conclusion stage, (4) Create production understanding stage, and (5) Learning community reflection stage. In the fourth stage of the unit, teacher determined the situation and assigned students in each group work together to planning and designing the projects about the solar calendar in their district location.

In the create production understanding stage, and learning community reflection stage, each student's group used the process involved in problem solving, creativity, designs that lead to the invention or creation of certain product and then presented their production understanding in their learning community for get some suggestions for further development. These process are classified as Technological process which in students' reflection and writing in worksheets could summarized their technological process into 6 steps as follows: (1) Identify and analyze the situations, (2) Gather more relevant information, (3) Data analysis And choose an alternative, (4) Plan and created, (5) Test, evaluate, improve and develop (6) Present a product in their learning community. Here was the student's interview in the following:

- *"First of all we read the situations and set the objective of the solar calendar architecture which we will make (1). Some ideas is not complete so we search more relevant information (2), when we have enough knowledge the member presented the ideas on the group and sometime debate (3). After that we choose the best ideas (3), planning together and created it (4). Before we send the product to teacher we check the model work (5) and presented perfect solar calendar architecture in front of 5 committees (6), and finally check and compare the result with others groups." (Group 4, interview)*

The six technological processes above is consistent with the STEM Education standards [9], and look like the Engineering process [14] and the stage of learning in the STEM Education that IPST promote for teaching and learning in Thailand. It consists of 6 stages as follows:

1) Problem Identification stage; understanding the problem or challenges, analyze conditions or limitations of problem situations, to determine the extent of the problem. It will lead to create the product or method to solve the problem.

2) Related Information Search stage; is a collection of information and concepts of science, mathematics, and technology which related to problem solving and evaluation the feasibility, the advantages and limitations.

3) Solution Design stage; is the application of information and ideas that related to the design of the product or solution designing which concern with the resource, limitations, and conditions situations.

4) Planning and Development stage; is determines the step for create the product and then create the product, work piece, or develop a solution to solve the problem.

5) Test, Evaluation and Design Improvement stage; is testing and evaluate the use of the product or the approach, and use the result to improve and develop it to the most effective product or solutions.

6) Presentation stage; is to present the production or effective step for solve the problems in the community, and get some suggestions for further development.

The author found that Sociocultural approach for teaching astronomy and architecture also supported students to learn through the STEM Education approach. In addition, in the create production understanding stage students also integrate conceptual approaches on STEM Education to create their product. In this study, the author called the integration of STEM education in celestial motion unit is A-STEM, because it include the archaeoastronomy concept in the units that include;

- A (Archaeo astronomy): Solar calendar
- S (Science): the apparent motion of the Sun (seasons concept)
- T (Technology): stellarium astronomical program, search information, integration of other knowledge for designing
- E (Engineering): inventing model of the solar calendar
- M (Mathematics): measurement the position of Equinox and Solstice on the horizon.

These assumed that the Sociocultural Astronomy and Architecture Unit could be enhanced students' technological process. They have the step for create the product and each product design base on their societal and cultural context. In addition, the process is consistent with the STEM Education standards [24] and the engineering design process [14] which IPST still promote for teaching and learning in Thailand. It aims this technological process will enable student to create innovation for Thailand.

5. Conclusion

Development for learning of Astronomy and Architecture through Sociocultural approach can promote not only student understanding, students' reasoning, but also the concept on STEM Education that students will integrate a knowledge between Science (apparent motion of the Sun and seasons concept), Technology (using the stellarium astronomical program), Engineering (Engineering design on the model of the solar calendar) and Mathematics (calculation and measurement) through societal and cultural context (archaeoastronomy). Finally, found that stage 4 (Create production understanding stage), and stage 5 (Learning community reflection stage) of the author Sociocultural approach could be enhanced students' technological process and the engineering design process.

6. Acknowledgements

The research was financially supported by the institute for the promotion of teaching science and technology (IPST) and the project for the Promotion of Science and Mathematics Talented Teachers (PSMT)

References

[1] Bernard J Sadusky 2012 *STEM standards of practice and STEM education definition* (Retrieved January 23, 2014 from www.marylandpublicschools.org/NR/rdonlyres/3ECF0379-2EE9-42AD-A5FCCA81E8F3FEEA/32260/MarylandStateSTEMStandardsofPractice_.pdf)

[2] Brown J S, Colin A and Duguid P 1989 *Situated Cognition and the culture of Learning* Educational Researcher, 18 (1): 32-42

[3] Bybee R W 2010 *Advancing STEM education : A 2020 vision.* Technology and Engineering

Teacher, 70 (1): 30-35

[4] De Jarnette N K 2012 *America's children: Providing early exposure to STEM (science, technology, engineering and math).* Initiatives Education, 133 (1): 77-84

[5] Driver R, Asoko H, Leach J, Mortimer E and Scott P 1994 *Constructing scientific knowledge in the classroom,* Educational Researcher, 23 (7): 5-12

[6] Goodman A, Fay J, Muench A, Pepe A, Udompraseret P and Wong C 2012 *WorldWide Telescope in Research and Education* (Retrieved May 16, 2016 from https://arxiv.org/pdf/1201.1285.pdf)

[7] International Astronomical Union [IAU] 2009 *Astronomy for the Developing World Building from the IYA2009 : Strategic Plan 2010-2020* (Retrieved January 23, 2014 from http://iau.org/static/education/strategicplan_091001.pdf)

[8] Institute for the Promotion of Teaching Science and Technology [IPST] 2008 *The Basic Education Core Curriculum B.E. 2551 (A.D. 2008) Science* (Bangkok-Kurusapha Ladpraw)

[9] Institute for the Promotion of Teaching Science and Technology [IPST] 2015 *STEM Education standard* (Bangkok-Success Publication)

[10] Jarman R and McAleese L 1996 *Physics for the star-gazer: pupils' attitudes to astronomy in the Northern Ireland Ireland science curriculum* (Physics education vol 31) chapter 4 pp 223-226 http://dx.doi.org/10.1088/0031-9120/31/4/020

[11] Leach J and Scott P 2003 *Individual and sociocultural views of learning in science education.* Science and Education, 12, 91-113

[12] McNally D 1982 *Astronomy at school* (Physics Education vol17) chapter 4 pp 157-160 http://dx.doi.org/10.1088/0031-9120/17/4/306

[13] Mollerup A 2007 *Solar-lunar events at prasat Phanom Rung in spring 2007 AD (2550 BE)* (Retrieved May 12, 2014, from http://www.sundial.thai-isan-lao.com/phanomrung2007.html)

[14] National Research Council [NRC] 2012 *A Framework for K-12 Science Education : Practices, Crosscutting Concepts, and Core Ideas* (Washington DC-National Academies Press)

[15] National Science Foundation 2007 *A National action plan for addressing the critical needs of the U.S. Science, Technology, Engineering, and Mathematics education system* (Retrieved June 1, 2017 from https://www.nsf.gov/pubs/2007/nsb07114/nsb07114.pdf)

[16] Osborne J 1991 *Approaches to the teaching of AT16-the earth in space: issues, problems and resources* (School Science Review vol 72) chapter 260 pp 7-15

[17] Siripattrachai P 2013 *STEM education and 21th century skills development* (Executive journal vol 33) chapter 2 pp 49-56

[18] Percy J R 2005 *Why astronomy is useful and should be included in the science curriculum* In J M Pasachoff and J R Percy (Eds) Teaching and Learning Astronomy: Effective Strategies for Educators Worldwide (New York-Cambridge University Press) pp 10-13 http://dx.doi.org/10.1017/CBO9780511614880.004

[19] Peterson C J 2000 *Astronomy.* (USA-IDG Books Worldwide)

[20] Plummer J 2006 *Students' Development of Astronomy Concepts Across Time* (Doctoral Dissertation The University of Michigan USA Retrieved from http://www.mapsplanetarium.org/maps-planetarium/Domain1/Julia%20Plummer%20Research/Dissertation-plummer.pdf)

[21] Posayajinda S 2013 *The Imagination platform: National star observational building O.S.T.* (Bulletin vol 53) chapter 7 pp 44-55

[22] Posayajinda S 2017 *Astronomy for STEM Education as the colloquium continues* (Retrieved from http://www.narit.or.th/en/index.php/news/611-itca-stem-education)

[23] Seattha P, Yuenyong, C. and Art-in, S. 2015. Developing STS Circular Motion unit for providing students' perception of the relationship between Science Technology Engineering and Mathematics. *Mediterranean Journal of Social Sciences*, 6 (3): 268-275

[24] Shepardson D P 1999 *Learning science in a first grade science activity: A Vygotskian perspective.* Science Education, 83 (5): 621-638

[25] Teemant A, Smith M E, Pinnegar S and Egan M W 2005 *Modeling sociocultural pedagogy in distance education.* Teachers College Record, 107 (8): 1675-1698

IOP Conf. Series: Journal of Physics: Conf. Series **1340** (2019) 012011 doi:10.1088/1742-6596/1340/1/012011

Promoting problem solving and thinking competencies for the middle school students using STEM activities in the Moderate Class, More Knowledge project.

P. Phaprom*[1], J. Nachanthong[1], A. Nachanthong[1] and S. Srihata[1]

[1]Science Department, Thakhanthowittayakarn School, Thakhanto, Kalasin, 46190, Thailand

*Corresponding author's e-mail: pimprom20@gmail.com

Abstract. This study aimed to investigate the problem solving and thinking competencies of the students using STEM activities in "The Moderate Class, More Knowledge" project of the Thakhanthowittayakarn School, Thailand. The participants were 20 middle school students who were **pleasured** to join the activities. The STEM activities applied in this study were adjusted from online resources and the STEM activities of the institute for the promotion of teaching science and technology (IPST). The five STEM activities were conducted in 20 hours, comprised of the following: young designer contest, mobility in the time of flood, balloon powered car, balloon rocket, and rubber band powered car. Multi-rater approach was used to investigate the students' problem solving and thinking competency. Self-checklists, observer-checklists and teachers-questionnaire with five rating scales were used as the study instruments. The analysis method is based on a handbook of a competency appraisal for Basic Education students developed by the OBEC. Percentage, mean and standard deviation were reported; the students' competencies were considered into three levels which were good, fair and unsatisfied. Findings revealed that 75% of the students were in good level, and the rest were in fair level for problem solving competency. According to thinking competency, 85% of the students were in good level and 15% were in fair level. The after action review (AAR) were also presented in this study. The result implied that STEM activities could be appropriate for fostering the students' competencies.

1. Introduction

1.1 Background and rational

In recent years, a different set of skills - 21st century skills – have been increasingly empowers the nations. Consequently, teaching and learning in K-12 education has focused on preparing students for globalization and the 21st century workforce. Moreover, the growth of the international trade needs Science, Technology, Engineering and Mathematics' (STEM) support. STEM has become the indicators that indicate the qualities of education in each country. It is necessary. [5, 12] Evidently, developing skillful and knowledgeable STEM graduates do not only support the growth of international trade but also enhance the qualities of social life. The knowledge and skills of STEM seem to be a crucial factor in order to develop the 21st Century skills which are 1) life and career skills 2) learning and innovation skills and 3) information, media and technology skills [16] Not surprisingly, the STEM education has played an important role in order to achieve the local educational system in each country and has spread continuously around the world such as the United States, United Kingdom, Australia, Taiwan and Singapore [12, 1, 8]

In order to improve students' competencies, in Thailand, there are students who constantly developing their skillful performance, higher order thinking, application of knowledge into the daily life and real life problem solving. They are eager to learn continuously. These objectives conform to the aims of Thai science teaching and learning of the Institute for the *Promotion of Teaching Science and Technology* (IPST) which highlights students' performance in order to construct their own knowledge and connect the process in order to create a productive result. Based on the real objects and evidences, students should be able to solve the problem, present skills in communication and able to make an appropriate decision to solve the various situations. [15]

To achieve these goals, many projects are initiated especially in K-12 education. As such, policymakers, researchers and educators are implementing and examining potential strategies for fostering students' efficiency. "The Moderate Class, More Knowledge" project is one of the educational policies led by the Office of the Basic Education Commission (OBEC) which is undertaken by the Thai Ministry of Education (MOE). This policy aimed to promote students learning through a variety of multi-media and styles in both inside and outside the classroom. The schools attending this project have to commit on a strict time management on teaching and learning for middle school students. The formal classroom schedule was reduced from 6-7 hours to 5 hours in each day, and the rest hour was replaced by learning through doing activities as the students' needs. The activities will be set to support students' efficiency in four domains as 4H Development which stands for *head, heart, health* and *hands* [17]

The Thakhanthowittayakarn School is one of the participants in this project. Therefore, 25 activities were set for students – learning four hours a week. As a significantly important STEM knowledge and the 21st century skills improvements, the researchers designed the STEM activity as a choice for the students who are interested in learning this area. According to activity-based learning, the 20 members of the activity have used these methods for five consecutive weeks and their skills in the form of problem solving and thinking competencies were assessed.

1.2 The purpose of the study

The purpose of this study is to investigate the problem solving and thinking competencies of the students using STEM activities in "The Moderate Class, More Knowledge" project of the Thakhanthowittayakarn School, Thailand.

2. Literature Review

To reach the purpose of the study, previous researches related to STEM education, activity-based learning and students' competency in terms of problem solving and thinking competencies were reviewed. The context of the Moderate Class, More Knowledge project in the school was described in following details.

2.1 STEM education

The STEM education is composed of four fields which include Science, Technology, Engineering and Mathematics. It is not only limited in those areas but also involves social, economics, psychology and politics [2, 4, 13] It seems that the main aim of the STEM Education is not only to emphasize one's performance or skills but also aims to enhance all performances of students to be able to adapt and adopt high technology for professional careers. That means students have to be knowledgeable and skillful in Science, Technology, Engineering and Mathematic [18].

In Thai context, there were many institutes that are using STEM education into high school classrooms. The IPST has established National STEM Education Centre (NSEC) in 2014 to promote STEM education as a new form for science education in Thailand. However, few researches have explored the current situation of STEM teaching and learning [10]. The example of research relating to STEM education in Thailand was the development of learning achievement and integrated science process skills of grade eight students through STEM education, entitled "Food and living" for the Faculty of Education, Mahasarakham University. The findings revealed that students reached 80 % of learning achievement and 79 % of integrated science process skills after the intervention using eight lesson plans [19]. Moreover, there was the development of STEM experiment laboratories based on the Royal-Initiated Projects for lower secondary students. The results showed that the students engaged with eight science activities could integrate scientific knowledge and science process skills related to the Royal-initiated projects through STEM education; and the students showed good attitudes in science [7]. Interestingly, there were few studies focusing on integrated STEM activity for outdoor classes.

2.2 Activity-based learning

Activity-based learning is a learning method through doing activity. This method allows students to learn and understand any concept from several activities such as games, hands-on activity or playing in groups. These activities were widely accepted as an effective method to encourage students to self-learning, construct their own unique ideas, support leadership and teamwork skills including social interaction [6]. The key features of the activity-based learning are: 1) Encourage students to be active in learning and thinking, 2) stimulate students to learn by themselves rather than listening or memorization. 3) Improve students' learning skills and keep continuous learning. 4) Support students' thinking and writing skills. 5) Foster students' satisfaction and their interest that (ABL) is better than passive learning. 6) Focus on students' responsibility for learning through reading, writing, discussion, and participation in problem solving [14]. There was a study anchored on the students' behaviors, discourse and learning in junior high school, related to science-based learning activity. The findings revealed more complex thinking and problem-solving skills both in their discourse and their responses on the follow-up learning probe [3]. In conclusion, the activity-based learning method is one of the effective methods that could be able to promote the student's competency.

2.3 Students' competency

Competencies are characteristics or individual abilities which are the key element in achieving success in works and also refer to knowledge or skills that lead to effective performance. According to David McClelland [9], the competency is composed of five elements which are skills, knowledge, self-concept, traits and motive. These students' competencies are, comprising hidden characteristics and surface characteristics. The hidden characteristics composed of three elements: self concept (representing positive attitudes and values), traits (personal characteristics, for instance respect or leadership), and motive (an internal motivation that drives a person towards a goal such as achievement orientation). The surface characteristics are comprised of specific knowledge and learning skills, such as coordination, effective communication, problem solving and decision-making [11].

In Thailand, according to the OBEC [17], students' competencies were defined into five competencies which are communication, thinking, problem solving, life-skills and technology skills. To achieve the purpose of this study, the problem solving and thinking competencies will be described as Thai curriculum definition. The students' problem solving competency has defined as the ability to solve problems or obstacles based on reasoning, moral and information. Moreover, this competency refers to understanding the relationship and changing in society, seeking new knowledge and applying in several situations to prevent and solve problems. It has also included effective decision makings, taking accountability on themselves, society and environment. Meanwhile, the thinking competency has explained as the ability to think synthetic thinking, creative thinking, critical thinking and systematic thinking. These thinking abilities are employed to construct knowledge, information for making self-determination related to their lives and community as well.

3. Methodology

To investigate the students' competency after learning with STEM activities in "The Moderate Class, More Knowledge" project for 20 hours. The methodology was employed as following;

3.1 Participant

The participants were 20 middle school students who were pleasured to join the activities. In doing the activity, the students were grouped into five. The students have to attend for four days in each week with one hour in each day, to learn through STEM activity. Two teachers are responsible in teaching STEM activity, and another two teachers are tasked to be observers and then later reflect and discuss the method.

3.2 Concepts of STEM activity

The five STEM activities were conducted in this study, comprised of the following: young designer contest, mobility in the time of flood, balloon powered car, balloon rocket and rubber band powered car. The STEM concepts behind these activities were shown in Table 1.

Table 1. The STEM concepts in each STEM activity.

STEM Activity	STEM Concepts			
	Science	Technology	Engineering	Mathematics
The young designer contest	Materials property	Drawing and sketching	Designing and problem solving in creating a skirt with the given constrains	Solving linear equation with one variable
The mobility in the time of flood	Sink and float, and Buoyant force of liquid	Drawing and sketching	Designing and problem solving in creating a float container with the given constrains	Calculating area, surface area, and volume
The balloon rocket	1. Action and reaction force 2. Motion and fraction	Technological system	Engineering and problem solving process	Measuring and calculating
The balloon powered car	Action and reaction force	Technological system and technology tools	Engineering and problem solving process	Measuring and calculating
The rubber band powered car	Elasticity	Technological system and technology tools	Engineering and problem solving process	Measuring and calculating

3.3 Instruments of the study

Multi-rater approach was used to investigate the students' problem solving and thinking competency. To cover all five elements of the competency, self-assessment as rating scale, observer (friends)-checklists and scoring rubric by teachers were used as the study instruments. Raters and the instruments are shown in Table 2.

Table 2. The instruments of the study

Element of competency	Rater/Instrument
Skills, Knowledge	Teachers/Scoring rubric Friends /Checklist Self-assessment/ Rating scale
Traits, Motives, Self-concepts	Teachers/ Checklist Self-assessment/ Rating scale

3.4 Data collection

As the purpose of the study is focused on two competencies, the indicator in each competency was described. According to Problem solving competency, comprised of two indicator which were 1) using problem solving process which are analyzing problems, planning, implementing and evaluating and concluding and 2) the result from the solution. For thinking competency, two indicators were defined as 1) basic thinking and 2) advanced thinking. All indicators were assessed with the multi-rater and specific instrument. After finishing each STEM activity, teachers and observers have discussed to assess the students competencies, including teachers' reflection. At the same time, the students assessed themselves and their friends after the activity as well.

3.5 Data analysis

There were two procedures in assessment the students' competency as following; 1) assessment in each instrument of the competency and 2) assessment in total of the competency. Firstly, in each indicator, the result from each assessment was concluded and transformed into percentage scores. After that, the percentage scores were processed with another instrument. In total, the competency assessment was estimated by finding the average score by weigh rating score (the weight rating proportion for teacher assessment: self-assessment: friend assessment was 2:1:1). The example for finding the average score for problem solving competency has shown as following;

The average score

$$= \frac{\left(2\times \text{ teacher assessment score } (\%)\right)+\left(1\times \text{ self-assessment score}(\%)\right)+\left(1\times \text{ friend assessment}\right)}{4}$$

The average score of problem solving and thinking competency was interpreted as the rubric defined by the OBEC; which was developed based on cut-off scores method [7]. The rubric was discussed by Thai educators, teachers and directors to validate its quality. The students' competency was defined into three level; good (more than 75%), fair (40% - 74%) and unsatisfied (less than 40%). The result from this study was reported in the form of the students' percentage in each level. The after action reviews were presented also in this study.

4. Results and Discussion

Upon conducting the STEM activity for 20 students of Thakhanthowittayakarn School as the promoting the students' competencies, the results are as follows. The students' competencies were considered into three levels; good, fair and unsatisfied. Findings revealed that 75% of the students were in good level, and the rest were in fair level for problem solving competency. And as for the thinking competency, 85% of the students were in good level and 15% were in fair level as showed in Table 3. As indicated in the table, there were no students who were in unsatisfied level since most of them fell under the good category.

Table 3. the problem solving and thinking competencies

Students No.	Problem solving competency		Thinking competency	
	Total percentage score (%)	The competency level	Total percentage score (%)	The competency level
1	78.0	Good	78.7	Good
2	75.6	Good	77.4	Good
3	76.7	Good	76.9	Good
4	77.3	Good	76.7	Good
5	77.6	Good	76.0	Good
6	76.6	Good	78.8	Good
7	76.0	Good	77.7	Good
8	75.3	Good	76.2	Good
9	76.1	Good	77.0	Good
10	76.2	Good	75.8	Good
11	73.8	Fair	77.6	Good
12	72.8	Fair	77.1	Good
13	71.2	Fair	72.8	Fair
14	75.1	Good	76.0	Good
15	75.6	Good	78.0	Good
16	73.5	Fair	72.9	Fair
17	75.6	Good	73.2	Fair
18	72.6	Fair	76.2	Good
19	76.8	Good	77.1	Good
20	75.2	Good	77.3	Good

It may be seen that the STEM activity which conducted in the study were doing great even without formal lecture. It is simply because they enjoyed what they were doing and did not feel any pressure about grading. The STEM activity allows them to do as they pleased but not secluding from its main objective, and of course, with visible results. This finding consequent with the study related to activity-based learning [14, 3] which mentioned that students' problem-solving and thinking skills will be promoted trough doing activity. In this research, both students and the teachers save a lot of time in leaning STEM concepts –the students can now easily construct the ideas that the teacher is trying to convey and the latter will not have a hard time imparting the knowledge to the students. This goes to show that the STEM activity, if applied properly can ease the burden of the students in learning and thus make their new experiences fun and exciting. Problem-solving and thinking skills could be created if given the right combination and this study proves it. Solutions and answers are easier to achieve and most especially, the students are having fun – which make it them more excited every day.

5. Conclusion

Based on the results, it is concluded that in using the STEM activity, results are evident and the application of knowledge in real-life situation is one step easier and faster. Students are learning in a fun and productive way. There is no doubt that this method is effective for both the learners and the educators.

References

[1] Baldi S, Jin Y, Skemer M, Green PJ , Herget D, and Xie H 2007 *Highlights from PISA 2006: Performance of U.S. 15-year-old students in science and mathematics literacy in an international context* (Jessup MD: ED Pubs)

[2] Breiner J M, Harkness S S, Johnson C C and Koehler C M 2012 *What Is STEM? : A Discussion About Conceptions of STEM in Education and Partnerships* (Science and Mathematics 112 vol 1) pp 3-11

[3] Gillies R M 2008 *The Effects of Cooperative Learning on Junior High School Students' Behaviors, Discourse and Learning during a Science-Based Learning Activity* (School Psychology International vol 29) chapter 3

[4] Green M 2007 *Science and engineering degrees: 1966-2004* (NSF 07–307) (Arlington VA: National Science Foundation)

[5] Kuenzi J J 2008 *Science, Technology, Engineering, and Mathematics (STEM) Education: Background, Federal Policy, and Legislative Action* (Congressional Research Service Reports) pp 35 http://digitalcommons.unl.edu/crsdocs/35

[6] Laksanaadisorn V 2007 *Theory of Knowledge Creation. (Constructivism) and learning through activity (Activity Based Learning: ABL)* (Retrieved from http://www.seedlesrning.com/LinkClick.aspx?fileticket=FenbQDYmzcc%3 d&tabid= 3 6 & mid=350 UploadClinic/RFID/A_ReflexRFID.pdf)

[7] Laloknaml S, Boonburapong B, Channamsai D, Pramote N, Konsaiya P, Nadoo P, Paingam W, and Pairoh S 2017 *From Activities of the Royal Initiated Project to STEM Education through Science Activity of Children's University Project at Srinakharinwirot University* (J Res Unit Sci Technol Environ Learning vol 8) chapter 2

[8] Marginson S, Tytler R, Freeman B and Roberts K 2013 *STEM: Country comparisons. Melbourne Victoria, Australia* (Australian Council of Learned Academies)

[9] McClelland D C 1973 *Testing for competence rather than for intelligence* (American Phycology, January. Retrieved from http://www.lichaoping.com/wp-content/ap7301001.pdf)

[10] Srikoom W, Deborah L, Hanuscin and Faikhamta, C 2017 *Perceptions of in-service teachers toward teaching STEM in Thailand* (Asia-Pacific Forum on Science Learning and Teaching vol 18) chapter 2

[11] Supamanee T, Krairiksh M, Singhakhumfu L and Turale S 2011 *Preliminary clinical nursing leadership competency model: A qualitative study from Thailand* (Nursing and health science vol 13) chapter 4 pp 433-439

[12] Tan J, Ngo B K, Chandrasekaran I and Cai Y 2013 *Cultivating STEM education Through 3D Sabbaticals.* (Springer Singapore vol 1) pp 71-77 DOI: 10.1007/978-981-4021-90-6_5

[13] Teaching Institute for Excellence in STEM (TIES) 2010 *What is STEM education ?* (TIES Retrieved from http://www.tiesteach.org/about/what-is-stem)

[14] The Bureau of Academic and Educational Standards 2015 *The guideline for teaching and learning in "The Moderate Class, More Knowledge" project* (Bangkok: The Agricultural Cooperative Federation of Thailand)

[15] The Institute for the Promotion of Teaching Science and Technology (IPST) 2002

[16] The Partnership for 21st Century Skills 2009 *Framework Definitions. The Partnership for 21st Century Skills* (Retrieved from www.p21.org/storage//P21_Framework_Definitions.pdf)

[17] The Thai Ministry of Education (MOE) 2015 *Handbook of "The Moderate Class, More Knowledge" project* (Bangkok)

[18] Vasques J A, Sneider C, and Comer M 2013 *Lesson Essentials Grades 3-8 : Integrating Science, Technology, Engineering, and Mathematics* (Postmouth, New Hampshire: Heinemann)

[19] Wanapathownl R and Nuangchalerm P 2017 *The Development of Learning Achievement and Integrated Science Process Skills of Mattayomsuksa 2 Students through STEM Education* (Journal of Educational Administration and Supervisor vol 8) chapter 3

IOP Conf. Series: Journal of Physics: Conf. Series **1340** (2019) 012012 doi:10.1088/1742-6596/1340/1/012012

A STEM Education project for training graduate students for STEM workforce

Narongsak Kodtharin[1], Phuvasa Chanonmuang[2] and UdomTipparach[1*]

[1]Department of Physics, Faculty of Science, UbonRatchathani University, Warinchamrab, UbonRatchathani, 34190 Thailand

[2]Thailand Institute of Scientific and Technological Research, Khlong Luang, Pathum Thani 12120, Thailand

*Corresponding author's e-mail address: udomt@hotmail.com.

Abstract. A thesis-based on STEM Education and STEM workforce has been carried out on the topic titled "Hydrogen production from cassava starch by a biological process." This work is a real world problem of an engineering project that needs to solve by the integration of knowledge and skills in Science, Technology, Engineering, and Mathematics (STEM). The project deals with social need which is hydrogen gas that will serve as environmentally friendly renewable energy in the future. This STEM Education project provides a way to train students to thinks and learn how to think about solving problems by using STEM Education. The work requires students getting involved this project to solve problems and the students have applied their STEM Education skills. The students not only learn STEM Education on the project but also teach others to learn through STEM Education.

1. Introduction

The Ministry of Science and Technology of Thailand has launched a STEM workforce program, the integration of Science, Technology, Engineering and Mathematics (STEM), to support graduate students who engage in the innovations and research on STEM workforce training program [5]. This integration program is considered a solution to the educational reforms skills to perform the high-tech knowledge based in research [4]. This STEM workforce project provides a way to train students to thinks and learn how to solve real problems by using STEM Education.

Our work is supported by the STEM workforce program. This work involves to do research on renewable energy. The work is intended to produce hydrogen from a biological process. Hydrogen is considered as a carrier for renewable and sustainable energy. This project also supports Thai government policy, Thailand 4.0: Biological/Biochemical fuel. Renewable energy received considerable attention because it can replace high-price fossil fuels and can reduce environmental pollution associated with their utilization. Both developing and industrialized countries consider biofuels for e.g. ethanol and hydrogen as relevant sources on the reasons of energy security. Hydrogen is recognized as a clean, renewable and promising future fuel [3]. Biological hydrogen production is classified into 4 categories, namely (1) biophotolysis of water using solar energy and algaecyano bacteria, (2) photodecomposition of organic compounds using light energy and photosynthetic bacteria, (3) fermentative hydrogen evolution to breakdown carbohydrate-rich substrates to hydrogen and other products such as acids and alcohols using anaerobic bacteria and (4) hybrid systems combining dark and photofermentation either directly or in a series-type [2]. Among these methods, fermentation hydrogen production is the simplest and cheapest method. However there is little information on the dark fermentation process for hydrogen production by controlling temperature and treating bacterial (Clostridia). In this paper, we present the method of biological hydrogen production method with STEM Education to prepare students for STEM workforce to support Thailand 4.0 policy. The purpose of this paper is to describe the method for hydrogen production with the use of STEM Education for STEM workforce.

2. Methods

An idea of STEM Education project for hydrogen is how to produce hydrogen with economical and engineering (E) feasibilities and the use of knowledge in Science (S), Technology (T), Engineering (E) and Mathematics (M). The science of a biological process is to ferment starch rich raw material to be sugar and acid and then turn into hydrogen gas by bacterial (Clostridia) as follows [1]:

- $C_6H_{12}O_6 + 2H_2O \rightarrow 2CH_3COOH + 4H_2 + 2CO_2$ (acetate fermentation)
- $C_6H_{12}O_6 + 2H_2O \rightarrow CH_3CH_2CH_2COOH + 2H_2 + 2CO_2$ (butyrate fermentation)

Technology used in this project includes lab instrument and tools as well as hydrogen analyzer. Engineering is used to design and control fermentation process for hydrogen production. Mathematics is used to calculate the concentration of raw materials, gas volume, and hydrogen yield.

Table 1. Knowledge and skills for STEM Education for training the students for STEM workforce

STEM	Cores knowledge and skills
Science (S)	fermentative bacteria for hydrogen production and fermentation process $C_6H_{12}O_6 + 2H_2O \rightarrow 2CH_3COOH + 4H_2 + 2CO_2$ (acetate fermentation) $C_6H_{12}O_6 + 2H_2O \rightarrow CH_3CH_2CH_2COOH + 2H_2 + 2CO_2$ (butyrate fermentation)
Technology (T)	Materials, methods, techniques, instruments and lab tools
Engineering (E)	Design, control and process
Mathematics (M)	Calculation related materials, process, and cost of investment

Teaching and learning suggests that each STEM Education project should be designed to include learning outcomes. However, STEM education as an instructional approach in which students participate in engineering design for a target or research meaningful learning through the integration and application of mathematics, technology and science. The students were assigned to learn by doing a project to solve real problems among STEM subjects in the real world for STEM workforce.

In the integrated applied, the STEM content are mixed and learnt as one subject concepts to solve real world problems. This learning process appears to activate the concentration and to increase the motivation in the topic titled "Hydrogen production from cassava starch by a biological process". While the students apply skills and knowledge to solve a real world problem, they learn and build up new knowledge and skills.

Procedure for Technical with STEM Education for biological hydrogen production was designed as follows:

- Step 1: The students were assigned to produce hydrogen for renewable energy by biological process of dark fermentation. The students were introduced to STEM project.
- Step 2: Raw materials, sources of bacteria, lab tools and apparatus were provided. The students were supervised to use of technology, methods and techniques and learned biological process for hydrogen production.
- Step 3: The student designed the experimental process and carried out the experiments under the supervision of an advisor.
- Step 4: The students measured volumes of biological gas, analyzed, and calculated amount of hydrogen gas.
- Step 5: The students made a conclusion and wrote a report including prepared for oral or poster presentation.

3. Results and Discussion

Graduate students learned and practiced. The students were able to produce hydrogen gas. They coached senior high school students of SciUs program, Science Classrooms in University-Affiliated School Project supported by the Ministry of Science and Technology of Thailand. The high school students could also generate hydrogen and earned medal awards in 2018 in national science competition. Biogas yielded in the experiments was composed of hydrogen, methane, hydrogen sulfide, water vapor, and others. A hydrogen analyzer was used to measure a percent of hydrogen as shown in Fig.1.

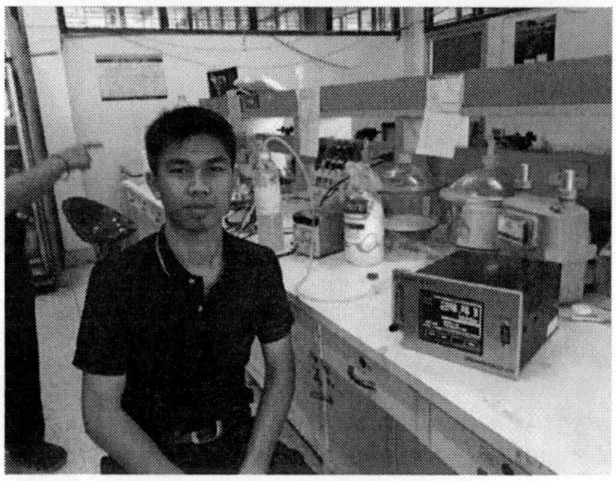

Figure 1. A graduate students with hydrogen analyzer used in our experiments

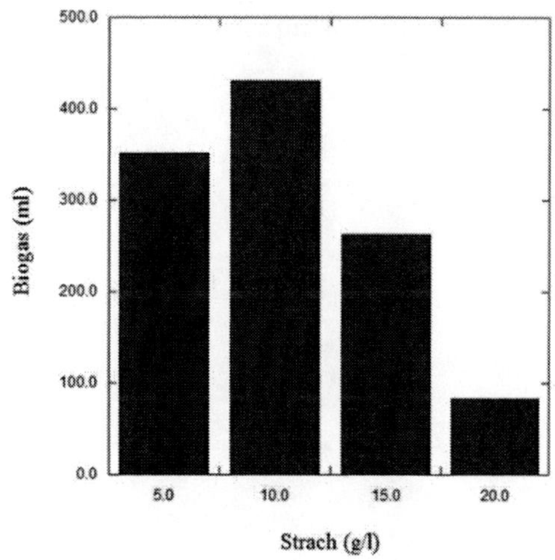

IOP Conf. Series: Journal of Physics: Conf. Series **1340** (2019) 012012 doi:10.1088/1742-6596/1340/1/012012

Figure 2. Amount of Biogas yielded in the experiments when starch ratio per liter (gram/l) were used for 50 milliliters of the fermentation bottle.

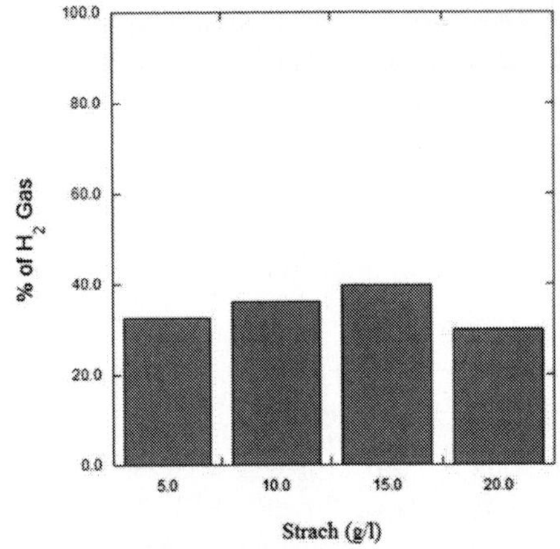

Figure 3. Percent of hydrogen gas yielded in the experiments when starch ratio per liter (gram/l) were used for 50 milliliters of the fermentation bottle.

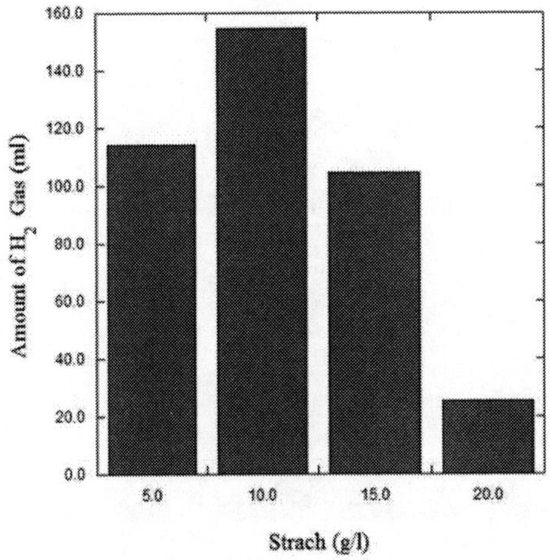

Figure 4. Amount of hydrogen gas yielded in the experiments when starch ratio per liter (gram/l) were used for 50 milliliters of the fermentation bottle

Figure 5 SciUs high school students award in national competition in 2018

The students who joined the project of hydrogen production from cassava starch by a biological process have actively learned by doing through STEM Education. The learn how to integrate their STEM knowledge and skills to carry out the experiments. They were also able to build up new knowledge and skills. They could achieve the target of hydrogen production. They very engaged in learning and doing STEM Education.

4. Conclusions

This work was successful in training graduate students and high school students to thinks, learn, and do their tasks to solve real problems by using STEM Education. This project-based STEM Education provided learning skills and building new knowledge effectively. The student could achieve their goals and enjoyed learning activities. The students were able to apply their STEM Education skills to solve problems and were prepared for STEM workforce to support Thailand 4.0 policy.

5. Acknowledgments

We would like to thank Thailand Institute of Scientific and Technological Research (TISTR) for financial supports under grant Partnership program in production of graduates in master and doctoral programs between Thailand Institute of Scientific and Technological Research and educational institutions.

References

[1] Bouallagui H, Torrijos M, GodonR J J, Moletta R, Ben Cheikh, Touhami Y, Delgenes J P, Hamdi M, 2004 *Microbial monitoring by molecular tools of a two-phase anaerobic bioreactor treating fruit and vegetable wastes* (J Biotechnology Letters lssus vol 10) pp 857-862

[2] Breiner J, Harkness S, Johnson C and Koehler C 2012 *What is STEM? A discussion aboutconceptions of STEM in education and partnerships School Science and Mathematics* 2012 vol 112) chapter 1 pp 3-11

[3] Le Xuan Quang, Le Huy Hoang, Vu Dinh Chuan, Nguyen Hoai Nam, Nguyen Thi, Tu Anh and Vu Thi Hong Nhung 2015 *Integrated Science, Technology, Engineering and Mathematics (STEM) Education through Active Experience of Designing Technical Toys in Vietnamese Schools* (J British Journal of Education, Society & Behavioural Science vol 11) chapter 2 pp 1-12

[4] Stylianos Sergis , Demetrios G, Sampson , María Jesús , Rodríguez-Triana Denis Gillet , Lina Pelliccione and Ton de Jong 2017 *Behavior Using educational data from teaching and learning to inform teachers' reflective educational design in inquiry-based STEM education.*

(J Computers in Human xxx) pp 1-15
[5] The Institute for the Promotion of Teaching Science and Technology 2014 *Strategies for Implementing STEM Education in Schools* pp 3-5

IOP Conf. Series: Journal of Physics: Conf. Series **1340** (2019) 012013 doi:10.1088/1742-6596/1340/1/012013

Opinions of Primary Science Teachers on 4C3RA+ Guidelines to Design STEM Lesson Plans

Worakanya Kaewklom[*], **Pinit Khumwong, and Chanyah Dahsah**

Science Education Center, Srinakharinwirot University, 114 Sukhumvit 23, Bangkok, 10110, Thailand

*Corresponding author's e-mail address: worakanya.swu@gmail.com

Abstract. The purposes of this study was to investigate the opinions of primary science teachers in 4C3RA+ guidelines. The sample consisted of sixty-two primary science teachers in Sra-Kaew Primary Education, Area Two, and fifteen Si Sa Ket Primary Education, Area Four, selected by purposive sampling and willing to participate in this study. The results found that primary science teachers had opinions on 4C3RA+ guidelines at a good level. Primary science teachers in Sra-Keaw Primary Education, Area Two, had opinions on 4C3RA+ guidelines at a good level ($\bar{x}$ = 3.05), and primary science teachers in Si Sa Ket Primary Education, Area Four, had opinions on 4C3RA+ guidelines at an excellent level ($\bar{x}$ = 3.76). They mentioned that 4C3RA+ guideline provided a clear view of STEM education. Then it made possibility to design STEM lesson plans using a six-step process based on 4C3RA+ guidelines.

1. Introduction

To promote the 21st century learning skills among learners and to be consistent with their career in the ASEAN Economic Community (AEC), The Institute for the Promotion of Teaching Science and Technology (IPST) developed the strategic plan to improve science, mathematics and technology learning in 2012-2016 through STEM education. [1,2]

STEM Education began to operate in Thailand in 2012, then it is the new topic for Thai teachers, and they lack the specific training and knowledge necessary to teach STEM subjects. Therefore, teachers must be properly trained to facilitate STEM lessons to their students and pique their interest, for increase quality of teaching and development their student by effectively.

I developed "4C3RA+ guidelines" to enhance in-service primary science teachers' understandings of STEM education, design problem situation and design STEM lesson plans. And study the opinions of primary science teachers with regard to 4C3RA+ guidelines.

2. 4C3RA+ Guidelines

The 4C3RA+ guidelines for designing STEM lesson plans consisted of a conceptual framework of 4C3RA+ and the steps in designing STEM lesson plans, were as follows.

2.1. Conceptual Framework of 4C3RA+

The conceptual framework of 4C3RA+ is a concept for design STEM problems situation, STEM activity and STEM assessment, adapted from "3C3R PBL" [3-4] and core feature of STEM Education.

Conceptual framework of 4C3RA+ comprises of three components (see fig. 1): 1) STEM Content: 4C including content, context, connections & integration, and challenge. 2) STEM Activity: 3R including researching, reasoning, and reflecting. 3) Authentic Assessment: A+ It means to include engineering design assessment, and integrated ideas assessment into normal assessment, in another word A+ mean beyond the normal authentic assessment.

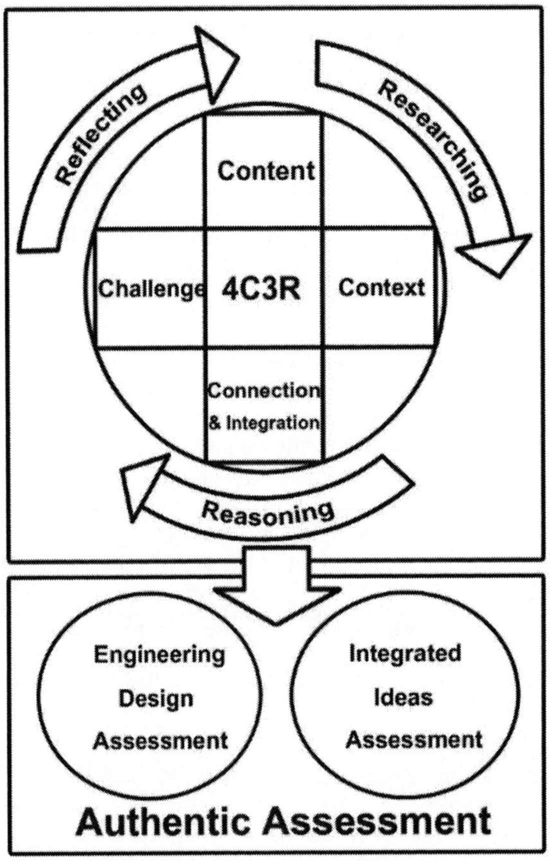

Figure 1. Conceptual Framework of4C3RA+.

2.2. The Steps in Designing STEM Lesson Plans

The steps in designing STEM lesson plansis a six-step process based on 4C3RA+ guidelines to design STEM lesson plans, were as follows;

Step1 Select STEM Content and Generate Problems Situation: 4C
Step2 Define Learning Objectives
Step3 Design STEM Core Activity: 3R
Step4 Select Learning approach/model
Step5 Activity Sequencing
Step6 Determine Authentic Assessment: A+

3. Purposes

The purposes of this study was to investigate the opinions of primary science teachers in 4C3RA+ guidelines.

IOP Conf. Series: Journal of Physics: Conf. Series **1340** (2019) 012013 doi:10.1088/1742-6596/1340/1/012013

4. Methodology

4.1. Participant
The sample consisted of sixty-two primary science teachers in Sra-Kaew Primary Education, Area Two, and fifteen Si Sa Ket Primary Education, Area Four, selected by purposive sampling and willing to participate in this study.

Table 1. Data of Participant.

	Sra-Kaew (N=62)		Si Sa Ket (N=15)		All (N=77)	
	f	percentage	f	percentage	F	percentage
1.Gendar						
1.1 Male	12	19.35	4	26.67	16	20.78
1.2 Female	50	80.65	11	73.33	61	79.22
2.Age						
2.1 Less than30 years old	16	25.81	4	26.67	20	25.97
2.2 31-40 years old	27	43.55	6	40	33	42.86
2.3 41-50 years old	12	19.35	4	26.67	16	20.78
2.4 51-60 years old	7	11.29	1	6.67	8	10.39
3. Science Teaching Experience						
3.1 1 – 5 years	28	45.16	3	20	31	40.26
3.2 6 – 10 years	9	14.52	3	20	12	15.58
3.3 11 – 15years	10	16.13	6	40	16	20.78
3.4 16 – 20 years	4	6.45	2	13.33	6	7.79
3.5 more than20 years	11	17.74	1	6.67	12	15.58
4. Graduate						
4.1 Science	34	54.84	13	86.67	47	61.04
4.2 Other	28	45.16	2	13.33	30	38.96
5. Teaching Science						
5.1 Only Science	34	54.84	15	100	49	63.64
5.2 All Subject in Class	11	17.74	0	0	11	14.29
5.3 Other Subject)Math, Another)	17	27.42	0	0	17	22.08

4.2. Instrument
To obtain the data for this study were by questionnaire after workshop for designing STEM lesson plans by using 4C3RA+ guideline, were as 15 questions by rating scale 4 level; strongly agree, agree, disagree, strongly disagree and one of open-ended question.

4.3. Data analysis
After the data was collected, researcher analyzing were as follow:

1. Statistic: Mean and divide into four levels

Excellent	3.35 – 4.00
Good	2.68 - 3.34

Poor 2.01-2.67
Very Poor 1.00–2.00
2. Content Analysis for additional comments.

5. Result

The opinions of primary science teachers with regard to 4C3RA+ guidelines showed as table 2;

Table 2. Opinions of primary science teachers with regard to 4C3RA+ guidelines.

List	Sra-Kaew (N=62)		Si Sa Ket (N=15)		All (N=77)	
	$\bar{X}$	S.D.	$\bar{X}$	S.D.	$\bar{X}$	S.D.
Opinion on Conceptual Framework of4C3RA+						
1. Concept image of 4C3RA+easy to understand?	3.08	0.64	3.87	0.35	3.23	0.67
2. Concept definition of4C3RA+easy to understand?	3.03	0.65	3.93	0.26	3.21	0.69
3. You can design STEM lesson plans by using Conceptual Framework of4C3RA+actually?	3.02	0.50	3.80	0.41	3.17	0.57
4. Conceptual Framework of4C3RA+make to design STEM lesson plans easier?	3.06	0.60	3.80	0.41	3.21	0.64
5. Overview of STEM Content: 4C	3.05	0.51	3.80	0.41	3.20	0.58
6. Overview of STEM Activity: 3R	3.06	0.50	3.64	0.49	3.17	0.55
7. Overview of Authentic Assessment: A	3.01	0.50	3.73	0.45	3.15	0.57
Mean	**3.06**		**3.75**		**3.18**	
Opinion on The Steps in Designing STEM Lesson Plans						
8. The steps in designing STEM lesson plansmade possibility to design STEM lesson plans?	3.15	0.60	3.73	0.46	3.26	0.62
9. The steps in designing STEM lesson plans make to design STEM lesson planseasier?	3.08	0.52	4	0	3.26	0.59
10. Step1 Select STEM Content and Generate Problems Situation: 4C	3.09	0.56	3.90	0.31	3.25	0.61
11. Step2 Define Learning Objectives	3.12	0.55	3.87	0.35	3.26	0.60
12. Step3 Design STEM Core Activity: 3R	2.90	0.54	3.60	0.51	3.12	0.59
13. Step4 Select Learning approach/model	2.97	0.53	3.57	0.52	3.09	0.57
14. Step5 Activity Sequencing	3.00	0.55	3.70	0.47	3.14	0.60
15. Step6 Determine Authentic Assessment: A+	3.04	0.57	3.87	0.35	3.20	0.62
mean	**3.05**		**3.77**		**3.19**	
Combined mean	**3.05**		**3.76**		**3.18**	
	Good		**Excellent**		**Good**	

From table 2: The primary science teachers had opinions on 4C3RA+ guidelines at a good level (=3.18). Primaryscience teachers in Sra-Keaw Primary Education, Area Two, had opinions on 4C3RA+ guidelines at a good level (= 3.05), and primary science teachers in Si Sa Ket Primary Education, Area Four, had opinions on 4C3RA+ guidelines at an excellent level (= 3.76),and found the notice as follow:

1. Conceptual framework of 4C3RA+:

The mean of opinions on Authentic Assessment: A (=3.15) were lowest. Because, both of engineering design assessment and integrated ideas assessment are the new for teachers.

2. The Steps in Designing STEM Lesson Plans:

The mean of Step 3 Design STEM Core Activity: 3R (= 3.12), Step 4 Select Learning approach/model (= 3.09), and Step 5 Activity Sequencing (= 3.14) were lowest.

Teachers did think these three steps were similar; they preferred to select learning approach/model first, then designing core activity and activity sequencing at the same time. Therefore 3R might be overlooked.

From additional comments; teachers mentioned that 4C3RA+ guideline provided a clear view of STEM education. Then it made possibility to design STEM lesson plans by using a six-step process based on 4C3RA+ guidelines.

6. Conclusion

The primary science teachers had opinions on 4C3RA+ guidelines at a good level. Primary science teachers in Sra-Keaw Primary Education, Area Two, had opinions on 4C3RA+ guidelines at a good level ($\overline{X}$ = 3.05), and primary science teachers in Si Sa Ket Primary Education, Area Four, had opinions on 4C3RA+ guidelines at an excellent level ($\overline{X}$ = 3.76). They mentioned that 4C3RA+ guideline provided a clear view of STEM education. Then it made possibility to design STEM lesson plans using a six-step process based on 4C3RA+ guidelines.

References

[1] Hung, W 2006 The 3C3R Model: A Conceptual Framework for Designing Problems in PBL *Interdisciplinary* **Journal of Problem Based Learning 1**

[2] Hung, W 2009 The 9-step problem design process for problem-based learning: Application of the 3C3R model **Educational Research Review 4** pp 118–141

[3] IPST 2015 *Teaching STEM Education* online: http://www.stemedthailand.org/wp-content/uploads/2015/05/STEM-Education2.pdf Retrieved 15 June 2015

[4] Montri Chulavatnatol 2013 November-December STEM Education in Thailand and STEM Ambassador **IPST Magazine 42** pp 14–18

IOP Conf. Series: Journal of Physics: Conf. Series **1340** (2019) 012014 doi:10.1088/1742-6596/1340/1/012014

Research about Metacognition in Science Education: A Case of Basic Education in Thailand

Soonthareeya Sanium[1] and Khajornsak Buaraphan[1]

[1]Institute for Innovative Learning, Mahidol University 999 Salaya, Phuttamonthon, Nakhon Pathom 73170 Thailand

E-mail: khajornsak.bua@mahidol.ac.th and soonthareeya.sai@student.mahidol.ac.th

Abstract. Metacognition has been regarded as a research variable more than four decades ago with its general definition of knowing about knowing. It is genuinely difficult to study about metacognition because it is an abstract process occurred in one's mind. A number of studies about metacognition have been conducted in an international context to derive more understanding about a metacognitive process and its benefits. There are some studies related to metacognition in the Thai context; so, it is a need to explore the current situation of metacognition research in Thailand. The metacognitive studies conducted during 2002-2016 have been intensively reviewed. The researchers found 22 research studies related to metacognition in Thailand; three, 10 and nine of them had been conducted in the primary, lower-secondary and upper-secondary levels, respectively. All metacognitive studies could be grouped into five categories: a) Impact of metacognition on other research variables (n = 9); b) Teaching strategies to enhance metacognition (n = 7); c) Developing measurement tool for metacognition (n = 3); d) Causal model of metacognition (n = 2); and e) Factors of metacognition (n = 1). Two dominant categories (16 of 22 or 72.73%) were the 'Impact of metacognition' (n = 9) and 'Teaching strategies enhancing metacognition' categories. Interestingly, there were five studies aimed to explore the impact of students' developed metacognition on problem-solving skill. The teaching strategies employed to enhance students' metacognition are diverse including 5E or 7E models, learning cycle, problem-based learning, open-ended problems, constructivist and Tri-Sik-Kha instruction.

1. Introduction

Metacognition is presently recognized as one research variable in science education context. The definition of metacognition is diverse. The most well-known definition of metacognition comes from Flavell the late 1970 that was cited by [5] as "cognition about cognitive phenomena" or "thinking about thinking" [6]. Other definitions of metacognition are "metacognition is the monitor and control of thought" [11], "Metacognition is essential for successful learning" [23] and "Metacognition is any knowledge or cognitive activity that takes as its object or regulates, any aspect of any cognitive enterprise" [22].

Flavell and his colleagues divided metacognition into two main components: *Knowledge of Cognition* (metacognitive knowledge) and *Regulation of Cognition* (metacognitive monitoring and self-regulation). Later, the third component was added, that is, *Metacognitive Experiences* [4]. Other researchers have addressed various components of metacognition depending on their theoretical or conceptual frameworks. Research about metacognition has taken rooted in science education context for five decades and there is one international journal specially devoted for it namely *Metacognition and Learning*. Metacognition is regarded as one desirable kind of thinking for students with its benefit in promoting students' thinking and learning achievement in various subjects and contexts.

In Thailand, metacognition is paid attention in educational context for two decades ago and there was a number of research studies related to metacognition. However, there is no study aimed to review the past studies of metacognition in the basic education context of Thailand. Consequently, the main objective of this study is to intensively review the research studies dealt with metacognition in the basic education context of Thailand during 2002-2016.

2. Literature review about review study in metacognition in science education context

From the review of literature related to metacognition in an international context. The review study about metacognition is rare. There are only two review studies and one book related to metacognition in science education. Then, we conclude the method of these studies to conduct the category in this paper.

The first review study is named "Metacognition: a literature review in science" [5]. In this study, the researchers focused metacognition as one goal in science education. It presented the definitions of metacognition stated by many researchers and the typology of metacognitive components. Also, the researchers presented the ways to assess metacognition and teacher development models regarding metacognition.

The second review study was entitled "Review of research on metacognition in science education: current and future directions" [29]. This review study presented metacognition in key trends in for 21st century and proposed a set of questions for future research. It highlighted the trends of metacognition, ways to categorize metacognition and discussion about the growth and expansion of metacognition in science education. The gap found from the review of literature is that there is no review study of metacognition in the Thailand basic education context.

3. Methods

3.1 Population and data collection

The population of this study is all research studies in metacognition reported and included in the Thailand national database so called *ThaiLIS* (Thailand Library Integrated System). After completed searching in ThaiLIS, we found 22 metacognitive studies conducted in Thailand during 2001 to 2016.

3.2 Data analysis

The researchers intensively read all studies in order to specify their major purposes. The major purposes of each research paper were compared with others. The research studies conducted with similar research purpose were grouped together and the category was subsequently created for them. This process was continued until the categories were exhausted.

4. Results and discussion

After review, the research about metacognition (n = 22) in the basic education content of Thailand.

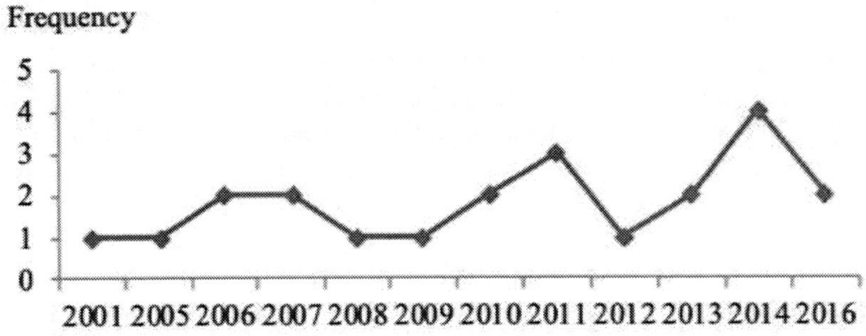

Figure1. Frequency of research studies in metacognition in Thailand during 2001 to 2016.

The metacognitive studies in Thailand could be grouped into five categories, that is, a) Impact of metacognition on other research variables, b) Teaching strategies to enhance metacognition, c) Developing measurement tool for metacognition, d) Causal model of metacognition, and e) Factors of metacognition.

Table 1. Categories of research about metacognition in Thailand.

Category	Educational level			Frequency	%
	Primary School	Lower-secondary	Upper-secondary		
a) Impact of metacognition on other research variables	1	5	3	9	40.9
b) Teaching strategies to enhance metacognition	1	1	5	7	31.8
c) Developing measurement tool for metacognition	0	2	1	3	13.6
d) Causal model of metacognition	0	2	0	2	9.1
e) Factors of metacognition	1	0	0	1	4.6
Total	3	10	9	22	100.0

From Table 1, there was two dominant categories of metacognitive studies in Thailand, that is, Impact of metacognition on other research variables (40.9%) and Teaching strategies to enhance metacognition (31.8%). Almost all studies about metacognition were conducted in the secondary school levels (86.36%). Interestingly, there was only three studies in metacognition conducted in a primary school level. The summary of each study related to metacognition in Thailand is presented as follows.

Category 1: Impact of metacognition on other research variables

The metacognitive studies in this category focused on the impacts of developed metacognition on other research variables.

Research 01: A study of relationship between metacognition and critical thinking of grade 6 students [20]. The purposes of this study were to study 1) the multiple correlation between subcategories of metacognition and critical thinking; and 2) the beta weight of metacognition which contributed to all subcategories and in each subcategory of critical thinking. The research instruments were the critical thinking test and the metacognition questionnaire. The results showed that 1) the multiple correlation coefficients between the subcategories of metacognition and the sum of all subcategories of critical thinking were significant at the .01 for all values and the multiple correlation coefficients between the subcategories of metacognition and each subcategory of critical thinking were statistically significant at the .01 level for all values. 2) The beta weight values in each subcategory of metacognition positively contributed to the sum of all subcategories of critical thinking for all values. There were some subcategories of metacognition that negatively contributed to critical thinking.

Research 02: Effect of a metacognition training with attribution training to effort on problem-solving skills of grade 7 students [27]. This research conducted with the three groups: the control group, the experimental group 1 received a training on metacognition and attribution to effort and the experimental group 2 receiving a metacognition training only. The result showed that, after training, the experimental groups 1 and 2 received higher problem-solving skill scores at the .05 level of significance; while the control group received lower problem-solving skill scores.

Research 03: Grade 8 students' scientific problem solving ability, metacognition development and scholastic achievement in work and energy topic by using metacognitive strategies [19]. The purposes of this research were: 1) to develop students' scientific problem-solving ability by using metacognition, 2) to develop students' metacognition by using metacognition development, and 3) to study student's scholastic achievement in work and energy topic by using metacognitive strategies. This research used four procedures: 1) analyzing problems, 2) planning to solve problems, 3) managing problems and 4) evaluating problems. The results were: 1) metacognition helped students develop scientific problem-solving ability as students analyzed and planned to solve scientific problems; 2) 4-procedure metacognition development was applied to solve situation that they received; and 3) teaching and learning based on metacognition development affected students' achievement.

Research 04: Learning management by using metacognition strategy to develop the scientific problem solving skills for grade 7 students [15]. The purposes of this research were: 1) to find out the efficiency of the metacognition learning plan in developing students' scientific problem-solving skills. 2) to compare the scientific problem-solving skills and the learning achievement of grade 7 students before, during and after the metacognition learning management. The results were: 1) the efficiency of the metacognition learning plan to develop the scientific problem-solving skill for grade 7 students was at 85.86/ 81.66; 2) when students used metacognition strategy, they had higher scientific problem-solving skill and learning achievement at the .05 statistically significant level; and 3) The students satisfied with the metacognition learning management at the highest level.

Research 05: The effect of metacognitive strategy on decision making and problem solving abilities in force and motion topic for grade 9 students [13]. The purpose of this research was to study grade 9 students decision making and problem-solving abilities in science after learning through metacognitive strategy. The results showed that, after learning through metacognitive strategy., the student's post-test average score of decision making and problem-solving abilities were significantly higher than pre-test at the .01 statistically significant level.

Research 06: Decision making and problem solving abilities in science learning through metacognitive strategy [24]. The purpose of this research was to observe and measure the decision making and problem-solving abilities of grade 9 students learned with metacognitive strategy in Force and Motion. The results presented that, after learned with metacognitive strategy, the students' pre-test average score of decision making and problem-solving abilities were significantly higher than pre-test at the .05 statistically significant level.

Research 07: The effects of cooperative learning activities management integrated with metacognitive thinking on learning achievement and metacognitive thinking ability of grade 10 students [25]. The purpose of this research was to compare students' learning achievement and metacognitive thinking ability after learned with cooperative learning only and learned with cooperative learning integrated with metacognitive thinking. The results revealed that the students who learned with cooperative learning integrated with metacognitive thinking has higher learning achievement and metacognitive thinking ability than those who learned with cooperative learning only at the .05 statistically significant level.

Research 08: Development of knowledge and understanding, critical thinking, awareness, environmental conservation behaviors of grade 12 students using the good science thinking moves with metacognition techniques [7]. The purposes of this study were: to compare knowledge and understanding, critical thinking, awareness and environmental conservation behaviors of the students who learned with good science thinking moves with metacognitive techniques and those who learned with traditional teaching method. The findings revealed that the lesson plans using the good science thinking moves with metacognitive techniques had an effectiveness index of 0.7290. The students in the experimental group developed higher knowledge and understanding, critical thinking, awareness and environmental conservation behaviors ($p < .001$). In overall, gender did not affect different learning outcomes.

Research 09: Using problem-based learning to promote metacognition and mathematics achievement of grade 10 students [10]. The purpose of this study was to study the use of problem-based learning (PBL) in promoting grade 10 students' metacognition and mathematics achievement. The result revealed that, after using the PBL approach, the students' metacognition mean score was increased from a moderate level to a good level. Also, most of the students (more than 80%) had mathematics achievement mean scores higher than the minimum criteria at 60%. The mathematics achievement mean score, in overall, was 14.9 of 20 points (74.5%).

In summary, the research studies in this category shows that developed metacognition in students could affect other desirable attributes as learning achievement, critical thinking skill, problem-solving skill, decision making skill, and awareness and environmental conservation behaviors.

Category 2: Teaching strategies to enhance metacognition

The metacognitive studies in this category focused on implementing several teaching strategies and models to enhance students' metacognition.

Research 10: Comparisons of effects of the 7E learning cycle using metacognitive moves on grade 5 students' alternative conception in biology and basic science process skills [2]. The purpose of this study was to examine the effect of 7E learning cycle using metacognitive moves on grade 5 students' alternative conception in biology and basic science process skills. The results showed that the 7E learning cycle using metacognitive moves helped students gain more understanding about biology concepts at the .05 level of significance. The students developed basic sciences process skills higher than the 50% criterion at the .05 level of significance. In addition, the female students evidenced more interning skills than the male students at the .05 level of significance.

Research 11: Effects on using open-ended problems to develop metacognition for grade 9 students (18). This study aimed to examine the effects of open-ended problems in helping promote problem solving ability with metacognition for grade 9 students. The results revealed that using open-ended problems could help 80.95% of the students passed the metacognition test criterion of 50%. The students also showed a "good" level of metacognition practice with mean score of 2.96.

Research 12: Development of physics learning activities to develop problem solving ability with metacognition for grade 10 students [26]. This study aimed to develop learning activities in physics subject to develop grade 10 students' metacognition and problem-solving ability. The results showed that 80.95% of the students passed the physics problem solving ability test with criterion of 70%.

Research 13: Effects of science learning management on grade 10 students through learning cycle with metacognitive reflection and awareness [28]. The research objective was to study: 1) the interaction between types of learning management and basic scientific knowledge, scientific problem-solving ability and metacognitive ability; and 2) the effects of different learning management and basic scientific knowledge on students' learning outcome, scientific problem-solving ability and metacognitive ability. This study presented that there was no interaction between types of learning management and basic scientific knowledge. The students who learned with the learning cycle with metacognitive reflection and awareness (LCMRA) performed the highest learning outcome, scientific problem solving ability and metacognitive ability, followed by those learned with the learning cycle with metacognitive reflection (LCMR) and learning cycle (LC) only. The degree of student basic scientific knowledge corresponded with the degree of achieved learning outcome, scientific problem-solving ability and metacognitive ability.

Research 14: Results of 5E instructional model with metacognitive reflection strategies on science learning achievement, science concepts and metacognition of upper secondary school students [12]. This study aimed to compare the result of 5E instructional model with metacognitive reflection strategies on student's science learning achievement. The result showed that, after learned with the 5E instructional model with metacognitive reflection strategies, the students in a bright, moderate and weak groups had higher science learning achievement, science concepts and metacognition than priori at the .01 of significance. Also, the students had science learning achievement higher than the criteria of 60%.

Research 15: Development of metacognition by using constructionist activities in the topic of evolution for grade 12 [21]. The purposes of this research were: 1) developing grade 12 students' metacognition by using the constructionist activities in the topic of evolution; 2) find the approach to develop grade 12 students' metacognition by using the constructionist activities. The result indicated that the constructionist activities could develop grade 12 students' metacognition in a high level. This was occurred from the characteristic of constructionist activities as working with multiple formats, a teacher allowing students to discuss and share ideas both in and outside the classroom.

Research 16: Effects of instruction based on Tri-Sik-Kha principle for developing grade 10 students' metacognition in the Ecosystems and Human and Sustainable Environment topic [17]. This study applied the Tri-Sik-Kha principles to design instruction to help develop grade 10 students' metacognition. The result showed that the instruction based on Tri-Sik-Kha principles could develop the students' metacognition. The student average score of metacognition was very high at 70.35%.

In summary, the researchers tried to apply several teaching strategies, that is, the 5E and 7E learning cycle, open-ended problems, constructionist activities, and Tri-Sik-Kha principle, in order to help promote students' metacognition. Also, the effect of those teaching strategies on students' metacognition was positive.

Category 3: Developing measurement tool for metacognition

The metacognitive studies in this category focused on the development of research tool for measuring metacognition in students.

Research 17: Development of metacognition test for Level 3 students [14]. This study aimed to construct a metacognition test for Level 3 students and investigate its quality. The results showed that the metacognition test for Level 3 students was consisted of six tests, that is, Test 1: A Metacognition Knowledge (Person), Test 2: A Metacognition Knowledge (Task), Test 3: A Metacognition Knowledge (Strategy), Test 4: A Metacognition Experience (Planning), Test 5: A Metacognition Experience (Monitoring), and Test 6 A Metacognition Experience (Assessment).

Research 18: Development of metacognition test in science subject for grade 9 students [9]. This study had two objectives: 1) to develop a metacognition test in science subject for grade 9 students; and 2) to examine the quality of a developed metacognition test. This study showed that the metacognition test in science subject for grade 9 students was consisted of 24 items. The metacognition test was consisted of four situations and each situation had six open-ended questions. The questions aimed to test six components of metacognition including Self-awareness, Task Awareness, Strategy Awareness, Planning, Monitoring, and Evaluating. The Index of Consistency of the metacognition test and its components was ranged from 0.67 to 1.00. The Index of Consistency of content and appropriateness of language was ranged from 0.67 to 1.00. Also, the difficulty and discrimination index of metacognition test were ranged from 0.22 to 0.44 and 0.44 to 0.88, respectively. Also, the reliability of metacognition test was very high at 0.93.

Research 19: Construction of metacognition test for Level 4 students [1]. The purposes of this study were to create a metacognition test and examine its quality regarding the discrimination index and construct validity by means of factor analysis and reliability. The results showed that the metacognition test for Level 4 students was qualified. That is, the Awareness, Planning, Self-evaluation aspects had discrimination index ranged between .402-731, .279-.561 and .326-.620, respectively. The construct validity of the metacognition test being analyzed by the factor analysis showed that the Awareness, Planning, Self-evaluation aspects had the factor loadings ranging between .480-.799, .353-.667 and .394-.712, respectively. The external reliability of Awareness, Planning, Self-evaluation aspects were .861, .816 and .842, respectively. The reliability of metacognition test as a whole was very high at .940. In this case, there was no interaction found between gender and class level of the students.

In summary, the researchers created the research tool for measuring metacognition in secondary students including Levels 3 and 4. The created tools were mainly quantitative and the researchers tried to established the quality of tools in quantitative way.

Category 4: Causal model of metacognition

The metacognitive studies in this category focused on the explanation of casual model of metacognition in students.

Research 20: Development of a causal model of metacognition of lower secondary school students [3]. This research aimed to study the metacognition level of lower secondary school students and develop a causal model of metacognition of lower secondary school students. The results showed that, in general, the lower secondary school students had high level of metacognition, especially in awareness and planning. The causal model of metacognition of lower secondary school students fit to the empirical data and accounted for 64% of variance in the student's metacognition.

Research 21: Causal model of self-efficacy and metacognition affecting efficacy and outcome expectancy of lower secondary students [8]. This research focused in developing and validating a causal model of self-efficacy and metacognition affecting efficacy and outcome expectancies of lower secondary school students. The results indicated that the causal model was consistent with the empirical data. The variables in the model accounted for 78% and 79% of the total variance in efficacy expectancy and outcome expectancy, respectively. Self-efficacy had the highest direct effect on students' efficacy expectancy and outcome expectancy.

In summary, the causal models of metacognition presented in the studies in this category fit to the empirical data. However, these studies focus on the causal models of metacognition of lower secondary school students.

Category 5: Factors of metacognition

The metacognitive study in this category focused on the factors of metacognition in students.

Research 22: Factors affecting metacognition of grade 6 students using the multilevel analysis [16]. This study aimed to study the relationship between the student and classroom levels and metacognition of grade 6 students' metacognition and to construct the predictive equation of the factors affecting the metacognition. The results were as follows. The student level factors positively affected students' metacognition were students' learning goals, achievement motivation, confidence in self-performance, parental support, parental educational expectations, teaching quality, learning aptitude, child-centered instruction, teacher personality and learning environment. The classroom level factor positively affected students' metacognition was child-centered instruction.

In summary, there are many factors that can affect students' metacognition. These factors may be internal factor (e.g. personality) and external factors (e.g. classroom environment).

5. Discussion

There are 22 research studies during 2001 to 2016 dealt with metacognition in the basic education context of Thailand. Almost all studies about metacognition were conducted in the lower- and upper-secondary school levels (86.36%) rather than the primary school level. The very small number of studies in primary school level may be dealt with the limitation of cognitive development of primary school students with their ages ranged from seven to 12 years old. According to Piaget's cognitive development, there are four stages of cognitive development: Sensorimotor stage, Pre-operational stage, Concrete operational stage and Formal operational stage. The students in primary school level are mainly in the Concrete operational stage (age ranged from 7 to 11 years old) that is the beginning of logical or operational thought. Consequently, a majority of research about metacognition in the basic educational context in Thailand deals with the secondary school students (ages ranged from 13 to 18 years old) because these students are mainly in the Formal operational stage (age 11 years and over). In this stage, Piaget stated that people develop the ability to think about abstract concepts and logically test hypotheses. Up to this, there is a need for more studies related to metacognition in primary school students (grades 1 to 6). One interesting research question is that what is the most suitable grade level for developing metacognition in students.

Another interesting issue is about the type of research participants. The major research participants are students more than pre-service and in-service teachers. Thus, there is a need for more studies with pre-service and in-service teachers. Particularly, there is a need of research aimed to prepare pre-service teachers or help in-service teachers in deriving in-depth understanding about metacognition and develop their ability to design and conduct the learning activities to enhance metacognition in their students in real classroom contexts.

The final issue is about the research methodology and tool in studying about metacognition. In Thailand, the metacognitive studies commonly employ the quantitative approach methodology and tools in conducting research and collecting data. The researchers usually employ the same or similar research quantitative tool and get the similar research findings. So, in Thai context, we need more metacognitive studies that employ different research methodology and tools. The researchers should consider to employ more qualitative research methodology and tools in their research. This will contribute more in-depth findings about metacognition in the Thai context.

6. Conclusions

The review of research about metacognition in the basic education context of Thailand during 2002 to 2016 shows some interesting issues as follows. First, there are five categories associated with metacognitive research in Thailand including a) Impact of metacognition on other research variables, b) Teaching strategies to enhance metacognition, c) Developing measurement tool for metacognition, d) Causal model of metacognition, and e) Factors of metacognition. Among these, two dominant categories are the 'Impact of metacognition on other research variables' and 'Teaching strategies to enhance metacognition' categories. Second, there is a lack of metacognitive research in a primary school level. Since 2002, there are only three studies on metacognition conducted in the primary school level. Third, the targeted participants of metacognitive research in Thailand focus on students rather than pre-service and in-service teachers.

7. Implications

This review study presents a need to develop more tools for measuring metacognition in different groups such as students, pre-service and in-service teachers. Regarding students, we need to develop tools for measuring metacognition in different levels ranged from primary, lower-secondary, upper-secondary to tertiary levels. Also, the research topic about causal model of metacognition and its associated factors is interested to study further. We hypothesized that the causal model of metacognition and its associated factors may be specific for different groups of participants. Particularly to research in students, there still be a need for more studies on metacognition for students in primary school level. In addition, there is a need for more research to enhance teachers' ability to design the lessons appropriated for promoting metacognition in their students. The teacher professional development on teaching to promote metacognition should be developed to suit specific educational contexts. More importantly, 'sooner is better', the pre-service teachers should be educated about metacognition and teaching to promote metacognition. Therefore, it demands the design of teacher education course as *'Metacognitive Teaching and Learning'* for pre-service teachers in university or teacher colleges.

8. Acknowledgment

This study was supported by the Royal Golden Jubilee (RGJ) Ph.D. Programme Scholarship under the Thailand Research Fund (TRF). The authors would like to express sincere gratitude for this support.

References

[1] Chalearmsuk N A 2007 Construction of Metacognition Test for Level 4 Students: A Case Study of Saraburi Province (Bangkok: Srinakharinwirot University)

[2] Chansana P, Suksringarm J and Parakham M 2008 Comparisons of effects of the 7–E learning cycle approach using metacognitive moves on alternative conception of some biology and basic science process skills of grade 5 students with different genders **Rajaphat Maha Sarakham University Journal 2**

[3] Duangamying C 2010 Development of A Causal Model of Metacognition of Lower Secondary School Students (Bangkok: Chulalongkorn University)

[4] Efklides A 2006 Metacognition and affect: What can metacognitive experiences tell us about the learning process? **Educational Research Review 1** 3

[5] Emily R Lai 2011 Metacognition: A Literature Review Available from: https://images.pearsonassessments.com/

[6] Flavell JH 1979 Metacognition and cognitive monitoring: A new area of cognitivedevelopmental inquiry **American Psychologist 34** p 906

[7] Ladawan C 2014 A Development of Knowledge and Understanding, Critical Thinking, Awareness, Environmental Conservation Behaviors of Mathayomsuksa 6 Students Using the Good Science Thinking Moves with Metacognition Technique (Mahasarakham: Mahasarakham University)

[8] Laprawat Y. A Causal Model of Self-efficacy and Metacognition Affecting Efficacy Expectancy and Outcome Expectancy of Lower-Secondary Students **Research Methodology & Cognitive Science 11** pp 75–85

[9] Latwong T 2016 Pruekpramool C, Phonphok N, Boonprakob M. A development of metacognition test in science subject for ninth-grade students **Journal of Education and Social Development 11** pp 293–305

[10] Lertchuwongsa J 2016 Using Problem-Based Learning to Promote Metacognition and Mathematics Achievement of Grade 10 Students (Chiang Mai: Chiang Mai University)

[11] Martinez ME 2006 What is metacognition? Phi Delta Kappan p 696

[12] Na Phatthalung N 2011 The results of 5E instructional model with metacognitive reflection strategies on science learning achievement, science concepts and metacognition of upper secondary school students (Songkhla: Thaksin University)

[13] Nonjuy D 2013 The Effect of Metacognitive Stratergy on Decision Making and Problem Solving Abilities in Force and Motion Topic For Mathayomsuksa III Students (Phitsanulok: Naresuan University)

[14] Oonseri P 2007 A development of Metacognition Test for the Third Level Students, Ubon Ratchathani Province (Ubon Ratchathani: Rajabhat Ubon Ratchathani University)

[15] Peungsaijai S 2011 The Learning Management by Using Meta cognition Strategy to Develop the Scientific Problem Solving Skills for Mathayom Suksa 1 Students (Chiang Rai: Chiang Rai Rajabhat University)

[16] Phochan T 2012 Factors Affecting Metacognition of Prathomsuksa 6 Students Under the Office of Loei Primary Education Service Arer 1 Using the Multilevel Analysis (Loei: Loei Rajabhat University)

[17] Phuaphan N 2014 The Effects of Instruction Based on Tri-Sik-Kha Principles for Developing

Grade 10 Students' Metacognition in the Topic of Ecosystems and Human and Sustainable Environment (Bangkok: Kasetsart University)

[18] Pitchayanurat S 2006 The effects of using open-ended Problems to Develop Metacognition For Mathayomsuksa III Students (Khon Kaen: Khon Kaen University)

[19] Prasertsook N 2011 Grade 8 Students' Scientific Problem solving Ability, Metacognition Development and Scholastic Achievement In Work and Energy Topic by Using Metacognitive Strategies (KhonKaen KhonKaen University)

[20] Ruangsri U 2001 A study of Relationship between metacognition and critical thinking of prathom six students (Bangkok: Srinakharinwirot University)

[21] Sangwong P 2013 The development of metacognition by using constructionist activities in the topic of evolution for grade 12 students (Bangkok: Kasetsart University)

[22] Schneider W 2002 The development of metacognitive knowledge in children and adolescents Edited by Timothy J. Perfect BLS (New York, NY, US: Cambridge University Press)

[23] Schraw G 1998 Promoting general metacognitive awareness **Instructional Science 26** p 123

[24] Senruang R 2006 Decision Making and Problem Solving Abilities in Science of Grade Level 3 Students Learning Through Metacognitive Strategy (Chiang Mai: Chiang Mai University)

[25] Teprattananan A 2014 The Effects of Cooperative Learning Activities Management Integrated with Metacognitive Thinking on Leaning Achievement and Metacognitive Thinking Ability of Mathayom Suksa IV Students of Suan Si Wittaya School in Chumphon Province **SKRU ACADEMIC JOURNAL 7** pp 31–43

[26] Thatsamee S 2009 The Development of Learning Activities in Physics Subject to Develop Problem Solving Ability With Metacognition For Mathayomsuksa IV Students (Khon Kaen: Khon Kaen University)

[27] Wongmak J 2010 Effect of A Metacognition Training Together with Attribution Training To Effort On Problem - Solving Skills Of Seventh Grade Students (Bangkok: Chulalongkorn University)

[28] Yatuuam D 2005 The Effects of Science Learning Management on Mathayom Suksa 4 Students Through Learning Cycle with Metacognitive Reflection and Learning Cycle with Metacognitive Reflection and Awareness (Bangkok: Srinakarinwirot University)

[29] Zohara A and Barzilaib S 2013 A review of research on metacognition in science education: current and future directions **Studies in Science Education 49** pp 69–121

IOP Conf. Series: Journal of Physics: Conf. Series **1340** (2019) 012015 doi:10.1088/1742-6596/1340/1/012015

Building a Professional Learning Community with Team Endeavors while Creating Elementary-focused STEM-integrated Lesson Plans

Krainara S[1*], Chatmaneerungcharoen S[2]

[1]Baankuku School, Phuket Educational Area District Office.
[2]Department of General Science, Faculty of Education Phuket Rajabhat University

*Corresponding author's e-mail address: drsiriwankief@pkru.ac.th

Abstract. This longitudinal study examined the development of pedagogical knowledge of teacher's participation in a 2-year professional development program that enhanced teacher's self-efficacy and prompted changes in science instruction in the elementary grades. The purpose of this research was to implement STEM activities which are developed by a professional team. A professional learning community (PLC) was used as a professional development tool helping the teacher increase her Pedagogical Content Knowledge. Collaborative design of STEM science lessons, implementation of the lessons and reflection on outcomes. Collaboration among the educator and teachers was offered in the form of online learning materials, exemplary lessons and the availability of an expert. The researcher participated for one year in the program. Results showed significant increases in teacher's overall self-efficacy in teaching science, personal efficacy, and outcome expectancy efficacy during the 2 years. The increase of self-efficacy was presented with changes in improved instructional practices, particularly student participation activities. The educators and experts were important in providing pedagogical support during the design and implementation, and reflected with teachers on how to improve on their teaching.

1. Introduction

Building individuals with good knowledge and understanding of Science and Technology have become the main goals of national education programs. Thai science education has advocated infusing 21st century skills into the school curriculum and several educational levels have participated in such efforts. Thai educational institutions that focus on STEM, such as the Office of Basic Education Commission (OBEC), The Institute for the Promotion of Teaching Science and Technology (IPST), Phuket Rajabhat University STEM Education Center (PKRU STEM Education Center) and Chevron Enjoys Science projects.

STEM is a curriculum based on the idea of educating students in four specific disciplines — science, technology, engineering and mathematics. Integrated approaches to STEM (science, technology, engineering, and mathematics) education are increasingly popular, but remain challenging and elusive. There is much hope that integrated approaches to STEM education can help the next generation of students to solve real-world problems by applying concepts that cut across disciplines as well as capacities of critical thinking, collaboration, and creativity [1]. However, most teachers have received training in only one discipline [3], and most schools and classes at all levels still have separate departments and class periods for the STEM subjects. Therein lies a significant challenge for educators and administrators interested in promoting integrated STEM.

Lesson design, Planning is an essential aspect of effective teaching. Planning begins with a design of units of study, and continues in the development of a sequence of lessons crafted to achieve the goals of the unit. Units are large chunks of the curriculum. Each unit contains multiple lessons and the lessons provide a pathway to mastery of the big idea. Each lesson should be an important step toward the goals of the unit.

Characteristics of a Good Lesson [2]

To plan effective lessons, you begin with the end – what are the outcomes you want from the lesson? What knowledge, skills, and attitudes will students possess if the lesson is successful? All lessons have the same basic structure: clear objectives, activities that provide students opportunities to meet the objectives, an introduction that activates prior knowledge and introduces questions that the teacher uses to check for understanding, and independent work in which students apply the concepts under varying degrees of difficulty.

So the essential elements of the lesson design are:

1. The objectives are set first and drive the design.
2. There is an opening that activates students' prior knowledge.
3. The lesson provides clear direction for students.
4. The teacher models or demonstrates the work to be done.
5. There is an opportunity for guide practice.
6. There are opportunities for teachers to check for understanding.
7. There is an opportunity for independent work.
8. There is a closing and summation of what was learned.

Goals identified in the framework include building STEM literacy and 21st century competencies; developing a STEM-capable workforce; and boosting interest and engagement in STEM. In terms of outcomes, the framework considers learning and achievement; and the ability to transfer understanding across STEM disciplines. Regarding the nature and scope of integration, the framework addresses which subjects are connected; which disciplines are dominant; and the duration, size, and complexity of an initiative. With respect to implementation, the framework focuses on instructional designs involving problem-based learning and engineering design; professional development and development of professional learning communities; and adjustments to the learning environment, such as extended class periods, extended lesson planning, team teaching, and partnering between STEM educators experts and teachers.

Watsuwankeereewong School is located in Patong, a major tourist attraction in the world. The students have migrated or have moved out from parents who come to work in Patong, and many of the student's behaviors have been observed. Students in school have lower academic achievement. Most lack the ability to think and solve problems without reason and are tired of learning, are lazy, and unable to connect their knowledge from the classroom to everyday life. As a result, students do not see the importance of science.

For the above reasons. If the teacher cooperates to adjust the teaching and learning and plans together the design of learning management, along with observation of teaching together and activities that focus on the students, they will be able to practice and think for themselves. Sharing the STEM activity design of teachers reflects and improves teaching the community of professional learning teachers. It develops students to their full potential, helps people in the 21st century and brings research results to effective teaching and learning through professional community processes. It will encourage students to recognize the importance and relevance of STEM (science, technology, engineering and mathematics)

Other paragraphs are indented (BodytextIndented style).

Purpose of study

To develop teachers in Watsuwankeereewong school to be able to design teaching and learning together.

To build a professional learning community in Watsuwankeereewong school Patong Phuket.

Research Questions

What processes of collaborative STEM lesson design enable teacher design teams to develop a STEM integrated lessons?

2. Methodology

2.1. Research Methodology

This Study is based on the classroom action research. The data analysis focused on processes of collaborative STEM lesson design enable teacher design teams to develop a STEM integrated lessons. This study was conducted for 1 year following the research framework which is presented in Picture 1.

2.2. Research Participations

The research participants were 4 science teachers and 2 science student teachers in major of General Science, Faculty of Education at Phuket Rajabhat University, themselves teach in Watsuwankeereewong School and educator from The Department of General Science Faculty of Education at Phuket Rajabhat University. The four science teachers and two science student teachers cooperative indicated their willingness to participate in the study choosing to focus their curriculum design within Force Lesson.

2.3. Research Instruments

In order to study the processes of collaborative STEM lesson design and enable teacher design teams with the Professional Learning Community (PLC). The researcher used data from, a design lesson observation meeting, classroom observation, individual interviews, a questionnaire, an inquiry-based lesson plan, reflection, and group discussions. In this research, classroom observations, interviews, questionnaires, case studies and documents were used as research methods to provide opportunities for participants and the researcher to generate an understanding for a particular situation. Research participants presented their lesson plan design through writing lesson plans, teaching, reflecting, meeting and answering or discussing topics in the interview process. Additionally an understanding of their PLC and STEM was transcribed and explained by the researcher who has used the inductive approach to generate the article.

2.4. Teacher design team descriptions

Team lesson design

The Team Lesson Design consisted of four teachers and two student teachers in major of General Science, the Faculty of Education at Phuket Rajabhat University, six of whom worked in the same schools. Two teachers participated in the Chevron Enjoy Science Project the previous year.

Team to the reflect

The Team to the Reflect consisted of an administrator and an educator from the Department of General Science Faculty of Education at Phuket Rajabhat University.

Table 1. Professional development overview and data collection timeline

Weeks	Activities	Data collected
1	-Design lesson. Collaborate with colleagues on designing and activity. How will it be assessed?	- Individual participant reflections - Field notes
2	-Teach Teach from the lesson plan design. -Assess How did your students do? -Share Share your assessments with your colleagues. -Reflect How did our students do?	- Individual participant reflections - Field notes
3	-Evaluate Are we making an impact? What changes do we need to make? -Inquiry What learning needs do our students have? How do we improve the lesson plan?	- Individual participant reflections - Field notes
4	Cycle repeat from wk2-3	

3. Data Collection

Classroom Observation was used in the ways of formal and informal observation of teaching while it was taking place in a classroom or other learning environment. It was conducted by fellow teachers, administrators, experts or instructional specialists.

Practice Reflective Journal by classroom teacher/ researcher.

Reflective Feedback Journal by PLC team.

4. Data Analysis

The researcher analyzed documents and underlying knowledge of participants in speaking and writing. Data from multiple sources such as teachers' journals and interviews, field notes from observations and card sorting were analyzed by the process of open coding to get the transcripts from the first interview, observation, reflection and card sorting, developing initial categories of the participant's technological pedagogical content knowledge and their practice. In developing categories, the researcher used a constant comparative method of analyzing multiple sources of data which served to triangulate the data in order to increase trustworthiness of the research findings and assertions made.

5. Result and discussion

The results of this research are as follows.

1. Form of community building, professional learning in schools. The use of STEM instruction on the strengths of Mathayomsuksa 3 of Watsuwankeereewong School consists of 4 main components and 10 sub-components as follows.

The main element is as follows

1 Organization for Learning consists of two sub-components:

1) Prepare the environment within the school with a focus on cleanliness, cleanliness, and health rate of personnel

2) Creating a corporate culture that is conducive to learning: being a good person, being open minded, listening to friends, enhancing the mental state of student teachers by organizing psychological activities.

The second element, the value creation and the shared vision, consists of two sub-components:

1) Co-ordinate the student's learning goals for understanding.

2) Define learning together based on the PLC concept: taking experience, exchanging experiences, visualizing success, exchanging expectations.

The Third element 3: Learning from the common practice consists of 3 sub-components: 1) Learning together. Collaborative Learning: Designing / Planning / Producing Exchanges. Reflect the teaching, and improve teaching

2) AAR (After action review): Successful exchange not less than one time a month.

The main component 3 evaluates the expected results.

1) Strong student achievement.

2) Student behaviour in STEM

3) Teacher Satisfaction The pattern of creating a professional learning community in schools.

The results of the evaluation of the community-based learning model in vocational schools in Wat Suwankeereewong School by contextual experts on the benefits. The possibility and the accuracy are at the highest level. The students in grade 7 - 9 had an average score of 73.14 percent. The teacher of Watsuwankeereewong School STEM teaching behaviours were at a higher level. Teachers were satisfied with the model of community building, professional learning in schools at a high level.

6. Conclusion

For one year with a collaborative STEM lesson design, and STEM teaching development by building a professional learning community, the results showed increases in teacher's overall self-efficacy in teaching science, personal efficacy, and outcome expectancy efficacy. The increase of self-efficacy was presented with changes in improved instructional practices, particularly student participation activities. This enabled the ability to build a professional learning community of teachers.

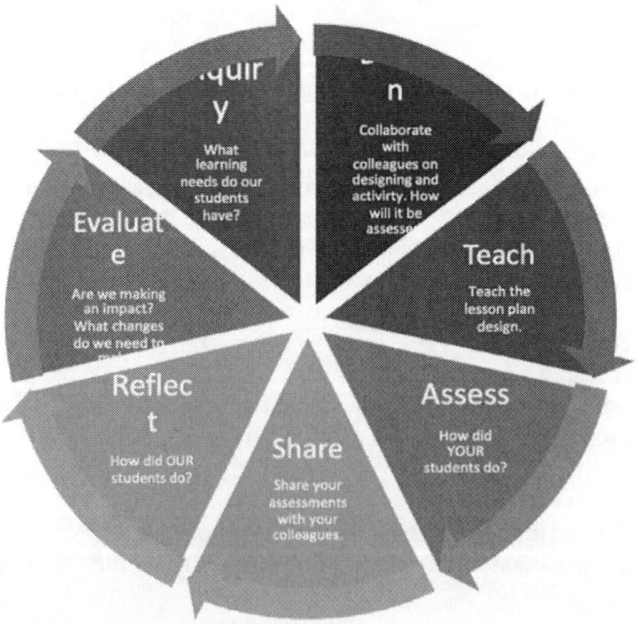

Figure 1. Research Frame work

IOP Conf. Series: Journal of Physics: Conf. Series **1340** (2019) 012015 doi:10.1088/1742-6596/1340/1/012015

Lesson Design consists of these elements:

- Understanding the prior achievement of the class.
- Clear learning objectives that specify the concepts or skills to be mastered;
- Modeling or demonstrating the skills or concepts;
- Guided practice to allow students to apply the skills or concepts; and
- Checks for understanding that permit the teacher to monitor student progress and when necessary adapt instruction to meet the needs of individual students.

In the Lesson design, teachers were asked to discuss:

- Group of Team Learning
  - ➢ Member's role; Group leader, supervisor, communicator, collectors, time controller.
- The Cognitive Demand of Classroom Tasks
  - ➢ Construction of knowledge, through disciplined inquiry, to produce discourse, products, and performances.
- Formative Assessment
  - ➢ Observations of student work, student responses to teachers' questions, or on the use of techniques that allow students to give continual feedback about whether they understand the material.

7. Research Implication

Results showed significant increases in teacher's overall self-efficacy in teaching science, personal efficacy, and outcome expectancy efficacy during the 2 years. The increase of self-efficacy was presented with changes in improved instructional practices, particularly student participation activities. The educators and experts were important in providing pedagogical support during design and implementation, and reflected with teachers on how to improve on their teaching.

University researchers/ experts are currently facilitating academic-year professional development for science coaches within the division to pilot STEM activities in classrooms. This continuing investment will build upon the school Professional development model for STEM integration and respond to the expressed needs of the participants. The sequential actions of experiencing, questioning, analyzing, adapting, envisioning, and reflecting on STEM lesson plans as a vehicle for STEM integration in science classrooms

References

[1] Burrows A and Slater T 2013 A proposed integrated STEM framework for contemporary teacher preparation **Teacher Education and Practice 28** pp 318–330
[2] Dougles L (2014) Teach like a Champion 2.0. (San Francisco, CA: Jossey-Bass)
[3] Honey M Pearson G and Schweingruber A 2014 STEM integration in K-12 education: Status, prospects, and an agenda for research (Washington: National Academies Press)
[4] John H 2013 Invisible learning for teachers
[5] McFadden J R and Roehrig G H 2017 Exploring teacher design team endeavors while creating an elementary-focused STEM-integrated curriculum (St. Paul, MN 55108, USA)

IOP Conf. Series: Journal of Physics: Conf. Series **1340** (2019) 012016 doi:10.1088/1742-6596/1340/1/012016

Bringing STEM Lessons into Thai Mathematics Classroom.

Nualkaew N.[1*], Chatmaneerungcharoen, S. [2]

[1]Mathematic Teacher at Kapongpittayakom School of the Office of the Region 14 Secondary Education Area

[2] Head of Department of General Science, Faculty of Education Phuket Rajabhat University

*Corresponding author's e-mail address: krumath124@gmail.com

Abstract. This study focuses on teaching and the learning development in mathematics through the collaborative design of technology-enhanced science lessons with STEM Education. The purposes of this research were as follows. The first purpose was to develop learning activities for the surface area and volume by organizing learning activities with STEM Education for Mathayomsuksa III students. The second was to compare achievements before and after for the surface area and volume by organizing learning activities with STEM Education for Mathayomsuksa III students. The third was to compare the ability to solve problems before and after on the surface area and volume by organizing learning activities with STEM Education for Mathayomsuksa III students. Finally, the fourth purpose was to study students' satisfaction in learning about the surface area and volume by organizing learning activities with STEM Education for Mathayomsuksa 3 students. The research samples consisted of 30 Mathayomsuksa III students at Kapong Pittayakom School, under the Jurisdiction of the Office of Region 14 Secondary Education Area. These students studied in the first semester of the academic year of 2017. 25 classrooms were acquired by a specific selection. The research instruments were: 1) a learning plan; 2) an achievement test; 3) a problem solving ability test; and 4) data that was analysed by percentage, mean, standard deviation and standard deviation. The results were as follows: 1) A plan for learning about surface area and volume for students of Mathayomsuksa III by organizing learning activities with STEM Education where efficiency was 86.31 / 89.47. 2) The learning of surface area and volume for students of Mathayomsuksa III by organizing learning activities with STEM Education had significant differences at a level of .05 after the study. 3) The ability to solve surface and volume problems for students of Mathayomsuksa III by organizing learning activities with STEM Education yielded significant differences at a level of .05 after the study. 4) The satisfaction of learning about surface area and volume for students of Mathayomsuksa III by organizing learning activities with STEM Education was observed. The learning activity was conducted based on the level of education.

1. Background

In today's world, society is changing rapidly with the advancement of technology and communication. It creates a phenomenal amount of information in a variety of sources, including the need to compete for economic advantage and to live and compete in the labor market with international civilization. So teaching 21st-century mathematics requires learners to learn with understanding, have a good foundation in mathematics and be used to solve problems with science and technology. As for observation behavior of school students, the observation found that most students lacked the ability to solve problems. Unreasonable technology is not beneficial for quality. Mathematics is of paramount importance to the development of human thinking and allows human creativity to think in a systematic manner, to analyze problems or situations carefully and carefully, to help them make predictions, plan, solve problems and apply them in their daily life properly. Therefore, teaching mathematics focuses on the students' thinking process. Practice in systematic study with various activities. Student learning takes place while students are directly involved in those activities. The ability to investigate and have the ability to solve problems help develop advanced thinking processes. As expected that the learning process enabled students to develop mathematical achievement.

From the context of Kapong Pittayakom School. The results of the National Basic Education Testing (O-NET) of Mathayomsuksa III students at Kappong Pittayakom School Learning Mathematics in the 2016 academic year are the average score of the school is lower than the provincial, sub-national and national levels, and from observation behavior of school students. The majority of the students lacked the ability to think and solve problems without reason and they were tired of learning, were lazy and could not connect the knowledge gained from the classroom to daily use. As a result, students do not see the importance of mathematics. In addition to negative addition, students look at other subjects as far away and as a result, students in Mathematics should expect to improve.

For the reasons mentioned above. The reporter is interested in introducing the concept of learning management in accordance with the study guide to use in teaching and learning of mathematics of surface area and volume, and it is expectation is that this will help improve students' ability to solve problems as well as to develop students to full potential. It allows for a skillful person in the 21st century and brings the results of the study to the development of effective teaching and learning. Students are aware of the importance and relevance of STEM (Science, Technology, Engineering and mathematics)

Purpose of study

1. To develop a plan for learning about surface area and volume for students of Mathayomsuksa III by organizing learning activities according to the study plan.
2. To compare learning achievement on surface area and volume for students of Mathayomsuksa III by organizing learning activities according to the study plan, before and after the study.
3. To compare the ability to solve surface area and volume problems for students of Mathayomsuksa III by organizing learning activities according to the study plan, before and after the study.
4. To study students' satisfaction toward learning activities on surface area and volume for students of Mathayomsuksa 3 by organizing learning activities according to the study plan.

Conceptual Framework

This study was conducted by studying STEM Education to develop learning achievement and problem solving ability of Mathayomsuksa III students on surface area and volume. By organizing learning activities according to the study, the framework summary is as follows.

IOP Conf. Series: Journal of Physics: Conf. Series **1340** (2019) 012016 doi:10.1088/1742-6596/1340/1/012016

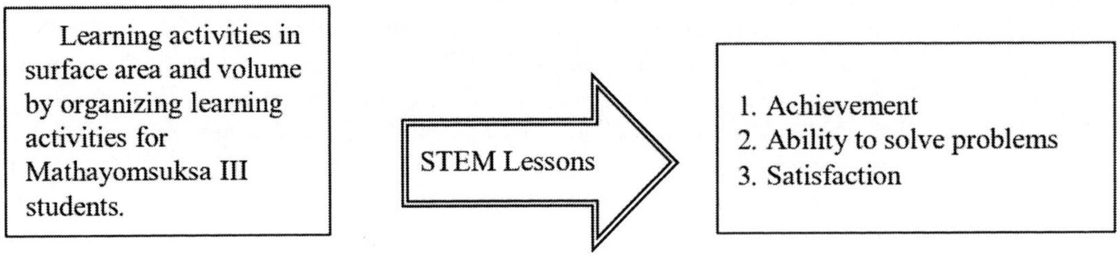

Figture 1: Research Framework

2. Method

To investigate the students' competency after learning with STEM activity in surface area and volume project for 14 hours. The methodology was employed as follows

Participants and Setting

The research participants were 25 secondary students. This purposive sampling method is to learn through STEM activity in surface area and volume and the teachers were tasked to be observers and then later reflect and discuss the method.

Research Instruments

The instruments used in this study included

1. A learning Plan for surface area and volume by organizing learning activities with STEM Education for Mathayomsuksa III students.
2. A learning Achievement Test on surface area and volume by organizing learning activities with STEM Education for Mathayomsuksa III students.
3. A problem solving ability Test on surface area and volume by organizing learning activities with STEM Education for Mathayomsuksa III students.
4. A student Satisfaction Assessment Form for Learning Activities about surface area and volume by organizing learning activities with STEM Education for Mathayomsuksa III students.

Data Collection

The reporter collected data in the first semester of the 2017 academic year. Data collection:

1. Experimental model this study Reporters used the One Group Pretest-Posttest Design

E	O_1	X	O_2

When O1 is pre-test
X is learning activities with STEM Education for Mathayomsuksa III students.
O2 is post – test

2. Data Collection Procedure

2.1 Understand how students learn. And the role of the student.

2.2 Pre - test the achievement test was used to measure learning activities with STEM Education for Mathayomsuksa III students, the study takes 2 hours.

2.3 Learning activities with STEM Education for Mathayomsuksa III students.

IOP Conf. Series: Journal of Physics: Conf. Series **1340** (2019) 012016 doi:10.1088/1742-6596/1340/1/012016

2.4 Post - test the achievement test was used to measure and the ability to solve problems on surface area and volume by organizing learning activities with STEM Education for Mathayomsuksa III students, the study takes 2 hours.

2.5The students make a satisfaction assessment on surface area and volume by organizing learning activities with STEM Education for Mathayomsuksa III students.

Data Analysis

The researcher analyzed documents and underlying knowledge of participants in speaking and writing. Data from multiple sources such as teachers ' journals and interviews;

1. Find the effectiveness of the learning plan on surface area and volume by organizing learning activities with STEM Education for Mathayomsuksa III students, by evaluating the efficiency of the process (E1) and the efficiency of the result (E2)

2. Compare learning achievement before and after learning about surface area and volume for students of Mathayomsuksa III by organizing learning activities according to the study plan. The independent sample t-test was used at the significance level of .05.

3. Compare the ability to solve problems before and after learning about surface area and volume for students of Mathayomsuksa III by organizing learning activities according to the study plan. The independent sample t-test was used at the significance level of .05.

4. Analyze students' satisfaction level on learning about surface area and volume for students of Mathayomsuksa III by organizing learning activities according to the study plan. With average and standard deviation, the criteria for interpretation are as follows:

4.51 - 5.00 means the highest level of satisfaction.
3.51 - 4.50 means high level of satisfaction.
2.51 - 3.50 means moderate satisfaction.
1.51 - 2.50 means less satisfaction.
1.00 - 1.50 means that there is minimal satisfaction.

3. Result and Discussion

The results were as follows;

1. Plan for learning about surface area and volume for students of Mathayomsuksa III by organizing learning activities with STEM Education were efficiency is 86.31 / 89.47.

2. Learning about surface area and volume for students of Mathayomsuksa III by organizing learning activities with STEM Education. Significant differences at a level of .05 were recorded after the study

3. Ability to solve surface and volume problems for students of Mathayomsuksa III by organizing learning activities with STEM Education. After the study, significant differences at a level of .05 were observed.

4. Satisfaction of students on learning about surface area and volume for students of Mathayomsuksa III by organizing learning activities with STEM Education, the learning activities were conducted according to the level of education.

Figture 2: Student's learning activities

4. Conclusion

Based on the results, it is concluded that by using learning activities with STEM Education about surface area and volume for students of Mathayomsuksa had results. The application of knowledge in real – life situations has a positive impact. The students are learning in a fun way and are happy. There is no doubt that by solving problems and finding answers that this method is effective for both the learners and the educators.

5. Research Implication

5.1 Suggestions to use the results of the study

Before incorporating the learning activities, Teachers should have an understanding of the process of organizing, learning by the way, prepare materials / materials for students and include problem-solving planning in the event that the learning activity does not follow the plan. In order to achieve maximum efficiency in learning, the teacher should clarify understanding with students before class to help students understand their roles by giving students the freedom to think fully. The teacher is a facilitator and a guide when students encounter problems, and teachers should be flexible in their time as appropriate.

5.2 Suggestions for future research

Study the guideline of learning management in accordance with the study plan and other related subjects and can do, and to study the effect on students. In addition to learning achievements and problem solving ability the results should be explored with other variables such as creativity, critical thinking and local relevance should be incorporated in the learning activities and focus on solving real problems in the student community.

References

[1] Barroody 1993 Problem Solving, Reasoning and Communicating, K-8: Helping Children Think Mathematically (New York: Merrill)

[2] Dossey and John 2002 Mathematics Methods and Modeling for Today's Mathematics Classroom: A Contemporary Approach to Teaching Grade 7-12 (Canada: Books/Cole Thomson Learning)

[3] Education: A Primer 1973 Washington DC: Congressional Research Service

[4] Good C V 1973 Dictionary of Education (New York: McGraw-Hill Book Company)

[5] Gonzalez H B and Kuenzi J J 2011 Science, Technology, Engineering, and Mathematics (STEM)

[6] Hynes, Morgan and others 2011 Infusing Engineering Design into High School STEM Courses. Utah: National Center for Engineering and Technology Education

[7] Krulik Stepphen and Rudnick Jesse 1996 The New Sourcebook for Teaching Reasoning and Problem Solving in Junior and Senior High School (Boston: Allyn and Bacon)

[8] Morse N C 1995 Satisfaction in the White Collar Job (Michigan: University of Michigan Press)

[9] National Council of Teachers of Mathematics (NCTM) 2000 Principles and Standards for School Mathematics (Virginia: NCTM Reston)

[10] Polya G 1969 How to Solve it (New Jersey: Princeton University Press)

[11] Polya G 1980 "On Solving Mathematical Problem in High School" Problem Solving in School Mathematics (Virginia: The National Council of Teachers of Mathematics)

[12] The Partnership for 21st Century Skills 2015 P21 Framework Definition. [Online]. Access form :http://www.p21.org/storage/documents/docs/P21_Framework_ Definitions_New_Logo_2015.pdf

Improving Thai Science Teachers' TPACK through an Innovative Continuing Professional Development Program

Chatmaneerungcharoen, S.

Department of General Science, Faculty of Education Phuket Rajabhat University

*Corresponding author's e-mail address: drsiriwankief@pkru.ac.th

Abstract. This study focuses on teaching and learning development in science through collaboration between science cooperative teachers and researchers. At the main goal was the ambition to integrate coaching strategy into internship course—with teaching practice, aligned with professional development on teacher's technological pedagogical content knowledge (TPACK). The phase where the collaboration moves from initial establishment towards a stable practice is investigated. The study aims to identifying the character of professional program with specific questions are "How does professional development program was identified as actions and arrangements impacting the development of teacher's TPACK and their emerging practice?" and "How does the teachers' TPACK change?" Therefore, the study focuses on developing 40 Cooperative Science teachers who were science cooperative teachers that participated in this study for 1 year. Data sources throughout the research project consisted of teacher refection, classroom observations, Semi-structure interviews, Situation interview, questionnaires and document analysis. Interpretivist framework was used to analyze the data.Research findings indicate that STEM Education Professional Development Model: STEMed- PD which was developed on the frameworks of coaching and mentoring, lesson study, after action reflection and professional development community can enhance the cooperative science teachers' understanding and practice on STEM lessons. These teachers conducted their classrooms with integrated science concept to other subject concepts. The lesson's learning objectives are related to the 21st century skills that encouraged students to apply their knowledge into their daily life. Different from the beginning, the teachers understand only the meaning of STEM Education but they do not know how to integrate STEM into their science classrooms and they thought STEM lesson should be taught through Science Camp, not in normal classroom. For 1 year of STEMed-PD progression, both Science cooperative teachers and their student teachers have shown their development of STEM understanding and practice in aspects of STEM definition, STEM background, learning goals based on STEM and the 21st century skills, STEM lesson plan development, classroom practice with STEM lesson, connecting STEM lesson into student's daily life, and evaluation and assessment of STEM classroom. The development of teachers' understanding and practice is presented in their lesson plans that are produced during STEMed-PD. Moreover, the study also indicates the importance of co-working between cooperative science teachers and university supervisors in lesson preparation through coaching and mentoring. Implication for further research, science educators should focus on teacher self-efficiency and teacher's background in technology knowledge which are factors to support or obstruct science teachers to develop their STEM practices.

1. Background

Building individuals with adequate knowledge and understanding of science and technology has become one of the main goals of national education. Current national documents are called for changes in K-12 science, technology, engineering, and mathematics (STEM) education to increase STEM literacy for preparing the 21st Thai citizenship. Will this happen without teachers? The answer is NO. Therefore, science teachers should have appropriated knowledge. Technological Pedagogical Content Knowledge or TPACK framework [7] has emerged as a professional development framework to describe required science teachers' knowledge for the 21st century science classroom. Emerging 21st century era, science teachers should conceptualize TPACK that is referred to a specific category of knowledge for teaching science. TPACK is a conceptualized blend of technological knowledge, content knowledge and pedagogical knowledge to design Information and Communication Technological lessons in science concept. Teachers' knowledge influences what they know, what they think, and how they act in the classroom. TPACK was extended from [13] who defines the knowledge of teaching a subject matter as pedagogical content knowledge (PCK): the distinctive bodies of knowledge for teaching. It represents the blending of content and pedagogy into an understanding of how particular topics, problems or issues are organized, represented, adapted to the diverse interests and abilities of learners, and presented for instruction. In emerging of the 21st century, people are taught to be familiar with technology because of the world communication and information access. Given this change, education must shift to incorporate computer based, electronic technologies integrating learning with these technologies within the subject area. Now the knowledge benchmark that teachers need in order to teach their subject matter with technology is more than just PCK, it needs the development of "Technology Pedagogical Content Knowledge (TPACK)" [6, 9]. From the National Research Council "The relationship between Science and Technology is so close that any presentation of science without developing an understanding of technology would portray and inaccurate picture of science" [8]. Similarly, the International Society for Technology in Education developed new technology standards for students and teachers that specifically confront teachers with integrating technology throughout education. These standards direct that electronic technologies become "an integral component or tool for learning of academic subject area" [5]. Therefore, the process of teacher education should focus to prepare the student teacher to have the basis of good teaching with technology and requires an understanding of the representation of concepts using technologies.

According to educational reform, teachers still have difficulties in implementing constructivist-based teaching and learning approaches. These approaches are seen to be radically new for the majority of science teachers and they are suspicious of their effectiveness [10]. Particularly, there are numbers of studies that document the problems in-service elementary science teaching. Thai elementary school teachers often do not have enough pedagogical content knowledge necessary to create a constructivist classroom. Most importantly, they lack an understanding about how to represent science content in ways that are personally meaningful and potentially accessible to students [11, 18] of pre-service teacher education [14], as it provides student teachers with essential bridge between theory and practice and the opportunity to define and refine their teaching skills. Similarly, the current literature supports the important of teacher practice, identifies student teaching as the most helpful part of professional education and comprises the first steps of a personal journey of becoming a teacher [15, 16, 17]. In school-based pre-service teacher education, mentoring refers to the supervision of a student teacher by an experienced teacher during the teaching practice. Student Teachers are assigned to teacher (Cooperative teachers) to be supervised throughout the practicum. Additionally, students are supervised by a university supervisor who occasionally visits the school. The university supervisor and the cooperating teacher serve as mentors to the student teachers, as both of them interact with the protégé and help him or her again the necessary professional knowledge and skills [4]. Cooperating science teachers are selected with care and with the knowledge that their experience will provide a nurturing environment for the science teacher education student. It is primary importance that the

IOP Conf. Series: Journal of Physics: Conf. Series **1340** (2019) 012017 doi:10.1088/1742-6596/1340/1/012017

clinical experience be a positive experience for student teacher. Therefore, the cooperating science teachers should carry with appropriate TPACK and supervisory skills.

Thus, effective professional development opportunities are important for helping science teachers to improve their TPACK.

Purpose of study

1. To develop science teachers' TPACK for integrating ICT into classroom.

2. To create professional development program to develop TPACK which move teachers' practices to more integrated way.

Research Questions

1. How do teachers develop their Technological Pedagogical Content Knowledge (TPACK)?

2. What is the relationship of TPACK development with the in-class use of new technologies such as the teachers' ICT integration and the emphasis on STEM education?

2. Method

2.1. Research Methodology

This study employs a phenomenological, interpretive multiple-case study design to develop science teachers' TPACK which promotes teachers to integrate STEM in their classes. The data analysis focused on assessing how science teachers develop their technological pedagogical content knowledge through Co-TPACK PD and STEMed PD. A multiple- case study design provides rich descriptions and interpretations of teachers' experiences relating to ICT and STEM integration. This study was conducted for 3 years.

2.2. Participants and Setting

Participants and The research participants were 40 science cooperative teachers and40 student teachers in major of General Science, Faculty of Education at Phuket Rajabhat University. This purposive sampling method is based on the criteria that the researcher would like to learn and find the answer for research questions. Forty science cooperative teachers indicated their willingness to participate in the study. In the end three criteria were used for choosing the participants and they were as follows:

2.1.1 Teachers who were teaching Science subject in the elementary and high school levels in school located in Phuket Province. They were cooperative teachers who worked with student teachers in 2015-2017 academic year.

2.1.2 The teachers were teaching Science in both semesters and could participate in long term professional development program. They showed a willingness to contribute to the profession by being open to classroom observations by the researcher, participate in follow-up interviews and be able to attend meeting of the TPACK PD for 2 years and STEMed PD for 1 year.

Research Instruments

In order to study the development of Thai science cooperating teachers' understanding of TPACK and practice through CO-TPACK PD and STEMed PD, multiple data sources have been used during the research process. Utilizing grounded theory in this research, the researcher used data from multi-sources to maximize flexibility and to help generate theory. Using data from a variety of resources or using more than one method has provided a fuller understanding of this study for the researcher. In interpretive case study, classroom observation, individual interview, questionnaire, inquiry-based lesson plan, written reflection, and group discussion are preferred to assess all teachers' TPACK with their thinking, actions, and reasons in the specific context and setting. In this research, classroom observation, interview, questionnaire, case study, card sort and documents were used as research methods to provide opportunities for participants and the researcher to generate an understanding for a particular situation. Research participants present their TPACK through writing lesson plans, teaching journals, meeting, card sorting, and answering or discussing topics in the interview process. An understanding of their TPACK was transcribed and explained by the researcher who has used the inductive approach to generate theory

Data Analysis

The researcher analyzed documents and underlying knowledge of participants in speaking and writing. Data from multiple sources such as teachers' journals and interviews; field notes and videotapes from observations and card sorting were analyzed by the process of open coding to get the transcripts from the first interview, observation, reflection and card sorting, developing initial categories of the participant's technological pedagogical content knowledge and their practice. In developing categories, the researcher used a constant comparative method of analyzing multiple sources of data served to triangulate the data in order to increase trustworthiness of the research findings and assertions made.

3. Result and Discussion

3.1. Science Teacher Program Innovation (STPI)

This study reports the finding of research on TPACK enhancement program effectiveness. TPACK PD and STEMed PD were parts of a large-scale Thai science cooperative teacher's professional development intended to improve in-service science teachers who work as cooperative teachers at elementary and junior high schools. The program consisted of a preparation stage, implementation stage, feedback exchange; and revising stage (Figure 1)

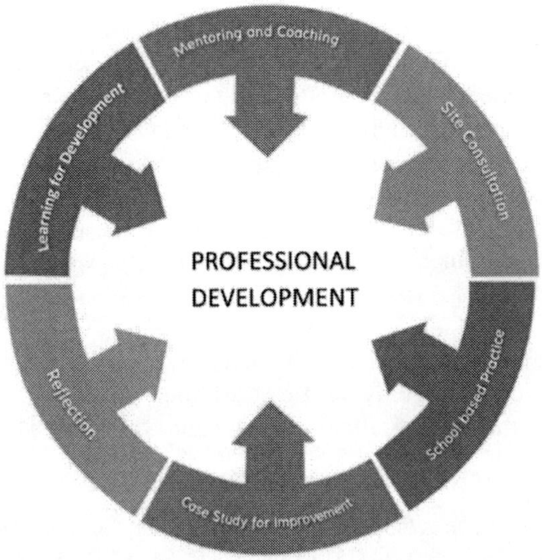

Figure 1. Professional Development Framework

The STPI program consisted of inquiry activities, teaching strategies (Explicit-Reflective and Content based approach) (Figure 2)

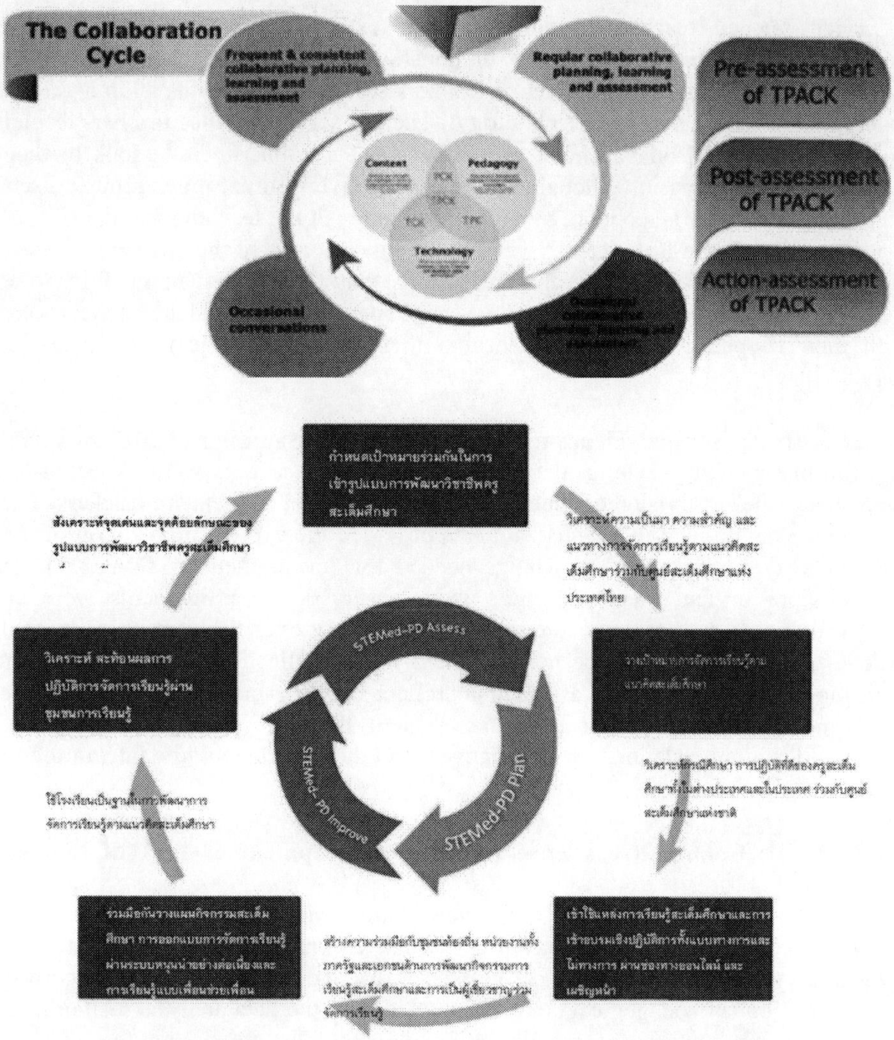

Figure 2. Continually Professional Development (Co-TPACK and STEMed PDs)

Co-TPACK 1 I: Integrating Technology in the Pre-service Science Teacher Preparation Program (Semester of 2016 Academic Year)

40 pre-service science teachers were attending Science Education course integrating ICT for one semester induration and emphasized learning through hands-on design of ICT lesson plans. The course was aimed at preparing pre-service science teachers to design student-cantered ICT-integrating science lessons that support the 21st century in Thai students. The pre-service science teachers' course was conducted 4 months (3 hours/lesson). The first two months were interactive lectures and workshops on Thai National Science Curriculum, science teaching methods, measurement and evaluation in science classroom. The last two months were about technological training, ICT lesson design in concepts of science, and microteaching lesson study. The specific technology section used science problem-based activities to guild the pre-service science teachers in learning about educational technologies, pedagogical considerations with these technologies and teaching/learning with these technologies. The first stage of the program, the student teachers explored a variety of science concepts integrating technologies that could be considered in the curriculum. Microteaching lesson study section focused

the pre-service science teachers gaining teaching experience, reflective skill on lesson plan development fours instructional methods: demonstrations, hands on/ laboratories, inductive and deductive activities. The pre-service science teachers were assigned to develop a science lesson for each model that has to integrate with technology. The pre-service science teachers teach (videotaping the instruction and upload on Facebook group by YouTube link) their lessons to their peers, and reflect (assessing and revision though written reflective journal and group discussion). The microteaching lesson study is recorded by videotapes to recall the teaching and debriefs of the lessons by their peer, instructor and ICT experts. The remaining 6 months of the program focused specifically on providing extended ICT workshops and professional classroom observations. Prior to be the student teachers starting their fulltime practicum at the school sites, the three student teachers were assigned to work with their cooperative teachers to design their sequence of lessons for an integration of technology.

Co-TPACK II: Pre-service science teachers' Practicum (Semester of 2017 Academic Year)

During fulltime student teaching, the pre-service science teachers were expected to adjust their lesson plans under the supervision of supervisor (researcher) and cooperative teachers. The technology integration theme highlighted in the shaded areas provided the explicit preparation of the pre-service science teachers' development of knowledge needed for development of TPACK. During teaching practicum, the pre-service science teachers were observed, all assignments were collected and analyzed, and the case studies were interviewed extensively over the various parts of the program. After each lesson, the pre-service teachers prepared written reflections that considered revised plans for succeeding lessons. They were assigned to reflect on their understanding of science concepts, instructional methods, learning and teaching assessment, the success of the integration of technology in the lesson and recommendations for the changes, and their teaching while integrating technology in teaching science.

Co-TPACK III: Cooperative science teachers' development of TPACK (Semester of 2017 Academic Year)

Data from teachers' journals and interviews, field notes and videotapes from classroom observations, videotaped transcriptions from group discussions, semi- structured interviews, and card sorting when using the CO-TPACK PD will be transcribed and developed to core categories of developing technological pedagogical content knowledge In the data analysis methods, the researcher attempted to find out patterns of growth or development by comparing the pre-service science teachers' understandings and practices of TPACK through CO-TPACK PD, The participants gained technological content knowledge in the use of both Weblogs and Facebook for sharing knowledge and experiences. The case studies are more confident of their TPACK perceptions. The finding presents that the pre-service science teachers tend to give more considerations to integrate technology into their lessons in three aspects (resource, teaching material, and student assignment). Through CO-TPACK PD and guided supervision within the school site enhances the student teachers development of TPACK based on educational theories and knowledge. This includes providing an effective way to convey the use of the technology to enhance student learning such as YouTube, Anusura, Dreamweaver, and Flash. By experiencing these different multimedia applications, they realized that they could represent many concepts to their students through technology. In an opposite way, students can learn science by creating project-based learning.

Sharing, reflecting, and discussing during co-planning, co-teaching, and co-evaluating of CO-TPACK PD were key activities in the CO-TPACK PD that enhanced Cooperative teachers' TPACK development. Through participating in the CO-TPACK PD, Cooperative teachers' understanding about student-cantered learning became clear. In their broadened ideas, student prior knowledge, and participating in hands-on activities were key aspects of student-cantered teaching. Three Cooperating teachers also had a chance to clarify his understanding of how to integrate knowledge bases for teaching particular content. These cooperative teachers reflected on their own teaching skills that helped them become aware of the importance of each knowledge base for teaching and its integration. In their second lesson plan, learning goals and purposes, learning activities sequence, instruction media, and assessment methods, had more detail and were more interrelated. They appeared focused more on teaching science by inquiry approach and enhanced students to learn science by integrated learning, PBL, or experimental learning as dimensions of student learning. They changed to use a variety of assessment methods such as asking questions of students, observing their behaviour, creating mind mapping and interaction with them. These student-student and student-teacher interactions appeared in their microteaching activity and their supervising to their student teachers. Three Cooperating teachers showed the CO-TPACK team can help their TPACK development and supervising skill because the CO-TPACK PD (during internship placement) contained a friendly and comfortable environment, interesting teaching and learning activities, and various types of assessment methods. The results of study suggest several aspects need to be addressed for science teacher education (pre-service and in-service science education) to be successful in integrating TPACK in their teaching of science, particularly in a classroom similar to the case study teachers'. First, the science teacher needs to hold the goals and purposes that focus on student learning with respect to science knowledge, science process skills, and scientific attitude. When the cooperating science teachers can teach science along with the reform-based science teaching, the student teachers will be trained to teach science as the same way of their cooperating science teachers. Second, strong pedagogical knowledge would make it easier for the science teacher to teach science through inquiry-based teaching and learning.

After 2 years with Co-TPACK, these teachers still continually worked with STEM project. Because the national proliferation of models of STEM that exist and the associated lack of practical advice creates confusion about how to integrate STEM in their classroom. Therefore, another 1 year of STEM professional development program was conducted in three stages: STEMed Plan; STEMed Assess; and STEMed Improve. The goal of STEM professional development is to increase student learning of science and mathematics by suing an engineering design-based approach for integrated STEM instruction to guide professional development and curricular design.

4. Conclusion

The study reported in this research is aimed at identifying how a Co-TPACK PD impacts on the development of cooperative science teachers' TPACK, and how does the Co-TPACK PD look like. The results focused on the teachers 'development of TPACK before and after they engaged in the Co-TPACK PD. As the CO-TPACK PD progressed during internship placement, Cooperative science teachers' TPACK knowledge base gradually broadened through learning activities in the CO-TPACK PD. Majority of cooperative science teachers have grown in their TPACK. Cooperating teachers were provided with many opportunities to broaden their understandings and practices about science concepts, pedagogy, assessment, technology integrated teaching and the nature of science. They had a chance to express their initial understandings and compare these understandings to constructivist understandings of teaching and learning science, proposed in the Basic Education Curriculum, and the Science Curriculum Framework. The Cooperating teachers were provided interesting ideas from the CO-TPACK members through sharing, reflecting, and discussion during their co-planning, co-teaching, and co-evaluating. Through these activities, three Cooperating teachers' understanding and practices of TPACK supporting teaching and learning science based on constructivism shifted to more constructivist understandings specifically, in the nature of scientific knowledge including the teacher 's supervising skill which was increased. The three stages of Co-TPACK PD provide guidelines for designing professional learning contexts for teacher development of TPACK knowledge. This study presents that effective professional development program that can help teachers developing their knowledge have the following these features:

1. Based on science teacher needs
2. Active engagements of science teachers, student teachers, and university supervisors.

3. Enhancement of technology knowledge into pedagogical content knowledge.
4. Ensuring science teacher's collaboration with other.
5. Provision opportunities for science teacher reflection and giving feedback throughout the professional development
6. Provision of experts and local supporters.
7. Along with STEM lesson plans were implemented, overall, these teachers reflected that their students enjoy the new activities which were the new way to represent science concepts or the new way of doing science through an integration of STEM unit.

5. Research Implication

According to the results of this study, the Co-TPACK PD and STEMed PD are productive in affecting changes of the teachers' TPACK in the classroom. The crucial components underlining these PDs that are likely to have an impact on the development of the science teachers might be: establishing common goals among program members, empowering teachers' leadership of the professional development program, providing opportunities for teachers to learn in their actual classrooms, giving time and support for teachers to plan, implement, observe, and reflect on their lessons, providing chances for teachers to learn through other teachers who are colleagues, having long-term assistance for continuous learning and practical change, and most importantly building and sustaining a trusting and respectful atmosphere among the teachers and the researcher. For readers and researchers who are interested in doing similar studies, it is important to remember that this study was conducted with a group of science teachers who taught at the elementary level or high school level. The findings from this study were not intended to generalize to all science teachers. Nevertheless, the description of how the Co-TPACK PD approach to professional development was implemented and the context surrounding the use of this approach may be useful to others who decide to use this as model for teacher professional development in their own context.

6. Acknowledgeable

My success would not be happened if I did not have helps from the National Research Council of Thailand and Phuket Rajabhat University for providing me the scholarship. Furthermore, this research would not have been completed without the support from the group of experts, research participants, and school administrators. I would like to thank all of the English editors and especially my husband, Robert Kief, for his patience and devotion in proof reading my writing.

References

[1] Beetham H and Sharpe R 2013 Rethinking pedagogy for a digital age: Designing for 21st century learning (New York: Routledge)
[2] Charmaz K 2003 Grounded theory In: J. A. Smith Qualitative Psychology: A Practical Guide to Research Methods (London, Sage)
[3] Chatmaneerungcharoen S 2008 The Elementary Science Teachers' Pedagogical Content Knowledge and Practices of Schools under the Bangkok Metropolitan Administration **KKU Research Journal 14** pp 1132–1142
[4] Healy C C and Welchert A J 1990 Mentoring relations: A definition to advance research and practice **Educational Researcher** pp 17–21
[5] International Society for Technology in Education 2000 National Educational Technology Standards for Teachers (U.S and Canada: Intel)
[6] Koehler M and Mishra P 2006 Technological Pedagogical Content Knowledge: A new framework for teacher knowledge **Teachers College Record 108** pp 1017–1054
[7] Koehler M and Mishra P 2009 What is technological pedagogical content knowledge (TPACK)? **Contemporary issues in technology and teacher education 9** pp 60–70
[8] National Research Council 1996 The National Science Education Standards (Washington, DC: Brooking Institution)

[9] Niess M L 2007 Mathematics teachers developing technological pedagogical content knowledge (TPCK) Paper presented at IMICT 2007 (Boston, MA)
[10] Office of the Education Council (OEC) 2004 Education in Thailand 2004 (Bangkok: Amarin Printing)
[11] Office of the National Education Commission (ONEC) 2001 The Research Report for Comparing of Educational Reform to Know in Thai Science Education Reform (Bangkok: ONEC)
[12] Office of the Rajabhat Institutes Council 2002 Quarterly report on World Bank project -The secondary education quality improvement (project IBRD Loan 4052 TH Bangkok: Author)
[13] Shulman L 1986 "Those who understand: Knowledge Growth in Teaching" **Educational Researcher, 15** pp 4–14
[14] Tanruther E M 1994 What does it mean to be a supervisory teacher? **College Journal** 167–171
[15] Thibeault J 2004 The relationship between student teachers and cooperating teachers as a foundation for the development of reflective thinking: An exploratory study based on student teachers' perceptions. Unpublished doctoral dissertation (McGill University, Montreal, Canada)
[16] Walkington J 2005 Becoming a teacher: encouraging development of teacher identity through reflective practice **Asia-Pacific Journal of Teacher Education 33** pp 53–64
[17] Williams C C 2001 Student teacher perceptions of university supervisor performance. Unpublished dissertation, Tennessee State University
[18] Yutakom N and Chero P 1999 Reporting of In-service Science Teacher Professional Development in Accordance with the National Education Act of B.E. (Bangkok: IPST)

IOP Conf. Series: Journal of Physics: Conf. Series **1340** (2019) 012018 doi:10.1088/1742-6596/1340/1/012018

Efficiency and Effectiveness of Universal Indicator from Native Plants in South of Thailand

Tatcha Sampim[1*],Salinkarn Phupa[2] and Supitcha Sampim[3]

[1] Faculty of Science and Technology, Nakhon Si Thammarat Rajabhat University, Thailand
[2] Watkuankaey School, Office of The Basic Education Commision, Thailand
[3] Phupripittayacom School, Office of The Basic Education Commision, Thailand

*the corresponding author's e-mail address : tatcha.sampim@mail.com

Abstract. A chemical universal indicator is generally the device of acid base indicator. But there are some limitations like environmental pollution, availability, and higher cost which lead to search for natural compounds as an acid base indicator was started. To solve these limitations based on STEM theory, the universal indicator from native plants in South of Thailand such as Melastoma, Phae, Mulberry, Sweet potato, Kuuam, Blue pea, Pandanus palm, Turmeric, and Roselle was developed to a useful pH indicator device. The objectives of this research are to compare the native plants colour extraction of the solution between distilled water and 95 % of ethanol solution, to identify the stable and clear colour drying paper, and to analyse efficiency and effectiveness of the universal indicator from native plants. The results showed that all of native plant can be extracted by ethanol solution 95 percent. There are six stable and clear colour paper including Melastoma, Phae, Mulberry, Blue pea, Turmeric, and Roselle that can be made to the universal indicator from native plants. The resolution efficiency can be divided the pH measurement into 7 ranges: 1-2, 3-4, 5-6, 7, 8-11, 12, 13-14. The 30 randomly effectiveness tests of the universal indicator from native plants has 100 percent of accuracy.

1. Introduction

The science standard 3, indicator 3.1 of grade 7 in The Thailand Basic Education Core Curriculum [7] specifies learners to experiment and explain acid-base properties. In generally, the simple method to test an acid-base property is to using a litmus paper. It is handy as a general acid-base indicator but cannot be used to determine the pH value. One way to indicate the pH value range from 1 to 14 is to use an universal indicator which composed of a solution of several chemical compounds such as water, propan-1-ol, phenolphthalein sodium salt, sodium hydroxide, methyl red, bromothymol blue monosodium salt, and thymol blue monosodium salt.

The world has become aware of environmental issue in recent year. Synthetic compounds are highly polluting, hazardous and much more costly [3]. For these reasons there has been an increasing interest in searching for alternative sources of indicators from natural origin [1]. Several authors have reported on the effectiveness of natural indicators in acid-base titrations [2] e.g. Nerium odorum, Thespesia populnea extract used as indicators [6]; Morus alba linn fruit extract indicator [5] and Ixora coccinea, Datura stramonium, Sun flower (Helianthus annus), pride of Barbados (Caesalpinia pulcher-rima) and rail creeper (Ipomoea palmate) flower petal extracts [4]. However, no evidence is available with regards to the pH ranges of the indicator, its optimum function and its possibilities of replacing some expensive commercial indicators [1].

IOP Conf. Series: Journal of Physics: Conf. Series **1340** (2019) 012018 doi:10.1088/1742-6596/1340/1/012018

On the other hand, as concern about the challenges of globalization and a knowledge-based economy, scientific and technological innovations have increased in theirs important [8]. One way to provide all citizens with the competences necessary to succeed in this new information-based and highly technological society, education in the fields of science, technology, engineering and mathematics (STEM) is becoming increasingly more important [8]. Therefore, this research further investigated the potentials of using native plant with extract and drying into paper as a source of natural indicator for simple pH indicator based on STEM education. The outcome of the study not only could promote the use of the native plant as a raw material to produce the natural pH indicator but also make learners to have the ability of problem solving, process thinking, project management, and STEM applying.

2. Materials Methods

2.1. Native Plants in South of Thailand

Different parts of native plant such as Melastoma, Phae, Mulberry, Sweet potato, Kuuam, Blue pea, Pandanus palm, Turmeric, and Roselle were collected from Moo 3, Sam Tambon sub district, Chulabhorn district, Nakhon Si Thammarat province, Thailand.

2.2. Chemical Reagents

The analytical grade of buffer solutions pH range 1 to 14 were prepared from The Center for Scientific and Technological Equipments of the Walailak University, Thailand.

2.3. Experimental Methods

2.3.1. Colour Extraction. 15 g f native plants were grinded by mortar and pestle. The grinded native plant were extract with two cases of 30 ml solvents (distilled water and 95 percent of ethanol). They were extracted and sieved though filter paper. The extracted solutions were observed the solubility and colour as shown in Table 1. The colour results of the solution was recorded by compare with standard 216 colour chart.

Table 1. The solubility and colour code of the native plant solutions.

Native plants	Solubility and Colour	
	Distilled water	Ethanol
1. Melastoma	Dissoluble #9900FF	Soluble #CC00FF
2. Phae	Dissoluble #CC0066	Soluble #660066
3. Mulberry	Soluble #990066	Soluble #663366
4. Sweet potato	Dissoluble #9900CC	Soluble #CC00CC
5. Kuuam	Soluble #FF0066	Soluble #FF3399
6. Blue pea	Dissoluble #0033FF	Soluble #330099
7. Pandanus palm	Dissoluble #666633	Soluble #336633
8. Turmeric	Dissoluble #FFCC33	Soluble #FF9933
9. Roselle	Dissoluble #CC0000	Soluble #CC00FF

2.3.1. Colour dyeing. The filter papers were dyed by extracted colour solutions and dried in the air. The only dyed papers which is stable and clear colour were collected to make the universal indicator from native plants as shown in Table 2.

Table 2. The dyed colour papers from the native plant solutions.

Stable and clear colour papers				Unstable and unclear colour papers	
1. Melastoma		4. Blue pea		1. Sweet potato	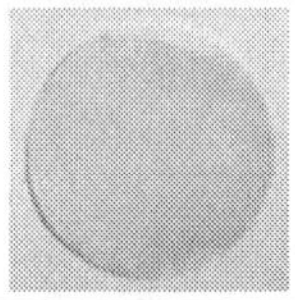
2. Phae		5. Turmeric	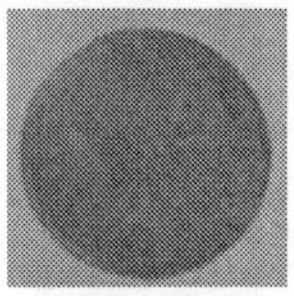	2. Kuuam	
3. Mulberry		6. Roselle	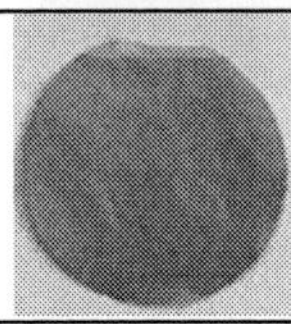	3. Pandanus palm	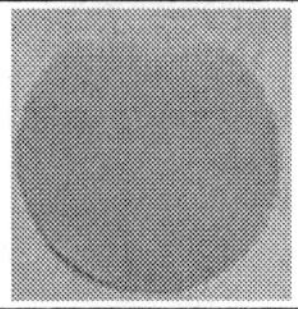

2.3.2. Efficiency of Universal Indication from Native Plants. The universal indicator from native plants was tested the colour change with buffer solutions pH range 1 to 14. The colour change patterns were analysed to make the table colour of the universal indicator from native plants which was used as measuring tool to indicate the pH value as presented in Table 3.

Table 3. Colour change and pH resolution efficiency of universal indicator with buffer solution pH range 1 to 14

	Buffer solution pH range 1 to 14													
	1	2	3	4	5	6	7	8	9	10	11	12	13	14
Colour change														
pH resolution efficiency	1-2		3-4		5-6		7	8-11				12	13-14	

IOP Conf. Series: Journal of Physics: Conf. Series **1340** (2019) 012018 doi:10.1088/1742-6596/1340/1/012018

2.3.3. Effectiveness of Universal Indication from Native Plants. The universal indicators were thirty sampling tests to analysed pH value with closed pH value. The pH test results were examined the accuracy and calculated the effectiveness as presented in Table 4.

Table 4. Thirty sampling tests of the universal indicator from native plants

Sampling No.	pH value test	pH value answer	Result	Sampling No.	pH value test	pH value answer	Result
1.	7	7	Correct	16.	8-11	8	Correct
2.	13-14	14	Correct	17.	5-6	6	Correct
3.	1-2	2	Correct	18.	13-14	14	Correct
4.	5-6	6	Correct	19.	1-2	2	Correct
5.	3-4	3	Correct	20.	13-14	13	Correct
6.	12	12	Correct	21.	3-4	3	Correct
7.	8-11	8	Correct	22.	5-6	6	Correct
8.	5-6	6	Correct	23.	1-2	2	Correct
9.	8-11	10	Correct	24.	13-14	14	Correct
10.	8-11	11	Correct	25.	8-11	9	Correct
11.	7	7	Correct	26.	5-6	5	Correct
12.	1-2	2	Correct	27.	3-4	4	Correct
13.	12	12	Correct	28.	7	7	Correct
14.	3-4	3	Correct	29.	12	12	Correct
15.	8-11	10	Correct	30.	5-6	6	Correct
100 % of pH measurement effectiveness							

3. Results and Discussion

From Experimental data, it is obtained that all native plants can be extracted to colour solutions by 95% of ethanol. However, in dyeing method, there are only six stable and clear colour dyed papers to make the universal indicator including Melastoma, Phae, Mulberry, Blue pea, Turmeric, and Roselle. The pH resolution efficiency of universal indicator from native plants can be divided into seven ranges: 1-2, 3-4, 5-6, 7, 8-11, 12, and 13-14. Additionally, its effectiveness to correctly indicate pH value is 100 %.

4. Conclusion

From the results obtained, it is confirmed that the universal indicator from native plants can be an alternative device to measure pH value based on STEM theory. It is also confirmed that the universal indicator is not only high resolution efficiency and pH measurement effectiveness but also green, available, and economical product.

Acknowledgement

I would like to express my sincere thanks to the Center for Scientific and Technological Equipments of the Walailak University for pH buffer solution preparation.

References

[1] Abugri D A, Apea O B and Pritchett G 2012 Investigation of a Simple and Cheap Source of a Natural Indicator for Acid-Base Titration: Effects of System Conditions on Natural Indicators **Green and Sustainable Chemistry 2** pp 117–122

[2] Ghatage T S, Killedar S G, Hajare M A and Joshi M M 2017 Extraction, Characterization, and Utilization of Different Plant Pigments as pH Indicators in Titrimetric Analysis **International Journal of Advance Research, Ideas and Innovations in Technology 3** pp 462–468

[3] Khan P M A and Faroooqui M 2011 Analytical Application of Plant Extract as Natural pH Indicator: A Review **Journal of Advance Scientific Research 2** pp 20–27

[4] Nwosu F O Adekola F A and Ihedioha K C 2004 Simple Titrimetric Colour Indicators from Some Natural Flower Petals **Centrepoint (Science Edition) 12** pp 74–89

[5] Pathade K S, Patil S B, Konda-war M S, Naik wade N S and Magdum C S 2009 Morus Alba Fruit-Herbal Alternative to Synthetic Acid Base Indicators **International Journal of ChemTech Research 1** pp 549–551

[6] Patil S B, Kondawar M S, Ghod-ke D S, Naik wade N S and Magdum C S 2009 Use of Flower Extracts as an Indicator in Acid-Base Titrations **Research Journal of Pharmacy and Technolog 2** pp 421–422

[7] The Ministry of Education 2008 The Thailand Basic Education Core Curriculum 124

[8] Thibaut L, Knipprath H, Dehaene W and Depaepe F 2018 The influence of teachers' attitudes and school context on instructional practices in integrated STEM education 71 190-205

Construct validity of attitudes towards mathematics questionnaire

S Srikoon[1*], P Thonjeang[2]

[1]Curriculum and Instruction, School of Education, University of Phayao, 19 Moo 2, Phahonyothin Road, Mae Ka, Mueang Phayao, Phayao, Thailand
[2]Student in Bachelor of Education (Education in Mathematics), School of Education, University of Phayao, 19 Moo 2, Phahonyothin Road, Mae Ka, Mueang Phayao, Phayao, Thailand

*Corresponding author's e-mail: sanit.sr@up.ac.th

Abstract. One goal of basic education in mathematics is to help students develop an attitudes towards mathematics. The purpose of this research was to validate of the attitudes towards mathematics questionnaire. Sample are 204 students for validating an attitudes towards mathematics questionnaire. The attitudes towards mathematics consists of the 14 items for self-assessment. Confirmatory factor analysis was used to confirm the construct validity. The results confirmed that the construct validity of these test an excellent fit. The results show that the fitness index of validating attitudes towards mathematics questionnaire were as follow: statistic of 68.77 (degrees of freedom = 53, p= 0.071), and the χ 2/df ratio having a value of 1.30 indicates a good fit. The comparative fit index (CFI) is 0.979, and Tucker-Lewis coefficient (TLI) is 0.964, Root mean square error approximation (RMSEA) is 0.028, Standardized Root Mean Residual (SRMR) is 0.030. All the indicators indicated that were a goodness of fit between the empirical data and the hypothetical measurement model.

1. Introduction

A growing body of literature suggests attitudes towards mathematics (AtM) are related to student's reflective thinking skills, problem solving skills, learning style, and achievement [8][10]. Consequently AtM are recognized as importance variation to student learning outcome and has become the focus of learning process [9].

2. Significant of the Study

Despite its wide-spread mention, little research has investigated the AtM, and only a few studies have defined the factor structure of AtM. For Example D S Memnun and R Akkaya refer AtM as teachers' assessment for enhancing students' learning outcomes but were undefined their construct [9]. Although J M arti-Parreno, A Galbis-Cordova, M J Miquel-Romero reported AtM in their research but were not clearly defined it, too [1]. To date, only a few studies have referred the factor structure of AtM [6].

3. Conceptual Framework

Attitude towards mathematics questionnaire (AtMQ) is the self-assessment inventory for diagnosing face validity. It consists of 14 items of three factor. These three are awareness (AW), perception (PE), and readiness (RE). AW is to know about the importance, the value, the useful of mathematics in daily life, studying in higher education, and developing in any case. PE is the felling after studying mathematics. RE is the readiness for studying mathematics [5]. Figure 1 show the conclusion of the three AtM. Each statement of AtM items is using a 5-point Likert scale namely 1 represents 'Does not agree', 2 represents 'Does not really agree', 3 represents 'Neutral, not sure', 4 represents 'Agree somewhat', and 5 represents 'Absolutely agree'

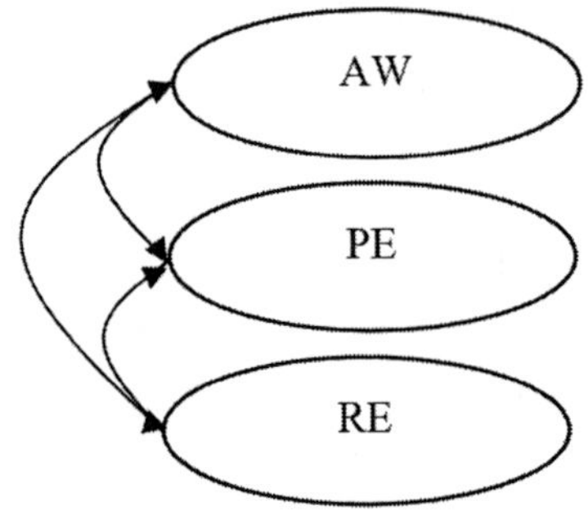

Figure 1: Conceptual framework

4. Aim of the Study

AtMQ should be precisely measured to examine the nature of construct and to make valid explanations for each individual's social activity. This study aimed to validate the AtMQ. The specific purposes or this study are to test the construct validity of the AtMQ with the Institute for the Promotion of Teaching Science and Technology [5].

5. Research Methodology

5.1. Sample of this Study

A total of 2,673 (N) samples of this survey research were draw from Grade 9 (n=472). Owing to incomplete response, a total of 204 cases.

5.2. Data Analysis

The rapid growing of advanced methodology would be provided the ability of researchers to analyze the construct validity. CFA was used to test the hypothesized of three-factor model consisted of 14 items in all samples. Covariance structures were fitted with the maximum likelihood method. The model fitted indicator was evaluated by means of several fit indices. Universally, the model is considered acceptable when probability value (p-value) >0.05, value of ratio between chi-square statistic and degree of freedom (χ 2/df) in 2:1 [3], the Comparative Fit Index (CFI) $\geq$ 0.90 and good when $\geq$0.95 [2][4] and Tucker-Lewis coefficient (TLI) $\geq$ 0.95 is good fit [11]. Moreover, Standardized Root Means Square Residual (SRMR) should not exceed 0.08 for a good fit [4]. Furthermore, Root Means Square Error of Approximation (RMSEA) value $\leq$ 0.06 are considered indicative of a good fit, $\leq$ 0.08 of fair fit, between 0.08 and 0.01 of mediocre fit and >0.01 of poor fit [4][7].

6. Research Finding

CFA was used to evaluate the goodness of fit. CFA were conducted using MPlus program. Model fit was assessed using χ2, χ2/df , CFI, TLI, SRMR and RMSEA. Results showed that all indicators indicated that there was a goodness of fit between the empirical data and the hypothetical measurement model for model. In other word, statistic of 68.77 (degrees of freedom = 53), the p-value in CFA models are not significant (.0.71), χ2/df values are fall in 2:1 (1.30), all CFI $\geq$ 0.95 indicate the good fit (0.979). Similarity all the TLIs $\geq$ 0.95 are good fit (0.964). Moreover all the RMSEAs are $\leq$

0.06 are considered indicative of a good fit (0.028). Finally all the SRMRs≤ 0.08 are accepted for a good fit (0.030) too. Detailed in Figure 2.

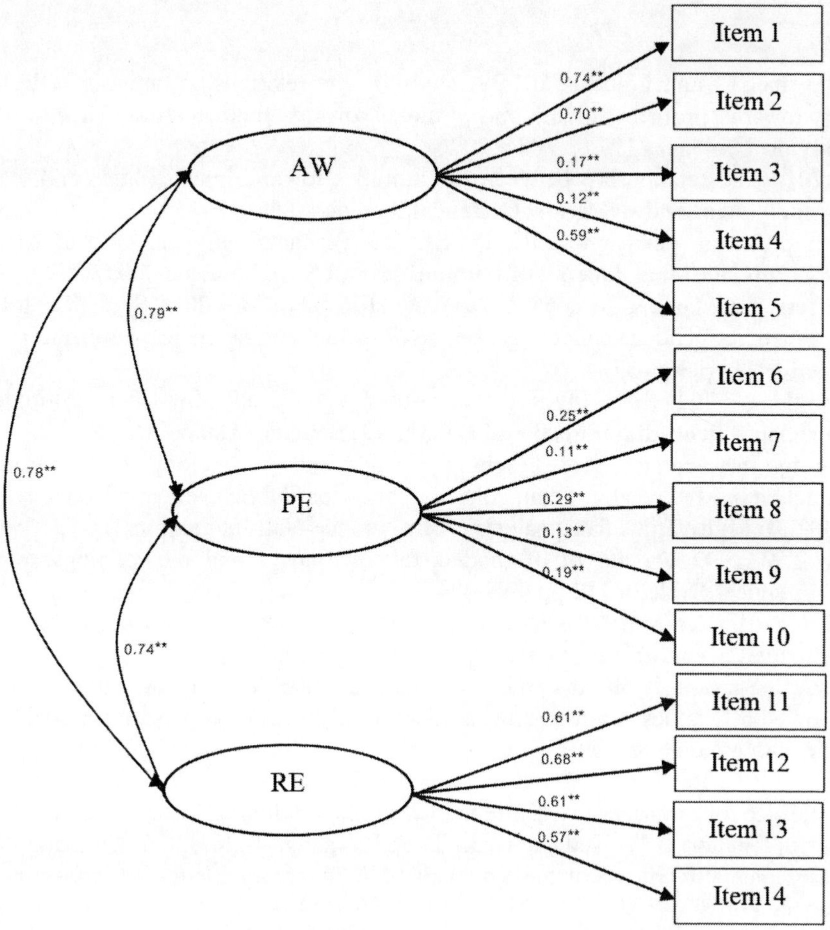

(**p<0.01)
[$\chi2$=68.77 , df=53 , p= 0.117 , $\chi2$/df = 1.30 ,CFI= 0.979, TLI= 0.964, RMSEA=0.028 ,SRMR=0.030]

Figure 2. Confirmatory factor analysis of AtMQ

7. Discussion

On this line of reasoning, AtM is considered as the most important factor because students who possessed high AtM are more likely to engage in mathematics learning outcome and achievement at high level [3]. In addition, investigating the construct validity provide the further understanding of how AtM can be measured and evaluated. This study was conducted purposively to examine the validity of the AtMQ with sample grade 9. The validity of the AtMQ are the well fit for all indicators. Results from this study provide evidence about the validity of the confirming AtMQ. It shows that AtMQ can measure the AtM and can give the valid information for classroom management. The validating information can be used for learning management in classroom ([12]). The AtMQ was measured the AtM in the learning classroom, was used the pre-post AtM measurement. These is one type of emotion which must occurred in classroom [13].

Acknowledgement
This work was supported by the Unit of Excellence in Research Methodology of Innovations and Learning Sciences based-on Educational Neurosciences, University of Phayao Research Fund, Thailand.

References
[1] Melek D, Ipek D and Edibe K 2015 A study on the relationship between reflective thinking skills towards problem solving and attitudes towards mathematics **Procedia - Social and Behavioral Sciences 197** pp 2086–2096
[2] Nur S 2010 The relationship between the attitudes towards mathematics and learning styles **Procedia Social and Behavioral Sciences 9** pp 644–648
[3] Memnun D S and Akkaya R 2012 Pre-Service Teachers'Attitudes towards Mathematics in Turkey **International Journal of Humanitiesand Social Science 2** pp 90–99
[4] Arti-Parreno J M, Galbis-Cordova A and Miquel-Romero M J 2018 Students' attitude towards the use of educational video games to develop competencies **Computers in Human Behavior 81** pp 366–377
[5] Kibrislioglu N 2015 An Investigation About 6th Grade Students' Attitudes Towards Mathematics **Procedia - Social and Behavioral Sciences 186** pp 64–69
[6] Institute for the Promotion of Teaching Science and Technology 2012 Measurement and Evaluation in Mathematics (Bangkok: SE-Education Public Company Limited)
[7] Hair J F 2006 Multivariate data analysis (N.J.: Prentice-Hall International)
[8] Bentler P M 1992 On the fit of models to covariances and methodology to the Bulletin **Psychological Bulletin 112** pp 400–404
[9] Hu LT and Bentler M 1999 Cutoff criteria for fit indexes in covariance structure analysis: Alternatives **Structural Equation Modeling 6** pp 1–55
[10] S Sharma, S Sharma, A Sharma and W R Dillon 2005 A simulation study to investigate the use of cutoff values for assessing model fit in covariance structure models **Journal of Business Research 58** pp 935–943
[11] MacCallum R C, Browne M W Sugawara and H M 1996 Power analysis and determination of sample size for covariance structure modeling **Psychological Methods 1** pp 130–149
[12] Srikoon S, Bunterm T, Nethanomsak T and Ngang T K 2016 Construc Validity and Measurement Invariance of the Research Skill Inventory *Mediteranean* **Journal of Social Sciences 7** pp 366–377
[13] Srikoon S, Bunterm T, Nethanomsak T and Ngang T K 2017 A comparative study of the effects of the neorocognitive-bases model an the conventional model on learners' attention, working memory and mood *Malaysain* **Journal of Learning & Instruction 14** pp 83–110

Construct validity of learning styles questionnaire

S Srikoon[1]

[1]Curriculum and Instruction, School of Education, University of Phayao, 19 Moo 2, Phahonyothin Road, Mae Ka, Mueang Phayao, Phayao, Thailand

Corresponding author's e-mail: sanit.sr@up.ac.th

Abstract. Learning styles is the importance goal for enhancing and focusing to help students develop learning outcomes. The purpose of this research was to validate of the learning styles questionnaire based on the left and the right brain theory. Sample are 201 students for validating the learning styles questionnaire. The learning styles questionnaire consists of the 40 items for self-assessment. Confirmatory factor analysis was used to confirm the construct validity. The results confirmed that the construct validity of these test an excellent fit. The results show that the fitness index of validating the learning styles questionnaire were as follow: statistic of 698.12 (degrees of freedom = 641, p= 0.058), and the $\chi 2/df$ ratio having a value of 1.089 indicates a good fit. The comparative fit index (CFI) is 0.977, and Tucker-Lewis coefficient (TLI) is 0.972, Root mean square error approximation (RMSEA) is 0.021, Standardized Root Mean Residual (SRMR) is 0.061. All the indicators indicated that were a goodness of fit between the empirical data and the hypothetical measurement model.

1. Introduction

Interdisciplinary educational neurosciences is increasingly important and valued in education [11]. Learning theory is developed with educational neuroscience research [2]. The left and the right theory (LRT) is one educational neuroscience approach to explain learning styles with neuroscience research. Herman (1996) reported that LRT comprised of thinker style (TS), organizer style (OS), humanitarian style (HS), and innovator style (IS) [6]. The learning styles (LS) has been used in education to explain individual differences in the ways students approach learning. It is assumed that instruction based in learning styles theory produces better achievement [1].

LS has been used in education to explain individual differences in the ways students approach learning. It is assumed that instruction based in learning styles theory produces better achievement [1]. Moreover LS relate with creativity, learning behavior and student's learning outcomes [4][10]. Consequently LS are recognized as importance part to enhance student learning outcome and has become the focus of student's learning process.

2. Significant of the Study

Despite its wide-spread mention, one of the critical problems with learning styles theory is the lack of clear, explanatory framework [8], and only a few studies have defined the factor structure of LS [5]. To date, only a few studies have referred the factor structure of LS.

IOP Conf. Series: Journal of Physics: Conf. Series **1340** (2019) 012020 doi:10.1088/1742-6596/1340/1/012020

3. Conceptual Framework

LSQ is the self-assessment inventory for diagnosing face validity. It consists of 40 items of four factor of the left and right brain theory [6]. These four are thinkers style (TS), organizers style (OS), humanitarians style (HS) and innovators style (IS). TS focus on working of left prefrontal lobe. The TS learning style student is who like to lecture and to critic content, to study with case study, to refer research, to set reasoning, to connect prior knowledge with new knowledge, to learn in any times, to do with strength criteria, and to have gifted ability. OS focus on left hemisphere lobe. The OS learning style student is who like to learn with thinking strategies, to learn with step by step on learning processes, to learn in any case study, to learn from figure or table, to create the new approach to group discussion, to solve problem on action plan, to believe in accountability knowledge, to be exhaustively person and systematic worker. HS focus on working on right hemisphere lobe. The HS learning style student is who like to learn with exchange experience and knowledge sharing, to learn with experience or tale or prover, to communicate with body language, to participate for all, to show music, to discuss in social or daily life. IS focus on working of right prefrontal lobe. The IS learning style student is who like to learn with brain storming, to compare cases, to create high imagination, to do freedom activity, to construct creative activity, to build creative problem solving, to learn interesting topics, and to show artist. Figure 1. show the conclusion of the four LSQ. Each statement of LSQ is using a 5-point Likert scale namely 1 represents 'Does not perform', 2 represents 'Does not really perform'. 3 represents 'Neutral, not sure', 4 represents 'Perform somewhat', and 5 represents 'Absolutely perform'.

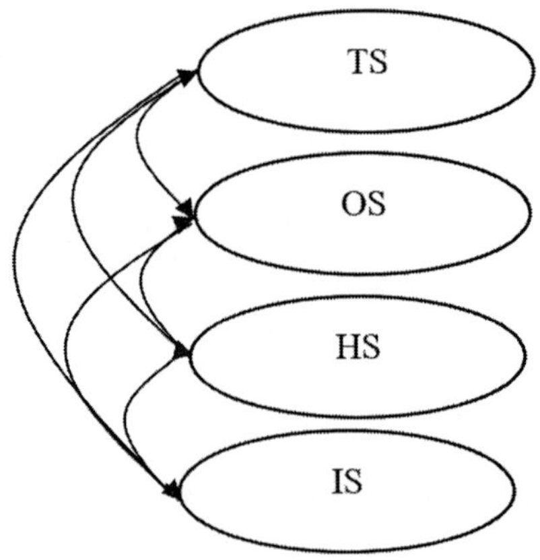

Figure 1: Conceptual framework of confirmatory factor analysis of LSQ

4. Aim of the Study

LSQ should be precisely measured to examine the nature of construct and to make valid explanations for each individual's social activity. This study aimed to validate the LSQ. The specific purposes or this study are to test the construct validity of the LSQ with the left and the right brain theory.

5. Research Methodology

5.1. *Sample of this Study*

A total of 2,673 (N) samples of this survey research were draw from Grade 9 (n=472). Owing to incomplete response, a total of 201 cases.

5.2. *Data Analysis*

The rapid growing of advanced methodology would be provided the ability of researchers to analyze the construct validity. CFA was used to test the hypothesized of four-factor model consisted of 40 items in all samples. Covariance structures were fitted with the maximum likelihood method. The model fitted indicator was evaluated by means of several fit indices. Universally, the model is considered acceptable when probability value (p-value) >0.05, value of ratio between chi-square statistic and degree of freedom (χ 2/df) in 2:1 [5], the Comparative Fit Index (CFI) $\geq$ 0.90 and good when $\geq$0.95 [2][7] and Tucker-Lewis coefficient (TLI) $\geq$ 0.95 is good fit [12]. Moreover, Standardized Root Means Square Residual (SRMR) should not exceed 0.08 for a good fit [7]. Furthermore, Root Means Square Error of Approximation (RMSEA) value $\leq$ 0.06 are considered indicative of a good fit, $\leq$ 0.08 of fair fit, between 0.08 and 0.01 of mediocre fit and >0.01 of poor fit [7][9].

6. Research Finding

CFA was used to evaluate the goodness of fit. CFA were conducted using MPlus program. Model fit was assessed using χ2, χ2/df , CFI, TLI, SRMR and RMSEA. Results showed that all indicators indicated that there was a goodness of fit between the empirical data and the hypothetical measurement model for model. In other word, statistic of 698.12 (degrees of freedom = 641), the p-value in CFA models are not significant (.0.058), χ2/df values are fall in 2:1 (1.089), all CFI $\geq$ 0.95 indicate the good fit (0.977). Similarity all the TLIs $\geq$ 0.95 are good fit (0.972). Moreover all the RMSEAs are $\leq$ 0.06 are considered indicative of a good fit (0.021). Finally all the SRMRs$\leq$ 0.08 are accepted for a good fit (0.060) too. Detailed in Appendix.

7. Discussion

LS is considered as the start point of learning processes because students who learned available with LS are more likely to engage learning outcomes or knowledge into long term memory at high level [11][14]. Moreover students who are supported with learning environment to comply with student's LS can possessed information processes at high level, too [2]. Therefore LS is the important factor of student's learning. In addition, investigating the construct validity provide the further understanding of how any tool can measure and evaluate any variable [13]. This study was conducted purposively to examine the validity of the LSQ with the sample grade 9. Result reports that the LSQ are the well fit for all indicators as LSQ can measure the LS and can give the valid information for learning management in classroom. Moreover the LSQ can used pretest measurement for diagnosis student's LS in classroom.

Acknowledgement

This work was supported by the Unit of Excellence in Research Methodology of Innovations and Learning Sciences based-on Educational Neurosciences, University of Phayao Research Fund, Thailand.

IOP Conf. Series: Journal of Physics: Conf. Series **1340** (2019) 012020 doi:10.1088/1742-6596/1340/1/012020

8. Appendix

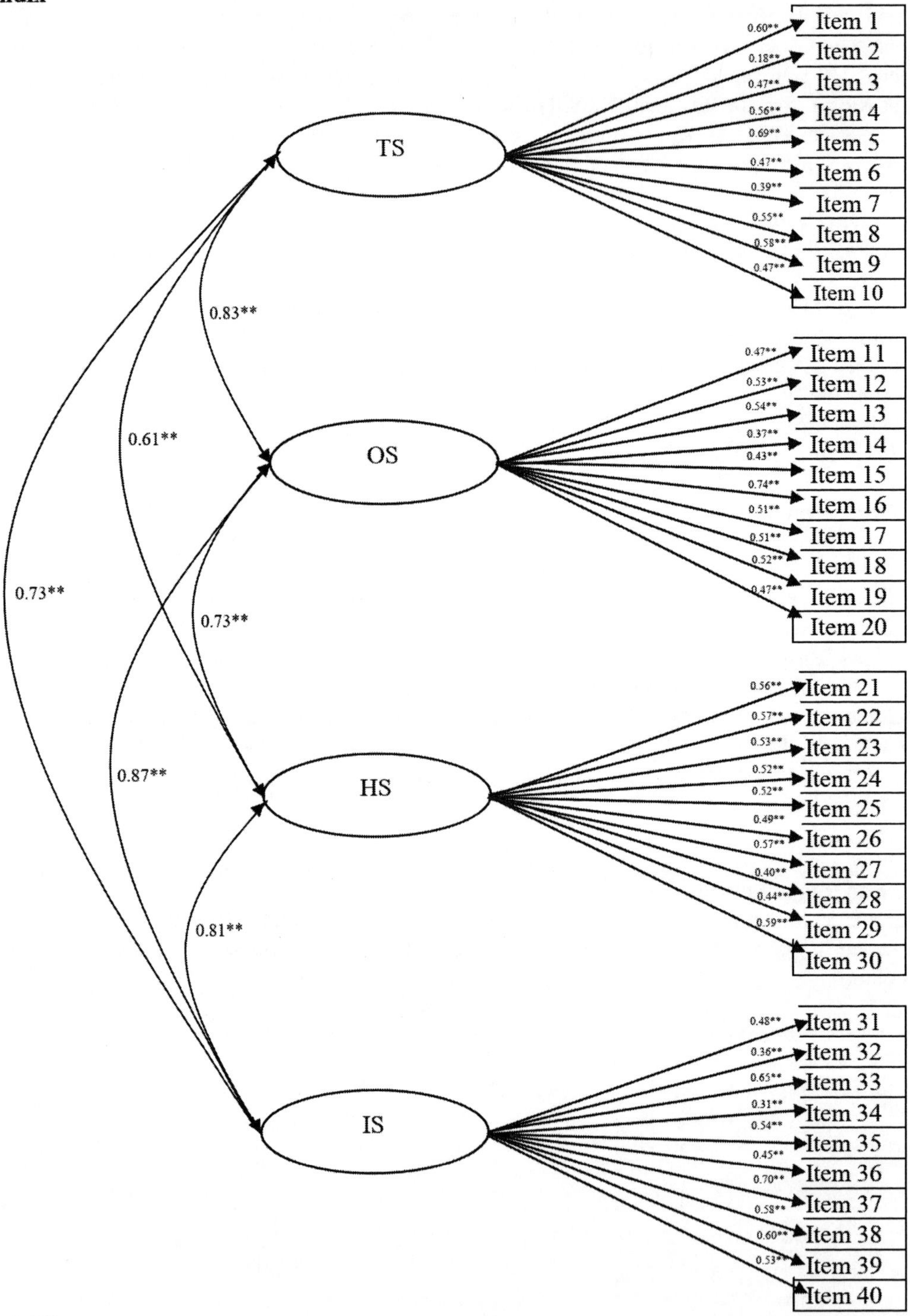

(**$p<0.01$)
[$\chi2$=698.12, df=641, p= 0.058 , $\chi2$/df = 1.089, CFI= 0.977, TLI= 0.972, RMSEA=0.021, SRMR=0.061]

Figure 2. Results of Confirmatory factor analysis of LSQ

References

[1] Palghat, J C Horvath and J M Lodge 2017 The hard problem of educational neuroscience *Trends in Neuroscience and Education* **6** 204-210
[2] Anderson 2009 Neurocognitive theory and constructionism in science education: a review of neoribiologycal, cognitive and cultural perpective. *Brunei Int. J. of Sci. & Math. Edu* **1** 1-32
[3] Herman 1996 *The Whole brain buriness book* (New York: McGraw-Hill)
[4] An and M Carr 2017 Learning styles theory fails to explain learning and achievement: Recommendations for alternative approaches *Personality and Individual Differences* **116** 410-416
[5] An and M Carr 2017 Learning styles theory fails to explain learning and achievement: Recommendations for alternative approaches *Personality and Individual Differences* **116** 410-416
[6] Eishami, E A Saad and Y Nami 2014 The Relationship Between Learning Style and Creativity *Procedia-Social and Behavioral Science* **114** 52-55
[7] Magdalena 2015 The relation of learning styles, learning behavior and learning outcomes at the romaninan students *Procedia-Social and Behavioral Science* **180** 1667-1672
[8] Jamil, F N Baharuddin and T S R Maknu 2015 Factors Mining in Engaging Students Learning Styles Using Exploratory Factor Analysis *Procedia Economics and Finance* **31** 722-729
[9] Hair 2006 *Multivariate data analysis* (N.J.: Prentice-Hall International)
[10] Bentler 1992 On the fit of models to covariances and methodology to the Bulletin *Psychological Bulletin* **112** 400-404.
[11] Hu and M Bentler 1999 Cutoff criteria for fit indexes in covariance structure analysis: Alternatives *Structural Equation Modeling* **6** 1-55.
[12] Sharma, S Sharma, A Sharma and W R Dillon 2005 A simulation study to investigate the use of cutoff values for assessing model fit in covariance structure models *Journal of Business Research* **58** 935-943.
[13] MacCallum, M.W. Browne and H M Sugawara 1996 Power analysis and determination of sample size for covariance structure modeling *Psychological Methods* **1** 130-149.
[14] Srikoon, T Bunterm, T Nethanomsak, T K Ngang 2017 A comparative study of the effects of the neorocognitive-bases model an the conventional model on learners' attention, working memory and mood *Malaysain Journal of Learning & Instruction* **14** 83-110
[15] Anderson 2009 Neurocognitive theory and constructivism in science education: a review of neurobiologyal, cognitive and cultural perspective *Brunei Int. J. of Sci & Math. Edu* **1** 1-32
[16] Srikoon, T Bunterm, T Nethanomsak, T K Ngang 2016 Construc Validity and Measurement Invariance of the Research Skill Inventory *Mediteranean Journal of Social Sciences* **7** 366-377

IOP Conf. Series: Journal of Physics: Conf. Series **1340** (2019) 012021 doi:10.1088/1742-6596/1340/1/012021

Effect of 5P model on mathematics achievement and mood

S Srikoon[1]

[1]Curriculum and Instruction, School of Education, University of Phayao, 19 Moo 2, Phahonyothin Road, Mae Ka, Mueang Phayao, Phayao, Thailand

Corresponding author's e-mail: sanit.sr@up.ac.th

Abstract. This study examined significant changes on the mathematics achievement and mood of Grade 9 students after the intervention of the 5P model and 5E model with an experimental group and control group, respectively. The 5P model is an integrated between educational neuroscience and research-based learning. In total, 77 out of 411 Grade 9 students were purposively selected from high school located in Buriram province, Thailand. The sample comprised of 38 students of the experimental group and 39 students of the control group for randomized pretest-posttest control group design. One-way analysis of variance is utilized for analysis of variance. The findings revealed that the effect of 5P model was found be greater compared to the 5E model.

1. Introduction

To date, teaching model is continuously developed for improved student learning outcomes. Neurocognitive-based Model (5P model) of teaching is design and is integrated research-based learning (RBL) and educational neuroscience (EN) for enhancing learning outcomes, too [4]. Moreover 5P model is the innovative teaching model and is produced for educational reform in 21st century. Consequently 5P model are recognized as important model of teaching to prove student learning outcome and has should become the new teaching model of student's learning process.

2. Significant of the Study

Despite its 21st innovative teaching model, little research has study the effect of 5P model, and only a few study have research the effect of 5P model. For example, S sanit, T Bunterm, T Nethanomsak, and T K Ngang study a comparative study of the effects of 5P model and 5E model on learner's attention, working memory, and mood [6], they study the effect of 5P model on academic achievement, creative thinking, and research characteristics [7]. Both research study the effect of 5P model on only science content knowledge. Moreover mood is the important factor as the effect on learning information processing for all phrase [1]. Consequently researcher should study the effects of 5P model on the mathematics achievement and mood. To date, this research aimed to examine the effect of 5P model on mathematics achievement and mood for proving the effect of 5P model.

3. Conceptual Framework

5E model is the instruction model used by control group. This 5E model is supported by the Institute for the Promotion of Teaching Science and Technology (IPST) in standard classroom, Thailand [3]. 5E model is consisted of 5 phases namely engage, explore, explain, elaborate, and evaluate. The 5E model allows teachers and learners to experience common activities, build on prior knowledge and experience, construct meaning, and continually assess their understanding of a concept.

5P model is constructed with integrated RBL and EN. RBL can enhance the academic achievement, promote learning on how to learn and construct the new knowledge by one's self. These capabilities are essential for education in the 21st century [5]. EN is interdisciplinary science and have a role to play in education reform [9]. Consequently 5P model is emerged from integrated RBL and EN for educational reform [5]. It covers 5 main phases namely persuasion, planning, performance, production, and presentation [6]. S sanit T Bunterm T Nethanomsak and T K Ngang report that 5P model can enhancing science achievement, attention, working memory, mood, creative thinking, and research characteristic but none research study the effect of 5P model on mathematics achievement. This research study to comparative the effects of the 5P model and 5E model on mathematics achievement and mood.

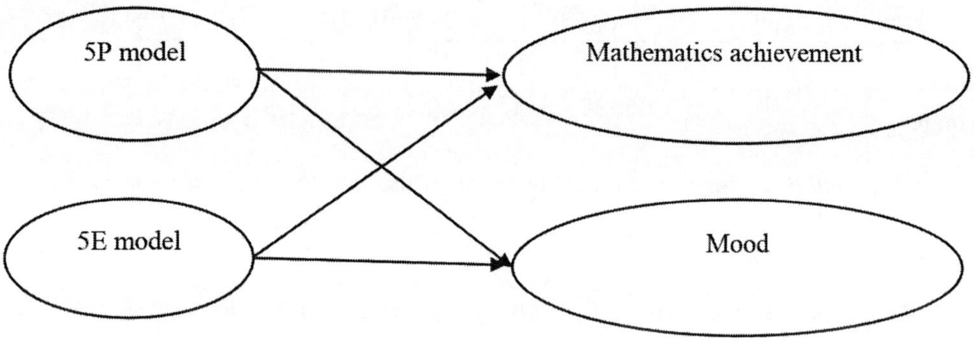

Figure 1 the effects of the 5P model and 5E model on mathematics achievement and mood

4. Aim of the Study

5P model should be generally used and studied the effects on the mathematics achievement and mood. This study aimed to examine the effect of 5P model on mathematics achievement and mood.

5. Research Methodology

5.1. Sample of this Study

Researcher utilized the randomized pretest-posttest control group design. The randomly select sample consisted of 77 from 411 Grade 9 students in twelve classes in a high school located in Buriram province, Thailand during the second semester of 2015 academic year. The 77 sample were divided into two group consisting of 38 student in the experimental group (5P model) and 39 student in the control group (5E model).

5.2. Research instrument

Research instruments were two type of tests used: mathematics content knowledge test (40 item) and Bond-Ladder visual analogue scale (16 item). Mathematics content knowledge test is developed to be multiple choice and was evaluated with 5 experts for content validity. Bond-Ladder visual analogue scale were evaluated the construct validity using S Srikoon, N Viriyapong, N Chutiman [8].

5.3. Data Analysis

The research use mean ($\bar{X}$) and standard deviation (S.D.) for analysed descriptive statistics. One-way analysis of variance (ANOVA) is utilized for tested the mean score of two group.

6. Research Finding

6.1. Descriptive statistics

The descriptive statistics of pre-test versus post-test mathematics achievement and mood for both 5P model and 5E model groups are presented in Table 1

Table 1. Descriptive statistics of mathematics achievement and mood.

Group	n	mathematics achievement		mood	
		pre-test	post-test	pre-test	post-test
5P model	38	23.79(0.76)	30.03(0.67)	111.76(2.38)	129.86(2.18)
5E model	39	22.44(0.66)	24.59(0.83)	111.41(2.43)	116.28(2.74)

Table 1 show descriptive statistics of mathematics achievement and mood for 5P model and 5E model group.

6.2. Significant F-test for univariate ANOVA follow up tests for the mathematics achievement and mood

6.2.1 Significant F-test for univariate ANOVA follow up pre-test for the mathematics achievement and mood

Table 2. Significant F-test for univariate ANOVA follow up pre-test for the mathematics achievement and mood

Effect	Sum of square	Mean square	df	F	p
mathematics achievement					
Between Group	35.26	35.26	1	1.82	0.18
Within Group	1455.91	19.41	75		
Total	1491.17		76		
Mood					
Between Group	2.40	2.40	1	0.01	0.92
Within Group	16726.30	223.02	75		
Total	16728.70		76		

Table 2 show that the univariate ANOVA follow up pre-test for the mathematics achievement and mood does not provide a specific mean difference. It shows that pre-test for the mathematics achievement and mood do not differ between 5P model and 5E model group.

6.2.2 Significant F-test for univariate ANOVA follow up post-test for the mathematics achievement and mood

Table 3. Significant F-test for univariate ANOVA follow up post-test for the mathematics achievement and mood

Effect	Sum of square	Mean square	df	F	p
mathematics achievement					
Between Group	568.86	568.86	1	25.95	0.00
Within Group	1644.41	21.93	75		
Total	2213.27		76		
Mood					
Between Group	3552.75	3552.75	1	14.94	0.00
Within Group	17830.24	237.74	75		
Total	21382.99		76		

Table 3 show that the univariate ANOVA follow up post-test for the mathematics achievement and mood does provide a specific mean difference. It shows that post-test for the mathematics achievement and mood do differ between 5P model and 5E model group.

Research conclude that pre-test of both the mathematics achievement and the mood do not differ between the control group and the experimental group but post-test of both the mathematics achievement and the mood do differ between the control group and the experimental group. Moreover researcher found that the post-test of the mathematics achievement ($\bar{x}$=30.03, S.D.=0.67) and the mood ($\bar{x}$=30.03, S.D.=0.67) greater than pre-test of the mathematics achievement ($\bar{x}$=129.86, S.D.=2.18) and mood ($\bar{x}$=116.28, S.D.=2.74), respectively. It concludes that the effects of 5P model was found be greater compared to 5E model.

7. Discussion

The findings of this study have revealed a greater understanding of 5P model on the improvement of learner outcomes (mathematics achievement and mood) and have the potential of making a significant contribution to education. The results of this study have supported the claim of effectiveness of the 5P model for enhancing learner mathematics achievement and mood. Researcher hopes that the 5P model will be applied in improving learner outcomes in the future.

The finding of this research imply that current teacher should focus on 5P model to design the instructional model in order to enhance and improve the mathematics learning outcomes of their student. Teachers should utilized the 5P model for sufficient opportunities for student to develop theses domain. The results of this study are similar to S Srikoon, T Bunterm, T Nethanomsak, T K Ngang [6], S Srikoon, T Bunterm, T Nethanomsak, T K Ngang [7], and M Hardiman [2]. Ultimately, 5P model is the linking of RBL and EN for creation the teaching model and is the innovative education for developing classroom practices in 21st century [2].

In conclusion, there is an innovative revolution of the teaching model in education. This findings concerning the influence of RBL and EN, in particular, will have develop teaching model, eventually leading to innovative paradigm in practice and policy. Ii suggested that 5P model can effectively utilized for enhancing the mathematics learning outcomes and can present the new approach of development for the implementation in classroom and for the design of learning instruction.

Acknowledgement
This work was supported by the Unit of Excellence in Research Methodology of Innovations and Learning Sciences based-on Educational Neurosciences, University of Phayao Research Fund, Thailand.

References
[1] Anderson O R 2009 Neurocognitive theory and constructionism in science education: a review of neoribiologycal, cognitive and cultural perpective **Brunei Int. J. of Sci. & Math. Edu 1** pp 1–32
[2] Hardiman M 2012 The brain targeted teaching model for 21st century school (California: CORWIN A Sage Company)
[3] Institute for the Promotion of Teaching Science and Technology 2012 Science measurement and assessment (Bangkok: V. Print Co., Ltd)
[4] Srikoon S 2015 Development of educational neurosciences and research-based instructional model (Thesis of doctor philosophy in currirulum and instruction of gratuate school Khon Kaen University: Khon Kaen University)
[5] Sanit S, Bunterm T, Samranjai J and Wattanathorn J 2014 Research synthesis of research-based learning for education in Thailand **Procedia-Social Behavioral Sciences 116** pp 913–917
[6] Srikoon S, Bunterm T, Nethanomsak T and Ngang T K 2017 A comparative study of the effects of the neorocognitive-bases model an the conventional model on learners' attention, working memory and mood **Malaysain Journal of Learning & Instruction 14** pp 83–110
[7] Srikoon S, Bunterm T, Nethanomsak T and Ngang T K 2018 Effect of 5P model on academic achievement, crative thinking, and research characteristics **Kasetsart Journal of Social Sciences** pp 1–8
[8] Srikoon S, Viriyapong N and Chutiman N 2018 Examining construct valodity and measurement invarince of mood across gendre and grade **Journal of Education Khon Kaen University 41** (inpress)
[9] Zadina J N 2015 The emerging role of educational neuroscience in education reform **Psicología Educativa 21** pp 71–77

The development of geography teacher training curriculum at Thai Nguyen university of education in Vietnam to meet the requirements in the new context

Nguyen Thanh Mai* and Pham Huong Giang

Department of Geography, Thai Nguyen University of Education, 20 Luong Ngoc Quyen Street, Thai Nguyen City, Vietnam

Corresponding author's e-mail: nguyenthanhmai.1812@gmail.com

Abstract. The development of higher education curriculum plays an important role in ensuring the quality of human resource training to meet the requirements of socio-economic development. Curriculum development at Thai Nguyen University of Education in Vietnam, especially the Geography teacher training curriculum, which is regularly reviewed and changed to respond the needs of the labor market and fit in the new context nowadays.

1. Introduction

In recent years, the development of curriculum in Vietnamese universities has been of great interest in improving the content, modification and adaptation to suit the country's needs and circumstances, the trend of international and regional integration. Renovation of training curriculum plays an important role in improving the quality of human resources. Thai Nguyen University of Education is a focal point in the higher education system of the midland and mountainous provinces of northern Vietnam and is one of the leading universities in the country in developing the curriculum. In particular, the geography teacher training curriculum is a prime example of developing a curriculum that is positive in response to new circumstances.

2. Content

2.1. The concepts of curriculum and curriculum development

Based on the study of domestic and foreign documents of curriculum development, the author finds that there are many different approaches to the terminology of the curriculum. However, in general, a curriculum is the set of modules offered by the institution depending on fields of the trainees. Training curriculum is a total package of learning activities designed to achieve the objectives of the training program. In a competency-based system, the objective, or desired end, is that trainees will acquire the specific knowledge and skills (competencies) they need to do their jobs. (Judith S.Rycis, 1994). Referring to domestic documents, foreign documents and from the current practice, according to the author, the higher education curriculum should be understood as all the modules and activities built by the school to equip for students the knowledge, skills and attitudes appropriate to the chosen majors.

Similarly, the terms of curriculum development have different perspectives among authors in many countries around the world. In general, the concept of developing curriculum is defined as to be a continuous process of refining a curriculum that involves compiling, developing a new curriculum, or improving a current curriculum. In addition, we use the term development of curriculum instead of "building", "designing" or "compiling" the curriculum, since "development" involves the change, continuous replenishment. Development is a cycle where the end will be the starting point, resulting in a new and better curriculum. Other concepts mean only one process and the result stops when we have a new curriculum.

2.2. Overview of Teacher Training Curriculum in Geography at Thai Nguyen University of Education

The teacher training curriculum in geography to train learners have the ability to teach geography in high schools, secondary schools. In addition, trainees can work in research institutes, ministries, committees ... related to the field of Geography or teach at the Universities and Colleges.

The goal of the Teacher Training Curriculum in Geography:

- Moral qualities. To be loyal to the Fatherland and with the people, to love the job, to have the qualities of teachers. Generally, seeing clearly the responsibility to the nation in education and particularly in building the socio-economic development of the country.
- Knowledge. Understand and be able to analyze, explain phenomena, natural and socio-economic processes. Knowledge of science, methods of research in the field of geography. Understanding basic knowledge of science education, teaching theory, teaching methodology geography.
- Skills. Have teaching and learning skills. Organizing educational activities, teaching geography in school. Have research skills, the ability to organize activities in the field of geography.

Training time and amount of knowledge accumulated: Training time: 4 years; Total amount of accumulated knowledge 135 Credits, not including the content of Physical Education and Defense Education.

Graduation requirements: accumulate sufficient amounts of knowledge including compulsory modules and elective modules; have a physical education certificate; have national defense education certificate; Certificate of English A2 (European Framework).

Table 1. Knowledge and number of credits of modules in the curriculum

Knowledge	Number of credits
General education, in which:	24
- Compulsory units	22
- Elective units	2
Professional education, in which:	69
- Basic knowledge	16
+ Compulsory units	12
+ Elective units	4
- Specialized knowledge	53
+ Compulsory units	49
+ Elective units	4
Pedagogy	35
- Compulsory units	33
- Elective units	2
Thesis/Equivalent unit	7
TOTAL	135

Table 1. shows the curriculum has a clear, concrete, logical structure that is systematically designed to meet the standards of knowledge, skills, attitudes, and human resources. quality force for practical

and academic ability to improve; designed to meet the needs of students. The curriculum consists of modules that are designed from the general knowledge block, the sectoral knowledge base to the subject area. Prerequisites are the foundation for the next module. The parallel modules in the sectoral knowledge and industry disciplines are designed to be integrated. Each module has a decisive role in meeting the standards of the program, regularly adjusting the curriculum to reflect the country's educational reform, learner and human resources requirements of the labor market.

The teacher training curriculum in geography is based on the knowledge of class 10, 11 and 12, and is divided into specialized knowledge units:

- Some Common Issues (Geography with Career Orientation for Students and Using Maps)
- General geography (General natural geography and general socio-economic geography)
- World Geography (Some socio-economic issues in the world and regional and national geography)
- Vietnamese geography (Natural geography, population geography, economic geography, economic geography, and local geography).

2.3. Developing teacher training curriculum in geography at Thai Nguyen University of Education in the new context

Based on the draft general education curriculum of the Ministry of Education and Training issued in January 2018 and based on the reference of construction materials of some pedagogical universities in other countries: Taiwan, Korea, China, the United States, the Philippines, etc. The development of the curriculum has been carried out in two main ideas.

2.3.1. Changes in the curriculum. About process and development*:* Firstly, survey managers (recruiters), oganizing seminars for geography teachers of some high schools in Thai Nguyen province. Then, building capacity profiles of students in the social sciences and geography. After that, develop the knowledge module and set up the subject matrix to determine the credit for each subject and compares coursework with competencies that need to be formed to identify required subjects, elective subjects ... Develop framework curriculum and curriculum details. Hence, experts have been appraised and corrected according to comments or saved opinions of appraisal experts. Finally, develop syllabus.

Implementation steps

- Step 1. Survey social needs and labor market.
- Step 2. Develop the learning outcomes of the training curriculum.
- Step 3. Determine the structure, modules of knowledge, the amount of knowledge of the training curriculum and the suitability of the graduates.
- Step 4. Comparison with other curriculum at the same level at other training institutions.
- Step 5. Design the syllabus.
- Step 6. Get feedback.
- Step 7. Finalize the program on the basis of feedback.
- Step 8. Review the program, update regularly.

Result: Elimination of some modules: Population and reproductive health, Methodology for research and Geography, History of geography science, Thematic map, Feng shui geography, Practice teaching... Reason: These subjects are no longer relevant due to the heavy content.

Put some new modules into the curriculum such as

General course of the World's History

General course of the Vietnam's History; World Civilizations

Sea Islands of Vietnam

Vietnam in the process of globalization

Pedagogical communication, Geography curriculum development

Teaching methods in social sciences; teaching local geography
Practice geography pedagogy 123
Assessment in geography teaching; Creative activity
Teaching skills in school
Climate and life
Water resources and life on Earth
Teaching methods by topic.

Table 2. Topics in the general education curriculum corresponding to new modules in the curriculum

Topics in the general education curriculum	New modules in the curriculum	Number of credits
Climate Change	Global Climate Change	2
Urbanization	Urban Geography	2
Some issues of world tourism	Tourism Geography	3
Some issues in Southeast Asia	Geography of East Sea	2
Mekong River Commission	Sea Islands of Vietnam	2
Peaceful Co-operation in the East Sea	Water resources and life on Earth	2
Natural disasters and preventive measures	Natural Disaster	2
Regional development	Regional development	3
Industrial Revolution 4.0		
Sustainable development and green growth	Environment and Sustainable Development	2
Globalization, economic globalization and global security.	Vietnam in the process of globalization	2

Table 2 shows new modules are associated with academic subjects in the curriculum of geography in the new context. Reason: The new subjects included subjects with useful content to meet the requirements of developing the capacity of Geography teachers.

The curriculum is designed in the form of credit training including a system of general knowledge subjects, specialized knowledge subjects, and specialized subject areas. Each module has a larger number of subjects than the required number of courses or credits; Students may refer to the teacher or academic counselor to select the ones that suit them, to fulfill the requirements for a degree and to serve their future careers. Students can actively choose the required subjects in a semester, while selecting the elective subjects that students find necessary for the accumulation of knowledge as well as interest in the subject. In addition, students can choose and enroll with qualified instructors and teaching methods.

At the same time, the formation of knowledge, professional skills, soft skills and ethical training to meet the challenges of professional development, to improve the quality and effectiveness of training. The improvement and renewal of the curriculum is based on the determination of the content and level of knowledge, skills and moral integrity for students.

Course syllabus for innovation has met the requirements and orientation of curriculum change from content approach to capacity approach, help students develop competencies: scientific research, lifelong learning capacity, problem solving capacity, creative capacity; Enhancing Discussion and Group Exercises.

Especially for the detailed objectives of the subject, students have the autonomy in preparing the lesson, self-assessment of their ability to determine the goal to reach the 3 levels.

In the syllabus of innovation, the content of the subject is very clear from the content, the form of teaching methods, the self-study time test, the place of study, so students can plan learning personal. Therefore, in addition to being equipped with specialized knowledge needed to apply in teaching, while explaining the natural phenomena, socio-economic happening every day, students can also expand their knowledge. Further develop and promote the ability of creative thinking, cooperation capacity in solving practical problems.

The new curriculum helps shape and develop student competencies, including general competencies and specific competencies.

General competencies:

- Communicative competence helps learners to use the appropriate communication in all situations, especially in the process of teaching a broad communication environment.
- The self-learning ability of professional fostering helps learners to develop self-study and fostering plans for each period, seeking for scientific research and training, and effectively implementing programs and learning resources (books, newspapers, magazines, equipment) for learning and career development.
- The capacity for scientific research increases the probability of identifying problems or research questions that need to be answered, expressing the research problem as a topic.
- Collaborative capacity is the capacity for teamwork, collaborative work that requires learners to leverage the power of teamwork, sharing, and collective collaboration, despite the critical opinions.

Specific competencies such as

- The ability to apply theoretical knowledge to explain natural geographic phenomena. For example: Recognize and explain the scientific basis for natural phenomena that occur in reality: seasonal phenomena, climatic and weather phenomena, the development of terrestrial, the rules of natural phenomena.
- The ability to apply theoretical knowledge to explain socio-economic geographic issues: Recognize and explain the scientific basis for socio-economic geographic issues that occur in reality.
- The capacity to establish charts and maps: Draw charts and maps according to criteria (depending on the nature and requirements of each article or topic): Science (accuracy); Visual (clear, easy to read); Beauty (Fine, beautiful).
- The ability to analyze spatial attributes and spatial relationships between things, territorial phenomena and integrative and differential teaching in the field of geoscience: Recognize and explain natural and socio-economic phenomena in a scientifically sound manner within certain spatial territories. Form the concept and derive the rules that give rise to the basic development of natural phenomena.

Compared with the old program, the new program has made fundamental changes, moving from the development of the curriculum towards the content approach to the capacity approach. As a result, many subjects have been built closer to practical needs in high school.

In the era of the industrial revolution 4.0, which took place strongly not only in developed countries but also in developing countries, the application of the achievements of this revolution was also significant in the development of the curriculum through E-learning and the use of information technology in teaching.

2.3.2. To develop and develop new training curriculum in the integrated and inter-disciplinary manner. In the trend of international integration, besides reviewing, modifying, supplementing and updating the teacher training curriculum in geography at Thai Nguyen University of Education, the development of new training curricula also play an urgent role in the current situation. In developed countries and in some countries in Southeast Asia, teachers not only teach one subject, but also teach many subjects in the natural sciences or social sciences. Therefore, the development of this curricula in the direction of integration, inter-disciplines are increasingly focused.

Specifically, Thai Nguyen University of Education has developed a curriculum of training social science teachers (130 credits) to train high school teachers who teach all three subjects: Geography, History and Education citizen; The Teacher Training Curriculum in Geography and History (130 credits) teachers can teach Geography and History. Expected, the school will enroll students in school year 2018 - 2019 to help students have more employment opportunities, meet the needs of the labor market.

For these curricula, employment opportunities are expanding, for exampe: Professional staff of Education, Education and Training Departments; Researcher at the Institute of Social Sciences and Humanities; Officials of state management agencies, mass organizations and social organizations: Department of Propaganda and Education, Committee for Ethnic Minority Affairs, Party Committee for Historical Research, Department of Culture - Sports - Tourism, Department of Information and Communications Communications, Districts, Provincial Youth Union, Labor Confederation, Fatherland Front from central to local; Reporters and radio and television news editors from local to central; Museums and interpreters at historical and cultural museums, historic monuments, traditional houses; tour guide.

3. Conclusion

Developing the teacher training curriculum in geography at Thai Nguyen University of Education has met the objectives of the curriculum in the learning outcomes of each subject and each block of knowledge; Ensure that the subjects are complementary, complementary and closely related; Combining professional career development with skills training, developing common capabilities and specific capabilities. Through this activity, the quality of learners is improved, contributing to helping them after graduation to work effectively in educational institutions or to adapt well to other working environment.

References

[1] Department of Geography, Thai Nguyen University of Education 2018 *Report building curriculum in a new context* (Thai Nguyen)

[2] Doan Thi Minh Trinh and Nguyen Hoi Nghia 2012 *Design and develop training programs to meet the output standards* (Ho Chi Minh City National University Press) pp 56-62

[3] Judith S, Rycus Ph D, and MSW IHSI 1994 *handout for "Training of Trainers on Development Curriculum"* (Ohio Child Welfare Program - December 2006) pp 1

[4] Ministry of Education and Training 2018 *The draft general education curriculum Geography* (Ha noi) pp 57

[5] Nguyen Thanh Son 2015 *A development on higher education curriculum to meet the output standards* (Journal of Sciene Vol 5) chapter 1 (An Giang University) pp 50-54

[6] Pham Hong Quang 2013 *Develop teacher training programs - theoretical and practical issues* (Thai Nguyen University Press) pp 22-35

IOP Conf. Series: Journal of Physics: Conf. Series **1340** (2019) 012023 doi:10.1088/1742-6596/1340/1/012023

The teacher's competence of integrated teaching at primary school science

Le Trung Hieu[1] , Nguyen Danh Nam[2]*

[1]Tan Trao University, Vietnam

[2]Thai Nguyen University of Education, Vietnam

*Corresponding author's e-mail: danhnam.nguyen@dhsptn.edu.vn

Abstract. The paper presents some research results on primary teachers' competence on integrated teaching in Vietnam. We have conducted a survey with teachers and educational managers in the context of renewing general education curriculum and texbooks as well as new teacher's professional standards. The research results have shown that most of respondents agreed with our proposed competencies for integrating teaching and their basic components. Based on these components, we have described indicators for each capacity and suggested some measures to design continuous professional development for primary teachers so that they could meet the requirements of general education renovation.

1. Introduction

Integrated teaching is an orientation in education reform, a radical change from content-based teaching approach to competency-based approach in order to train dynamic and creative people who are capable of applying knowledge when solving problems in real life. Integration is an approach to teaching and learning based on a variety of fields, sectors and practice. It involves the acquisition of knowledge, skills, attitudes and values within or outside the subjects to develop a stronger understanding of crucial ideas. Integration occurs when components of the curriculum are linked and related in meaningful ways by students, teachers, specialists... accordingly, many of the areas or topics of the subject are linked together. Integration is not a new way to organize teaching at the primary school. Educators first explored the notion of integrated curriculum in the 1890s. Over the years, there have been many educational researchers, such as Susan Drake, Heidi Hayes Jacobs, James Beane and Gordon Vars, who have described different understandings of the integrated curriculum, referring to interwoven, connected, thematic, multi-disciplinary, related, linked and holistic curricula. Some educators, such as Robin Fogarty, go beyond a single definition of integrated curriculum and consider it as a continuum. How does the term "integration" show its relation and linking between subjects in the curriculum, especially the topics of the school program in general and primary school program in particular [1,6,13].

Integration in teaching is a trend of modern teaching approach, but now in Vietnam it is not widely applied in schools. Integrated teaching is the suitable orientation and a means by which students can develop deep knowledge structures that are connected with one another. However, there is no consensus on the definition of integrated teaching. Some researchers have argued that explicit definition is very difficult, so many terms are often used to indicate this coordination [2,8,3]. These

IOP Conf. Series: Journal of Physics: Conf. Series **1340** (2019) 012023 doi:10.1088/1742-6596/1340/1/012023

research have showed that integrated teaching helps students shape competencies including the ability to apply knowledge to solve problems, especially the use of knowledge in practice because problems arising in life. Business are rarely related to a certain field of knowledge but usually require the integrated application of knowledge in a number of different disciplines. This is to say that primary education must help students have an overview of the world in its inherent complexity, not divided into disciplines or sectors too early.

In order to have an integrated teaching result as expected, teachers are required to have suitable integrated teaching competencies. Therefore, in this research, we have explored and identified the characteristics of integrated teaching and some basic components of integrated teaching competence for primary teachers in Vietnam.

2. Literature Review

We examined the primary teacher competency framework of some countries in the world. This international experiences suggest us to propose teaching competencies for primary teachers in my country.

Columbia and British Primary School Teacher Profiles

Comlumbia primary school teacher profiles include three areas: knowledge, skills, values and attitudes. For the knowledge, they refer to curriculum, teaching content, pedagogical profession, theoretical teaching and updating knowledge of contemporary society. For the skills, they refer to practice, emotion, pedagogical skills, life skills, management skills, research and reflection skills, social skills, cooperation and teamwork skills, communication skills. For the values and attitudes, they refer to personal qualities and ethics, ethics and professional responsibility.

British primary school teacher profiles include two areas: teaching capacity; qualities, personal and professional ethics.

Australian Primary School Teacher Profiles

Australian primary school teacher profiles include the following basic components: Understanding of students and how they learn; understanding of the teaching content and how it is taught; plan and implement effective teaching activities; establish and operate a safe, friendly teaching environment; evaluate, give feedback and report the results of the evaluation of the learning process; practice and improve professional skills; promote effective relationships with colleagues, parents and other educational stakeholders.

Thus, in comparison with the primary teacher competency framework of some countries in the world, it is generally recognized that the capacity of primary teachers is focused on three main areas such as: Political qualities, ethics, lifestyle; knowledge; pedagogical skills.

Characteristics of Integrated Teaching at Primary Schools

Primary school is the first level of the education system that provides basic and life-long education for children who continue to study at the continuous levels, helping children to form initial foundations of personality. Thus, primary education has special pedagogical characteristics, with its own nature. In integrated teaching, it was found that teaching content is related to concepts in a way that is not bound by the boundaries of particular disciplines.

We can see more clearly about the characteristics of integrated learning through a number of studies by many authors describing its requirements. Brown describes integrated teaching in four main characteristics [2]: (i) the unity of all knowledge that integrated science has a holistic view of basic knowledge as a whole; (ii) the conceptual unity of science that make up the defined frameworks; (iii) the unified scientific process emphasizes the distinction between methodological and scientific disciplines; (iv) in integrated teaching, a problem can be solved in many areas of coordination.

In general, the purpose of integrated teaching at primary schools is to shape and develop students' abilities, enable the students to apply mathematical knowledge and related fields to solve real life problems. The core of competence is the ability of the subject to combine the knowledge, skills with attitudes, values, motives in a flexible way in order to meet the complex requirements of an activity and ensuring that the activity is successful in a certain teaching context.

3. Research Results and Discussion

The research aims at finding the basic competency components of teachers for integrated teaching at primary schools. This research has used cross-sectional surveys. The sample consisted of 80 primary school teachers, 20 educational administrators at some primary schools in 4 provinces in Vietnam. The research tool includes a structured questionnaire with 100 copies released and 97 copies collected with full responses. After conducting the survey, we have compiled the answers, calculated the percentage for each question in the questionnaire. From the data collected, we have analyzed and evaluated the necessary level of components of integrated teaching competence, then suggested the teacher's components in integrated teaching that should be trained at the teacher training institutions.

In order to identify the basic components of the integrated teaching competence of primary school teachers, as well as to explore the need for developing and fostering those components, we considered the following scientific bases: (i) a theoretical study of the teacher competence framework; (ii) international studies of integrated teaching competence; (iii) new school curriculum and teachers' professional standards.

In addition, we have conducted a survey of 97 teachers and educational managers at some primary schools. The research results have shown that teachers and educational managers are interested in the components of integrated teaching (see Table 1).

Table 1. The necessity of competency components for integrated teaching

Basic components	Very necessary (%)	Necessary (%)	Normal (%)	Unnecessary (%)
Capacity to define teaching objectives	52.58	32.99	11.34	3.09
Capacity to plan integrated teaching	48.45	39.18	9.28	3.09
Capacity to design integrated teaching lessons	63.92	35.05	1.03	0.00
Language competence in integrated teaching	46.39	48.45	5.15	0.00
Capacity to cooperate in integrated teaching	39.18	34.02	19.59	7.22
Capacity to design and create an integrated teaching environment	47.42	42.27	10.31	0.00
Capacity to connect lesson content and practice	45.36	46.39	8.25	0.00
Capacity to organize integrated teaching	52.58	43.30	4.12	0.00
Capacity to evaluate students' learning outcomes	51.55	39.18	8.25	1.03

Through the survey, more than 52% respondents said that the ability to determine the objectives of a lesson was very necessary. Identifying instructional goals for integrated teaching is important because it involves the contents, the objects, the objectives to be achieved, and finally the integrated learning outcomes of students. Only 11% of the respondents considered it as normal. This was because, when the research team distributed the questionnaires to elderly teachers in remote areas, they did not accept modern teaching methods. More specifically, they said that, due to retirement, there was not much motivation for them to change the form of teaching, however, they also acknowledged that the change in the direction of modernity is better. Meanwhile, only 3% of respondents did not care about the teaching objectives. Nearly 49% respondents said that integrated teaching planning capacity was very necessary. Only about 9% was normal and 3% unnecessary. Thus, we have found that teachers and educational managers appreciated the need for a fully integrated teaching plan.

From Table 1, it could be seen that most teachers and educational managers are very interested in the basic components (e.g., integrated teaching design; language competence in integrated teaching; cooperative capacity in integrated teaching; design and create an integrated teaching environment; connect lesson content and practice; organize integrated teaching; evaluate students' learning outcomes) and would like to have continuous professional development courses for fostering these competencies.

We have examined the teachers' need for developing and fostering the competency components for integrated teaching. Table 2 showed that most of the primary schools and teachers urgently had demands for teacher training so that they had the capacity for effective integrated teaching.

Table 2. Primary teacher's need for integrated teaching competencies

Contents	Very necessary (%)	Necessary (%)	Normal (%)	Unnecessary (%)
Capacity to define teaching objectives	44.33	30.93	21.65	3.09
Capacity to plan integrated teaching	39.18	36.08	20.62	4.12
Capacity to design integrated teaching lessons	76.29	22.68	1.03	0.00
Language competence in integrated teaching	42.27	48.45	9.28	0.00
Capacity to cooperate in integrated teaching	38.14	31.96	23.71	6.19
Capacity to design and create an integrated teaching environment	50.52	34.02	15.46	0.00
Capacity to connect lesson content and practice	48.45	46.39	4.12	1.03
Capacity to organize integrated teaching	49.48	43.30	7.22	0.00
Capacity to evaluate learning outcomes	51.55	39.18	9.28	0.00

Table 2. showed that the proportion of the need for the basic components at the level of very necessary and necessary was 70%. In particular, over 98% of teachers and educational managers considered it very important to train the capacity to design integrated teaching lessons. However, in some schools, where the socio-economy is still low, investments in education reform are still difficult, some teachers do not pay much attention to integrated teaching methods.

The teachers' need for capacity building in integrated teaching at primary schools is urgent. This is a task that contributes to improving the quality of primary education towards a modern, appropriate education in the 21st century.

3.1 Capacity to Define Teaching Objectives

Identifying the lesson objectives is based on the knowledge, skills and requirements of the program. This step is set by defining the goal of the lesson as an indispensable role in each lesson plan. The goal (requirement) is both the destination and the need for learning; in other words, it is the measurement for the result of teaching process. It helps teachers identify what tasks they will have to do (instruct students to learn, apply knowledge, skills, scope and extent, and then withdraw lessons to learn).

When defining instructional goals, teachers must point out the process in which students can find, identify, occupy, and develop knowledge, skills, and attitudes. That is, point out the way students think to achieve those results. For example, with the topic "measure the length of straight line" in Math Grade 1, the goal that teachers expect students to achieve is: Students have understandings of "shorter, longer", using the things that are around them as a tool to help with their measurement. Students can then measure length in practice (which can be done by hand, foot, arm, etc.). Thus, in order to set appropriate teaching objectives, teachers need to have the capacity to define teaching goals.

3.2 Capacity to Plan Integrated Teaching

Effective lesson planning requires time, dedication, and understanding of student abilities and goals. In primary education, every teacher tries to encourage students to acquire knowledge as much as possible during class and apply it. The teacher performs the lesson as stated in the plan to achieve the goal of the course. Measuring a successful lesson plan is often proven through the use of a board, homework, or test.

To plan a lesson is to know your students, understand who the teacher is going to educate and predict different students' learning styles (visual, auditory, tactile, or combination, etc.). Teachers adjust their lesson plans to incorporate all learning styles through independent and group exercises. If teachers are familiar with student group learning, teachers may choose to plan ahead to increase interaction. The lesson plan should be specific to the activities, the content of the learning, the role of the participants, etc. In order to have a fully integrated teaching plan, the teacher should have the capacity for the task.

3.3 Capacity to Design Integrated Teaching Lessons

The teacher's preparation for a teaching hour is often reflected in the preparation of the lesson plan. This is the activity of developing plan for a specific lesson, demonstrating the interaction between teacher with students, students with students in order to achieve the objectives of the lesson. Based on the syllabus, both teachers' professional qualifications and pedagogical skills can be assessed, reflecting their perceptions of educational issues such as educational objectives, how to use teaching methods, teaching aids, forms of teaching organization, and how to evaluate students' learning outcomes in relation to relatively stable factors like: plan, time, facilities and target students. Therefore, the preparation for one teaching hour has a decisive role and significance to the quality and efficiency of teaching hour.

For students to have good academic results, teachers need to have good lesson plan design, which is only possible when the teacher is competent to integrated teaching design.

3.4 Language Competence in Integrated Teaching

Language competence is a broad term that includes linguistic or grammatical ability, presentation capacity, social or cultural linguistics capacity and what can be termed literary competence. Language competence refers to language knowledge and the ability to use that knowledge to interpret and produce meaningful texts appropriate to the situation in which they are used. Language competence is best developed in the context of activities or tasks where language is used for practical purposes, in other words, in practical applications.

Taking a holistic approach means educating a child as a whole, and taking all aspects of development into consideration. It is to gather every aspect of the individual, emotional, cognitive, social, and linguistic development in an integrated way. Teachers with good language ability can help students to transform real world problems with mathematics and vice versa. Integrated teaching aims at applying many skills, especially STEM skills, so the language skills of teachers are very necessary.

3.5 Capacity to Cooperate in Integrated Teaching

John Dewey, a pragmatic American educator, was considered the first person to initiate a cooperative teaching trend in the early 1900s. Previously, the concept of education was the transmission of ideas and experience, or an enlightening process that helped people effectively use their knowledge. John Dewey always emphasized the role of education and considered education as a means of teaching people how to live cooperatively in a democratic society.

Some researchers argued that collaborative learning is a structured, systematic, instructional strategy that is implemented together in small groups to achieve common tasks. Collaborative learning is the whole learning activity that students perform together in groups, within or outside the classroom [4,6,12].

The learning environment should be designed to promote active learning and teaching, teaching in classroom, individuals, small groups, and whole group learning. Integrated learning is an example of a teaching arrangement that can be used to motivate students to learn positively, which is an important aspect of learning mathematics and is validated by mathematics educators and researchers. In integrated teaching activities, students may be given the task of discussing, solving problems, and undertaking collaborative learning activities that can be used to supplement instruction by providing students the opportunity to practice newly introduced skills or review skills and concepts. Teachers can use cooperative activities with other teachers or scientists to help them integrate contents that are not their strengths to help students make the connection between math content and other related fields carefully, reaching the required goal. Therefore, teachers need to have the capacity to cooperate in the integrated teaching process.

3.6 Capacity to Design and Create an Integrated Teaching Environment

Teachers need to have a clear idea of the topics they are integrating so that they can point out and explain the connection to learners and encourage them to link together. For example, while buying food we may ask for the price of something; buy it if we can afford it, but decide not to buy it because it does not fit the culture; therefore, language, mathematics and cultural studies skills are integrated and applied in this situation. In addition, after school, learning often takes place in a context and is associated with other languages that are part of our lives. For example, we could read about a special offer, using the internet, a phone or asking a friend to learn more about it, using what we know to compare it with other offers and make decision if we want to the offer. Teachers help students use different skills to learn and place education in context.

Although timetables and time distribution for each topic need to be included in the schedule in a flexible manner, teaching and learning at primary schools sometimes cannot follow a rigid schedule. For instance, it does not allow learner-centered approaches or integrate teaching in education. Therefore, designing and creating an environment for integrated learning is very important. In order to have an integrated teaching environment consistent with the lesson content, students' psychology and in accordance with general education objectives and subject objectives, teacher needs to be competent for this task.

3.7 Capacity to Connect Lesson Content and Practice

In integrated teaching, choosing meaningful connections between subjects will help students develop upon their diverse knowledge and experience on a topic. This provides the students with a comprehensive view of the world and makes learning more meaningful.

When teaching is organized in a holistic way, it is a better reflection of the real world and learning path at home and in the community. It puts learning in context and allows students to see the scene and make it easier to learn and memorize information. The integrated curriculum allows students to develop a unified view of the curriculum and extend their learning context beyond the single subject area. It also allows the transfer of learning from one area of knowledge to another. Through integrating the curriculum, teachers can plan the development of important skills and knowledge that transcend the content and subject of the course. Students may also work with many sources of information. Thus, connecting the content of the lesson with practice needs the role of teacher, who helps students approach problems with the knowledge of the subject.

3.8 Capacity to Organize Integrated Teaching

As mentioned above, the integrated teaching process includes the learning activities of students insideand outside classroom, at home and in the community. Classroom instruction is primarily for activities aimed at identifying issues, proposing solutions to problems, developing problem solving plans, presenting reports, and discussing the outcomes of the problem solving process. The process of learning and research outside the classroom should be monitored and supported regularly to ensure success and effectiveness. Therefore, the organizing integrated teaching topics need to strengthen the application of information technology to perform that function of the teacher.

3.9 Capacity to Evaluate Learning Outcomes

Assessment is a continuous process of teaching and learning to determine what learners can do in areas where they need improvement and whether learning objectives have been met. It is a tool that measures the knowledge and skills of learners and shows where they have weaknesses and strengths. Assessment is also important in giving feedback to parents and is used to change teaching methods, ensuring that all learners have the best chance of acquisition. Also, it is associated with self-evaluation of teachers for their teaching and the methods used. Thus, in order to assess the learning outcomes of students in the integrated learning process, teacher needs to be competent to perform this function.

Based on the above research results, we have also found and determined that the primary school teacher competencies for integrated teaching include two crucial following areas: (i) *professional knowedge and understanding*; (ii) *professional skills and practice*.

For professional knowledge and understanding, it includes the knowledge to teach different ages, the different stages and appropriate contents. It is also necessary to understand how students learn and how they can be taught effectively in the major learning areas. The basis for all competency standards is the knowledge of the primary education program. Table 3 below describes in details minimum requirements and indicators for teachers' professional knowledge and understanding.

Table 3. Indicators for professional knowledge and understanding

Minimum requirements	Indicators
Qualification A1: Know how students learn	
A1.1. Demonstrate an understanding of how students learn related to their ages.	- Give examples of student perceptions, psychological, social, emotional and moral development that can affect their learning. - Prepare integrated learning activities that are consistent with the student's level of cognitive, linguistic, social, emotional and physical development.

IOP Conf. Series: Journal of Physics: Conf. Series **1340** (2019) 012023 doi:10.1088/1742-6596/1340/1/012023

Table 3 (Cont')

Minimum requirements	Indicators
A1.2. Demonstrate an understanding of the different ways in which integrated teaching methods can meet the individual learning needs of students.	- Identify different teaching methods to help different students (gender, ethnicity, culture) have a better and specific learning needs. - Prepare integrated learning activities, explain and support students to link new concepts with their prior knowledge and experience.
Qualification A2: Know how to use information technology for integrated teaching	
A2.1. Demonstrate an understanding of a wide range of teaching strategies, and resources.	- Make plan and provide opportunities for learning;student interactions, investigations, problem solving, and creativity. - Use teaching methods, strategies, and materials as required in textbooks and additional low-cost materials to support students' learning.
A2.2. Demonstrate appropriate use of information and communication technologies (ICT) in integrated teaching and learning.	- Describe the strategy to support responsible, safe and relevant use of information technology for integrated teaching and learning, including at school or personal use.
Qualification A3: Know how to communicate well with students and their families	
A3.1. Demonstrate understanding of the role, and expected tasks of primary school teachers.	- Describe the role and duties of teachers in accordance with the standards of the Ministry of Education and Training, accepted by society.
A3.2. Demonstrate an understanding of the social, linguistic and cultural diversity of students and their communities.	- Identify the social culture of students, parents, community elders, and educational administrators when interacting with them during integrated teaching. - Give examples of strategies to assist all students and participate in integrated learning activitiesin classroom.
Qualification A4: Know about the curriculum	
A4.1. Demonstrate an understanding of the structure, content and expected outcomes of primary education programs.	- Describe the key concepts, content, learning objectives, and outcomes of the core curriculum for target learners. - Prepare lesson plans that reflect the requirements of the curriculum and include relevant integrated teaching and learning materials and activities. - Describe the assessment principles that underpin the core curriculum.

Table 3 (Cont')

Minimum requirements	Indicators
Qualification A5: Know the content integrated topics	
A5. 1. Demonstrate understanding of issues to be taught in integrated subjects in classroom.	- Describe key concepts, skills, techniques and applications for the subjects mentioned in classes. - Give correct and appropriate information in each lesson, examples and exercises to assist students in learning core content and developing skills. - Describe the methods used to promote integrated learning in the core areas: mathematics, science, social studies for classroom and link to the real life.
A5. 2. Demonstrate understanding of how content is conveyed depending on the student's response, needs, and learning context.	- Describe how to integrate the learning context in terms of age, language, ability, and culture for students to develop understanding of issues related to principles, ideas, and concepts. - Explain how lessons are contextual and give examples relate to content, concepts and topics. - Describe the approach to develop student fluency in reading and doing mathematics.

For professional skills and practice, we described what teachers can do, the knowledge and expertise of teacher, teaching strategies for different educational contexts to meet the needs of each student. Table 4 below describes in details minimum requirements and indicators for teachers' professional skills and practice.

Table 4. Indicators for professional skills and practice

Minimum requirements	Indicators
Qualification B1: Coordinate various approaches in integrated curriculum	
B1.1. Demonstrate the ability to teach programs related to concept and topic content.	- Explain clearly the content of the curriculum and the expected learning outcomes. - Choose the guide to link learning with previous knowledge of students, concerns, daily life and local needs. - Help students be aware of their ideas to build new understanding.
B1.2. Ability to apply different strategies in integrated teaching and learning.	- Use appropriate strategies for integrated teaching and learning in classroom. - Use mathematics to support students learning in different areas of expertise. - Provide opportunities for students to investigate into subjects of related content and concepts through practical activities.

Table 4. (Cont')

Minimum requirements	Indicators
B1.3. Demonstrate a well-integrated lesson plan and preparation that fits the students' learning and experience.	- Plan and structure the lessons to ensure that all lessons are used effectively. - Provide and introduce lessons to link new learning content with previous knowledge, to attract students' interest and encourage them to learn. - Prepare teaching experience, continually integrate learning areas and be enthusiastic with the interests and experienceof students. - Use techniques and interrogative examples to introduce and illustrate acquired concepts.
Qualification B2: Assessing, monitoring and reporting students' learning results	
B2.1. Ability to keep track of students' progress and evaluate their learning results.	- Plan and use tasks to assess, integrate with learning activities. - Use assessment to enable students to demonstrate achievement in a variety of ways. - Use questions and discusstechniques to test students, and provide feedback.
B2.2. Ability to record, evaluate and use assessment information to guide the learning process of students.	- Record students' progress in a consistent and accurate way. - Use different validations to track student progress and report the next plan of the curriculum. - Talk about the learning progress and achievement to students, parents, and other educators.
Qualification B3: Create a supportive learning environment for students	
B3.1. Ability to create a safe and effective learning environment for all students.	- Use classroom space, materials and resources to ensure the participation of all students in integrated learning. - Encourage students to interact with each other and work independently and in groups.
B3.2. Demonstrate strategies for managing student behavior.	- Encourage students to interact with each other with respect and safety. - Learn about the needs of each student and interact regularly with all students. - Encourage the well-adjusted behaviors of students by working in collaborative and independent learning groups.
Qualification B4: Work with teachers, parents and the community	
B4.1. Demonstrate strategies to work with other teachers, parents and local communities to improve the integrated learning environment for students.	- Talk positively to parents about academic culture and curriculum to promote their understandings. - Describe strategies to encourage parents to engage in their children's learning at school, at home and in the community. - Consider colleagues' perspectives in an effort to cope with learning problems and to accept feedback positively.

4. Conclusion

Based on the research results, we can conclude that the proposed components for teachers at primary schools are in line with the orientation of Vietnam's general education renovation and the goal of evaluating primary teachers' competencies in the 21st century. These teachers' competencies and their basic components for integrated teaching would provide a teacher competency profiles for training curriculum framework as well as for continuous professional development curriculum framework at the teacher training universities and colleges. Additionally, the research results have also confirmed that primary teachers need to update and foster new knowledge and skills for meeting the demands of general education reform in Vietnam. In this research, we have also identifed professional knowedge, professional skills and practice for primary teachers so that they could develop their professional competencies on their own and even by participating some online courses.

References

[1] Blum A. Towards a rationale for integrated science teaching. In P. E. Richmond (Ed.), *New trends in integrated science teaching*, 2, 29-51. Paris: UNESCO, 1973.

[2] Brown S. A. A review of the meanings of, and arguments for, integrated science. *Studies in Science Education*, 4, 31-62, 1977.

[3] Frykholm J., Glasson G. Connecting science and mathematics instruction: Pedagogical context knowledge for teachers. *School Science and Mathematics*, 105 (3), 2005.

[4] Johnson, D. W., & Johnson, R.T. *Cooperation and competition: Theory and research.* Interaction Book Company, 1989.

[5] Hieu L. T., Nghia D. D. Integrating mathematics and science in Vietnamese primary schools, *Journal of Education Management*, 8 (8), 2016.

[6] Huntley M. A. Design and implementation of a framework for defining integrated mathematics and science education, *School Science and Mathematics*. Hoboken, 1998.

[7] Kıray S.A., Gök B., Çalışkan İ., Kaptan F. *Perceptions of science and mathematics teachers about the relations between what courses for qualified science mathematics education in elementary schools*, 2008.

[8] Lonning R. A., DeFranco T. C. Integration of science and mathematics: A theoretical model. *School Science and Mathematics*, 97(4), 212-215, 1997.

[9] Maria Åström. Defining integrated science education and putting it to test, *Studies in Science and Technology Education*, 26, Sweden, 2008.

[10] Mark David Otarigho, Djagbo Daniel Oruese. Problems and prospects of teaching integrated science in secondary schools in Warri, Delta State, Nigeria. *Techno LEARN*, 3(1), 19-26. New Delhi Publishers, 2013.

[11] Ministry of Education and Training. *Training materials for integrated teaching at junior and senior high school.* Hanoi University of Education Publishing House, 2014.

[12] Nam N. D. Modelling in Vietnamese school mathematics, *International Journal of Learning, Teaching and Educational Research*, 15(6), 2016.

[13] Zollman, A. Learning for STEM literacy: STEM literacy for learning, *School Science and Mathematics*, 112(1), 2012.

IOP Conf. Series: Journal of Physics: Conf. Series **1340** (2019) 012024 doi:10.1088/1742-6596/1340/1/012024

Introducing the process of developing open education program for universities of information technology in Vietnam

Cao Thi Hao*, Nguyen Thi Bich, Nguyen Thi Hai Anh

Thai Nguyen University, Vietnam

corresponding author's e-mail address: caohaokv@gmail.com

Abstract. The great achievements made in every field so far have given educators many challenges. In Vietnam, following the guidelines of fundamental and comprehensive education reform, the development of "open education" program is an urgent requirement contributing to the successful implementation of the education reform. The program will provide access to education for all people in need, regardless of age, education level, occupation, social status, gender, and can be implemented at all times and places. In the 4.0 era, the role of IT universities has been well defined. In order to maximize their role, they need to develop an open program. To establish and develop an open education program for Universities of Information Technology, the article proposes a process of eight steps. Step 1: analyzing context and assessing education needs (international context, national, and university context). Step 2: identifying six criteria and four objectives (knowledge, skills, ethics, capability) when developing open programs. Targets must be specific, detailed, easy to review and evaluate. Step 3: reviewing the current program, and designing new program contents. Step 4: selecting the form of training organization. Step 5: organizing examinations and accreditation of education quality, ensuring accuracy, objectivity, seriousness and flexibility in accordance with the training methods, subjects and contents. Step 6: verifying the training programs. Step 7: implementing the program (including two steps: Pilot implementation, evaluation and adjustment; mass implementation applied throughout the university). Step 8: evaluating the program (after implementation, the university collects, analyzes, and processes data to determine the value of the program). With the above process, the open education program will enhance the quality of education, in line with the trend of higher education in the world today. By attaching training goals to real life, to the development of learners' capacity, IT universities will affirm their "brand", their advanced and active role in the digital era, and the globalization of education today.

1. Introduction

The great achievements made in every field have given educators many challenges. In line with the trend of popularization, marketization, diversification and globalization, Vietnam's education is also changing from a knowledge-based to a capacity-based education. The outcomes of the new education must be capable of acting with the highest level of four competencies: professional competence, methodological competence, social competence, and individual competence corresponding to the four pillars of education identified by UNESCO: "learn to know, learn to do, learn to live together, and learn to be." The Party and State of Vietnam have paid great attention to education and proposed the

IOP Conf. Series: Journal of Physics: Conf. Series **1340** (2019) 012024 doi:10.1088/1742-6596/1340/1/012024

guideline: "Reform the education system in an open, flexible and interdisciplinary manner between different levels and modes of education; Standardize and modernize the education and training" (Central Resolution 8, Session XI). Hereby, the Party has set the important task of "renovating the program to develop the capacity and qualities of learners, as well as virtue, mind, body and manner; teaching people, knowledge and profession"[5]. Therefore, developing an open program is an urgent task to contribute to the success of education reform. In order to obtain achievements in education reform in Vietnam, aiming at quality development for information technology universities, we need to consider the process of developing an open educational program for these schools.

2. Content

2.1. The concept of open education

Open education is a concept that refers to one type of education or the nature of an education. In the world, the concept of "open curriculum" was initiated in the early 20th century and developed strongly in the seventies of this century. Open education, however, has made a real breakthrough in the last 10 years since when the MIT Institute published its "open" learning materials online. The general understanding of open education is "Education in which unnecessary barriers on the path to education are broken down. Learners have reasonable opportunities to succeed in an educational system that is diverse in learning scopes and meets different needs of learners" [1]. Thus, developing an open curriculum is a prerequisite of the flexible education system. (Because the open curriculum will link the components together such as program contents, methods, forms of training, time, teaching and learning space, subjects, educational environment, etc). This will provide access to education for those in need, regardless of age, education level, occupation, social status, sex at all times and places.

In the 4.0 era, the role of IT universities has been confirmed. In order to maximize their role, universities need to establish and develop an open curriculum. "Curriculum development is the whole process of developing an educational program for a certain period of time: from reviewing the need for innovation to undertaking basic research; from proposing directions for the innovation (objectives, contents, methods of teaching, testing and evaluation) to the implementation"[2]. Thus, open education creates conditions for "teachers, scientists, administrators, parents and students to access, use, exchange, comment, supplement and finalize the program better and better" [6]. Here, we mention the development of an open educational program for IT universities towards the formation and development of quality and capacity for learners. This is reflected in the design, compilation, modification and adjustment of the curriculum in order to renovate and improve the quality of the program, in line with the context and educational needs of the country and the world.

2.2. Following process

To build and develop an open program, IT universities can follow the following process:

2.2.1. Step 1: Context analysis, educational needs assessment. Step 1: Context analysis, educational needs assessment. First of all, it is necessary to analyze the international context. Today, the Industrial Revolution 4.0 has been creating not only new opportunities but also many challenges faced for mankind. With the "unprecedented breakthrough" and "exponential rather than linear speed progress" (Klaus Schwab, Executive Chairman of the World Economic Forum), it is breaking down most industries in every country, including Vietnam. It can be said that the magnitude of these changes has signaled the transformation of the entire production, management and administration system. The Deputy Prime Minister Vu Duc Dam (at the High-Level Policy Dialogue on Human Resource Development in the Digital Age within the framework of the Second APEC Senior Officials Meeting in Hanoi) said: "The Fourth Industrial Revolution and the digital age not only help increase productivity, economic growth but also open up a human-to-human connection". As a result, it is imperative to learn about the international experience of developing open educational programs at IT universities in developed countries in the region and around the world. When studying international experience, curriculum development projects we should focus on such issues as higher education regulations, common points in the education system of the IT universities, the cycle of innovation, the goal of higher education, the structure of the program, the program implementation process and the cost of higher education. When studying the domestic curricula, the research team needs to explore the pros and cons in the current curricula of the same field in terms of theoretical and practical basis; then, studying the orientation of education reform, the point of view in developing the open program and other fields of education and training. Based on this, each university will develop an open educational program that is consistent with the conditions and characteristics of their establishments.

Next, the program development team needs to analyze the narrow context - the context of the university - to develop the program accordingly. The research team needs to collect general information about:

- The needs for socio-economic development, geographical characteristics, population and cultural identity of the area where the university is located.
- The educational management viewpoint of the university: What specific training goals does the university have to meet the educational needs of the country and the region?
- Conditions and facilities of universities: State funding sources, their own revenues, sources of funding for the curriculum development, infrastructure (classrooms, teaching facilities, other conditions for teaching and learning activities ...).
- Staff involved in teaching, management and scientific research: Do survey and provide specific data on the number of lecturers involved in the program: male/female; minority/religion; qualification; age; job skills; years of experience; achievements. The following sample can be used:

Table 1. Information checklist (for staff, lecturers)

No.	Full name	Age	Gender	Minority/ religion	Qualification/ major	Years of experience	research and teaching results
1							
2							

- Students / trainees: The educators conduct surveys and give specific figures on the number of students / trainees; minority/religion; age; hobby; career orientation; achievements; then, generalize the characteristics of learners. The following sample can be used:

Table 2. Information checklist (for students, trainees)

No.	Full name	Age	Gender	Minority/ religion	Career orientation	Major	Learning results
1							
2							

- The rank of the university in the university system of the country, the region, and the world: Observe, collect opinions, and make statements about the role of the university in the development of the country.

For example, when researching and analyzing narrow context for the curriculum development of the Information and Communication Technology University - Thai Nguyen University - the research team had to collected necessary data as basis. The University of Information and Communication Technology was established under Decision No. 468 / QD-TTg dated on 30th October 2011 of the Prime Minister. The mission of the university is "a human resources training center with undergraduate and postgraduate degrees; scientific research and technology transfer in the field of information and communication technology for the cause of industrialization and modernization of the country". About the organizational structure, the university has: Board of Directors (Principal, 2 deputy principal); 10 functional departments; 6 faculty of specialization; 2 divisions; 8 centers/institutes. In addition, there are organizations of Party Committee, Labor Union, Youth Union, Student Association, Veterans Association. The school has a total area of 86,795 m2, 54 classrooms (33 lecture halls and 21 labs) with a total area of 10,044 m2. There are 681 sets of computers, of which 485 are for practice /traineeship; 196 sets for the office work, test data management.

The hostel has 11 three-storey buildings with a total of 168 rooms to accommodate 1,344 students. In addition, the school has a stadium for sports activities. The area of the library is over 1000m2; the library has a reading room with 500 seats. Total number of books: 1,455 with 20,893 copies. Digital documents: 4,862; 6,000 files of exogenous resource, with 29 collections. Number of electronic data visits: 279,000 / year. At present, the university has undergraduate 17 majors, 2 master's degree majors and 1 doctoral degree major. About the staff (up to March 2016): the total number of full-time lecturers: 337, of which 23 lecturers (7.4%) have doctoral degree or above; 215 (71.2%) have master's degree. On such basis, the open educational program of the university will be develop to be "open", flexible, facilitating cross-vertical training, supporting learners of all ages at different times and places.

Context analysis to build an open curriculum is essential to ensure that the curriculum is feasible.

2.2.2. Step 2: Define the goals and objectives for the development of the open curriculum. First of all, it is necessary to define criteria to ensure that the training program is "open". In our opinion, the "open" education program of the IT University must ensure the following criteria:

Criterion 1: The university's training program is developed according to current regulations promulgated by the Ministry of Education and Training; with reference to the training programs of prestigious universities in the country and in the world; with the participation of professional scientists, lecturers, managers, representatives of social-professional organizations, employers and graduates.

Criterion 2: The training program has clear and specific objectives, logical and systematic structure, meeting the knowledge and skills requirements of undergraduate training and flexibly responding to labor market needs.

Criterion 3: The regular training programs are designed according to prescribed regulations, ensuring the quality of training.

Criterion 4: The training program is periodically supplemented and adjusted on the basis of reference to international advanced programs, feedback from employers, graduates, organizations, other educational and social organizations in order to meet the demand for human resources for socioeconomic development of the locality, the country and the world.

Criterion 5: The training program is designed to ensure inter-linkage with other training levels and programs.

Criteria 6: The training program is periodically assessed and implemented quality improvement based on evaluation results.

On that basis, the curriculum development team needs to identify the following objectives:

- Knowledge: equipping learners with general education knowledge; basic knowledge of the sector; specialized knowledge;
- Skills: develop soft skills and career skills for learners to meet the diverse needs of jobs;
- Moral quality: the training program aims to promote the noble qualities: honesty, courage, self confidence, etc; be responsible, enthusiastic, serious in work; respect the law, actively participate in community activities to contribute to building strong community;

- Competencies: the program also aims to develop competencies for learners to meet their corresponding positions.

Therefore, it is necessary to base on the results of context analysis and the needs assessment of learners to determine the objectives of the curriculum development. Targets must be specific, detailed, easy to review and evaluate. In this section, educators need to identify common goals for the whole curriculum and the individual goals for each module.

2.2.3. Step 3: Review existing programs, and design new program content. When choosing and arranging contents, educators should consider such issues as the value, significance, and the suitability of those contents with the social reality. The selection of program content is often based on the basic criteria: Significance (content must be meaningful to the needs and interests of learners and the society); Utility (content must be really useful in the life of learners); Effectiveness (content must be accurate and timely); Appropriateness (content must be consistent with the level of cognitive and psychological development of learners); Feasibility (content must be appropriate with the actual context of the educational environment, socio-economic conditions of the country).

Thus, to design the content of the open educational program, first of all, it is necessary to thoroughly study the current training program. Thereby, the Universities of Information Technology shall develop a highly practical training program that meets the requirements of labor recruitment; open more disciplines and specializations to meet learners' needs. The training program shall be designed to ensure deep profession, also have reasonable interlinking, increase the number of optional modules for all levels of students. The training program is developed in the direction of application, increasing the practice time, linking training at the university with training at enterprises; developing a number of high-quality programs, advanced training programs; applying some training programs of prestigious international universities so as to improve and upgrade their own programs towards the international trends. These universities also need to develop highly qualified degrees, increase the number of graduate students and postgraduates; develop vocational training certificates to serve training needs of the society.

2.2.4. Step 4: Training implementation. The Universities of Information Technology (IT) need to develop a modern, friendly management and training process to become an open training service that meets various levels (PhD, Master, Bachelor) inside and outside the universities. Establishments have decentralized management functions, clearly defined roles and responsibilities of functional departments, faculties and departments throughout the system. Universities should have a software system to manage the online training, decentralization from the management board to departments, sectors, divisions, lecturers, students and modern resource services system (library, warehouse data). Training institutions should prioritize investment in upgrading equipment for training in practice rooms; improving the organization of practical learning to ensure practical efficiency; enhancing business cooperation, allowing enterprises to combine the training of a number of modules in the optional modules, replacement modules, graduation thesis; assign professional units to actively implement the process of training and evaluating students, enhancing the activeness and creativeness in studying and research for learners.

In addition, the organization of teaching methods should be flexible according to the requirements of each module such as inside classes, outside classes (clubs, competitions, seminars, exchanges, etc); interactive classes (teachers and students exchange online via the internet); dynamic classes (enhance the practical content); friendly classes (highly emphasizes democracy in the teacher-student relationship), open classes (learning anytime, anywhere through computers, telephones, televisions), etc.

With such an orientation, the open educational program will help learners to actively explore, exploit and accumulate open learning resources; Form learners professional competence, methodological competence, social competence, individual competence to prove and assert themselves in life and work.

2.2.5 Step 5: Organize examinations and accreditation on educational quality. The program aims at building a modern system of examination and evaluation to ensure accuracy, objectivity, seriousness and flexibility in accordance with the training methods, subjects and contents; establish evaluation centers in each university. Universities and training programs are accredited and meet the requirements of the Ministry of Education and Training; Standardize the stages of issuing, implementing documents, organizational structure, management process, facilities and human resources. To achieve this, each university needs to improve the evaluation methods suitable with the training programs, learners and reality. The quality assessment must be done right in the learning process, attaching importance to the dynamic and creative elements of learners; develop a modern evaluation technology system to increase efficiency, objectivity, comprehensiveness and accuracy; increase skill evaluation criteria rather than knowledge. The system of criteria and procedures for quality assurance should meet the objectives of modernization of each university

2.2.6 Step 6: Program verification. Training programs for the development and adjustment should be assessed. The appraisal committee includes reputable scientists. They are experts from the Institute of Information Technology - Vietnam Academy of Science and Technology, Institute of Information Technology - Hanoi National University, Institute of Technology, Post and Telecommunications, Academy of Management Education. Employers and alumni, etc. The program appraisal focuses on key issues such as objectives, curriculum content, methods and forms of training, examinations and accreditation of education and facilities. Then, the minutes of the appraisal meetings are recorded. The minutes can be written as the figure 1.

Annually, the Information Technology Universities should conduct surveys on learners, graduates, employers and other stakeholders on training programs and training quality. Based on the analysis and evaluation of the survey results, universities shall adjust the content of the curriculum, content of the subjects to meet the needs of learners and social reality.

2.2.7 Step 7: Program implementation. After the verification, the program will be implemented in two steps: 1) Piloting, then checking, evaluating, adjusting; 2) large-scale application.

2.2.8 Step 8: Program assessment. After the implementation of the program, the university will collect, analyze and process data to determine program values (whether the program meets the educational objectives and the practical needs; the process of program implementation creates outcome that fit the goal or not; advantages and shortcomings of the program; the impact of the program on learners). Review the program as follows: 1) Preparation for the assessment (define the purpose and scope of the evaluation, select evaluators, the aspects to be evaluated); 2) Developing an evaluation plan (objectives, data to be collected, method and form of assessment, time); 3) Data collection (through various forms and methods such as observation, test, study and research products, feedback, etc); 4) Data analysis and processing (classification in multiple levels with clear criteria); 5) Verifying the results (confirm test results, compare with given objectives, comments and recommendations).

IOP Conf. Series: Journal of Physics: Conf. Series **1340** (2019) 012024 doi:10.1088/1742-6596/1340/1/012024

MINISTRY OF EDUCATION AND TRAINING SOCIALIST REPUBLIC OF VIETNAM

UNIVERSITY ………….. Independence - Freedom - Happiness

MINUTES OF TRAINING PROGRAM VERIFICATION

At the ... day ... month ... year ... the university verification council ... has met

I / Composition of the Verification Council

1……………………………………………………………………………………………..

2……………………………………………………………………………………………

II. Verification contents

1. The university report summarizes the progress of the development of the open education program

2. Opinions of members on the program:

-Objectives:

- Contents:

- Training organization:

- Examinations and quality accreditation:

- Infrastructure conditions:

3. Answers from the Council

4. The council opens separate discussion

5. Results of the evaluation by votes

- Number of votes issued:
- Number of votes collected:
- Valid votes:
- Invalid votes:
- Number of votes required:
- Number of failing votes:

6. Conclusion of the Council

The meeting ended at: ... hours ... day ... month ... year ...

Secretary of the Council Presidents of the Council

(sign and full name) (sign and write full name)

Figure 1. Ministry of education and training socialist republic of vietnam University.

IOP Conf. Series: Journal of Physics: Conf. Series **1340** (2019) 012024 doi:10.1088/1742-6596/1340/1/012024

3. Conclusion

In the current context of education reform, the development of open curriculum is essential for the university system in general and the universities of information technology in particular. Depending on the actual situation of each university, the program may be developed in accordance with different "open" levels. A dynamic, flexible program will improve the quality of education, consistent with the trend of higher education in the world today. By attaching the training objectives to life, to the development of learners' competencies, IT Universities will affirm their "brand", advanced and positive role in the digital age, and the trend of educational globalization. The open curriculum of IT universities creates learning opportunities for all people of all ages at any times and places, for information technology always offer the hidden challenges that raise aspiration and passion to all people.

References

[1] Cao Van Phuong 2015 *Journey to open education: Open to learn - Learn to open- To become a responsible person* (Literature Publishing House Hanoi)
[2] Ministry of Education and Training 2013 *Training materials for pilot curriculum development of education program at school* (Education Publishing House Hanoi)
[3] Ministry of Education and Training 2014 *International experience in curriculum development* (workshop materials Hanoi)
[4] Nguyen Ky 2007 *Open Educational Model - Learning Society Model* (Education Publishing House Hanoi)
[5] Resolution No 29-NQ / TW dated on 4th November 2013 *Central Resolution 8, Session XI on fundamental and comprehensive education and training reform* (http://baochinhphu.vn/Tin-noi-bat/Ban-hanh-Nghi-quyet-ve-doi-moi-can-ban-toan-dien-giao-duc/184826.vgp.)
[6] N d 2018 (from Website http://vietnamnet.vn/vn/tuanvietnam/giao-duc-viet-nam-mo-nhu-the-nao 106454.html March 31st 2018)

Developing critical thinking of students through STEM educational orientation program in Vietnam

Nguyen Quang Linh [1], Nguyen Mau Duc[1], and Chokchai Yuenyong[2]*

Thai Nguyen University of Education, Thai Nguyen, Veitnam, email: nguyenquanglinh@dhsptn.edu.vn
Faculty of Education, Khon Kaen University, Khon Kaen, Thailand

*Corresponding author's e-mail address: ychok@kku.ac.th

Abstract. One of the requirements in teaching process is to develop the thinking of learners. Thus, how an educator could organize teaching activities in order to enable the critical thinking ability of students in high school? Based on the stages of the solving process in the learning process of the student, the expression of critical thinking in each stage, the author would like to suggest educational orientation program following STEM system (STEM activities) in order to develop critical thinking of students. This article analyses the impact of using STEM activities on the development of students in some high schools in mountainous regions in the northern part of Vietnam. The results show the development of critical thinking of students through STEM activities and some superior suggestions for using STEM activities to motivate critical thinking of students in high school.

Keywords: STEM education, Critical thinking, Physics teaching

1. Introduction

Critical thinking is one of the important skills that each individual in the 21st century should know in order to work and live effectively in the global society. Critical thinking is a high-level skill, not only analyses, identifies and evaluates information or arguments from different sources but also thinks of private thought. It is complicated to accomplish; however, it is not impossible in teaching and one of the most effective ways is to take students into problem-solving activities [1]. In the world, many authors have been studied about critical thinking of students. However, there are not any specific studies concerned critical thinking of students in high school through STEM educational orientation program [7].

The new curriculum in Vietnam (July, 2017) indicates that the importance and enhancement of STEM activities is a radical innovation of the new curriculum. In particular, STEM educational orientation program is known as an effective solution in the development of critical thinking, which every high school with a new curriculum is aiming for. Based on STEM activities, students not only perceive scientific knowledge but also improve necessary abilities, creative thinking, critical thinking, and solving abilities [6]. STEM activities help the school incorporate academic knowledge in books with life and society; help students balance physical and mental, develop thinking, especially critical thinking.

On the other hand, the demand for jobs related to the STEM field is increasing and skills of labour in the 21st century are required, and STEM program can support for necessary skills. Therefore, applying for STEM educational orientation program in teaching in high schools can bring effective results and meet the requirements of society.

The issue is how to create STEM educational orientation program so that students can develop critical thinking. Aspects of critical thinking of studens during STEM educational orientation program were analyzed. Definitions as to what Critical thinking is vary but are often rather involved, complex and difficult to understand. Perhaps a more useable de finition for current affairs classes might be Fisher and Scriven's "Critical thinking is skilled, and active interpretation and evaluation of observations and communications, information and argumentation" [2].

According to Lipman (2003) , "Critical thinking is involved in all responsible interpretation (the production of meaning), and all responsible translation (the preservation of meaning)". These views have been summarized in "Developing Thinking Developing Learning" [5]. According to the study of the aspects of critical thinking by Luyen (2009) and Lipman (2003), and the study of problem-solving characteristics, we have identified the basic expressions of critical thinking of students during STEM activities as below:

(1) Identifying new matters;
(2) Giving a reliable prediction;
(3) Suggesting alternative options;
(4) Successfully executing chosen options;
(5) Improving the chosen options;
(6) Selecting the optimal option;
(7) Applying knowledge to tackle problems in every situation.

Before each problem encountered, the learner will consider how to find the reasonable and unreasonable points, which means critical thinking. STEM activities are built in order to create more opportunities for students when solving problems.

2. Methodology

The research methodology of this study is qualitative research. Based on targets of development thinking of students (critical thinking), researchers figured out some initial observations: STEM activities can improve critical thinking skill of learners. According to this result, researchers built a survey to find out the difficulties as well as actual demands of students about STEM educational orientation program, built a system consisting of 6 STEM activities and accomplished this system to improve critical thinking for students. The effect of this impact will be measured based on critical thinking scale during the time students tackle problems in the assigned tasks. Analysing results will help answer the questions below:

- What is the effect of STEM education on the development of critical thinking?
- How does critical thinking relate to learning outcomes or to creative thinking?
- In term of demographic factors, how do they influence the development of critical thinking ability of students?

2.1 Target group

Target group were randomly selected, including 52 students grade 10, divided into 10 groups of 4 schools in some mountainous provinces at the northern part of Vietnam. At the present, the pressure and expectation from examinations and grade points in Vietnam are remarkably high. Parents want their children to gain high academic achievements (grades), and they are not concerned about other factors such as the ability to solve problems or to develop critical thinking ability of their children. That is the reason why the authors could not invite all students in one class taking part in activities which we have created. Besides, with a big number of students taking part in at the same time would be a big difficulty for the research in observing and evaluating critical thinking level of each individual. Instead of this, the person in charge of this paper had volunteers under the permission of the school and parents. Each member of this research has worked with a group of students for 2 to 3 months, 1 to 2 days per week including STEM activities. Every group, the researchers spent about 8 days working with them, sometimes the process lasted about 2 months as students had to join other activities of the school.

The pedagogical experiment process was carried out with students at Grade 10 chosen randomly, in some high schools in the northern part of Vietnam as showed in Table 1.

Table 1. Research subjects and time

Group	**Research time**	**Research subjects**	**Number of students**	**Areas**
1-2	From 12/10/2016 to 07/12/2016	Students in class 10A1 of Vung Cao Viet Bac High school – Thai Nguyen province	10 Students	Moutainous area
3-4	From 12/10/2016 to 14/12/2016	Students in class 10A2 of Vung Cao Viet Bac high school – Thai Nguyen province	10 Students	Moutainous area
5-6	From 15/10/2016 to 22/12/2016	Students in Physics class 10 of Thai Nguyen gifted high school – Thai Nguyen province	10 Students	Urban area
7-8	From 16/10/2017 to 28/12/2017	Students in class 10A9 of Luong Ngoc Quyen high school – Thai Nguyen province	12 Students	Urban area
9-10	From 10/10/2017 to 18/01/2018	Students in Physics class 10 of Ha Long gifted high school – Quang Ninh province	10 Students	Urban area

IOP Conf. Series: Journal of Physics: Conf. Series **1340** (2019) 012025 doi:10.1088/1742-6596/1340/1/012025

2.2 Intervention of STEM activities executing process for develop students' critical thinking

In order to nourish and develop students' critical thinking and creative thinking, it is required that the teaching process needs to boost students' creativeness, such as: encouraging them to ask questions, exchanging ideas, discussing logically, problem-solving. Students are allowed and provided the best conditions to carry out the option that they suggested or to defend their ideas and results. Hence, we propose a step-by-step process which is illustrated in Figure 1 [8]; [9].

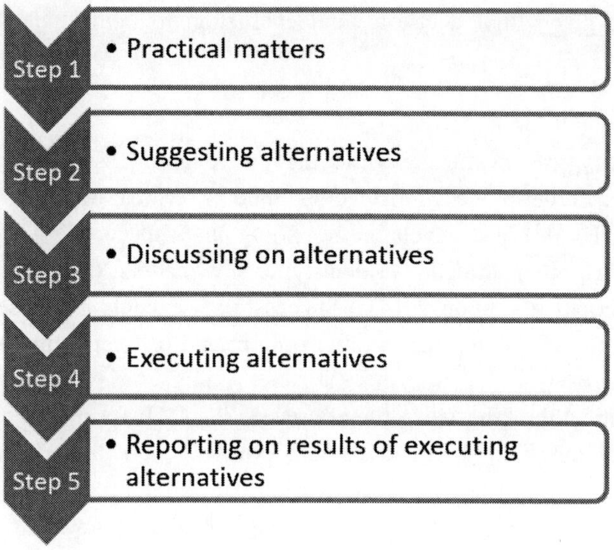

Figure 1. Steps for organizing STEM activities

Step 1 Practical matter: From real-life problems, the teacher and students should come up with a specific problem with the need to tackle it. For example: questioning them to measure their houses' height. Or how to create a basic boat? Is it possible to design anti-flood/earthquake-resistant houses?...

Step 2 Suggesting alternatives: From the assigned task, students will have discussions in groups to suggest options to tackle the task they got from Step 1.

Step 3 Discussing on altrenatives: After working in teams, students could bring out two or more options. Then they will present their ideas in front of the class to analyze and collect feedback from other teams. At the end of this step, teams will decide their one or more options.

Step 4 Executing alternatives: After making a decision from the third step, student teams will start to carry out their selected options. Here students could do only one or more alternatives. However, students are required to collect important data to prove their results and discussions.

Step 5 Reporting on results of executing alternatives: From the above steps, all groups will present the results of their tasks. With the evidence and data gained, the groups will discuss with each other to find the optimal solution. This discussion will add to the conviction of the conclusions after the first discussions in Step 3. Although previous opinions might be correct, might be not, there is a lack of the persuasiveness since they have not tested and proved (most of the opinions are suggested based on personal experience, theories). Nevertheless, in this step students will prove their findings/researches.

IOP Conf. Series: Journal of Physics: Conf. Series **1340** (2019) 012025 doi:10.1088/1742-6596/1340/1/012025

At the end of Step 3, the teacher should ask students to plan their tasks, to collect needed information for reports (data collecting, photo illustrating, report presenting, ...). Also, teachers should provide and instruct students to make their own plans and settle the working timetables with them. During outdoor research activities, teachers should always supervise students' safety. Moreover, this could help teachers to evaluate each individual better.

During discussing time (Step 3, Step 5) teachers play the role as the host, encourage students to generate and defense for their own ideas, and avoid unnecessary intervention. At the end of Step 5, teachers become the person who gives final decisions and evaluation to both students and their performances.

2.3 Data Collection and Analysis

Students critical thinking ability was evaluated while they were doing assigned tasks in the intervention of STEM activities executing process for develop students' critical thinking. The tool of the critical thinking evaluation (CTE) form was developed regarding on 7 aspects of basic expressions of critical thinking [3]; [4]. These aspects included (1) Identifying new matters; (2) Giving a reliable prediction; (3) Suggesting alternative options; (4) Successfully executing chosen options; (5) Improving the chosen options; (6) Selecting the optimal option; and (7) Applying knowledge to tackle problems in every situation.

To collect data of students' critical thinking when they on tasks, the CTE form was completed as the table 2 and 3. Data points obtained from the three main points: observation points (a); self-assessment (b); peer-to-peer review (c) and teamwork evaluation (d). After that, the final scores of each student for each STEM activity will be calculated as the average of all the points.

Table 2. Critical thinking evaluation form based on the observation chart

Evaluated student's full name: ....

Levels / Criteria	Level 1	Level 2	Level 3	Evaluation points
1) Bring out questions about a process, a thing or a phenomenon; a technical application, ... that they encounter in daily life				
2) Suggest solutions to solve the newfound problem				
3) Have proper basis points during discussing process about a specific problem				
4) Can evaluate other viewpoints and prepare for arguments				
5) Adjust their own opinions and activities when a new information is found				
6) Select the optimal options in problem-solving process				

Table 2 is an example of an observation form of a teacher for each student. In which, the form will clearly state what students can achieve after learning or doing an activity. Each student will receive their own assessment via the form. This is also the reason why the researchers did not choose a large group of students since that would make the observation and follow-up difficult. To achieve the wanted results, the research group has chosen to record a full video when students were working and discussing. Then the videos will be reviewed, students' performances will be evaluated and given points based on criteria in the above table. Level 1 means 1 point, while level 3 means 3 points. As mentioned earlier, this practice will only work with small group of students (in this study, the authors only allow 5 to 6 people per group) since other factors such as: discussing noise or recording difficulties, should not affect students during their activities to bring the most realistic reflection of thinking or working process of all students.

Table 3. Self-assessment and peer-to-peer review form

Reviewer's full name: ....

Number	Review criteria / Full name	1) Bring out questions about a process, a thing or a phenomenon; a technical application, ... that they encounter in daily life	2) Have proper basis points during discussing process about a specific problem	3) Have valuable contribution to other members' opinions	4) Listen and receive feedback from other members	5) Select the optimal options in problem-solving process	**Total**
1	Nguyen Van A						
2	Nguyen Van B						
3	....						

Table 3 is the self-assessment and peer-to-peer review form and it will be distributed to each student. When the reviewer (student A) evaluates himself, the points in the column with his name (Nguyen Van A) will be his own assessment, other points are from peer-to-peer reviews. Student A also will not be allowed to evaluate the whole class but his team members to ensure the reliability since he only works with 4 other members of the group (in a group of 5). Of course, before distributing the forms, teachers are expected to explain clearly every criterion in it and instruct them the evaluating steps to get the highest reliability. This form will be given to the class at the end of the activity (Step 5).

Besides, the researchers also collect other data through in-depth interviews during and after each STEM activity (if necessary) to contribute to the evaluation process in the above forms.

3. Research Findings

One of the most important targets in the teaching process is to create and develop thinking of learners, especially critical thinking. Enormous ways could be utilized to achieve this target, and STEM educational orientation program is one way. The paper will clarify the general view of organizing STEM activities and motivating critical thinking of students in high school and then examine students' critical thinking ability in the intervention of STEM activities

3.1. General view of STEM activities in high schools

Researchers have carried out the survey from July, 2015 to August, 2015 on 42 teachers teaching Sciences and 442 students from different high schools such as Thai Nguyen High School, Pho Yen High School, Chu Van An High School, Phu Luong High School, Dai Tu High School of Thai Nguyen Province and some neighboring provinces. The result of the survey presented as follows:

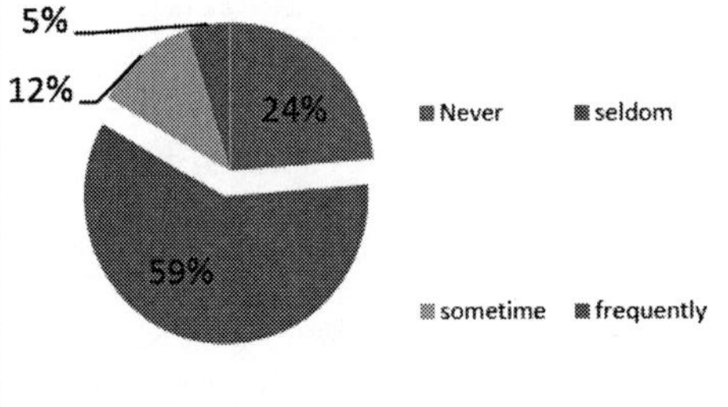

Figure 2. The proportion of organizing STEM activities of teachers

+ Sciences' teachers: They used to try organizing STEM activities in teaching, however, the rate is still low (Figure 2), the frequency rate is extremely small (5%). The reason is that teachers have to follow the teaching plan of the school, meanwhile, STEM activities would take a lot of time. Another reason comes from the evaluation process, examinations in Vietnam are not concerned about ability evaluation. To deeply understand about this situation, researchers have had a meeting with teachers and known that many teachers were not trained or introduced the way to organize STEM activities and the way to evaluate students through those activities as well.

+ Students: The result finds out that during studying time, students were absolutely excited about observing one physics experiment and they would like to do a specific experiment by themselves. They also had many questions need to be answered. Most students enjoyed lessons with STEM activities. This is one of the factors that urge the need to conduct this study.

3.2 Students' critical thinking ability in the intervention of STEM activities

Students' critical thinking ability was examined when they on tasks in the STEM education activities. The findings about students' critical thinking ability will be calified as following; a) General evaluation about the development of students' critical thinking ability, b) Comparing the development of critical thinking ability between male and female students, c) the development of critical thinking ability for students with different background, and d) relationship between learning outcomes and critical thinking development.

a. General evaluation about the development of students' critical thinking ability

During the process of designing STEM activities, the researchers decided that the difficulty level, as well as the opportunities, should be the same in the first activity (Activity 1) and the last one (Activity 6) to let students express themselves equally. The analysis result shows that, by participating in STEM activities, most students increased their points of critical thinking ability (based on the comparison between Activity 1 and Activity 6), however, the change is not similar among students in researched groups. Some students noticeably rose their points (about 1 – 2 points), but some almost remained unchanged (Figure 3). Results also specified that no students experienced unusual increase/decrease (2 or more points). In conclusion, STEM activities can change students' critical thinking ability.

Figure 3 The change in critical thinking points of students

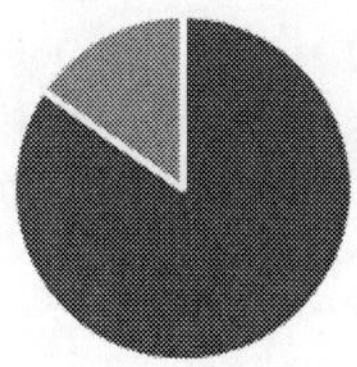

b. Comparing the development of critical thinking ability between male and female students

The researchers were also curious about if there is a difference between genders in term of developing critical thinking ability or not. Here is the final outcome: The critical thinking average points of female students are lower than that of male students; the increase in critical thinking points of male also higher than that of female students (Table 4). This indicates that male students have better critical thinking ability, though the difference is not remarkable. This also allows the researchers to come to the conclusion: the psychological barriers in communication and the lack of self-confidence have hindered female from the development of critical thinking ability. Besides, it could be that STEM field was not the strengths of them. So it is recommended that each student group should have the same numbers of male and female numbers in order to balance out the differences. That could let students with higher critical thinking points help ones with lower, and they can learn from each other since male students tend to do heavy work while female is much better in doing activities required sophistication, skilfulness. By combining these two suggestions, each student could do their best in STEM activities.

Table 4. Critical thinking points of male and female students

	Average	Average increase
Male	7.2	1.1
Female	6.4	0.5

c. The development of critical thinking ability for students with different background

From Figure 4, everyone could see that students in mountainous, remote and isolated areas have low critical thinking scores, low academic achievement, whereas students in the cultural and administrative centers of the city did get higher critical thinking ability and academic results. The main reason could be that students coming from mountainous areas are limited to have experiments, extracurricular activities, science clubs, learning activities related to practical tasks. In addition, these groups are required by the school to stay in their dormitories and could only go outside with permissions. While city-based students are more likely to live and work independently, receive better practical knowledge, their opportunities of developing critical thinking ability may be higher. Thus, it can be said that the living environment of a student can influence his ability to develop critical thinking.

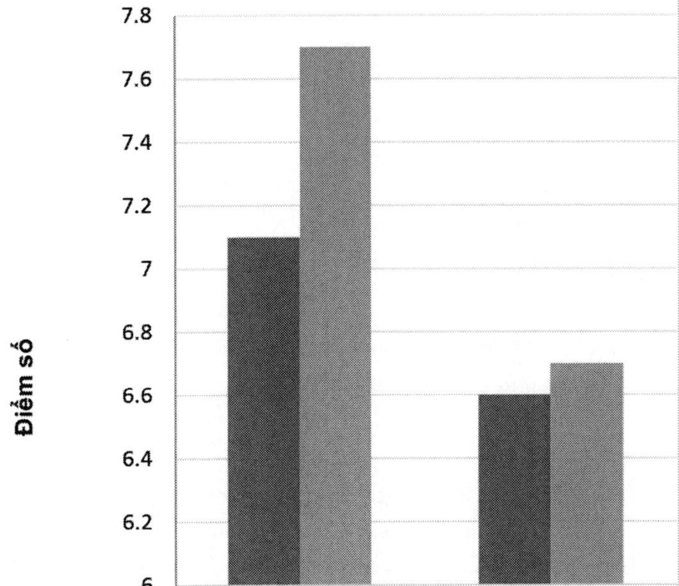

	group of students city area	group of students in rural areas
critical thinking point on average	7.1	6.6
point average academic results	7.7	6.7

Figure 4 critical thinking point, point averge academic results,

d. Relationship between learning outcomes and critical thinking development

To evaluate the correlation between academic performance and critical thinking ability, the research team has taken the most recent grade point average for students in Maths, Physics, Chemistry, Biology, Computer Science and Technology to compare (Figure 4). The results show that students with high academic scores also develop critical thinking better. This made the researchers quite unexpected compared to the team's prediction before doing pedagogical experiments. This fact suggests that classroom assessments have already included comments about critical thinking assessments, or, in other words, good assessments. This could be considered as a model for evaluating student learning outcomes in the future.

4. Conclusions

The research results obtained from pedagogical experiments showed that:

- Critical thinking ability of students can be developed through STEM activities
- Students with high level of critical thinking also have high academic achievement and vice versa
- Teachers' assessments at the experimental schools have paid close attention to the student's critical thinking
- Gender can influence the development of critical thinking and the ability to solve problems in STEM activities
- Organizing STEM activities is suitable for all students from highland and mountainous areas to the urban, metropolitan places; from average to good academic performance, from normal school to gifted high school students, both male and female
- To facilitate the development of critical thinking, teachers should select and design practical STEM activities that are close to their daily lives in order to stimulate interest and motivation for their students. Also, teachers have to create a safe and friendly learning environment, always encourage students to ask questions, criticize and contribute ideas, ...
- Research results should be tested on a broader range of subjects, longer timeline so that conclusions could be more persuasive

Acknowledgement

This article is published from the research results of the topic ĐH2017-TN04-09" University of Science and Technology Program "in a critical critical research paper" University of Science and Technology ". This article is by Nguyen Quang Linh, Nguyen Mau Duc, Chokchai Yuenyong.

References

[1] Ennis, R. (March, 1992). Critical thinking: What is it? Proceedings of the 48th Annual Meeting of the Philosophy of Education Society (pp. 1-2). Colorado: Denver.

[2] Fisher, A., & Scriven, M. (1997). Critical Thinking: Its definition and Assessment. Edgepress and Centre for Research in Critical Thinking University of East Anglia. Cambridge: Critical Thinking An Introduction.

[3] Lipman, M. (2003). Thinking in Education, 2nd edn. Cambridge: Cambridge University Press.

[4] Luyen, P. T. (2009). Biện pháp rèn luyện tư duy phê phán cho HS trung học phổ thông qua dạy học chủ đề phương trình bất phương trình. Tạp chí GD, số 209, 25-27.

[5] McGregor, D. (2007, 4 10). Developing Thinking Developing Learning. London: open univercity Press. Retrieved from hhhhh.

IOP Conf. Series: Journal of Physics: Conf. Series **1340** (2019) 012025 doi:10.1088/1742-6596/1340/1/012025

[6] Nga, N. T. (2018). Thiết kế và tổ chức hoạt động dạy học theo chủ đề STEM cho học sinh THCS và THPT. Hồ Chí Minh: Đại học Sư phạm thành phố Hồ Chí Minh.
[7] Sternberg, R. J. (1985). Teaching Critical Thinking, Part 1: Are We Making Critical Mistakes. JSTOR: Phi Delta Kappan International.
[8] Tong, P. H. (2004). Dạy học Vật lí ở trường phổ thông theo định hướng phát triển hoạt động tích cực, chủ động, sáng tạo và tư duy khoa học. Ha Noi: NXB ĐHSP Hà Nội.
[9] Tra, D. H. (2002). Một số vấn đề dạy học vật lí theo tiến trình nghiên cứu khoa học. Tạp chí Giáo dục, số 23, 35-38.

IOP Conf. Series: Journal of Physics: Conf. Series **1340** (2019) 012026 doi:10.1088/1742-6596/1340/1/012026

Enhancing scientific literacy – research the case of teaching the topic Water in life

Do Huong Tra[*1] , **Nguyen Thi Thuan**[2]

[1]Do Huong Tra: Hanoi National University of Education
[2]Nguyen Thi Thuan: Hanoi Metropolitan University

*Corresponding author's e-mail address: dhtra@hotmail.com

Abstract: Scientific literacy has been a strong target in many countries for many years, including Vietnam. Educating students to become future citizens has been emphasized in the science curriculum, with the goal of providing students with the tools so they need to be able to adapt and solve practical issues in a responsible way. The article mentions the proposal for structure of scientific literacy and research the case of enhancing scientific literacy through activities to explore while learning the topic Water in life.

1. Introduction

The nature of science is humanitarian, and so science activities are responsible for society. After all, scientist is a member of society, and cannot stand outside or above society, so their scientific research must contribute to the community. Therefore, teaching science must towards bigger goals not only the scientific knowledge and skills but also the application of scientific knowledge in the problems of reality and society.

Over the years, educating students to become future citizens has become a major challenge for science education in many countries, especially in Europe and in developing countries. [8][12], including Vietnam. Scientific education for all should be directed to scientific literacy, including attitudes of learners to the community.

There are many organizations and programs in the world which are interested in researching scientific literacy such as the Organization for Economic Co-operation and Development (OECD), such as Australia, Singapore and Germany [10]. However, there are certain differences in the structure of the scientific literacy. In the general, education program of Germany, the competence model of natural sciences is divided into four groups of competency components: specialist knowledge, knowledge discovery, communication and assessment. In the US and Singapore programs, the structure of scientific literacy includes elements: questioning and problem identification; development and use of the model; planning and conducting surveys; analyze and present data; use mathematical thinking; solution design and solution design; gathering evaluations and exchanging information. PISA is initiated and directed by the OECD. It recognizes scientific literacy as a basic ability of students to study science and proposes a suitable structure to evaluate secondary school students [2]. In Vietnam, scientific literacy is one of the goals of the general education program after 2018, so many authors have been interested in research. In our study, we emphasize that the knowledge that is taught in schools should refer to the social issues that the learner lives in. We use the term Scientific Literacy but the term Scientific Competency in order to emphasize the attitude of using their scientific knowledge.

The research topic is how to select the topic of teaching and how to organize learning activities to meet the scientific literacy goal of the learner?

2. Research content

2.1. Scientific literacy

2.1.1. Concept. The concept of competency has been mentioned in many domestic and foreign documents with different approaches, so there are many different definitions of competence. At present, there are three different terms that are translated into Vietnamese as "năng lực": competency, ability and literacy.

The term Competency emphasizes the ability to effectively implement actions, issues related to a certain area is based on knowledge, skills and willingness to act [9], [13]; Perrenond. P [7] and consider competency as such as qualities, psychology, personal attributes [1].

In the term Literacy - this term when translated into Vietnamese means academic level, the ability to read and write. The PISA's documents use the term Literacy. The authors translated into Vietnamese as "năng lực" [2]. According to PISA, Literacy includes knowledge, skills and cognitive processes. PISA and some other countries emphasize the use of knowledge and skills accumulated in schools in the context of real life situations in order to contribute to changing attitudes of learners.

In our opinion, Literacy refers to understanding in general and applying that knowledge into real-world contexts to emphasize the responsibility of future citizens for the problems of society, community. The term (literacy) is different from the word competency. In terms of competency, competence consists of a fairly complex set of elemental powers. In terms of literacy, it cannot be divided into elements but only express their manifestations, It emphasizes understanding in order to solve problems in a real context as responsible citizens with the future, with the society.

From the two terms of competency, it leads to two different ways of describing scientific competency.

- The first group is the term "Scientific competency". It is the ability to perform scientific activities to create new concepts, new methods, new tools of a scientist or a future scientist. This concept emphasizes the scientific ability associated with the scientific process. Therefore, enhancing scientific competency for students emphasizes the fostering process of scientific research. Therefore, teaching science according to scientific process has become a central issue in educational reform in some countries such as the United States, France, and Canada.
- The second group is the term "Scientific literacy". PISA 2006 [5] uses the concept of scientific literacy - when translated into Vietnamese, the authors have called it năng lực khoa học (scientific competency). According to PISA, scientific literacy is the ability to use scientific knowledge, analyze questions and draw reasonable conclusions to make sound decisions about the natural world and human changes. Created for the natural world. This concept emphasizes that teaching aims are attitudes toward science, to the problems of society.

OECD has defined the scientific literacy as follows: Scientific literacy is the individual's scientific knowledge and the ability to use knowledge to identify questions, acquire new knowledge, explain scientific phenomena, and draw conclusions; the ability to identify problems and the ability to draw conclusions based on issues related to science; Scientific literacy is the ability of the individual awareness of the impact of science and technology on human life, material and cultural life. Scientific literacy is also expressed in the willingness to solve issues such as a citizen with scientific knowledge and thinking [2]. In this view, a person with scientific ability should have the following elements:

Have scientific knowledge.

Understand the characteristics of science as a form of human knowledge and discovery activities of human.

Use knowledge to identify, acquire new knowledge, identify scientific issues, explain scientific phenomena and draw conclusions on the basis of evidence on scientific issues.

Aware of the role of science and technology in shaping the cultural, spiritual and material environment.

Willingness to participate as an active citizen, applying scientific understanding to solving related problems.

Thus, the manifestations of scientific literacy do not overemphasize the process of scientific research and the scientific knowledge that attaches importance to the use of knowledge and attitudes to the use of scientific knowledge. This is very suitable for students in secondary schools.

2.1.2. Structure of scientific literacy. According to PISA, scientific literacy is structured by three components: (a) Explaining the facts in a scientific way; (b) Evaluation and design of scientific research; (c) Present the data and evidence in a scientific way [5]

PISA has a broader approach to assessing students' knowledge, skills and attitudes, which is aimed at assessing the capacity to use knowledge in addressing everyday tasks and challenges. This broad and practical approach helps PISA to better reflect changes in the curriculum, more specifically, a model based on a lifelong learning model in which the accumulation of knowledge and skills is needed to adapt in an ever-changing world and a lifelong process, not limited to school.

However, this structure only evaluates the students after the study but not interested in the research activities of learners during the study. On the one hand, based on the structure of PISA's scientific literacy, on the other hand, based on the point of view that learning activity is in fact the learning activity in the scientific process, we add activities on scientific discoveries and propose the structure of scientific literacy in teaching science in junior high school.

Table 1. Proposing structure components of scientific literacy

Components/ Behavior index	
1. Explain the phenomenon in a scientific way	1.1. Recognize (recall) and apply scientific knowledge appropriately
	1.2. Create models to explain the phenomenon in a scientific way
	1.3. Make trusted predictions/ Provide hypotheses to explain
	1.4. Explaining the meaning of scientific knowledge to life, society.
2. Evaluation, design and implementation of scientific discovery	2.1. Ask questions to discover a scientific mission.
	2.2. Propose how to carry out research tasks (proposed solutions)
	2.3 Select research solution
	2.4 Do the research
3. Present data and evidence in a scientific way	3.1.Analyze and process data
	3.2. Review and compare the results
	3.3. Show the results

2.1.3. Enhancing scientific literacy. In the world, there are many authors interested in researching the literacy of science by various means such as [4, 3, 6, 11]... Many of these studies emphasize:

First of all, enhancing scientific literacy must be through the learner's own activities. The research by Holbrook. J & Rannikmae. M indicates that the nature of science teaching based on the theory of enhancing scientific literacy should be linked to the goals that individual learners need to attain, including personality and attitudes. positive, civic responsibility with goals in the field of social education [4].

Secondly, teaching should put learners in a practical context so that problem solving can change the attitude and responsibility of learners to society. Derek Hodson believes that it is time for the school's science curriculum to move towards social responsibility. The author believes that in order to solve the current social and environmental issues, a generation of citizens with scientific understanding is needed. From there, the author proposes a scientific program focusing on seven areas of concern (human health, food and agriculture, land, water and mineral resources, energy and consumption, public communications, transport, ethics and social responsibility) with the participation of the community. Therefore, it is necessary to focus on how to choose teaching content to direct learners to social responsibility and community [3].

Thirdly, attention should be paid to the role of social interaction in enhancing scientific literacy. Janice D. Gobert and Barbara C. Buckley shows that learning science is effective when learners develop and evaluate models of their own, others ones, and scientists. The study also found that group discussion and collaboration were measures to enhance students' cognitive and metacognitive skills. [6]

Fourthly, Care should be taken with the role of the learner's conception and the teaching of science with changing the conceptions of students in explaining the phenomenon in a scientific way. Reinders Duit & David F. Treagust believe that one of the ways to dramatically improve teaching and learning science is to change the perceptions of learners, which discuss the research that changes our perception. analyze, explain the practical phenomena with scientific knowledge [11].

Putting students into activities, organizing group interaction, accessing the process of exploring science comes from life issues, linked to individuals, local, national and global issues. are of interest to researchers worldwide. Research will mention discovery activities in the process of building knowledge as well as in the application of knowledge to solve issues that are linked to the real context of society to change attitudes and responsibilities of the learner.

2.2. Develop the topic of enhancing scientific literacy for students

2.2.1. Principles of selecting topics. With the connotation and structure of the concept of scientific literacy as presented, teachers need to choose practical situations that are linked to science. This focus on situations is also consistent with the emphasis on "life context" in PISA surveys. The context of life related to health, natural resources, environment, environmental and technology risks will be an opportunity for learners to understand and be responsible for individuals, local, global and society.

With the topic of health: at the local level, it is the matter of preservation and food safety, nutritional issues. At the social level, that is production and distribution of resources. And at the global level, that's the problem of the growth of the world population and the sustainable use of resources.

Thus, in order to gain scientific literacy, learners must have scientific knowledge, knowledge of scientific process to explain the practical phenomena in a scientific way. Local, social, and global contexts are the top priorities for developing teaching topics. This may lead to inconsistencies in the choice of teaching content. On the one hand, it emphasizes the scientific nature of the scientific knowledge itself (science teaching is primarily prepared for the training of scientists rather than future citizens). On the other hand, it emphasizes the "social" aspect of scientific knowledge. The emphasis on the social aspect of scientific knowledge leads to the appearance of alternative subjects instead of traditional subjects (such as physics, chemistry, biology). These subjects are characterized by the integration of knowledge content at different levels.

However, in our opinion, it is important to establish a balance between these two tendencies, in addition to emphasizing the nature of scientific knowledge, it also needs to concern the social nature of the knowledge. This means that in selected topics, discovery activities to build knowledge with explaining the phenomenon in a scientific way always support one another. If the social nature of knowledge is concerned so much, it can lead to the risk of teaching "beyond the subject".

In order to enhance the scientific literacy of the students, the selection of subject content requires the teacher to follow the following steps (Diagram 1):

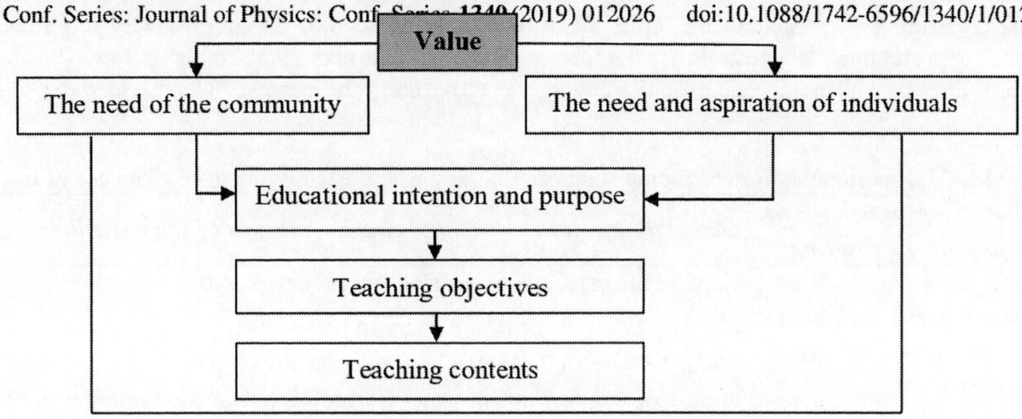

Diagram 1. Steps to select teaching content

Analyzes the social, cultural and economic needs of the community as well as the needs and aspirations of their students or parents.

Carry out a detailed analysis to clarify the educational intentions and purpose, and to identify values and priorities that need to be taught in order to meet aspirations to encourage action to solve the issues of the event. living, serving the society.

Establishing teaching goals and specific objectives of the situations revolves around a central theme that relates to the content of knowledge and methodology in the various subjects.

Social responsibility is associated with the need to solve complex situations of social life, requiring the use of knowledge and methods of many subjects. In other words, integrative teaching has created a favorable opportunity for enhacing social responsibility for students.

2.2.2. Teaching the topic Water in life

Water is one of the topics chosen for enhancing scientific literacy because of its importance to every aspect of life. Topics include the following contents:

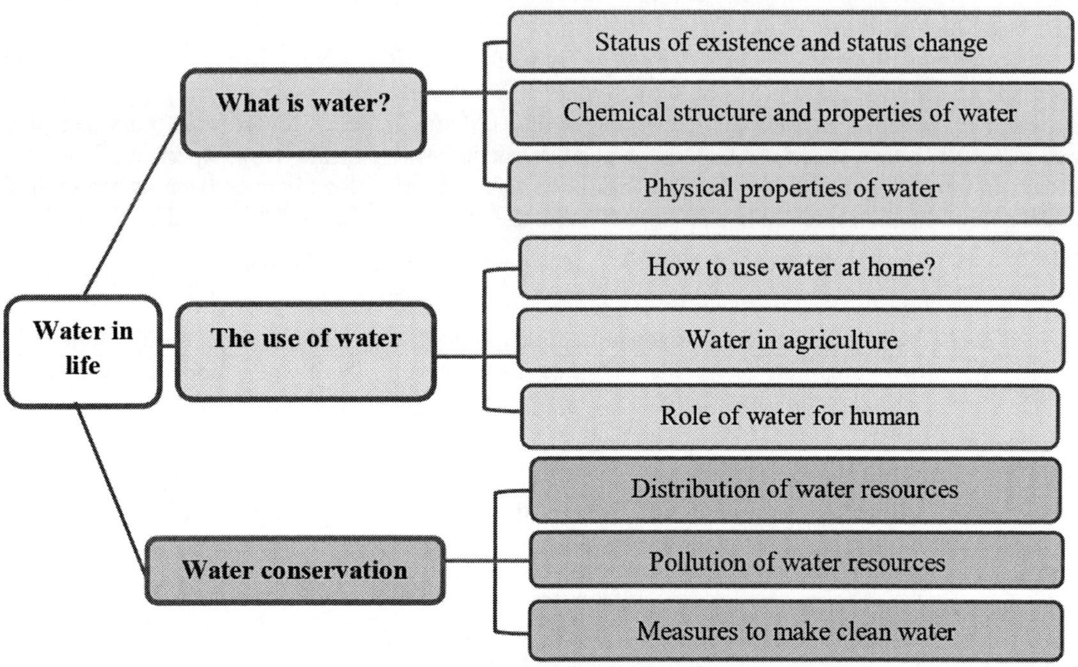

Diagram 2. The topic Water in life

After defining the content of the topic, the next issue is the need to design discovery activities that ensure opportunities for fostering scientific capacity in learners. The table below shows the opportunities for enhancing scientific literacy in two activities of the content "The use of water".

Table 2. Opportunities for enhancing scientific literacy when teaching the topic The use of water

The process of exploring, discovering	*Opportunities for enhancing behavior indexes of scientific literacy*
How is water distributed to the family and used at home?	Explain the principle of distribution to households Create models to distribute water to households Make research questions: how do you know how water is used at home? Assess how much water is used? What causes wasted water?
	Implement research solution - Learn how to use water at home.
	Solution Review: Learn how to use water at home Analyze and interpret data to draw conclusions about the amount of water used in the studied families. Propose and adjust solutions to save water in your family.
What is the role of water in agriculture?	Make research questions: How is water used in agriculture? What is the role of water in plants? Implement the solution - Understand the role of water in agriculture Describe the reseatch results on the role of water for crops.
	Propose solutions: Smart and saving irrigation model Evaluate and adjuste solutions on smart and saving irrigation model

2.3. Assessing the scientific ability

2.3.1. Use the observation checklist (Rubic). To assess scientific literacy in the process of learning each specific knowledge unit of the topic Water in life, the research makes description table of the structure of scientific literacy and levels of quality criteria. In order to preserve the value of the measurement tool, a description of the capacity structure was consulted by expert opinions. After processing expert opinions, the checklist has been adjusted, revised and concretized as shown in the table below. In the topic "What is water?", the checklist of behaviors of scientific literacy is as follows:

Table 3. Scientific literacy assessment checklist in the topic **"How is water used at home?"**

1. Explain the phenomenon in a scientific way	Explaining the meaning of scientific knowledge to life, society.	Explain the important role of water for self, community and social life
	Ask questions to discover a scientific mission.	How is the water used in your home (or the surrounding family) reasonable? How do you adjust the right amount of water use in your home? Proposed water quality test plan is reasonable? - What is your family's and your child's daily use?
	Propose how to carry out research	- Find out how much water is used daily, monthly by tracking water meters and charting, reading water bills, collecting data

IOP Conf. Series: Journal of Physics: Conf. Series 1340 (2019) 012026 doi:10.1088/1742-6596/1340/1/012026

	tasks (proposed solutions)	and drawing water use charts in other activities: each other at home for a month. ... - Assess the amount of water used by your family and their families. - Devise measures to regulate the use of reasonable water in your home such as automatic spill prevention design, water-saving plant watering design.
2. Evaluation, design and implementation of scientific discovery	Select research solution	Analyze, evaluate, select the optimal solution for the proper use of water in the family.
	Do the research	- Do you have a detailed plan for researching and checking water usage? - Make a detailed plan for the design of water-saving appliances and equipment in the home + Assignment of specific tasks in the group + Select and prepare raw materials.......... - Successfully follow the steps of the plan
3. Present data and evidence in a scientific way	Analyze and process data	Transfer daily water usage data to other charts. Select and arrange the appropriate information to present the report
	Review and compare the results	From the chart, does the collected data compare the amount of water consumed or less, saved? Comment on the results of water use after the implementation of water saving measures Review comments on the operation of automatic spill prevention devices and watering tools save.
	Show the results	Choice of scientific presentation: poster, apphiche ...

Using this checklist, the study confirmed the behavioral manifestation of the student's scientific literacy while learning the subject matter.

From the checklist that has been developed, observations have been made, the research has collected

evidence on the behavior of scientific literacy. Specifically, with the task of discovering **"How to use water at home"**, the behavioral manifestation of the student's scientific literacy has been demonstrated:

- Explain the important role of water for self, community and social life
- Ask questions to discover a scientific mission.
- Proposed water quality test plan.
- Evaluate the amount of water used in a single family day. Students track the amount of water used in the household by reading the water bill, tracking the water meter…

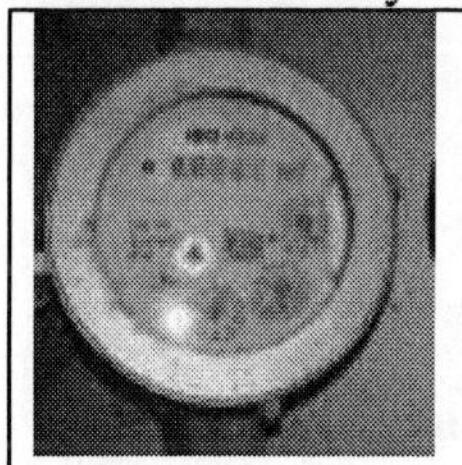	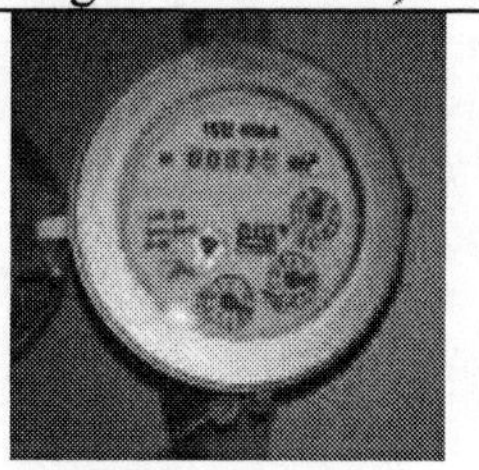	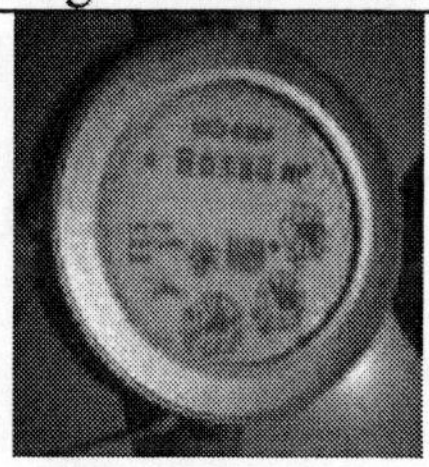	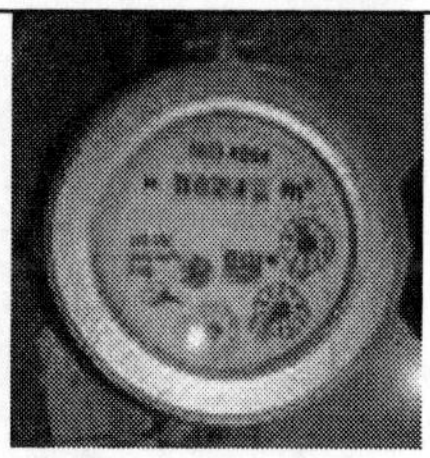
First day - 1	seventh day - 1	First day - 2	Seventh day - 2

IOP Conf. Series: Journal of Physics: Conf. Series 1340 (2019) 012026 doi:10.1088/1742-6596/1340/1/012026

Students evaluate the amount of water used in the home for various needs such as bathing, washing, toilet use, etc. From that set up evaluation.

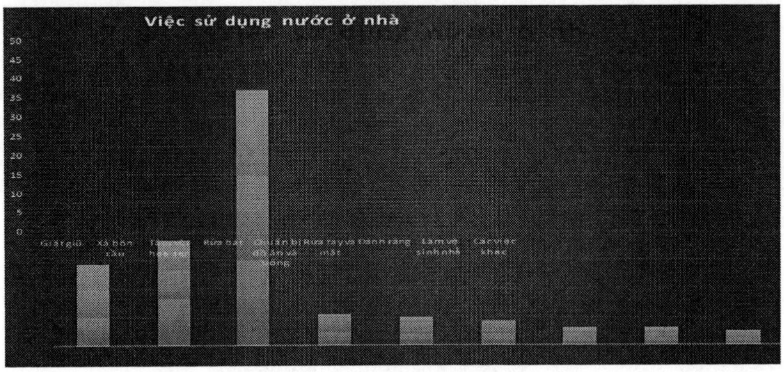

Figure 1. Learn how water is used in various home activities such as laundry, showering, hand washing and face cleaning.

- Devise measures to regulate the use of reasonable water in your home such as automatic spill prevention design, water-saving plant watering design.

With solutions designed to save water, students also assign each other to search, purchase tools to assemble and operate the equipment. The individual and team working out the plan have outlined:

Preparing instruments:

Assemble the equipment according to the design plan

Check the layout and assembly of test instruments

- Analyze, evaluate and select the optimum design options with the use of simple, easy-to-find materials.
- Present, discuss and identify the stage of designing product.

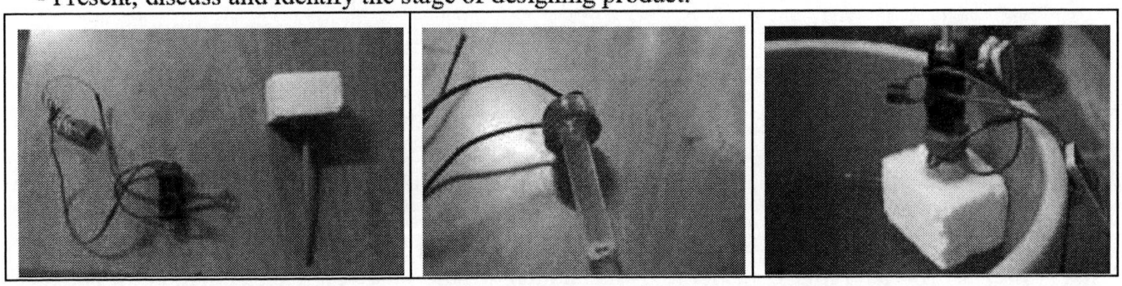

Figure 2. Fabrication of equipment to avoid wastage of water: "Automatic anti-spill equipment"

*2.3.2. Scientific literacy test.*Test is a fairly popular assessment tool that allows researchers to collect data through answers or describe the way students think about a system of questions or tasks that require people. must demonstrate knowledge of knowledge and skills required. Therefore, designing tests must follow a strict processthat includes the following steps: clarifying the purpose of the test, identifying the object to be evaluated, determining the variables to be measured, determining the method and condition take quizzes, set up test matrices, compile items and score guides, discuss quizzes and items; test and evaluate, complete the test.

After the student finishes studying the topic Water in Life, the research uses the test to assess the scientific literacy. The test was standardized with Conques and SPSS software to test the tool.

• Assess the reliability of the test

The Cronbach Alpha coefficient considers the correlation of the performance of a question with the whole test. Calculation of the coefficient Cronbach alpha using SPSS software gives:

IOP Conf. Series: Journal of Physics: Conf. Series 1340 (2019) 012026 doi:10.1088/1742-6596/1340/1/012026

Reliability Statistics	Cronbach's Alpha	N of Items
	.749	16

In the table above, the Cronbach alpha coefficient is 0.749. This coefficient is at average (from 0.6 to 0.8). So the test used to measure the scientific literacy is relatively good.

• Information curve of the test subject

For this test, the information curve is given by the following figure:

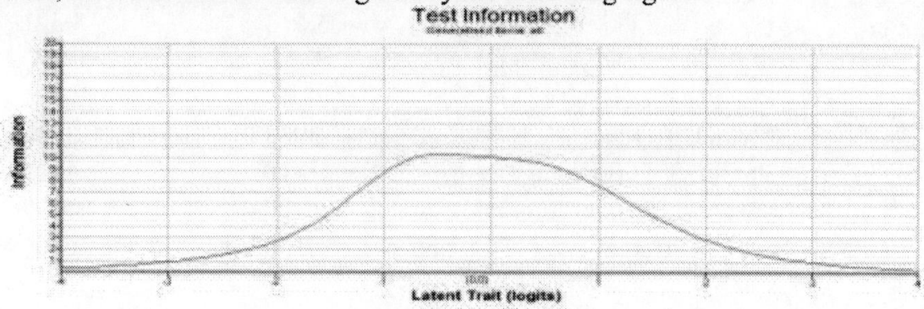

Figure 3. Information curve of the test subject

Looking at the information curves of the test, which can be seen as an average test for students, and this test has the most information in the range from -1 to 0.5.

• *Map of balance between the difficulty of the task and the competency of the learner*

Figure 2 shows the balance between the difficulty of the 16 questions and the competency of 43 students in the same Logit scale, with "0" is the average value.

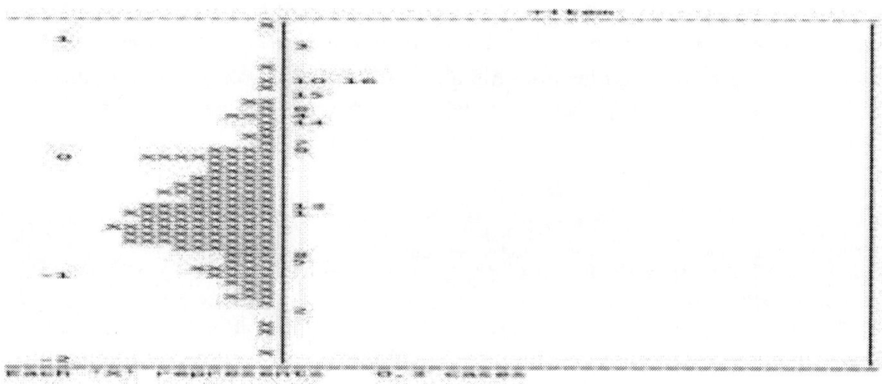

Figure 4. Difficulty balance map of 16 questions and student's competencies.

The outer left of the map is Logit scale. Next is the distribution of students' competency along the logit scale, for every 0.3 children denoted by an 'x'. The right side of the map is the difficulty level of 16 questions. Students in the same position as the question, the probability of answering the question correctly is 0.5. When the student is in higher or lower position, the probability of answering the question correctly will be higher or lower than 0.5.

The map shows that the test has a degree of difficulty that is distributed fairly to the student's competency (spread evenly over the scale). Some questions are easy such as the 2nd, 5th question (most students are able to answer correctly), and some difficult questions like the 3rd, 10th, 16th question (few students have the correct answer). However, there are some students who has competency with the same difficulty level of questions in the subject.

After normalization, determination of the difficulty and reliability of the test, the test was conducted in two control and experiment classes with 43 students in grade 6 - the first grade of lower secondary school in Ha Noi. Independent Samples Test – T is used to test the difference in the average value between the experimental and control groups shown in the table below.

Table 4. Independent Samples Test.

		Equal variances assumed	Equal variances not assumed
Levene's Test for Equality of Variances	F	.001	
	Sig.	.978	
t-test for Equality of Means	t	3.513	3.501
	df	87	84.702
	Sig. (2-tailed)	.001	.001
	Mean Difference	.39333	.39333
	Std. Error Difference	.11197	.11234
	95% Confidence Interval of the Difference Lower	.17078	.16996
	95% Confidence Interval of the Difference Upper	.61587	.61669

Based on the equality test of the two variances, it shows that:

The sig. value in the Levene test = 0.978 (> 0.05), so the variance between the two experimental and control classes remains unchanged, we will use the results of the Equal variances assumed.

Sig. value in Equal variances assumed in the t-test < 0.05 so it can be concluded that there is a significant difference in the average values between the two experimental and control classes. This shows that the goal of enhancing scientific literacy for students when teaching the topic Water in life has been initially confirmed.

3. Conclusion

The research results in the teaching of the theme Water in Life have shown that the construction of a scientific literacy structure model and the design of learning activities have responded to the need to demonstrate the behavior of the Literacy, it is suitable for junior high school students.

The results of these learning activities not only allow students to have a common understanding of water, but also help them to apply this knowledge in real life situations, with a focus on the responsibility of future citizens with the problems of individuals, local and global - here is Water in Life.

References

[1] Bich N N 1998 Personality psychology - Some theoretical issues (House of education publication, Ha Noi)

[2] Hoa N T 2015 PISA and some issues of Vietnamese education (University of Education Publishing House) p 69

[3] Hodson D 2013 Time for action: Science education for an alternative future. International Journal of Science Education, 25:6, 645-670, DOI: HYPERLINK "https://doi.org/10.1080/09500690305021" 10.1080/09500690305021

[4] Holbrook J and Rannikmae M 2007 The Nature of Science Education for enhancing Scientific Literacy **International Journal of Science Education, Taylor Francis (Routledge) 29** pp 1347–1362

[5] Http://www.oecd.org/.../pisa-2015-assessment-and-analytical-framework-978

[6] Janice D G and Barbara C B 2000 Introduction to model-based teaching and learning in science education **International Journal of Science Education 22** pp 891–894, DOI: HYPERLINK "https://doi.org/10.1080/095006900416839" 10.1080/095006900416839

[7] Perrenoud P 1997 Construire des compétences de l'école. Paris. ESF (4e éd. 2004)

[8] Robine F 2009 L'enseignement des sciences, France et Europe. Communication présentée aux journées d'études S-TEAM. Grenoble. Disponible sur Internet à l'adresse : http://iufm.ujfgrenoble.fr/index.php/accueil/188.html?start=5 (consulté le 26 janvier 2011)

[9] Romainville M 1996 L'irrésistible ascension du terme «compétence» en éducation. In Enjeux, n°37/38, pp 132–142

[10] Thai L V 2011 (Head of project) Scientific research project: Development of general education program following the directions of developing learner competency (Code: B2008-37-52 TD, Hanoi)

[11] Treagust D and Duit R 2011 Conceptual change: A powerful framework for improving science teaching and learning **International Journal of Science Education, 25** pp 671–688

[12] Virginie A 2011 Finalités socio-éducatives de la culture scientifique. Revue francaise de pédagogie. Janvier – Mars 2011. HYPERLINK https://journals.openedition.org/rfp/2605 Évaluation, développement professionnel et organisation scolaire

[13] Weinert F E 2001 Concept of competence: a conceptual clarification. In D. S. Rychen & L. H. Salganik (Eds.), Defining and selecting key competencies pp 45–65 (Seattle: Hogrefe & Huber Publishers)

IOP Conf. Series: Journal of Physics: Conf. Series **1340** (2019) 012027 doi:10.1088/1742-6596/1340/1/012027

Online training in regular fostering of school science teachers to meet education innovation in Vietnam towards the development of cross-culture teachers

Do Vu Son

Thai Nguyen University of Education, Thai Nguyen, Vietnam

Corresponding author's e-mail address: dovoson@dhsptn.edu.vn

Abstract. Regular fostering teachers is an annual imperative task for the school teachers. This has been stipulated in the Government document, the Ministry of Education and Training (MoET), the (provincial) Departments of Education and Training in Viet Nam. The form of concentrated and periodic fostering teachers as implemented in Vietnam has many limitations, those are: for millions of teachers need fostering and retraining in order to meet the time requirements of education reform, the concentrated fostering is not feasible; professional contents do not meet each local or individual learner; It is costly to spend on classrooms, teaching equipment; costs for teachers and experts; travel, accommodation for concentrated stay are also costly, and time consuming. The article provides regular training form for geography teachers by online training to address the limitations of concentrated regular training while promoting the outstanding advantages of online teaching. This article has developed a number of special subject contents for fostering the social sciences teachers to meet the requirements of current education reform in Vietnam. The article also mentions the possibility of using online training to foster teachers that are qualified to teach in the region and the world.

1. Introduction

Regular training (BDTX) in general and of social science subjects in particular for high school teachers is a mandatory annual task as stipulated in Circular No. 26/2012 / TT-BGD ĐT dated 10/7/2012 of The Ministry of Education and Training of Vietnam promulgating the Regulation on regular training of pre-school, [2],[3] has clearly pointed out:

"Teachers study and are regularly trained to: 1/ Update their professional knowledge, improve their political qualities, professional ethics, develop their teaching and education capacity and other competencies according to the teachers' job standards, requirements of the school year tasks, educational levels, requirements of educational development of the province, requirements for renewal and improvement of the quality of education in the entire branch; 2 / Development of self-learning ability, self-improvement of teachers; capacity for self-assessment of the effectiveness of regular training; Capacity of organization, management of self-learning, self-improvement activities of teachers of the school."

The periodically-concentrated form of teacher training as implemented in Vietnam has shown many limitations, such as: the number of millions of teachers need training and retraining to be able to timely meet the requirements of education reform, the concentrated training is not feasible;

professional content does not meet the local or individual learners; costly for classrooms, teaching equipment, teachers and experts; traveling, accomodation and food supply is expensive, time consuming; ...

Research on regular fostering high school social sciences teachers by online training in addition to improving the professional qualifications of teachers to meet the fundamental and comprehensive reform of education in Vietnam today, it also open an opportunity for training, that is, online training teachers to reach regional and world standards, towards the development of cross-cultural teacher resources.

2. Research methods

System Analysis: Use the systematic analysis method to review and evaluate regular fostering social sciences teachers with those in other subjects in the high school curriculum, and with the overall of fundamental and comprehensive reform of education process in Vietnam.

Modern teaching technology methods: Online teaching using advanced technology such as the Internet, computer and computer software, together with modern teaching methods and management of learners that contribute to improving the quality and effective training.

Methods of survey, experiment, summation of experience to ensure the contents and regular fostering forms close to the actual requirements of each locality, subject, and subject learner.

3. Results and discussion

3.1. An overview of online training

According,[1] Teaching Online Geography textbook, "E-learning is an advanced, comprehensive training method that is able to connect and share knowledge effectively. It aims to achieve good teaching and learning objectives, including convenient direct interaction between learner and trainee as well as learning community through Information and Communication Technology"(p. 3).

Compared with traditional teaching, online training has the following outstanding advantages:

- Flexibility: Learners can study anytime, anywhere (Just-In-Time) as long as they have an Internet connection or other means of communication. This is very suitable for those who have a stable job wanting to further improve their knowledge level. Under the current conditions, the Internet has been widely spread throughout the territory of Vietnam, the Internet connection in teaching is quite possible.
- Personal adaptability: learners can complete the course quickly or slowly according to their cognitive ability and timetable.
- Reduced training costs: With the initial cost of designing online learning content, the training is evaluated as with much lower training costs than traditional classrooms.
- The rich learning materials, always updated: Online training can help learners to access the vast knowledge treasure of humanity with the support of the software to search on the Internet. On the other hand, learners can also share their learning resources with others. In particular, the materials are regularly supplemented, updated with new knowledge.
- To contribute to the training learning attitude skills: To learn in an online environment requires the learner to have good learning habits, self-learning skills and self-management of time. This gives the learner the ability to work independently and creatively. Unlike traditional classrooms, many students are hesitated that they have not yet bravely raised their objections and criticisms about teaching contents and teaching methods, online training creates discussion forums openly and straightforwardly, and at the same time eliminate the concerns of learners that with the same content, but each teacher has a different approach to the issue. This helps to train serious attitudes and develop critical thinking in learners.
- Enhance the ability to remember in the learner: Thanks to the outstanding advantages of multimedia technology, online training affects learners through many information channels such as text, images, charts, audio, video, animation, simulation, ... The learning environment has integrated

multimedia technology helps the learners to increase the ability to memorize. In addition, the learning content is designed, transmitted and linked to the online repository of data to help students easily review the knowledge that for the first time was not very clear to them.
- Easy management of learner learning: Through the LMS / LCMS remote learning management system, managers, faculty, families and other learners can easily follow the learning process. and results of the learner in each module. As a result, managers and faculty can identify what content the trainees need to be retrained or further trained to help them achieve the best results at the end of the course, and this helps the teaching to well differentiate and individualize learners.

3.2. Regular fostering social science subjects for high school teachers in the form of online training

3.2.1. Building some content for regular training of high school social science subjects to meet the educational reform.
In order to meet the requirements of education reform, by surveying the needs of a number of high schools, the author has built some special subjects on high school social sciences for regular training with new subject matter content and is necessary for high school social sciences education:
Core subject 1: "Fostering and Developing Social Science Subjects Program in High School"
a) The objective of the core subject
* Theory
- Understand and analyze the content of social sciences knowledge innovation in general program of the current education reform in Vietnam.
- Determine the scientific basis of social sciences knowledge content in the general curriculum
- Assessment of social sciences knowledge content in integrated and differential teaching of the general curriculum.
* Skills
- Ability to analyze and propose knowledge on social sciences in the general school curriculum in the direction of innovation.
- Skills to renovate the program
b) Summary of thematic contents
Providing learners with innovative models of education from some countries in the world that can be applied in the conditions of Vietnam; Basic concepts and methods of program development in general and of social sciences in particular: basis for the development of the program, the relationship between scientific knowledge and teaching knowledge in general schools with professional knowledge. The subject provides learners with a complete set of curriculum content related to social sciences in general schools, the degree of social sciences knowledge integration in the various disciplines. On this basis, the teacher can actively develop the school curriculum suitable for the target education group.
Core subject 2: "Teaching social sciences in a multimedia environment"
a) The objective of the subject
* Theory
To provide a basic understanding of multimedia teaching, about E-learning. Learn software applications in online course construction. The process of building and the method of using E-Learning in teaching social sciences.
* Skills
- Forming and practicing skills for teachers in building and using electronis lectures in social sciences teaching;
- Provide the skills to build courses and organize online teaching social sciences.
b) Summary of thematic contents
The subject provides the contents of the need for innovation of teaching methods in a multimedia environment; learn some application softwares; process and manipulation of e-lesson design, web-

based online teaching design, the development of a number of model lectures and the organization of e-learning.

Core subject 3: "Training social sciences teaching skills in High School"

a) The objective of the subject

* Theory

- Strengthen the practical skills and skills to exploit knowledge from different sources.
- Improve knowledge about the organization of educational activities for the learner.
- Strengthening skills in handling and dealing with pedagogical situations in the education process.

* Skills

Through the course, learners are well aware of the importance of fostering professional knowledge, practicing skills, teaching techniques, enhancing teaching capacity and quality, and professional ethics to meet the requirements of society in the current educational innovation cause.

b) Summary of thematic contents

The subject provides learners with a systematic view of the teacher's basic skills, reinforces learners with the skills they need during classroom instruction as well as organization of educational activities.

Core subject 4: "Organizing creative experiential activities for social science subjects in High School"

a) The objective of the subject

* Theory:

Identifying, analyzing and evaluating skills of appling theories learned through social sciences into problem solving in reality with the formation of competencies for learners; Points to note for teachers when guiding creative experiential activities; Implementation methodologies for teaching creative experiences in social sciences.

* Skills

Set up a quiz questionnaire system / game organization, contest organizing skills, experience organizing skills, etc.

b) Summary of thematic contents

The content of the subject equips learners with common issues about creative experiential activities; Identify creative experiential activities in social sciences teaching through the topics. Design of some specific creative experiential activities for social sciences teaching in general schools. Creative experiential activities in social sciences are diversed and integrated contents, bringing together knowledge and skills on a wide range of topics, fields of study and education that are designed as open topics.

3.2.2. Building the website 'regular fostering of high school social sciences.

The software used in website building:

The main tool for building online learning websites is the software E-Learning XHTML Editor (eXe) and the Modular Object-Oriented Dynamic Learning Environment (Moodle). eXe is an authoring tool designed to run on the web environment to help faculty in designing, developing, and publishing teaching and learning materials on the web. EXe software aims to provide an intuitive, easy-to-use tool that allows lecturers to publish professional web pages for teaching. Moodle is an open-source learning management (LMS) system that enables the creation of courses on the Internet or remote learning websites. Moodle has the advantages: flexible and customizable, each person can build Moodle activities on his/ her own and share them; Moodle has a global community that can coordinate and help each other during the work; Moodle is free to download and use.

Interface of the homepage (see figure 1).

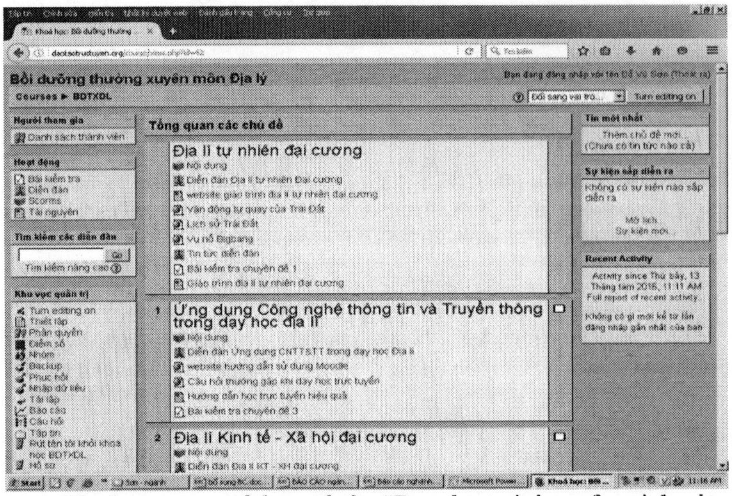

Figure 1. The homepage of the website "Regular training of social sciences".

* Main contents and teaching activities on the website:

- The course information: includes the name of the course, the requirements of the course, summary of the course, starting time, ending time of the course, ...
- Learner management: including student enrollment or exclusion, student attendance management (class or interest groups).
- Course materials: Contains the full content of the course including learning resource packages, video lectures, PowerPoint presentations, video clips, online reference materials, and more.
- Study Forum: is a place to raise and discuss topics. Learners can participate in one or more forums, find information in the forum, read and reply, send E-mail, chat with interested people. Trainers monitor and conduct the forum to the right direction, as well as review and evaluate learners.
- On-line testing and evaluation: On the website there is a tool to check online learners. Learners can participate in objective tests (biennial, right-wrong, multi-choice), in self-examination, or do exercises. After submitting the answers to the test, the learner receives the feedback including scores and comments from the teacher. Learners can self-assess their learning ability, thus adjusting their learning accordingly. The faculty member also grasps the acquirement of the learner to make timely adjustments to individual instruction (see Figure 2).

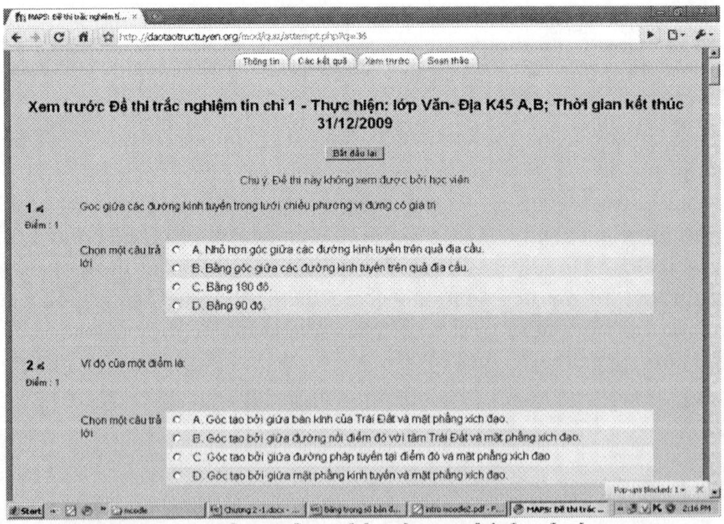

Figure 2. Interface of an objective multiple-choice test.

3.2.3. Design models for regular fostering courses in social sciences.

Online teaching and face-to-face teaching have their advantages and disadvantages. In order to maximize the effectiveness of each form of teaching, the author selects a model that combines these forms called blended learning.

According to [1] Teaching Online Geography textbook: "Combined teaching is the co-ordination of content, methods and ways of teaching and learning organization, between different forms of learning to optimize the strengths of each form, ensuring the effectiveness of education achieved is the highest" (p. 98).

There are several levels of combination in regular fostering of social sciences as follows:

- Level 1: Online teaching parallel to concentrated regular training. In level 1, learners will learn a part of the content in the form of face to face, often the hard part that needs to be explained and exchanged directly between the teacher and the learner, the rest they will learn on their own via the Internet. This type of teaching reduces the time spent in class while still ensuring the quality of teaching. For example, when teaching Core subject 4: "Creative experiential activities for social sciences in High School," it is possible for learners to pre-read, watch video tutoring experiential activities on the Internet, time in the classroom is for hands-on activities.
- Level 2: Online learning part of the refresher program. In level 2, the regular training program will be subdivided into modules, based on the content and relevance of each module to apply the online or offline teaching form accordingly. Level 2 instruction will provide an active, independent way for learners to be part of the program, thereby improving their online learning skills completely. Example of teaching Core subject 2: : "Teaching social sciences in a multimedia environment" can be devided into modules (as chapters of the subject), modules such as the theory of online teaching, the introduction of software used in online teaching,... students will self-study on the Internet, others parts will be taught in the form of face-to-face.
- Level 3: Online teaching the whole program. For learners who have improved their online learning skills while selecting appropriate regular training content, this training can also be conducted completely on the Internet.

3.3. Ability to use online training in teacher fostering to develop cross-cultural human resources

In the integration period, the training of teachers who have enough knowledge and skills to undertake international teaching is very important and is of particular interest. In order to meet the training and fostering of teachers capable of cross-cultural teaching, using online training is the best solution for the following reasons:

- With the ability to connect without borders, learners have access to cutting-edge programs with enormous human knowledge. They learn and exchange with many teachers, colleagues, people of the same interests from different nationalities.

Learners themselves can choose the appropriate knowledge for their expertise, determine the learning process accordingly.

Learners do not have to pay for accommodation, transportation, living expenses or other expenses such as tuition and lecture fees, etc.

- Learners can study anywhere, at any time, as long as they are suitable for the individual learner, to: "Learn any time, anywhere, and study for life."
- English is a barrier to the development of cross-cultural teachers. Through online training, they will improve their English skills through online English courses, through study and exchange in English with foreign lecturers and students.
- Online training is a new modern form of teaching that is only initially implemented in Vietnam. Therefore, the use of online training in teacher fostering is an opportunity to bring online teaching into the schools in Vietnam, helping Vietnamese teachers have the opportunity and conditions to reach other countries in the region, and the world.

4. Conclusion

The content of the article focuses on studying and analyzing the advantages and disadvantages of online training methods. From this, the combined teaching (Blended learning) model proposed is an effective and suitable option for social sciences teachers regular fostering at present. Introduce the regular training website for social sciences in high school with selected topics to meet the education reform in Vietnam; online teaching, exchange activities, discussion, evaluation, ... aim to achieve high efficiency teaching. The article also mentions the use of online teaching system in cross-cultural teacher training in the near future. The success of social science teacher training and retraining will be the basis for applying for other subjects.

In order to promote the effectiveness of research results, the author would like to make some notes:

1. It is necessary to maintain a team of highly qualified professional trainers to use and develop the regular fostering website. Regularly renovate regular training content in accordance with Vietnam MoET's orientation, suitable to the actual conditions of each locality and learners.

2. To create funding for the maintenance and development of websites, funding for building and development of the curriculum of regular training subjects and teaching expenses for lecturers.

3. General school teachers need to raise awareness about self-improvement, self-training, and improve their English. General schools should create conditions on funding, time and facilities for teachers to carry out their own self-refreshing tasks and regularly check and evaluate their regular training quality.

References

[1] Do Vu Son 2016 Online Geography Teaching Textbook (Thai Nguyen University Publishing House Thai Nguyen)

[2] Ministry of Education and Training 2011 Circular No.30/2011 / TT-BGD DT, Curriculum framework for regular fostering for kindergarten, general and regular training teachers (Hanoi)

[3] Ministry of Education and Training 2012 Circular No. 26/2012 / TT-BGD ĐT, Regulation on regular training of pre-school, general and regular training teachers (Hanoi)

Implementation of a primary sciencecurriculum designed in accordance with a social constructivist appoarch for Vietnamese Confucian heritage culture

Ngô Vũ Thu Hằng[1] , Astrid M. W. Bulte[2] , Albert Pilot[3]

[1] Hanoi National University of Education, 136 Xuan Thuy, Cau Giay, Hanoi, Vietnam
[2] Freudenthal Institute for Science and Mathemactics Education, Utrecht University, The Netherlands
[3] Freudenthal Institute for Science and Mathemactics Education, Utrecht University, The Netherlands

Email: [1]hangnvt@hnue.edu.vn , [2]A.M.W.Bulte@uu.nl , [3] A.Pilot@uu.nl

Abstract : This paper describes the implementation of a social constructivism-based primary science curriculum, which was designed culturally appropriate with Vietnamese Confucian heritage culture. It focuses on analysing a specific science lesson best functioning in classroom practice. Both achievements and shortcomings of the designed science curriculum were uncovered and provided an adequate evaluation to the designed curriculum. This creates a basis for appropriate adjustment and improvement to the curriculum design for a further better application in primary science education in Vietnamese Confucian heritage culture.

Keywords: Confucian heritage culture, design, science curriculum, social constructivism

1. Introduction

In spite of reforms, science education in Vietnam Confucian heritage culture (CHC) still adheres a teacher-centered approach with passive students in the classrooms in which reproductive teaching and learning is more dominant than a constructivist approach. Confucianism remains its influences in convetional education in the Asian countries such as Korea, China, Taiwan, and Vietnam because of its deep and long generic existence. It provides both hindering and fostering factors on the implementation of a social constructivist approach in Vietnam primary science education due to the similarities with and divergences from social constructivism [3]. To address the problems in primary science education in Vietnamese CHC and to foster the application of a social constructivist approach in teaching and learning science for better education quality, a science curriculum was designed to be culturally appropriate [4]. The curriculum design takes similarities and the divergences between social constructivism and the Vietnamese CHC into consideration and provides new educational guidelines promising for educational progress. Such a design-based approach aligns with perspectives on Asian learning approaches, that is to focus on applying and refining educational theories and make appropriate with Asian context [6] to avoid a cultural mismatch. The designed curriculum was implemented and accompanied by a program of teacher professional development, which requires interaction between Vietnamese teachers and the new designed curriculum [5], as a way to bridge the ideal curriculum and the operational curriculum [8]. The program of teacher professional development helped Vietnamese teachers, who have been influenced by CHC and accustomed to the traditional teaching of science, adapt to and accommodate the designed curriculum [5]. This paper focuses on analyzing the implementation of the designed social constructivism-based science curriculum by zooming into a specific lesson, which is assessed to be best functioning in the classroom practice. This aims to provide details about both the achievements and shortcomings of the designed curriculum. In this way, the designed curriculum could be adequately evaluated, providing a basis for appropriate adjustment and improvement for a further better application. This study contributes to the knowledge base about designing and applying a social constructivism-based curriculum for primary science education in Vietnamese CHC.

2. A social constructivism-based primary science curriculum designed culturally appropriate with Vietnamese CHC

The formal curriculum was designed based on 5 key features of a social constructivist approach, including: (i) Learning is social; (ii) Knowledge is experience based; (iii) Knowledge is constructed by learners; (iv) All aspects of a person are connected; and (v) Learning communities should be inclusive and equitable [1]. The total curriculum design consists of three interrelated parts: 1) educational goals, 2) the instructional framework, and 3) exemplary curriculum units. The educational goals are determined to develop scientific knowledge that is relevant and useful for students, to foster scientific thinking and scientific skills, and to develop critical attitudes towards science [4]. These goals are recommended by science educators [2] and consistent with the constructivist principles [7]. They also align with the goal of primary school education in Vietnam, that is to help students know how to develop self-directed learning and to be cooperative in learning, to remain proactive and active in finding and solving problems so that they can master new knowledge. Derived from these goals, a curriculum framework was constructed with the alignment of learning functions, learning settings, and educational expectations corresponding to each of the learning phases, which were labelled *Engagement*, *Experience*, *Exchange*, and *Follow*-[4]. The framework of the designed curriculum is presented in Table 1.

IOP Conf. Series: Journal of Physics: Conf. Series **1340** (2019) 012028 doi:10.1088/1742-6596/1340/1/012028

Table 1. The designed framework of the social constructivism-based science curriculum for Vietnamese CHC

PHASE	FUNCTION	LEARNING SETTING		EDUCATIONAL EXPECTATION
		Activities	Layouts	
Engagement	A. To provide students with a motivation to learn	1. Doing a small hands-on task with a relevant example related to scientific subject matter 2. Answering questions about *What*, *How*, and *Why* on a relevant example related to scientific subject matter	1. In small groups and/or in the class as a whole	a. Students are interested in scientific subject matter
Experience	B. To evoke attitudes towards science C. To acquire procedural knowledge D. To acquire conceptual knowledge E. To acquire argumentative skills	3. Predicting: Observe and discuss in order to answer questions: *What do you observe? What will happen if...? Why do you think so?* 4. Hands-on: Do experiment and discuss in order to answer questions: *What did you observe? How to explain? Why do you think so?* 5. Questioning: Formulate questions related to scientific subject matter	2. In small groups	b. Students are curious about learning representative examples of scientific subject matter c. Students are active in learning representative examples of scientific subject matter d. Students use their intuitive knowledge to learn about scientific subject matter e. Students argue with each other to attain consensually agreed knowledge on representative examples of scientific subject matter

Table 1 (Cont')

PHASE	FUNCTION	LEARNING SETTING		EDUCATIONAL EXPECTATION
		Activities	Layouts	
Exchange	F. To build on attitudes towards science G. To build on procedural knowledge H. To build on conceptual knowledge I. To build on argumentative skills	6. Presenting results to other groups 7. Discussing results with other groups 8. Answering formulated questions related to scientific subject matter	3. In the class as a whole and/or in combined groups	f. Students are interactive in learning scientific subject matter g. Students argue with each other to attain consensually agreed knowledge on scientific subject matter
Follow-up	J. To acquire cognitive flexibility K. To develop motivation for further learning	9. Providing answers and/or solutions for questions and/or problems related to scientific subject matter	4. In the class as a whole	h. Students can provide proper answers and/or solutions on applying attained knowledge i. Students show their desire to learn more about scientific subject matter

Based on the curriculum framework, three curriculum units (or lessons), including *Air pressure*, *Plant roots*, *CO_2 reactions*, were designed to exemplify the ideas of the formal curriculum [4] and conducted in succession by 3 Vietnamese teachers in classroom practices through the program of teacher professional development [5]. The lesson *Plant roots* in the sixth class (in total 9 implemented classes) was assessed to be best functioning and used for an in-depth analysis. The overall design of lesson *Plant roots* is presented in Table 2.

Table 2. The design of lesson *Plant roots*

Phase	Learning activity
Engagement	1. Drawing a complete plant 2. Answering: *What did you draw? Why did you draw the plant roots like that? How could you know the plant has such a root system?*
Experience	3. Predicting (Exercise 1): Choose a wild plant in the school garden to observe. Discuss in the group the answers to the following questions: *a. What do you think the plant root looks like? Draw them.* *b. Why do you think they look like this?* 4. Hands-on (Exercise 2): Pull out the wild plant in the school garden. Discuss in the group the answers to the following questions: *a. What does the plant root system look like? Draw it.* *b. Why does this plant have a root system like that?* *c. What are the functions of the plant root system? Why do you think so?* 5. Questioning (Exercise 3): Write down questions or ideas related to the subject matter that you want to discuss.
Exchange	6. Presenting results to other groups 7. Discussing results with other groups 8. Answering formulated questions related to subject matter
Follow-up	9a. Answering questions: - *What did you learn from the lesson today?* - *Can you provide some examples of root types and explain why you think those plants have such root types?* 9b. Determining type of root for some plants

The above pre-designed lesson *Plant roots* was later used for the teachers to rely on in order to co-design and develop the detailed lesson plans before they conducted the lesson in their science classrooms.

3. Methodology

3.1. Participants

The sixth class with the lesson *Plant roots* was conducted by Vietnamese teacher T1, whose pedagogical capability was appreciated by the school board of a public school located in an urban area in the North of Vietnam. Teacher T1 was 38 years old and had got 18 years of experience teaching in primary school. She was considered as a CHC teacher because her observed teaching characteristics, as analyzed in a previous empirical study [3], aligned with those of traditional CHC [5]. She was willing to experience a new approach of teaching as in the designed curriculum and stimulated by the board of the school.

Thirty Vietnamese students aged 10 were involved in the class of the experimental lesson *Plant roots*. The numbers of male and female students were relatively equal. These students had no experience with a social constructivist approach as in their conventional science curriculum [3].

3.2. Data Collection

A. Classroom observation. The lesson practice was observed by the first author and the other teachers of the team. The classroom observation was implemented with the activities of note taking, video recording, and audio recording. One video camera was located in a convenient place either in the classroom or in the school garden, depending on where the teaching and learning took place in order to have the best overview of the lesson. All of the discussions of the student groups in the class were audio recorded to help the authors obtain data which cannot be grasped by observing student learning from a distance. To record discussion of the groups, audio recorders were held by one of the students in each group (for outdoor activities) or set in the middle of the student tables (for indoor activities). The videotape and the audiotapes were later watched and listened to carefully, and transcribed verbatim.

B. Student questionnaires. Pre-questionnaires and post-questionnaires were employed for the study. The pre-questionnaires were distributed two months before the lesson took place; the post-questionnaires were distributed after the lesson. In both of the questionnaires, the students were asked to illustrate lessons in drawings (about a conventional science lesson and about the experimental lesson on *Plant roots*). In the post-questionnaires, the students were asked to answer closed-ended questions and open-ended questions that aimed to elicit their reflections on and evaluations of the lesson.

C. Post-interview. Three students were randomly selected by the first author to get involved in a face-to-face semi-structured group interview for about 30 minutes after the lesson. All of them were encouraged to be free in answering the questions, which were aimed at attaining their impressions of the lesson and testing their acquisition of knowledge on plant roots as introduced in the lesson. Students' answers were elaborated in more detail in the interview. The interview was audio recorded and transcribed verbatim afterwards.

D. Learning materials. All of the plant drawings and learning cards completed by the students were collected and used for the analysis of the lesson.

3.3. Data Analysis

The educational expectations (Table 1) were used as the organizing elements to analyze the lesson and provide the leading themes for presenting the findings. Classroom observation was used as the main data source, the findings of which were verified and triangulated by data from the other sources. The analysis of the lesson led the first author to formulate the main conclusions about the attainments of the expectations. Then the analysis and the corresponding conclusions to the attainments of the expectations of the lesson were discussed with the teachers in the team. The team discussion provided the consensus on the first cycle for the analysis and the conclusions about the lesson. In the second cycle, the in-depth analysis was carried out by the authors of the research team with several sub-cycles. In this way, the analysis and the corresponding conclusions about the attainments of the expectations were recorded in the format of a thick description of the lesson *Plant roots*.

4. Findings

Below is the detailed description of the lesson implementation and its achievements in the classroom practice.

4.1 Engagement Phase

Expectation a: *Students are interested in scientific subject matter.* In this phase, all five groups of six students were attentive in listening to the teacher's instruction. All of the students in the five groups were highly involved and concentrative in doing the practical task of drawing a complete plant as the main activity of this phase (Source A). This could be recognized by the students' actions, speech, and expressions, such as knitting the brows and furrowing the forehead. Five drawings of plants were presented by the five groups (Source A & D). The students remained rather timid in presenting their work in front of the teacher and the class, and in providing answers for questions (Source A). Nevertheless, they looked curious and attentive in attitude while listening to answers and observing drawings. Based on their concentrative and attentive attitudes and their high involvement in learning in this phase, the classroom observation showed that the students were interested in learning about plant roots. This observation is consistent with the views of the students on the lesson shared in the post-interview. The interviewed students expressed that they enjoyed the activity of drawing a complete plant because it made them want to know what the other groups drew for a complete plant and why they drew the plant with the root system they did (Source C). The conclusion therefore is that the students were interested in plant roots.

4.2 Experience Phase

Expectation b: *Students are curious.* In this phase, all students of the five groups were highly involved in doing the two learning tasks: (i) providing predictions about the root systems of the pre-determined plants grown in the school garden and providing explanations of their predictions, and (ii) pulling the real plants out, observing their root systems, and providing explanations of the observed root systems (Source A). Fifty-six questions regarding plant roots were posed by the students personally during their discussions in the five groups (Source A). Twenty-two of the 56 questions were "Why" questions which were often elaborated from previous answers and focused on the specific plant root examined by each group (Source A). In addition to individual questions, eight questions were formulated cooperatively by the five groups and presented on the completed learning cards, such as "*Why do big old trees have their root systems rising to the surface?*" (Group 1); "*How many root types are there in total?*" (Group 2); "*How will the tree be if it has no root*" (Group 3); "*Why do small plants often have fibrous roots and big plants have tap roots*" (Group 4); "*Why can plants live in the water?*" (Group 5) (Source D). In addition, practical actions, such as touching the plant, spreading the soil out at the foot of the plant to observe the root part better, bringing the root part closer to the eyes to observe details, and looking thoroughly at the root part, often took place while the students discussed the plant and its root system (Source A). The students' questions and actions show that the students' curiousity was evoked when they were learning about plant roots. We therefore conclude that the students were curious in learning about the root systems of the real plants.

Expectation c: *Students are active in learning.* All five groups looked highly enthusiastic and excited in their learning. This could be recognized through their voices, their body language, their smiles, and their laughter (Source A). The various scientific activities, such as observing real plants, predicting what their root systems will look like, pulling the plants out of the ground, describing the root part, discussing and explaining why the plants have such root systems were observed in all five groups (Source A). All of the five groups of students wrote about and provided drawings of the plant roots they examined (Source D). This active learning was also obvious as presented in the students' feedback on the lesson (Source B & C). All thirty students referred to the lesson in a positive way (Source B). According to the students, they had opportunities to work with real plants, stay in small groups, learn outdoors, discuss and cooperate with each other, and speak freely regardless of wrong answers (Source B & C). Twenty-two of 27 drawings in the student pre-questionnaires illustrated static, individualistic learning or teacher-centered learning. This was shown by the images presented in the drawings, such as that of a student sitting neatly, alone or separately beside her peer(s), looking at the blackboard or reading a textbook (20 of 22 drawings), or that of only the teacher teaching (2 of 22 drawings). Such static learning can be represented by drawing *a* in Figure 1 (Source B). In the post-questionnaires, a very different picture of learning emerges: 20 of the 28 drawings illustrate student learning in action and cooperation in various learning activities, as represented by drawing *b* in Figure 1 (Source B).

a. In a conventional science class b. In the (sixth) class with lesson *Plant Roots*

Figure 1. The students' drawings of their learning activities (Source B)

Expectation d: *Students use their intuitive knowledge.* Intuitive knowledge was used by the students when they worked on the predicting and hands-on tasks on plant roots (Source A). Since the students could not at first see the root systems of the given plants (because the root systems were under the ground), they often referred to their prior knowledge to describe the plant roots. When interacting with the real plant roots for the hands-on task, the students in the five groups used everyday knowledge to describe the forms of the root systems and to provide explanations, such as *spreading into many sides*, *falling down like a bunch*, *like a willow tree*, *like leaves of the willow tree hanging down*, *small as a finger*, *like thinly threaded meat*, and *like an octopus* (Source A). All of these expressions show their intuitive knowledge regarding plant roots. Similar comparative phrases and expressions were also used in the other discussions, in which students compared the root to a snake, determined sizes of roots by comparing them with the span of one group member, and referred to their non-scientific prior knowledge (Source A). In the post-interview, the interviewed students also acknowledged that many "funny" words and expressions were used by their peers when they were discussing the root systems of the given plants (Sources C). The "funny" expressions noted by the interviewed students reflected intuitive knowledge that the students applied to learning about plant roots in this lesson. The conclusion, therefore, is that the students used their intuitive knowledge to learn about the plant roots.

Expectation e: *Students argue with each other to attain consensually agreed knowledge.* Various personal scientific arguments and explanations about plant roots were raised by the individual students in the discussion in all five groups for answering the given question *Why do you think the plant has the root system like that?* (Source A). All of these personal arguments and explanations were simple in general and often used the relation-indicating linking word *because*, for example, *because this plant is small; because its leaves are large; because its shape is bent*; *because its foot has many small bodies growing up; because the plant is smaller than the others;* and *because it has serrated leaves* (Source A). Ten scientific arguments were made cooperatively in the five groups and written on the learning cards (Source D). Both of the personal arguments and the groups' arguments were constructed by the linking words *because, thus* or *therefore*, and supported by data and evidences, which often relied on the outer characteristics of the plant, such as small body, large leaves, bent shape, serrated leaves, or plant foot with small bodies (Source A). Despite the explanations and arguments, the students hardly argued with each other. They mainly provided single explanations and arguments, most of which were not evaluated, judged, qualified, or rebutted by their peers. The groups' final answers were often found to have been decided by the individuals who were the group leaders or advanced students in the groups (Source A). The conclusion, therefore, is that the students argued with each other to a small extent and they attained imposed knowledge rather than consensually agreed knowledge on plant roots.

4.3 Exchange Phase

Expectation f: *Students are interactive.* The students took up the roles of the questioner and the answerer by reusing the questions introduced for the hands-on task in their exchanging dialogues (Source A). At the beginning of this phase, the students exchanged ideas with rather passive attitudes (Source A). In the later part of this phase, the students became more active and involved in exchanging ideas (Source A). The activity of questioning was taken up by one or two of the students in the combined groups. These students used the questions given in the Experience phase as the leading questions. All of the students in the combined groups in turn provided their answers for the questions asked by their peers (Source A). In this way, the students presented the results of the inquiries carried out in the former phase to those from different groups. The conclusion, therefore, is that the students were interactive in learning in this phase.

Expectation g: *Students argue with each other to attain consensually agreed knowledge.* The style of argumentation in this phase was repeated what took place in the Experience phase. The students from two different groups played two roles: one as a questioner and the other giving answers. Many personal explanations were expressed by the students not evaluated, judged, qualified, or rebutted by their peers as in their discussion (Source A). Although the students provided many answers and presented results for students from different groups, the knowledge the students achieved in this phase remained separate rather than being consensually agreed. The conclusion is that the students argued with each other to a small extent in this phase.

4.4 Follow-Up Phase

Expectation h: *Students can provide appropriate answers and/or solutions.* Overall, the students provided appropriate answers for new questions asked by the teacher. Nevertheless, some answers given by the students were unsatisfactory when the teacher asked the students to determine the root type of a prepared plant (Source A). This happened again in the post-interview, in which the students were asked to answer some questions posed by the first author (Source C). 4 of 13 answers given by the students were unsatisfactory. Given the inappropriate answers of the students in the Follow-up phase and in the testing, the conclusion is that the expectation about students providing satisfactory answers by applying attained knowledge was not completely satisfied.

Expectation i: *Students show their desire to learn science.* Despite unsatisfactory answers, the students were highly active and involved in applying the knowledge of plant roots conveyed in the lesson to answer questions asked by the teacher (Source A). Twenty-nine out of 30 students were attentive and concentrative about learning in this phase (Source A). More than half of the class actively volunteered to provide answers to the teacher's questions (Source A). Only one student, whose seat was backside to the blackboard, was inattentive during the majority of this learning time (Source A). The students' desire to learn can be partly revealed by their answers in the questionnaires, in which all of them showed high extent they liked the lesson (Source B). This also could be found in their answers to the open-ended question asking them to reflect on the lesson (Source B). All three of the interviewed students revealed their wishes of having more opportunities to learn science as they had with the experimental lesson (Source C). We therefore conclude that the lesson was effective in inspiring students to have a great desire to learn science.

5. Conclusions

This study shows that the lesson (or curriculum unit) of the designed social constructivism-based curriculum can encourage Vietnamese students to be interested in scientific subject matter, curious in learning about scientific subject matter, and active in learning science. Also, the findings uncover that the curriculum unit could help students use intuitive knowledge to learn science, be interactive in learning, and have desire to learn science. With these educational attainments, the authors suggests that the designed curriculum can be a possibility for applying into primary science education in Vietnamese CHC in order to overcome the analyzed problems associated with primary science education.

Nevertheless, the study also showed that the designed curriculum was not effective enough in encouraging Vietnamese students to practice scientific argumentation in classroom practice. Due to the signinificance of argumentation to science and possible influences of Vietnamese CHC to it, the study suggests that the designed curriculum should be adjusted and improved to foster argumentation for Vietnamese primary students so that they can achieve better scientific knowledge and skills.

References

[1] Beck C and Kosnik C 2006 *Innovations in teacher education – A social constructivist approach* (Albany NY: State University of New York Press)
[2] Duschl R A, Schweingruber H A and Shouse A W (Eds) 2007 *Taking science to school – Learning and teaching science in grades K-8.* Washington, D.C.: National Research Council of the National Academies; The National Academies Press.
[3] Hằng N V T, Meijer M, Bulte M W A and Pilot A 2015 *The implementation of a social*

constructivist approach in primary science education in Confucian heritage culture (the case of Vietnam Cultural Studies of Science Education vol 10) chapter 3 pp 665-693

[4] Hằng N V T, Meijer M, Bulte M W A and Pilot A 2017 *Designing a primary science curriculum in a globalising world: how do social constructivism and Confucian heritage culture meet?* (Cultural Studies of Science Education vol 12) chapter 3 pp 739–760

[5] Hằng N V T, Bulte M W A and Pilot A 2017 *Interaction of Vietnamese teachers with a social constructivism-based primary science curriculum in a framework appropriate for a Confucian heritage culture* (Asia-Pacific Science Education vol 3) chapter 2 DOI 10.1186/s41029-017-0013-0

[6] Örtenblad A, Babur M and Kumari R 2012 *Learning in Asia.* (Asia Pacific Journal of Education vol 32) chapter 2 pp 131-136

[7] Sunal S C and Haas M E 2002 *Social studies for the elementary and middle grades.A constructivist approach Boston* (MA: Allyn and Bacon)

[8] Van den Akker J J H 2003 *Curriculum perspectives: An introduction.* In J J H Van den Akker, W Kuiper, and U Hameyer (Eds) (Curriculum landscape and trends Dordrecht, Netherlands: Kluwer Academic Publisher) pp 1-10

IOP Conf. Series: Journal of Physics: Conf. Series **1340** (2019) 012029 doi:10.1088/1742-6596/1340/1/012029

Implement of STEM education in Vietnamese high school: unit of acid-base reagent from purple cabbage

Nguyen Mau Duc [1*], Nguyễn Quang Linh [1], and Chokchai Yuenyong[2]

[1] Thai Nguyen University of Education, Thai Nguyen, Vietnam
[2] Faculty of Education, Khon Kaen University, Thailand
*Corresponding author email: nguyenmauduc@dhsptn.edu.vn

Abstract. This study aimed to clarify implement of STEM education in Vietnamese high school in the unit of acid-base reagent from purple cabbage. Methodology regarded experimental design. The samples included 84 students of experimental class grade 11 of 4 high schools: 02 schools in Hanoi are Hai Ba Trung high school and Van Lang high school; 02 schools in Thai Nguyen are Thai Nguyen high school and Gang Thep high school, in the term of 2017-2018. The intervention of STEM topic of acid-base reagent from purple cabbage was organized for the samples. Students' problem solving capacity in learning about the STEM topic of acid-base reagent from purple cabbage will be evaluated through their performance on conducted activities. Students' problem solving capacity score was compared between before and after impacts. The findings showed that the students' performances on conducted activities revealed their problem-solving capacity. The findings will clarify 1) summary of students' problem-solving capacity comparing between before and after intervention, and 2) the development level of students' problem-solving capacity. It found that average score of problem solving capacity of experimental class before impacts = 2.47. And, average score of problem solving capacity of experimental class after impacts = 3.13. The scores of the criteria for assessing the environmental impact of the experimental class after impact are always greater than before the impact, the value of $p < 0.05$, the level of ES effect is 0.78. The value of ES indicates that the experimental results have a moderate effect.

Keywords: STEM education, acid – base reagents, purple cabbage

1. Introduction

A boarded view of perceptions, beliefs and actions on STEM, Yuenyong [7] suggested consensus five characteristics distinguish integrated STEM instruction from other teacher pedagogy: (a) the content and practices of one or more anchor science and mathematics disciplines define some of the primary learning goals; (b) the integrator is the engineering practices and engineering design of technologies as the context; (c) the engineering design or engineering practices related to relevant technologies requires the use of scientific and mathematical concepts through design justification; (d) the development of 21st century skills is emphasized; and (e) the context of instruction requires solving a real-world problem or task through teamwork. This conceptualization of STEM is grounded in learning research [7].

Vietnam is moving to the industrialization and modernization to become a country which has the modern economy and get on well with international economy. Before the fourth industrial revolution, education has a vital role in the development of society. According to the direction 16/TC – TTg of the

Prime Minister on May 4th 2017, the solutions of education were presented: "...Strongly change the policy, program, method of education and training in order to create the qualified resources to deal with new technology, furthermore, we need to focus on pushing the training in Science, Technology, Engineering, Mathematics (STEM), foreign language and informatics in general education system. The tasks were also presented: Conducting education focusing on Science, Technology, Engineering, Mathematics (STEM) in general education system, organize to experiment at a few high schools from the term 2017-2018..." [4]

STEM education appeared in Vietnam a few years ago. It has just been the stage of media announcement and experimental. It has not been an official education activity in high school. However, STEM education with the tasks of providing crucial knowledge and skills for students in 21st Century will become a widespread education model in the future [1]. Therefore, STEM education is highly needed the concern and awareness of the whole society.

Nowadays, in Vietnam, STEM and STEM education have not been researched deeply. Those articles and materials of STEM education in Vietnam are just information, comments and applications in teaching through integrated topics [6]. Especially, STEM education topics in Chemistry are limited.

In the process of teaching Chemistry and determining soil environment of farmers, using reagents causes some problems: the lack of reagent paper because of the poor facilities; expired and broken reagents, unclear identification signals because of the poor quality, the size of reagent paper makes observing difficult. In this case, making reagent solution is a practical method. Today, some ingredients are easy to find in life to use as a reagent such as: hibiscus petals, gompherna globose, purple cabbage, hydrangeas flowers, turmeric, geranium petals, blueberries [5].

The method of making these solutions from available ingredients provides fate, excitement and students' learning efficiency. It also provides the purposes of innovating teaching method and improving education quality. Besides, this method makes students, farmers determine the soil environment to cultivate, using familiar products: vinegar, lemon, laundry soap without using harmful chemicals that makes Chemistry become closer to students and farmers lives, putting knowledge in reality.

2. Methodology

Methodology regarded experimental design. Students' products on the Intervention of the STEM topic of "acid-base reagent from purple cabbage" will be evaluated between before and after impacts in order to clarify students' problem-solving capacity.

2.1 Sample

The sample is randomly selected, including 84 students of grade 11, which is divided to 4 classes in 4 different schools (2 of them are in the city and the others are from the countryside of the Northern mountainous of Vietnam. Nowadays, the pressure of examinations and marks at high school is very high. The parents of students always want their children to have a good learning achievement but they do not care about other factors such as: problem solving capacity, creative capacity... of their children. Among with the number of students at the same time, we also feel difficult to observe and evaluate the development of those capacities of each and every student. Instead of that, we examine and evaluate each capacity using the observing form and take the average score for each capacity in the experimental class before and after the impact. We have worked with each group for about 1-2 months, 1-2 times a week with the education activities according to STEM orientation. We worked with the total of 6 periods.

2.2. Methods of Inquiry

The crucial method is impact research. Basing on the purpose of developing problem-solving capacity for students, the research group has presented some initial assessments: STEM activities are able to develop problem-solving capacity of students. Therefore, the research group builds practical inquiring forms in order to find out the difficulties and the needs of students about STEM education activities and problem-solving capacity. Using the result, the research group proposes the process of teaching Science using STEM education orientation to develop problem-solving capacity of students and design a full detail of one STEM activity with the topic: "Invent acid-base reagent from purple cabbage". The result of the impact will be evaluated and measured by problem-solving capacity criteria while students are solving these problems from the topic. Analyzing the result will help the research answer the questions: How STEM education activities affect on problem-solving capacity of students? How will these capacities be developed before and after conducting the STEM topic of "acid-base reagent from purple cabbage".

2.3 Intervention of the STEM topic of "acid-base reagent from purple cabbage"

The STEM education intervention was developed regarding on the general process of teaching Science in STEM education orientation including 5 steps [3] as the figure 1.

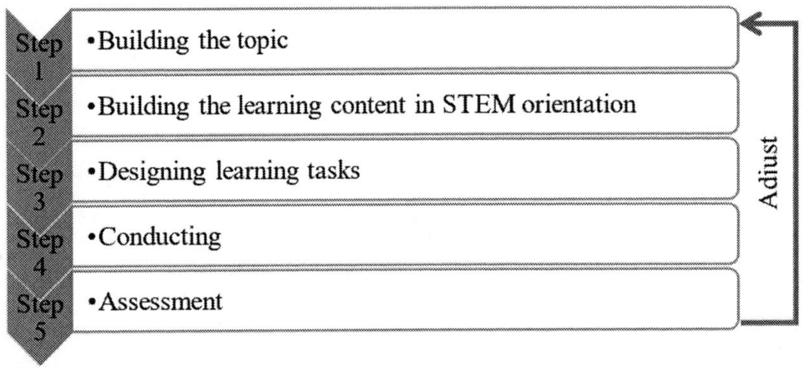

Figure 1. STEM Process of teaching Science in STEM education orientation

The learning activities in STEM topic of "acid-base reagent from purple cabbage" will be clarified through the following aspects: reason of choosing the topic, STEM knowledge in the topic, aims of the topic, and conducting of activities.

2.3.1 Reason of choosing the topic

Purple cabbage is a popular ingredient and easy to find in life. The main pigment extracted by purple cabbage is *cyanidin 3.5-diglucoside* of *the anthocyanin color* (purple) and it changes clearly based on the pH of the environment [2]. In acid environment, it has strong red color, in base environment, it changes to blue and maintains for a long time. Using purple cabbage solution to recognize acid and base and soil environment with simple methods, to choose fertilizer and plants which are suitable. Besides, making litmus paper helps students know about the environment of matters, helps farmers know about soil environment without purple cabbage.

2.3.2 STEM knowledge in the topic

- Science (S): How to determine the environment of matter, soil, the pH of the solutions
- Technology (T): Use safe and available ingredients: purple cabbage, lemon, baking soda, vinegar, soap, bleach, and so on.
- Engineering (E): The process of making reagent solution form purple cabbage
- Mathematics (M): Determine the pH of the solution

2.3.3 Aims of the topic

- Knowledge:
  + Students know how to use Science knowledge to solve problems.
  + Students are able to distinguish the environment of familiar substances in life using acid-base reagent.
  + Students analyze the acid-base soil environment, use it to give the solution to improve the soild and choose the suitalbe plants for the soil.
  + Students apply knowledge to make litmus paper from purple cabbage.
- Skills:
  + Students know the soil environment using acid-base reagent.
  + Students make litmus paper for farmers and high school laboratory
- Attitude
  + Improve the excitement to explore science, develop and create new things
- Capacities:
  + General capacities: Problem-solving capacity, cooperation capacity
  + Particular capacities: Chemistry experimental capacity, problem-solving capacity through Chemistry, capacity of apllying Chemistry knowledge in life.

2.3.4 Conducting of activities

The STEM topic of "acid-base reagent from purple cabbage" provided seven learning activities. These included a) activity 1: Building process of extracting purple cabbage solution, b) activity 2: Practice making an acid-base indicator extracted from purple cabbage, c) activity 3: Applying test set of indicators to identify environmental substances, d) Activity 4: Making indicator paper from purple cabbage solution, e) Activity 5: Test the indicator paper with substances need indentifying environment, f) Activity 6: Quantify pH of substances, and g) Activity 7: Complete the content of evaluation form of the whole process

a) Activity 1: Building process of extracting purple cabbage solution
- Aim: Students know how to extract purple cabbage solution
- Conduct: Students could be able to present the processes of extracting purple cabbage solution in the following 3 steps.
  - Step 1 included 1) Finely chopped purple cabbage, 2) put into a puree machine, and pressed, filtered for water.
  - Step 2 included 1) Finely chopped purple cabbage, 2) Soak in hot water for about 20 minutes, and 3) take the purple water.
  - Step 3 included 1) Separate layers of purple cabbage, 2) boil for 5 minutes. After heating at different times, the color of the solution is extracted from the purple leaf that is gradually darkened by purple pigmentation. Anthocyanin in the leaf is separated from the cell into a solution, and 3) filter the purple water.

b) Activity 2: Practice making an acid-base indicator extracted from purple cabbage
- Aim: Students know how to do some progress of extracting purple cabbage indicator for reagent.
- Conduct: Students can choose one in three processes to complete the task with the condition of the class

c) Activity 3: Applying test set of indicators to identify environmental substances.
- Aim: Students know who to determine environment using purple cabbage reagent.
- Conduct:
  - Take samples of the solution into the named cup. Squeeze in the cups of purple

cabbage solution that each group has prepared in advance.

- Observe and comment on the color change of indicator solution
- Make the following conclusions about the environment: acidic solutions are vinegar, lemon, chloroacetic acid, orange juice, yogurt; Basic solutions are backing soda, calcium hydroxide, ammonia solution, potassium permanganate, Vim water, Javen water. These could be seen as the figure 2.

Figure 2: Photos of students identifying environmental substances using purple cabbage solution

d) Activity 4: Making indicator paper from purple cabbage solution

- Aim: Students succeed in making indicator paper from purple cabbage solution
- Conduct:
  - Propose the problems: How long can purple cabbage solution be? How to preserve purple cabbage solution? If you have not kept long or no purple cabbage for a long time, how do you get the indicator?
  - Solve the problem: Depending on the temperature conditions to preserve the purple cabbage solution, usually at normal temperatures, purple cabbage solution can be kept for 3 days if not tightly closed, for longer will be damaged. To preserve the purple cabbage solution longer, it is possible to add alcohol 90o, or as a color indicator for long term use, in the absence of purple cabbage.
  - Making indicator paper from purple cabbage as following steps. Step 1: Prepare purple cabbage solution (process similar to the above, can use alcohol 90o). Step 2: Soak the filter paper in about 60 minutes (Figure 3). Step 3: Dry, filter paper (Figure 4). Step 4: Cut the 1 × 5 indicator (Figure 5. Step 5: Store in a clean zip lock or glass jar.

IOP Conf. Series: Journal of Physics: Conf. Series **1340** (2019) 012029 doi:10.1088/1742-6596/1340/1/012029

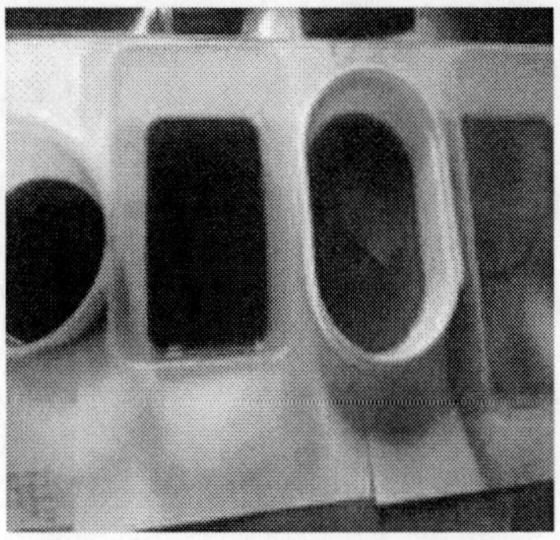

Figure 3: Soak filter paper

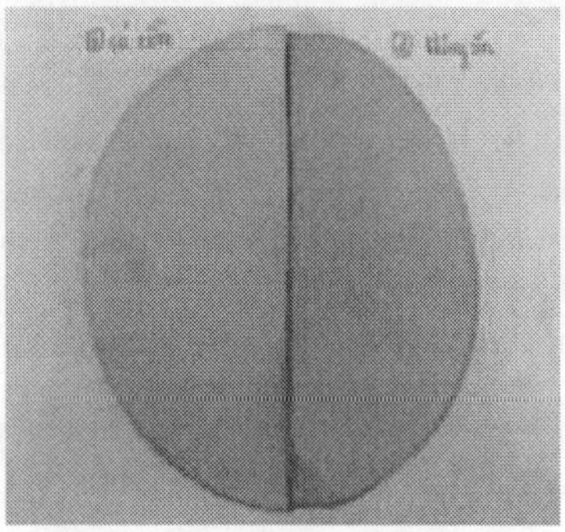

Figure 4: Dry filter paper

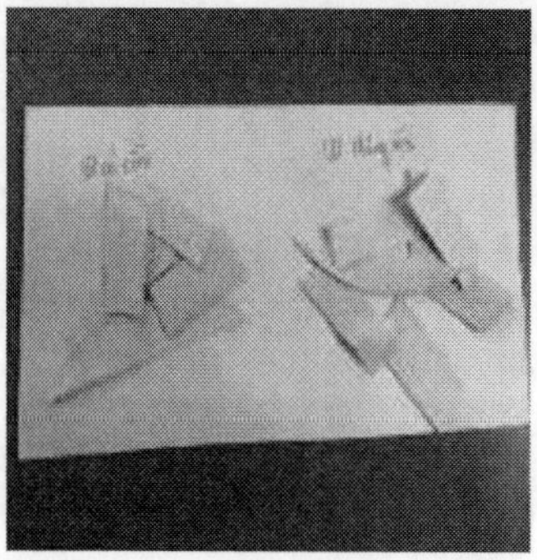

Figure 5: Cut the filter paper

e) Activity 5: Test the indicator paper with substances need indentifying environment

- Aim: Students identify environmental substances using indicator paper
- Conduct: Test paper produced with the substances that need to determine the environment as above. It was found that the paper had the character of purple litmus paper, so it is possible to use this method to make lintels for the identification papers in the laboratory at school or to test the soil environment in the season without cabbage. purple (Figure 6).

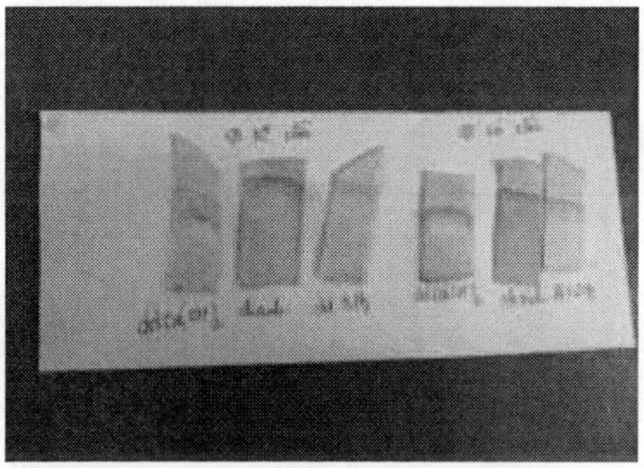

Figure 6. Testing indicator paper

f) Activity 6: Quantify pH of substances

- Aim: Students quantify pH of substances.
- Conduct: Indentify pH of substances
  - Based on the pH determination formula of the solutions: 0.1M hydrochloric acid, 0.1M ammonia solution, 0.1M NaOH solution, 0.01M Ca(OH)2 solution.
  - Based on the pH-determining scale of the lemon, vinegar, nabica salt, Vim water, Javen water, orange juice, yogurt.

g) Activity 7: Complete the content of evaluation form of the whole process

- Aim: Students complete evaluation form of products
- Conduct:
  - Teachers design self-evaluation form, question form for students' excitement after conducting this topic. This is a meaningful and necessary activity to finish STEM topic
  - Products score of each group will be counted by average score of student self-evaluation and teachers' evaluation

IOP Conf. Series: Journal of Physics: Conf. Series **1340** (2019) 012029 doi:10.1088/1742-6596/1340/1/012029

3. Findings

The 84 students of experimental class grade 11 of 4 high schools: 02 schools in Hanoi are Hai Ba Trung high school and Van Lang high school; 02 schools in Thai Nguyen are Thai Nguyen high school and Gang Thep high school, in the term of 2017-2018; designed lesson plans and conducted activities, evaluated products of students. The students' performances on conducted activities revealed their problem-solving capacity. The findings will clarify 1) summary of students' problem-solving capacity comparing between before and after intervention, and 2) the development level of students' problem-solving capacity.

3.1 summary of students' problem-solving capacity comparing between before and after intervention

All of groups have their own products with good quality. The result is calculated by the average score of students' self-evaluation and teacher's assessment. We used the observing form for teacher as a tool to evaluate problem-solving capacity of students in experimental class before and after impacts. The result was handled by mathematical statistics, we have synthesized the results as follows:

Table 1: Summary table of parameters evaluating problem-solving capacity of students before and after intervention.

Criteria	*Experimental class after impacts*					*Experimental class before impacts*				
	Number of students				*Average score of each criteria*	*Number of students*				*Average score of each criteria*
	1.0	*2.0*	*3.0*	*4.0*		*1.0*	*2.0*	*3.0*	*4.0*	
1	4	8	31	41	3.30	19	19	18	28	2.65
2	4	15	32	33	3.12	22	20	29	13	2.39
3	5	7	34	38	3.25	18	23	28	15	2.48
4	9	13	32	30	2.99	20	23	31	10	2.37
5	6	11	32	35	3.14	18	19	17	30	2.70
6	6	14	30	34	3.10	19	23	30	12	2.42
7	9	12	30	33	3.04	19	23	30	12	2.42
8	4	7	37	36	3.25	18	24	17	25	2.58
9	8	14	32	30	3.00	20	22	30	12	2.40
10	4	17	31	32	3.08	24	20	31	9	2.30
Average score of problem solving capacity of experimental class after impacts = 3.13						Average score of problem solving capacity of experimental class before impacts = 2.47				
Average score difference = 0.66										
Standard deviation of experimental class after impacts = 0.88						Standard deviation of experimental class before impacts = 0.85				
Independent t-test verification $p = 1{,}71.10^{-4}$										
Impact Level ES = 0.78										

Name the criteria in numerical order: 1) Find and identify problematic situations in chemistry; 2) identify and address problematic situations in chemistry related to practical applications; 3) Analyze problematic situations in chemistry and related practical applications; 4) Planning and solving some simple problems in studying chemistry and in practice; 5) Identify information, collect and clarify information relevant to the issue to be addressed in the STEM topic; 6) Proposing some solutions for resolving problems; 7) Analyze and select the most suitable solution; 8) Successfully implemented the solution according to the selected plan; 9) Evaluate the effectiveness of the chosen solution accurately and briefly; 10) Apply the solution in the same context or situation.

3.2 The development level of students' problem-solving capacity

The development level of students' problem-solving capacity could be represented as the figure 7. From the results of the problem-solving capacity criteria in students after the topic evaluated by the teacher, the scores of the criteria for assessing the environmental impact of the experimental class after impact are always greater than before the impact, the value of $p < 0.05$, the level of ES effect is 0.78. The value of ES indicates that the experimental results have a moderate effect.

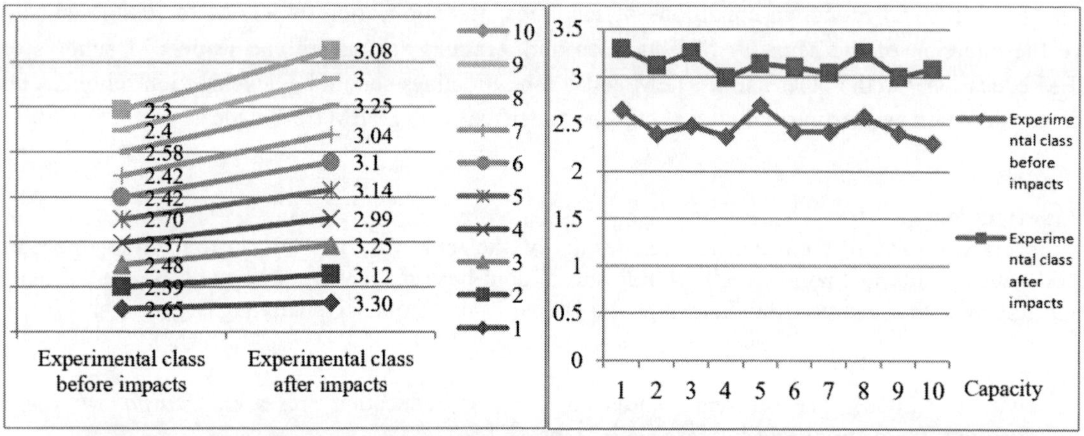

Figure 7: Chart progress of problem solving capacity of experimental class before and after impacts

Criteria for the higher grade of the experimental class were criteria 1 (3.30), criteria 3 (3.25) and criteria 5 (3.14), criteria 8 (3.25). This means that the students has realized that after completing the topic, students will be able to analyze the problematic situations in studying chemistry and related to practical application in studying Science, successfully implemented the solution according to the selected plan. The ability to identify information, to collect and clarify information relevant to the problem solved in the STEM topic is very good. However, the criteria for having a lower grade is that of criteria 7 (3.04), criteria 9 (3.0) indicates that students have not shown good performance in analyzing and choosing the right solution. Incorporate problematic situations in the study of Science and related to practical application, have not evaluated the effectiveness of the solution has been selected precisely and briefly. As such, STEM education orientation has created a very natural environment for students to express themselves in situations where they need to apply knowledge to solve practical problems.

4. Conclusion

It is possible to develop problem-solving capacity of students through STEM activities. Teaching in STEM education orientation provides the needs of innovating teaching method in high schools. Not only it completes the aim at teaching, it also make the opportunities for students to self-study, self-research, develop problem-solving capacity of students. Problem-solving capacity of students in experimental class after impacts always better than before impacts which is shown clearly by the observing form of teachers. Research result need to be experiment in wider objects, in longer time so that the conclusion given will be more realiable.

Nowadays, STEM education is an education orientation which is necessary in Vietnam. It encourages training in Science, Technology, Engineering and Mathematics, improve creativity, desire to explore science. With the solution of making indicator from available ingredients, it creates belief, excitement and efficiency in learning for students. This makes Chemistry become closer to students' life, connect the knowledge to real life, apply the development of science. This is a practical activity that helps students build experimental capacity, the ability to apply knowledge in reality, to solve problems and give more motivation in learning Chemistry.

The exploitation and application of the strengths of STEM education is extremely beneficial and necessary for general education in Vietnam. STEM education creates people who can meet the needs of the 21st century, responding to the nation's economic and social development and can positively impact the changing economies. knowledge in the context of globalization. But in the real situation of the country in terms of qualification and capacity of the managerial staff of teachers; material facilities of the school; Socio-economic conditions of localities, the application of STEM to schools should have the attention of the Ministry of Education and Training, managers and leaders of schools to STEM education. On the other hand, STEM education guidelines should be developed in schools, and testing, evaluation and testing activities should be consistent with STEM's basic ideas.

Acknowledgement
This article is published from the research results of the topic CS-2019-05" in the " Implement of STEM education in Vietnamese high school: unit of acid-base reagent from purple cabbage " article, this article by author Nguyen Mau Duc, Nguyen Quang Linh, Chokchai Yuenyong.

References

[1] Brown J. (2012). The current status of STEM education research. *Journal of STEM Education: Innovations and Research*, 13(5), pp. 7-11.

[2] Chigurupati, N., Saiki, L., Gayser, J. C., Dash, A.K. (2002), "Evaluation of red cabbage dye as a potential natural color for pharmaceutical use", *International Journal of Pharmaceutics*, 241, pp. 293–299

[3] Lantz Jr H. B. (2009), *Science, technology, engineering and mathematics (STEM) education. What form? What function*, CurrTech Integrations, Baltimore.

[4] Minister of Education and Training (2018), *The overall general education program in new general education program,* January 19th 2018.

[5] Phuong, N.T.L. (2015), *Describe the concept and standard output of problem-solving capacity with high school students,* Institute of Educational Sciences, Vietnam.

[6] Quang, L.X. (2017), *Teaching general Technology in STEM education orientation,* Doctoral thesis of Educational Sciences, Hanoi National University of Education.

[7] Yuenyong, C. (2019). Lesson learned of building up community of practice for STEM education in Thailand. *AIP Proceeding*, 2081: 020002-1- 020001-6.

Situation of organizing STEM activities in Vietnamese Schools

Nguyen Mau Duc [1*], Nguyễn Quang Linh [1], and Chokchai Yuenyong[2]

[1] Thai Nguyen University of Education, Thai Nguyen, Vietnam
[2] Faculty of Education, Khon Kaen University, Thailand
[*]Corresponding author email: nguyenquanglinh@dhsptn.edu.vn

Abstract. This study aimed to clarify situation of organizing STEM activites in Vietnamese schools. Methodology regarded the survey research and document analysis The study aimed to clarify 1) situation of organizing STEM activities in Vietnamese schools, 2) Existing ideas of problem-solving capacity as STEM education evaluation framework, and 3) developing framework of STEM education orientation. Samples inclued total of 58 teachers and 628 students of these schools: Hai Ba Trung high school, Nguyen Tat Thanh high school and Van Lang high school in Hanoi; Thai Nguyen high school, Dong Hy high school, Gang Thep high school, Luong Ngoc Quyen high school, Luu Nhan Chu high school, Dai Tu high school in Thai Nguyen and other nearby high schools. The findings revealed that some of teachers' awarenss on STEM activities including STEM, STEM education, STEM festival, and STEM club. And, Low teachers' awareness on STEM activities included STEM occupation, STEM resources, and robotics competition. The reality of developing problem-solving capacity in schools was provided in small frequency. It found that the the Ministry of Education and Training (Vietnam, 2018) provided some ideas of problem-solving capacity as STEM education evaluation framework. And, the MOET framework and literatures suggested an STEM Process of teaching Science according to STEM education orientation.

Keywords: STEM education, activities, Vietnamese schools

1. Introduction

Increasingly, educators, policy developers, non-government organization, and business and industry organization are stressing the urgency for improving STEM skills to meet current and future social and economic challenges. There are many concerning with the STEM education in shaping innovation and development. This push the policy of many countries continues under the orientation of STEM education [10].

Internationally, there is a boarded view of perceptions, beliefs and actions on STEM. However, the leaders and educators in each country held the varied beliefs, focuses, and actions of STEM education. They organized various kinds of STEM activities STEM educational reform, preparing for STEM workforce and STEM literacy, structuring STEM as a track of school curriculum, providing a broad range of content and practices, exploration of different teaching strategies for STEM, teacher preparation for STEM education teaching approach, and so on [1]; [2]; [3]; [4]; [6]; [8]; [12]; [13]; [14]; [15].

STEM education appeared in Vietnam a few years ago. It has just been the stage of media announcement and experimental. It has not been an official education activity in high school. However, STEM education with the tasks of providing crucial knowledge and skills for students in 21st Century will become a widespread education model in the future [3]. Therefore, STEM education is highly needed the concern and awareness of the whole society. To move further on STEM education in Vietnam, the studying of what they organize the STEM activities may provide baseline of developing STEM education in Vietnamese schools.

2. Methodology

Methodology regarded the survey research and document analysis The study aimed to clarify 1) situation of organizing STEM activities in Vietnamese schools, 2) Existing ideas of problem-solving capacity as STEM education evaluation framework, and 3) developing framework of STEM education orientation.

2.1 Samples

Samples inclued total of 58 teachers and 628 students of these schools: Hai Ba Trung high school, Nguyen Tat Thanh high school and Van Lang high school in Hanoi; Thai Nguyen high school, Dong Hy high school, Gang Thep high school, Luong Ngoc Quyen high school, Luu Nhan Chu high school, Dai Tu high school in Thai Nguyen and other nearby high schools.

2.2 Data collection and analysis

The research team conducted the survey from July 2016 to August 2017. The survey form of organizing STEM activities was developed to examine the reality of organizing STEM activities in Vietnamese schools, 2) the reality of developing problem-solving capacity in schools. Then, statistics descriptives were provided to describe the reality of organizing STEM activities at high schools and developing problem-solving capacity for students.

Document analysis was also carried out. The literatures were categorized and find pattern to develop existing ideas of problem-solving capacity as STEM education evaluation framework and STEM Process of teaching Science according to STEM education orientation.

3. Findings

3.1. Reality of organizing STEM activities at high schools

For teachers: The result of the awareness of teachers in Science subject about STEM shows that a few teachers know about STEM and STEM education activities, issues related to STEM education is presented in figure 1. High teachers' awarenss on STEM activities inclued STEM, STEM education, STEM festival, and STEM club. Low teachers' awareness on STEM activities included STEM occupation, STEM resources, and robotics competition.

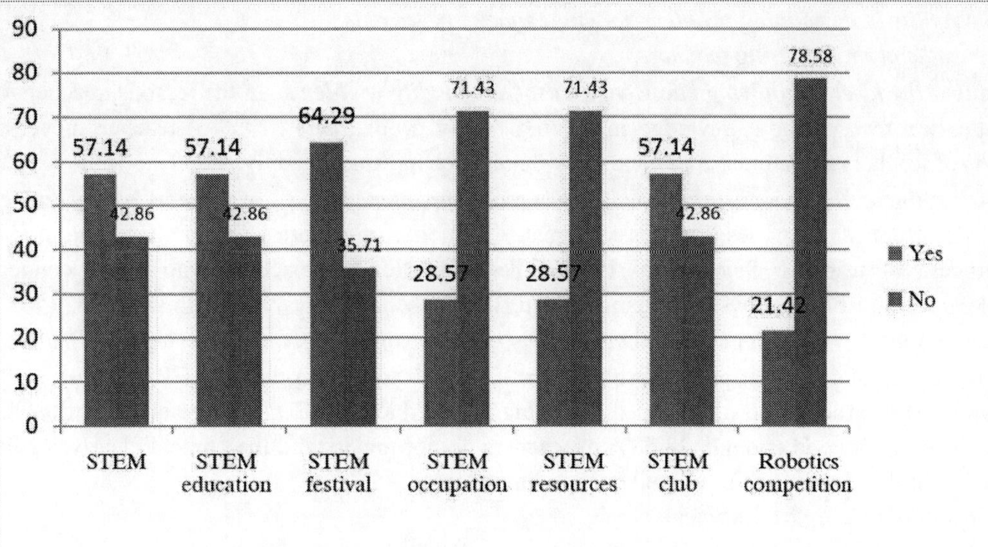

Figure 1. *Graph of STEM awareness*

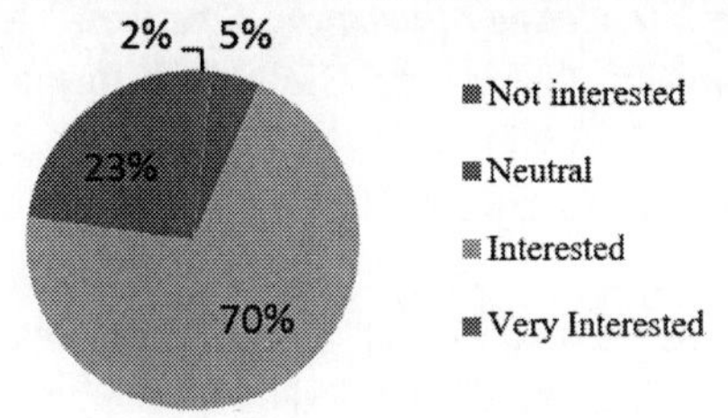

Figure 2. Graph of STEM activity excitement of students

Exchanging with teachers gives these comments: integration and problem-solving capacity related to reality are difficult problems with teachers. Although it has been focused for years, the reality of conducting is not as expected. Applying the advantages of STEM education is very beneficial and necessary with general education in Vietnam but STEM education is just conducted in high school in order to attract a number of students attending the competition, for example, the annual Intel ISEP.

With students: The result shows that in the learning process, students are interested in learning STEM education orientation (Figure 2). STEM education highly recommends the improvement and the development of problem-solving capacity for students. In each lesson with STEM topic, students have to deal with a practical situation which needs to be solved and relates to scientific knowledge. To solve the problem, students need to research the knowledge of the subjects related to the problem (through the text book, learning materials, experiments, technology) and use them to solve the problem. These knowledge and skills have to be integrated and connected to each other to help students know about the principles, practise and make products in daily life.

3.2 The reality of developing problem-solving capacity in schools

a) Results of investigating teachers

- *About the level of applying knowledge in solving reality problems:* In the lessons, teachers do not pay attention to *applying* knowledge in solving reality problems, 14,29% of teachers never apply, 62,84% of teachers sometimes apply and 22,87% of teachers usually apply.

- *About the level of connecting knowledge in teaching Science:* Most of teachers hardly connect the *knowledge* from Mathematics, Physics, Chemistry, Biology, Informatics, Technology, Engineering in the process of teaching Science (57,19%). The proportion of teachers who never connect the knowledge and just teach Physics, Chemistry and Biology separately is very high: 14,26%.

- *About the level of difficulty that teachers have while using problem solving teaching method*: The majority of teachers say that problem solving method takes too much time (85,71%) and it is difficult to evaluate the participation of students in solving problem (50%). Besides, the quality of teachers in this method is not good enough, 28,58% of teachers do not understand the content clearly. 42,84% of teachers find that the capacities of students are limited.

b. Result of investigating students.

- *About the level of applying knowledge in solving reality problems:* In the lessons, students do not pay attention to *applying* knowledge in solving reality problems, 14,29% of students never apply, 42,84% of students sometimes apply.

- *About the level of connecting knowledge in learning Science:* Most of students hardly connect the knowledge from Mathematics, Physics, Chemistry, Biology, Informatics, Technology, Engineering in the process of teaching Science (77.38%). The proportion of teachers who never connect the knowledge and just *teach* Physics, Chemistry and Biology separately is very high: 12.62%.

- - *About the attitude when recognize the conflict between the exercises and the questions:* Most of students do not care about strange problems (56.2%), a small number of students pays attention and find out how to solve a conflict (17.3%). This could be caused by teachers who do not focus on the situations in teaching.

3.3. Existing ideas of problem-solving capacity as STEM education evaluation framework

3.3.1. The concept of problem-solving capacity

Problem-solving capacity shows the ability of an individual (working by himself/herself or working in groups) to think about the situation, research and conduct the solution for the problem. Therefore, we can know as: *Prblem solving capacity is the ability to use effectively the awareness process, actions and attitudes, motive and emotion to solve problems that there is no availabe process and normal solution* [10].

3.3.2. Expression of problem-solving capacity

According to general education of the Ministry of Education and Training [9] general capacities of students were determined. Problem-solving capacity of high school students has these expressions:

- Find out and clarify problems: Analyze the situation in learning and life, determine and propose the situation in study and life.
- Propose and select solutions: Collect and clarify the information related to the problem; propose and analyze some solutions to solve the problem; choose which one is most suitable.
- Conduct and evaluate the solutions: conduct and evaluate the solutions of the problem, think about the method and process of solving problems to adjust and apply in new scenes
- Aware of new ideas: Determine and clarify information, new ideas and different sources of information; analyze the information independently to see the new ideas.

- Form and improve new ideas: Propose new ideas in learning and life; think in a new way; créate new factors from different ideas; form and connect the ideas; research to change the methods before the change; evaluate the risk and have back up plan.
- Think independently: Ask valuable questions, do not take one way information easily; do not prejudice after evaluating; care about the evidence, ready to evaluate the problem again.

From the expressions of problem-solving capacity developed in Vietnamese high school students, we find the similarities with the process of STEM education, so that we use this expressions to organize learning activities for students, to build the evaluation tools for problem-solving capacity development through STEM education for students in teaching Science.

3.4 STEM Process of teaching Science according to STEM education orientation

From the international experience in approaching STEM education [7] (Process 5E, process of approaching scientific research method, process of approaching engineering design) and the reality of the program and Science text book in the new general education in Vietnam, we research and propose the general process of teaching Science in STEM education orientation including 5 steps (Figure 3):

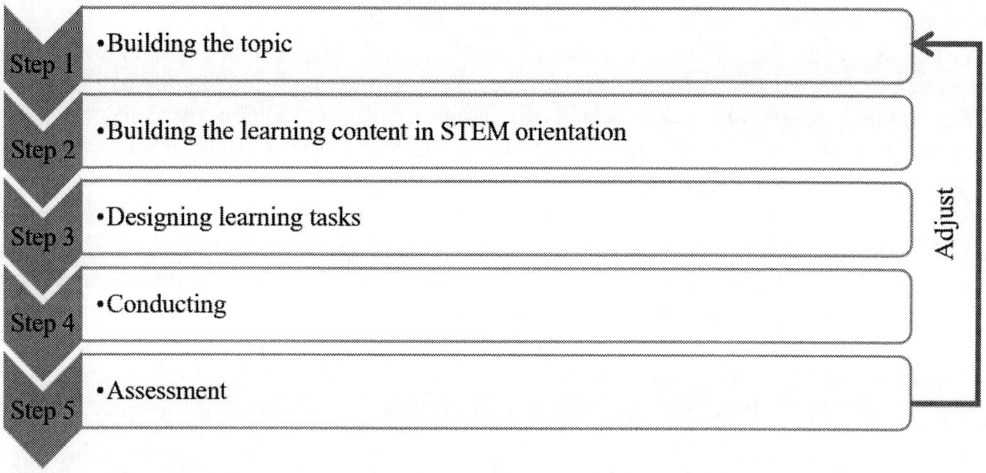

Figure 3. STEM Process of teaching Science in STEM education orientation

5. Conclusion

This study provided some beliefs and actions about STEM Education in Vietname. The teachers' awareness on STEM education revealed the reality of organizing STEM education in Vietname. According to the findings, it seems that STEM resources and robotics competition should be strongly supported. And, based on the MOET, it have showed that MOET provided alreay STEM education orientation in school setting. Therefore, Vietnamese teachers may develop STEM education learning activities regarding on the 5 steps. These included (1) Building the topic, (2) Building the learning content in STEM orientation, (3) Designing learning tasks, (4) Conducting, and (5) Assessment.

Acknowledgement

This article is published from the research results of the topic CS-2019-05" in the "Situation of organizing STEM activities in Vietnamese School" article, this article by author Nguyen Mau Duc, Nguyen Quang Linh, Chokchai Yuenyong.

References

[1] Barlex, D. (2009, August). The STEM Programme in England—Help or hindrance for design and technology education? Paper presented at the PATT22 conference, Delft, Netherlands. Retrieved from http://www.iteaconnect.org/Conference/PATT/PATT22/Barlex.pdf.

[2] Bell, D. (2012). Perceptions of STEM, within the context of teaching D&T in secondary schools: A phenomenographically inspired study. In T. Ginner, J. Hallstro¨m, & M. Hulte´n (Eds.), *Technology education in the 21st century* (pp. 74–79). Linko¨ping: Linko¨ping Electronic Conference Proceedings.

[3] Brown J. (2012). The current status of STEM education research. *Journal of STEM Education: Innovations and Research*, 13(5), pp. 7-11.

[4] Bybee, R. W. (2013). *The case for STEM education: Challenges and opportunities*. Arlington: NSTA Press.

[5] Herschbach, D. R. (2011). The STEM initiative: Constraints and challenges. *Journal of STEM Teacher Education*, 48(1), 96–122.

[6] Kim, J. S. (2011, November). New national curriculum of technology education in Korea. Paper presented at the 2011 international conference on technology education in the Pacific-Rim Countries, Nagoya, Japan.

[7] Lantz Jr H. B. (2009), *Science, technology, engineering and mathematics (STEM) education. What form? What function*, CurrTech Integrations, Baltimore.

[8] Marginson, S., Tytler, R., Freeman, B. and Roberts, K. (2013), STEM: country comparisons: international comparisons of science, technology, engineering and mathematics (STEM) education. Final report. Australian Council of Learned Academies, Melbourne, Vic.

[9] Minister of Education and Training (2018), *The overall general education program in new general education program,* January 19th 2018.

[10] National Science and Technology Council. (2013). *A report from the committee on STEM education*. Washington, D.C: National Science and Technology Council.

[11] Phuong, N.T.L. (2015), *Describe the concept and standard output of problem-solving capacity with high school students,* Institute of Educational Sciences, Vietnam.

[12] Ritz, J. M. and Fan, S. C. (2015). STEM and technology education: international state-of-the-art. *Int J Technol Des Educ* (2015) 25:429–451

[13] Sanders, M. (2009). STEM, STEM education STEMmania. *The Technology Teacher*, 68(4), 20–26.

[14] Williams, P. J. (2011). STEM education: Proceed with caution. *Design and Technology Education*, 16(1), 26–35.

[15] Yuenyong, C. (2019). Lesson learned of building up community of practice for STEM education in Thailand. *AIP Proceeding*, 2081: 020002-1- 020001-6.

IOP Conf. Series: Journal of Physics: Conf. Series **1340** (2019) 012031 doi:10.1088/1742-6596/1340/1/012031

Teaching and Learning about Magnetic field and Electromagnetic Induction Phenomena integrated Science, Technology, Engineering and Mathematics (STEM) Education in Vietnamese high schools

Thai Quoc Bao, Cao Tien Khoa, Ngo Tuan Ngoc, Nguyen Thi Thu Ha, Vo Quang Hoan, Pham Hong Quang and Chu Viet Ha*

Thai Nguyen University of Education, 20 Luong Ngoc Quyen, Thai Nguyen, Vietnam

*E-mail: chuvietha@tnu.edu.vn

Abstract. This work presents the study on teaching and learning about magnetic and electromagnetic induction phenomena integrated science, technology, engineering and mathematics (STEM) education in Vietnamese high schools for addressing issues of relevance and epistemic practice. We have explored STEM education with the creative and experiential activity in magnetic and electromagnetic induction phenomena. The STEM experiments and technical toys about magnetic and electromagnetic induction phenomena have been designed to be applied for active learning for students. The integration of STEM education will increase motivation and achievement in physics subjects and enhance capacity to improve student learning.

1. Introduction

STEM stands for Science, Technology, Engineering, and Math. STEM education is essentially meant to equip learners with the necessary knowledge and skills in the areas of science, technology, engineering and math. Teaching STEM has helped to "flatten" the global economy in the 21st century. STEM education has evolved into a meta-discipline, an integrated effort that removes traditional barriers between these subjects, and instead focuses on innovation and the process of applying design solutions to complex context issues using existing tools and technologies [5]. STEM education projects and plans have been developed all around the world. The US has launched the STEM education strategic plan that was used multiple strategies to make progress on improving STEM education including: Making STEM a priority in more of the Administration's education efforts, getting ambitious but achievable goals and challenging the private sector, and even Capitalizing on the President's deep interest and leadership [4]. There is even a project to enhance the access of female students to STEM by developing a bridging program for female students at the University of Malawi (Africa) [8]. The survey of IRIS project (Interests and Recruitment in Science; http://iris.fp-7.org/about-iris/) in Austria and Germany has reported that the STEM studies choice and preference of first-year students depend on high school experience and good teachers [3].

STEM education activities show that STEM education applies a learning approach based on practice and experiential activities. The most progressive, flexible educational methods such as project-based learning, learning through the games, and especially learning through practice are fully applied STEM integrated subjects. Students who follow the STEM approach have outstanding advantages such as: firm knowledge of science, technology, engineering, and mathematical; creativity, logical thinking, excellent performance in learning and working. Also, they have the opportunity to develop soft skills more comprehensively while not feeling overloaded [1-7]. For high school students, taking STEM courses also has a positive effect on future career choices. When learning a variety of knowledge in an integrated way, students will actively enjoy their learning rather than being afraid of or avoiding a certain area. This will, thereby, encourage them to have a better orientation when choosing majors for higher education and certainty for future careers.

To date in Vietnam, STEM education is not just studied and developed in universities but has been already progressed in high schools [9]. Research on STEM education has been carried out in the Faculty of Physics, Thai Nguyen University of Education. First of all, STEM education is developed to teach students how to apply STEM teaching in high schools [7]. Physics lessons will be interesting, engaging, and help students understand better if teaching is linked to visualizations and active learning. The skills and knowledge in the areas of science, technology, engineering and math (called STEM skills) must be integrated, complemented with each other to help students not only understand the principles but also be able to apply them in practice and create products in daily life. Compared with natural sciences, physics is a subject that is closely related to mathematical, technological, technical and scientific contents, so STEM-oriented education in teaching Physics has a very favorable condition. STEM education is aimed at developing following specific capabilities of each subject for students:

i. Critical thinking and problem solving skills
ii. Communication and collaboration skills
iii. Creativity and discovery skills
iv. Culture, information and communications technology
v. Ability to perform project-base tasks
vi. Presentation capacity

In this work, we designed the process of teaching some STEM-oriented contents of "Magnetic field" in Physics Grade 11 in order to enhance the activeness, train skills and develop capabilities for students

2. Methodology

2.1. The process of teaching physics under STEM education

STEM education applies practice-based and experiential approaches; therefore, when following STEM method, students will be able to use materials to generate products that will enable them to develop their competences in science, engineering, technology and mathematics to acquire knowledge. Through studying the theory of active teaching, and STEM teaching, we have proposed a process of teaching physics related to STEM education which includes the stages showed in figure 1.

The stages of teaching physics integrated STEM are described in detail following steps:

Stage 1: Select STEM topics

STEM topics in teaching are issues associated with the designing and manufacturing of products related to high school physics knowledge. STEM topics should have a close relationship with reality, solving practical problems, serving the daily life or entertainment of learners.

Stage 2: Define STEM tasks

Based on the STEM topics, students are asked to identify STEM tasks that need to be addressed

Stage 3: Study STEM documents and materials

After receiving STEM assignments, students study STEM documents so that they can clearly identify the ideas and methods of generate outcomes as well as learn about the contents of relevant physics knowledge in from the STEM topics.

IOP Conf. Series: Journal of Physics: Conf. Series **1340** (2019) 012031 doi:10.1088/1742-6596/1340/1/012031

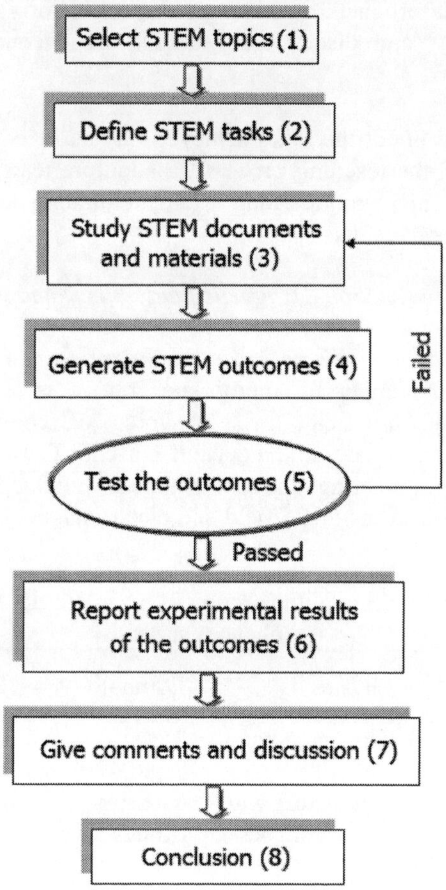

Figure 1. Procedure of the teaching process under STEM education.

Stage 4: Generate STEM outcomes
- Divide classes into groups
- Groups actively assign specific tasks to each member of the group according to their capacity
- Investigate the tools and materials distributed
- Generate STEM outcomes

This is a difficult step so teachers need to keep track of the activities of each group and give timely help and guidance if necessary. On the other hand, product manufacturing also requires students to meet safety rules so teachers should prepare safety rules and disseminate them to learners. Some complex details may need processing, and then teacher can help learners in this process.

Stage 5: Test the outcomes

During this phase, there are two possibilities for the groups that test the outcomes to access the relevant knowledge of of the topic: (1) If the product has defects in its manufacturing process or not consistent with the given knowledge, those groups should review the STEM documents to further refine the product; (2) If the product meets the requirements, those groups will use it to collect the results, make conclusion on the knowledge of magnetic field.

Stage 6: Report experimental results

Groups report the results obtained in stage 5 in form of tables, graphs or qualitative conclusions.

Stage 7: Comments and discussion

Groups review and evaluate products of other groups in the form of peer review. Then, teacher evaluates and gives comments and discussion. Students are encouraged to review the teacher's assessments.

Stage 8: Conclusion

Teacher makes conclusions about the tasks achieved, the knowledge students have acquired, the positive and limited aspects in the teaching process. In addition, teacher gives rewards to members, groups with good sense of learning, good learning results, reminding and giving advises to individuals and groups who do not have good results.

2.2. Analyze the knowledge of the section "Magnetic field - electromagnetic induction phenomenon" under STEM education

After studying the magnetic field - Electromagnetic induction phenomenon - and the facilities (experiments) are equipped, a number of appropriate content is selected to teach under STEM education including the magnetic field lines of force of conductors with special shapes, electromagnetic interaction, Lenz's law, and Foucault current. These contents are dissected firmly based on science, technology, engineering, and mathematical knowledge. Table 1 presents the content of STEM teaching in the section of magnetic field and electromagnetic induction phenomena that are performed into 4 parts.

Table 1. The content of STEM teaching in the section of magnetic field and electromagnetic induction phenomena.

No.	Contents	Outcomes	Physics	Technology	Engineering	Maths
1.	The magnetic field lines of conductors with special shapes	Set of experiments of the magnetic field lines of 3 conductors: straight wire, carrying loop of wire, solenoid	The magnetic field of conductors with special shapes	Pressed paper, copper wire, iron fillings	Do experiments to figure out the magnetic field lines caused by special conductors	- Measure to choose appropriate copper wire - Calculate the number of the wire rings, the diameter of the ring, and the suitable distance between the rings.
2.	Electroma-gnetic interaction	- Magnetic interaction experiments between the magnet and the current - Car driven by magnetic interaction	- Characteristics of the magnetic field - Magnetic interaction	Copper wire, magnet, trash timber, plastic bottle, pressed paper	Manufacture products	- Calculate, measure to make good moving vehicles. - Measure and determine the length of the moving car's track.
3.	Lenz's Law	The experimental set used to study the Lenz's law.	The magnetic line of force of conductors with special shapes	Copper wire, magnet, trash timber, pressed paper	Manufacture products	Measure and calculate to create experimental setup

4.	Foucault current	Experiment of Foucault current	Foucault current	Magnet, trash timber, metal rings	Manufacture products	Calculate to create inclined plane; measure the distance between the magnets on the inclined plane

The experimental products of the contents of magnetic field and electromagnetic induction phenomena are manufactured from simple, easy-to-find and inexpensive materials. All students can do it themselves under the guidance of teachers. These products are showed in table 2 corressponding to each part of content.

Table 2. The experimental products corressponding to each part of content in the section of magnetic field and electromagnetic induction phenomena.

No.	Contents	Experimental products
1.	The magnetic field lines of conductors with special shapes	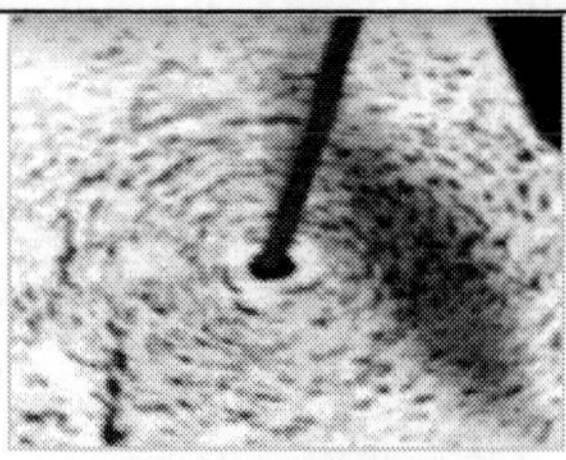 *a. A straight wire* 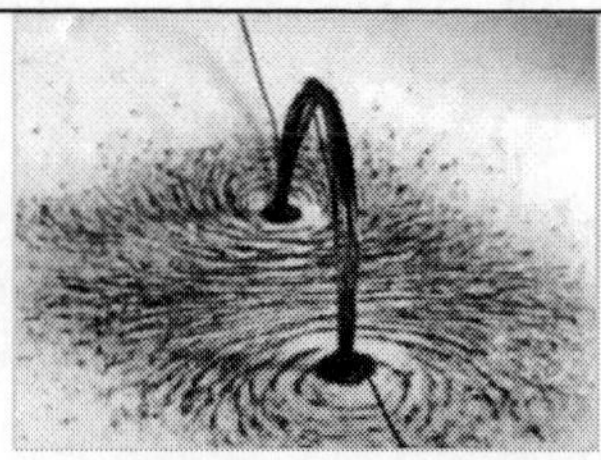 *b. A carrying loop of wire* 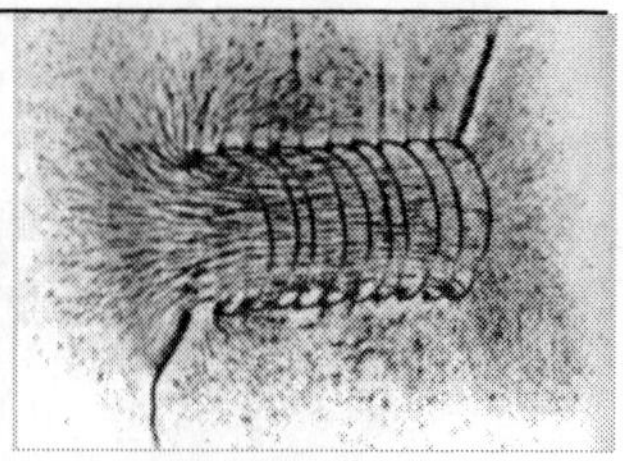 *c. A solenoid*
2.	Electroma -gnetic interaction	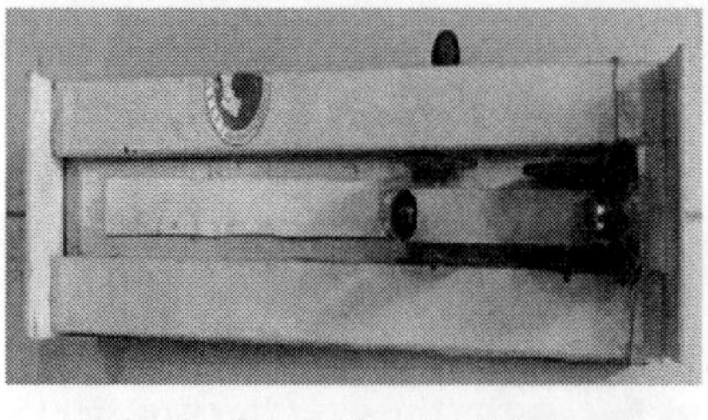 a. *Vehicle driven by the interaction between electric current and magnet* 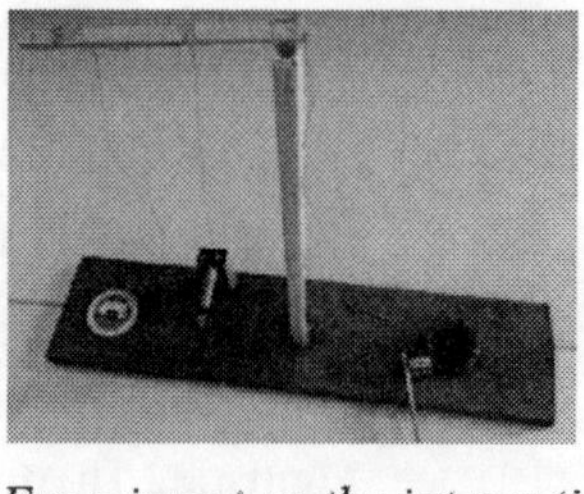 *b. Experiment on the interaction between electric current and magnet*
3.	Lenz's Law	 *Lenz's Law Experiment*
4.	Foucault current	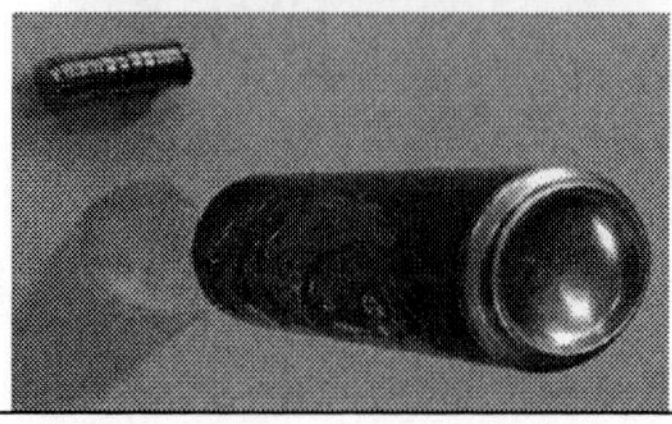 *Foucault current experiment*

IOP Conf. Series: Journal of Physics: Conf. Series **1340** (2019) 012031 doi:10.1088/1742-6596/1340/1/012031

Experimental preparation of STEM products is guided by the teacher so the students can do as following suggestions:

1. Figure out the magnetic field lines of conductors:

- A straight wire:

Use a piece of straight copper wire (5 mm in diameter) piercing through the paper, sprinkle the iron flings on the paper around the wire and then make the current of 3-4A flow in the wire, tapping will get the image of the magnetic field lines.

The magnetic field lines of a straight wire can be obtained as the image showed in figure 2.

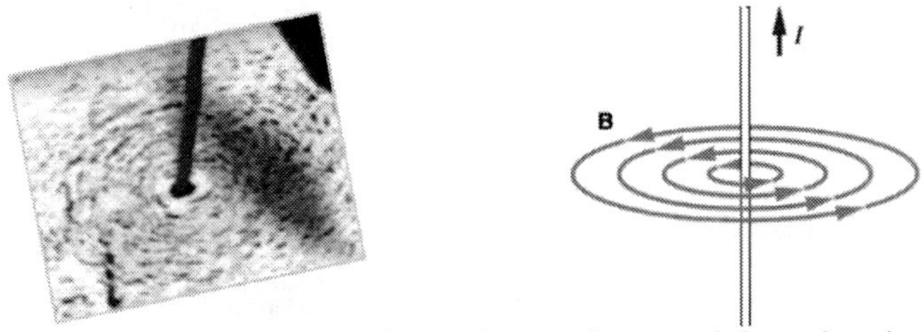

Figure 2. The magnetic field lines of a straight wire in the experiment (on the left) and in illustration (on the right).

- The current loop:

+ Using a square paper (20cm in diameter), cut two small holes in the middle of the paper with the distance between the holes of about 5-6 cm. Use copper wires through round the holes to form a loop (about 30-50 rounds), sprinkle the iron filings around the wire into the paper.

+ The current flow in the wire is controlled at 2 – 3 A. tap lightly on the paper to observe the line of iron filings. The magnetic field lines of a current loop can be obtained as the image showed in figure 3.

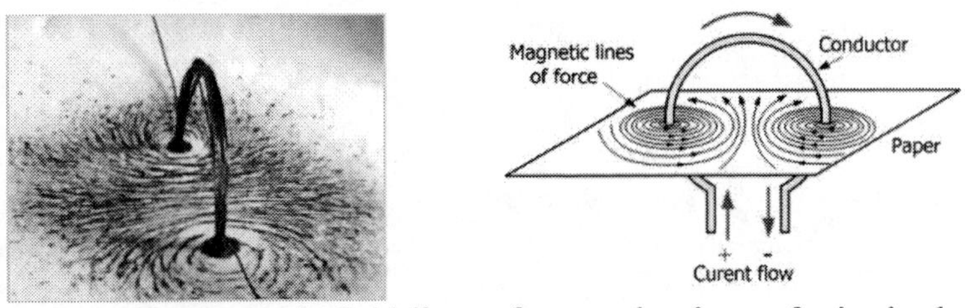

Figure 3. The magnetic field lines of a carrying loop of wire in the experiment (on the left) and in illustration (on the right).

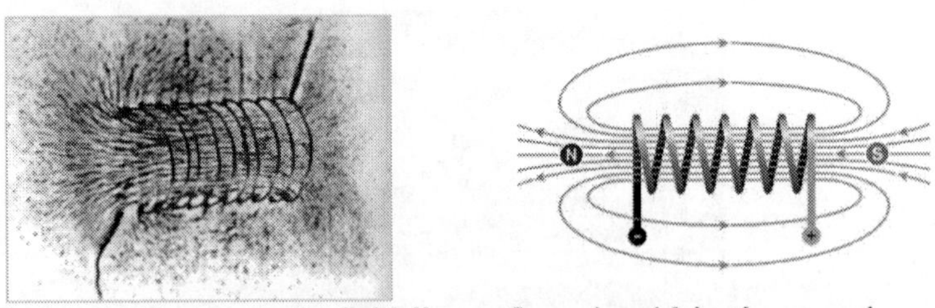

Figure 4. The magnetic field lines of a solenoid in the experiment (on the left) and in illustration (on the right).

- The solenoid:

+ A solenoid is a coil of many circular loops wrapped up in the shape of a cylinder. Wrap the copper wire around a can of coke then cut the can to form a solenoid.

+ Use a piece of paper and cut 2 slits on the paper with the distance between 2 slits equal to the diameter of the solenoid. Put the solenoid matched to slits and fix the solenoid with glue then sprinkle the iron filings in the paper around the solenoid.

+ Make the current flow in the solenoid, tap lightly on the paper to observe the line of iron filings. The iron filings on the paper show the magnetic field lines of the solenoid. The magnetic field lines of a solenoid can be obtained as the image showed in figure 4.

2. Electromagnetic interaction: Vehicle driven by the interaction between electric current and magnet

+ The vehicle (a small car) is made from 2 wood spatulas, 2 straws, and 4 plastic caps. The end of the car is attached with a rare earth magnet.

+ Manufacturing gutters (for the car moving on it): An inclined plane made from wood with many trenches. The lower end of inclined plane is designed a coil (the larger the round number, the better). Connect the wires to the power supply and switches to form a closed circuit.

+ Principle of operation: When the vehicle is at the bottom, the force of the magnetic field of the coil acts on the magnet so that the car can move slowly on the top of the inclined plane, then the car moves down.

This experiment is very dificult to control the strength of magnetic force so that the car can move up on the inclined plane (figure 5). The main objective of this experiment is calculating and measuring to make good moving vehicles, also determining the length of the moving car's track. The moving of the car on inclined plane by electromagnetic interaction is normally for good students. We will report it detail in another paper.

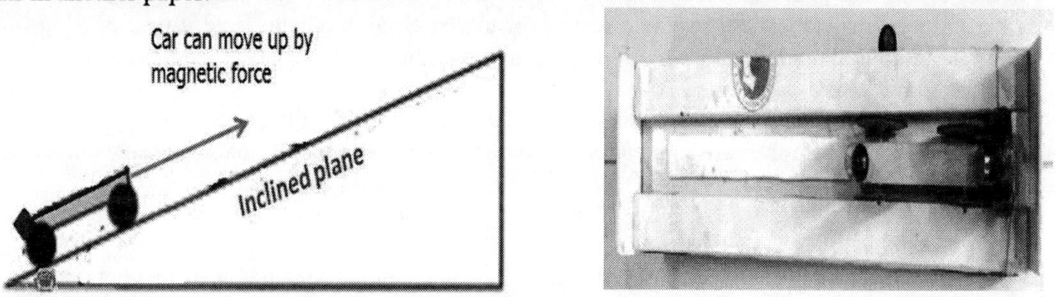

Figure 5. Illustration of the moving of the car on inclined plane by electromagnetic interaction (on the left) and the experimental product by students (on the right).

3. Lenz' law experiment

The materials including cans, sponges, candle stick, ink ball pen out of ink, plastic bottles are set up and showed in table 2. Let the magnet move close or away from the wire frame we can determine the movement of the wire frame. Observing the direction of motion of the magnet determines the direction of the induced current. The detail of this experiment will be reported in the teaching process of lesson plan of Lenz's law.

4. Foucault current experiment

Foucault currents are loops of electrical current induced within conductors by a changing magnetic field in the conductor due to Faraday's law of induction. The experimental setup is carried out by using a can of coke as conductor and permanent magnet.

Let the magnet move around the conductor, the moving of the can can be observed. This implied that there is one induced curent in the can. This curent is attributed to Foucault current.

3. Pedagogical experiment

We have designed four teaching plans based on products: magnetic interaction experiments, cars driven by magnetic interaction, magnetic line of force caused by conductors with special shapes, experiments on Lenz's law, experiments on Foucault current to organize teaching under STEM education.

The number of products in the section of magnetic field and electromagnetic induction phenomena taught under STEM education at least as following:

- Magnetic interaction: 2 products
- Magnetic line of force of special wire shapes: 3 products
- Lenz's Law: 2 products
- Foucault current: 1 product.

In this paper we present in detail the teaching plan for the lesson "Lenz's law of inductive current" in Physics Grade 11 for a period of 45 minutes.

3.1.1. Objectives of lesson plan: "Lenz's Law"

- Knowledge: the content of Lenz's law
- Skills: Do experiments with Lenz's law according to STEM documentation; learn how to exploit experiments to draw the relationship between external magnetic field and electromagnetic induction; apply the Lenz's law to determine the direction of the inductive current.

3.1.2. Materials of lesson plan: "Lenz's Law"

- STEM teaching materials: copper wire, rare earth magnets, plastic bottles, cans, scissors, sponges, paper knives, glue guns, 502 glue, tin and welding torches.
- STEM manual of "Lenz's Law"

a. *Lenz's law experiment*: This is a set of experiments that help us determine the inductive current directly or indirectly. In this lesson, we introduce a set of experiments that help determine the direction of the induced magnetic field by the direction of the external magnetic field, then applying the right hand rule to determine the direction of the inductive current.

b. *Materials of the experiment*

- Rare earth magnets: help generate magnetic fields with defined direction
- The coils cut from the cans and attached to the two ends of the light balance scales that can rotate around a fixed axis.
- The stand of the rotary shaft is attached to the plastic bottle

c. *Theoretical background and the experimental procedure*

- Theoretical background for the experimental design: Interactions between magnets and magnets, electromagnetic induction and magnetic fields caused by a round wire.
- Experimental procedure:

-> Move the magnet near the wire; we observe the movement of the wire. If the wire moves in the same direction as the magnet, we can infer that direction of the induced magnetic field is opposite to that of the magnet. It is easy to determine the direction of the induced magnetic field.

-> Move the magnet away from the wire, we observe the movement of the wire. If the wire moves in the same direction as the magnet, the magnet is attracting the wire. Then, we can infer that direction of the induced magnetic field must be in the same direction as the magnetic field of the magnet. It is easy to determine the direction of the induced current.

- Collect data and give conclusion.

d. *Experiment of the Lenz's Law*

i. Purpose: to study the relationship between the directions of the induced magnetic field and the external magnetic field when the flux sent through the wireframe increases or decreases. Then, make conclusions about the Lenz's Law.

ii. Tools and materials

In order to do Lenz's Law experiment, we need to use the following materials and tools:

+ Materials: magnets, plastic bottles, cans, candle glue, ink pen balls, paper pins.
+ Tools: Glue gun, scissors, pliers, ruler, and paper knife.

iii. Processing, assembling experiment

Step 1: Make round wire from cans

Step 2: Make scale stand from sponge

Step 3: Make rotary shafts from plastic bottles, paper pins
Step 4: Assemble, complete the test kit
Step 5: Operation, testing
Step 6: Perform the experiment, collect the results and write the report
iv. Evaluate and discuss the design of experiment
The evaluating Lenz's law experiments are evaluated via 5 criterias presented in table 3.

Table 3. Criteria for evaluating Lenz's Law experiments

No.	Criteria	Maximum scores	Evaluation scores
1.	Successfully implement the experiment as scheduled	10	
2.	Clear observation; the test results have high reliability; determining the given requirements.	40	
3.	Demonstrate the principle of structure, principle of operation of the experiment clearly, coherently.	30	
4	Find out answers to questions	10	
5.	Active and enthusiastic participation in the experiment.	10	
Total		100	

3.1.3. Teaching process of lesson "Lenz's law
a. *Introduction and question*
Teacher (T): When the flux is sent through the variable wire frame, the induced current appears in the frame. So, how to determine the direction of induced current flowing in the wire frame?
Students (Sts): Consider the direction of induced current through experiment.
T: Orient students to do experiments to determine the direction of induced current with simple, easy-to-find materials.
Sts: Explore methods to design the experiment
b. *Processing, manufacturing and conducting experiment: students work in groups*
Groups in the class receive instructions and conduct experiments to study; then, receive materials and tools from teacher to begin processing, manufacturing and conducting experiments. Teacher monitors the activities of groups, provides timely support, and advises them to make sure that groups can complete the task within the required time period.
c. *Report the activity results*
Groups send representatives to report the results of their activities: Product is the experiment kit and the results of experiment are showed in the table 4.

Table 4. Report on experiment results of student group

Rounds	Motion of magnet	Motion of round wire	Magnetic flux variation	Relationship between the directions of inductive current and the motion of magnet
1.	Magnet approaching the round wire			
2.	Magnet moving away from the wire			
3.	Magnet approaching the round wire			
4	Magnet moving away from the wire			

d. *Conclusion*

- When the magnetic flux increases, the inductive magnetic field is .... direction with external magnetic field.
- When the magnetic flux decreases, the inductive magnetic field is .... direction with external magnetic field.
- Infer that the inductive magnetic field always has a tendency .... the process of varying the magnetic flux sent through the round wire

e. *Conclusion on Lenz's Law of induced current direction*

3.1.4. Conclusion on the lesson content

T gives conclusion to the lesson content.

+ Lenz's Law: The induced current that appears in round wire produces an induced magnetic field that tends to resist what generates it.

+ Steps of applying Lenz's law to determine the induced current direction:

Step 1: Determine the direction of the external magnetic field

Step 2: Determine the increasing or falling variation process of the flux sent through wire frame

Step 3: Determine the direction of the magnetic field

- If the magnetic flux falls, the direction of the external magnetic field is the same as that of the induced magnetic field
- If the magnetic flux increases, the direction of the external magnetic field is opposite to that of the induced magnetic field

Step 4: Apply the right hand rule to determine the direction of the induced current

+ Application: T gives some cases for students to determine the direction of the induced current.

3.1.5. Results

Through the implementation process in class 11A2 at Thai Nguyen High School with 28 students divided into 4 groups, we have obtained the following results:

+ The process of manufacturing, experimenting: Students were very excited, actively involved in the process of doing experiments on the basis of guidance from STEM documents. However, there were many students confused in the use of tools: fine art saw, glue gun, drill, etc. Some groups did not know how to make rotary shafts, quite embarrassed to roll copper wire tube, frame. However, when instructed by teacher, they acquired the basic skills of using the tools in the manufacturing process and completing the tasks assigned.

+ Reporting Results: Groups reported the results of their work. Generally they had good reporting skills, full presentation of the structure, effect of the components, the principle of product operation, advantages - limitations of the product.

Through the teaching process under STEM orientation, students are facilitated to develop competencies that are expressed through the table 5.

Table 5. Report on experiment results of student group

Competencies	Manifestation
Studying STEM documents	Students actively seek information about the tools, materials, principles of manufacturing, principles of operation of the product in STEM documents.
Stating and solving problems	The groups have proposed and set out solutions for the given tasks. Some groups recommend excellent solutions.
Reviewer	The students are quite enthusiastic in group discussions, giving a lot of arguments and discussion about the manufacturing process, as well as the principles of operation.
Practice	Groups know how to manufacture products, how to divide specific tasks and successfully complete the products as required.
Communication	Students in groups have had discussions and exchange of ideas to complete the tasks.
Calculating	Know how to calculate the collected data, utilize materials in optimal way, and arrange time to make products.
Group work	Know how to divide specific tasks, solve problems together.
Creativeness	Recommend solutions to the task; some groups also offer a variety of solutions to choose the optimal one.
Presentation	Students have good way to present their group products and results.

4. Conclusion

This work demonstrates that the STEM-oriented teaching process is highly feasible. This process not only meets the 45-minute requirement but also creates excitement for students, enabling them to develop the necessary skills and abilities: presentation, teamwork, critical thinking, problem solving, creativity, etc. Especially, this process ensures that students are equipped with the basic knowledge of the subject and the knowledge of other disciplines such as technology, science, mathematics. It can be seen that STEM-oriented teaching process should be studied and experimented to gradually apply in the process of high school teaching to meet the change of the secondary education reform.

References

[1] Appianing J, Richard N and Eck V 2018 Development and validation of the Value-Expectancy STEM Assessment Scale for students in higher education **International Journal of STEM Education 5** p 16

[2] Blotnicky K A, Franz-Odendaal T, French F and Joy P 2018 A study of the correlation between STEM career knowledge, mathematics self-efficacy, career interests, and career activities on the likelihood of pursuing a STEM career among middle school students **International Journal of STEM Education 5** p 15

[3] Elster D 2014 First-Year Students' Priorities and Choices in STEM Studies – IRIS Findings from Germany and Austria **Science Education International 25** pp 52–59

[4] Feder A L 2013 Science, Technology, Engineer Ing, A Nd M Athem At Ics (STEM) Educat Ion 5-Year Str ategic pl a n, A Report from the Committee on STEM Education National Science and Technology Council

[5] Kennedy T J and Odell M R L 2014 Engaging Students In STEM Education **Science Education International 25** pp 246–258

[6] Knaub A V, Henderson C and Fisher K Q 2018 Finding the leaders: an examination of social network analysis and leadership identification in STEM education change **International Journal of STEM Education 5** p 14

[7] Linh N Q, Suong H T H and Khoa C T 2017 STEM Contents in Pre-service Teacher Curriculum: Case Study at Physics Faculty, International Conference for Science Educators and Teachers (ISET) 2017 **AIP Conf. Proc.** 1923, 030071-1–030071-8

[8] Mbano N and Nolan K 2017 Increasing Access of Female Students in Science Technology, Engineering and Mathematics (STEM), in the University of Malawi (UNIMA) **Science Education International 28** pp 53–77

[9] Quang L X, Hoang L H, Chuan V D, Nam N H, Anh N T T and Nhung V T H 2015 Integrated Science, Technology, Engineering and Mathematics (STEM) Education through Active Experience of Designing Technical Toys in Vietnamese Schools **British Journal of Education, Society & Behavioural Science 11** pp 1–12

IOP Conf. Series: Journal of Physics: Conf. Series **1340** (2019) 012032 doi:10.1088/1742-6596/1340/1/012032

Realistic mathematics education (RME) and didactical situations in mathematics (DSM) in the context of education reform in Vietnam

Nguyen Tien Trung[1], **Trinh Phuong Thao**[2], **Tran Trung**[3]

[1] Ph.D, Editor, Vietnam Journal of Education, 4 Trinh Hoai Duc, Hanoi, Vietnam
[2] Ph.D, Teacher, Thai Nguyen University of Education, Vietnam
[3] Assoc. Prof., Ph.D, Senior Teacher, Vietnam Academy for Ethnic Minorities, DreamTown-COMA6 urban area, road 70, Nam Tu Liem district, Hanoi, Vietnam

E-mail: [1]nttrung@moet.gov.vn, [2]trantrung@cema.gov.vn, [3]trinhthao.sptn@gmail.com

Abstract. Mathematics Education plays a key role in the ongoing education reform in Vietnam, which commenced with the renewal of the curriculum and textbooks (from primary to secondary and high schools). In the world, Didactical Situations in Mathematics and The Realistic Mathematics Education have been widely and effectively applied in the Netherlands, America, France, Indonesia, etc. This article presents some cultural characteristics of Mathematics Education in Vietnam and the results of initial research on these two theories, providing some models of teaching situations and examples for pilot implementation, initial survey the ability to apply Didactical Situations in Mathematics in Vietnam, providing some suggestions for the renovation of Mathematics curriculum and textbooks in Vietnam at present.

1. Introduction

Mathematics is a science that arises, develops from real life and serves as a practical tools to solve reality problems. Therefore, it is necessary that mathematics teaching be linked with the reality in society as well as in learner's life. Hence, how mathematical teaching is linked to practice, what extent is appropriate in the context of mathematical education reform in Viet Nam, what adjustments are required in mathematics teaching are the questions that need investigating and resolving.

At present, the Ministry of Education and Training of Vietnam is organizing a major educational reform, commencing with the renovation of the curriculum and textbooks (from primary to secondary and high school). It is projected that in principle, there will be only one general education program in the whole country and possibly a range of textbooks, approved under the regulations of the Ministry of Education and Training. The general education curriculum after 2018 will be designated in the aim of developing learners' competencies. As explained by experts and the Ministry of Education and Training, the current program was designed in the direction of content development. In the context of general education curriculum reform, Mathematics curriculum also needs to be renovated towards learners' competency development. The researchers believe that this innovation should be implemented comprehensively in terms of objectives, programs, textbooks and teacher training.

This study focuses on the Theory of Didactical Situations or the Didactical Situations in Mathematics (DSM) and the Realistic Mathematics Education (RME) in order to propose some implications for the current reform of the mathematical education program in Vietnam.

2. Mathematics teaching in Vietnam in the context of education reform

2.1. Mathematics Curriculum and Mathematics Textbook after 2018 (MoET)

Maths Cirriculum Outcomes [10]: Mathematics Education aims to develop students core qualities; general competences, and mathematical competence with the core components of including competences of mathematical reasoning, mathematical modeling, Maths problem solving, mathematical communication, competence to use mathematical tools and media. Mathematics education also develops key knowledge and skills and creates opportunities for students to experience and apply Mathematics to real life. Mathematical education connects mathematical ideas, Mathematics and other disciplines, and Mathematics and real life. Mathematics education is conducted in many subjects such as Mathematics, Physics, Chemistry, Biology, Technology, Informatics, Experimental Activities, etc., in which Mathematics is the core subject.

In this study, we focus on proposing some implications for the development of the Mathematics curriculum in general, in the context of education reform in Vietnam. However, it is advisable to discuss the current teaching context in Vietnamese schools.

2.2. Systems of Vietnam Education

Vietnam education system is shown in the following diagram:

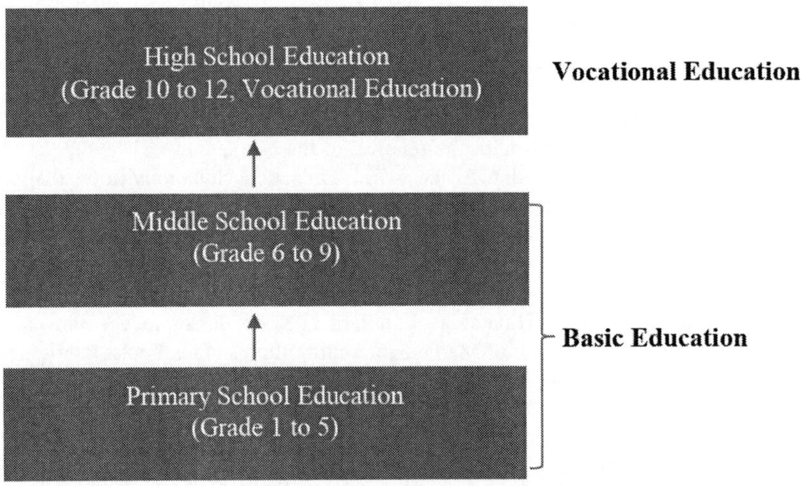

Figure 1. Vietnam education system

2.3. The role of Curriculum and Textbook

Teachers and students mainly use textbooks as the core materials in the classroom in teaching Mathematics

Mathematics textbook are divided into several chapters containing lessons and a mathematics lesson often has some formal mathematical definitions, theorems, regulations or formulae [8]. There is one series of mathematical textbooks in school despite the fact that students live in different regions [8]. However, there are some sets of books that are used exclusively in some elementary schools, or in some provinces, such as books for ethnic minority students (translated into ethnic minority languages), Maths book for students in some international schools (based on the international curriculum, in parallel with the current textbook).

Recently in the Math textbooks, there have been more real life related exericses, which is just somehow related to real life problems rather than actual real life ones. It is common that problems in Math textbooks are generated by modelling real life problems, in stead of examining the actual real life ones.

2.4. Classroom Organization

Teachers often spend much less time on teaching concepts and theories, mainly focusing on teaching rules, methods, and problem solving skills. This is primarily because current assessment tools mainly focus on testing basic knowledge and skills of the lessons in textbooks through solving exercises. At present, in order to approach the new cirriculum, in the test of some schools, some Departments of Education and Training have included Math questions that require students' application to other subjects, to reality or real life related situations (in end-of-term, end-of-year or entrance exams at all levels).

According to Le Tuan Anh [8]. students usually have to take part in some extra-lessons to keep up with the curricula or pass examinations.

Mathematics lessons in the textbooks as well as examinations in schools offer few examples or applications relating to real life or real world [3]. Vietnamese students often struggle to apply mathematical knowledge in reality [8].

In spite of some changes in the content of mathematic textbook, a formula or a theorem is often presented as follow: +) Step 1: Content of a formula or a theorem; +) Step 2: A proof of the formula or theorem; +) Step 3: Application of the formula or theorem in some pure mathematics examples. Similarly, a concept is often performed as follows: +) Step 1: Definition; +) Step 2: Some examples of the concept; +) Step 3: Characteristics [8].

Although there have been various changes in the practice of mathematics teaching these days, some following situations are still common: 1) The teacher mainly imparts content and knowledge, pupils learning by the examples; 2) The majority of teachers generally prefer explanations, lectures and samples with frequent incidental questioning; 3) The teachers are not monitored, they do not help pupils to create problems and 'occupy' new knowledge [3]. 4) Mathematics teachers usually use 'training for exams' method to help their students carefully practice forms of problems which usually appear in examinations; 5) One 'real goal' of teaching mathematics in school is to help students score higher on examinations [8]. The second situation greatly affects Mathematics teaching and learning.

A session usually lasts 45 minutes, while a lesson from one to three or four hours. It is up to teachers to appropriately adjust the teaching content (up to 30% different to the original subject teaching schedule) and lesson plans. However, the motivation for innovating teaching content and techniques is not high and popular so teachers tend to closely comply with the prescribed teaching content distribution (according to the regulations of the Ministry of Education and Training).

Mathematics teachers usually use pieces of chalk, one blackboard, ... and students have the specific desk in the classroom for semester or school year [8]. Teachers sometimes use overhead projectors, computers, beamers, videos and other tools for their teaching in case of the good teacher contest.

In the primary classroom, contrary to the description of the secondary classroom that "*students do not have the chance to play games, which can help them to learn mathematics*" [8], students often play at least one game in each Mathematics lesson.

At present, more and more primary and secondary students want to participate in the Maths Competition in English language organized by national and international organizations, for example, International Kangaroo Math Contest (Contest-kangaroo-math.vn), Violympic online (violympic.vn), International Mathematics Assessments for schools (imas.ieg.vn), etc.

3. Research questions and method

The research questions are:

- Is it possible to combine and apply the DSM and RME theories in the process of reforming Mathematics education in Vietnam?
- Can Vietnamese students learn Mathematics based on models and examples of applying these two theories in Mathematics teaching?

To answer the two above questions, we studied about DSM and RME, developed some models of Maths teaching situations by combining the research results on the two theories, proposed some examples (some teaching situations) relevant to the Vietnamese cultural and practical context. The research was then conducted using the proposed teaching situations in order to give some initial recommendations for the application of the two theories in the process of reforming Mathematics education in Vietnam. These situations were employed to teach some groups of students (140 students at 12th grade in some high schools) in some provinces including Bac Giang, Thai Nguyen, Hai Duong, Hanoi. The findings help us to make an initial evaluation of the effects or obstacles in promoting mathematical education in practice in the context of Vietnam.

To access the student's compentence and readiness to learn Mathematics in practice and by practice, in each situation, we provided a work sheet with the given tasks for students, which were then collected and evaluated in terms of: ability to recognize and identify the situational tasks and Maths problems; Maths modeling compentence, the ability to solve the modelled problems and compentences to apply knowledge and make the conclusion in the real life contexts. The statistics of the number of students who could solve the tasks and the degree of completion of the tasks allow us to find out the outstanding tendencies and difficulties in teaching Maths in connection with real life among Vietnamese high school students. The analysis of some typical answers to the assigned tasks partly helps to determine the level of students' task achievement and the difficulties in certain contexts. Subsequently, it helps to identify the cause of those difficulties. Hopefully, we then may propose suitable recommendations for teaching Mathematics in the current context of Vietnam.

4. A brief overview about Didactical Situations in Mathematics and Realistic Mathematics Education

4.1. Some basic concept in Didactical Situations in Mathematics

The theory of Didactical Situations is stated by the psychologist, educationalist Guy Brousseau. Subsequently, a variety studies created a school called Didactical Situations in Mathematics.

In this school, knowledge is constructed in various forms. According to Guy Brousseau [6], stated knowledge is usually "hidden the "true" functioning of science, which is impossible to communicate and describe faithfully from the outside, and replaces it with an imaginary genesis", and, "to make teaching easier, it isolates certain notions and properties, taking them away from the network of activities which provide their origin, meaning, motivation and use. It transposes them into a classroom context. Epistemologists call this didactical transposition.", and to have effective teaching, the production and teaching of mathematical knowledge requires an effort to transform this knowledge into institutionalized knowledge, a depersonalization and a decontextualization that tend to blot out the historical situations which had presided over their appearance.

There are two main important concepts: adidactical situation and didactical situation. In the process of teaching, the teacher need to provoke the expected adaptation in her students by a judicious choice of problems that she puts before them. These problems, chosen in such a way that students can accept them, must make the students act, speak, think, and evolve by their own motivation. Between the moment the student accepts the problem as if it were her own and the moment when she produces her answer, the teacher refrains from interfering and suggesting the knowledge that she wants to see appear. Not only can she do it, but she must do it because she will have truly acquired this knowledge only when she is able to put it to use by herself in situations which she will come across outside any teaching context and in the absence of any intentional direction. Such a situation is called an adidactical situation. Each item of knowledge can be characterized by a (or some) adidactical situation(s) which preserve(s) meaning. In the adidactical situation, teacher's specific intentions are hidden and students can function without teacher intervention. We can say that in the learning process, students face to face the adidactical situation with the support of teacher is the context of didactical situation.

Discussing the learning process, Brousseau said that: Knowing mathematics is not simply learning definitions and theorems in order to recognize when to use and apply them [6]. And the work of teacher is imagining and presenting to the students situations within which they can live and the knowledge will appear as the optimal and discoverable solution to the problems posed.

And these are three types of situations called situation of action; situation of formulation (or situation of communication); situation of validation. These situations is presented clearly in the famous example the game "the race to twenty" which created by Guy Brousseau. The Situation of action lays the essential foundation for the explicit models and formulations which follow. The Situation of action provide feedback to the student on which to base, and against which to test, his models. In general, formulation occurs in Situations where the student has a certain amount of information, but either needs more information than she can come up with on her own or does not have the means of taking action on her own, and in order to proceed must communicate with other members of the class. If the groups become argumentative, the next Situation may develop while the group planning sessions are going on. In any case, it will do so in the following one. This is the Situation of validation.

4.2. Realistic Mathematics Education

The Realistic Mathematics Education (RME) developed by the Freudenthal Institute is also known as "real-world mathematics education" [15]. RME aims at enabling students to apply mathematics. In RME, this connection to reality is not only recognizable at the end of the learning process in the area of applying skills, but also reality is conceived of as a source for learning mathematics. Just as mathematics arose from the mathematization of reality, so learning mathematics has to originate in mathematizing reality. Van den Heuvel-Panhuizen, M. [17]. Even in the early years of RME, it was emphasized that if children learn mathematics in an isolated fashion, divorced from their experiences, it will quickly be forgotten and the children will not be able to apply it [4]. "In RME, mathematics is viewed as a human activity which connects mathematics to the reality. Reality refers to mathematics that is relevant to everyday situations and problem situation that are real in student's mind" [9]. And according to Lu Pien Cheng, the *real-life context problem* refer to problems embedded in real life situations that have no ready-made algorithm [9].

According to Freudenthal, mathematics was not the body of mathematics knowledge, but the activity of solving problem and looing for problems, and, more generally, the activity of organizing matter form reality or mathematical matter – which called "mathematizing" [4]. And he clarified what mathematics is about: "There is no mathematics without mathematizing" [4]. So the teacher need to find out the context, create the context which support student to construct mathematics knowledge. There are some suggestions for teachers find and create contexts for mathematics teaching: context in history of mathematic; context in real life (primary students' life: games, shopping, saving and using money, film,...; social issues: traffic, weather forecast, lottery, ...); integrated education (mathematics in Physical, Chemistry, Informatics Technology, etc.).

It is possible to point out some crucial principles in the studies of Mathematics teaching in the light of RME:

- +) *Activity Principle*: The learner is considered as an active subject in the teaching process whose their activity is the key factor to the outcome of this process. Therefore, the best way to learn Maths is by *solving* Maths problems;
- +) *Reality Principle*. Learners must be able to apply Mathematics knowledge to solve practical problems and mathematics education *should start* from meaningful practical situations with learners to give them opportunities to cooperate those meanings into the mathematical structures in their minds.
- +) *Level principle:* emphasizes cognitive development through various levels of mathematical learning: from non-mathematical contexts involving knowledge, through symbols, diagrams, to pure mathematics content of knowledge. *Models* are very important as a bridge between informal experiences, the mathematical context involved, and pure mathematical knowledge. The models here can be understood as mathematical modeling.

- *+) Intertwinement principle:* Learners are placed in a variety of situations in which they may perform various types of tasks intertwined (reasoning, calculating, statistics, algorithms conducting, etc.), using a lot of Mathematics knowledge and tools from different disciplines, even other sciences.
- *+) Interactivity Principle:* encourages interpersonal and group activities to create opportunities for individuals to share their skills, strategies, discoveries, ideas, etc. with other learners. In return, they can benefit from others for cognitive and personal development.
- *+) Guidance Principl*e is described as a process of guided reinvention in mathematics instruction. Specifically, teachers need to design scenarios or situation (or context) that are potentially rich in activity, of which the implementation can create meaningful cognitive leaps for learners.

5. Some recommendations for teaching mathematics in the direction of RME and DSM

5.1. Renovating the maths teaching program in direction of RME and DSM by designing appropriate teaching situations

From the conceptual and theoretical analyses of the above-mentioned mathematical schools, we believe that in their classrooms, teachers can innovate the teaching programs by designing teaching situations as follow.

The teaching situations employed here are not necessarily derived from the DSM perspective. Every teaching situation, designated by combining the approach of the two theories, fulfils the following requirements:

Each situation is a dual context consisting of two interlocking contexts: a knowledge context designed by the teacher (the context in which the students handle a performance task, and the knowledge discovering requirement may not be explicit); and the second context is a classroom setting with teacher - student interaction with cultural relevance and adaptability.

Each situation must include all the basic types of situations as described in the DSM: situation of action, situation of communication and situation of validation.

Each situation is designed to be either started or terminated in practice, in one of the four following type of situations.

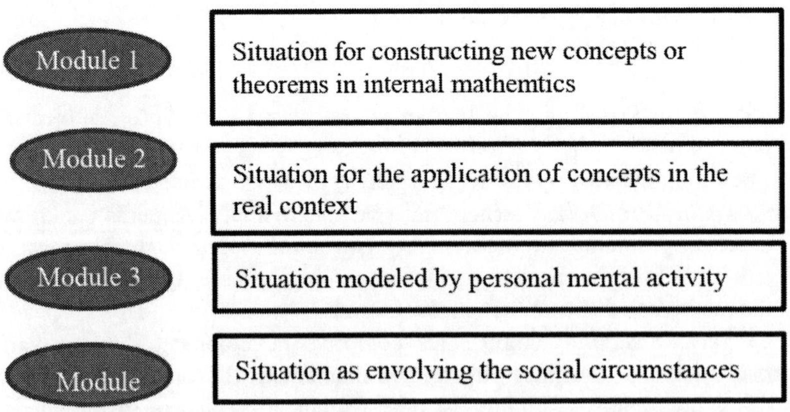

Figure 2. Some types of Maths teaching situations

IOP Conf. Series: Journal of Physics: Conf. Series **1340** (2019) 012032 doi:10.1088/1742-6596/1340/1/012032

5.2. Examples of some teaching situations and assessments of students' compentences through teaching situations

5.2.1. Some example of teaching situations.

Situation 1 (Model 2). Your school is organizing a sport games, and you are assigned to work as a coach of triathlon. The athletes are required to swim and run from point A to point B is as shown in the figure below. The swimming speed is 1.6 m/s, the running speed is 4.8 m/s and the distance between the two banks AH = 200 m, HB = 1000 m. Advise the athlete the best strategy for the situation.

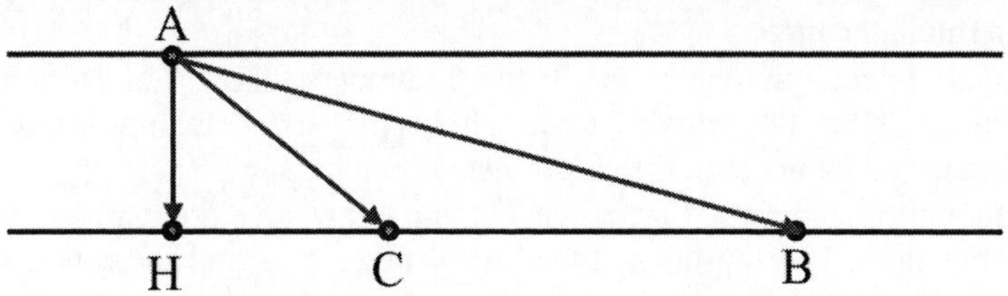

Figure 3. The athletes are required to swim and run from point

Perform the two following tasks:

- *Task 1.* Describe possible ways to perform the athlete's task.
- *Task 2.* What is the best way to perform the task? Why?

Situation 2 (Model 2). The water wheel is a device of Vietnamese farmers in the mountainous North Westhern region. Its shape is formed by a reel, which is a large wheel with the five meter diameter. The rotating axis is made of solid bamboo trunk with hundreds of "spokes", forming a stable frame for the water reel. The reel rim is about 80 cm wide, with the nests in place to halt the water to creat a thrust of the treadmill, and the inclined tentacles tied up with an angel of about 30 degrees. Connected to the water wheel is often the gutters leading back to local houses or fields, often made from the half-spilt bamboo trunk. Each water has 40 bamboo tentacles, each tube is 60 cm in length, 20 cm in diameter, of which the shell thickness is about 1 cm.

Figure 4. water wheel

Let's do the following two tasks:

- *Task 1:* Explain how the water wheel in the above figure can collect water?
- *Taks 2:* How many rounds does the water wheel rotate to collect enough water for the pool with 3 m^3 in capacity?

IOP Conf. Series: Journal of Physics: Conf. Series **1340** (2019) 012032 doi:10.1088/1742-6596/1340/1/012032

Situation 3 (Model 4). Here is the rules and prizes of Vietlot (a new form of digital lottery in Vietnam).

Rules:

1. Playing one line of numbers: Players select 6 out of 45 numbers (from 01 to 45) to form a line of numbers to play.
2. Playing multiple lines of numbers:
   - Players choose 5 numbers from 01 to 45 (bao 5), the sixth number will be selected randomly by the software system from the remaining 40 numbers, which will form 40 lines of numbers. Compare your lines with the winning one in the selected draws to determine the prize.
   - Players select 7 numbers (bao 7) to 15 numbers (bao 15) or 18 numbers (bao18) from 01 to 45. Then, the software system helps players create all 6-digit combinations from the selected numbers to form the lines of numbers to play.
3. Playing in multiple draws: Players select the draw and are entitled to play in up to 6 consecutive draws. In each draw, there are 4 prize tiers and there is 1 draw in each prize drawing session to select a winning line of numbers composed of 6 number from 01 to 45.

Prizes

- Playing one line of numbers: (Players select 06 numbers to form a line of numbers). Each draw ticket costs 10,000 VND.
- Playing multiple lines of numbers: Draw ticket cost: (10,000 VND per play, multiplied with the number of selected lines).

Table 1. Rules and prizes of Vietlot (a new form of digital lottery in Vietnam)

MEGA 6/45			
Prize Categories	Matching numbers	Prize) VNĐ(	Projected prize proportion
Special Prize	●●●●●●	Minumun 12 billions VND and added value	41.31%
First Prize	●●●●●	10,000,000	2.87%
Second Prize	●●●●	300,000	4.09%
Third Prize	●●●	30,000	6.73%
	Total		55.00%
Note :● is the matching number with the winning line, in any order.			

In the case that a winner's tickets win multiple prizes, the winner will only be awarded with the highest prize. In case there are multiple winners of special prize, the special prize is shared in proportion with the winners' ticket values. As for the First to the Third Prize, the prize value is calculated by the number of draws (each draw ticket costs 10,000 VND) multiplied with the prize value in each draw.

Question 1. Calculate the probability of winning special prize and other prizes when playing Vietlot.

Question 2. Analyze the advantages of the two current lottery types (the traditional National and Provincial Lottery and Digital Lottery Vietlot). Advise lottery players the more favourable type of lottery and the optimal playing.

Question 3. Study some international types of lottery that you know and compare with the lottery types in Vietnam in terms of prizes, winning ratio and other aspects.

Question 4. Why do people like playing lottery? Are playing lottery and lotto betting (Lô-Đề) is an important and even indispensable human's need? Why did the lottery company launch the digital lottery Vietlot in the South of Vietnam?

Question 5. Select and do one of two following tasks:

1. If you are a lottery advocate, write a paragraph that analyzes its advantages reminds players about the pitfalls objectively and scientifically;
2. If you are an opponent of lottery playing, write a paragraph that analyzes the disadvantages and advantages objectively and scientifically.

Question 6. If you were an investor, which lottery company would you choose invest between these two type of abovementioned lottery?

5.2.2. Findings and discussion

Situation 1

Task 1. All the students were able to describe some of the strategies to carry out the given task. However, all the proposes solutions included: swimming from A to H then running to B; swimming from A to C and then running to B, in which the point C was identified mainly on the drawing without any special features; or swimming straight from A to B.

Task 2. All the test takers failed to fulfil this task. Most of the assignments submitted were either incompleted or incorrect using their backdground knowledge of planar geometry (Figures 5., 6., 7., 8. below). In other words, many of the student's assignments focused on composing the images to work out the shortest distance. As a result, students neglected the spped when calculating the swimming or running time. Thus, students can not perform the modelling task, which means they failed to transform the situational task into a Math problem.

Figure 5.

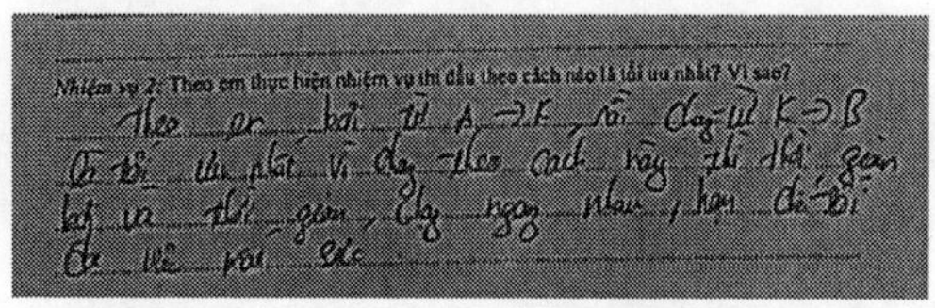

Figure 6.

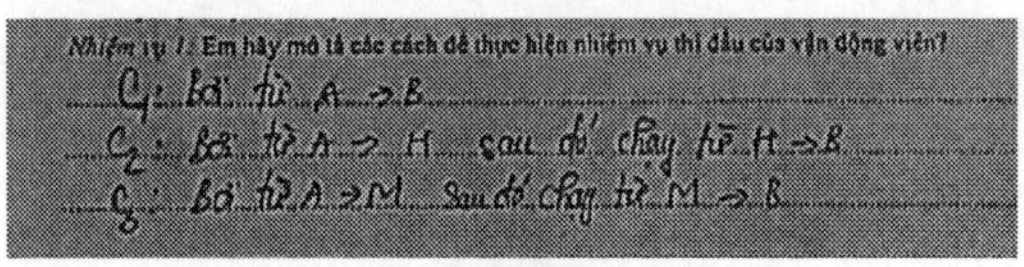

Figure 7.

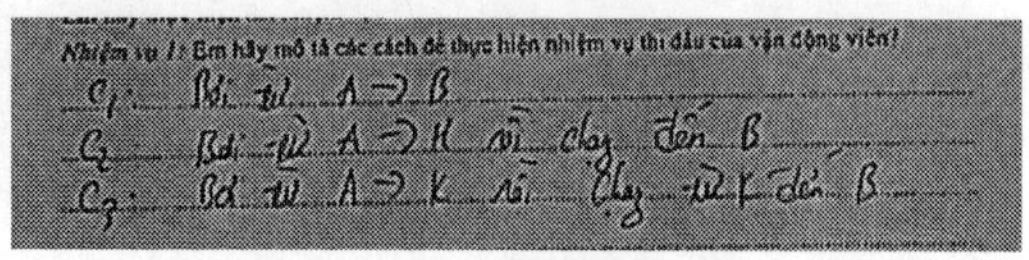

Figure 8.

Moreover, two proposed solutions to the task were revealed included: either figuring out the shortest path by drawing the diagram first and then calculating the required time in that scenario (Figure 9.), or pointing out some possible scenarios in the diagram, then calculating the time under each circumstance, then selecting the option with the shortest time (Figure 10.).

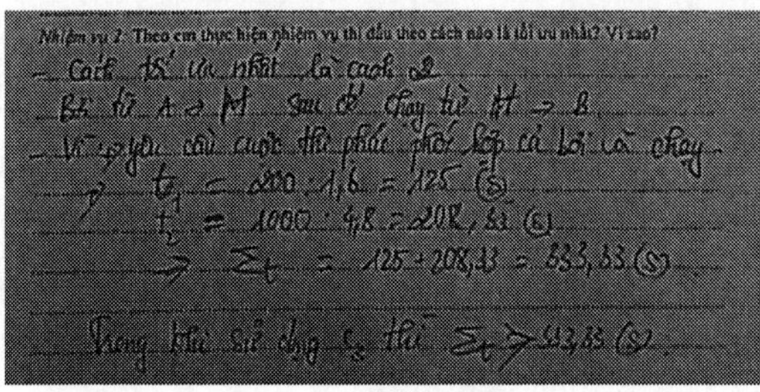

Figure 9.

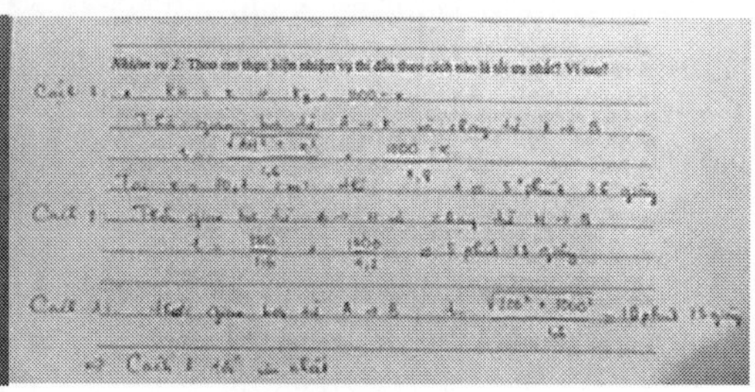

Figure 10.

It is shown that all 12th graders in this survey were not be able to model the situations, and of course failed to solve the problems in the given context. In addition to the above mentioned reasons, their teachers also said that students were not familiar with such Maths tasks due to the lack of proper practice with that task type. Moreoever, students are not motivated enough to solve those practical Maths problems as these problems are not included in important examinations.

Situation 2

Task 1. Many students could not explain how water could be collected. A clear and well-structured explanation as in figure 11. below is quite rare.

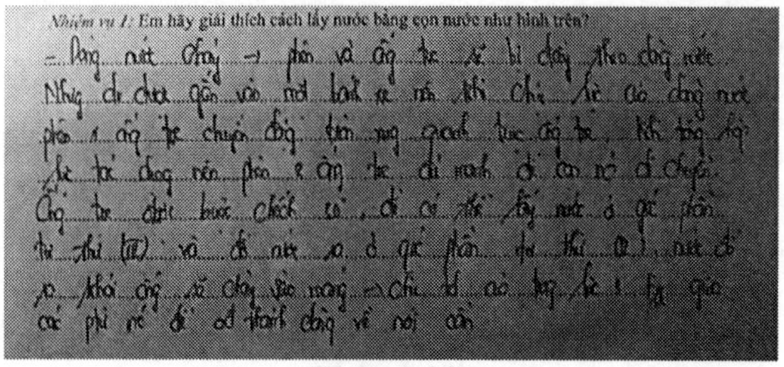

Figure 11.

Task 2. Many students can not relate or refer to calculating the capacity of water in the inclined tube rather than the filled tube. Thus, many students couldn't explain how to collect the water because they made mistakes by calculating the full capacity of the tube rather than considering all the factors in the real situation (Figure 12, and Figure 13).

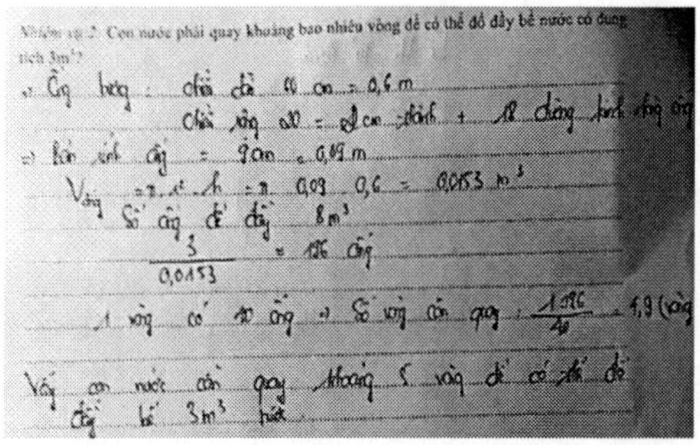

Figure 12.

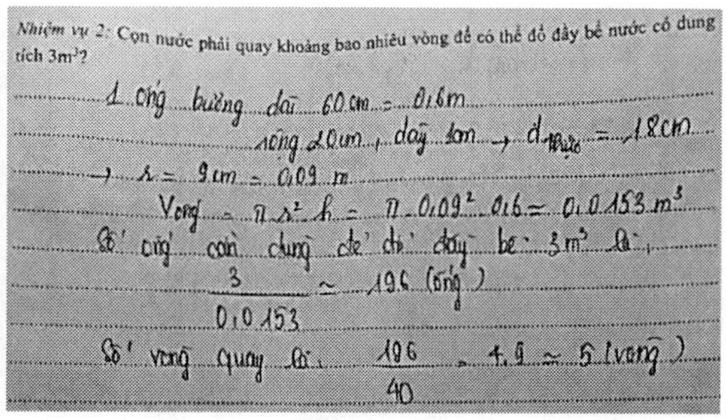

Figure 13.

The given task is considered an interesting, yet challenging one. Unfortunately, in reality, it turned out to be such a huge challenge with all student participants in the research that all of them left the problem completely unsolved or failed to transform the situational task into a Maths problem. If they couldn't even understand the questions, it's certain that they couldn't answer it.

Situation 3

Question 1. Most students could not answer this question (by writing 'I don't understand.') or answer incorrectly. The reason identified is that although the instruction is very specific, the students still could not understand the rules of the game. Also, despite being given plenty of time, they still did not try asking any lottery agents or players to understand the rules of the game. Thus, they failed to model the situation in the task using their own understanding of the lottery rules, which affects the subsequent questions.

In all the students' answer sheets, only one worksheet was found with the correct answer to question 1 as shown below (Figure 14). However, as the answer was too brief, we conducted an interview to ask the student about the way he did it. It was found out that the student understood the question, the rules of the lottery game as well as the basic concepts such as accordance, combinations or permutations and managed to apply them.

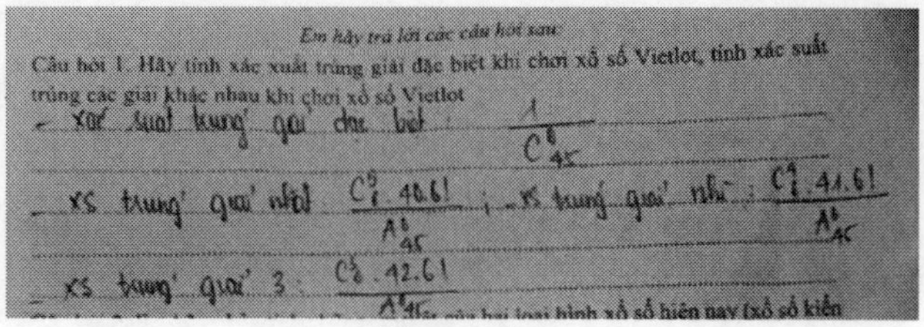

Figure 14.

Question 2. Because many students failed to answer Question 1, very few students succeeded in answering this question correctly. Some students still answered question 2 without answering Question 1 (Figure 15.). Interestingly, they all gave the same advice which was not to play, because the winning ratio of the lottery in Vietnam was too small. Another interesting finding is that when asked about the playing strategy, they all replied "do not play" (Figure 15.), and if you play, you should choose Vietlot (Figure 16.).

On the other hand, we also see the unclear rules of the game in the answer of a student to the first question when he wrote that Vietlot's special prize is 12 times worthier than the special prize of the tradition Northern Lottery (Figure 16.). This is not true. The students could not project the cases of multiple winners and did not know how to calculate the unstated prize of Vietlot.

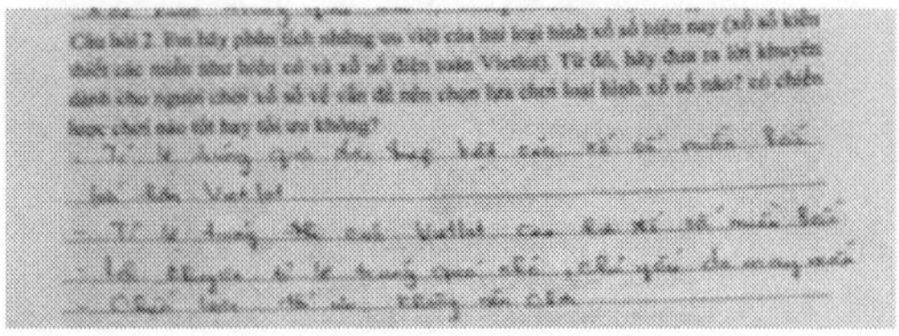

Figure 15.

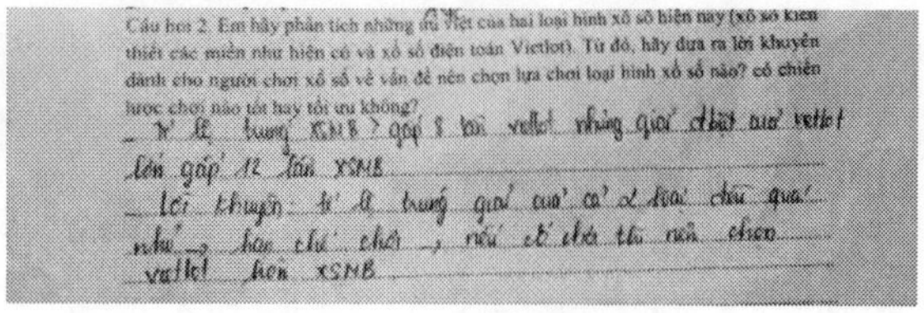

Figure 16.

Question 3. Collecting some answers like "I do not know", "Most lottery types are similar because the winning chances are very small", ... (Figure 17., Fig. 18.), we followed up by asking the students in person what lottery types they refered to with "most". It was revealed that the students were aware of neither the rules of both lottery games mentioned in the task nor those of other types. When being asked, "Did you use the Internet to search for the rules?", all students answered similarly "no" or "I did not think of that way". This points out the lack of practical knowledge and skills and the habit of using the internet as an important searching engine. Although Internet-connected PC and smart phones are widely available with many students, they still haven't got this habit and competence.

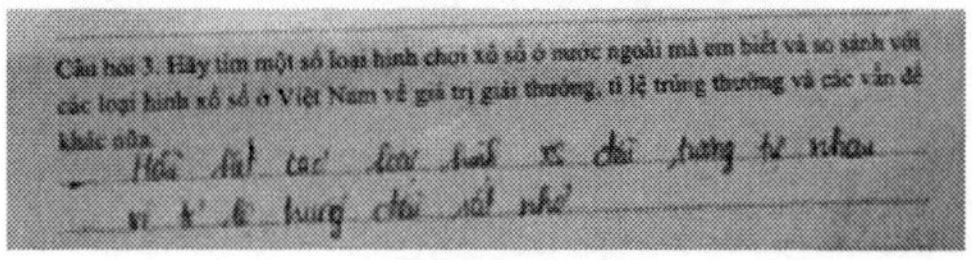

Figure 17.

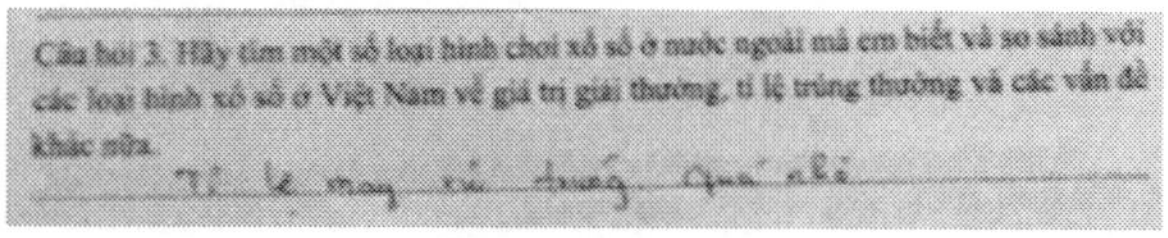

Figure 18.

Question 4. With this question, although students could not provide complete answers, the submitted ones were quite satisfactory. They are expected to study and discuss the belief in "good luck" as well the the subtle distinction in the culture and lifestyle between Northerners and Southerners in Viet Nam regarding playing lottery. Students were able to partially explain the reasons for playing the lottery (such as "want to spend less, get more" (Figure 19., Figure 20.) and the difference in lottery players' behaviour in the North and the South. For example, they mentioned that Southerners do not have the habit of saving as the Northern people (Figure 20.). All the answers we collected stated that playing the lottery is not an essential need.

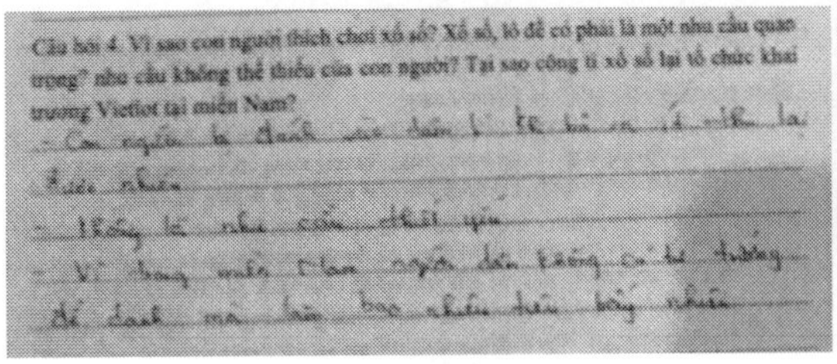

Figure 19.

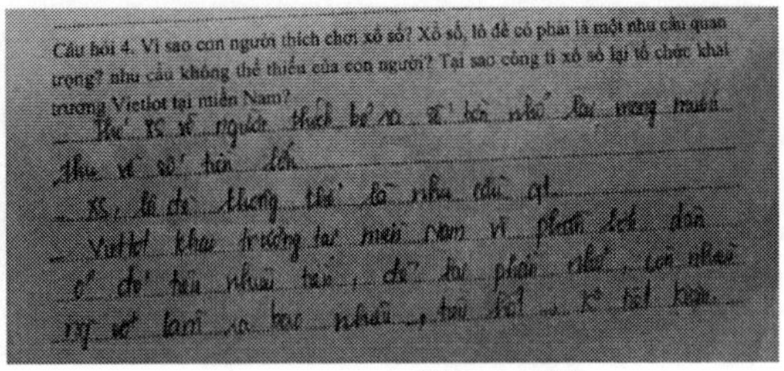

Figure 20.

Question 5. There were very few answers to this question. All the submitted answers warned people against playing lottery given the explanation that "the chance to win is too small" (Figure 21., Figure 22.) and the money spent would be much more than the amount earned (Figure 21.). Interestingly, there was a student who compared the lottery with lotto betting - a form of illegal gambling based on the daily results of the official lottery draw- and stated that the chances to win in lotto betting might be much stronger!

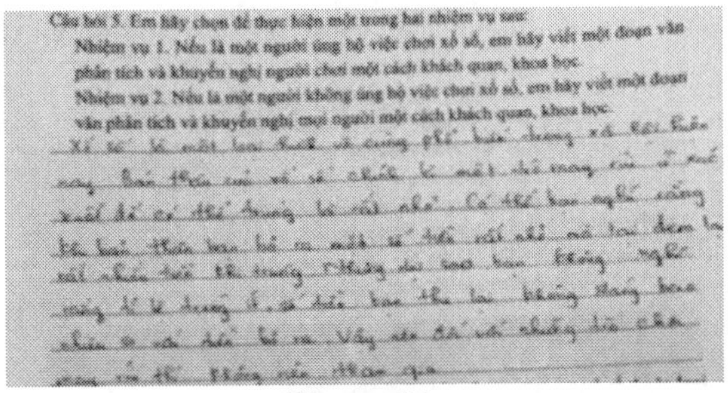

Figure 21.

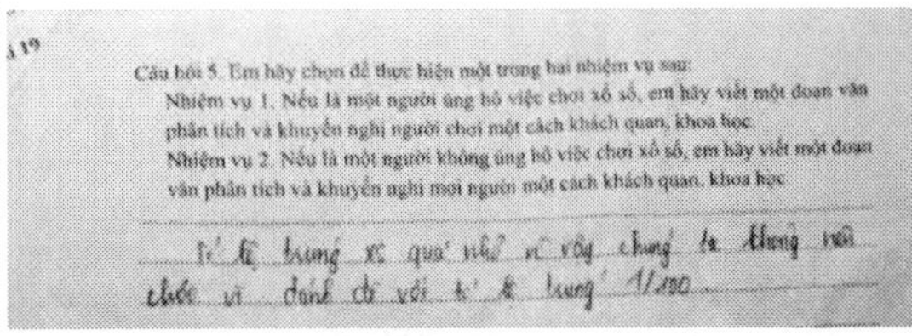

Figure 22.

Question 6. Most of the students choose Vietlot as the investment plan but they could not explain their choice. When we interviewed some students, they still did not point out specific or convincing justifications for their option (Figure 23.). Their answers even implied the misunderstanding between investing in lottery companies and investing in playing lottery. (Figure 24.).

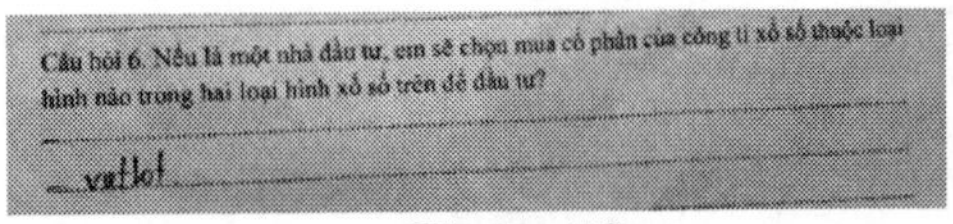

Figure 23.

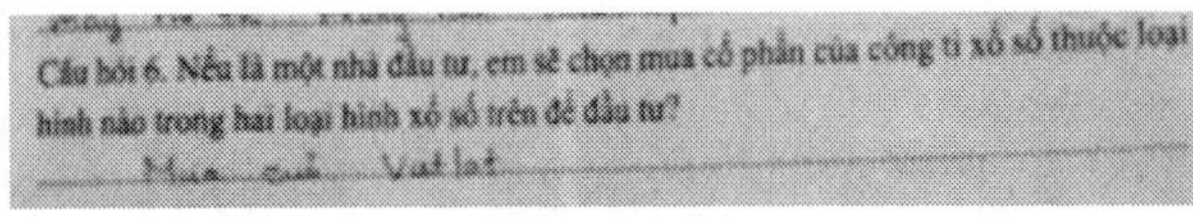

Figure 24.

6. Conclusions and recommendations

In all the situations mentioned, the students were always given plenty of time, not only in class but also independent self-study time. The knowledge we asked is just basic knowledge rather than complex knowledge, which requires complex mathematics transformational techniques or belongs to the field of probability and requires derivatives to identify the maximum and minimum values of the functions. However, the findings of the initial research point out various concerns about the students' competence to apply mathematics into practice including:

1) Students have not yet adapted to the proposed Maths task types as well as the organising methods of practical learning activities.
2) Students are not proactive in exploit relevant information in Maths problems, in life when encoutering new task type and knowledge they are familiar to in school.
3) Students' reading comprehension is generally limited;
4) The given contexts have not been fully and profoundly exploited and examined by the students.

5) Student's modelling competence is not adequate.

Therefore, we would propose some recommendations for developing the mathematics cirriculum as follow:

1) It's crucial that the Maths cirriculum take into account and introduce the requirements and regulations for developing learners' modeling compentence in a more practical way. In the context of Vietnam, the Mathematics program needs to be designed in a more open and practical way. One solution would be learning from the effective international Maths cirricula in other countries such as the Netherlands, USA, Singapore or Indonesia;
2) In the Vietnamese context, the implementation of school curriculum and classroom programs needs to be more flexible to further promote the teachers' creativity in designing classroom or school programs more effectively and progressively;
3) It is necessary to consider and systematically innovate assessment methods in major mathematics examinations throughout the school year or provincial and national examinations in the aim of assessing learners' modeling competence and reading comprehension ability because assessment form and content play an important role in exam preparation and teaching and learning methods.
4) Maths teachers need to update their teaching content in the direction of enhancing the application and connection of Maths knowledge and real life in order to to help students use Mathematics and realise the significance of Maths;
5) It's necessary to develop of students' reading comprehension ability in teaching Mathematics.

References

[1] Baker E L, O'Neill H F and Linn R L 1993 *Policy and Validity Prospects for Performance-Based Assessment* (Am Psychol vol 48) pp 1210-1218

[2] Bennett J, Lubben F and Hogarth S 2007 *Bringing Science to Life :A Synthesis of the Research Evidence on the Effects of Context-Based and the Science-Technology-Society Approaches to Science Teaching* (Science Education vol 91) pp 347-370

[3] Do Dat 2000 *Mathematics teaching and learning in Vietnam* (International Journal for Mathematics Teaching and Learning) Retrieved on December 15, 2005 from : http//:www.cimt.plymouth.ac.uk/journal/

[4] Freudenthal H 1968 *Why to teach mathematics so as to be useful* (Educational Studies in Mathematics vol 1) pp 3-8

[5] Gilbert J K 2006 *On the nature of "Context "in Chemical Education* (Intenational Journal Science Education vol 28) pp 957-976

[6] Guy Brousseau)2002 .(Theory of Didactical Situations in Mathematics (Kluwer Academic Publisher)

[7] Kamaliyah, Zulkardi and Darmawijoyo 2013 *Developing the Sixth level of PISA-Like Mathematics Problems for Secondary School Students* (Indonesian Mathematics Society, Journal on Mathematics Education vol 4) chapter 1 pp 9-28

[8] Le Tuan Anh 2006 Applying Realistic Mathematics Education in Vietnam :Teaching middle school geometry (Dotoral Thesis, Institutional Repository of the University of Postdam)

[9] Lu Pien Cheng 2013 *The disign of Mathematics Problem Using Real-life context for Young Children* (Journal of Science and Mathematics, Education in Southeast Asia vol 36) chapter 1 pp 23-43

[10] MoET 2017 *National general Education Curriculum –Overall curriculum.*

[11] Ngo Vu Thu Hang 2016 *Context-Based Education :An Advanced Approach* (Journal of Science, Educational Research vol 32) chapter 3 pp 11-17

[12] Nguyen Tien Trung, Hoang Ngoc Anh and Nguyen Duong Hoang 2017 *Constructivist teaching in Vietnam mathematics class :A case study of teaching equations of straight lines in the plan*

)*Geometry Grade 10th – High School, Vietnam* ((The European Journal of Education And Applied Psychology vol 4) pp 41-51

[13] Reudenthal H 1973 *Mathematics as an Educational Task.* (Riedel Publishing Company, Dordrecht, The Netherlands)

[14] Van den Heuvel -Panhuizen M 1996 *Assessment and Realistic Mathematics Education* (Utrecht :CD-Beta Press)

[15] Van den Heuvel-Panhuizen M 2000 *Mathematics education in the Netherland :A guided tour* (Freudenthal Institute Cd-rom for ICME9 .Utrecht :Utrecht University)

[16] Van den Heuvel -Panhuizen M 2003 *The didactical use of models in Realistic Mathematics Education :An example from a longitudinal trajectory on percentage* (Educational Studies in Mathematics vol 54) chapter 1 pp 9-35

[17] Van den Heuvel-Panhuizen M 2005 *The role of contexts in assessment problems in mathematics* (For the Learning of Mathematics vol 25) chapter 2 pp 2-9; 23

[18] Vu Dinh Phuong 2014 *Using Video Study to Investigate Eighth-grade Mathematics Classroom in Vietnam* (Dotoral Thesis, Institutional Repository of the University of Postdam)

IOP Conf. Series: Journal of Physics: Conf. Series **1340** (2019) 012033 doi:10.1088/1742-6596/1340/1/012033

STEM education in non-formal learning institutions in ethnic monority and mountainous areas: a look from Ha Giang city's community learning centre

Ha Thi Thuy, Hoang Dieu Thuy – Ha

Thai Nguyen University of Education Giang City's Community Learning Centre

*Corresponding author's e-mail address: hathuyduc2002@gmail.com

Abstract. Community Learning Centre (CLC) is an non-formal learning institution working at grassroots level of the society. It derives from the initiative of UNESCO (the United Nations Educational, Scientific and Cultural Organization) launched in 1988, as part of the Education for All Target Program in the Asia-Pacific region. The idea of building a lifelong learning society to promote the economy and appeal to participation of all social groups through education and the role of the CLC as a special autonomous community is the spirit and reason of CLCs' existence [7]. In Vietnam, As of 2015 there have been more than 11,900 CLCs established in 10,600 communes and wards nationwide [4], working as informal and non-formal educational institutions in the commune level. STEM education has been a new trending method of teaching and learning in the last decades originated from US under the demands of educating and training new labour generations with expertise in practical knowledge and skills. STEM education can be in forms of formal and informal learning, in schools and in lifelong learning institutions. STEM education has been imported to Southeast Asian countries including Vietnam in the last recent years and become a trending word in educational discourse. In Vietnam, STEM education has been piloted in a few public schools of big cities of designed in international schools and private institutions of the nation. Yet it is a luxury educational product and strange in remote areas of Vietnam where most people are poor and live under disadvantaged conditions. While, CLCs as part of continuing educational institutions of the country have exited in all provinces, but struggling with the crisis of operation. In this context, there has been a model of CLC choosing STEM education as its main operating component and working with mostly children in informal ways such as supporting STEM clubs in local schools, running mobile STEM in villages and connecting STEM experts with local teachers. This is a groundbreaking pilot model in the ethnic and mountainous context, with close link between science - education - research and private sector participation with local government budget and government-led activities: Ha Giang City's Community Learning Centre (HCLC) in Vietnam. It can probably be considered as a good example for other places to learn. The STEM education program of the HCLC is designed with the purpose of promoting formal and informal STEM education for ethnic minority students within the city. The three main aspects for changes after operating this STEM element are (1) improvement of learning environments and accessing to hands-on learning materials for children through a formal and informal science-integrated education system; (2) teachers' and education managers' professional development, and the enhancement of their provision of teaching learning material and through capacity building programs; And (3) opportunity of accessing better quality education for community, especially poor and ethnic minority children within remote suburbs of the city through free of charge enrollment for the community STEM program. In line with the content of the workshop, the following analysis focuses on the third aspec the STEM program that has been contributing to the improvement of better quality education for poor ethnic minority children in the 3 rural communes of Ha Giang city. This presentation will also provide some suggestions to the operation of STEM integrated education in CLCs or other community-based learning institutions in ethnic minority areas.

1. Introduction

1.1. The functions of CLCs

Community learning centre as defined by UNESCO is a local educational institution. It is theoretically established and operated by the local communities with the support of different sectors such as local government, private and Non-Profit/Government Organisation to deliver non-formal learning programs. The programs normally are designed and operated based on needs of communities and the conditions of locality.

Since CLCs aims at promoting lifelong education for all people in the community, they benefit mostly vulnerable groups who have less chances to access formal education like poor children, pre-school children, dropouts, women and similar others .

In Vietnam, the CLC model has been supported by the Vietnamese government with strong commitments and resources for development. Vietnamese Government has issued specific resolutions to guide the implementation and operation of this model, with the responsible agencies including the Ministry of Education and Training – MOET (Continuing Education Department - CED which is similar to Non-Formal Education Departments in other countries), Vietnam Association for Promoting Education (VAPE) and Viet Nam Institute of Educational Sciences (VNIES). At the local level (communes, wards and towns), the CLCs are approved by the People's Committee (districts, counties, towns and cities in the province), and are managed and professionally directed by the commune People's Committee and the Department of Education and Training.

As of 2015, Ha Giang province established 195 CLCs at 195 communes/wards/towns in over the province. Ha Giang city has 8 CLCs, being respective to 8 wards/villages. However, according to the report on results of CLSs' activity implementation in Ha Giang (509/BC-SGDDT on 20th July 2016) in the academic year 2015-2016 Ha Giang city is one of the districts that have the lowest number of classes being held in CLCs of the 11 districts of the provinces (142 classes/25,820 times of participation). Even, in the fields of strength as development tendency of urban areas, such as technical sciences, there was only 1 class a year . Or no classes was held and no particular activities were implemented in the fields that are considered as features of disadvantaged areas, such as illiteracy eradication and primary education univerlisation.

STEM education or STEM integrated education.

STEM education stems from science education as it stands for four disciplines of sciences studies Sciences, Technology, Engineering and Math. Therefore, since last two decades, STEM education has been considered to be the main target of developed nations where there was a prediction of lacking well-educated human resources in economic and technological areas. Along with this, notion of STEM integrated-education also occurs based on the division of STEM branches such as STEM career, STEM robotics, STEM professional, ...

STEM integrated-education is defined as approaches focussing on integration or combination of two or more areas of STEM subjects or a STEM subject and other subjects in order to design a lesson or curriculum showing connections between the subject areas and the real world issues. [5, 8]. Thus, all knowledge and skills of the teaching programs will concern the connections between the real life problems and seeking to find the ways to solve them.

However, on the other side, it is also defined as an approach of teaching STEM disciplines bounded up with learning practices within contexts to enhance student learning [9]. Therefore, it seeks to locate connections between STEM subjects and provide a relevant context for learning the content. Educators should remain true to the nature in which science, technology, engineering, and mathematics are applied to real-world situations. Hence, the more students practice, the more they reach authentic knowledge and skills. Regarding to this point of view, the authors create a conceptual framework for STEM integrated-education that put the community of practice into the assisting position to support the learning process. It helps the students to observe, discuss and participate into sharing and learning (p.4).

According to [2], in recent years, there has been a disconnection between schools and students communities, especially in urban environments where there are diversity of culture and linguistic. This disconnection leads to the situation that the school knowledge can not be used to solve the real-world problems. The authors also found out a form of connection called "connected science" using community-based knowledge and school –based knowledge to bridge the students to the community and provide them opportunities for meaningful and effective science learning programs.

There is also a rethinking of science education role in lifelong learning in our society. [8] state that, science education for community- participatory approach can not only prepare the life outside the school but also allows children "to participate in legitimate ways in community life and therefore provides a starting point for uninterrupted lifelong learning across the presently existing boundary separating formal schooling from everyday life outside schools". The relation between science learning and cultural practices are also cited in studies in both rural and urban areas. Particularly, states the efficacy of community-based science education programs conducted in Native American Communities [1].

Given that, science education particularly STEM integrated education is a flexible approach in which the teachers and learners can apply not only Western science knowledge but also local cultural knowledge. It cannot solely be taught in school curriculums but in informal programs too. Its contents are more than the four disciplines as its name and can be added by multi-subjects.

Ha Giang City's Learning Centre and Its STEM education

As in 2008 UNESCO Assessment Report [3], and the secondary data source survey in Vietnam, there are three groups of issues that a CLC model is facing, including: (i) management and operation (including infrastructure, specialized human resources and leadership); (ii) coordination mechanisms and engagement with relevant stakeholders (citizens, agencies join function); and (iii) strategies of CLCs (to solve specific problems locally, or linked with more macro issues).

Data analysis from professional reports and web-based surveys also indicates the importance and impact of CLCs on the implementation of socio-economic development goals at the grassroots level (illiteracy elimination, anti- illiteracy, knowledge and technology transfer, community engagement, livelihood support and health care). However, the performance of CLCs is uneven, and in many localities this model exists only in nominal terms but not in actual activity, leading to a situation where people do not approach and abandon this model.

In order to continue to exist and to maximize the effectiveness of lifelong learning promotion, and to help economic model transformation and socio-economic development in both rural and urban areas through learning activities, the model of CLC needs to change. In that circumstance, the HCLC is designed and advocate to pilot in Ha Giang City, a mountainous city grade 3 in the North part of Vietnam.

IOP Conf. Series: Journal of Physics: Conf. Series **1340** (2019) 012033 doi:10.1088/1742-6596/1340/1/012033

The Setting:

HCLC is first community - based non-formal education initiative in Vietnam, under the direct governance by Ha Giang City's People's Committee. It is entitled to own stamp and financial account to implement assigned tasks in compliance with the Law. It is the first model of CLC in the city level that cooperating between local government and a non-profit organisation working in sciences-technology fields named Ha Giang Centre for Community Assistance (HCA).

ORGANISATION CHART OF HCLC

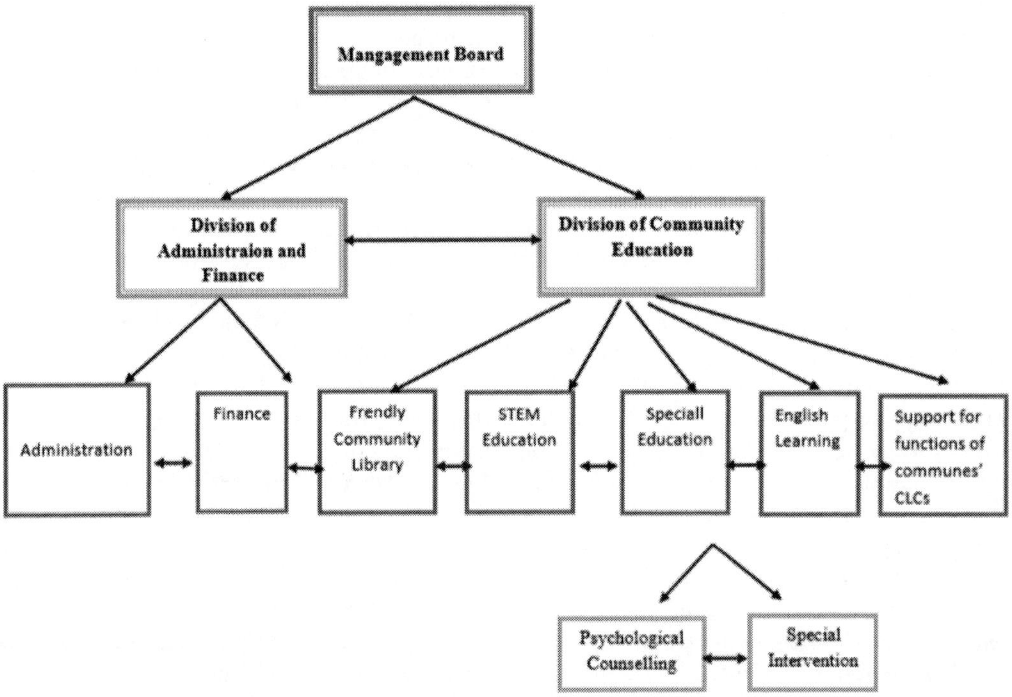

Diagram 1. Organisation chart of HCLC

According to Ha Giang City's People's Committee's decision, HCLC's functions and missions include:

Playing a role as a master Centre to support the activities of the district - level community learning centres within the city; Counsels and orientates community learning centres plans; Support the important events of the district - level community learning centres regarding assistants.

Promoting and attracts the participation of the community in non-formal activities through STEM education programs, English language learning, Community library, special and inclusive education, mental health care. The place for community activities are modern and friendly where individuals and the community are beneficial and have access to opportunities of continuing and lifelong education and comprehensive care.

Promoting the attraction of social resources through public and private cooperation, gratis assistance, non-interest loans, fund raising...in order to reinforce and extend the Centre's activities and improve its workforce quality (soft skills, foreign languages, knowledge of multi cultures…) in fields of economy, culture and society within the city.

Based on that, HCLC is responsible to:

Assist the activities of district - level community learning centres regarding techniques and human resource in within Ha Giang city and Vi Xuyen district in implementing annual tasks

Implement such education services as: STEM education courses, special education, foreign language teaching, art education....

Organises events, counselling activities; Assists psychological counselling and mental health care for children, parents and other groups within the community.

Organise the courses that enhances foreign language levels for the community tourism within the city and the province's districts.

Implement free social benefit programs as part of such modules as: library, STEM education , special education, local knowledge education and community mental health care

Cooperate, connects different resources within and beyond the community so as to practise and develop update, modern teaching and learning methods in compliance with the Law.

Hence, HCLC has demonstrated at the same time the parallel functions of (1) a model CLC which is able to deliver programmes and activities in response to the learning needs of the community and formal schools and (2) a resource center which is capable to provide technical assistance for other communes and formal schools in organizing practical classes.

HCLC's STEM education Programs

The STEM education Module is one of the 4 main modules of the Centre, being responsible for receiving the transfer of STEM education models from other countries thus serving different local children groups. It plays a role as a bridge and assists local schools in general education curriculum innovation. Also, it contributes to create opportunities of accessing the information about education and creates creative learning spaces for children within Ha Giang city through high quality outdoor education activities. This is the first STEM education model implemented in Ha Giang and is the first model in Vietnam implementing education programs through a community-based and non-profit centre.

The vision of HCLC in implementing this Module is, till 2020, the centre will actively cooperate with high quality secondary schools to improve the province's education, creating opportunities for developing high quality education in the province, serving local people over the province. This means the HCLC plays as a coordinating and mornitoring role in building STEM education ecosystem of the city and the province as well.

Given that, community-based approaches is considered as the main strategy of developing STEM Education of the centre. Since, all part of planing, designing and conducting activities of STEM Education program focus on the integration and harmony between sciences knowledge and local cultural elements and local traditional knowledge.

Community STEM Education

1. Infrastructure, teaching facilities and human resources of HCLC:

Ha Giang is a mountainous and one of the poorest provinces in Vietnam with nearly 90% of the population are ethnic minorities. Ha Giang city is the province's centre of economy, culture and politics with 45% of ethnic minorities and 8 administration units at ward/village level. 3 villages lie in the especially - disadvantaged group. Therefore, the implementation of STEM education can not be as robotic and technical as in developed cities and private STEM education centres in Vietnam. Due to characteristic features of the teaching materials and the orientation of teaching method towards the nature, agriculture, and local knowledge application, the centre just spent a little part of the budget on teaching facilities (about 5000 USD), but take advantage of material from the locality and nature.

IOP Conf. Series: Journal of Physics: Conf. Series **1340** (2019) 012033 doi:10.1088/1742-6596/1340/1/012033

Figure 1. Community STEM Education

The centre has 5 STEM education teachers. Most of them are teachers of science subjects, and have been trained by domestic experts under the coordination of the US experts. After being trained, the teachers are in person conduct STEM programs at HCLC and work with villages/wards in order to implement STEM activities to the community's children.

Figure 2. Implementation STEM activities

2. STEM integrated programs

2.1. After school programs for children in 5 inner-city wards

Figure 3. STEM activities to the community's children

The students have participated in programs of support STEM literacy. These programs include hand-on activities that involve active learning, such as making icream without refregerator, creating a model of wooden stilt house, designing a bridge, planting in hydroponic systems, and putting together a science fair. All after-school components integrate asset building activities that promote learning, positive values, social competencies, and positive identities, cooperation around areas of STEM education. These programs have been held at the elementary and middle schools on Friday after 10 am and at the HCLC on Saturdays (Fantastic Saturdays). The programs have complemented initially with demonstration in class, hands-on experiment, field trips, with the number expanding in the following years to include more rural areas in Ha Giang province. Lots of parents came to the center to observe the positive activities in which their children have been involved, and witness their child receiving an award for participating.

Since December 2017 to present, the HCLC has organised 32 events like STEM fairs, STEM after-school sessio in the centre. There were approximatly from 20 to 30 children joined in every single activities, of which 55% is girls.

2.2. STEM Club Network

In collaboration with communities and schools, STEM clubs have been promoted and engaged in a STEM Clubs network in vision of development up to 2010. This network supposes to provide the forum where students and teachers will share experiences focusing on STEM learning and teaching during their regular meetings. Activities will include STEM showcases, science fairs, science contests, and career exploration activities. These activities have been facilitated by trained teachers who will act as change agents and partners to HCLC. Up to July of 2018, 3 STEM clubs in 3 villages (Tien Thang, Tha, Thai Hai) of 3 moutainous communes of Ha Giang City have been set up and running effectively. Regarding to this activity, the HCLC's STEM teachers have gone to the villages to conduct educational programs for the clubs weekly. The clubs' programs are varous from teaching STEM experiments to reading science books, watching science movies and learning life-skills. Recently, there have been about 700 children benefited from this network. 400 of those are girls.

Figure 4. STEM Club Network

2.3. Community-Free of Charge Programs:

- STEM Fair:

STEM fair held on November 2017 was a playstation for knowledge sharing and learning among teachers and students. The fair was given at 2 places, Ha Giang city and Bac Quang district, with the participation of 120 teachers from Ha Giang and 95 teachers and education administrators from Ha Noi and from nearby provinces. The 16 stalls in the fair already attracted more than 3600 students to participate in experience activities. The show contents and experience classes included Robotics, Reaction engines, Motorcycles….

Figure 5. Community-Free of Charge Programs

- STEM on the Move

This is a free program periodically serving the community, being held on every Sunday. The teachers will establish an integrated programs consisting of Animation movie shows, book reading, science games and STEM based experiment practice at CLSs in suburban villages to ethnic minority children. More than 1000 ethnic minority children in 8 villages benefit from the project. This is an on-going-mobile program and working with children at various ages, different background and social status. Therefore making curriculum in for this program based on the traditional local knowledge and real-life knowledge rather than formal education programs. The project is funded by US Embassy to Vietnam and will be finished in Jun 2018.

Figure 6. STEM on the Move

3. Conclusion

Lessons from HCLC:

STEM education in CLCs contributes to lifelong learning and in the community and develops learning society.

STEM education in CLCs contributes to close the gap between the city and the country, mountainous areas in high quality education access.

STEM education is not an costly teaching method (despite the requirement of teaching facilities).

"No one fits all". Effective STEM education programs should be flexible and characterised with local features, being well - matched with the perceptions and culture of local students and being suitable with local culture- economy conditions.

References

[1] Bang M and Medin D 2010 Cultural processes in science education: Supporting the navigation of multiple epistemologies **Sci. Ed. 94** pp 1008–1026
[2] Bouillion L M and Gomez L M 2001 Connecting school and community with science learning: Real world problems and school–community partnerships as contextual scaffolds **J. Res. Sci. Teach. 38** 878–898
[3] Haddad C 2008 Community Learning Centres: Country Report from Asia (Bangkok)
[4] Hossain Z 2016 Towards a lifelong learning society through reading promotion: Opportunities and challenges for libraries and community learning centres in Viet Nam **International Review of Education 62** pp 205–219

IOP Conf. Series: Journal of Physics: Conf. Series **1340** (2019) 012033 doi:10.1088/1742-6596/1340/1/012033

[5] Moore T, Stohlmann M, Wang H, Tank K, Glancy A and Roehrig G 2014 Implementation and integration of engineering in K-12 STEM education. In S. Purzer, J. Strobel, & M. Cardella (Eds.), Engineering in Pre-College Settings: Synthesizing Research, Policy, and Practices pp 35–60 (West Lafayette: Purdue University Press)

[6] National Research Council 2011 Successful K-12 STEM education: Identifying effective approaches in science, technology, engineering, and mathematics. National Academies Press.

[7] PHAM D 2010 Community Learning Center

[8] Roth W and Lee S 2004 Science education as/for participation in the community **Sci. Ed. 88** pp 263–291

[8] Sanders M 2009 STEM, STEM education, STEMmania **The Technology Teacher, 68** pp 20–26

[9] Todd R K and Knowles J G 2016 A conceptual framework for integrated STEM education **International Journal of STEM Education** pp 3–11

IOP Conf. Series: Journal of Physics: Conf. Series **1340** (2019) 012034 doi:10.1088/1742-6596/1340/1/012034

The needs assessment of internal supervision of teachers in the department sciences and mathematics of Khon Kaen University demonstration secondary school (Suksasat)

Metta Marwiang

Department of Mathematics, Faculty of Education, Khon Kaen University, Thailand

corresponding author's e-mail address: metmaw@kku.ac.th

ABSTRACT. The research procedures in the study of the needs assessment of internal supervision of secondary school teachers in the department sciences and mathematics of Khon Kaen University Demonstration School (Suksasat) consisted of three following steps: 1) to identify the concept to study the actual and desired model of internal supervision in six dimensions; including (i) internal supervision planning, (ii) direct support for teachers, (iii) group-working skill development, (iv) professional development, (v) curriculum development, and (vi) learning assessment and evaluation, 2) to investigate the actual and desired model of internal supervision of secondary school teachers in the department sciences and mathematics of Khon Kaen University Demonstration School (Suksasat) in all dimensions. The result showed that the overall actual internal supervision was in a high level ($\bar{x}$= 3.64, S.D. = 1.09) while the desired internal supervision was in a strongly high level ($\bar{X}$ = 4.78, S.D. = 0.49). Therefore, the needs assessment of internal supervision of secondary school teachers of Khon Kaen University Demonstration School (Suksasat) in all dimensions. The results indicated that the participants need internal supervision with the between 0.29 and 0.34, and with the overall of 0.30. The was found at the highest (0.34) when the direct support for teachers was independently considered.

1 .Background

According to National Education Act B.E. 2542 [14], the educational administration or learning administration aims to enhance knowledge for the learners to further obtain self-learning and self-development by focusing on the learners as priority .Thus, teachers must have strong command in the educational administration to reach the standards of curricular and national education .Also, the school must maintain its potential to manage the development of the quality assurance and educational evaluation system as well as upgrading the leading schools under the Office of the Basic Education Commission to become the World-Class Standard School, where all personnel must be engaged in the management collaboratively .Internal supervision is an important tool in building partnerships between teachers and educational personnel .Internal supervision is a collaborative process for teachers and administrators to develop curriculum and teaching and learning in order to solve learners 'learning problems and broaden their skill .The information concerning quality of education in the school is used as an indicator for the development of school management work in all aspects . However, the access to information for non-school personnel is limited. As a result, it is necessary that the personnel must cooperate closely in identifying the problems to solve and develop in accordance with the existing data and educational information according to Office of the Basic Education Commission [15].

One of the important tasks for school administrators, executives and staff in the school is to supervise, especially in school internal supervision is widely used to improve the quality of education . The school internal supervision aims to solve academic deficiencies and encourage teaching and learning to achieve its goals; as a consequence, the school curriculum is responsive to the needs of all school personnel .When the internal supervision is implemented precisely and appropriately, the school management is accepted by the school staff resulting in the positive internal and external evaluation.

The supervision in the school is a collaborative work process among administrators and school personnel to guide, support, and promote teaching and learning procedures to meet school standards [3]. The objectives of internal supervision are to develop staff in the organization, motivate teachers to obtain wider the knowledge and skills to develop the curriculum, teaching and learning management, coordinate and collaborate to ensure good understanding among related staff and departments . Moreover, it creates confidence and job security in the profession as a teacher, which in turn, encourages developing teaching and learning to reach the greater quality of the educational administration as expected .Hence, the internal supervision should be highlighted in all schools and complied with the STEM Education management .Particularly, sciences and mathematics learning holds fundamental structure of knowledge in any subjects that facilitates teachers to teach and combine knowledge in other fields since learners are encouraged to utilize problem-solving skills, seeking to meet the rapid pace of world's change .Therefore, the collaboration among teachers in different departments is key success in learning management .This helps develop skills required in the globalized world in the 21st century [6], [17], [4]

The commitment of Khon Kaen University Demonstration Secondary School (Suksasat) is to develop the academic abilities of learners, life skills and desirable attributes to produce graduates, and provide academic services and educational innovations while maintaining the national culture to be accepted in the international arena .The school is a leading demonstration school in ASEAN that does not only develop innovative education and research but also co-produce graduates with the Faculty of Education of Khon Kaen University .As a learning society, the school and university collaboration mainly supports the excellence of learners 'academic identity, and encourages learners to have confidence to express their vision [7].

Furthermore, school internal supervision is essential in learning management and the quality of education development to achieve the school goals .Therefore, the school administrators and all personnel in the educational institution are obliged to cooperate in all development tasks of the educational institution .The Education Service Area Office, as a main agency who is responsible for educational management, seek ways to ensure that the importance of supervision in the school has been addressed as imposed in the standards of the Education Service Area Office .In the statement, the school internal supervision is suggested to be used as a guideline for the establishment of standards for educational institutions for greater effectiveness in practice .

Nevertheless, in the actual practice, the implementation of internal supervision has faced with numerous obstacles. For example, Kunnatham School Center under the operation of Yuwa Sathiratham Foundation [10] tested follow-up internal supervision to voluntarily support the counseling and share educational vision among the staff which helps to establish the unity in team building process leading to sustainable success in the future.

In addition, Ratanawitoon [16] studied the setting academic events in elementary school which is under the jurisdiction of the Office of Nan Provincial Primary Education, and noted that 'the internal school supervision is often found with the lack of supervision staff and understanding in internal school supervision management.' The study also investigated that the current situation of the internal supervision and found that personnel development is insufficient for the current issues and needs. These results were consistent with the findings of the Office of the National Primary Education

Commission [12] mentioning that the school has limited competent personnel to supervise the school and supervision skills.

The needs assessment is considered as the first critical step in the assessment cycle to help diagnose problems and solutions .In the context of the school, research needs assessment can identify what is needed by the administrators and teacher, which is further used in the design of teaching and learning development .The learning assessment and evaluation are used to point out weaknesses and strengths that should be developed and promoted .They are also used to determine the direction that school administrator leads the school. In addition, the assessment and evaluation is beneficial to seek the most appropriate development methods, [18], [11] leading school development to a clearer and more direct to the goals. Due to the importance of the needs assessment, certain schools are required to report the assessment results in the school financial sheet as critical information used in supporting financial activities or projects of the school.

According to the importance of internal supervision in the department of sciences and mathematics for developing operational tasks of Khon Kaen University Demonstration Secondary School (Suksasat) and its needs assessment, the researcher team is aware that it is essential to identify the actual internal supervision in the current situation and the desired internal supervision that the administrators, teachers in the department of sciences and mathematics, and staff would preferred in order to develop teaching and learning management in line with STEM Education approach .Since education in the secondary level is strong foundation for students to further education in the higher level, the school must ensure that prospective graduates are fully equipped with knowledge to perform in their future career and serve the society in various fields .On the other hand, if the education in the secondary level is poor, the learning outcomes in the higher level will undoubtedly dissatisfied .

Thus, the effective implementation of internal supervision in the department of sciences and mathematics is one of the key factors to success influencing the quality of teaching and learning in school, especially in students 'learning outcomes .In this study, the needs of internal supervision of Khon Kaen University Demonstration Secondary School (Suksasat)are assessed to develop the model of internal supervision in the department of sciences and mathematics in accordance with the particular context of Khon Kaen University Demonstration Secondary School (Suksasat)

2 .Purpose of Study

1. To develop a conceptual framework in the study of the actual and the desired internal supervision development in the current circumstances
2. To investigate the current development of the internal supervision model of the teachers of Khon Kaen University Demonstration Secondary School (Suksasat)
3. To assess the needs for internal supervision of the teachers of Khon Kaen University Demonstration Secondary School (Suksasat)

3 .Concept of Study

The purpose of the research is to reach the scope of the study to

1) identify the needs assessment of internal supervision of teachers and staff of Khon Kaen University Demonstration Secondary School, and

2) develop the model of actual and desired internal supervision in six dimensions, including (i) internal supervision planning, (ii) direct support for teachers, (iii) group-working skill development, (iv) professional development, (v) curriculum development, and (vi) learning assessment and evaluation.

The internal supervision model development is designed as a conceptual framework of the study as shown in Figure 1.

Independent Variables

Attributes of teachers and staff in the department of science and mathematics of Khon Kaen University Demonstration Secondary School (Suksasat)

Dependent Variables

Dimensions in the internal supervision 6 .

1. Internal supervision planning
2. Direct support for teachers
3. Group-working skill development
4. Professional development
5. Curriculum development
6. Learning assessment and evaluation

Figure 1. Conceptual framework of the study

4 .Population and Samples

The population was 80 secondary school teachers of Khon Kaen University Demonstration School (Suksasat). A total of 28 samples were selected by purposive sampling .

5 .Research Tools

In this study, the questionnaires were used as research tool designed by the author from the theoretical frameworks and concepts to obtain the information of the internal supervision .The questions are divided into 3 parts .The first part is mainly asked to obtain general information of the respondents : gender, age, education background, work position, work experience, and the questions regarding model internal supervision using checklist technique .In the second part. The questions are concerning the current issues and the needs assessment of internal supervision by using 5-Likert's scale, cited in Kaiyawan [8]. The final part of the questionnaire focuses on the needs assessment and suggestions for internal supervision by using open-ended questions.

6 .Research Tool Testing

1. The validity of research tool was used to indicate that the content is related to the scope of the study. The questionnaire was developed and revised by the recommendations of experts in all dimensions with the objective congruence (IOC) values of 0.6 - 1.00.
4. The reliability of the questionnaire was used to try out with the respondents, namely teachers and staff at the Khon Kaen University Demonstration Secondary School (Mordindaeng). The reliability were 0.97.

7 .Results

The result of needs assessment of internal supervision in the department of sciences and mathematics of secondary school teachers of Khon Kaen University Demonstration School (Suksasat) is shown in Table 1.

Table 1. Results of overall needs assessment of internal supervision in the department of sciences and mathematics of secondary school teachers of Khon Kaen University Demonstration School(Suksasat)

Dimensions in Internal Supervision	Actual implementation		Desired implementation		$PNI_{Modified}$	Rank
	$\overline{X}$	*S.D.*	$\overline{X}$	*S.D.*		
1. Internal supervision planning	3.69	1.02	4.74	0.59	0.29	6
2. Direct support for teachers	3.61	1.05	4.82	0.52	0.34	1
3. Group-working skill development	3.61	1.13	4.80	0.50	0.33	2
4. Professional development	3.73	1.09	4.84	0.39	0.30	5
5. Curriculum development	3.59	1.12	4.74	0.46	0.32	4
6. Learning assessment and evaluation	3.62	1.13	4.76	0.51	0.33	2
Total	3.64	1.09	4.78	0.49	0.32	-

8 .Discussion

Based on the results of the research, the needs of internal supervision in the department of sciences and mathematics of the teachers of Khon Kaen University Demonstration Secondary School)Suksasat(, the significant data were found as follows:

Part 1: The descriptive study was used to identify current issues of internal supervision of teachers in the department of sciences and mathematics of Khon Kaen University Demonstration Secondary School (Suksasat) in order to indicate its needs for future development. This research method allows an insight into the key issues of actual internal supervision occurred in the school as described below.

First, educators are required to assist the teachers directly by providing a summary of the teachers' performance to solve their problems or obstacles in teaching and learning. This is a key factor to improve teaching activities development and evaluation. However, the validity of the evaluation is varied from different evaluators [9] Amphol [1] also pointed that human resources development should be implemented based on the actual practice in a specific context in order to facilitate and encourage the staff effectively.

Second, the overall result the actual practices of internal supervision in the department of sciences and mathematics of teachers of Khon Kaen University Demonstration Secondary School (Suksasat) showed the following statistic data: mean (3.64), and SD (1.09). The three highest mean of all six dimensions of the internal supervision was professional development ($\overline{X}$= 3.64, S.D. = 1.09), followed by internal supervision planning ($\overline{X}$= 3.73, S.D. = 1.09), and assessment and evaluation ($\overline{X}$ = 3.69, S.D. = 1.13), respectively.

Third, the overall desired practices of internal supervision in the department of sciences and mathematics of teachers of Khon Kaen University Demonstration Secondary School (Suksasat) showed the three highest mean of 4.80, and S.D. of 0.49. When each dimension of the internal supervision was independently considered, it was found that three highest mean of all six dimensions was professional development ($\overline{X}$= 4.84, SD = 0.39), followed by direct support for teachers ($\overline{X}$= 4.82, SD = 0.52), and group-working skill development ($\overline{X}$= 4.80, SD = 0.50), respectively.

Part 2: The results of needs assessment of internal supervision of teachers of Khon Kaen University Demonstration Secondary School)Suksasat (are described below.

First, the needs of internal supervision of respondents of Khon Kaen University Demonstration Secondary School (Suksasat) were found in all dimensions with the $PNI_{Modified}$ between 0.29 and 0.34 with the average of 0.32 .The dimension of direct help for teachers was found with the highest $PNI_{Modified}$ of 0.34, followed by group-working skill development and assessment and evaluation with $PNI_{Modified}$ of 0.33, and internal supervision planning with $PNI_{Modified}$ of 0.29, respectively .

The results in this study were consistent with the strategy of the Office of the Public Sector Development Commission [14], mentioning about the national development plan of Thai public sector in the span of 2008 – 2012. The purpose of the strategic plan for Thai public sector development section III was established to lead government organizations to become a high-performing with talented and competent staff so that they are able to learn, innovate, adapt and adjust to different circumstances.

In the context of the educational institute assessment of Khon Kaen University Demonstration Secondary School (Suksasat), the Balanced Scorecard (BSC) in the perspective of internal education management was used as an assessment tool. The implementation of internal supervision was consistent with the approach because it is a process of developing instructional effectiveness. The actual teaching was observed and found with the innovation using 'Lesson Study', which requires the educators to directly assist the teachers, to develop curriculum and learning assessment and evaluation.

According to Chindanurak [5], teachers must have sufficient knowledge and skills to develop learners with the appropriate qualifications that help develop country under the national policy, called 'Thailand 4.0 Policy'. This means that teachers must be able to analyze and apply accumulated knowledge to improve the teaching and learning process.

The results reflected that the respondents were developed by the Faculty of Education to develop the understanding toward bureaucracy which was in compliance with government's quality management guidelines that highlighted the awareness of globalization under the educational policy of STEM Education. This approach is anticipated to solve the current bureaucratic issues while improve the test results e.g. PISA which lead the citizens to gain wider knowledge [2]. In the era of information-driven society, the pursuit of knowledge becomes common that leads teachers to share knowledge, understanding and awareness of internal supervision. Thus, the curriculum of the Faculty was aligned with the national development strategies.

In summary, it can be concluded that the results of needs assessment of internal supervision of Khon Kaen University Demonstration Secondary School (Suksasat) are consistent with the preliminary actual and desired implementation of inter supervision. As a consequence, the results can be used as a guideline for the model development of internal supervision. The results are also in accordance with the concept of quality management the development of government's quality management

9. Recommendations

Based on the results and discussion in the study, the practical recommendations for future internal supervision practices and further academic research are presented as follows.

9.1 Practical implications

9.1.1 Practical implications for policy plaining

- The results showed that supervisors had limited time to supervise teachers in certain topics. Moreover, there was no clear evidence showing that the school administrators have seriously applied the knowledge of internal supervision into the development of teaching and learning in practice. Hence, the school administrators should play a key role in taking internal supervision application from the empirical results into account as part of an internal supervision approach.

- From the findings, it can be suggested that the internal supervision model should be implemented in line with the country's sustainable strategic development. Also, the model should enable teachers to be aware of the importance of internal supervision; as a result, teachers can improve and develop their practices constantly.
- Schools should impose systematic and long-term plan and policies to develop human resources of Khon Kaen University Demonstration Secondary School (Suksasat) to maximize their effectiveness while promoting their morale.

9.1.2. Practical implications for implementation

- Each learning department should schedule meeting to supervise the staff in every semester or establish training for internal supervision skill development.
- Develop a project to study the model of supervision in accordance with the government's sustainable management of education.
- Support and promote the visit to external agencies that have similar characteristics, at least one time per academic year to allow human resources to understand the internal supervision of other agencies and apply it appropriately in their organization.

9.2 Suggestions for future research

- There should be an analysis of the needs assessment, guidelines, and solutions designed for a particular department of teacher group; as a result, teachers can implement accordingly.
- It is recommended that all participants should collaboratively design the model of internal supervision in order to acquire the accurate data from different positions and meet the needs of organization as a whole.

10. Acknowledgment
The dedication of Khon Kaen University Demonstration Secondary School in this study was deeply appreciated.

References

[1] Amphol R 2015 *Practice of Internal Supervision of School Administrator under the Office of 1st Education Service Area Ayutthaya* (Krungkao Rajabhat Journal 2nd year vol 1) pp 25-32
[2] Blackley S, Sheffield R, Maynard N, Koul R and Walker R 2017 *Makerspace and reflective practice :Advancing pre-service teachers in STEM education (*Australian Journal of Teacher Education Online vol 2) chapter 3 pp 22
[3] Boonya C 2003 *Principles and Concepts of New Age of Supervision* (Supervision Department and Education Management Office of General Education Educational Service Area Office Area 8)
[4] Breiner JM, Carla CJ, Harkness SS and Koehler CM 2012 *What is STEM? A discussion about conceptions of STEM in education and Shelly Sheats Harkness Partnerships* (School Sci and Math vol 112) chapter 1 pp 3-11
[5] Chindanuluk T 2017 *Education 4.0 Management and Sustainable National Development*, retrieved on 12 December 2017, from http//:adacstou.wixsite.com/adacstou/single-post/*01/09/2017/*
[6] Dejarnette NK 2012 *America's children :providing early exposure to STEM)science, technology, engineering and math* (initiatives Edu vol 133) chapter 1 pp 77-84
[7] Demonstration School Khon Kaen University 2015 *Annual Report* (Secondary Department (Suksasat)Khonkaen University)
[8] Kaiyawan Y 2002 *Research Foundation* 4th edn (Bangkok :Suweeriyasarn)
[9] Kanjanawasee S 2007 *Theories of Evaluation* (Bangkok :Chulalongkorn University)
[10] Kunnatham School Center under Yuwa Sathiratham Foundation 2017 2nd edn Bangkok :

Sahamit Printing and Publishing Public Company Limited
[11] Marwiang M 2013 *The Development of A Model for Mathematic Classroom Assessment : Collaborative Assessment Pyramid* (Philosophy Doctoral Thesis of Education Assessment and Evaluation Faculty of Education Khon Kaen University)
[12] Office of the National Primary Education Commission 1991 *Evaluation Manual for Education* (National Education Standards Bangkok :MPT)
[13] Office of the National Education Commission 1999 *The National Education Act* (Bangkok :Prig wan Communication)
[14] Office of the Public Sector Development Commission 2009 *Good Governance Rating* (Bankok: Premier Pro Company Limited)
[15] Office of the Basic Education Commission 2010 *Principles of Quality Assessment for General Education Standards :Internal Quality Assessment of Educational Institutes* (Bangkok : Office of Educational Testing .Department of Education)
[16] Rattanawitoon S 2007 *Supervision Behavior of Administrators of Model Schools :Recognition of Teachers* (Office of Nan Provincial Primary Education)
[17] Wayne C 2012 *What is S.T.E.M and why do I need to know?* (Retrieved 28 July 2018 from https//:issuu.com/carleygroup/docs/stem12online)
[18] Wongvanich S 2007 *Research of Needs Assessment* 2nd edn (Bangkok :Thammada Place)

IOP Conf. Series: Journal of Physics: Conf. Series **1340** (2019) 012035 doi:10.1088/1742-6596/1340/1/012035

Compose the topic of integration to foster the competency of integrating natural science knowledge for students of Can Tho University's Physics Pedagogy with Webquest _method

Thu K. T. Tran

Department of Physics Education, CanTho University, Can Tho, VietNam

*Corresponding author's e-mail address: ttkthu@ctu.edu.vn

Abstract. Description integrated science competency of physics education students is important work. Lecturers may be evaluating their students based on level of integrated science competency description table exactly. The paper presents result about structure and expression competency model integration science and give examples of integrated teaching to students through the X-rays topic to develop their skills and knowledge.

1. Introduction

Improve the quality of comprehensive education, attach importance to the application of knowledge into practice, develop creativity, self-study, encourage lifelong learning in students [5].

In Vietnam, The new general education program has been implemented since 2019. Whereby, "science" subjects will stars to teach to students (grades 4-5) in elementary, junior high school with "natural sciences" subjects (grades 6-9) and elective subjects in natural sciences: physics, chemistry, biology for high school students (grades 10-12) to direct their career.

The teacher should be equipped and develop skills teach integrated knowledge or coordination, organize interdisciplinary teaching [4]. The best way to learn and perceive natural phenomena of the real world in science should be based on an effective interdisciplinary teaching [8].

The integrated teaching ability of pedagogical students include the knowledge, skills and attitudes [1, 2], the capacity to integrate scientific knowledge is a core competence of integrated teaching capacity in high school. The steady integration of knowledge helps students develop integrated ideas and the concepts of the environment so that they can compile and teach the subject better integrate in high school.

Integrated teaching to develop higher-order thinking, teaching physics for education student should approach competency. "Method of development is progressive. This means, the assessment must be required at the level [6].

Based on the research for the Bloom'capacity model, we have the given table to describe for the integrating science five level during general physics training, this help for the rating is specified and science.

IOP Conf. Series: Journal of Physics: Conf. Series **1340** (2019) 012035 doi:10.1088/1742-6596/1340/1/012035

2. Literature Review

2.1. The Three Domains of Learning
The committee identified three domains of educational activities or learning [2]
Cognitive: mental skills (knowledge)
Affective: growth in feelings or emotional areas (attitude or self)
Psychomotor: manual or physical skills (skills).

2.2. Webquest _ method
Defining a webquest

A WebQuest is an inquiry-oriented activity in which some or all of the information that learners interact with comes from resources on the internet [3].

Producing a webquest

An introduction that sets the stage and provides some background information.
A task that is doable and interesting.
A set of information sources needed to complete the task.
A description of the process the learners should go through in accomplishing the task
A conclusion that brings closure to the quest.

3. Findings

3.1. The integrated science competency of physics education student in teaching general physics show in the table 1

Table 1. Description integrated science competency of physics education student in teaching general physics.

Competence component	Standards
Knowledge	General physics knowledge
	Physical research method
	Integration science theory
Skills	Integrated skill in the physics subject
	Integrated skill in the science (physics - chemistry and biology)
	Study skill groups the general physics of students
Attitudes	Positive, self-reliant in the study of general physics
	Recognizing the importance of knowledge in general physics, the knowledge of integrating the subject and the integration of theoretical knowledge with other sciences to have integrated teaching ability in high school

3.2. Each standard description in 5 level, they are from the lowest level of competency in the figure 1.

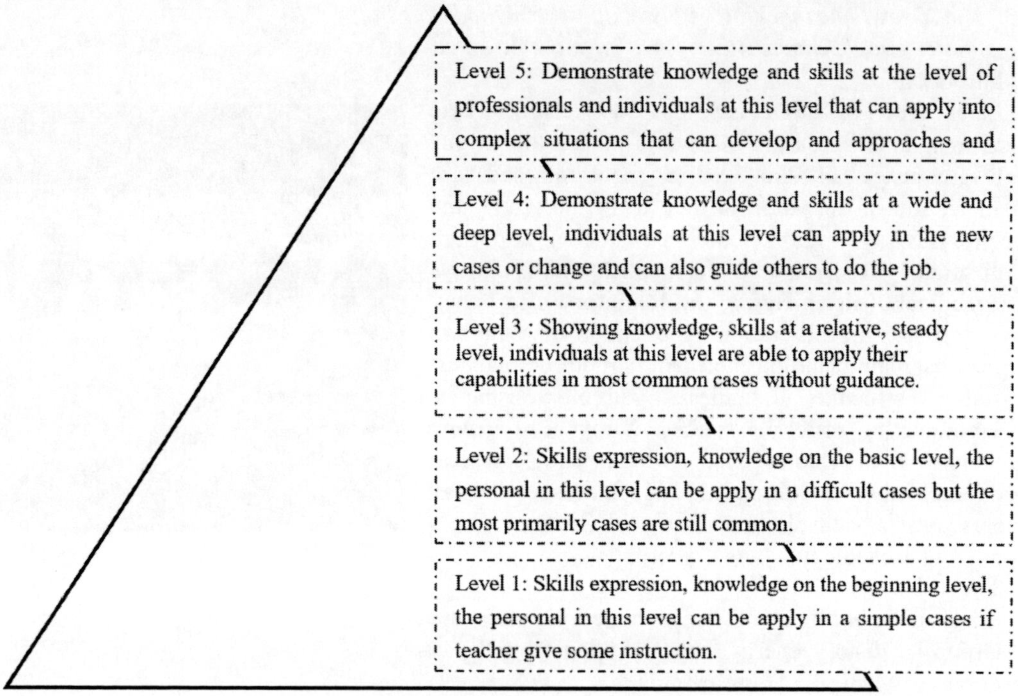

Figure 1. Capacity is described in five levels.

Thus, the new matrix would look similar to this

	Level 1	Level 2	Level 3	Level 4	Level 5
General physics knowledge					
Physical research method					
Integration science theory					
Integrated skill in the physics subject					
Integrated skill in the science (physics - chemistry and biology)					
Study skill groups the general physics of students					
Positive, self-reliant in the study of general physics					
Recognizing the importance of knowledge in general physics, the knowledge of integrating the subject and the integration of theoretical knowledge with other sciences to have integrated teaching ability in high school					

IOP Conf. Series: Journal of Physics: Conf. Series **1340** (2019) 012035 doi:10.1088/1742-6596/1340/1/012035

3.3. The X-ray topic teaching with webquest _ method Website: https://sites.google.com/a/ctu.edu.vn/x-rays-theme/home

Introduction

Each type of atom has its own characteristic electromagnetic spectrum. X-rays is at the top of the high frequency of the atomic spectrum and are characteristic of the atom. In this theme, we study the specific X-rays and some of their important applications. Research into the application of X-rays into medical X-ray imaging, the inhibitory effect they have during cell proliferation. Application of X-rays in airport baggage handling, protein structure analysis, crystals structure studies, atomic shapes, superconducting materials studies. X-rays can irritate the atom then fluorescence (which emits typical electronic wave radiation), making fluorescent X-ray fluorescence a valuable analytical tool in many fields from art to archeology.

The task

Study the applications of X-rays in the following aspects

Group 1: History, energy ranges, properties of X_rays

Group 2: Medical X_rays imaging

Group 3: Protein and crystals structure analysis, atomic shapes, superconducting materials studies

Group 4: Art to archeology

Group 5: Astronomy

Information sources

https://en.wikipedia.org/wiki/X-ray

https://www.fda.gov/radiation-emittingproducts/radiationemittingproductsandprocedures/medicalimaging/medicalx-rays/default.htm

https://www.orau.org/ptp/PTP%20Library/library/Subject/Early%20Publications/A_Text_book_of_radiology__X_rays_.pdf

https://proteinstructures.com/Experimental/Experimental/protein-crystallography.html

https://www.cambridge.org/core/services/aop-cambridge-core/content/view/8D6FAD3025CCAA979E533D03BC0522A7/S0885715600008563a.pdf/xrays_in_art_and_archaeology_an_overview.pdf

https://en.wikipedia.org/wiki/X-ray_astronomy

https://www.youtube.com/watch?v=hTz_rGP4v9Y

https://www.youtube.com/watch?v=T1WwHh4b__M&pbjreload=10

The process

Create 4 student/group. Request they to the tasks and reports with powerpoint software.

Conclusion

Understanding X-rays is an interesting topic because it is related to many disciplines, an interdisciplinary topic, student want to do task in this theme, they must use knowledge from other disciplines together. Although hard work, but certainly this will help students in real life.

4. Conclusions and Recommendations

In conclusion, this article presents the structure and expressions of integrated science capabilities for students of physical education in CanTho University. Lecturers who compose these topics can provide opportunities for students to develop skills and improve their knowledge of integration, meeting teacher standards after they graduate. The above description is a useful tool for instructors in evaluate the progress of students.

In addition, through the web-based learning activities, students are also trained in self-study, group work and information technology skills.

Integrated teaching process should be done regularly, which can be accomplished by instructors giving students a short essay test after each topic, aiming to assess their integration competence, if passed this test, instructors tasked with continuing with a new topic, and so on.

References

[1] Anderson L W, Krathwohl D R and Bloom B S 2001 A taxonomy for learning, teaching, and assessing: A revision of Bloom's taxonomy of educational objectives (Allyn & Bacon)

[2] Bloom B S (Ed.), Engelhart M D, Furst E J, Hill W H and Krathwohl D R 1956 Taxonomy of Educational Objectives, Handbook I: The Cognitive Domain (New York: David McKay Co Inc)

[3] Dodge B 1995 *"Some Thoughts About WebQuests"* (San Diego State University)

[4] Ministry of Education and Training The summary record of a conference, Raising the capacity of training and fostering teachers and educational administrators of pedagogical schools to meet the requirements of renewing the general education (Đa Nang, VietNam)

[5] Resolution 2013 On "Fundamental and comprehensive innovation in education, serving industrialization and modernization in a socialist-oriented market economy during international integration" ratified in the 8th session (HaNoi, VietNam)

[6] Roegiers X 1996 Khoa sư phạm tích hợp, hay làm thê nào để phát triển càc năng lực ở nhà trường The pedagogy of integration, or how to develop competences at the school (Hanoi, Vietnam: Nhà Xuât Bản Giáo Dục Vietnam Education Publishing House)

[7] www.ndt.org

[8] You H S 2017 Why teach science with an interdicciplinary approach: history, trends, and conceptual frameworks **Journal of Education and Learning 6**

Assessment of learning progression on mathematical problem solving of students using open approach

Metta Marwiang[1] , Kitipat Dechdan[2] , Chakkrit Sunyalak[3] , Putcharee Junpeng[4]

[1,2,3,4]Faculty of Education , Khon Kaen University, Thailand

Email [1]metmaw@kku.ac.th, [2]kitipat_d@kkumail.com, [3]lpnp_pae@hotmail.com, [4]jputcha@kku.ac.th

ABSTRACT: The purpose of this research was to assess the progress of mathematical problem solving skills of students at Bannamphrae School in Chiang Mai in which the Open Approach was employed for class management. This research study was a quantitative research. The scope of the study included 23 fifth-grade students at Bannamphrae School in Chiang Mai, who had been selected through the purposive random sampling from the population covering 520 students in total. All of the students were in the fifth to sixth grades and were from 18 schools nationwide, which were participating in the Project on Advanced-Thinking Skills Development on Mathematics in the Northeast. The instrument used in the study was the mathematics problem solving test, which was comprised of 6 main items and 9 minor items. The data was analyzed via Conquest 2.0 software to discover the progress of their mathematical problem solving abilities through the developed framework of learning progress including 4 levels ranging from 0 to 3. The research revealed that the students at Bannamphrae School had shown different levels of progress with respect to mathematics problem solving as follows: 10 students were at Level 0 (Need Improvement), 6 students at Level 1 (Fair), 2 students at Level 2 (Good), and 6 students were at Level 3 (Very Good).

1. Background

In the 21st century, the world is dramatically changing. Mathematics, therefore, plays an important role in the development of human thinking. Learning mathematics focuses on learners. Through lesson study and the Open Approach, the learners become the center of learning by providing students with creative and systematic thinking and updated knowledge. The students become able to analyze problems or situations thoroughly and carefully. Moreover, they develop skills that will allow them to make predictions, plans, and decisions, as well as properly solve problems in their daily lives.[8] What was needed for the students to become more advanced was basic knowledge, which included memorizing basic rules; understanding the principles or theories, which are used to solve simple problems and more complex ones; and appropriately understanding and interpreting problems and situations so that when seeking to find the answers, a distorted perception of the problem could be avoided. Moreover, it was necessary for them to write explanations and justifications for the answers they had discovered by themselves, as well as to validate the answers and the accuracy of their calculations. In order to find the answers, they also needed to use numbers, mathematical symbols, or other means, such as figures and diagrams.[5]

Problem solving was an important focus of mathematics instruction. This has resulted in educators around the world turning to mathematical problem solving at all levels of the mathematics curriculum.[4] In Thailand, the Core Curriculum for Basic Education (BE 2551) was set up with the aim that learning management is to provide learners with knowledge, skills, competence, and desirable characteristics according to Learning Standards, which are based on the principle that learners are most important given the belief that everyone is able to make progress in learning and developing himself/herself. In learning management, students should be encouraged to naturally develop themselves to reach their full potential. Most importantly, individual differences and brain development should be taken into account.[6] However, what has happened in the classroom has not matched the purpose of this course; that is, learning management generally focuses on learning achievement, while the importance of the learning process is overlooked. This does not help the learners develop their competence.[1] Regarding the statistics of the results of the National Primary Education (O-NET) on mathematics in academic year 2016, sixth grade students, ninth grade students, and twelfth grade students had average scores of 43.47, 32.40, and 26.59, respectively, all of which were below the standard of 50 percent.[7] Moreover, according to the International Student Assessment 2015 (PISA 2015), in which 72 countries participated, Thailand received the 54th highest rating among the participating countries and had an average score of 421 points in mathematics, which is significantly lower than the international average of 500 points.[3]

Most mathematics teachers in Thailand use Direct Instruction as a teaching method, and as a result, students cannot express their own mathematical solutions.[1] Maitree Inprasit, et al. [5] argued that in order to promote effective learning reform, not only is there the need to change the teaching methods, but innovation is also needed. At present, Open Approach consists of a new teaching innovation called "Lesson Study" in which teachers need to understand the concepts or ideas of the students as much as possible in order to assist the students to discover ways or ideas to solve problems under the supervision of the teachers. As a result, learning management was developed using the concept of the Open-Ended Problems developed by Professor Nobuhiko Noida of the University of Tsukuba in Japan. It is used as a guideline for the development of the instructional activities of the learning management plan in the manner of creating a situation for mathematical problems. Students solve the open-ended problems through an investigative process and problem-solving. The students can create and develop new mathematical knowledge including the development of their thinking skills based on the different answers. Thereby, the different abilities of the students can be supported. The students can also collaborate in solving the open-ended problems because each student's response to the activity will be different depending upon each individual's ability. In addition, this method, can empower the students to reach their full potential. Hence, the students will be led to discover mathematical formulas and rules of their own. The emphasis of the Open Approach does not focus upon the correct answer, but instead, focuses on different ways of searching for the answers.

Therefore, the researcher is interested in examining how to assess the students' progress in learning mathematical problem solving in the classroom. Furthermore, in order to analyze the data received as a guideline to further develop learning management in the field of mathematics, the researcher is also interested in assessing the progress of the students, who have received Open-Approach instruction in learning mathematical problem solving.

2. Purpose of Study

To assess the progress of mathematical problem solving skills of fifth grade students on numbers and procedures using Lesson Study innovation by means of the Open Approach.

3. Concept of Study

The Investigation of the Progress of Mathematical Problem Solving of Fifth grade students using Lesson Study innovation by means of the Open Approach

The concept was as follows:

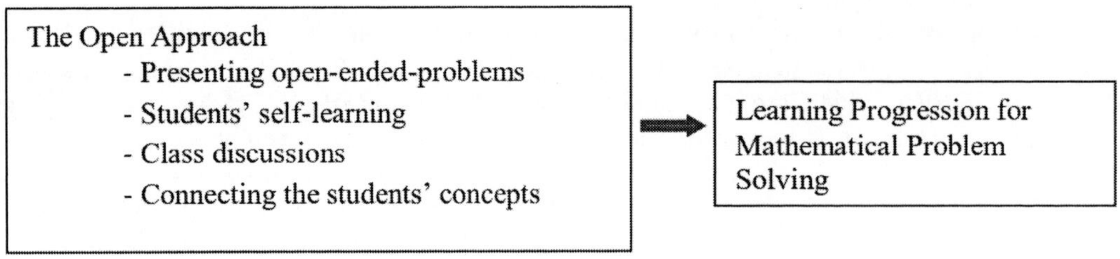

Figure 1. Conceptual Framework

4. Population and Samples

The sample consisted of 23 fifth grade students of Bannamphrae School in the Hangdong District of Chiang Mai.

The population was 520 students nationwide from 18 schools who were participating in the Project on Advanced-Thinking Skills Development on Mathematics in the Northeast.

5. Terms and Definitions

5.1 The study on learning progression on mathematical problem

Refers to the process of showing students' methods or strategies about problem solving.

5.2 The assessment on the learning progression of mathematical problem solving

Means the process of investigating the students' learning progression beginning from the lowest level to the uppermost level through the assessment form on solving open-ended questions.

5.3 Mathematical procedures

Mean the progress in the following areas: 1) problem solving learning, 2) the connections between mathematics and mathematics or between mathematics and other sciences, 3) giving reasons, 4) presentations and communication, 5) creativity, 6) skills, and 7) the problem solving process. In other words, it is a process of applying mathematical knowledge, the steps of the problem solving process, the strategies, and the students' experiences to solving mathematical problems. The process ranges from Unrecalled Memory, Basic Memory and Reproduction, Simple Skills and Concepts, to Strategic / Extended Thinking through open-ended questions.

5.4 Open Approach

Refers to the process of setting up situations with open-ended problems and then encouraging the students to solve the problems by themselves. The Open Approach focuses on developing the learners' mathematical potentials through open-ended questions.[2] The mathematical thinking process, used in mathematics class through the Open Approach, is as follows:

1) Presenting open-ended problems
2) Self-learning
3) Promoting class discussions and extended thinking
4) Connecting all concepts which were touched upon in class

6. Mathematical problem solving

Means solving open-ended questions or mathematical situations.

6.1 Methodology

This was a quantitative research study. An assessment form on mathematical competence was employed to collect the data from fifth grade students at Bannamphrae School. The data was analyzed together with the data derived from the 17 other schools. The learning progression on mathematical problem solving was assessed in terms of the mathematical process and the mathematical strategies on the basis of the Structural Model.[3]

6.2 Instrument and Instrument Quality

The Research Instrument was the assessment form on learning progression, which focused upon mathematical problem solving on numbers and procedures comprising 6 major items and 9 minor items for fifth grade primary students

- Quality of individual items: the data was analyzed using ConQuest 2.0 program developed by Margaret L. Wu, Raymond J. Adams, Mark R. Wilson, and Samuel A. Haldane in 2007. Item difficulty, discrimination, and suitability were considered in relation to MNSQ and T in Weight Fit. It was found that all open-ended questions had been at a moderate level, and the MNSQ was in an acceptable range. Subsequently, all items were accepted to be used.

According to the Threshold at each level of each response in the Wright Map, it was found that Items 1, 2, 4, 5, 6, 8, and 9 had a higher difficulty value, which corresponded to the Difficulty Level or Progress Map and had a range level from Easy to Difficult (the easiest level was scored 0; the hardest level was scored at 4). However, Item 3 and Item 7 did not have any increase in difficulty in relation to Difficulty Level or Progress Map. In addition, it was found that the MNSQ at each level of response on the Wright Map was within an acceptable CI range. The data indicated that this assessment form could be used as a tool to distinguish learner's learning progression based upon the assessment level in the created Progress Map.

- Quality of the test as a whole: According to the ConQuest 2.0 program developed by Margaret L.Wu, Raymond J. Adams, Mark R. Wilson, and Samuel A. Haldane in 2007, the test information has the maximum point of the y-axis at zero, which indicates that the difficulty level of the whole test is moderate to easy. The Internal consistency reliability (Cronbach's alpha coefficient; α) was 0.83, which meant that the quality of the whole test has internal consistency. Regarding the quality analysis for the whole test, it revealed that EAP/PV reliability was 0.827, indicating that the quality of the whole test had exhibited a high degree of reliability.

7. Results

According to the analysis of learning progression on mathematical problem solving of students which is based on the set criteria, it can be concluded that 23 fifth grade students at Bannamphare School showed different levels of progression as follows: a) 10 students were at Level 0 (Poor), b) 6 students were at Level 1 (Fair), c) 3 students were at Level 2 (Good), and d) 4 students were at Level 3 (Very Good). This indicated that the students were progressing in learning mathematical problem solving at a poor to fair level. The following table shows 4 different levels of learning progression involved with mathematical problem solving.

Table 1: The Levels of Progress Assessment in the Mathematical Process

Level	Process	Progression Level Description/Example
3 (Very Good)	Strategic/ Extended Thinking	Students can demonstrate a complex multi-step solution to find the right answer and/or can extend existing knowledge to other knowledge. For example, from the problem, *"Find the sum of 3 consecutive numbers of odd number."*, students can find the sum and write it in the following mathematical equation: $n + (n + 2) + (n + 4)$.

Table 1: (Cont')

Level	Process	Progression Level Description/Example
2 (Good)	Simple Skills and Concepts	Students can apply basic knowledge to solving both single and double-stage math problems by solving more complex problems. The problem solving process used by the students mostly reflects that the students have an understanding of the principles or theories. However, the solution has not been totally completed. There may be some misconceptions. For example, students can apply the principle of addition, subtraction, multiplication and division to solve problems, such as $2 + 3 x 8 \div 4-3 =?$ The multiplication and division have to be done first before proceeding to the addition and subtraction. However, the results may not be accurate due to some inappropriate or incorrect calculation.
1 (Fair)	Basic Memory and Reproduction	Students can apply basic knowledge to solve simple math problems. This is a one-step solution. It is a problem that students are familiar with, or a problem, which is similar to exercises in the textbook. For example, students can tell the properties of addition, subtraction, multiplication, or division. They can also remember the multiplication table or basic multiplication concepts from Sections 2 to 12 (2 x 3 = 6; 12 x 3 = 36. However, students lack the basic understanding of how to apply this principle to higher levels. It can be considered from a mathematical process that is not suitable for problem solving. This cannot lead to further discoveries because there are misconceptions about basic mathematical concepts.
0 (Poor)	Unrecalled Memory	Students can respond to some knowledge, but the responses reflect that the learner could not memorize important or necessary content as the basis for solving problems, such as using inappropriate mathematical processes or procedures. Students have no fundamental knowledge about the properties of addition, subtraction, multiplication, or division. Neither is it possible for them to solve simple problems because they cannot recognize multipliers or basic multiplication concepts from Sections 2 to 12. Subsequently, they have no mathematical basis to solve the problem at a higher level. Students do not give answers, nor do they answer questions, which are not related to open-ended questions.

8. Discussion

In regard to the assessment of the learning progression on mathematical problem solving skills of fifth grade primary students at Bannamphrae School where Open Approach innovation was used in the instruction, the raw score and the the full raw score was 36 points. The highest and the lowest scores were 19 and 1 points, respectively. The average score was 7.65 points of 23 students in total. Considering the individual student's progress in learning the math problems (EAP), there were four levels of grading based on the Wright Map at Level 0 to Level 3 as follows: a) 10 students were at Level 0 (Poor), b) 6 students at Level 1 (Fair), c) 3 students were at Level 2 (Good), and d) 4 students were at Level 3 (Very Good).

The researcher considered the raw score and the actual progress score of mathematical problem solving. Next, these were compared with the progress map, and then one step at a time, what students needed to know in order to enhance their mathematical competence was presented. Based on the actual competency level of the students, most students were found to be at a Poor or a Fair Level. This indicated that the students had been unable to remember important or essential content as a basis for solving problems, such as using inappropriate mathematical processes or procedures. No fundamental knowledge about the properties of addition, subtraction, multiplication, or division had been found. Furthermore, the students had been unable to find simple solutions to solve problems. Therefore, when carried out in proper order, the assessment based on the Progress Map will be a guideline for the development of the students' potential. This process helps students learn important points that should be improved for effective self-development.

9. Suggestions

According to the analysis of learning progression on mathematical problem solving for fifth grade primary students, it was found that a large number of students had been at a poor or a fair level. Teachers should provide students with extra classes or methods to adjust their knowledge base, especially those having misconceptions about the lessons. For example, divisibility is one of the problems that students encounter. When the students' notes had been analyzed, it was found that each of them had had different misconceptions. If these research results are implemented in math classes, students will know their levels, as well as their weaknesses, which need to be improved.

- The research results indicated that the assessment on learning and the lesson study had reflected the actual problems that had occurred in classes. Both of these should be conducted together with the Open Approach in order to successfully develop the learners' potential to its highest.

- Based on the test results gained from fifth grade primary students at Bannamphrae School, the researcher believes if the results from this study were applied to junior primary students, it would be more effective given that these students would have more time to develop their self-learning than the senior primary students.

References

[1] Inprasit M et al. 2003 *Reforming of the Learning Processes in School Mathematics with Emphasizing on Mathematical Processes* (Khon Kaen: Khon Kaen Print)
[2] Inprasitha M 2004 *Teaching by Open-Approach Method in Japanese Mathematics Classroom* (Journal of Mathematics in Thai 2547 vol 2) chapter 5 pp 30–35
[3] Junpeng P, Wilson M, Inprasite M and Marwiang M 2016 *Developing the framework for assessing cognitive process of Thai students' mathematical problem solving* (Unpublished manuscript California Graduate School of Education University of California at Berkeley)
[4] Krulik Stephen 1980 *Problem Solving in School Mathematics*: *Yearbook of the National Council of Teachers of Mathematics* (Reston Virginia: National Council of Teachers of Mathematics)
[5] Marwiang M, Klaharn J, Saree L and Junpeng P *2017 Assessing Students'Mathematical Problem Solving kill Through The innovative Lesson Study and Open Approach* (The

IOP Conf. Series: Journal of Physics: Conf. Series **1340** (2019) 012036 doi:10.1088/1742-6596/1340/1/012036

Turkish Online Journal of Educational Technology Special Issue for INTE 2017 October 2017) pp 385-395

[6] Ministry of Education. 2017 *The Implementation of the Basic Education Core Curriculum B.E. 2551* (AD 2008) (Retrieved August 8 2017 from http://www.ibe.unesco.org/curricula/thailand/th_befw_2008_eng.pdf)

[7] National Institute of Educational Testing Service 2016 *Quality Assurance in Thailand* (Retrieved August 5 2017 from http://www.onesqa.or.th/en/download/1004/)

[8] Panich W 2014 *Open Approach – Applying Active Learning for 21st Century Learning* (Retrieved 11 February 2017 from https://www.gotoknow.org/posts/568714)

[9] The Institute for the Promotion of Teaching Science and Technology (IPST) 2015 *PISA Results in Focus* (Retrieved August 7 2017 from https://www.oecd.org/pisa/-2015-results-in-focus.pdf)

STEM education Project -Based Learning Activities impacting on teamwork skills and Satisfaction of grade 11 Students in KhonKaenWittayayon School.

Nakin Satchakett and Aduldej Thana

KhonKaenWittayayon School, the Secondary Educational Service Area office 25

*Corresponding author's e-mail address:annakin_s@hotmail.com

Abstract This research aims to 1) To study the teamwork skills of the students 2) To study the effects of the project STEM education-based learning activities on the students' satisfaction. The target group was the Glade 11 students of KhonKaenWittayayon School. The study was conducted in the first semester of the academic year 2017. The 92 students were selected by purposive sampling. Research instruments included 9 lesson plans STEM approach, Teamwork assessment and satisfaction questionnaire. The result found that 1. The ability of teamwork skills of students who have been provided with the project STEM education-based learning activities was found that to Level 4 include 1) Listening to the opinions of others (125 points) Collaboration (122 points) and the reflection of their knowledge (122 points) all that means the students are attentive Listen to the proposals and ideas of the group members before presenting their ideas. Students can work with other members of the group well and share responsibilities with all team members. Students present or reflect a positive and creative concept. Level 3 include students Participation(116 points) Working behavior (107 points). Level 2 include Leadership (84 points) 2. Satisfaction with project STEM education-based learning activities was high level (4.70 S.D. = 0.2).

1. Introduction

The world of education has changed dramatically in the 21st century. Knowledge acquisition tools are more important than knowledge content. The advancement of communication technology enables learners to find their own knowledge from various sources [12]. The key to facing change and the challenge is the quality of people. In the present age of education, children are encouraged to develop their lifelong learning abilities [16]. According to the aforementioned problem, it leads to a crisis which is not how bad the students score on the exam but that they are not capable of analyzing[8]. Students do not truly understand the lessons since they try to remember. But have no idea how the lessons can be used in real life. Students are not able to connect their knowledge to the big picture[9]. When students encounter problems that are different from the classroom or have to do tests that have never encountered such problems, and can not solve the problem. However if the students have learned from the actual experience, practice, doing, think, have a continuous learning and discover self-knowledge. which can be applied or solved in real life. This will lead to self-development. Project-based learning is a process of seeking knowledge or search for the answer to what the students want to know or doubt in various ways. It is a way of learning that students choose to study according to their own interests or groups. It's a mutual decision, until the production can be used in the real life [11]. STEM education is Integration of integrated teaching across disciplines. Science (S), Technology (T), Engineering (E), and Mathematics (M), with the emphasis on nature and interdisciplinary teaching approach that are supported by a large number of research.

According to the problem and importance, researchers as teachers are interested in STEM learning styles to integrate in the course Knowledge Creation and Independent Education (IS*1-2)* in order to study Teamwork skills and Satisfaction of Grade 11 students, Khon Kaen Wittayayon School.

2. Purposes

1. To study the ability of teamwork skills of the students who have been provided with the STEM education project-based learning and Robotic activities.

2. To study the students' satisfaction of the STEM education project-based learning and Robotic activities.

3. Target group

Target group used in this study are Grade 11 students in KhonkaenWittayayon School. The *92* students in the first semester of the academic year 2017 were selected by purposive sampling.

4. Research Plan

One Group Pretest Posttest Design [10]. was used as a guideline for conducting the research. The pretest was administered by researcher and the treatment was provided during the 1st semester After the treatment section the post test was administered to evaluate the understanding with the following pattern.

Group	**Pre-test**	**Technique**	**Post-test**
E	**T1**	**X**	**T2**

When E represents the target audience
X represents the learning activities of the researcher created by the researcher.
T*1* represents the student's grades before class.
T2 represents student's grades after class.

5. Research tools

The instruments used in the study were the research instruments.

1) A plans of the learning activities of the project of the full school education project of the Grade 11 students and the test of learning achievement evaluating the basic knowledge on research methodology.

2) Teamwork assessment was used form to study the skills of teamwork. The results of the study were as follows.

3*)* Satisfaction Questionnaire on STEM Educational Activities The questionnaire consisted of *5* sets of questionnaires.

Research Framework

independent variable		dependent variable
Learning Management for Grade 11 students is as follows. Step *1:* Intro STEM education. Step *2:* Improve Critical thinking from model. Step *3:* Review the concept Step *4:* Create and develop Step *5:* Presenting than improvements.	➡	*1.* Teamwork Skill 2. Satisfaction

6. Data collection

The researcher collected the data manually as follows:

1) Pre-test Grade 11 students using the basic achievement test, knowledge, research methodology. A total of *10* items were taken and checked for grade point average.

2) Conduct learning activities in full education with Grade 11 students. There are *10* plans for *18* periods using the creative assessment record in every lesson

3) At the completion of the learning management of 18 class periods, the test of achievement(posttest) was administered and had students answer their attitude and satisfaction questionnaires were used to collect the data.

4) All data were analyzed using statistical methods. As discussed in the statistical section used for data analysis.

5) Summary of data analysis.

7. Data analysis

1) Analyze Teamwork Skill using the Teamwork assessment form to the group, then summarize the results in the table.

2*)* Analyze students' Satisfaction using answer in the assessment of individual Satisfaction Questionnaire then summarized in the table.

8. The analysis of data

Teamwork Skill

Table 1. Teamwork Skill

Evaluation Item	Number of students showing behavior				Score Sum	Rating
	4	3	2	1		
1. Collaboration	22	10	2	0	122	4
2. Participation	18	12	4	0	116	3
3. Listening to others	25	7	2	0	125	4
4. Reflecting knowledge	21	12	1	0	122	4
5. Leadership	6	6	20	2	84	2
6. Working behavior	14	11	9	0	107	3

From Table *1*, shows the total score. Listening to the opinions of others. Overall, students have a maximum score of *125*, which is a *4*-point scale, which means that students are attentive and listen to the proposals and ideas of the group members before presenting their idea. The second is the student participation. Overall, a total score of *122*, which scores *4* points scale shows that the student is able to work with other members of the group well and is responsible for all members of the team. The reflecting knowledge score is *122* points, which is a score of *4* points, indicating that the student is offering or reflecting on the positive and constructive concept the third aspectis student participation displaying a total score of *116*, which is a *3*-point scale, indicates that the student is involved in most work and is interested in their work. The fourth aspect is working behavior scoring 107 points which is a 3 point scale indicating that students paid attention to what they had been learning. Leadership with a total score of *84* points, compared to a score of *2* point scale, indicates that students are involved in the work, but find that they wasted time with unprofitable work and finding problems with sharing. Show that little students show leadership.

Figure 1. Teamwork Skill

Satisfaction of students toward the project activities.

Table 2. The students' satisfaction towards the project-based learning activities.

	Evaluation Issues	**Number of students (n) = 92**		
		($\bar{x}$)	**(S.D.)**	**Assessment**
Part 1 The nature of the activity.				
1.	Activities Activities	4.93	0.25	The Most
2.	Steps to Organize Activities	4.91	0.29	The Most
3.	Time spent on activities.	4.96	0.21	The Most
4.	Documents and media for activities.	4.91	0.29	The Most
5.	Appropriate activities.	4.90	0.25	The Most
	Average of Part 1	**4.92**	**0.26**	**The Most**
Part 2 Teacher.				
6.	The teacher has good manners.	4.16	0.21	The Most
7.	Teachers have the ability to care and solve problems.	4.78	0.15	The Most
8.	Teachers are ready to facilitate the provision of learning materials.	4.90	0.21	The Most
9.	Teachers take care of every teaching activity.	4.91	0.21	The Most
10.	The teacher can explain it comprehensively.	4.54	0.15	The Most
	Average of Part 2	**4.65**	**0.19**	**The Most**
Part 3 Measurement and Evaluation				
11.	How to measure the evaluation of teachers.	4.35	0.21	The Most
12.	Evaluation is difficult.	4.00	0.15	The Most
13.	Activity sheets are appropriate for the time spent learning.	4.96	0.21	The Most
14.	Activities to promote a variety of thinking.	4.21	0.32	The Most
15.	Encourage students to practice data collection. And self-knowledge.	4.98	0.15	The Most
	Average of Part 3	**4.50**	**0.21**	**The Most**
Part 4 Aspect of success in learning				
16.	This activity allows students to apply knowledge to their daily lives.	4.41	0.25	The Most
17.	Students become more curious and committed to their careers in this profession.	5.00	0.00	The Most
18.	Students are happy to learn from this activity.	4.91	0.29	The Most
19.	Student activities have a positive attitude towards science and math.	4.24	0.32	The Most
20.	A series of activities enables students to succeed in their studies.	4.98	0.15	The Most
	Average of Part 4	**4.71**	**0.21**	The Most
	Average Total	**4.70**	**0.20**	**The Most**

From Table 2, it was found that Satisfaction with Learning Activities Project The average of 4.70 (S.D. = 0.2) was at the highest level followed by the assumptions set. When considering satisfaction, it was found that The highest mean of satisfaction was in the first aspect of the activity (" $\bar{x}$ " = 4.92, SD = 0.26). The fourth aspect of learning achievement (" $\bar{x}$ " = 4.71, SD = 0.21). 2) Teachers (" $\bar{x}$ "=4.65, SD = 0.19) and the third aspect of measurement and evaluation (" $\bar{x}$ "‾= 4.50, SD = 0.21) respectively.

Figure 2. student Satisfaction

9. Discussions

According to the study, the STEM education project-based activity and Robotic activities affects creative thinking and attitude of Grade 11 Students, KhonKaenWittayayon School. The results can be discussed as follows.

1) Grade 11 students who have been provided with project-based learning activities have skillfulness in teamwork shown by the students, attention by listening to the proposals and ideas of the group members before presenting their ideas. Also, they work with other members of the group well and share responsibilities with all team members offering or reflecting a positive and creative concept. Students pay attention to their work by being reminded and engage in work, but find that wasting time with unprofitable work and finding problems with sharing the findings also show that students show leadership. This is consistent with the assumption that The students in the Grade 11 receive a full-scale learning project. And have the ability to work as a team after learning more than before. Since the project learning activities, each step has a group process, Students will need to adjust their behavior. Step *1:* Understand the basics of STEM education. Step *2:* Develop creative ideas from the third stage. Step *4:* Create and develop the ideas. Step *5.* The results of the research project of the Faculty of Science, Chulalongkorn University said that integrated teaching across disciplines. Science, technology, engineering and mathematics. The nature of the teaching as well as the teaching methods of each field to blend perfectly. To teach the students to use the knowledge to solve problems, research and development. As in the actual work or in daily life, a large amount of knowledge is needed in the process of group work throughout the project, students are encouraged to learn through real-life activities or projects that enhance the experience of life skills, creativity leading to innovation.

2*)* Grade 11 students received the project-based learning activities. The satisfaction of the students on the project-based learning activities was *4.70 (*SD = *0.2).* At the highest level Followed by the assumptions set. When considering satisfaction, it was found that The highest mean satisfaction were the aspect of activity, the success of learning teachers in the assessment and evaluation respectively, which is consistent with the assumption that students in the Grade 11 are satisfied with the project learning activities in full education. Students are expected to perform at all stages with fewer lectures in the activity. Use of high technology integration is used in this STEM project. Students use a variety of thinking processes. Teachers have the knowledge and ability to organize activities. There are assessment models that make students successful in learning. Students are satisfied with the level of activity.

10. Suggestion

1) The learning activities of the project should focus on how to access and evaluate each activity before class in order to give students a sense of the importance of making a portfolio, group activities, answering questions, and projects.

2) Teachers play an important role in encouraging them in Learning activities of the project. Students should be encouraged by question or problematic situation and receiving continuing suggestions to result in curious and self-learner.

3) Teachers should create a learning environment that allows learners to be free in their learning, thinking and practice to stimulate their thinking processes in many way.

4) The learning activities of the project focus on group work. To achieve the purpose Teachers must focus on the importance of working in a very group. Some activities may cause students to have different opinions. Teachers need to guide the students to think about the value of the good and the bad. Then share the ideas and choose the best idea.

5) In project-Based Learning Activities, there are many activities in the learning plan. Teachers will see the ability of different students, try to let the students express themselves. And instruct the students to see their abilities. However, in some activities, some students did not always collaborate especially brainstorming. Teacher should ask students to brainstorm their ideas in a paper and then organize them into groups. Teachers may motivate to grade each group. In the manner of accumulating points. And award the highest scoring group in each learning activity.

6) The media and research resources are important to organize learning activities. Teachers should provide sufficient media, materials and resources. This will result in effective activities.
To effectively

7) Presentation will give the student to show their potential and be confidents in presentation which will make them proud of themselves and develop a learning habit.

11. REFERENCES

[1] Chulavatnatol, M. (2015). *Preparing Thai STEM Workforce with 21st Century. Skills by STEM Education.* Paper presented in the 2nd International Conference on Innovation in Education (ICIE 2015).

[2] Csikzentmihalyi, M. (1988). *Society, culture and person: A system view of creativity.* In R. Sternberg (Ed.), The nature of creativity. Cambridge, NY: Cambridge University Press.

[3] Falk, J. H., & Needham, M. D. (2013). *Factors contributing to adult knowledge of science and technology.* Journal of Research in Science Teaching, 50, 431–452. doi:10.1002/tea.21080.

[4] Falk, J.H., Dierking, L.D., Staus, N.L., Wyld, J.N., Bailey, D., and Penuel, W. (2016). *The Synergies research–practice partnership project: a 2020 Vision case study.* Cult Stud of Sci Educ. 11:195–212.

[5] Guilford, J.P. (1967). *The Nature of Human Intelligence.* McGraw-Hill, Book Company.

[6] Gupta, B. N. ; Singh, R. B. ; Chatterjee, D., (1981). *Chemical composition and nutritive value of rice bean (*Phaseolus calcaratus *Roxb) hay.* Indian Vet. J., 58 (9): 527-530

[7] Hathcock, S.J., and Dickerson, D.L. (2015). *Scaffolding for Creative Product Possibilities in a Design-Based STEM Activity.* Res Sci Educ 45:727–748.

[8] IPST (2003). *Science evaluation guide.* Bangkok : Printing of Teachers Council of Ladprao

[9] Kittibhum Lertkittikulyothin.(2007).*The Study of Problem Solving Ability and scientific creative Thinking by Practical Science Project and Technology of Mathayom Suksa II Students.* Master's Project, M.E.d.(Secondary Education). Bangkok: Graduate School, Srinakarinwirot University. Advisor Committee: Assoc. Prof. Dr. Chutima Wattanakeeree.

[10] Loun Saiyos and Angkana Saiyos, (1995). Educational research techniques. 4th. Bangkok : Printing of Suweeriyasan.

[11] Office of the Secretary of the Education Council, (2007). *Framework for educational development During the 10th National Economic and Social Development (2007-2011).* Bangkok: Printing of Chulalongkorn University.

[12] Pornthip Siripatrachai, (2013). *STEM Education for development of skills in the 21st century.* journal (April-June 2013)

[13] Seattha, P., Tupsai, J., Sranamkham, T., and Yuenyong, C. (2016). *Students' view on STEM in learning about circular motion through STS approach.* AIP Conference Proceedings 1775, 030063; doi: 10.1063/1.4965183.

[14]Sompanya Sripakananon.(1992). *The study of the ability to create inventions and ideasCreation of Mathayom Suksa 2 students using a set of constructions to create artifactsKarma, repairing scientific inventions.* Master's Project, M.E.d.(Secondary Education). Bangkok: Graduate School, Srinakarinwirot University.

[15]Tassanee Pruksatan(1974).*The creation of scientific creativity test of lower secondary school students.* Master's Project, M.E.d.(Secondary Education). Bangkok: Graduate School, Chulalongkorn University.

[16] Wachara Loareindee, (2010) . *Development of teaching and learning styles to promote critical thinking abilities Science learning strand of grade 5 students.* Journal

IOP Conf. Series: Journal of Physics: Conf. Series **1340** (2019) 012038 doi:10.1088/1742-6596/1340/1/012038

STEM education project-Based and Robotic Learning Activities impacting on creativity and Attitude of grade 11 Students in KhonKaenWittayayon School.

Siwa Pinasa[1*] and Luksanawadee Srisook[1]

KhonKaenWittayayon School, the Secondary Educational Service Area office 25

*Corresponding author's e-mail address:Krusiwa@kkw.ac.th

Abstract. This research aims to 1) To study the creativity abilities of the students who have been provided with the STEM education project-based learning and Robotic activities. 2) To study the effects of the STEM education project-based learning and Robotic activities on the students' attitudes. The target group was the Grade 11 students of KhonKaenWittayayon School. The study was conducted in the first semester of the academic year 2017. The 92 students were selected by purposive sampling. Research instruments included 9 lesson plans STEM approach, Evaluate creativity, attitudes questionnaire. The result found that 1. The creativity of the students in the STEM education Project-Based Learning and Robotic Activities was highest level in the Fluency (average 26.80, SD = 2.56) and Flexibility (average 25.04, SD = 2.72). At the high level in the Initiatives (average = 23.82, SD = 2.04), and thoughtfulness (average = 21.78, SD = 2.42) 2. The attitude of glade 11 Students After Learning STEM education Project-Based Learning and Robotic Activities was all high Level

1. Introduction

The world of education has changed dramatically in the *21*st century. Knowledge acquisition tools are more important than knowledge content. The advancement of communication technology enables learners to find their own knowledge from various sources [13]. The key to facing change and the challenge is the quality of people. In the present age of education, children are encouraged to develop their lifelong learning abilities [16]. According to the aforementioned problem, it leads to a crisis which is not how bad the students score on the exam but that they are not capable of analyzing [9]. Students do not truly understand the lessons since they try to remember But have no idea how the lessons can be used in real life. Students are not able to connect their knowledge to the big picture [9]. When students encounter problems that are different from the classroom or have to do tests that have never encountered such problems, and can not solve the problem. However if the students have learned from the actual experience, practice, doing, think, have a continuous learning and discover self-knowledge. which can be applied or solved in real life. This will lead to self-development. Project-based learning is a process of seeking knowledge or search for the answer to what the students want to know or doubt in various ways. It is a way of learning that students choose to study according to their own interests or groups. It's a mutual decision, until the production can be used in the real life. [12]. Office of STEM education is Integration of integrated teaching across disciplines. Science (S), Technology (T), Engineering (E), and Mathematics (M), with the emphasis on nature and interdisciplinary teaching approach that are supported by a large number of research.

According to the problem and importance, researchers as teachers are interested in STEM learning styles to integrate in the course Knowledge Creation and Independent Education (IS*1-2)* in order to study creativity and Attitude of Grade 11 Students, KhonKaenWittayayon School.

1.1. Purposes

To study the creativity abilities of the students who have been provided with the STEM education project-based learning and Robotic activities.

To study the effects of the STEM education project-based learning and Robotic activities on the students' attitudes.

1.2. Target group

Target group used in this study are Grade 11 students in KhonkaenWittayayon School. The *92* students in the first semester of the academic year 2017 were selected by purposive sampling.

1.3. Research Plan

One Group Pretest Posttest Design [11]. was used as a guideline for conducting the research. The pretest was administered by researcher and the treatment was provided during the 1st semester After the treatment section the post test was administered to evaluate the understanding with the following pattern.

Group	Pre-test	Technique	Post-test
E	T1	X	T2

When E represents the target audience
X represents the learning activities of the researcher created by the researcher.
T*1* represents the student's grades before class.
T*2* represents student's grades after class.

1.4. Research tools

The instruments used in the study were the research instruments.

1. A plans of the learning activities of the project of the full school education project of the Grade 11 students and the test of learning achievement evaluating the basic knowledge on research methodology.
2. Creativity assessment form of Grade 11 students were evaluated in four aspects: fluency, flexibility, Initiative, and thoroughness.
3. The Attitude Test of Grade 11 Students.

Table 1. Research Framework

independent variable	dependent variable
Learning Management for Grade 11 students is as follows. Step *1:* Intro STEM education. Step *2:* Improve Critical thinking from model. Step *3:* Review the concept Step *4:* Create and develop Step *5:* Presenting than improvements.	*1.* Creativity 2. Attitude

2. Data collection

The researcher collected the data manually as follows:

1. Pre-test Grade 11 students using the basic achievement test, knowledge, research methodology. A total of 10 items were taken and checked for grade point average.
2. Conduct learning activities in full education with Grade 11 students. There are 10 plans for 18 periods using the creative assessment record in every lesson
3. At the completion of the learning management of 18 class periods, the test of achievement(posttest) was administered and had students answer their attitude and satisfaction questionnaires were used to collect the data.
4. All data were analyzed using statistical methods. As discussed in the statistical section used for data analysis.
5. Summary of data analysis.

IOP Conf. Series: Journal of Physics: Conf. Series **1340** (2019) 012038 doi:10.1088/1742-6596/1340/1/012038

3. Data analysis

1. Analyze creativity using the creative evaluation form to the group, then summarize the results in the table.

2. Analyze students' attitudes using answer in the assessment of individual attitudes then summarized in the table.

The analysis of data

Creative ability

Table 2. Creative ability

Creativity	Average	SD	Level
Fluency	26.80	2.56	highest
Flexibility	25.04	2.72	highest
Initiative	23.82	2.04	high
Thoughtfulness	21.78	*2.42*	high

From Table 2, it was found that the creativity scores of the students who received the project-based learning activities were as follows. The average score of the first aspect was *26.80 (*SD = *2.56)* which is at the highest level. The average score of the second aspect was *25.04 (*SD = *2.72)*, when considering the criteria, which is at the highest level. The 3rd aspect is initiative. The average score was *23.82 (*SD = *2.04)* which is at high level and the 4th aspect is thoughtful. The average score was *21.78 (*SD = *2.42)* when considering the criteria. It is at a very high level. The results show that the students have creative ability have been found as high and very high.

Figure 1. student Achievement.

Student Attitude

Table 3. Attitude of students towards learning activities in the project.

Student Attitude	Number of students						
	Most agree	Agree	Not sure	Disagree	Strongly disagree	Total score	Average
1. Attitude of interest							
1.1 Research is a new pursuit of knowledge.	78	12	2	0	0	444	4.83
1.2 Experimenting, I always like to find new knowledge and solve problems.	60	26	6	0	0	422	4.59
1.3 Friends spend time in the library, I feel tired.	3	14	25	32	18	324	3.52
1.4 Keeping up with the news is a good thing.	80	8	4	0	0	444	4.83
1.5 I like to observe while experimenting.	70	16	6	0	0	432	4.70
Average						413.20	4.70
2. Rational attitude							

Student Attitude	Number of students Most agree	Agree	Not sure	Disagree	Strongly disagree	Total score	Average
2.1 We should accept the description of others when there is evidence or information to support.	74	12	6	0	0	436	4.74
2.2 If so many people leave comments, it will be confusing and confusing.	0	0	23	26	43	388	4.22
2.3 If the leader of the group allows members to comment, it will make the job faster.	65	14	12	1	0	419	4.55
2.4 I like to work alone without criticizing or commenting on my work.	63	25	4	0	0	427	4.64
2.5 I always follow the teacher's recommendations.	70	14	6	2	0	428	4.65
average						419.60	4.65
3. Appropriate attitude to accept the conclusion.							
3.1 Before working on a group, I had to plan with my friends first.	85	5	2	0	0	451	4.90
3.2 People who accept the opinions of others are unreliable.	0	0	12	39	41	397	4.32
3.3 The reason of others, despite their feelings, should be heard.	70	16	6	0	0	432	4.70
3.4 I am willing to work with others.	70	16	6	0	0	432	4.70
3.5 When it comes to doing something, it must take into account the benefits shared over their own interests.	70	16	6	0	0	432	4.70
average						428.80	4.70
4. Attitude towards generosity and generosity.							
4.1 Be willing to publish what they have or know and seek more.	65	16	11	0	0	422	4.59
4.2 I like to talk or exchange ideas with those who have knowledge.	84	5	2	1	0	448	4.87
4.3 Attending lectures from lecturers makes it very time consuming.	5	16	25	20	26	322	3.50
4.4 Telling yourself to others will make others imitate.	0	16	25	25	26	337	3.66
4.5 Telling new stories to your friends is a proud thing.	70	19	3	0	0	435	4.73
Average						392.80	4.73

From Table3, it was found that the mean of attitude of students in Grade 11 after learning project was full. The fourth highest mean of attitudes toward generosity and generosity was *4.73* on the basis of the criteria. The average level of satisfaction was *4.70 (*mean *4.70).* Acceptance of reasoned conclusion (mean of *4.70)* was based on the criteria. This is a very good level and reason (average *4.65).* Holds a very good level.

Figure 2. student Activities

4. Discussions

According to the study, the STEM education project-based activity and Robotic activities affects creative thinking and attitude of Grade 11 Students, KhonKaenWittayayon School. The results can be discussed as follows.

1. Grade 11 students who have been provided with project-based learning activities. Have a very high level of creativity. The creativity with the highest scores was the average of *26.80*, SD = *2.56*, and the flexible idea (average *25.04*, SD = *2.72)*. Initiatives (mean = *23.82*, SD = *2.04)* and thorough thinking (mean = *21.78*, SD = *2.42)* were based on the assumptions set by the researcher. This is probably because of project learning in each lesson plan, there are *5* steps to develop creative thinking students, especially in the second step. This is a remarkable step in creativity because students must create their own unique works and show their originality. In creating and developing, we can express our creativity in every aspect. Since students have to take every step of the way to create a piece of work in accordance with the given problem, Gupta *(1981, 29)* [7]. states that creating a problematic situation for a teacher. It encourages students to use their ideas and interests. Tassaporn Prutcholatan *(*1974; From [10]). in other words, show that scientific creative is initiative fluency in thinking and flexibility of problem solving by using scientific process skills and Sompunya Sripakananon *(*1992; from [1]*.)*. that the ability of the person to express many ideas which is to create something that exists to the new model and that is not recreated with others. In thinking, basic knowledge and scientific process will be used. This ability consists of a streamlined thinking, flexible thinking and initiative used to solve the problem or find answers. So, it can be seen that scientific creativity has the same meaning as general creativity. That is the thought process. And the effect is different. Scientific creativity requires scientific process skills effectively to create intellectual development, problem solving and new knowledge discovery. Which corresponds to the meaning of science, the body of knowledge and process of science inquiry. The process of creative thinking. Kitiphum Lertkittikulyotin *(*2007*)* studied the ability to solve problems and creativity in science using the science and technology project of the students in the eighth grade. Students taught by using the science and technology projects have the ability to solve scientific and creative sciences. Is .01 significantly higher.

2. Student Attitude After the project was completed, the highest mean of attitudes is generosity. The acceptance of the rational conclusion is based on the assumption that the project-based learning activity can improve the attitudes of students in the Grade 11 to be better. This is because each step of the project learning activity is fully encouraged students to complete the group process. Students are required to exchange knowledge within the group. The attitude of generosity. In the second and third stages of project-based learning, there are so many searches which cause the attitude of interest of knowing and when the exchange of learning occur. Therefore, the attitude of accepting the rational conclusion and reasoning.

5. Suggestion

1. The learning activities of the project should focus on how to access and evaluate each activity before class in order to give students a sense of the importance of making a portfolio, group activities, answering questions, and projects.
2. Teachers play an important role in encouraging them in Learning activities of the project. Students should be encouraged by question or problematic situation and receiving continuing suggestions to result in curious and self-learner.
3. Teachers should create a learning environment that allows learners to be free in their learning, thinking and practice to stimulate their thinking processes in many way.

4. The learning activities of the project focus on group work. To achieve the purpose Teachers must focus on the importance of working in a very group. Some activities may cause students to have different opinions. Teachers need to guide the students to think about the value of the good and the bad. Then share the ideas and choose the best idea.
5. In project-Based Learning Activities, there are many activities in the learning plan. Teachers will see the ability of different students, try to let the students express themselves. And instruct the students to see their abilities. However, in some activities, some students did not always collaborate especially brainstorming. Teacher should ask students to brainstorm their ideas in a paper and then organize them into groups. Teachers may motivate to grade each group. In the manner of accumulating points. And award the highest scoring group in each learning activity.
6. The media and research resources are important to organize learning activities. Teachers should provide sufficient media, materials and resources. This will result in effective activities.
To effectively
7. Presentation will give the student to show their potential and be confidents in presentation which will make them proud of themselves and develop a learning habit.

References

[1] Buason Tumma 2011 The study of science and creative learning achievement Scientific Of Mathayom Suksa 3 students Who have been teaching science projects and science inquiry activities (Thesis, Srinakharinwirot University)
[2] Chulavatnatol M 2015 Preparing Thai STEM Workforce with 21st Century. Skills by STEM Education Paper presented in the 2nd International Conference on Innovation in Education
[3] Csikzentmihalyi M 1988 Society, culture and person: A system view of creativity In R. Sternberg (Ed.), The nature of creativity. Cambridge (NY: Cambridge University Press)
[4] Falk, J. H., & Needham, M. D. (2013). *Factors contributing to adult knowledge of science and technology*. Journal of Research in Science Teaching, 50, 431–452. doi:10.1002/tea.21080
[5] Falk J H, Dierking L D, Staus N L, Wyld J N, Bailey D and Penuel W 2016 The Synergies research–practice partnership project: a 2020 Vision case study **Cult Stud of Sci Educ 11** pp 195–212
[6] Guilford J P 1967 The Nature of Human Intelligence (McGraw-Hill, Book Company)
[7] Gupta B N, Singh R B and Chatterjee D 1981 Chemical composition and nutritive value of rice bean (*Phaseolus calcaratus* Roxb) hay **Indian Vet. J. 58** pp 527–530
[8] Hathcock S J and Dickerson D L 2015 Scaffolding for Creative Product Possibilities in a Design-Based STEM Activity **Res Sci Educ 45** pp 727–748
[9] IPST 2003 Science evaluation guide (Bangkok : Printing of Teachers Council of Ladprao)
[10] Kitiphum Lertkittikulyotin 2007 The study of problem solving ability and scientific creativity by using science project training (Thesis, Srinakharinwirot University journal)
[11] Loun Saiyos and Angkana Saiyos 1995 Educational research techniques. 4th (Bangkok : Printing of Suweeriyasan)
[12] Office of the Secretary of the Education Council 2007 Framework for educational development During the 10th National Economic and Social Development (2007-2011) (Bangkok: Printing of Chulalongkorn University)
[13] Pornthip Siripatrachai 2013 STEM Education for development of skills in the 21st century journal
[14] Seattha P, Tupsai J, Sranamkham T and Yuenyong C 2016 Students' view on STEM in learning about circular motion through STS approach **AIP Conference Proceedings 1775** 030063; doi: 10.1063/1.4965183
[15] Torrance E P 1966 Thinking Creatively with Words: Verbal Booklet A. (Benseville, IL: Scholastic Testing Service, Inc.)
[16] Wachara Loareindee 2010 Development of teaching and learning styles to promote critical thinking abilities Science learning strand of grade 5 students

IOP Conf. Series: Journal of Physics: Conf. Series **1340** (2019) 012039 doi:10.1088/1742-6596/1340/1/012039

Design Framework of Constructivist Mobile Application Learning Environment to Foster Creative Thinking on Basic Photography Skill for High School Students

Thanaban Kwanman[1] and Charuni Samat[2*]

Master degree student of Educational Technology, Faculty of Education, Khon Kaen University, Khon Kaen, Thailand, Thanaban.kwanman@gmail.com

[*]Assistant Professor of Computer Education, Faculty of Education, Khon Kaen University, Khon Kaen, Thailand, thaibannok@hotmail.com

Abstract. This research study aimed to synthesize theoretical framework and designing of constructivist mobile application learning environment to foster creative thinking on basic photography skill for high school students. The target group consisted of 3 expert reviewers for content, mobile application and learning environment designing. Research methodology is developmental research; developmental research consisted of 3 processes: (1) designing process, (2) developing process, (3) evaluate process. The procedures were as follows: (1) to examine and analyze the principles and theories, (2) to synthesize theoretical framework, and (3) to synthesize designing framework. The result revealed that: 1) to synthesize theoretical framework comprise of 6 components as following that. Contextual base, Psychological base, Pedagogies base, Creative Thinking base and Technologies and media base. 2) To synthesize design framework of mobile application learning environment to foster creative thinking of 6 components i.e., (1) Problem base, (2) Resource, (3) Creative Thinking Photo Lab, (4) Collaboration, (5) Coaching, (6) Scaffolding. The efficiency of this learning environment was evaluated by expert review. It was found that the learning environment is appropriate on 3 aspects: content, mobile application learning design, and learning environment design..

1. Introduction

Thailand society stepped into the digital world and economic social activities are carried out quickly, Competition was increasing for more data access of information through the online world, so Human Development to prepare to face the change is important. Education is an important tool in improving the quality of human resources in the country. The important mechanism for economic development, in the economic arena national and international through education. Creativity is a required skill in the global society in 2020. Enhancement of creative thinking on learner based on the web-based learning environment was achieved using the principles and theories for synthesizing the theoretical framework and the environmental design which promote creative thinking [9]. The theories and web-based characteristics were brought into the design of instruction that utilized the learning environment media and methods with important components of the Constructivist Theory. Mobile application technology can be used for promoting learner's creative thinking in order to have meaning learning. This accordingly happens from interaction with the learning environment. That can create virtual images that appear in a 3D animation, sound, and hypertext hypermedia. In this paper, are presenting the principles related to the basis of creative thinking and innovation skills for producing students in

Thailand 4.0 and the basic context of students training in Thailand that would lead to the development of innovation the enhance creative thinking for students in the world economic 2020.

Thus,this research was aimed at designing framework of constructivist augment reality mobile application learning environment to enhance creative thinking, from synthesizing of the theoretical framework and learning environment. In order to obtain the basis for constructing the appropriate and efficient learning environment models for the learners.

2. Literatuer review

A.Mobile Learning Environment

The rapid growth of information technologies along with the increasing flexibility in communications among users have provided new modalities of learning as well as innovative ways to deal with the limitations of traditional learning. For instance, due to the advent and evolution of technology, allied to the ubiquitous computing, a new modality of education based in mobile computing, referred to as mobile learning, has emerged [1].

3. Theoretical framework

The results show that the theoretical framework and designing of constructivist mobile application learning environment comprised of 5 theoretical bases. (1) Contextual base is following: basic education curriculum in Thailand, Course content. (2) Psychological base is following: Constructivist theory; cognitive constructivist [9] and social constructivist [8] and Cognitive Theory; information processing theory. (3) Technologies and media base are following: Web-based learning [10], Augmented Reality [1], the system of media [4], (4) Creative Thinking base are following: creative thinking theory [3] consisted of 4 abilities to think as follows; Fluency, Flexibility, Originality and Elaboration. (5) Pedagogies base Model learning environments are following: OLEs Model [5], SOI Model [6], Situated learning [7], Cognitive apprenticeship [2]. Fig.1. Showed theoretical framework of constructivist augmented reality web-based learning environments to enhance creative thinking on topic design and create three-dimensional grade, 9 students.

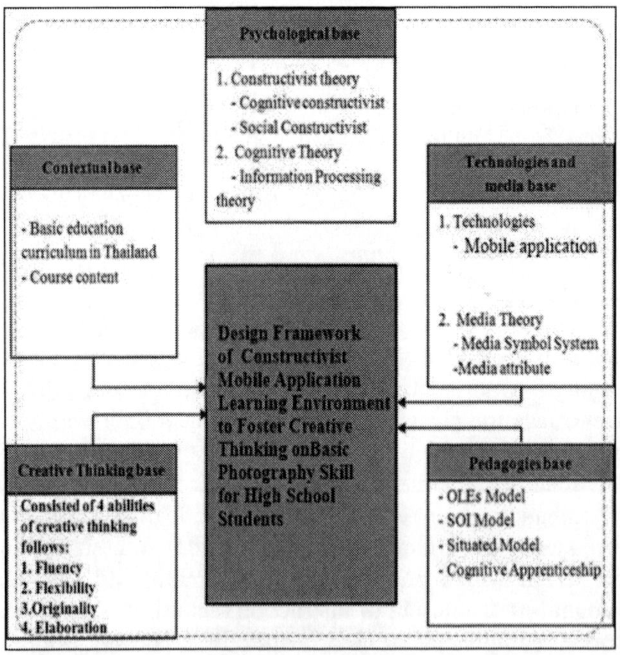

Figure 1. The theoretical framework of the learning environment for enhance creative thinking.

TABLE I. The results of an expert on learning content, the expert assessment. Theoretical framework synthesis and design framework.

No.	Lists of preconception towards the Constructivist web-based learning environments	results of the expert (Percentage)
Learning Content		
1.	Appropriate learning content	80
Synthesis of theoretical framework		
2.	Theoretical framework	80
Web-based learning environments components		
3	Problem base	72
4	Resource	75
5	Creative Thinking Center,	80
6	Collaboration	74
7	Coaching	82
8	Scaffolding	80
Total		77.87

4. Method and result

4.1. Data collection

The researchers collected the data as follows: (1) Synthesis of theoretical framework and Components of the learning environment. The data were collected by analyzing principles, theories, related research of the constructivism theory, cognitive theory, media and technology theory, pedagogy and contextual study, (2) Synthesis of Designing framework of the learning environment: The above synthesized theoretical framework was taken into this process. The underlined theories base such, Contextual base, Psychological base, Technologies and media base (AR: technology and media symbol system), Creative thinking base, and Pedagogies base. (3) Designing and developing of the learning environment based on the foundation of creating designing framework was adopted. (4) Evaluate of the learning environment by experts. The analytical description, summarization, and interpretation were used to analyze data.

4.2. Resaech resault

The instruments in this study as following details: (1) The document examination and analysis recoding form to synthesize a theoretical framework, (2) The recoding form for synthesis of the design framework to learning environment to enhance creative thinking, and (3) The evaluation form for synthesize theoretical framework and designing of constructivist augmented reality web-based learning environments to enhance creative thinking.

References

[1] Basaeed E I, Berri J, Zemerly M J and Benlamri R 2007 "Webbased context-aware m-learning architecture," **Int. J. Interact. Mob. Technol. 1** pp 5–10
[2] Brown J S, Collins A and Duguid P 1989 Situated cognition and the culture of learning **Educational Reaearch 18** pp 32–43
[3] Chijaroen S 2009 Education Technology: Priciples, Teories, and Implementation (Khon Kaen: Khlung Nanawittaya)
[4] Donald D M 2014 Augmented Reality on Mobile Devices to Improve the Academic Achievement and Independence of Students with Disabilities (Doctoral Dissertations, University of Tennessee, Knoxville)
[5] Guilford 1967 The Nature foHuman Intelligence (New York: McGraw-Hill BookCompany)
[6] Hannafin M 1999 Open Learning (Enviroment: Foundation, Method)

[7] Mayer R E 1996 Designing Instruction for Constructivist Learinig Instructional Design Teories and Model: A New Paradigm of Instructional Teory 2 (Newjersy: Lawrence Erlbaum Associaes)
[8] Piaget J 1992 Judgment and reasoning in the child Translated by Marjorie Warden (London: Roultedge & Kegan Paul)
[9] Samat C and Chaijaroen S 2015 Design and development of learning environment to enhance creative thinking and innovation skills for teacher training in the 21st century **Proceedings of the 23rd International Conference on Computers in Education** pp 667–672
[10] Vygotsky L 1962 Thinking and SApeaking (Cambridge, MA; Mit Press)

IOP Conf. Series: Journal of Physics: Conf. Series **1340** (2019) 012040 doi:10.1088/1742-6596/1340/1/012040

The Framework Developmentfor Interpretation Of Students Reading Literacy In Teaching And Learning Through Open Approach

Chayanit Lahnwong

Faculty of Education , Khon Kaen University, THAILAND

corresponding author's e-mail address: chayath@kku.ac.th

Abstract. The PISA 2000, PISA 2003 and PISA 2006 scores indicate that Thai students have read ability level 2 and 1. Students are able to read and read the text only when the text is readable. Can be referenced or compared or linked to what has been known at low levels. Assess and analyze at a basic level, linking the readings to the things that are related to their lives. No reference or comparison can be made if further analysis is required. This paper comprehensively Reading literacy includes the ability to extract the relevant information from texts and also to understand, use and reflect on written texts. The ability and incentive to engage with texts and address their content is also examined. In order to measure the diverse dimensions of reading literacy, in PISA it is defined as a complex competency with multiple facets. The exercises therefore include different situations and text formats. Solving reading exercises also calls for various cognitive processes of reading. A new feature of the 2018 PISA study is interactive exercises with several texts to be read in a simulated web environment. Aspects of reading literacy in the digital age are thus examined, including the ability to find, relate and assess information. This framework incorporates two approaches: 1) promoting students Reading Literacy instructions; and 2) and Learning through Open Approach. The teaching about problem solving process which students can be improved the cognitive skill and develop several skills of students in the reasonal thinking, creative thinking, analyzed thinking , self-study and systematically problem solving ability. However, this framework emphasizes which grade 9 students in Demonstration School of Khonkaen University, Thailand.

1. Introduction

The goal of education has shifted its emphasis from the collection and memorization of information only, to the inclusion of a broader concept of knowledge " :The meaning of knowing has shifted from being able to remember information, to being able to find and use it"[16]. The ability to access, understand and reflect on all kinds of information is essential if individuals are to be able to participate fully in our knowledge-based society .The PISA framework for assessing the reading literacy of students towards the end of compulsory education, therefore, must focus on reading literacy skills that include finding, selecting, interpreting and evaluating information from the full range of texts associated with situations that reach beyond the classroom .

According to Holloway [9] reading skills are essential to the academic achievement of middle -and high school students. Olson [15] claimed that in today's society, reading literacy introduces a bias because it provides advantages to those who acquire the necessary skills .As the currency used in schools, literacy provides access to literate institutions and has an impact on cognition, or thinking process [11]; it also shapes the way in which we think. Achievement in reading literacy is not only a foundation for achievement in other subject areas within the educational system, but also a prerequisite for successful participation in most areas of adult life [5].

PISA 2000, PISA 2003, and PISA 2006 were used for evaluation .The Thai students were able to read at level 2 or 1 .Thai students could read and write the text .The reading is quite straightforward . Referenced or compared or linked to what has been taught at low levels .Assess and analyze at a basic level, linking the readings to the things that are related to their lives .No reference or comparison can be made .Further analysis is required.

The results reading skills from PISA 2000 to PISA 2006 of Thai children, the average score has dropped .The number of students being able to read below the basic level)Level 2 (increased from 37) %PISA 2000 (to 44.5) %PISA 2006 (and the proportion of students being able to read at Level 4 and Level 5 was reduced. This exposes the flaw of Thai education in reading skills, so it is essential to find a way to develop students in reading skills. Students are encouraged to read .Reading is an activity that requires perseverance .Motivation is another major factor that helps such activity to succeed .Because motivation is what motivates and influences the students and increase reading durability .Motivation is something that can make the student try to reach their goals as well.

The open-ended approach [1] or open approach [14] is a flexible, student-centered method which has recently gained popularity in the field of mathematics education .Here, students, working individually or in groups, are expected to apply their own unique methodology to solve given problems .These problems are so designed, that there may be more than one correct answer or there may be more than one way of arriving at an answer, thus they are able to challenge students at various levels of cognitive development .The problems used may be created from student's experience or by modifying regular classroom questions .Questions may also be created to satisfy the expectations or demands of the teacher.

A good open questions is difficult enough to challenge high performing students, yet simple enough so that the slowest member of the class can find at least one solution [14]. The use of open questions in the teaching of mathematics is promoted as they stimulate critical thinking, unearth idiosyncrasies in students, have practical applications to student's daily life.

As mentioned above, the researchers were interested in applying their solutions "The development framework for interpretation of students reading abilities and teaching and learning through open approach ."This framework incorporates two approaches: promoting students reading abilities instructions and learning through open approach .As noted above, the framework set forth in this paper consists of two approaches Reading Literacy and Open Approach :promoting reading instruction and learning strategies; and creating opportunities for engaging in reading instruction and learning strategies.

2. Literature review :The framework of Reading Literacy

Definitions of reading and reading literacy have changed over time in parallel with changes in society, economy, and culture. The concept of learning, and particularly the concept of lifelong learning, have expanded the perception of reading literacy. Literacy is no longer considered an ability acquired only in childhood during the early years of schooling. Instead it is viewed as an expanding set of knowledge, skills and strategies that individuals build on throughout life in various contexts, through interaction with their peers and the wider community. Cognitively-based theories of reading literacy emphasize the interactive nature of reading and the constructive nature of comprehension, in the print medium [6] and to an even greater extent in the electronic medium [4]. The reader generates meaning in response to text by using previous knowledge and a range of text and situational cues that are often socially and culturally derived. While constructing meaning, the reader uses various processes, skills, and strategies to foster, monitor, and maintain understanding. These processes and strategies are expected to vary with context and purpose as readers interact with multiple continuous and non-continuous texts both in print and (increasingly) when using digital technologies [4].

Reading literacy includes sufficient items in the 2000 or 2009 PISA assessments to report on each of the five aspects as a separate subscale, for reporting on reading literacy these five aspects were organized into three broad aspect categories: access and retrieve, integrate and interpret, reflect and evaluate.

Retrieving information tasks, which focus the reader on separate pieces of information within the text, are assigned to the access and retrieve scale.

Forming a broad understanding and developing an interpretation tasks focus the reader on relationships within a text. Tasks that focus on the whole text require readers to form a broad understanding; tasks that focus on relationships between parts of the text require developing an interpretation. The two are grouped together under integrate and interpret.

Tasks addressing the last two aspects, reflecting on the content of a text and reflecting on the form of a text, are grouped together into a single reflect and evaluate aspect category. Both require the reader to draw primarily on knowledge outside the text and relate it to what is being read. Reflecting on content tasks are concerned with the notional substance of a text; reflecting on form tasks are concerned with its structure or formal features. An elaboration of the three broad aspect categories is given below.

2.1 Access and retrieve

Accessing and retrieving involves going to the information space provided and navigating in that space to locate and retrieve one or more distinct pieces of information. Access and retrieve tasks can range from locating the details required by an employer from a job advertisement, to finding a telephone number with several prefix codes, to finding a particular fact to support or disprove a claim someone has made.

In daily life, readers often need to retrieve information. To do so, readers must scan, search for, locate and select relevant information from some information space (for example, a page of continuous text, a table or a list of information). The required information is most frequently found in a single location, though in some cases the information may be in two or more sentences, in several cells of a table or indifferent parts of a list.

In assessment tasks that call for retrieving information, students must match information given in the question with either identically worded or synonymous information in the text and use this to find the new information called for. In these tasks, *retrieving information* is based on the text itself and on explicit information included in it. Retrieving tasks require the student to find information based on requirements or features explicitly specified in questions. The student has to detect or identify one or more essential elements of a question, such as characters, place/time and setting, and then to search for a match that may be literal or synonymous.

Retrieving tasks can involve various degrees of ambiguity. For example, the student may be required to select explicit information, such as an indication of time or place in a text or table. A more difficult version of this same type of task might involve finding synonymous information. This sometimes involves categorization skills, or it may require discriminating between two similar pieces of information.

2.2 Integrate and interpret

Integrating and interpreting involves processing what is read to make internal sense of a text. Integrating focuses on demonstrating an understanding of the coherence of the text. It can range from recognizing local coherence between a couples of adjacent sentences, to understanding the relationship between several paragraphs, to recognizing connections across multiple texts. In each case, integrating involves connecting various pieces of information to make meaning, whether it be identifying similarities and differences, making comparisons of degree, or understanding cause and effect relationships.

In the fixed-text display space, information might be located in a single paragraph, across different paragraphs or sections of text, or across two or more texts. In the dynamic-text display space, integration can be more complex.

Interpreting refers to the process of making meaning from something that is not stated. It may involve recognizing a relationship that is not explicit or it may be required at a more local level to infer (to deduce from evidence and reasoning) the connotation of a phrase or a sentence. When interpreting, a reader is identifying the underlying assumptions or implications of part or all of the text. A wide variety of cognitive activities is included in this approach. For example, a task may involve inferring the connection between one part of the text and another, processing the text to form a summary of the main ideas, requiring an inference about the distinction between principal and subordinate elements, or finding a specific instance in the text of something earlier described in general terms.

Both interpreting and integrating are required to form a broad understanding. A reader must consider the text as a whole or in a broad perspective. Students may demonstrate initial understanding by identifying the main topic or message or by identifying the general purpose or use of the text. Examples include tasks that require the reader to select or create a title or assumption for the text, explain the order of simple instructions, or identify the main dimensions of a graph or a table. Others include tasks that require the student to describe the main character or setting of a story, to identify a theme of a literary text, or explain the purpose or use of a map or figure.

As mentioned above, interpreting signifies the process of making meaning from something that is not explicitly stated. In recognizing or identifying a relationship that is not explicit, an act of interpretation is required: thus interpretation is perhaps always involved somewhere in the process of integration as described above. The relationship between the processes of integration and interpretation may therefore be seen as intimate and interactive. Integrating involves first inferring a relationship within the text (a kind of interpretation), and then bringing pieces of information together, therefore allowing an interpretation to be made that forms a new integrated whole.

2.3 Reflect and evaluate

Reflecting on and evaluating the content of a text requires the reader to connect information in a text to knowledge from outside sources. Readers must also assess the claims made in the text against their own knowledge of the world. Often readers are asked to articulate and defend their own points of view. To do so, readers must be able to develop an understanding of what is said and intended in a text. They must then test that mental representation against what they know and believe on the basis of either prior information, or information found in other texts. Readers must call on supporting evidence from within the text and contrast it with other sources of information, using both general and specific knowledge as well as the ability to reason abstractly.

The outside knowledge to which textual information is to be connected may come from the student's own knowledge or from ideas explicitly provided in the question. In the PISA context, any outside knowledge required is intended to be within the expected range of 15-year-olds' experiences. For example, it is assumed that 15-year-olds are likely to be familiar with the experience of going to the movies, a context that is drawn upon in the items related to the stimulus Macondo, discussed below.

Reflecting on and evaluating the form of a text requires readers to stand apart from the text, to consider it objectively and to evaluate its quality and appropriateness. Implicit knowledge of text structure, the style typical of different kinds of texts and register play an important role in these tasks. These features, which form the basis of an author's craft, figure strongly in understanding standards inherent in tasks of this nature. Evaluating how successful an author is in portraying some characteristic or persuading a reader depends not only on substantive knowledge but also on the ability to detect subtleties in language – for example, understanding when the choice of an adjective might influence interpretation.

3. Literature review :The Open Approach Theoretical

Teachers are student's "guides" to academic achievement. Irrespective of geography, ethnicity or social status, the purpose of teaching remains the same: to produce intelligent, self-sufficient learners who are able to apply their knowledge to solve future problems and make valuable contribution to the society in which they live. Pedagogical methodology has been categorized as either teacher-centered or student-centered, with the latter deemed more appropriate for developing critical thinkers, self-sufficient learners, and problem solvers[8]. In the teacher-centered approach, the student is a passive learner who tries to "absorb" and "reproduce" the prescribed body of knowledge presented by the teacher. In contrast, student-centered pedagogy focuses on students being active participants in the teaching – learning process. The student is in control of the pace and, in some cases, the content of their learning. "Student-Centered or "Learner-Centered" education has its origins in constructivist developmental theory [8].

Constructivism refers to the idea that individuals "construct" their own understanding of the world as a product of their interactions with their environment. With respect to mathematics, rather than giving students a prescribed method of interpreting and calculating (as is seen in teacher-centered approaches), student are encouraged to apply their experience to solve routine and no routine problems in their own unique way. May be more than one correct answer or there may be more than one way of arriving at an answer, thus they are able to challenge students at various levels of cognitive development. The problems used may be created from student's experience or by modifying regular classroom questions Questions may also be created to satisfy the expectations or demands of the teacher.

Unlike in the usual teacher–centered approach this method eliminates the use of patterns, heuristics and generalizations, in solving problems. Here, the teacher facilitates and guides discussion and summarizes the lesson by recapitulating the strengths and weakness of each presented method [1] suggested that students learn in different ways, and therefore, there is the need for a teaching method that will allow students to demonstrate their knowledge in a variety of ways. An open question is set so that students at various levels within the class can solve the problem based on his or her ability, experience and interpretation of the problem given.

3.1 The Open-Ended Approach Framework

In other words, the process of organizing activities or situations is open-ended, encouraging students to think in terms of three scenarios.

Situation A is about defining a situation that is a problem or the source of problem(s). The role of the teachers is to simulate a situation that is a problem or the source of problem(s) and presenting the classes to stimulate a variety of solutions or foreseeing the situation. The role of the learner is to think of an answer or a solution. Or create another situation in response to the given situation which will motivate or influence an extensive learning. Situation B is a session of investigation to find solutions to their problems. Based on the encountered experience. The role of the teachers is to guide and spark a discussion about connection with other ideas. A variety of solutions can be acquired to integrate the responses that may be seen as unrelated which will generate new knowledges. Situation C is a new and more complicated situation than ever before. Students will try to create more common problems in response to given situation. Based on the activities of Situation B and from the solution of the problem. Students will be able to find more general responses.

3.2 The problems presented in the "open method" have the following characteristics

"Process is open" means the process that students use to solve problems. Students are free to use any ways of solving problems based on their current ability or knowledge. When the students enters a problem situation the teachers given. Students will be able to solve problems in a variety of different solutions. Some students may find the words to be divided into 5 categories, such as compound words, duplicated words, repeated words, Thai words, foreign languages words. Some students may find the criteria to be divided into seven words: compound words, repeated words, Thai words, foreign languages words, one syllable words, two syllables words, three syllables words. Some students may find the words to be divided into 10 words: compound words, repeated words, Thai words, foreign words, slang words, independent words. Depending on the view and the student's interests or abilities, even if the circumstances in Examples A and B are the same. It can be seen in Example B that the situation is open. But students can have a variety of thinking processes. As in Example B, where the situation is a closed situation. Students have limited thinking processes. End products are open. When defining students, the problem must be a problem with many correct answers. One of the key aspects of open-source problem solving is the open-source approach to problem solving.

In other words, after the student has solved the problem. Students can develop their problem-solving process into new problems or solve new problems by changing the conditions or elements of the original problem [10].

Considering that in the "open-end" method proposed by Shimada focus on managing the classes that benefit from having many correct answers. But in the "Open Approach Method", the meaning of openness is considered in a wider perspective than the open-ended method. The "open-ended approach" accepts that the problem is an open-ended problem, and that there is a wide range of answers here are many problems in the problem.

Inprasitha [10] has introduced "open approach" as a teaching method that focuses on problem solving. The teaching sequence is as follows.

1. Posing open-ended problem is an open-ended problem to students. The focus of the open problem is to make the problem a student's problem.
2. Students' self-learning is a way for students to learn in person while solving their problems. The aim is to make students learn from one's own learning to learn together with others.
3. Whole class discussion and comparison is the introduction of a variety of concepts for students to discuss the class to other students to have the opportunity to learn from the ideas of their fellow classmates.
4. Summarize by linking students' concepts that occur in the classroom. The concept of mathematical ideas emerges in the classroom.

It is connected by the concept of students in the classroom. At the student's self-learning stage and discussion stage, the whole class can go back and forth, as in the case of student ideas, sometimes without waiting for ideas to emerge. Teachers can bring ideas to debate at the moment in these two steps there are overlapping parts and can be reversed.

As mentioned, this concludes that open learning management uses open-ended problems to motivate students to think through situations. The problem solving process of the 4-step students in Inprasitha [10] is as follows: 1) Presentation of open-ended problems 2) Self-study of students 3) Class discussion and comparison 4) Summary by linking the concept of students that was discussed in the classroom.

4. The framework of students Reading Literacy in Teaching and Learning through Open Approach

The researcher studied document, theory and research related to Reading Literacy and teaching style using "Open Approach" as a process that promotes reading literacy development. Therefore, the researcher is interested in introducing an "Open Approach" to develop reading literacy skills among students. The research framework is summarized as follows.

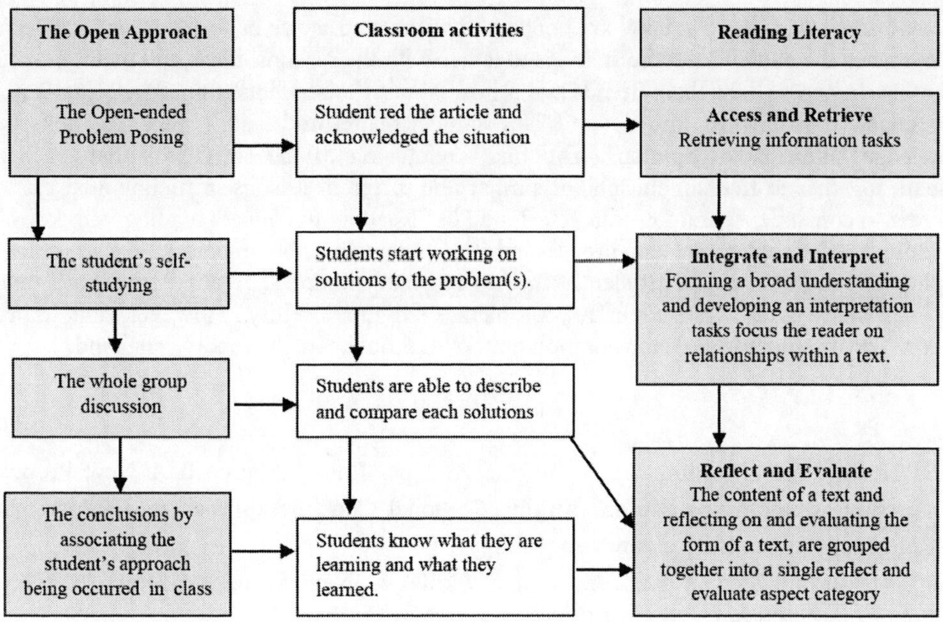

Figure 1. The Develop framework for interpretation of students Reading Literacy in Teaching and Learning through Open Approach

5. Conclusion

From the study we can conclude that "The Develop framework for interpretation of students Reading Literacy in Teaching and Learning through Open Approach" has gained the ability in aspects guided the development of the reading literacy assessment tasks:

Step 1 : The Open-ended Problem Posing has gained the ability in aspects guided the development of the reading literacy assessment tasks Retrieving information tasks, which focus the reader on separate pieces of information within the text, are assigned to the access and retrieve scale. Because in This s step, the instructor must present the problem situation as follows: Open process is the correct solution. There are several open-ended approaches. The answer is correct. The open development approach is a solution to problems that can be developed into new problems. With the pre-reading. Pre-Reading to introduce and stimulate interest in the topic. Encourage the learner to reason for reading. Prepare the language for the students before reading the story.

Step 2 : The students' self- studying has gained the ability in aspects guided the development of the reading literacy assessment tasks is integrate and interpret, forming a broad understanding and developing an interpretation tasks focus the reader on relationships within a text. Tasks that focus on the whole text require readers to form a broad understanding; tasks that focus on relationships between parts of the text require developing an interpretation. The students in the group will help each other to figure out how each one is done, then discuss the group in order to find a conclusion and the reason for this answer is because there is a way to finish it. Read the concept of the group together with the reading stage by making the reader understand the purpose of reading according to the structure of the writing or the body to read.

Step 3: The whole group discussion and Step 4: The conclusions by associating the students' approach being occurred in class has gained the ability in aspects guided the development of the reading literacy assessment tasks reflect and evaluate *the content of a text* and *reflecting on and evaluating the form of a text*, are grouped together into a single *reflect and evaluate* aspect category. Both require the reader to draw primarily on knowledge outside the text and relate it to what is being read. *Reflecting on and evaluating content* tasks are concerned with the notional substance of a text;

reflecting on and evaluating form tasks are concerned with its structure or formal features. Because in step 3 and 4, when the students get the answers together with the reasons, ideas and methods, they will present the class page so that their friends will know about the students' thinking. Next With Post-Reading, students transfer their knowledge of the subject matter and read a story that links them to their knowledge, interests, or opinions. The final conclusion of the activity is that teachers and students learn together to find conclusions of similar and different lessons in finding answers to each group in order to conclude a common concept. Read Post-Reading to connect reading with knowledge.

the teaching about problem solving process which students can be improved the cognitive skill and develop several skills of students in the reasonal thinking, creative thinking, analyzed thinking , self-study and systematically problem solving ability. However, this framework emphasizes 9th grade students in Demonstration School of Khonkaen University, Thailand.

References

[1] Anon Becker, J .P., & Shimada, S) .1997 .(The Open-Ended Approach A New Proposal for Teaching Mathematics .Reston, Virginia National Council of Teachers of Mathematics - . References -Scientific Research Publishing

[2] Anon Measuring Student Knowledge and Skills :A New Framework for Assessment - Publications 2000 – OECD

[3] Anon The World Is Flat :A Brief History of the Twenty-first Century :Thomas L .Friedman : 9780374292881 :Amazon.com :Books

[4] Britt M A and Rouet J-F 2012 Learning with Multiple Documents Enhancing the Quality of Learning ed J R Kirby and M J Lawson)Cambridge :Cambridge University Press (pp 276–314

[5] Cunningham A E and Stanovich K E 1997 Early reading acquisition and its relation to reading experience and ability 10 years later Dev .Psychol .**33** 934–45

[6] Dole J A, Duffy G G, Roehler L R and Pearson P D 1991 Moving From the Old to the New : Research on Reading Comprehension Instruction Rev .Educ .Res .**61** 239–64

[7] European Commission and European Commission 2001 European report on the quality of school education :sixteen quality indicators :report based on the work of the Working Committee on Quality Indicators)Luxembourg :Lanham, Md :Office for Official Publications of the European Communities ; Bernan Associates)distributor((

[8] Fosnot C T 2005 Constructivism :theory, perspectives, and practice)New York :Teachers College Press(

[9] Holloway J H 1999 Improving the Reading Skills of Adolescents Educ .Leadersh .**57** 80–1

[10] Inprasitha M 2011 One Feature of Adaptive Lesson Study in Thailand :Designing a Learning Unit J .Sci .Math .Educ .Southeast Asia **34** 47–66

[11] Kern M L and Friedman H S 2008 Early educational milestones as predictors of lifelong academic achievement, midlife adjustment, and longevity J .Appl .Dev .Psychol .**30** 419–30

[12] Kirsch I S 2001 The Internationnal Adult Literacy Survey)IALS :(Understanding What Was Measured ETS Res .Rep .Ser .**2001** i–61

[13] Loipha, Suladda N Inprasitha Lesson Study :An Innovation for Development of Thai Language Teaching Profession Journal of Education J .Educ .Khon Kaen Univ .**2007**

[14] Nohda N 2000 Teaching by Open-Approach Method in Japanese Mathematics Classroom

[15] Olson D R 1977 From utterance to text :The bias of language in speech and writing Harv .Educ . Rev .**47** 257–81

[16] Simon H A Observations on the Sciences of Science Learning

Examining existing ideas about sustainable leadership of private school science principals

Pha Agsonsua [1], Ekkarad Kositpimanwech [2,*], and Chokchai Yuenyong[3]

[1]North Eastern University, Khon Kaen, Thailand
[2] Eksuksa School, Khon Kaen, Thailand
[3] Khon Kaen University, Khon Kaen, Thailand
[*]Corresponding author email :ekkarad2552@gmail.com

ABSTRACT. The objectives of this research were :to examine existing ideas about sustainable leadership of private school science principals in jurisdiction of Khon Kaen Provincial Education Office. The quantitative data of population were the private school principals and teachers working in private schools under jurisdiction of Khon Kaen Provincial Education Office from 127 principals, 2,425 school teachers, total of 2552 persons .The samples were 92 school principals selected by Purposive Sampling, and 331 teachers selected by Simple Random Sampling .The research instruments included 5 Level Rating Scale, the content validity was 0.84, and the Reliability of total issue was 0.96 .Data were analyzed by using the computer program to calculate the frequency, percentage, and standard deviation . The research findings found that the sustainable leadership of private school administrators under jurisdiction of Khon Kaen Provincial Education Office, in all, was at "High "level .The mean values were ranked in order from high to low as follows :1 (the in-depth body of knowledge, 2 (the steadfast standpoint and far reaching stretch, 3 (the breadth, comprehesiseness and balance 4 (the construction of cooperative network and 5 (the various techniques for success and goal achievement .The study may have implications for professional development of administrators' sustainable leadership in private science school.

Key words :Existing ideas, Sustainable Leadership, Private School Principals

1. Introduction

Hargreaves and Fink recognize that the key forces influencing change and continuity in the long term are leadership, leadership sustainability and leadership succession. They also note the urgent need to rethink the constant drives in educational policy and practice—the obsession with reaching higher and higher standards—without regard to social justice. The authors take the position that to solve problems effectively and ensure meaningful changes, good leaders need to pursue and model sustainable leadership, and identify and define what matters. Across the pages of the book, there are interesting insights, which will be useful for reflective practitioners. The book provides a core vocabulary for leaders to talk and debate further about change and the authors offer this Last Word as conclusion:

... Sustainable educational leaders promote and practice sustaining learning. Sustainable leaders sustain others as they pursue this cause together. Sustainable leaders also sustain themselves, attending to their own renewal and not sacrificing themselves too much as they serve their community. Sustainable leaders stay the course, stay together, stay around and stay alive...Sustainability is the first and final challenge of leadership. (p. 272)

The private educational institutions, the educational organizations under jurisdiction of Office of Private Education Commission (OPEC) , were implemented under National Education Act 1999, and Revised Issue (the Second Issue) 2002, and the Third Issue 2010. Since then, the private educational institution was a part of Thai Educational System throughout the time, and took part in national educational management for a long period of time. In addition, it played an important role in helping shouldering the national burden and responsibility in public sector. Therefore, the role of private educational institutions in Thailand, the private

IOP Conf. Series: Journal of Physics: Conf. Series **1340** (2019) 012041 doi:10.1088/1742-6596/1340/1/012041

educational institutions could not ignore refuse the educational changing trend. On the contrary, they had to be aware of the significance of those changes for educational institutes including: the high competition among private schools as well as public schools. It was necessary for private schools to include leadership for administrators so that the schools would be effective and sustainable although there were many changes that have an impact on. (Boonrod Leau-ngam, 2017)[6]

It was to be based on the administrators' leadership, the school development, the educational quality development, and student development so that the quality would reach standard at universal level. The problems of private schools were: the private schools had to invest by themselves. Consequently, they had to be careful in developing their own schools in order to be survival and, to be secure, sustainable, and recognized by the parents. Since there were a lot of students each year, the sustainable leadership of school administrators were necessary for secure and confident administration in order to have secure administration, sustainable and sound immunity, confidence, and be ready to face with changing in time, policy, law, politics and government, society, economic, environment, and globalization trend. As a result, the private schools had to run their business continuously and effectively, and be recognized by society. (Kingkaew Sri-la-sri-kul-rat, 2015)[8]

A variety of academics concluded that the leadership was an important ingredient in promoting the efficiency of schools. Likewise in Thai society, the educational crisis was cased by lack of effective leadership. Since the administrators often played their role as the administrators rather than leaders. They focused on their routine work to be successful in time without considering the future goal and direction. They had no of interest in bringing new innovation to be used in organization. They didn't improve their working technique since they always clang to traditional one. The sustainable leadership was interested by various organizations around the world. Because every leader had the same goal that they would administer for successful organization according to the organizational objective. More important thing was the technique how the organization would be successful sustainably and securely throughout the time, even to work under the new administrators. (Ti-wa-kan Sri-sa-wad, 2015)[9]

Realizing the significances of the problems slaked above, the researcher team working in the private schools in Khon Kaen Province, were interested in studying the sustainable leadership of private school administrators, under jurisdiction of Khon Kaen Provincial Education Office so that the utility from this research study would be applied for determining policy into practice, and guiding direction of school development .

2. Objectives

1. To study the sustainable leadership of private school principals, under jurisdiction of Khon Kaen Provincial Education Office.

3. The Related Theoretical Approach and Research Literature

According to the approach of Brent Davies,2012[2] in textbook called "Developing sustainable leadership," the foundation factors of leadership were presented: the administrators had to have self-development for obtaining their leadership by focusing on the Outcome instead of Output. They had to integrate the working balance among mission, goal, short term plan (Annual Action Plan), and Long Term Plan to be harmonious. The working process should be emphasized instead of planning. In addition, how to make their organizational staffs to have trust, belief, faith, was the focus of the leadership. It was necessary to possess personal humility humbleness, courage, sincerity, strategies timing and detachment, building rapacity and Creating involment, developing strategic measure for continuously success organization.

According to research study of Ti-wa-kan Sr—sa-wad (2015) [9] in " Sustainable Leadership of School Administrators under The Office of Secondary Educational Service Area 19," the findings found that the sustainable leadership of school administrators, in rall, was at "High" level. The values were ranked in order from high to low as follows: 1) the development of organizational working culture, 2) the mission and commitment in promoting the community and society, 3) the in-depth body of knowledge, 4) the various techniques for success, 5) the innovation, 6) the cooperative network development, respectively. The guidelines for developing the sustainable leadership of school administrators included: 1) the school administrators should have sharing experience, develop clear goal of network, 2) the school administrators should analyze the situation for improving the suitable and relevent services for current changes, 3) the school administrators had to use their competency and skill in communication, providing supplementary budgets suitable, 4) the school administrators had to create the counseling and coaching system in schools, and really putting into practice until they became their organizational culture, 5) the public relation should be provided for community to know and understand so that there would be cooperative in enhancing the academic performance, and 6) the teachers should be encouraged to have continual self-development by furthering their studying or being on promotion of academic standing, providing sharing platform or stage among school staffs, and collaboration with various outside organizations as learning source for the staffs.

According to the above related approaches and research literatures, the significance of guidelines for developing the private school administration, under jurisdiction of Khon Kaen Provincial Education Office, could be viewed as useful in improving, and developing the efficiency in work administration of private schools to meet the standard, quality reguired and to reach school goal accordingly.

4. Research Methodology

This research was Mixed Methods Research.

4.1 Population and Samples

1. The population of this research consisted of 127 school principals and 2,425 teachers, total of 2,552 persons.

2. The samples of this study consisted of the school principals selected by Purposive Sampling. The key informants included: 1) 92 school principals selected by Simple Random Sampling, and 2) 331 teachers selected by Simple Random Sampling.

4.2 The instruments using in this study

The quantitative data were collected by using the questionnaire with 3 Parts. Part 1, the questionnaire asking the respondents' demographic data, was the Checklist. Part 2, the questionnaire asking the sustainable leadership of private school administrators, under jurisdiction of Khon Kaen Provincial Education Office, was 5 Level Rating Scale. Part 3, the recommendations of guidelines for sustainable leadership of school administrators, under jurisdiction of The Office of Khon Kaen Provincial Education Office, was the open-ended questionnaire.

4.3 Data Analysis

The computer program was used for data analysis by analyzing the following statistics: 1) the respondents' demographic data were analyzed by using the frequency and percentage, 2) the data of sustainable leadership of private school administrators under jurisdiction of Khon Kaen Provincial Education Office, was the Checklist, analyzed by using the mean and standard deviation, and 3) the content was analyzed from recommendations in order to determine guidelines for developing the sustainable leadership.

5. Research findings

1. The sustainable leadership of private school principals, under jurisdiction of Khon Kaen Provincial Education Office, was shown in Table 1.

Table 1. The sustainable leadership of private school principals

Number	Sustainable Leadership of Private School Administrator connectivity	Mean	Standard Deviation	Interpretation
1.	Development of Connectivity and Supportive Network	4.22	.74	High
2.	breadth, comprehensiveness and Balance	4.25	.64	High
3.	In-depth Body of Knowledge	4.44	.70	High
4.	Steadfast standpoint and far reaching stretch	4.38	.68	High
5.	Variety of Technique for Success and Overall Goal	4.15	.75	High
	Overall	4.29	.70	High

According to the number respondents of questionnaires, there were 120 school principals (89.13%), 331 teachers (100%), their work experiences were less than 5 years (10.43%), more than 5 years but less than 10 years (37.45%), more than 10 years but less than 20 years (43.24%), more than 20 years (8.88%).

According to Table 1, found that the sustainable leadership of private school principals, under jurisdiction of Khon Kaen Provincial Education Office, in overall, was in "High" level มาก ($\bar{X}$ = 4.29). Considering each aspect, found that every aspect was in "High" level. The mean values were ranked in order form high to low as follows: 1) the intensive body of knowledge ($\bar{X}$ = 4.44) , 2) The steadfast standpoint and for reaching stretch ($\bar{X}$ = 4.38), 3) the breadth, comprehensiveness and balance ($\bar{X}$ = 4.25), 4) the creation of cooperative and supportive network ($\bar{X}$ = 4.22), and 5) the various techniques for success and goal ($\bar{X}$ = 4.15).

6. Conclusion and Discussions

According to research study, found that the sustainable leadership of private school administrators under jurisdiction of Khon Kaen Provincial Office, in fall, was at " High" level. It was indicated that Khon Kaen Provincial Office recognized the significance of school administrators who needed to be competent in developing the cooperative climate, conducive working climate providing an opportunity for organizational staffs to express their viewpoint freely. They recognize the significance of flexibility. It was supported by the experts' viewpoints during Focus Group Discussion. It was also relevant to viewpoint of Abdul Ka-lam (2012)[1] that the administrators had to have leadership and be able to adjust themselves, collaborate with, and connect with boundless association. They should develop firm standpoint and far stepping with extensiveness, covering, and completeness. They should create good relationship with their colleagues, understand the other persons, be able to communicate efficiently. They should have creative thinking as well as competency in managing the changes. They had to build relation network in both of personal and workplace for increasing the fluency and flexibility in collaborative work in both of inside and outside the schools. They should search for various technique in order to accomplishing the goal. Considering each aspect, found that the highest level of mean value was in the in-depth body of knowledge. It might be due to current private school administrators tried to search for body of knowledge throughout the time in order to gain self-development as professional administrators. It was supported by viewpoint of Tee-ra Roon-cha-roen, 2013) [7] that the necessity of professional administrators was owing to various problems in school management during the present time. The outcome of educational management had no quality. In addition, the law specified that the school management had to be of high professional school as well as professional standard. The aspect with second order of mean value, was the steadfast standpoint and for reaching stretch, and readiness for lasting stance. It was supported by viewpoint of Brent Davies (2012)[2] that the Lencth or sustainability, persistence, and for reaching stretch, would emphasize on well constructed foundation, not be unwarming or shaker easily. It consisted of clear steadfast standpoint and far reaching stretch being great or powerful alway. The third order of mean value was the breadth, comprehensiveness and balance by being the persons who had good relationship with different work units related in every aspect. It was supported by viewpoint not be unwarming or shaker easily. of Gayle,c. Avery (2014)[4] that the sustainable leaders had to obtain the following characteristics: 1) the continuous staff development, 2) the good relationship with both of the members and subordinates, 3) the long period of staff maintenance, 4) the value for staffs, 5) the administrators with leadership, 6) the moral and work practice, 7) the shared decision making, 8) the listening to every level of colleagues, 9) the

References

[1] Abdil Kalam. (2012) . Creative leadership in the global knowledge economy. Retrieved January II,2015 From http// www.india.Jbs cam.ac.uk/new/events/2009/09063_abdul_kalam.htm/.
[2] Brent Davies.(2012). Developing Sustainable Leadership, London:ECIY JSP.
[3] Fullan,M.(2013).The six Secrets of Change.New York:John wiley&cons
[4] Gayle,C.A.(2014).Leadership for Sustainable Future. Massacchusetts. Edwaerd Elgar Publishig.
[5] John P.kotter.(2016).Model Planning for Higher Education. Journal of Higher Education,52,470-489.
[6] Leau-ngam, Boon-rod. (2017). Strategy of Private School Administration based on Effective Sufficiency Economy Philosophy, under The Office of Nongkai Primary Educational Service Area 1. Doctor of Philosophy Dissertation in Educational Administration, Faculty of Education, North Eastern University.
[7] Roon-cha-roen, Tee-ra. (2013). Professional in Educational Administration and Management during The Educational Reform. Bangkok: L.P.Press Co. ltd.
[8] Sri-la—li-kul-rat, Kingkaew. (2015). Strategic Leadership. Searching on the 19th November 2015 From http://gotoknow.0rg/blog /kingkaew_aems/206245.
[9] Sri-sa-wad, Ti-wa-kan. (2015). Sustainable Leadership of School Administrators under The Office of The Secondary Educational Service Area 19. Doctor of Philosophy Dissertation in Educational Administration, Khon Kaen University.

IOP Conf. Series: Journal of Physics: Conf. Series **1340** (2019) 012042 doi:10.1088/1742-6596/1340/1/012042

Suggested guidelines for developing sustainable leadership of private science school principals

Pha Agsonsua [1], Ekkarad Kositpimanwech [2,*], Chokchai Yuenyong[3]

[1]North Eastern University, Khon Kaen, Thailand

[2] Eksuksa School, Khon Kaen, Thailand

[3] Khon Kaen University, Khon Kaen, Thailand

[*]Corresponding author email :ekkarad2552@gmail.com

ABSTRACT. The research aimed to study the guidelines for developing the sustainable leadership of private school principals, under jurisdiction of Khon Kaen Provincial Education Office. The participant included Ten experts for focus group discussion selected by Purposive Sampling including the Lecturers teaching Leadership Course in the university, the Provincial Education Office administrators, the school director, the expert, and the academic receiving doctoral degree in Educational Administration. Methodology regarded qualitative paradigm. The guidelines of developing the sustainable leadership of private school principals were interpreted through Focus Group Discussion from the experts selected by Purposive Sampling. Data were analyzed by using the Content Analysis based on conceptual framework and other issues being studied, and concluded into main issues. The findings revealed that the guidelines for developing the sustainable leadership of private school principals, under jurisdiction of Khon Kaen Provincial Education Office were provided into 9 categories. These included 1) outcome based oriented, 2) strategic administration, 3) the process based oriented (instead of planning), 4) Ethical and moral based administration, 5) professional administrors, 6) strategic abandonment or detachment, 7) cooperative development potentiality, 8) strategies measure for continual success, 9) building up sustainability. The paper will discuss about these guidelines. The paper may have implications for developing professional development of sustainable leadership in Thailand.

Key words: Guidelines, Administration, Sustainable Leadership, Private School Principals

1. Introduction

There is a growing body of evidence that school leadership has the greatest impact on teachers in the classroom and is the key factor for successful achievement of a school's organizational goals (Barber et al. 2010; Day et al. 2010). This is because leadership is a process of convincing others to follow the leader to act and achieve the goal envisioned by the leader (Samriangjit et al. 2016). To be an effective leader, a school principal must influence his teachers in a positive way to achieve the goals of the organization (Hall et al. 2015). Since teacher quality is the most important factor related to student achievement (Barber et al. 2010), the extent to which school leaders are successful in driving teachers' professional development is another major measure of their success as leaders. There is no single, all-embracing theory of educational leadership. This reflects the fact that school leaders have adopted different leadership styles to match the astonishing diversity of educational institutions, ranging from small, rural primary schools to extralarge secondary schools, across the widely different national contexts in Thailand (Sirisooksilp et al. 2015). This is also related to the varied nature of the problems encountered in schools, requiring different approaches and solutions. As a result, several perspectives may be valid simultaneously (Bush 2011). There are broad compilations of the main theories of educational leadership based on systematic review of literature and research (Bush 2011).

The private educational institutions, the educational organizations under jurisdiction of Office of Private Education Commission (OPEC), were implemented under National Education Act 1999, and Revised Issue (the Second Issue) 2002, and the Third Issue 2010. Since then, the private educational institution was a part of Thai Educational System throughout the time, and took part in national educational management for a long period of time. In addition, it played an important role in helping shouldering the national burden and responsibility in public sector. Therefore, the role of

IOP Conf. Series: Journal of Physics: Conf. Series 1340 (2019) 012042 doi:10.1088/1742-6596/1340/1/012042

private educational institutions in Thailand, the private educational institutions could not ignore refuse the educational changing trend. On the contrary, they had to be aware of the significance of those changes for educational institutes including: the high competition among private schools as well as public schools. It was necessary for private schools to include leadership for administrators so that the schools would be effective and sustainable although there were many changes that have an impact on. (Boonrod Leau-ngam, 2017)[6]

It was to be based on the administrators 'leadership, the school development, the educational quality development, and student development so that the quality would reach standard at universal level .The problems of private schools were: the private schools had to invest by themselves .Consequently, they had to be careful in developing their own schools in order to be survival and, to be secure, sustainable, and recognized by the parents .Since there were a lot of students each year, the sustainable leadership of school administrators were necessary for secure and confident administration in order to have secure administration, sustainable and sound immunity, confidence, and be ready to face with changing in time, policy, law, politics and government, society, economic, environment, and globalization trend .As a result, the private schools had to run their business continuously and effectively, and be recognized by society) . Kingkaew Sri-la-sri-kul-rat, 2015[8] (A variety of academics concluded that the leadership was an important ingredient in promoting the efficiency of schools. Likewise in Thai society, the educational crisis was cased by lack of effective leadership. Since the administrators often played their role as the administrators rather than leaders. They focused on their routine work to be successful in time without considering the future goal and direction. They had no of interest in bringing new innovation to be used in organization. They didn't improve their working technique since they always clang to traditional one. The sustainable leadership was interested by various organizations around the world. Because every leader had the same goal that they would administer for successful organization according to the organizational objective. More important thing was the technique how the organization would be successful sustainably and securely throughout the time, even to work under the new administrators. (Ti-wa-kan Sri-sa-wad, 2015)[9]

Realizing the significances of the problems slaked above, the researcher team working in the private schools in Khon Kaen Province, were interested in studying the sustainable leadership of private school administrators, under jurisdiction of Khon Kaen Provincial Education Office so that the utility from this research study would be applied for determining policy into practice, and guiding direction of school development.

2. Objectives

To study guidelines for developing the sustainable leadership of private school principals, under jurisdiction of Khon Kaen Provincial Education Office.

3. The Related Theoretical Approach and Research Literature

According to the approach of Brent Davies,201[2]2 in textbook called "Developing sustainable leadership," the foundation factors of leadership were presented: the administrators had to have self-development for obtaining their leadership by focusing on the Outcome instead of Output. They had to integrate the working balance among mission, goal, short term plan (Annual Action Plan), and Long Term Plan to be harmonious. The working process should be emphasized instead of planning. In addition, how to make their organizational staffs to have trust, belief, faith, was the focus of the leadership. It was necessary to possess personal humility humbleness, courage, sincerity, strategies timing and detachment, building rapacity and Creating involment, developing strategic measure for continuously success organization.

According to research study of Ti-wa-kan Sr—sa-wad (2015)[9] in "Sustainable Leadership of School Administrators under The Office of Secondary Educational Service Area 19," the findings found that the sustainable leadership of school administrators, in rall, was at "High" level. The values were ranked in order from high to low as follows: 1) the development of organizational working culture, 2) the mission and commitment in promoting the community and society, 3) the in-depth body of knowledge, 4) the various techniques for success, 5) the innovation, 6) the cooperative network development, respectively. The guidelines for developing the sustainable leadership of school administrators included: 1) the school administrators should have sharing experience, develop clear goal of network, 2) the school administrators should analyze the situation for improving the suitable and relevant services for current changes, 3) the school administrators had to use their competency and skill in communication, providing supplementary budgets suitable, 4) the school administrators had to create the counseling and coaching system in schools, and really putting into practice until they became their organizational culture, 5) the public relation should be provided for community to know and understand so that there would be cooperative in enhancing the academic performance, and 6) the teachers should be encouraged to have continual self-development by furthering their studying or being on promotion of academic standing, providing sharing platform or stage among school staffs, and collaboration with various outside organizations as learning source for the staffs.

According to the above related approaches and research literatures, the significance of guidelines for developing the private school administration, under jurisdiction of Khon Kaen Provincial Education Office, could be viewed as useful in improving, and developing the efficiency in work administration of private schools to meet the standard, quality required and to reach school goal accordingly.

4. Research Methodology

Methodology regarded qualitative paradigm. The guidelines of developing the sustainable leadership of private school principals were interpreted through Focus Group Discussion

4.1 Participants included ten experts for focus group discussion selected by Purposive Sampling including the Lecturers teaching Leadership Course in the university, the Provincial Education Office administrators, the school director, the expert, and the academic receiving doctoral degree in Educational Administration.

4.2 The instruments using in this study

The qualitative data were collected by using purposive Focus Group Discussion. The key informants were 10 experts selected by Purposive Sampling. The instrument of Focus Group Discussion Record.

4.3 Data Analysis

Data from Focus Group Discussion were analyzed by using the Content Analysis based on conceptual framework and other issues being found. The main issues were concluded.

5. Research findings

The guidelines for developing the sustainable leadership of private school principals, under jurisdiction of Khon Kaen Provincial Education Office, were as follows:

5.1 Outcome-oriented

1) The work practice was emphasized on in economic, spirit, and soul aspects, 2) the stakeholders had trust and confidence and faith, 3) the work implementation focused on sustainability and security, 4) the creative and valuable common benefit of schools were focused on, 5) the academic leaders in developing the schools up to universal standard, and 6) the quality, value, knowledge, wisdom, and profound attitude were emphasized, on.

5.2 Strategic Administration

1) The annual action plan was established to be relevant to the School Development Plan to serve for National Education Act, 2) the Strength, Weakness, Opportunity, and Threat (SWOT), were analyzed from both of inside and outside for determining the appropriate development strategy, 3) the new works were regularly initiated, and being Trice creative leaders in every aspect, 4) the progressive and for advanced school administration was performed, 5) the new approaches for development to outperform the trading competitors, and new and original strategic technique always, 6) the school administrators could think out of the box, be courageous to think and do new things, and estaftish new values and beliefs creatively, and 7) the intelligent administrative strategies, improved and modified change appropriately.

5.3 Process-oriented instead of Planning

1) The process and valuable activity were emphasized instead of just planning oriented, 2) the participative collaboration in promoting success, advancement, and pride were developed, 3) the consultation was performed in order to propel the process to accomplish goal, 4) the PDCA quality cycle was used for the best benefit, and 5) the cooperation to have consensus and understanding were developed for problem solving.

5.4 The Ethical and moral-based Administration

1) The administrators played their role as good examples with principle in behaving themselves, 2) they had awareness of principle in behaving themselves, 3) they practice good governance and used their wisdom in leading life, 4) they believed in moderate practice or middle way, and were ready to accept change throwing what someone was up to in both of the globalization trend as well as present occurrence, 5) the encouragement for persistence and competency in administering with reasonableness, and 6) the justice as well as carefulness in decision making.

5.5 The Professional Administrators

1) They should promote the staff development to meet professional standard with service-mindedness, 2) they have Professional Will including to be persistent, earnest, and conscientious, 3) they are persistent in dedicating their performance to be successful, secure, and sustainable for schools, 4) they develop their staffs' potentiality for developing as well as transcending and promoting their competency for future generation, 5) they preferred to improve or behave themselves without blaming the others or outside factors, and 6) they honored and gave credit to others rather than themselves.

5.6 Detachment or strategic abandonment

1) The administrators had to know time and place to have detachment or leave their position, 2) they plan the strategy for developing future leaders or administrators for inheriting, 3) they are ok sincerity and justice in assigning their work transparently, 4) they provide opportunity for the other persons to work nobly, and 5) they should not be grudge in their position by offering opportunity for the others.

5.7 Development of Participative Potentiality

The stakeholders had unity, and everyone participate in their work creatively, 2) they recognized and honored with each other, they paid respect to each other, respected in seniority and were trustworthy, 3) they participated in cooperation in order to search for the best alternative in school development, 4) they collaborated in thinking, doing, listening to their co-workers' opinion and accepting their co-workers' competency for the highest successful development of their schools, and 9) they shared their viewpoint in seeing their success, adjusted their thought with their colleagues, and were one part of friendly team workers.

5.8 The Measurement and Evaluation of Work Performance

They focus on intensive and extensive measurement covering every dimension , 2) they supervise and follow up their staffs' performance, 3) they encourage their colleagues to analyze the rising problems in their work for developing the better alternatives in their group, 4) they compare their work performance with the criterion for indicating their work success, 5) they evaluate their strategies periodically so that in case of obstacles, they would be able to solve the problems in time, and 6) they use their evaluative findings to be guidelines for developing the modern strategies which would lead to more efficient school development.

5.9 The Development of Sustainability and Security in Organization

1) They had to create cooperative and sustainable culture for success, 2) they enhance, fulfill, and transfer regularly and continuously, 3) they develop the team leaders by not waiting for being dependent on the administrators only, 4) They are administrators with great and grand vision they were transformational leaders, 5) they are flexible to propriety as well as sustainable and secure situation recued. They were of great spirit in persisting, stretching far wide with comprehensiveness and balance.

6. Discussions

The Guidelines for Developing Sustainable Leadership of Private School Administrators, under jurisdiction of Khon Kaen Provincial Office, were as follows:

According to content analysis in recommendations and experts' focus group discussion, there were following guidelines for developing the sustainable leadership: 1) the Outcome-oriented, 2) the Strategic Administration and Management, 3) the Process-oriented instead of Planning, 4) the Ethical and moral baud Administration and Management, 5) the Professional Administrators, 6) the Detachment, 7) the Participatory Development of potentiality 8) the Measurement and Evaluation of Work Performance, 9) the Development of Sustainability and Security in Organization. It was supported by viewpoint of Brent Davies (2012)[2] that there were 9 factors of sustainable leadership including: 1) the Outcome-oriented instead of Product, 2) the short term and long term Integrated Plan for developing the balance of goal, 3) the Process-oriented instead of Planning-oriented, 4) the passion and Faithfulness, 5) the personal Humility and professional will, 6) the the strategic Timing and abandonment as detachment 7) the development of Potentiality, 8) Developing strategic measures of success, and 9) Building in sustainability. It was supported by viewpoint of John.P.Kotter (2016) in transformational leadership which included the following factors: 1) the Participation and Recognition by persons in environmental context. 2) the Comprehension and Knowing of the standpoint of organization, and Reasonableness, 3) the Knowing of Goal to be accomplished, and What to be measured for persons to know and to reach the goal, 4) the Outcome-oriented, how to lead the organization to accomplish its goal or objective, 5) the Planning for goal accomplishment, and 6) the Provision for an Opportunity for organizational staffs to express their opinion or viewpoint.

7. Conclusions and Recommendations

7.1 Recommendations for The Application of Findings

7.1.1 The school administrators should develop the cooperative network with various high competent organizations such as higher educational institutions and other institutions.

7.1.2 The school administrators should develop the innovation in both of process and of structure in order to improve the service system in school to be more convenient and time saving.

7.1.3 The school administrators should provide opportunity for community to participat more by reducing the steps and regulations in work process, promoting the flexibility so that the community would process sufficient resources in educational management.

7.1.4 The school administrators should create the counseling and coaching system systematically, and be committed in actual practice in schools.

7.1.5 The school administrators should develop the cooperation with community in conducting research and academic product for the public.

7.1.6 The school administrators should create new instructional innovation for society.

7.2. Recommendations for further research

7.2.1 The qualitative research as case study in sustainable leadership of successful school administrators who had made great performance and were recognized by community as well as society, and could develop their school effectively and successfully.

7.2.2 The study of variables of sustainable leadership of school administrators in other factors which were not considered yet suchar, toughness, justice, should be studied so that the study of sustainable leadership would be more comprehensive, and also of wide variety.

References

[1] Abdil Kalam.(2012).**Creative leadership in the global knowledge economy.** Retrieved January II,2015 From http// www.india.Jbs cam.ac.uk/new/events/2009/09063_abdul_kalam.htm/.
[2] Brent Davies.(2012). **Developing Sustainable Leadership, London:ECIY JSP.**
[3] Fullan,M.(2013).**The six Secrets of Change.New York:John wiley&cons**
[4] Gayle,C.A.(2014).Leadership for Sustainable Future. Massacchusetts. Edwaerd Elgar Publishig.
[5] John P.kotter.(2016).**Model Planning for Higher Education.** Journal of Higher Education,52,470-489.
[6] Leau-ngam, Boon-rod. (2017**). Strategy of Private School Administration based on Effective Sufficiency Economy Philosophy, under The Office of Nongkai Primary Educational Service Area 1. Doctor of Philosophy Dissertation in Educational Administration, Faculty of Education, North Eastern University.**
[7] Roon-cha-roen, Tee-ra. (2013). **Professional in Educational Administration and Management during The Educational Reform.** Bangkok: L.P.Press Co. ltd.
[8] Sri-la—li-kul-rat, Kingkaew. (2015**). Strategic Leadership.** Searching on the 19th November 2015 From http://gotoknow.0rg/blog /kingkaew_aems/206245.
[9] Sri-sa-wad, Ti-wa-kan. (2015). **Sustainable Leadership of School Administrators under The Office of The Secondary Educational Service Area 19. Doctor of Philosophy Dissertation in Educational Administration, Khon Kaen University.**

Grade 11 Student's Mental Model of Fluid and Analytical Thinking in Science Teaching Through Science Technology and Society (STS) Approach

Amorn nongkhunsarn, Chokchai Yuenyong*, Jiraporn Tupsai, and Tawee Sranamkam

Science Education Program, Faculty of Education, Khon Kaen University 4002, Thailand

Corresponding author's e-mail address here: ychok@kku.ac.th*

Abstract. This study reported Grade 11 student's mental model of fluid and analytical thinking in science teaching through science technology and society (STS) approach. The participants were 38 Grade 11 students in Moeiwadee, Roi-et, Thailand. The teaching and learning about fluid through science technology and society (STS) approach had carried out for 12 hours. The fluid through STS approach was developed based on framework of Yuenyong (2006). Students' mental model of fluid and analytical thinking were collected during their learning by participant observation, open questionnaire, semi-structured interview, student' task, analytical thinking test, and journal writing. The findings revealed that students' mental model of fluid were represented as verbal, equation, drawing. Data were analyzed identify patterns that offer insight into aspects of student understanding and discussed in the scientific concept. The student's analytical thinking could behave the characteristics for analytical thinking such as thinking for classifying, compare, contrast, reasoning, interpreting, collecting data and decision making. This study will discuss implication of these for science teaching and learning through STS in Thailand.

1. Introduction

In recent world, the knowledge of Science and Technology was an important instrument in preparing young people to be able to live in society with Science and Technology as a foundation. It affected every level of life, both in person, occupation, and socio-culture of every life. Consequently, people could be able to perceive and make decision in social problems occurring from the effect of Science and Technology with knowledge and understanding, participation in society in community level, national level, and global level with full esteem. [4]

For Science instructional management, the students should be focused on knowledge and understanding in boundary, nature, and limitation of Science, components of Science as the content collecting facts, concepts, hypotheses, principles, theories, rules of Science until it could be utilized as well as the technique in searching foe Science Process Knowledge, obtaining understanding and awareness of the importance of Science toward participation in society so that the students would be developed for higher order thinking , problem solving ability, communication and decision making, Scientific attitude [4]. Therefore the teachers had to provide learning facilitated in developing human beings to have characteristics for wide vision, far thinking, love to know, and focusing on process for the students to develop their analytical thinking, opinion expression, and knowledge search by themselves, as well as participation in thinking and practicing [1]. So, teaching was very important thing in instructional management for developing quality of students.

One approach the Scientist around the world encourage and applied in developing Science instructional in recent time, was Science, Technology, and Society (STS) Approach which was an approach of Science instructional management in context of human's experience by focusing on situation or problem situation regarding to Science an Technology that truly occurred, and attempt to encourage the students to find answers for that situation. It was the encouragement for students to analyze and apply their approaches and processes in real situation of students' society [5]. It was supported by Yeunyong (2008) [6] statement that the objective of instructional management based on Science, Technology, and Society (STS) Approach was : the students could be able to have Science concept integrated by themselves with knowledge in Science, Technology, and Society Approach, competency in applying the studied knowledge in daily life, with reasons, analytical thinking, self development, could work with the others well and be good members, responsibility for themselves, community, local, and society.

When the students obtained experience from knowledge management based on Science, Technology, and Society (STS) Approach, they could construct their body of knowledge by themselves from experience they obtained by knowledge management. Each student had different concept model from learning for explaining some things which based on their characteristics including general competency, prior knowledge, motivation and environment, culture, language, and communication [2]. Therefore, if we wanted to know the students body of knowledge constructed from their experience by knowledge management, could be studied from mental model as their representation.

Although there was a policy for instructional reform focusing on developing analytical thinking, it was in limited boundary and didn't achieve the ultimate goal as wanted. It could be viewed by the findings of assessment 7,273 schools passed the assessment by The Office of Accreditation and Educational Quality Assurance (Mass Organization) 2004, found that the standard of students, Standard 4, only 18.12% of them had analytical thinking ability, synthetic thinking, critical thinking, creative thinking, and vision, as the standard that the students should be improved and developed. It was supported by the second round of assessment of Moeiwadee Pittayakom School in 2007, found that the standard of students, Standard 4, it was in "Fair" level. According to the problems, the researcher was interested in developing the students' analytical thinking ability since it would help us to know the facts, consider the appropriateness, know fundamental reasons, understand the incidence. Consequently, the facts as basis for applying in making decision for solving problems, assessment, and making decision in various things correctly [3].

So, the researcher was interested in studying the Mental Model and Analytical Thinking in "Fluid" of Grad 5 Students of Moeiwadee Pittayakom School, Roi-et Province, under jurisdiction of Roi-et Educational Service Area 3, by using instruction based on Yuenyong's (2006) Science, Technology, and Society (STS) Approach. The characteristics of fluid to analyze Mental Model of this study included "Density and pressure of Liquid."

1.1. The scientific concept of density and pressure of liquid.

Density is a term that refers to a measure of mass per volume. The symbol of density is "ρ"(rho). It is a Greek symbol. Its SI unit is the kilogram per cubic meter or kg/m^3.

$$\rho = \frac{m}{v}$$

Pressure (symbol :P) is the force per unit area applied in direction perpendicular to the surface of an object.

$$P = \frac{F}{A}$$

The pressure exerted by a static fluid depends only upon the depth of the fluid, the density of the fluid, and the acceleration of gravity.

The pressure from the weight of a column of liquid of area A and height h is

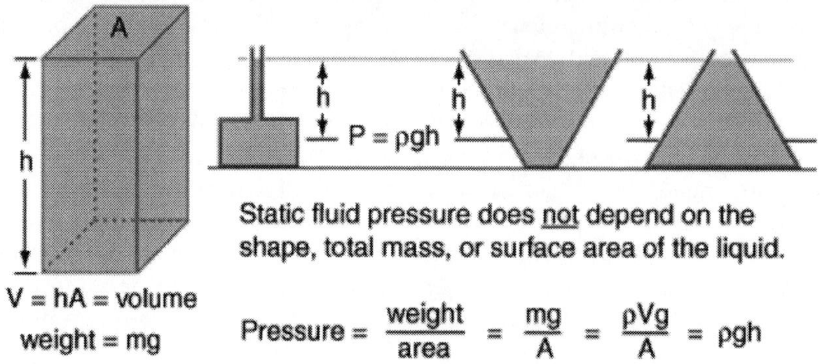

Figure 1. Pressure of liquid

1.2. Science Technology and Society (STS) Approach of Science Learning

The STS Approach of teaching was Science instruction focusing on student centered. Consequently, they viewed that Science and Technology were things around them. They could value in Science on living and be able to apply the studied knowledge for usefulness. As a result, they were more knowledgeable in various issues of Science and Technology affecting on students and whole society. They could use Science process in improving problems, searching for information. They would understand and learn how to adjust their needs with each other. The teaching based on STS Approach didn't emphasize on students' knowledge or memorization of concepts while they were searching for information to solve problems and answer questions. They would directly learn various concepts from their work (Myer, 1996 cited in [4]).

Many Scientist Educators constructed and presented instructional management model based on Science, Technology, and Society Approach. It was broadly used in organizing different activities. The STS Approach of instruction based on Yuenyong's (2006) approach was another using in Thailand (STS Approach) based on Yuenyong's (2006) approach including 5 stages as follows [8]:

(1) Identification of social issues stage. This stage is designed to focus student attention and attitudes on learning about energy. The STS instruction has to begin in the realm of society. Instruction will be begun by posing issues related to energy use in society. These questions or problems of social issues need to be solved by citizens. For energy concept, the issued of saving energy and the social problem related generating power should be brought into classroom by various strategies such as informing situation related these issues from posing in newspaper; posing the social questions related these issues for allowing students to participate in public decision-making; seeing social problem by taking field trip.

(2) Identification of potential solutions stage. Students plan to solve the social problem related to energy use. This stage supports students to concern with the technological aspects for find the possible solutions. Technological aspects are skills to support student decision making. Students need to think of what, why, and how ideas, design, systems, volition of application scientific knowledge work for that social problems. Teaching strategies may be used discussion among students' group, role-play, brain storming, searching information via internet, and discussion with expert (e.g. engineers or scientists).

(3) Need for knowledge stage. This stage involves developing scientific knowledge. Social questions and technological knowledge create the need to know some science content. Energy concept was formulated in many strategies to help students to understand the technology and social issues. The strategies, for examples, include reflection reading document provided by teacher, and lecture. To give

feedback students' understanding about energy concept, the short quiz will be taken after class of this stage.

(4) Decision-making stage. This stage involves student in making a decision on how to use energy knowledge and technology as solution of the social problem. This aspect public rhetoric about energy related technological and societal issues becomes dominated by dichotomies like 'chances and problem', 'advantages and disadvantages', or uses and abuses'. Student will be given chance to learn to choose between alternatives and in a thoughtful way systematically comparing as many relevant pro's and con's as possible. Teaching strategies may be used discussion among students' group, role-play, and brain storming.

(5) Socialization stage. Students need to act as people who are a part of society by reporting their proposal for solving problem. Student might exhibit their solution in public by produce a poster, a newspaper article or a plan, present science project.

2. Methodology

The research design of this study was Qualitative Research emphasizing on interpretative paradigm for studying the Mental Model and analytical thinking of Grade 11 Students studying Science titled "density and liquid pressure" based on Science, Technology, and Society (STS) Approach.

2.1 The Target Group

31 Grade 11 Students, Moeiwadee Pittayakom School, Moeiwadee District, Roi- et Province, during the first semester, 2009 school year.

2.2 Data Collection

The researcher collected data by using 8 Science Knowledge Management Plans based on Science, Technology, and Society (STS) Approach titled "fluid" with the target group, total duration of 16 hours, during the first semester of 2009 school year. Students' conceptual model of fluid and analytical thinking were collected during their learning by participant observation, open questionnaire, semi-structured interview, student' task, analytical thinking question, and journal writing.

2.2.1 Preparation for Students before teaching. The researcher oriented the students by informing the Expected Learning Objectives , learning and teaching activity management based on Science, Technology, and Society (STS) Approach by using Yuenyong's (2006) STS Approach for their understanding whereas the open questionnaire making in "density and fluid pressure" asking their conceptual model and analytical thinking question before learning .

2.2.2 Learning and Teaching Implementation

The target group was taught by using 3 Science Knowledge Management Plans titled "density and fluid pressure" by the researcher, and instructional model based on Yuenyong's (2006) Science, Technology, and Society (STS) Approach, duration in implementing the instructional activities for 6 hours.

(1) Identification of social issues stage: The teachers presented the news issue of "Dam Bursts in Indonesia, Killing Dozens in Flood". Then, the students were asked for setting questions they interested in the news of dam bursts in Indonesia, that be able to get answers by applying knowledge in Density and Pressure in liquid.

(2) Identification of potential solutions stage: The teachers and students collaborated in grouping the questions with nearly the same kind which could be find common answer. Then, they were assigned into groups to brainstorm their thought and guidelines in finding for the answers.

(3) Need for knowledge stage : The students explored, searched for Density, Pressure in Fluid, relationship between Pressure, Depth, and Density of liquid, the force of water on the water gate or dam, the teacher suggested supplemented knowledge in calculating quantity, by assigning each group of students to study the construction of dam outside their class schedule by studying from the officers

who took care of Wangnong reservoir, Moeiwadee District, civil officers in Sub-district Administrative Organization, or scholars in constructing weir in community.

(4) Decision-making stag: The students each group brainstormed the construction technique of dam.

(5) Socialization stage: The students proposed their ideas they designed for constructing dam in their classroom. The teachers and classmates expressed their opinion and provided additional suggestions.

2.2.3 After Learning and Teaching

When the knowledge management process was ended, the researcher analyzed students' mental model and analytical thinking findings from participant observation, open questionnaire, semi-structured interview, student' task, analytical thinking question ,and journal writing Questionnaire of decision making, Interview Form and Observation Form, Worksheet.

3. The Findings and Discussion

This research was Interpretive Paradigm focusing on explaining Mental Model and analytical thinking in "density and Pressure" of students before and after instructional management based on Science, Technology, and Society Approach. The students had to respond the open ended questionnaire. Then, all students' answers were read for viewing overall view in concept of fluid titled" Density and Pressure. The data were analyzed by grouping the students' concepts and analytical thinking. The findings would be presented as follows:

3.1Mental Model in 'Density and Pressure."

3.1.1 Mental Model "Density". When the students were asked that: "If the cylinder shape iron, with its amass 1,000 kilograms, dropping into the sea, do you think that the iron would be sunk or floated in the sea water?"..

If the same lump of iron was built the boat, then it was dropped into the sea again, do you think that it would be sunk or floated in the sea water?........................

Why does it happen?

The analysis of findings from the students' responses. According to the students' answers, the responded that if the hollow round cylinder shape of iron with its amass 1,000 kilograms, dropping into the sea, they thought that it would be sunk in the sea water. Besides, if the same lump of iron is built for the boat and dropped into the sea again, they think that the iron built for the boat would be float in the sea. The model of concepts in "Density" could be grouped into 3 groups including: Science Concept, Physics Concept, and Object Shape Concept as follows:

Mental Model of student, Group 1: Scienctific Concept. For the concepts from the students' answer explaining the sinking of iron or floating of the boat including concept of density as the examples of following answers:

"because of its hollow round cylinder shape was more than density of sea water as well as the density of boat is less than the sea. So, the boat could be floated above the water." (Suwat)

"because the density of the hollow round cylinder is more than the density of water. But, its area acted on water is les, So, it is sunk in the water. The boat is floated because it had the area acted on surface of water more, and inside of the boat if shallow. So, its density is less than water." (Tewarit)

"the lump of iron is sunk because its mass and density were more than water. If it is built for the boat, although its weight is much, its shape and form would include the width, length, depth. So, the water couldn't be inside. As a result, the boat is floated." (Anon)

According to the answers and reasons of this group of students, found that they had their concepts the same as Science Concept. The researcher thought that this students group understood concepts of this topic very well since they had knowledge by instructional activity management in schools as well as the content they had already studied from Secondary School.

Mental Model of student, Group2: Force Concept. According to the answers of students in this group, there were explanations of answer by explaining The Force in Physics, which was the expression of the understanding in Science Concept relating to sinking and floating of object. The researcher interpreted that the students had Physics Concept by providing their reasons for the Force and Pressure because they had studied in Grad 10 in "Force and Movement". Then, they might apply the knowledge of Force connecting and explaining the phenomenon that would occur as: if the object and water had gravity force, the object would be sunk. But, it the water and object had impulsion force, the object wouldn't be sunk. The molecule of water would hold together by molecular force, as the students' following answers:

"because when the same lump of iron is built as a boat and dropped into the sea, the molecule of water would hold with each other. So, the liquid wouldn't separate from each other. Therefore, the iron built as a boat wouldn't be sunk." (Prisna)

"when the limp of iron which isn't built, it is only a kind of object. When it is dropped into water, this lump of iron would be sunk. But, it is built as a boat, this lump of iron would be floated because there is pressure on boat. So, the boat is floated." (Wipawan)

Mental Model of student, Group 3: Shape of Object concept. According to the answers of student in this group, there were explanation of shape of object affecting object to be sunk or float. So, the researcher interpreted that the students had experience by observing the shape of object with good balance, the object had to be flat with square. If the good balanced object was put into the water, the object would have balance the same as when it was put on general ground. The students didn't have concept of changing of volume of object, when the object changed its shape, its volume would also be changed. Consequently, the lump of iron had the same weight as 1,00 tons. When it was built as a boat, the volume of this object would increase, its density would decrease. As a result, the boat would be floated as the following answers:

"because the boat has its surface area more and is shallow, not hollow as the piece of iron before it is built as a boat. The iron is flat, so it could be floated" (Supapon).

"the boat is floated because for building boat, we couldn't built as a circle. We have to build it as a square shape. If we build it as a square, the iron would have more balance than when it is a round shape. When the iron is a round shape, it doesn't have a balance." (Laddwan)

"because the iron that would be built as a boat, it has to be shaped. Then, the iron would be floated." (Supachai)

3.1.2 Mental Model in "Pressure of Liquid" When the students were asked that "the diver explored the variety of fish living in 7 meters depth, the first person dived to survey freshwater fish. The second one dived to survey sea fish. Do you think that the pressure of water acted on two divers would be equal? Why? Explain.

The analysis of findings from the students' responses. According to the students' answers, the researcher interpreted the model for understanding in "Pressure in Liquid" for 3 groups including: Science Concept, Concept Model of Process based on Depth, and Academic Concept Model as follows:

Mental Model of students, Group 1: Scientific Concept. The researcher interpreted the students' answers that the had Science Concept in "Force in Fluid", the pressure in any point in static fluid, would be vary to the depth and density of liquid when the temperature is stable, and in static liquid at any point with the same depth, the pressure would always be equal. If the students could correctly answer the question and explain as: Although both of divers would dive at the same depth, the pressure of water acted on them wouldn't be equal since beside the depth, the pressure was based on density of liquid as the following answers:

"not be equal, the sea level has more pressure than fresh water. The force acted on the divers is different. The diver surveys freshwater fish would have less pressure than the diver surveying sea fish because the sea water has more density than fresh water." (Nawarat)

" not be equal because although the divers equally dive 7 meter depth, it doesn't mean that the pressure acts upon them would be equal which could be compared with the following examples: floated egg, sunk egg, when the egg is dropped into water, it appears that the egg would be sunk. When we add salt into water, the salt would be dissolved, the dropped egg would be floated because of different density levels." (Prisna)

Mental Model of students, Group 2: Concept of Pressure based on Depth. According to the answers of students in this group, found that they had concept in "Pressure in Liquid only some Part," The students' answers were that the pressure of water acted on both divers would be equal since both of them dived into the same depth of water as following answers:

"be equal, both surveyors dive into the same depth. As a result, the pressure on both of them isn't equal since the sea water has more density than fresh water, and the pressure isn't based on the density of water. But, it is based on the depth of water." (Titima)

Mental Model of students, Group 3: Academic Concept. The students in this group could answer that even the two dicers dived into the same depth of water. But, the pressure acted on both of them wasn't equal. The students in this group explained the differences between some kinds of substances in sea water and fresh water: sea water tasted salty. So, they gave reasons that the concentration of some substances in sea water was more than in fresh water. The researcher interpreted the answers of students in this group that the students had academic concept because their explanation in concentration of substance as the following answers:

"I think that the pressure of water acted on both divers is different. The diver in sea water would have more density than in fresh water since the sea water has more density. Consequently, its density is more. But, if the sea water is very salty, the object of man would be floated without swimming" (Anon)

"not be equal because the sea water has many substances. These substances would have reaction on the diver. But, the fresh water even it has the same pressure, different kinds of substance and mineral are not equal to the sea, As a result, the sea water has more pressure than fresh water." (Patpan)

"not be equal since the fresh water has pressure, the sea water has more pressure because it includes salt in it. Each kind of water would have different pressures. Besides, the divers dive into 7 meters depth, the pressure would act on the explorers. If the sea water, it has more pressure because the concentration of salt is more than the fresh water." (Kanyarat)

3.2 The study of students' analytical thinking

IOP Conf. Series: Journal of Physics: Conf. Series **1340** (2019) 012043 doi:10.1088/1742-6596/1340/1/012043

The data collected from situation of questions measuring the students' analytical thinking before knowledge management based on Science Approach, Technology, and Society Approach, were analyzed and interpreted according to the measurement of analytical thinking of Luaen Saiyod and Ankaan Saiyod (1996) as follows:

Analytical Thinking, Aspect 1: Significant Analysis. Question from situations measuring analytical thinking ability in analysis of aspect 1 including: What is the main idea of this topic?

According to the students' answers, found that they could diagnose by searching for main idea or thesis statement from the read statement only by some students who used their skill in classifying situations into sub-topic which was early understood and explained in details as following answers:

'the body of Titanium Ship had an accident crashing with the huge iceberg. As a result, the water flowed into it and other cabins. So, the ship was wrecked. There were many people were killed by this incidence." (Supaopon)

"the titanium ship was wrecked in the sea because of the careless of the captain causing the tragedy." (Mongkontip)

"the titanium ship crashed with huge iceberg until its body was broken and the sea water flowing the ship. So, the ship was wrecked and there were many losses." (Tipawan)

But, most of students couldn't be able to classify the incidences into parts to be easily understood as well as explain its details as the examples of following answers:

"Luxurious and biggest titanium ship in the world crashed with a huge iceberg" (Rampai)

"for this tragedy caused Pooyai Lee and Mrs Ma to death" (Rachada)

"the size of iceberg under the ocean" (Kanyarat)

Analytical Thinking in Aspect 2: An Analysis for relation. Question from situations measuring the analytical thinking in aspects 2, What caused Titanium ship be wrecked?

The analytical thinking to find relationship of different major parts that How did it related together? such as relationship between cause and effect. According to the students' answers, found that only some students could relate relationship of information in order to analyze cause and effect as the following answers:

"because it was the captain's careless since he thought that it could pass this huge iceberg. But, it wasn't the case, there was a very huge iceberg under the ship. So, the body of it crashed with the iceberg. Consequently, the ship was wrecked."

But, found that most of students couldn't cause and effect by using skill in associating relationship of different information in situations.

"the crash of Titanium ship with the iceberg." (Nawarat)

"the body of ship was crashed with the huge iceberg." (Natapong)

'because the body of ship was crashed with a hige iceberg under water." (Kamonrat)

Analytical Thinking in Aspect 3: An Analysis for Principle. Question, this incidence could cause any experiences. Did the students think that the sailors could be able to apply this accidence to prevent the accident on the wreck ship according to crashing with iceberg? How?

The analytical thinking fir finding principles was based on ability to summarize or find the main idea from situations, and apply knowledge from the situation in conjecturing, expanding, and applying to the future evidence. The students summarized, integrating experience from the passage, and concluded principle in applying the lessons as the students' following answers"

"planning in advance would make us safe, without danger. The captain's careless and no planning until there were losses were received as a motto. The lesson could be applied since it caused the sailors' lesson for warning them that the should make plan before starting their journey which would be benefit for themselves and other as well." (Sasiton)

"critical thinking, observing, problem solving, should be applied so that we could have critical thinking with more careful in solving problems." (Kamonrat)

"We shouldn't be careless on our lives. But, we should be conscious throughout the time and able to apply lessons from this situation for preventing accident of ship wreck from crash with the iceberg since we should have consciousness and make good decision because every life is valuable. So, we ought to be thoughtful and critical thinkers by considering real principles." (Jakrish)

"the weather condition and route should be investigated before stating journey. In addition, they should follow the rule or cancel the schedule when they knew that there is an incidence which they couldn't sail from the radio warning for danger from the station on the shore. The lesson of this incidence should be applied for preventing an accident of shipwreck by crashing an iceberg by designing ship with endurance. The accident should be prevented and warning for their crews." (Supapon)

"should be careful, conscious, and careful. This lesson should be used because we would be careful and not be careless." (Napapon)

According to the answers and interpretation from the analysis of Analytical Thinking in 3 aspects including: the analytical thinking for importance, analytical thinking for finding relationship, and analytical thinking for principles, found that each student had different analytical thinking abilities. Most of them still lacked of analytical thinking in classification skill, grouping skill, associating skill, concluding skill, and application skill which might be because each of them had different characteristics regarding to their competency, experience, or knowledge background. They were not trained or practiced before. As a result, they showed different analytical thinking abilities.

4. The Findings and Discussion

According to the study of model of concept and analytical thinking ability in "Fluid" of Grad 11 Students before managing the instruction based on Yuenyong's (2006) Science, Technology, and Society Approach, found that:

The model of concept of Density of students could be classified into 3 groups. The Scientific Concept, Force Concept, and Shape of Object Concept. Moreover, the concept of Pressure in Fluid could be classified into 3 groups: Scientific Concept, Concept of Pressure based on Depth, and Academic Concept.

The analytical thinking ability could be classified into 3 groups including; the analytical thinking in the importance, analytical thinking for finding relationship, and analytical thinking for finding the principle, found that each student had different analytical thinking ability. Most of them still lacked of analytical thinking skill including the classification skill, grouping skill, summarizing skill, and application skill.

Ref References

[1] Areerat W 2007 *The development of learning management medel focusing on thinking skill in basic education curriculum* Research in Science Education, Khon Kaen University.
[2] Hubber P 2009 *Year 12 students' mental models of the nature of light* Retrieved May 10 2009 (from http://springer.com)
[3] Jaroenwongsak K 2003 *Analytical Thinking (*Bangkok: Success Media Ltd*)*
[4] Klainil S Pramojanee A and Dechsri P *2008 The knowledge and scientific capability for future: the project evaluates report of PISA* (Bangkok: Seven printing group)
[5] Portjanatanti N 2005 *Document for Teaching through science technology and society (STS)* Faculty of Education *(*Songklanakarin University*)*
[6] Yuenyong C 2006 *Teaching and Learning about Energy: Using STS Approach* Bangkok Thailand Thesis of Doctoral Degree in Science Education (Kasetsat University)
[7] Yuenyong C 2008 *Document for Training on The Application of Sufficient Economy Philosophy in Physics Knowledge Management* Faculty of Education *(*Khon Kaen University*)*
[8] Yuenyong C and Narjaikaew P 2009 *Scientific Literacy and Thailand Science Education* International Journal of Environmental and Science Education, 4 (3): 335 – 349

IOP Conf. Series: Journal of Physics: Conf. Series **1340** (2019) 012044 doi:10.1088/1742-6596/1340/1/012044

How to Apply Technology in STEM Education Lesson By Project Based Learning

Artnarong Manosuttirit

Department of Innovation and Educational Technology, Faculty of Education, Burapha University, Chonburi, Thailand, 20131

Corresponding author's e-mail address: m_artnarong@hotmail.com

Abstract. STEM Education has been playing a part in Thailand for quite some time. STEM teaching and learning is originated from the US. Later, many countries have applied it to their own education systems. Since the arrival of STEM Education, ministry of education, under the supervision of teaching science and technology by Institute for the Promotion of Teaching Science and Technology, has used this teaching model. National Legislative Assembly of Thailand has also proposed STEM policy in order to gain tangible results. Until now, there is still no clarity and there are confusions with the organization of STEM teaching and learning. This article aims to propose the idea of the application technology in project-based STEM Lesson. The content of this article contains concept of STEM, meaning of STEM, policy and principle of STEM, STEM teaching and learning condition, and case study of applying technology with project-based Learning in STEM lesson idea.

1. Introduction

1.1. STEM Education in U.S.

"[Science] is more than a school subject, or the periodic table, or the properties of waves. It is an approach to the world, a critical way to understand and explore and engage with the world, and then have the capacity to change that world..."— President Barack Obama, March 23, 2015

The United States has developed as a global leader, in large part, through the genius and hard work of its scientists, engineers, and innovators. In a world that's becoming increasingly complex, where success is driven not only by what you know, but by what you can do with what you know, it's more important than ever for our youth to be equipped with the knowledge and skills to solve tough problems, gather and evaluate evidence, and make sense of information. These are the types of skills that students learn by studying science, technology, engineering, and math—subjects collectively known as STEM.

Yet today, few American students pursue expertise in STEM fields—and we have an inadequate pipeline of teachers skilled in those subjects. That's why President Obama has set a priority of increasing the number of students and teachers who are proficient in these vital fields

IOP Conf. Series: Journal of Physics: Conf. Series **1340** (2019) 012044 doi:10.1088/1742-6596/1340/1/012044

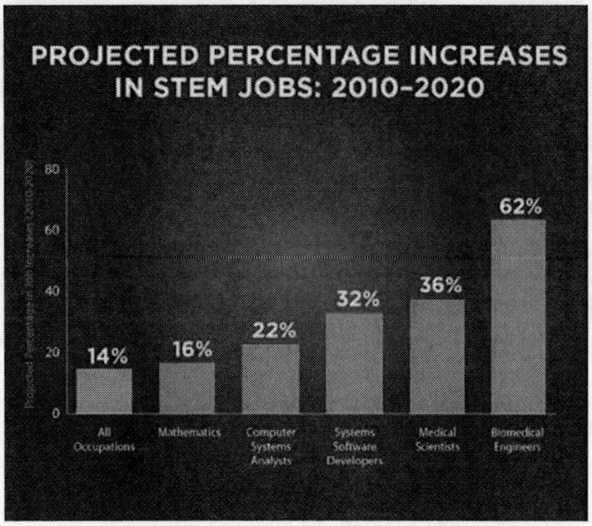

Figure 1. Source: U.S. Department of Education https://www.ed.gov/stem

1.2. The need

All young people should be prepared to think deeply and to think well so that they have the chance to become the innovators, educators, researchers, and leaders who can solve the most pressing challenges facing our nation and our world, both today and tomorrow. But, right now, not enough of our youth have access to quality STEM learning opportunities and too few students see these disciplines as springboards for their careers.

For example, we know that only 81 percent of Asian-American high school students and 71 percent of white high school students attend high schools where the full range of math and science courses are offered (Algebra I, geometry, Algebra II, calculus, biology, chemistry, and physics). The access to these courses for American Indian, Native-Alaskan, black, and Hispanic high school students are significantly worse. Children's race, zip code, or socioeconomic status should never determine their STEM fluency. We must give all children the opportunity to be college-ready and to thrive in a modern STEM economy.

We also know that only 16 percent of American high school seniors are proficient in math and interested in a STEM career. Even among those who do go on to pursue a college major in the STEM fields, only about half choose to work in a related career. The United States is falling behind internationally, ranking 29th in math and 22nd in science among industrialized nations. What's more, a recent survey revealed that only 29 percent of Americans rated this country's K-12 education in STEM subjects as above average or the best in the world. In our competitive global economy, this situation is unacceptable.

1.3. The goals

President Obama has articulated a clear priority for STEM education: within a decade, American students must "move from the middle to the top of the pack in science and math." The Obama Administration also is working toward the goal of fairness between places, where an equitable distribution of quality STEM learning opportunities and talented teachers can ensure that all students have the chance to study and be inspired by science, technology, engineering, and math—and have the chance to reach their full potential.

Specifically, the President has called on the nation to develop, recruit, and retain 100,000 excellent STEM teachers over the next 10 years. He also has asked colleges and universities to graduate an additional 1 million students with STEM majors.

These improvements in STEM education will happen only if Hispanics, African-Americans, and other underrepresented groups in the STEM fields—including women, people with disabilities, and first-generation Americans—robustly engage and are supported in learning and teaching in these areas.[13]

2. Concept of STEM in Thailand

Apisith Thongchai and et al [1] concluded that STEM is an acronym for Science, Technology, Engineering and Mathematics. The concept of STEM is originated in the US from the gathering of the representatives from every prominent sector in the country with the aim to improve the quality of the population by enhancing their skills in order to be able to compete with other nations. The government had the education policy to support learning and teaching by integrate these four subjects together. Up to now the word STEM is popularized as the government wished to promote this kind of teaching and learning. Nevertheless, there were still quite many misunderstandings toward STEM. Many people were more familiar with Science and Mathematics. General Education also concentrated on these two subjects. Moreover, many also misunderstood STEM as an invention or development of a way of learning that promote each individual subject. In fact, STEM is an integration of four subjects, namely Science, Technology, Engineering and Mathematics, which each of the subject holds the same importance. STEM engages learners to use all subjects to solve the problems, research and create or develop things in the present. The teaching is formed by teachers which come from many areas of expertises. In the real working world, knowledge from many areas needs to be combined and not separate into parts. This supports the essential skills of development in globalization. However, from the past to present, each subject is still being taught independently.[1]

Rakpol Thananuwong [8] wrote that STEM Education is learning of content and the skills of Science, Mathematics, Engineering and Technology. These four cores are the subjects which promote learners to gain knowledge and has a quality living in the 21st century, a fast-pace, globalized and full of technology era. The four subjects are essential to enhance the capabilities of economic competition, the development of the life quality and the nation's security.

In the past, the learning of these four subjects in STEM are separated and independent, for example STEM teaching and learning in Thailand or in the US had courses of Science, Mathematics, Engineering and Technology separately with the lesser focus on the engineering. Educators and academics in the US have compared this separation of the four subjects as "silo", which lies independently in the field [8]

It was found that at present the US was not the first when measuring nation's capacity like it used to be in the past. Many countries around the world have made progresses. The PISA results also showed that the US was worse than many countries. Moreover, the population of science, technology and engineering also dwindled. Thus, the government has implemented a policy which supports the development of STEM. They hoped to raise the PISA results. Furthermore, the swift changes of technology, especially in the communication, logistics, trading, etc, makes the world globalized so it is utmost importance to bring STEM teaching and learning to support the necessary skills in the 21st century. [1]

Pornthip Siripatracha [6] said the necessary skills for the 21st century began from the assembly of academics from many expertises in the US. The government wanted to improve to quality of the people in order to elevate the capacity of the nation so it can compete with others. They also wanted their people to has quality and ability in the society so that they can live in the swiftly changing world. Thus, equipping learners with the 21st Century Student Outcomes, which contains knowledge, skills and expertise is essential [11]. The outcomes are as follows, [6]

IOP Conf. Series: Journal of Physics: Conf. Series **1340** (2019) 012044 doi:10.1088/1742-6596/1340/1/012044

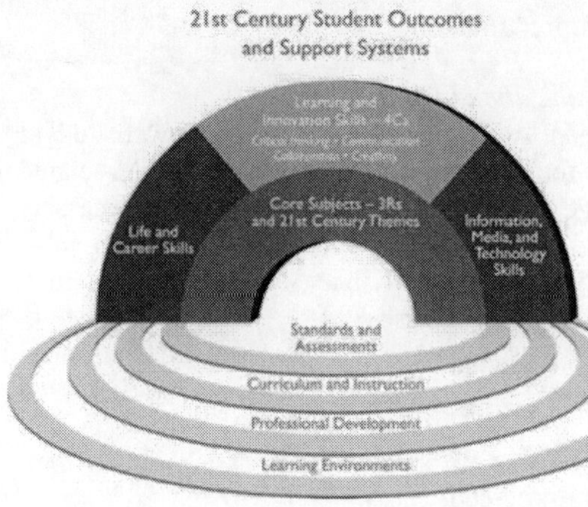

Figure 2. Source :The Partnership for 21st Century Skills)2009 (http//:www.p21.org/about-us/p21-framework

2.1. Core Subjects and 21st Century Themes are English, Reading,
Art of language using, foreign languages, mathematics, economics, science .Art, Geography, History, Citizenship, Political Science .These subjects also need to cover the new areas of content which will influence work and community .Education institutes also have not given enough importance toward subjects such as awareness toward the earth, fundamental course of finance, economics, administration, entrepreneurship, civil and awareness for health and welfare .

2.2. Learning and Innovation Skills, as follows

- Creativity and Innovation which covers creative design, collaborate creatively with others and
- application of creativity.
- Critical Thinking and Problem Solving which includes rational thinking, systematical
- thinking, decision making and solving problems.
- Communication and Collaboration which concentrates on communication by using various
- types of media and collaboration with others effectively and clearly.

2.3. Information, Media and Technology Skills.
In the 21st century, which many technological advancements, learners need to know the these following skills .

- Information Literacy
- Media Literacy
- ICT (Information, Communications & Technology) Literacy

2.4. Life and Career Skills .In our daily and working lives,
We need not only the people who have knowledge and thinking skills, but we also need the people who are able to work in a complex context .These necessary skills are as follows.

- Flexibility and Adaptability
- Initiative and Self Direction
- Social and Cross-cultural Skills
- Productivity and Accountability
- Leadership and Responsibility [6]

3. STEM: Meaning and its importance

3.1. The importance of STEM education in U.S.

All of this effort is to meet a need. According to a report by the website STEMconnector.org, by 2018, projections estimate the need for 8.65 million workers in STEM-related jobs. The manufacturing sector faces an alarmingly large shortage of employees with the necessary skills — nearly 600,000. The field of cloud computing alone will have created 1.7 million jobs between 2011 and 2015, according to the report. The U.S. Bureau of Labor Statistics projects that by 2018 [13], the bulk of STEM careers will be:

- Computing – 71 percent
- Traditional Engineering – 16 percent
- Physical sciences – 7 percent
- Life sciences – 4 percent
- Mathematics – 2 percent

STEM jobs do not all require higher education or even a college degree. Less than half of entry-level STEM jobs require a bachelor's degree or higher. However, a four-year degree is incredibly helpful with salary — the average advertised starting salary for entry-level STEM jobs with a bachelor's requirement was 26 percent higher than jobs in the non-STEM fields, according to the STEMconnect report. For every job posting for a bachelor's degree recipient in a non-STEM field, there were 2.5 entry-level job postings for a bachelor's degree recipient in a STEM field.

This is not a problem unique to the United States. In the United Kingdom, the Royal Academy of Engineering reports that the Brits will have to graduate 100,000 STEM majors every year until 2020 just to meet demand. According to the report, Germany has a shortage of 210,000 workers in the mathematics, computer science, natural science and technology disciplines.[3]

3.2. Meaning and Status of STEM in Thailand

The Institute for the Promotion of Teaching Science and Technology, Thailand (2018) STEM originated from the US, the concept and principle of STEM was announced to be its national policy. After that many countries received and adapted STEM to their own nations. Especially in Thailand, STEM has been applied to teaching and learning of Science and Technology. There are many definitions to STEM such as

STEM is a way of organizing education that integrates four subjects, which are Science, Engineering, Technology and Mathematics. STEM concentrates on practicality of the knowledge to be used in real life, including the development of processes or products, which benefit the working and living. [11]

Wasinee Itsarasena Na Ayutthaya [14] gave definition to STEM as a way of organizing education that is influencing the changes of human in the 21st century. STEM Education or STEM is derived from the integration of four subjects as S stands for Science, T for Technology, E for Engineering and M for Mathematics. STEM also means root, which can imply to the foundation that causes development and changes to human. These four subjects are necessary skills to human of the present and the future. The teaching of STEM is different from the past that required learners to recite. STEM Education is an integration of Science, Technology, Engineering and Mathematics which allows learners to experience, practice, and think creatively of how to solve the problems. STEM Education should be taught from kindergarten level as it trains learners to be curious for new knowledge and know to solve problems. Consequently, they grow up having a systematic thinking and they are aware of the changes and the advancement of technology in the world. Teaching STEM to learners from their young age will also create good attitude toward Science and Mathematics, which often seen from the learners as difficult subjects and being dislike. In the long term, this could influence the young generation to be interested in Sciences, Technology, Engineering and Mathematics. Consequently, more people will be working in the areas of Science-Mathematics which allows the number to be met with the needs in economic and technological markets at present and in the future. STEM Education is therefore very important for human development. Teaching and learning STEM with other subjects such as language, arts, social science and morality will create holistic knowledge and skills for people in order to improve the quality of our society[14].

3.3. Policy of STEM education

3.3.1. The plan. The Committee on STEM Education (CoSTEM), comprised of 13 agencies—including all of the mission-science agencies and the Department of Education—are facilitating a cohesive national strategy, with new and repurposed funds, to increase the impact of federal investments in five areas : 1) improving STEM instruction in preschool through 12th grade; 2) increasing and sustaining public and youth engagement with STEM; 3) improving the STEM experience for undergraduate students; 4) better serving groups historically underrepresented in STEM fields; and 5) designing graduate education for tomorrow's STEM workforce.

Coordinated efforts to improve STEM education are outlined in the federal, 5-year Strategic Plan for STEM Education and concentrate on improving the delivery, impact, and visibility of STEM efforts .Additionally, the Department of Education, the National Science Foundation, and the Smithsonian Institution are leading efforts to improve outcomes for traditionally underrepresented groups.

3.3.2. Supporting Teachers and Students in STEM. At the Department of Education, we share the President's commitment to supporting and improving STEM education .Ensuring that all students have access to high-quality learning opportunities in STEM subjects is a priority, demonstrated by the fact that dozens of federal programs have made teaching and learning in science, technology, engineering, and math a critical component of competitiveness for grant funding .Just this year, for the very first time, the Department announced that its Television grant competition would include a priority to promote the development of television and digital media focused on science.

The Department's Race to the Top-District program supports educators in providing students with more personalized learning—in which the pace of and approach to instruction are uniquely tailored to meet students' individual needs and interests—often supported by innovative technologies. STEM teachers across the country also are receiving resources, support, training, and development through programs like Investing in Innovation (i3), the Teacher Incentive Fund, the Math and Science Partnerships program, Teachers for a Competitive Tomorrow, and the Teacher Quality Partnerships program.

Because we know that learning happens everywhere—both inside and outside of formal school settings—the Department's 21st Century Community Learning Centers program is collaborating with NASA, the National Park Service, and the Institute of Museum and Library Services to bring high-quality STEM content and experiences to students from low-income, high-need schools. This initiative has made a commitment to Native-American students, providing about 350 young people at 11 sites across six states with out-of-school STEM courses focused on science and the environment.

And in higher education, the Hispanic-Serving Institutions-STEM program is helping to increase the number of Hispanic students attaining degrees in STEM subjects.

This sampling of programs represents some of the ways in which federal resources are helping to assist educators in implementing effective approaches for improving STEM teaching and learning; facilitating the dissemination and adoption of effective STEM instructional practices nationwide; and promoting STEM education experiences that prioritize hands-on learning to increase student engagement and achievement. [13]

3.3.3. The Policy of STEM Education in Thailand. The policy proposal of STEM education which aims at integrating knowledge of science, technology, engineering and mathematics emphasizing on problem solving in real life to promote additional experiences, life skills and creativities as well as to prepare the students to be ready for any practices that require scientific, technological and mathematical knowledge and process that lead to future innovation .It is an enjoyable learning and teaching process for students and benefits their future careers .STEM education helps establish manpower who has skills on problem solving, creativity and new creative innovation .Hence, it is the important foundation of innovative skill development and the significant mechanism in improving economy of Thailand to higher income level in the future and being able to integrate knowledge from real life and working.

Social stimulation of STEM educational significance would cause developmental mechanism for scientific, technological and innovative manpower for society .Public hearing of organizations, institutes and associations, especially those who use the products of STEM education, plays a role in developing manpower and pushing forward national policy to support Thailand's innovative systems [4].

4. Organizing STEM teaching and learning

4.1. Suggestion of STEM Education curriculum in U.S.

4.1.1. Blended learning. What separates STEM from the traditional science and math education is the blended learning environment and showing students how the scientific method can be applied to everyday life .It teaches students computational thinking and focuses on the real world applications of problem solving .As mentioned before, STEM education begins while students are very young:

4.1.2. Elementary school. STEM education focuses on the introductory level STEM courses, as well as awareness of the STEM fields and occupations. This initial step provides standards-based structured inquiry-based and real world problem-based learning, connecting all four of the STEM subjects. The goal is to pique students' interest into them wanting to pursue the courses, not because they have to. There is also an emphasis placed on bridging in-school and out-of-school STEM learning opportunities.

4.1.3. Middle school. At this stage, the courses become more rigorous and challenging. Student awareness of STEM fields and occupations is still pursued, as well as the academic requirements of such fields. Student exploration of STEM related careers begins at this level, particularly for underrepresented populations.

4.1.4. High school. The program of study focuses on the application of the subjects in a challenging and rigorous manner. Courses and pathways are now available in STEM fields and occupations, as well as preparation for post-secondary education and employment. More emphasis is placed on bridging in-school and out-of-school STEM opportunities.

Much of the STEM curriculum is aimed toward attracting underrepresented populations. Female students, for example, are significantly less likely to pursue a college major or career. Though this is nothing new, the gap is increasing at a significant rate. Male students are also more likely to pursue engineering and technology fields, while female students prefer science fields, like biology, chemistry, and marine biology. Overall, male students are three times more likely to be interested in pursuing a STEM career, the STEMconnect report said.

Ethnically, Asian students have historically displayed the highest level of interest in the STEM fields .Prior to 2001, students of an African-American background also showed high levels of interest in STEM fields, second only to the Asian demographic .However, since then, African-American interest in STEM has dropped dramatically to lower than any other ethnicity .Other ethnicities with high STEM interest include American Indian students]3[.

5. STEM teaching and Learning in Thailand

Apisit Thongchai and et el (2012) elaborated the above mentioned of the necessary skills for the 21st century and the nature of the its science as follows,

- Science is a study of natural phenomena by using the process of Scientific Inquiry. Science is tools that make human understand more about nature. In May 2012, the US promoted the new Science K-12 Framework and published for online public hearing. The concept was to combine Technology and Engineering together with raising the importance of engineering design to be equal to scientific inquiry.
- Technology is a study of working process to solve, adjust and develop in order to serve the needs of human. The processes of solving the problems or work, in the technological term is called Engineering design or Design process, which has similar process to scientific enquiry. Organizing this learning pattern is a problem-based or project-based learning. Nevertheless, most of the people usually misunderstood that Technology is computers and ICT equipment. In fact, Technology means the process of solving the problems or work in order to create things that serve our needs. In the US, there is International Technology and Engineering Educators Association or ITEEA, who sets the standard of Technology so teachers can teach in the same direction.
- Engineering is a study of creation or innovation or other things that facilitates human's needs by applying the knowledge of Science, Mathematics and Technological process. However, even in the US, the subject of Engineering is not clearly observable in mandatory education but rather is being covered in the same umbrella as Technology.
- Mathematic is an important subject. It has a clear nature of theory. Mathematics is a subject which can be well-combine with the other three subjects. [1]

For organizing teaching and learning, Rakpol Thananuwong (2013) proposed that the integration of STEM Education lies in the learning that link Science and Mathematics explicitly in Engineering Connection. Or it can be said that STEM learning is a design-based learning, which explicitly shown in Engineering knowledge. Integration of the four subjects in STEM teaching and learning requires group practice, discussion and communication skills in order to present the work and result. It is similar to project-based learning, which has been studied that help learners to acquire "deeper learning". Deeper Learning contains five aspects as follows;

- Mastering Core content
- Think critically and Solving complex problems
- working collaboratively
- communicating effectively
- Self-directed and Incorporate feedback Apart from that, integration of STEM Education is also combine two aspects of teaching and learning.
- Context integration which integrates four subjects together with learner's daily life in one context.
- Content integration which integrates four subjects together with one "Big Idea". [8]

6. How to Apply Technology in STEM Education.

Tomoki Saito and et al [12] researched about The Problem about Technology in STEM Education: Some Findings from Action Research on the Professional Development & Integrated STEM Lessons in Informal Fields , Since 2013, the authors' Japanese team in the Department of Science Education at Shizuoka University has held trials of STEM Education in informal fields as participatory action research (e.g., Science museum in Shizuoka, Lifelong Learning Center in Fujieda City, and STEM Summer camp) for the preparation for implementing STEM education in public schools and for proposing science education reform in a Japanese context. Problems in preparing STEM lessons include numerous new instructional materials and programs and emerging specialized schools. In addition, while most of these initiatives address one or more of the STEM subjects separately, there are increasing calls for emphasizing connections between and among the subjects (Honey, Pearson and Schweingruber, 2014). Unfamiliar problems for Japanese teachers are, What is Engineering? What is Design? and How can they be implemented in lessons? While gathering STEM learning materials to implement in their STEM Summer Camp, the authors noticed a pattern with which to develop a STEM lesson and developed a template "T-SM-E" in reference to prior STEM studies. After the STEM Summer Camp, the authors introduced the model in the pre-service teacher preparation program. As a result, the authors received suggestions about how teachers can develop integrated STEM lessons, how undergraduate (UG) teachers can implement it in their lessons, and how teachers can assess student learning in their STEM lessons. From standard based student assessments and reflections written by the UG teachers, the authors found that it was difficult for the UG teachers to include technology in their lessons, and their assessment also indicated that the students did not show performance proficiency in technology. The authors discuss this existing problem in the Japanese education system.

After research found, Teachers experienced difficulty including "Technology" in the lessons. The Japanese UG teachers understood that they could not provide enough of a chance for their students to learn about Technology and it might have influenced the learning result. The authors discussed with the UG teachers why they believed they could not provide enough information about technology. The teachers replied that they thought the definition of technology was a little different from their experiences. In Japan, there is a subject in the curriculum called "Technology" in the junior high school level, in which students learn about fixed themes, such as "Technology related to material and processing", "Technology related to energy transformation", "Technology related to biological development", and "Technology about information" with woodworking, electrical working, metalworking, or cultivation, and so on. Therefore, though its objectives as a subject (Table 7) are similar to STEM, UG students thought of "Technology" as involving vocational skills. In other words, there was a gap between the historical recognition and the objectives of the Course of Study. As a professional development subject, there needs to be more study about the nature of technology. On the other hand, an UG teacher who put "A" on Viewpoint 1 (I) prepared a story about technology of the water purification system (See teacher's reflection about the Engineering Problem). This may show that a good description or illustration about technology leads students to the engineering design processes naturally, an important point to remember when preparing and implementing a STEM lesson [12].

Priya Chacko and et al [7] researched about Integrating Technology in STEM EDUCATION ,For the last two years, two groups of students in 9th through 12th grade from various New York City schools participated in a paperless summer science program in which technology was fully integrated into science education. This program aimed to educate high school students about health related topics including diabetes, cancer and HIV/AIDS while encouraging them to conduct independent research. The four week program ran from July-August for a group of an average of 18 students (the number of students who participated in the program varied from year to year, but overall attendance never wavered from week to week). Each topic was introduced by a technical lecture. Once the students were given the background information, laboratory experiments were conducted in small groups. Afterwards, students and instructors discussed the results. Using what they learned from the lecture and lab, students were then given classroom activities to complete based on the course curriculum. For our curriculum, classroom activity is defined as experiences involving students manipulating their knowledge by participating in discussions, creating presentations, assessing case studies, watching videos, and topic related games/activities. At the end of every week, students were given an evaluation of the module to determine whether or not they had increased their understanding in the STEM field covered and their interest level in pursuing a STEM field as a career after high school. In this way, the modules could be evaluated by data provided by students [7].

7. Introducing project-based learning

NSW Department of Education [5] Project-based learning (PBL) is an approach to teaching and learning that engages students in rich and authentic learning experiences. PBL can be transformative for your teaching practice but requires strong, supportive leadership and a commitment to innovation and contemporary pedagogies.

In a PBL environment, students gain knowledge and skills by investigating and responding to an engaging question, problem or challenge.

A PBL learning and teaching framework addresses cross-curricula content and learning dispositions through rigorous, authentic, hands-on, interactive learning experiences.

7.1. Why should I implement project-based learning?

PBL is interdisciplinary (cross-curricula) and focuses on active, student-directed learning. It gives students an authentic, real-world context for learning where student voice matters.

Students face complex challenges in a post-school environment where problem-solving, collaboration and creativity are highly valued skills.

Key reasons for using PBL include:

- well-scaffold PBL engages students in their personal learning journey
- offers students an opportunity to build confidence, solve problems, work in teams, communicate ideas, and manage themselves more effectively
- encourages students to use technology in authentic ways
- Connects students and schools with communities locally and globally.

7.2. How do I implement project-based learning?

In order to implement PBL effectively it is important to understand the essential elements of PBL and consider how the roles of student and teacher are redefined.

This guide will help develop an understanding of the processes of PBL and point to a range of resources that can be used to support its implementation.

PBL does not need to be implemented in its entirety when first starting. A 'project slice' is a good starting point, and as students and teachers become more familiar with the elements of PBL, further elements and strategies may be implemented. [2]

8. STEM teaching and learning condition in Thailand

Artnarong Manosuttirit [2] researched STEM teaching and learning condition. It was found that there are many factors that require supports and modification. The details are as follows.

1. Educational policy and curriculum: Each school cannot determine the direction and format of STEM instruction because it lack enough information and understanding in STEM education management and there is no clear guidelines provided. This makes each school worried that STEM instruction may affect the teaching and learning of other subjects, especially in terms of limited studying time which is not conducive to any other activities.
2. Integration and collaboration: As there is no clear curriculum and policy, it is not possible to determine details of STEM teaching and activities in term of time, credit, place, and budget. Many schools still have not received financial support or have to use their own budget. This is an ongoing problem that needs to be solved.
3. Responsible teachers: STEM education is a new concept that needs time to understand and adopt. Normally, the teachers already have a lot of responsibilities to take care of. If they are assigned to handle STEM teaching, they will have more burdens and it may have an effect on their daily jobs. Therefore, nobody are willing to take responsibility in this respect. This point should be clearly understood and personnel tasks should be certainly defined.
4. Support from responsible organizations: Many schools carry out their own STEM instruction and activities without correct guidance and support so they are not certain whether what they

did are right or wrong. This situation leads to some mistakes in STEM teaching and has an impact on related factors, including place, budget, teaching personnel, teacher training and development, which are all vital to STEM education advancement.

8.1. STEM instructional guidelines

The data obtained from the experts and schools in STEM education network can be concluded as follows.

1. Open approach and horizontal learning should be applied. The teachers should control the situation, listen to the students' opinions, and allow the students to speak, think, practice, and research on their own.
2. Any surrounding topics can be applied to STEM activities, especially in application field as everything is already integrated. STEM activities should be opened without any limitation and should focus on problem-solving methods, concepts, and processes.
3. The teachers of each subject should design the activity together in holistic manner. All involved aspects should not be split and inserted with irrelevant ones. The teachers should emphasize understanding on integration, skills, and ways of thinking.
4. Activities need to be challenging. They should not be too difficult or too easy. The challenge should be moderate and appropriate with environment and restriction.
5. STEM should be used with other instructional activities and teaching methods in order to respond to skill and ability requirements such as using STEM together with project-based teaching to practice 21st century skills, critical thinking, and project executing.
6. Activities should be flexible and suitable for educational environment. They should neither have rigid format nor add burden to the teachers and students. All of the activities should be blended in routine work and inserted in regular class content concerning problem-solving in daily life and practical situation in the community. [5]

From the circumstances and the problems occurred, there were still difficulties that make STEM teaching and learning and activities possible. With many factors that need to be integrated within the four subjects, it takes a huge amount of time in preparation, set up the right time and organize activities for STEM teaching and learning. A case study presents here is in a Computer class in Lower Secondary Level, called "Arduino STEM" which is a project-based Learning.

9. Example of Technology application to STEM teaching and learning

Table 1. STEM LESSON IDEA Department of Career and Technology Subject: Computer.
Topic: Embedded System (Arduino) Grade: 8 Time: 2 hrs./week

Lesson Title	Automatic Fan
What will you teach in this lesson	1.Computer Programming 2.Arduino's Project 3.STEM Education or STEAM 4. Project Based Learning
Situation	The situation is when living in a house. Outside temperatures are higher than inside. Every time go home have to turn on the air conditioner with a manual control or a remote control. This situation must be solved by automatic control. Using computer technology and programming to control automatic opening and closing in various ways.

IOP Conf. Series: Journal of Physics: Conf. Series **1340** (2019) 012044 doi:10.1088/1742-6596/1340/1/012044

Table 1. (Cont')

Lesson Title	Automatic Fan
What will the class do in this lesson	1. Solve problems with process 2. Programming language and can be analyzed by using the computer. 3. Could understand the working principles of Arduino design and function. 4. Responsible, Creative and local resources to apply them appropriately. 5. Designs created by making fantasy computer language properly.
Media and Example	1. www.youtube.com 2. e-Learning(Moodle) 3. Arduino's Project example 4. Project: Smart home 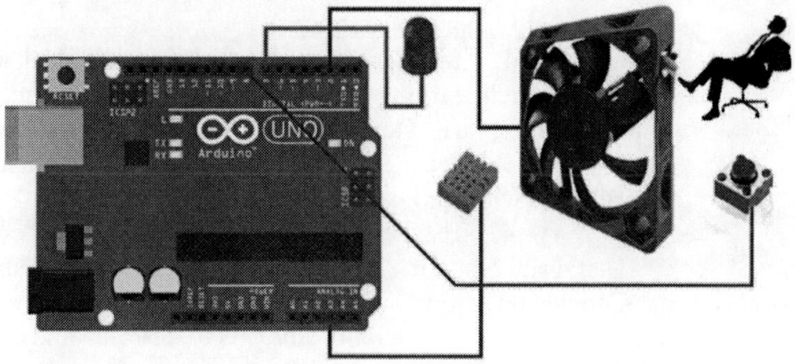 **Figure 3**. Source : Arduino STEM (Automatic Fan)_www.acsp-cs.com
Curriculum subjects Covered	21st Century Core Subject 1. Sciences(Temperatures, Lighting) 2. Technology(Computer Programming, Arduino, Electronics Device, Circuit) 3. Engineering(How to assemble circuit, How to choose device) 4. Mathematics(Timing, Lighting Gain) Additional: 5. Art/Architecture (Designing, Materials)
Skill being taught	Sciences Process Skill and 21st century skills 1. Learning and Innovation Skills - Creativity and Innovation Skills - Critical Thinking and Problem Solving skills - Communication and Collaboration skills 2.Information,Media and Technology Skills - Information Literacy - Media Literacy - ICT(Information, Communication &Technology Literacy) 3. Life and Career Skills - Flexibility & Adaptability - Initiative & Self-Direction - Leadership and Responsibility

Table 1. (Cont')

Lesson Title	Automatic Fan
Learning Contents	- Lecture by media E-learning. - Reporting group. - Write a program and control the Arduino board Working with Functions on Arduino IDE Using LEDs. -Write a program (To control Automatic Fan) - Create jobs Project Learning Activity: <u>Starter activity</u> 1.The students meditate before class and learn about morals in daily life for 2 minutes. Brain Gym 2. The teacher states the learning objectives to the students. <u>Main and Elaboration activity</u> 1. Teacher did the analysis of individual student results to improve their learning and performance. 2. The student did login to Moodle learning platform 3. Teachers make a plan to prepare students to analyze and solve problems with application programs. 4. Teachers teaching about Recommended program " Moodle system and Installing Arduino Using LEDs and App Inventor" <u>Conclusion activity</u> 1. Ask random students to describe and demonstrate teaching content information from the lecture 2. The teacher separates the students into a team for group Activity projects and show Conception Teachers and students to share a summary discussion.
Materials	1. e-Learning(Moodles system) 2. Hand Out 3. Arduino board and electronics device
Assessment	1. Peer assessment and self assessment via e-Learning 2. Quality work project. 3. Oral Presentation 4. Quiz
Classroom Setup	1.Computer Lab 2.Pair or participants work in group of 4-5 3.Should be able to access the internet(for Moodles system)

10. Conclusion

Although STEM Education has reached Thailand for quite some time and the relating government organizations have tried to push STEM Education's teaching and learning, in term of policy, personnel and other issues, it is still quite an abstract concept. Thus, it is necessary to study STEM and plan how STEM in Thailand should be. As Sutheera Prasertsil [10] suggested evidences in the US shown that in 100 working people, 96 is hired by 4 people and those 4 people are educated in STEM, which is Science-Technology-Engineering-Mathematics. So, the US has tried to propel the education system by using STEM in its fundamental education and encourage STEM to be used in university level. The problems, when compare to the US, that stops STEM Education from succeed in Thailand, is Thai context, teachers' development and the ability of the teachers to deliver knowledge to learners. Sutheera added that "STEM teachers" need to equip themselves with these necessary skills. [10]

1. Awareness of the patent
2. Knowing various kinds of technology
3. Ability to see Mathematics as representative of Science and Simulation
4. Ability to see assumption
5. Understanding the subject thoroughly
6. Seeking opportunities
7. Understanding systematically
8. Creating reasoning charts which can be explained by each subject's content
9. Application of knowledge
10. Five mantras of inventions [9]
    - What are the previous inventions?
    - What are the disadvantages of the previous inventions? What needs to be solved? What solution did we choose and why?
    - What principle do we applied to the disadvantages? Why do we choose these principles?
    - How do we apply with our invention?
    - How can we prove that our inventions are better than the previous ones? [9]
11. Understanding liberal arts
12. ability to sketch
13. ability to write flow chart of the process with feedback
14. Computational thinking

These skills might only be some parts of the suggestion that STEM teachers should possess, further study and application to each and individual context also need to be done. This, however, needs assistance from many organizations in order to concretely improve Thailand STEM.

References

[1] Apisith Thongchai and et al 2012 *Composer. Special Lecture Handout in Science, Technology, Engineering, and Mathematics Education: Preparing students for the 21st Century* (by Prof. Dr. Edward R. Reeve.15 May 2012, at Sanan Sumit Hall The Institute for the Promotion of Teaching Science and Technology)

[2] Artnarong Manosuttirit 2016 *A Study of Teaching STEM Education in Thai High School* (Thammasart International Journal of Science and Technology Vol 21) No 5 pp 1-6

[3] Elaine J Hom 2014 *What is STEM Education?* (Online. https://www.livescience.com/43296-what-is-stem-education.html[22/10/2018])

[4] National Legislative Assembly 2015 *Report of Policy Recommendation STEM Education Proactive policy Youth development and manpower in science, technology, engineering and mathematics* (Online, http://www.senate.go.th/w3c/senate/pictures/comm/1547 STEM.pdf [19/6/2018])

[5] NSW Department of Education 2018 *Introducing project-based learning* (Online https://education.nsw.gov.au/teaching-and-learning/curriculum/learning-for-thefuture/future-focused-learning-and-teaching/project-based-learning-resource-guide/introducing-project-based-learning [25/10/2018])

[6] Pornthip Siripatrachai 2013 *STEM Education and Development in 21 Century* (Bangkok University, Journal of Administration Yrs vol 33) ep 2 April-June 2013 pp 49-56

[7] Priya Chacko, Sarah Appelbaum, Heejoo Kim, Jinhui Zhao and Jin Kim Montclare 2015 *integrating technology in stem education* (Journal of Technology and Science Educaton Vol 5) chapter 1 pp 5 On-line ISSN 2013-6374 – Print-ISSN 2014-5349 DL: B-2000-2012 – http://dx.doi.org/10.3926/jotse.124

[8] Rakpol Thananuwong 2013 *Composer . Training Report in STEM Education by Prof. Mitchell Nathan* pp 10-11 Jan 2013

[9] Sutheera Prasertsil 2015 *STEM in Invention : Development of Vision and Imagination* (Poh Pan

Punya Project Songkhla 2015) Online https://kruyaivijai.wordpress.com/ [9/7/2018]

[10] Sutheera Prasertsil 2016 *Decode STEM Teaching. Poh Pan Punya Project* Online https://kruyaivijai.wordpress.com/ [9/7/2018]

[11] The Institute for the Promotion of Teaching Science and Technology 2018 *Knowing STEM* (Online http://www.stemedthailand.org [24/6/2018]

[12] Tomoki Saito Yoshiyuki Gunji and Yoshisuke Kumano 2015 The *Problem about Technology in STEM Education* (Some Findings from Action Research on the Professional Development & Integrated STEM Lessons in Informal Fields. K-12 STEM Education Vol 1) chapter 2 Apr - Jun 2015 pp 85-100

[13] U.S. Department of Education 2015 *Science, Technology, Engineering and Math: Education for Global Leadership* Online https://www.ed.gov/stem [23/10/2018]

[14] Wasinee Itsarasena Na Ayutthaya 2017 *Knowledge of STEM Education* (Chulalongkorn University Press, Bangkok)

Tri-Patterns of Community Network for Carbon Emission Reduction in Thai-Isan Urban

T Chamaratana[1*], L W J Knippenberg[2], E B P De Jong[2], P Prabnok[1], K Phatchaney[3] and R Dasri[3]

[1]Assistant Professor, Faculty of Humanities and Social Sciences, Khon Kaen University, Thailand
[2] Assistant Professor, Department of Anthropology and Development Studies, Radboud University Nijmegen, the Netherlands
[3] PhD. Student in Development Science, Faculty of Humanities and Social Sciences, Khon Kaen University, Thailand

*Corresponding author, E-mail: thanacha@kku.ac.th

Abstract. In 2014 more people lived in urban areas than in rural areas. In the energy sector that is the highest proportion of all carbon emission sectors, urban area release carbon 71-76% of global carbon emission. The first priority solution for solving climate change is carbon emission reduction in urban areas. The low carbon city is an area base management to meet the target of city carbon emission reduction. One of means for develop the carbon emission behavior in everyday life of the people is collaborative action in form of networking. The social network is the one of major concept for urban study. It used to focus in individual behavior, but now extended to study group of urban people as community network. The objective of this research was to study the pattern of community network for carbon emission reduction in Urban of Northeastern Thailand as Khon Kaen city. Data were collected via in-depth interviews as well as by participatory and non-participatory observations with 24 key informants who were the community heads and members from 4 communities in Khon Kaen city. Data analyses were content analysis with ATLAS.ti program and social network analyzed by UCI-net program. The results of the study indicated that there are three patterns of community network for carbon emission reduction in Khon Kaen city; One-center network, Dual-center network, and Full network.

1. Introduction

At the present time, various nations give precedence to solve the climate change problem which as a result of global warming [21]. They succeeded in the "Paris Agreement" to hold the increase in the global average temperature to well below 2 °C above pre-industrial levels [6]. The increase in the global averaged temperature since the mid-20th century results from the increase in anthropogenic greenhouse gas concentrations [7]. In 2014 more people lived in urban areas than in rural areas, with 54% of the world's population and will be 66% in 2050 [6]. In the energy sector that is the highest proportion of all carbon emission sectors, urban area release carbon 71-76% of global carbon emission. [8] May, conclude that the first priority solution for solving climate change is carbon emission reduction in urban areas.

The low carbon city is an area base management to meet the target of city carbon emission reduction. There are three solutions for low carbon city driving, technology development, infrastructure investment, and lifestyle change [12]. The lifestyle change is the significance solution for low carbon city driving at present and future [13]. There are two main solutions for low carbon city driving, technical solution and lifestyle change solution. The first depends on science or technology knowledge and high budget and also relate to many stakeholders. But the second does not depend on knowledge and budget like the first or depend on only a few budgets moreover, it can proceed immediately [22].

The almost all activities in human's everyday life concern to carbon emission more or less so that the individual behavior changing can reduce carbon emission. Even though human's everyday life is dominated by consumerism cause of individual losing but this dominance is not complete. A human has the potential and strategy that scrambles the benefits from structure of social [3]. Therefore, human can change carbon emission behavior in their everyday life and these changes will be the significance solution for low carbon city driving at present and future.

One of means for develop the carbon emission behavior in everyday life of the people is collaborative action in form of networking. The social network is the one of major concept for urban study. It use to focus in individual behavior [4],[5] but now extended to study group of urban people as community network. Also, the growth urban in the Northeast of Thailand, "Isan", such as Khon Kaen city.

Khon Kaen is a center of commercial, investment, and transport in the Northeastern of Thailand where the part of Greater Mekong Sub-region are. It is determined to be a MICE city and ICT city so that economic growth and population in urban areas are more and more. Besides the goal of the Khon Kaen Municipality in their master plan (2015-2018 and 2018-2021) is a low carbon city [10]. One of their main solutions is promoting citizen to improve their knowledge how to reduce carbon emission. They expect that when citizen gain the knowledge, they will change their behavior in everyday life to reduce carbon emission.

The carbon emission reduction network of each group may like or different, it depends on their context. Thence, if we understand the carbon emission reduction network and the factors that influence the carbon emission reduction network strengthening these will be an important tool for Isan Urban as Khon Kaen Municipality policy maker to drive the city to be a low carbon city.

From the reasonable above lead us to solve the research's questions as "How to develop the strengthen community network for carbon emission reduction in Thai-Isan Urban? However, this article is the first phase result of the research project with the objective as to study the pattern of community network for carbon emission reduction in Thai-Isan Urban.

2. Conceptual Framework

In order to understand social network relationships of Isan urban community, it was necessary to apply an appropriate concept, i.e., the social network concept, which has an assumption that the social network consists of persons or actors who are related with one another according to the roles or functions of each individual or partners. Each person does not take only one role but many in the daily life. The relation among people in the social network may follow the Exchange Theory because a person does not only act according to his role, expected social function or according to the transferred standard; but relation among individuals also depends on perception and decision for exchange between the pair [2], [5], [18]

It can be said that the social network concept focuses on the presence of social relation web among individuals which extends and covers all of the society. The analysis of social network in terms of sociology, on the contrary, focuses on social relation among individuals in the social network, to see if and how this influences the behaviors expressed to others. The patterns and characteristics of the social network must be used to explain an individual's behaviors. In this respect, [17]applied the concept of Boissevain (1974) to develop the framework for studying and analyzing the social network by dividing into 4 dimensions [5] as follows:

The first dimension is the roles of social network members. They are the factors behind multiplicity of relations, which can be explained by the Role Theory. A social network consists of individuals who have relationships with one another according to their roles or functions. Each person does not have only one role but many roles to act each day. Therefore, two persons can have relationships both in one role or many roles, which can be referred to as Uniplex or Single-Relation and Multiplex or Multi-Relation, respectively. The role of each case will have the norms or expectation as the lead for behaviors acted towards one another.

The basis of relationships in the social network was considered to be the second dimension. This usually was hidden in the Exchange Theory. It was believed that social relation among individuals in the social network is based on the Benefit Exchange or Transactional Contact. An individual does not act his role or does according to the norm being transferred to him, but the relation also depends on perception and decision to exchange between the related pair in terms of materials such as money and things, of the mind such as assistance that brings about indebtedness to be returned in the future, or personal admiration and respect. The person would decide what to select, what behavior, what to exchange with the other through consideration of appropriateness and satisfaction obtained from the exchange.

The third dimension is the directional flow. It was the factor causing social relation which could have equal and unequal characteristics since social relationship among individuals on the exchange basis can result in cooperative and competitive relationship and bring about balance reciprocity as well as negative reciprocity. Here, it means obtaining benefits from each other both equally and unequally with one party obtaining more than the other.

Finally, the frequency and duration of relationship was the factor leading to quality of relation, especially the attachment and influence on one another in terms of behaviors. In other words, when a person is frequently related to the other and the duration of relation is long, then the relation between the two will influence and establish the behaviors between each other owing to a lot of attachments. However, frequency of relation is not only sufficient to predict the influence of personal relation or individual relation. There must be the factors of time and roles of relation, too.

According to thesis statements above, a concept of social network is implemented in this paper as a framework to investigate the pattern of community network for carbon emission reduction in Thai-Isan Urban. The form of relationship within social network and embed within the network until it becomes the pattern of community network for carbon emission reduction, which is mentioned in the following chapter. Moreover, social network can be considered as another type of collaboration among people within the network, and it is useful to analyze behavior of people. Social network is a social phenomenon that expresses social relations among individuals [5]. Therefore, the concept of social network is applied to conceptual framework as a figure below (see figure 1).

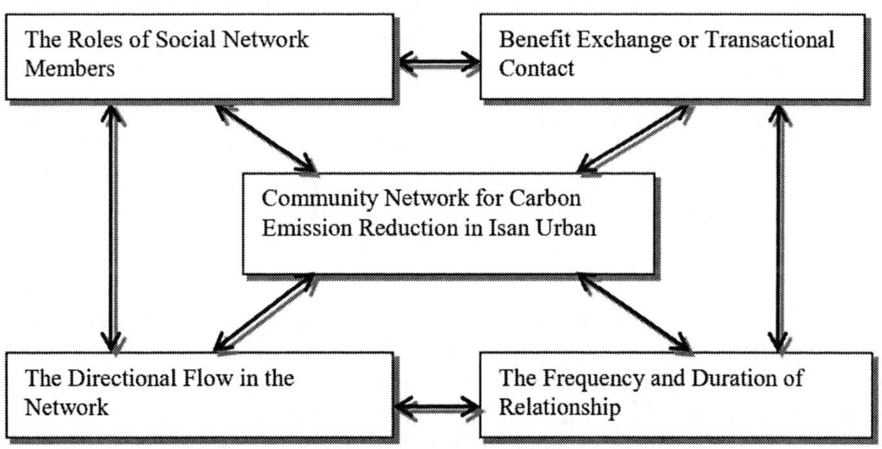

Figure 1. Research Conceptual Framework

3. Research Methodology

The study applied qualitative methodological approach to investigate the pattern of community network for carbon emission reduction in Thai-Isan Urban as Khon Kaen city. Key informants consisted of twenty four community heads and members from 4 communities in Khon Kaen city; Non Nong Wat, Bann Bakham, Karn Keha Khon Kaen and Bann Hua Tanon (see figure 2). The community heads and members in each community were selected by purposive sampling technique.

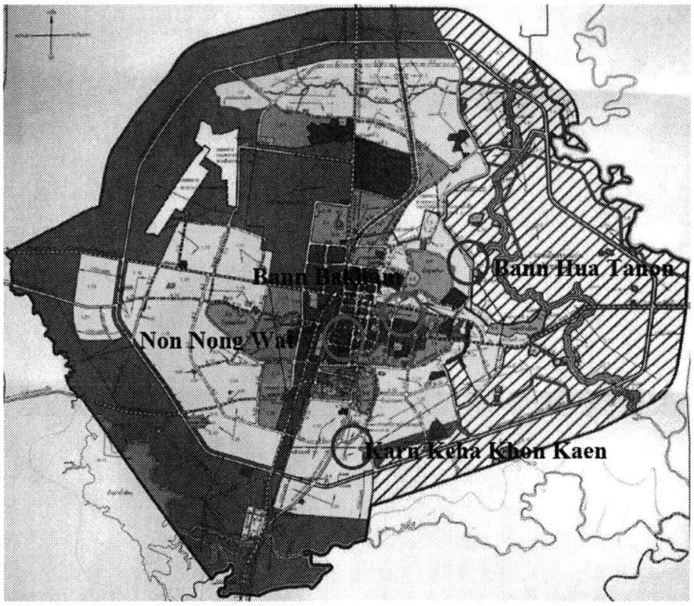

Figure 2 Research Area

For data collection, the study applied in-depth interview with interview guideline to collect field data from the key informants in community mentioned above. Participatory observation and non-participatory observation were also applied to the study to observe characteristics and scopes of relationship with the networks, elements of the social phenomenon, and behavior of individuals [4]. We conducted interviews in research sites as mentioned earlier, and observed general contexts surrounding the targeted key informants in order to obtain more information. In addition, we used field note, voice recording, and camera to collect the data with permission from key informants and related stakeholders. In case the permission was denied, field note would be used instead. Data from the interview and the observation would be categorized and edited everyday in order to update the data completely.

Regarding validity and reliability of qualitative data in this study, triangulation method was introduced. Data was collected from different sources among four research sites. Validity and reliability of data was cross-checked several times in each day. Questions on the interview guideline were asked at different times, and questions would be updated all the time in order to meet interviewee's manners. In addition, data was checked by traceability method with the interviewed farmers to prevent data discrepancies, and to ensure that the data was reliable [6].

Data was updated every day in order to check the completion of data. Later on, data was categorized by topics investigated. Data was comparative analyzed and constructed the summary. Data from in-depth interview was arranged by the transcription of audio recordings and the field note editing. Then, data was typed with Microsoft Word Program, and was prepared for qualitative analysis by the ATLAS.ti program. We then assigned quotations, coding, categorizing, and networking data. Once the data was ready for analysis, content analysis method was implemented to interpret data to find relationships and patterns within social phenomena in order to construct the inductive conclusions [4].

Finally, we applied simple social network analysis method by the UCI-net program, and construct the virtual network modeling through the program code directly [20. Data analysis began with the matching of relationships among key informants, and then ran a program to display the relationship. In addition, the social network scenario was displayed via the Netdraw Program. Analytical description was used for data presentation together with diagrams and photos in order to make a clear presentation.

4. Khon Kaen City : The Rapid Growth City and Determination to the Low Carbon City

Khon Kaen city established in 1797, or estimated 200 years ago. Khon Kaen city was established by a Laotain leader, named Pea Mueng Pan ("Pea" represented a governing position in the ancient Lao kingdom). Pea Mueng Pan and his people about 330 people moved from Ban Chee Loan (currently located in At-Samat district, Roi-Et province) to Khon Kaen city. In 2442, they had a second movement and settled community at the north of Kaen Nakorn pond, where located in Khon Kaen city at present. Khon Kaen city nowadays is a major city of northeastern region of Thailand. The city is the fifteenth largest city of Thailand by area, and it is the fourth populated city of Thailand. In addition, Khon Kaen have a Gross Domestic Product (GDP) of 145,275 million Baht, which is the city with the second highest GDP in northeastern Thailand [16].

The rapid growth of Khon Kaen city has been caused by many factors. For instance, the city is a hub of finance and banking in northeastern region. The city also is an education hub of northeastern region especially Khon Kaen University, where is the biggest university in northeastern region. These factors boost rapid growth to Khon Kaen city. The city currently has become center of transportation, industrial sector, healthcare, and administration of the region. Moreover, the city located on a strategic area where Mittaphab Road (a main route of northeastern region) and the East-West Economic Corridor (EWEC) where connects Indo-China region, cross to each other. This advantage is a highly focus and includes in the Strategic Development and International Cooperation Projects such as the Greater Mekong Sub-region (GMS) and the Ayeyawady-Chao Phraya-Mekong Economic Cooperation Strategy (ACMECS). Due to those positive factors, it contributes rapid expansion to the city, and the city is expanding align with Mittraphap Road and other main roads of Khon Kaen city such as Maliwan Road, Srichan Road, and Khon Kaen - Yangtalad Road [15]

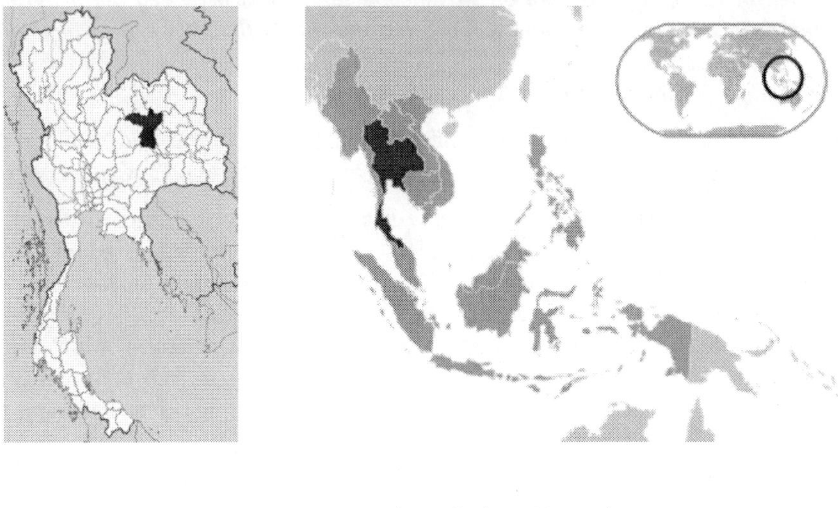

Figure 3. Location of Khon Kaen city.

5. Implementing Community Network into the Low Carbon City

Based on data collected from 24 key informants in four communities of Khon Kaen Municipality, community networks represented in forms of community committees and village health volunteers. In general, those two groups shared the same members and worked under supervision from governmental agency, which was Khon Kaen Municipality Office. In 2015, however, Khon Kaen Municipality Office decentralized its authority on waste management and environment to local community under the strategy "Small municipality under the big municipality". The strategy convinced communities in Khon Kaen city to have more roles in environmental management. Therefore, community networks were constructed with an efficiency of environmental management, and the networks contributed significantly to the reduction of carbon emission in community through several activities of community network.

Although the programs of "Khon Kaen: Low Carbon City" and "Zero Waste Office" had initiated by Khon Kaen Municipality Office, the programs were rarely to be success without participation of people in community especially since the authority decentralization. According to a study in four communities, it was found that all the communities had problem on too much waste. Waste disposal, by Khon Kaen Municipality Office, was not effective enough to get rid of waste at all communities, and it resulted in waste pollution in communities. Moreover, communities had to pay an expense on waste disposal to Khon Kaen Municipality Office a lot. In 2013, situation of waste disposal in Khon Kaen city was more serious when villagers who lived in Kham Bon village, where Khon Kaen Municipality Office used as the landfill of waste disposal, started protesting because they did not want to receive waste from Khon Kaen city anymore. The landfill was closed for two weeks, and it resulted in community in Khon Kaen city had a huge waste waiting for disposal. Moreover, it was reported that there were no waste separations within households. Organic waste was mixed with other kinds of waste, and it produced waste pollution disturbing people in community.

Communities, therefore, were increasingly aware of waste management in community. Many communities, represented as community network, applied those two programs as a pilot to initiate waste management activities under financial support and training from Khon Kaen Municipality Office and the Khon Kaen Environmental Office. Background of the activity was told by Mr.Kritpas, the community leader, as quotation below.

> *"In 2013, the high-ranked municipal administrators began to pay attention to waste problem because the municipality paid a lot for waste disposal expense. Therefore, they started the training program on waste separation, and provided the program to communities in the city. The Khon Kaen Environmental officers were responsible to train us by firstly started with community committee and then extended to the village health volunteers"* (interviewed on August *15, 2018)*

Waste management programs in each community had different patterns and methods depended on community network's decision and agreement. However, at least three main activities were included.

1. Household waste separation; this activity was considered as the most difficult activity of community network, since the activity was an attempt to change behavior of people in community. Some households might not concern about waste separation because they did not want to have any difficulties on waster separation, or they were not aware of waste separation. The community network worked hard on this issue at the beginning stage of activity. The village health volunteers spent much time to provide knowledge on waste separation to all the households in four communities. However, only one third of the households decided initiating waste separation as a statement below from *Ms.Ura*, leader of Karn Keha Khon Kaen community.

> *"The activity at early stage was quite difficult because people did not expect to succeed the activity. Someone thought that only paid the waste disposal fee for 10-20 THB per month to Khon Kaen Municipality Office was enough to get rid of waste from their*

> *houses. It was hard for the village health volunteers to convince people concerning this issue and started separating waste in households"* (interview on August 15, 2018)

The activity was not easy to succeed because only a few households were aware of waste separation. Therefore, the community network tried to find ways to motivate the households to focus on waste separation. All the four communities used a same way by showing an example of waste separation to the households. For instance, Bann Hua Tanon community assigned responsibility to a village health volunteer to manage waste in public areas. Each volunteer covered 15 households in community. The volunteers started asking collaboration from households to separate waste. The community network would support a garbage bag to all households, and then the volunteers brought all the garbage bags to the disposal site in community. While Bann Bakham community and Non Nongwat community had volunteers to collect waste in community. The volunteers collected the household waste and separated waste every day until the community members realized on an advantage of waste separation. When many households saw advantage, they turned themselves to start waste separation as the volunteers did.

2. Organic waste disposal; in the past, communities disposed organic waste by putting in the garbage bag and waited the Khon Kaen Municipality Office for disposal, or they burned waste in house area so that it caused pollution in community especially smoke or dust pollution. When smoke or dust flow to water source, people in community could not use water for household consumption. However, after the success of household waste separation, the success contributed more convenient to the community network in order to run waste disposal activity. It was to say that they were able to identify types of waste. They knew which type had to dispose out of community and which type could be recycled or used for value added. The community realized that organic waste was able to utilize as the bio-fertilizer. Therefore, all the communities asked their community members for a help in separating organic waste. The households or community networks would put organic wastes in the "organic waste stall" they provided. The stalls were mostly made by bamboo. The bottom layer was used for organic waste, such as food, while the top layer of the stall was used for dry waste such as leaf. This method prevented stinky smell and the fly, which might be a carrier of disease.

When organic waste was stalled, the community network processed organic waste as fertilizer in both liquid form and solid form, and gave to the households to use in their farms. Some communities, such Non Nongwat community, had an effective production until they were able to produce commercially as a below statement from *Mr.Jaroon,* community leader of Non Nongwat community.

> *My colleagues and I picked the waste up every day. Each households helped us separating waste. Waste such as foods, fruits, or vegetables from the market were brought to our stalls as well. We processed the waste into fertilizers and liquid compost. I firstly used at my farm, or gave fertilizers to the households in community. When we had a lot of production, we decided doing commercially. We contained it into bottles and sold. Now it becomes community product"* (interviewed on August 17, 2018)

3. Waste banking; the activity was a part of the Khon Kaen: Low Carbon City program, initiated by Khon Kaen Municipality Office. Later on, the program was sent to all 95 communities in Khon Kaen Municipality under supervision from governmental agencies. At present, President of the program is *Mr.Jaroon*, Non Nongwat community leader. Secretary is *Ms.Ura*, Karn Keha Khon Kaen community leader, and registration officer is *Ms.Toy*, Bann Bakham community leader

Brief concept of the activity was that each community sent invitation to all the households to sell their recycle waste to the program. Then, the program sold all the recycled waste to waste disposal operators. Households would have accumulated saving accounts from the sale of the waste on a regular basis for6 months. When the money was accumulated up to300 Baht, they were able to withdraw the money. In case member of the household died, the program would give a funeral fee of 2,000 Baht, deducted from the account. It is possible to increase the cremation fee if there are more project members in the future. This concept was a good choice to motivate households attending the program. Households would separate waste and took recycled waste to the community network on monthly basis in order to generate household income. It was beneficial to all the households, and it was better to attend the program than leaving waster without any benefits. *Ms.Banyen*, a member of Karn Keha Khon Kaen community, stated as below.

> *"At first, they were not interested in the program. When we applied the funeral fee to the program, many households started separating waster and sold to us. They would not get the money at time, but the money would be accumulated as the saving account. Moreover, I thought it was an additional income. If someone in the households died, we still had 2,000 Baht to manage the funeral"* (interviewed September 5, 2018)

Besides those three waste management activities, four community networks in Khon Kaen city also had activities that contributed reduction of carbon emissions in urban area. It was not quite successful. However, it was a beginning step to become low carbon city. For instance, the campaign to save energy by turning off air conditioner at least one hour per day, run by Bann Bakham community, the campaign to clean the public road at Karn Keha Khon Kaen community, and the campaign to stop using plastic bag in which the activity operated by Khon Kaen urban community network. This activity received a positive feedback from major markets in Khon Kaen city.

Based on the overall success of carbon emission reduction in urban areas which operated by, researchers analyzed patterns of community network by considering contexts in each community. Contexts of community included roles of member in the network, particularly exchanging relationship, frequency and duration of interaction, and direction of relationship, that were effective to contribute those three community networks to become a pioneer of the low carbon community in Khon Kaen city.

6. Patterns of Community Network for Carbon Emission Reduction

Considering all four community networks, the study analyzed the network members who played major roles in the community network. Each network has six members, which is a proper size for social network analysis in terms of working and decision-making [11]. Researchers implemented center of network in the study in order to identify patterns of community network for carbon emission reduction in urban areas. The results identified 3 patterns of community network included 1) one-centered network, 2) dual-center network, and 3) full network. Details of each patterns are as follows.

6.1 One-center Network

One-center Network is a pattern of community network with one centered-individual who linked social network in community. This pattern of community network was found in Non Nongwat community and Karn Keha Khon Kaen community. Model of those two community networks were presented as in figure 4 and 5 respectively.

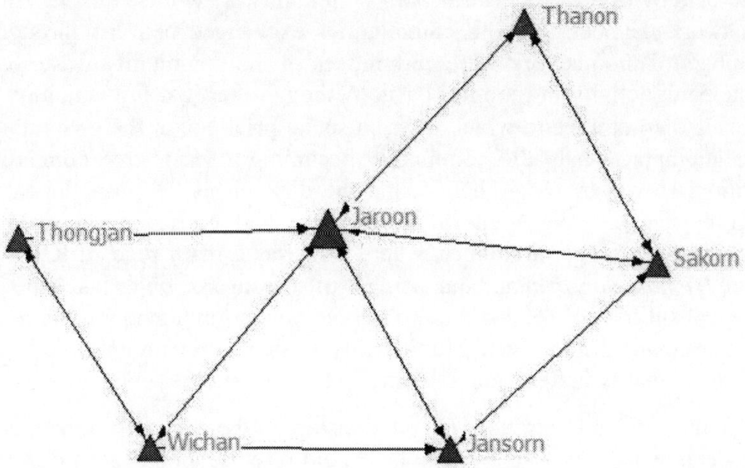

Figure 4. model of Non Nongwat community network.

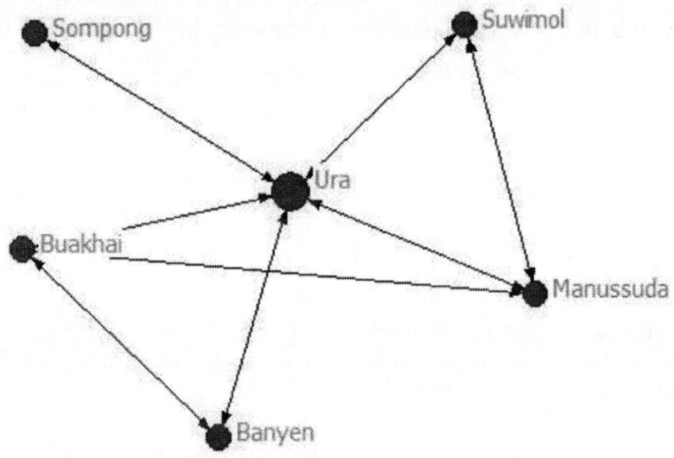

Figure 5. model of Karn Keha Khon Kaen community network.

Considering both community networks, the center of Non Nongwat community network was *Jaroon*, and the center of Karn Keha Khon Kaen community network was *Ura* respectively. When considered relationship pair between centered-persons and their network members, both of them interacted closely with all members within the network. Considering the role within the network, found that they were community leaders who had an outstanding role in implementing low carbon emission activities in community especially the Low Carbon City program and the Zero Waste Office program. In 2018, the two communities qualified to the final round of the national waste-free community competition, which sponsored by the Ministry of Natural Resources and Environment. Both communities won the prizes. Non Nongwat community won the 1st runner up of the big-size community category, and Karn Keha Khon Kaen community won the 1st runner up of the medium-size community category.

Regarding the benefit exchanges within both community networks, the benefit exchanges were different. The network members in both communities exchanged benefits through the waste bank activity. All members of the network were committees of the community waste bank, and they got paid for the waste bank activity in monthly basis. Although they got few amounts of payment, they had benefits of being a committee in which it was a social privilege in the community. This privilege represented their attempts to help the community becoming a waste-free community that won the national competition. However, researchers found the differences between the two communities in terms of benefit exchange. Non Nongwat community had entrusted its members to maintain cleanliness on main road of the community, where also was a main road of Khon Kaen city. Three network members, *Wichan*, *Jansorn*, and *Sakorn*, got trust from *Jaroon* to take care of the road. Their roles were to keep cleanliness of the road and observed general incidents on the road. They received monthly payment from the Khon Kaen Municipality Office in accordance with a decentralization policy on "Small municipality in a big municipality" which mentioned in the previous topic.

In terms of frequency and duration of relationship, although both urban communities were comprised of residents who came from many places, those residents had stayed in communities for a decade. They met each other and always participated in community, so relationship of network members was quite strong. These relationships of the network members were linked by the network center. Some members had to use the center to link relationship to other members. For instance, *Sompong*, the community philosopher of Karn Keha Khon Kaen community, who had close relationship and worked with *Ura*. When he wanted cooperation from other network members, he always asked *Ura* for a help. It was considered as the cut point of the network [20] (see Figure 5).

Regarding the direction flow of relationship, direction was clearly spread out from the center. Both *Jaroon* and *Ura* were leaders who commanded and asked the network members for cooperation, but they did not command the network by using priority of leader position. They used a relationship as "neighbors" or "siblings" in developing their communities. It was reflected by a statement of *Banyen*, a village health volunteer, as below.

> *"We came from different places. It was difficult to find Khon Kaen born-people in our community. Since we have lived together, relationships were getting closer. When Ura volunteered to work as community leader and leader of the village health volunteer, we all agreed with her. We were happy to do things when Ura needed our help"* (interviewed on August 15, 2018)

Moreover, patterns of both community networks were performed as the density knit, since relationships between members in the network related to each other. At least 2 relationship pairs were presented between the center and the members (except a case of *Sompong*). This reflected the strength of the social network which correlated to the number of individuals who had relationship to each other [18],[20].

6.2 Dual-center Network

Dual-center Network is a pattern of community network with two centered-persons who served as a link within social network. This pattern of community network was found in Ban Bakham community. Model of community network was presented as figure 6.

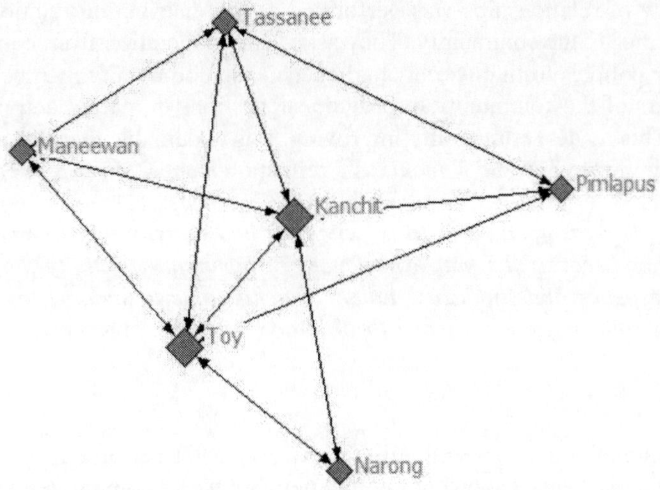

Figure 6. model of Ban Bakham community network

Considering pattern of community network, the center of network was *Toy*, who was community leader for more than 10 years, and *Kanchit*, who acted as the right-handed man of *Toy*. Considering the number of relationship pairs within the network, both the centered-persons had closely relationships with all the members in the network. They helped each other to run activities of the community. *Toy* was those who run the Zero Waste Office program. In addition, Ban Bakham community was a pilot community since 2013. The community was able to develop bio fuels from organic waste before they transferred technology to Khon Kaen Municipality Office in order to avoid any dangers from the production process within community.

Considering the benefit exchanges within community network, the exchanges were similar to Non Nongwat community and Karn Keha Khon Kaen community. The network members got benefits from the waste bank program. In addition, some network members, Narong and Tassanee, were assigned to oversee main road in community. They received monthly payment from the Khon Kaen Municipality Office as well.

In terms of frequency and duration of relationship, Ban Bakham community was originally a suburban community, where people had rural lifestyle. When Khon Kaen city grew, urbanization invaded a community until it became an urban community. Community members were those who have lived for decades and the "new comer" who mostly were young students from various provinces, studying in Khon Kaen city. There are 24 student dormitories in community. Although they came from different places, they have stayed long enough until they had closely relationships to each other. They participated in community activity and worked together. It resulted in the relationships between community members were quite strong. It is noticed that members of this community network had more than 2 relationship pairs. *Kanchit* stated about this matter as below.

> *"I came to this community as a son in law twenty years ago. I firstly worked alone, and Toy later invited me to work for community. I had a good feeling when I did something good to others....The people here were the elderly. They knew each other very well. The community committee here worked together for a long time. They never had a serious fight between them"* (interviewed on July 10, 2018)

The direction flow of relationships was performed as the distributing relationship. Both *Toy* and *Kanchit* worked together in the community. They asked for cooperation from community members by distributing orders or policies from governmental agencies to community members. At the same time, they allowed members of the community to provide suggestions or started activities that were useful for the community. This made a community improving. It is undeniable that those who played a major role in decision-making were *Toy* and *Kanchit* as a reflection from *Maneewan* below.

> *"... Normally Toy assigned roles to us when we had activity or community project. We trusted her because she had experiences in community activity. She made our community in peace and solidarity. Later, Kanchit came to support toy's work. So, works within community were more flexible and smooth"* (interviewed on July 10, 2018)

6.3 Full network

Full network is a community network without the network center because all the members have good relationships and solidarity. Hua Tanon community network was a community with full network. The community was a suburban community of Khon Kaen city, where had been widely recognized as a waste-free community. The community won the King Rama 9th national championship in 2015 which brought the most proud to people in community. Model of community network was as shown in Figure 7.

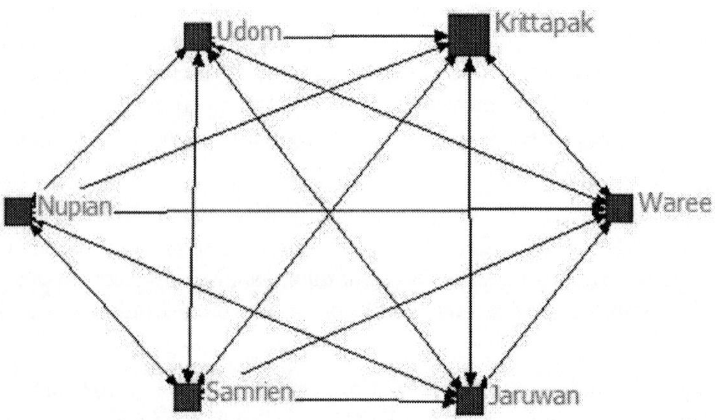

Figure 7. model of Hua Tanon community network.

This pattern of community network assigned equally role to members in community network. All the members played important role in driving carbon emission reduction in urban areas especially the development of a waste-free community program until the community achieved at national level. Udom, the ex-community leader who worked along with *Kritapak*, a current leader, gave statement below.

> *"... We were working together like brothers. We all knew well about our roles and functions. When our tasks finished, we would help others. We did everything together..."* (interviewed on July 18, 2018)

Considering the benefit exchanges within community network, the benefit exchanges were done through the waste bank program as same as other communities. Hua Tanon community developed the program, and they were able to develop a waste management learning center, funded by various agencies. The community became a learning source for those who interested over the country. On average, there were at least 20 visits per month from agencies over the country. This stimulated community members to be active and always welcomed guests. Community members also had chance to get small incomes from being a lecturer at the learning center and souvenirs. These incomes became a motivation for community members to work for community.

Considering frequency and duration of relationship, the community was a suburban community with those who have lived together for decades. So, they were very close friend. Community members worked together for a long time, and there were community activities every week. They also had to take care of visiting groups from over the country. This caused relationship of network members was very strong. it was noticed that all the network members had relationships to each other as a full network [18], [20].

According to above pattern, direction flow of relationship of Hua Tanon community network was dynamic. Relationship was not depending on any community network members. *Krittapak* was those who represented as a community leader, and he did any actions based on the position he took.

In addition, when considering the connection among four community networks, it was found that the centers of the community network serve as a network connecting point, whether *Jaroon* of Non Nong Wat community network, *Ura* of Karn Keha Khon Kaen community, *Toy* and *Kanchit* of Ban Bakham community network, and *Krittapak* of Hua Tanon community network. Model of Khon Kaen urban community network is presented as in Figure 8.

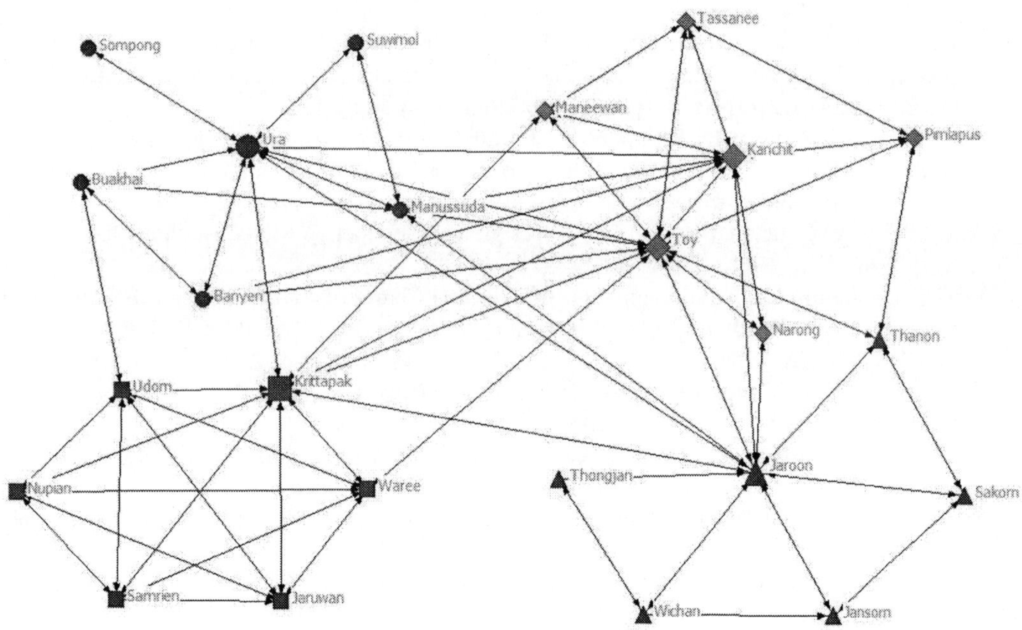

Figure 8. model of Khon Kaen urban community network.

In conclusion, patterns of community network in carbon emission reduction are identified into 3 patterns including 1) one-center network-Non Nongwat community and Karn Keha Khon Kaen community, 2) dual-center network-Ban Bakham community, and 3) full network-Hua Tanon community

7. Suggestions

The result points out that the social network of the Thai-Isan communities is a key tool to carbon emission reduction in terms of recycles practices, waste management, and quality of life upgrading. The social network is also an important mechanism that provides benefits to community members through green environment and some benefit from activities. Therefore, if the Khon Kaen Municipality and other Thai government agencies apply functions of the social network to empower the social network of community through the supporting and decentralization, it will be opportunity to improve Thai-Isan Urban to carbon emission reduction. In addition, empowering social capitals through the social network is beneficial to urban people who can construct and extent the network to other community, and are able to live together in society sustainably.

8. Acknowledgements

This research article is a part of research project entitled "Community Network for Carbon Emission Reduction in Isan Urban". The research project has been funded by the 2018 Research Supporting Fund (Cooperative Research with the Top 400 University of the World), Faculty of Humanities and Social Sciences, Khon Kaen University, Thailand.

References

[1] Ayuwat D and Chamaratana T 2013 The Role of Labour Broker Networks in Setting the Price of Working Abroad for Thai Migrant Workers. **Asia-Pacific Population Journal 28** pp 51–68

[2] Boissevain J 1974 Friends of Friends: Network, Manipulators and Coalitions (Oxford: Basil Blackwell)

[3] Certeau M D 1984 The Practice of Everyday Life (Berkeley: University of California Press)

[4] Chamaratana T, Ayuwat D and Chinsri O 2013 "Plural characteristics of social network of Isan labor brokers" **Kasetsart Journal (Social Sciences) 34** pp 1–16

[5] Chamaratana T, Ayuwat D, Knippenberg L and Jong E D 2010 Connecting the Disconnected: Background, practices and motives of labour brokers in Isan, Thailand - an explorative study **The International Journal of Interdisciplinary Social Sciences 5** pp 359 – 372

[6] Creswell J W and Plano-Clark V L 2007 Design and Conducting Mixed Methods Research (New York: SAGE)

[7] IPCC 2007 Summary for Policymakers. In: Climate Change 2007: The Physical Science Basis. Contribution of Working Group I to the Fourth Assessment Report of the Intergovernmental Panel on Climate Change [Solomon, S., D. Qin, M. Manning, Z. Chen, M. Marquis, K.B. Averyt, M.Tignor and H.L. Miller (eds.)] (Cambridge University Press, Cambridge, United Kingdom and New York, NY, USA)

[8] IPCC 2014 Summary for Policymakers. In: Climate Change 2014: Mitigation of Climate Change. Contribution of Working Group III to the Fifth Assessment Report of the Intergovernmental Panel on Climate Change [Edenhofer, O., R. Pichs-Madruga, Y. Sokona, E. Farahani, S. Kadner, K. Seyboth, A. Adler, I. Baum, S. Brunner, P. Eickemeier, B. Kriemann, J. Savolainen, S. Schlömer, C. von Stechow, T. Zwickel and J.C. Minx (eds.)] (Cambridge University Press, Cambridge, United Kingdom and New York, NY, USA)

[9] Jong E D 2008 Living with the death: The Economics of Culture in the Torajan Highlands, Indonesia (Nijmegen: CIDIN)

[10] Khon Kaen Municipality 2016 **Khon Kaen Municipality Strategy Plan.** Accessed on 23 October 2017 at http://www.kkmuni.go.th/2017/admin/file/files/1486108398_1269816638.pdf

[11] Kilduff M and Tsai W 2003 **Social Networks and Organizations** (London: SAGE Publication)

[12] Milner J, Davies M and Wilkinson P 2012 Urban energy, carbon management (low carbon cities) and co-benefits for human health. **Current Opinion in Environmental Sustainability 4** pp 398–404

IOP Conf. Series: Journal of Physics: Conf. Series **1340** (2019) 012045 doi:10.1088/1742-6596/1340/1/012045

[13] Moriarty P and Wang S J 2014 Low-carbon cities: Lifestyle changes are necessary **Energy Procedia 61** 2289–2292

[14] Nooy W D, Batagelj V and Mrvar A 2005 Exploratory Social Network Analysis with Pajek (Cambridge: Cambridge University Press)

[15] Phatchaney K 2017 Livelihood Security of Labor Households in Khon Kaen Peri-Urban Master of Art Thesis in Social Development, Graduated School, Khon Kaen University

[16] Promphakping B 2018 The Origin and Evolution of Fresh Marketand Petty Traders in Khon Kaen City (Khon Kaen: Print sri sood suay)

[17] Public Health and Environment office 2013 Performance Khon Kaen People Heart free Project Report (Khon Kaen: Khon Kaen Municipality Printing)

[18] Ritzer G 1992 Sociological Theory 3rd ed (New York: McGraw-Hill, Inc)

[19] United Nations 2015 World Urbanization Prospects: The 2014 Revision (New York: Economic and Social Affairs)

[20] Wasserman S and Faust K 1999 Social Network Analysis: Methods and Applications 4th ed (New York: Cambridge University Press)

[21] World Bank 2010 Eco2 Cities: Ecological Cities as Economic Cities (Washington: The world bank)

[22] Zhang X Q 2016 The trends, promises and challenges of urbanization in the world **Habitat International 54** pp 241–252

Developing the Garbage Problem in Iligan City STEM Education Lesson Through Team Teaching

Ivy Claire Mordeno[1*], Algin Michael Sabac[1], Angel Jane Roullo[1], Hazel Dwight Bendong[1], Amelia Buan[2], and Chokchai Yuenyong[3]
[1]Integrated Developmental School, College of Education, MSU – Iligan Institute of Technology, Iligan City, 9200 Philippines
[2] Department of Science and Mathematics Education, College of Education, MSU – Iligan Institute of Technology, Iligan City, 9200 Philippines
[3] Faculty of Education, Khon Khaen University, Thailand
*Corresponding author's e-mail address: ivyclaire.mordeno@g.msuiit.edu.ph

Abstract. The paper will clarify the learning activities which provide through Sutaphan nad Yuenyong [11] context based STEM education learning approach . This approach consists of 7 stages including (1) Identification of social issues, (2) Identification of potential solution, (3) Need for knowledge, (4) Decision-making, (5) Development of prototype or product, (6) Test and evaluation of the solution, and (7) Socialization and completion decision stage. This will be implemented using team teaching approach by the Earth Science, Physics, Chemistry, and Biology teachers. Identification of social issues, the earth science teacher raises the social issue about the garbage problem affecting the residents of Brgys. Bonbonon and Digkilaan, Iligan City, Philippines. Students may answer that the probable source could be the leachate from the sanitary landfill and from there the teacher will put the problem causes problem to the people in the locality where the landfill is located and to the City's tourism as well. Identification of potential solution, the Earth Science teacher facilitates the class discussion about thinking of possible solutions to help solve the problem on garbage in Iligan City. The need for knowledge stage, the input of the scientific knowledge necessary for the learners' decision in designing their prototype or product in solving the problem on garbage in Iligan City will be done separately in Physics, Chemistry, and Biology class. Physics teacher may provide scientific inquiry about force and compression to help the students design a prototype of a garbage/trash compactor. Chemistry teacher may provide learning activities about classification of mixtures according to particle size helps the students think of a design for a water treatment process. Biology teacher may provide learning activities related to macroinvertebrates and a bioindicator of the water quality of the streams/rivers. The decision making stage, lets the students decide on how they are going to make the required prototype for Physics, Chemistry and Biology aspect as possible solutions to the problem. The development of prototype or product stage, students are given enough time to create the compactor, water treatment process and the poster-infographic about macroinvertebrates. Test and evaluation of the solution stage, students will apply their knowledge related to their prototype or product for test and evaluation of the solution. The socialization and completion decision stage, the students' will have an exhibit of their output. The paper will discuss how to provide students chance to applying STEM knowledge through these activities.

1. Introduction

The Philippines had just implemented the K to 12 Basic Education Program. It aims to create a functional basic education system that will produce productive and responsible citizens equipped with the essential competencies and skills for both life-long learning and employment. Today, the education system tries to impose on students the attitude of lifelong learning. Life-long learning (LLL) as viewed involves all strategies that are put in place to create opportunities for people to learn throughout life by providing both the individual needs and that of the relevant community [1]. Lifelong learning can be associated to the following notions: (1) individuals take ownership of the need for and the content of learning, (2) learning focuses on how students think rather than on what to think about, (3) teachers become role models and mentors for lifelong learning rather than distributors of information, (4) evaluation encourages students for self-development and becomes a facilitator in the process rather than classifying them according to a norm, and (5) learning is considered as an enjoyable and essential part of the individual [2]. Methods such as problem-based learning, conscious learning, mutual teaching and cognitive apprenticeship, and such naturally motivating activities and similar teaching strategies as cooperation, reflection and student autonomy are employed by individuals in order to remain updated in today' world [5]. It is also important to note that problem solving, critical thinking, and reasoning skills are essential to lifelong learning for these skills allow learners to acquire the knowledge, skills and attitudes necessary not only for academic success but also for their overall well-being [7]. In order to help students become lifelong learners, the schools should be organized in such a way that the students will have the opportunities to develop these skills.

The science curriculum framework for basic education in the Philippines has three primary themes: maintaining good health and living safely; utilizing energy and coping with changes; and conserving and protecting the environment which are used in various real life contexts across grade levels. The framework embeds science concepts in the context of students' daily life to help them develop a more clear understanding of the concept or process they are learning. Connecting learning to events and activities in students' lives enables them to develop understandings that will last a lifetime. In order for learners to apply science into context, teaching science should not be compartmentalized so that they will have opportunities to make connections for a deeper understanding of the importance and application of science and technology in everyday life. As the interaction among science, technology and society are learned, students begin to appreciate that technological design and problem solving involves many factors besides science. Teachers then need to communicate that science alone cannot solve human problems or meet human needs but it has to be integrated with other disciplines [9].

Studies have noted that science process skills are effective on teaching and learning about Science [3]. Teaching strategies such as inquiry teaching, problem solving, problem based learning and project based learning relies heavily on the effective use of the science process skills by students to complete in an investigation [4]. Problem solving is considered as the most important constructivist learning method [6]. Constructivist course goals should provide realistic problems that elicit social cognition, facilitate student application of external knowledge sources, and encourage students to utilize scientific reasoning. Students are more likely to persist in pursuit of problem solutions if they feel efficacious, that is, if they judge themselves as capable of solving the problem [10]. Self-efficacy can be promoted by encouragement, but is best promoted by actual problem solving success. Students, as well, can apply their scientific knowledge when trying to find a solution about the problem found in the society.

2. Developing STEM Education Learning Activities

The developed Lesson Plan is a collaboration among Physics, Chemistry & Biology teachers to address the issue on the garbage problem in Iligan City. All of the concepts were based from the Grade 12 Curriculum Guide of the three specified subjects. The concept on Pressure and Compression, Separating Mixtures, and Macroinvertebrates all aims that students learn the subject on the relation between concepts in science, technology, engineering and mathematics. The lesson plan was developed on concept of Context-based STEM education learning approach which the STEM

IOP Conf. Series: Journal of Physics: Conf. Series **1340** (2019) 012046 doi:10.1088/1742-6596/1340/1/012046

education learning activities should provide not only the ways of investigation and solving problem but also a real world problem solving. Regarding on Sutaphan and Yuenyong [11], the context-based STEM education learning approach included (1) Identification of social issues, (2) Identification of potential solution, (3) Need for knowledge, (4) Decision-making, (5) Development of prototype or product, (6) Test and evaluation of the solution, and (7) Socialization and completion decision stage.

Regarding on the 7 stages of context based STEM education teaching approach, the activities could be started from social issues and/or human needs such as disasters, pollutions, environmental issues, biotechnology, health, cosmetics, ecosystem, energy, market, commerce, and designing some technological products. The 7 stages will allow students using applying scientific and other knowledge for designing the solutions, and provided the context of instruction requires solving a real-world problem or task through teamwork. And, students will have also chance to apply their scientific and other knowledge for problem solving in context of engineers, technology, or entrepreneurship [11].

2.1 Purposes of the Lesson Plan

1. Explain the scientific principles about Pressure & Compression, Separating Mixtures, and Macroinvertebrates.
2. Explain the factors affecting the compression of objects.
3. Investigate ways of separating dissolved solids from a mixture classification.
4. Identify the macroinvertebrates which could be present in streams/rivers and explain what it tells about the quality of the river/stream.
5. To design a manual garbage/waste compactor (Physics), an infographics about macroinvertebrates (Biology), and a water treatment process (Chemistry).
6. Select suitable materials for creating the model.
7. Explain the process of creating the prototype/product.

2.2 The activities of Garbage Problem in Iligan City STEM Education Lesson

The garbage problem in Iligan City STEM Education was developed through team teaching as provided in the table 1.

Table 1. Lesson Plan on Garbage Problem in Iligan City adapting the Context-based STEM education learning approach

Stage	Activity
1. Identification of Social Issue	1. Ask students to discuss hazards of improper waste segregation and disposal. 2. Teacher raises the issues, "How can students minimize the size of the garbage sent to sanitary landfills?", "What pollutants are present in the landfill and how can it affect humans and the local ecosystem?", "What methods can be used to treat waste water or leachate?" 3. The teacher raises the social issue about the garbage problem affecting the residents of Brgys. Bonbonon and Digkilaan, Iligan City. Due to improper waste management and design of the Compost Material Recovery Facility (CMRF), the sanitary landfill in the area is being filled fast and the leachate contaminates the river where the local people source their livelihood. Also, the Dodiongan Falls is located downstream and is also affected which could affect the local tourism.
2. Identification of Potential Solution	1. Students and teachers discuss about a. the design of a simple garbage/waste compactor considering the use of materials that are easy to find, its cost & functionality. **[Physics]** b. the sequence of processes that can be used to clean waste water. **[Chemistry]** c. the design of an poster-infographics showing the different macroinvertebrates. **[Biology]**

IOP Conf. Series: Journal of Physics: Conf. Series **1340** (2019) 012046 doi:10.1088/1742-6596/1340/1/012046

Table 1. (Cont')

Stage	Activity
	The topics will be discussed separately to grade 10 leaners in their Earth Science Class as a requisite discussion about Surface and Groundwater. Each of the lesson will be conducted for an hour and the class will be divided into 4 groups with 8 members each.
3. Need for Knowledge	**PHYSICS (1 hour)** 1. The related knowledge that will be raised includes Pressure & Compression. 2. Students may do some activities about pressure to explain that solids are hard to compress because the molecules are closely packed together – they have high density, thus it requires greater pressure in order to reduce its size/volume. 3. Do Activity 1 Pressure 3.1 Give the students push pins. Let them think of these two situations. 4.1.a. Pushing a drawing pin into the table with pointy ends towards the table. 4.1.b. Pushing a drawing pin into the table with pointy ends towards your thumb. 3.2 What result did they get? Students will write down their answer on a paper. 3.3 The teacher and students discuss the reason for the difference in pressure according to the activity. Assuming that the same force is applied, each case would have a different pressure acting on the thumb. In the first situation, the thumb pushes on a large area so the force is spread out and the pressure is low. In the 2nd situation, the force is concentrated on a small area so the pressure is much higher. Given the relationship $\boldsymbol{P} = \frac{\boldsymbol{F}}{\boldsymbol{A}}$, it can be deduced that Pressure is directly proportional to the Force applied and inversely proportional to the surface area.
	CHEMISTRY (1 hour) 1. Students will review their previous knowledge on classification of mixtures according to particle size (solution, colloids, suspension). 2. Students will think of ways to separate the dissolved solids from each mixture classification. 3. Do Activity 2 separation of mixtures 3.1 Group the class into 5 members. Give each group 3 containers each with milk, salt solution, and muddy water. 3.2 The students will be tasked to answer pertinent questions on their worksheets. Questions to be answered will be as follows: (1) how will you classify the three mixtures presented? (2) What are the properties of each classification (solution, colloid, suspension)? (3) What possible methods can be used to separate the water from the other component(s)? (4) Is there a process that can be used to separate mixtures regardless whether it's a solution, colloid or suspension? Why or why not?

Table 1. (Cont')

Stage	Activity
	3.3 Provide each group with filter paper, layer of rocks and sand in a container. Let the students use the above-mentioned materials to separate the 3 mixtures previously given to them. Students write down their conclusion on worksheets. 3.4 Teacher and students discuss their answers and have an open discussion on why particle size of the dissolved or suspended component matters in determining the separation method to be used. 3.5 Teacher will explain that solutions have suspensions contain solids that are big enough to be affected by gravity. The suspended solids in this type of mixture can be separated from the liquid component by simply allowing it to stand for a certain amount of time. Filtration can also be employed to separate the components of a suspension. Colloids, on the other hand, have dispersed particles that are much smaller than suspensions and hence, do not settle out. The colloidal particles can be removed from the mixture by adding coagulants or flocculants, i.e. alum, which help the particles to form aggregates that precipitate out of the mixture. 3.6 Among the three, solutions have the smallest particle size. Components of a solution mixture are usually separated through distillation.
	BIOLOGY (1 hour) 1. The students will review their previous knowledge about macro invertebrates present in the river 2. Students will use the Biotic Index of Water Quality and will familiarize which group or level these macroinvertebrates belong.
4. Decision-making	**PHYSICS** 1. The students design a manual garbage/waste compactor. They should apply scientific principle, force, pressure and other knowledge for presentation to the class. 2. Let them draw the draft of their design on a paper and indicate the materials that they are going to use and the reason behind using those.
	CHEMISTRY 1. So as to simplify the problem, the students will only be tasked to design a water treatment process. Students will apply scientific knowledge in classifications of matter and separation techniques. 2. Students will work on a draft of the waste water treatment process on paper. They would also indicate the materials that they will be using and justify why they will use those materials.
	BIOLOGY 1.The teacher will group the students with 5 members each. They will design a Digital Infographics about Macroinvertebrates as Bioindicators of Water Pollution in streams/rivers and will post it to their social media account (e.g. Facebook, Instagram, etc.) to maximize the spread of information about the importance of these macroinvertebrates.

IOP Conf. Series: Journal of Physics: Conf. Series **1340** (2019) 012046 doi:10.1088/1742-6596/1340/1/012046

Table 1. (Cont')

Stage	Activity
	2. The students will apply all their knowledge about Macroinvertebrates present in the streams/rivers as well as the classifications of these organisms (Pollution-Sensitive, Moderately Pollution-Sensitive, Pollution- Tolerant) 3. Students will think of their design and content of their Digital Infographics in a piece of paper.
5. Development and prototype or product	**PHYSICS** The students will create a simple waste compactor using the materials that they have proposed.
	CHEMISTRY Students will design a water treatment process. They will be given freedom to choose their materials.
	BIOLOGY The students will create a Poste-Infographic that will present the information about the Macroinvertebrates as a Bioindicator of Water Quality in the streams/rivers. The students must use the following: Laptop with Adobe Photoshop/Microsoft Office and Internet connection.
6. Test & Evaluation of the solution.	**PHYSICS** Let the students develop ideas of how their compactor will work. They need to test the following issues: 1. What maximum volume/size of waste material can the compactor compress or reduce in size? 2. By what factor does of the volume of the trash/waste reduced (compaction ratio)?
	CHEMISTRY The teacher will test their prototype by letting them treat provided muddy water. The students will be asked to conduct turbidity and BOD tests of their influent and effluent streams. They need to test for the following issues: 1. Does the water quality of the effluent pass the minimum requirement required by the Department of Environmental and Natural Resources (DENR)? 2. If the prototype is implemented in real life, how much would it cost?
	BIOLOGY The teacher will rate/ critique the Infographics done by the students based on the rubrics provided.
7. Socialization and completion decision stage	1. The class will exhibit their final prototype – the compactor, a water treatment process and poster-infographic. 2. Present how to test and evaluate the solution. 3. Present how the prototype can be explained as the real one. 4. Share what they learn from the comment and what they will revise for the completion solutions.

This lesson is designed for 10th graders who are taking up Earth Science. The Earth Science class is conducted for one (1) hour daily. This will be implemented using team teaching approach by the Earth Science, Physics, Chemistry and Biology teachers. The Earth Science and students discusses first about Ground and Surface Water. The Earth Science teacher raises the issue on the garbage problem in Iligan City which could affect surface and Ground Water. With the help of the Physics, Biology &

Chemistry teachers, the students identify possible solutions to the problem using scientific concepts. This will be implemented to two (2) Earth Science classes comprising of 33 students per class.

The table below shows the class schedule and activities per day during the implementation period. Each class will be divided into 4 groups and each group will be tasking themselves in the development of the product/prototype for the solution in the Physics, Chemistry & Biology aspect. It is also expected that the students will make a write-up about the results of the testing & evaluation of their prototype.

Table 2. Class activities during the lesson implementation period.

	Monday	Tuesday	Wednesday	Thursday	Friday
Week 1					
7:30 – 8:30 8:30 – 9:30	Stage 1 & 2 Identification of the Social Issue & the Potential Solution	*Physics* Stage 3 & 4 (Need for Knowledge and Decision Making)	*Chemistry* Stage 3 & 4 (Need for Knowledge and Decision Making)	*Biology* Stage 3 & 4 (Need for Knowledge and Decision Making)	Stage 5 Development of prototype or product.
Week 2					
7:30 – 8:30 8:30 – 9:30	Stage 6 Test and evaluation of Prototype or Product	Stage 6 Test and evaluation of Prototype or Product	Stage 7 Exhibit of their prototype or product		

Stage 1. Identification of Social Issue

It is assumed that the discussion of ground and surface water is already done the previous meetings. In this stage, the Earth Science teacher is going to ask first the students about the possible sources of contamination/pollution in the ground and surface water in general. The students may answer that the probable source could be the leachate from the sanitary landfill and from there the teacher will put the problem causes problem to the people in the locality where the landfill is located and to the City's tourism as well. The teacher and students will discuss the big issue of the garbage problem of the city.

The Problem

The residents of Brgy. Bonbonon and Brgy. Digkilaan felt that their livelihood, food security and safety are threatened by the landfill leachate coming from the Central Materials Recovery Facility in Sitio Bangko, Bonbonon, Iligan City. This is because of the excessive amount of garbage that are collected from the whole vicinity of the city itself. Since then, residents started to see black and brown liquid being carried by the streams into the falls and nearby rivers which were approximately a couple of kilometers below the landfill.

Residents complained that the contamination has caused skin diseases to some and it would contaminate their drinking water in the long run and they fear this would discourage tourists and affect their means of living because they rely primarily on springs and deep wells. The city government has yet to issue a statement on their long- and short-term solution to the leak, and the rehabilitation.

Figure 1. Central Recovery Facility of Iligan City

Upon identifying the social issue, questions such as "How can students minimize the size of the garbage sent to sanitary landfills?", "What pollutants are present in the landfill and how can it affect humans and the local ecosystem?", "What methods can be used to treat waste water or leachate?" will be asked which leads the students to think of possible solutions to the social issue identified.

Stage 2. Identification of Potential Solution

In this stage, the Earth Science teacher facilitates the class discussion about thinking of possible solutions to help solve the problem on garbage in Iligan City. The students may think of solutions in the personal level, in the households and as a community. The possible solutions that may arise could be that the garbage to be brought to the landfills be reduced in size using a compactor, the leachate from the landfill be treated before throwing to the river or let the people be aware that they can determine the quality of the water in the river or streams using macroinvertebrates and many others.

Stage 3. Need for Knowledge

The input of the scientific knowledge necessary for the learners' decision in designing their prototype or product in solving the problem on garbage in Iligan City will be done separately according to the time allotment shown in Table 2.

Physics

The discussion on force & compression helps the students design a prototype of a garbage/trash compactor. It is assumed that the students already have learned the concept of force in the previous grade level. The discussion starts with Pressure which is produced by applying a force on a certain area. It is important to note also that the application of force will cause a deformation on the object – changes the size and form of the object. It must be emphasized that the compression is dependent on whether the object is solid, liquid or a gas. A simple activity (see Table 1) to enhance the understanding of the learners as well giving them the opportunity to think will be given. The activity lets the learners describe what they have observe when trying to push the drawing pin against the table with the pointy end on top or the other way around. The teacher leads the students to explaining that when a force is applied to a small area it results to a bigger pressure and when the same amount of force is applied to a bigger area it results to a smaller pressure. The bigger the pressure the greater the strain (deformation) and vice versa. These concepts help the students decide how to design the compactor and choose the cost-effective materials.

Chemistry

Discussion on classification of mixtures according to particle size helps the students think of a design for a water treatment process. It is assumed that the students have had prior knowledge of solutions, colloids and suspensions from their previous grade level. The discussion starts by reviewing the classification of matter before focusing on mixtures. It will be emphasized to the students that what sets mixtures apart from pure substances is that the components of the former can be separated through physical means.

A simple activity (see Table 1) to enrich the understanding of the students and encourage critical thinking will be given. The activity lets the students observe the properties exhibited by each type of mixture and think of ways on how the separate the suspended or dissolved solids from the liquid component. For the learners to demonstrate the physical methods they suggested, additional materials will be provided.

The teacher leads the students to realize that suspended solids in a suspension, having the largest particle size, can be separated with the help of gravity by allowing it to stand for a certain amount of time or through filtration. Colloidal particles are much smaller than suspensions and can be removed from the mixture by adding coagulants or flocculants which help the particles to form aggregates that precipitate out of the mixture. Lastly, solutions have the smallest particle size and are usually separated through distillation or evaporation.

Biology

The previous knowledge of the students about Macroinvertebrates will serve as a guide for them in the identification process knowing that by definition, macroinvertebrates are organisms that are large enough to be seen with unaided having no backbone. Further discussions about macroinvertebrates will be about the qualities that made these aquatic organisms a bioindicator of the water quality of the streams/rivers, how will these aquatic organisms be collected in terms of their morphology and classified in terms of their tolerance level (pollution-sensitive, moderately pollution-sensitive, and pollution-tolerant).

The teacher will present the Macroinvertebrate Biotic Index chart and will lead the students to think and make conclusions about how the presence of a certain macroinvertebrate indicates the health of the stream/river.

The information about these macroinvertebrates as a bioindicator will not be caged in the four corners of the classroom and so is the reason why the students will be given an activity to make a poster infographic to further disseminate the information on how these macroinvertebrates function in the ecosystem aside from being a food source from other higher trophic organism such as fish. By having all of those ideas and information, the students will then think of an infographic design that is powerful enough to allow the people to easily understand and digest the information about Macroinvertebrates as a bioindicator of water quality with the proper use of color, words, attention-grabbing and persuasive graphics and images.

Stage 4. Decision-making

This stages lets the students decide on how they are going to make the required prototype for Physics, Chemistry and Biology aspect as possible solutions to the problem. The class is divided into 4 groups and the groups must have a clear tasking to ensure that each of the members has a part to perform.

Physics

To help the students design their prototype, the things to be considered must be presented. It is important to stress that the materials to be used must be readily available and at low cost, the design should be sturdy or with strong support and it should have a big compaction ratio (compressed volume/original volume). The learners must give the rationale of the design and the use of materials; or that it must be supported with scientific concepts. A draft of the design is to be presented first and will be approved by the Physics teacher.

Chemistry
The teacher will remind the students of the things to be considered in designing their prototype. It is necessary to emphasize that in choosing materials, those that are readily available is preferred. The students are also reminded to consider the cost and efficiency of each material. The effluent of the water that they will be treating must pass the minimum requirements of the DENR and DOH. The students must explain and justify their design and materials chosen through sound scientific reasoning.

Biology
To help the students design their poster infographic, they need to consider the materials needed for it to actualize the project. The students must first put everything into writing all the information they will present and how the design would look like. The students must give highlight to the scientific information about the Macroinvertebrate and not just merely focus on the design itself.

Stage 5. Development of the Prototype or Product

The student are given enough time to create the compactor, water treatment process and the poster-infographic about macroinvertebrates. Each group should have a clear tasking of who is going to be in-charge of the development of each of the required prototype or product. They already have the design and it is now ready for development. It is understood that as scheduled (see Table 2), the students bring with them the materials needed for the task.

Stage 6. Test and Evaluation of the Solution

In this stage, students will apply their knowledge related to their prototype or product for test and evaluation of the solution. This will help the students improve their prototype or product and make the necessary actions before it will be exhibited.

Physics
The garbage/trash compactor will be tested according to its functionality – its ability to compress. Its ***compaction ratio*** will be determined by dividing the new volume by the original volume of the compressed object. The maximum volume that the compactor can compress will also be determined. Each group's compactor will be tested to similar materials like carton, sheets of paper, soft plastic containers, and trash available in the trash bins of the school premises. Based from the output of testing the compactor, necessary improvements will be made.

Chemistry
The students' design of the water treatment process will be tested by determining if the effluent's turbidity and BOD tests passed the Department of Health and Department of Environment and Natural Resources' requirements.

Biology

The students' poster-infographic will be evaluated using an adapted rubric. The poster-infographic will be posted on a certain location in school and comments from the evaluators will be solicited for the improvement of the product. Sample Rubric for the poster-infographic is shown in Table 3.

Table 3. Rubric for Evaluating the Poster-Infographic.

Category	4	3	2	1	0	Total
Content Accuracy	At least four accurate facts about macroinvertebrates are displayed on the poster-infographic	At least three accurate facts about macroinvertebrates are displayed on the poster-infographic	At least two accurate facts about macroinvertebrates are displayed on the poster-infographic	Less than two accurate facts about macroinvertebrates are displayed on the poster-infographic	No accurate facts about macroinvertebrates are displayed on the poster-infographic	
Graphics	All graphics are related to the topic and make it easier to understand.	All graphics are related to the topic but there are those that cannot be understood.	All graphics are related to the topic.	Some graphics are related to the topic.	No graphics are related to the topic.	
Layout/ Design	Makes excellent use of font, color, graphics, effects, etc. to enhance the presentation.	Makes good use of font, color, graphics, effects, etc, to enhance the presentation.	Makes some use of font, color, graphics, effects, etc. to enhance the presentation	Makes little use of font, color, graphics, effects, etc. to enhance the presentation.	Makes no use of font, coolor, graphics, effects, etc. to enhance the presentation	
Grammar	No misspellings or grammatical errors.	Two or fewer misspellings and/or mechanical errors.	Three misspellings and/or grammatical errors.	Four or more spellings or grammatical errors.	Piece so filled with errors that meaning is undetectable.	
Mechanics	Capitalization/punctuation are current and consistent throughout the poster infographic	Two or fewer capitalization/ punctuation errors exist in the poster infographic	Three capitalization/ punctuation errors exist in the poster infographic	Four capitalization/ punctuation errors exist in the poster infographic	Piece so filled with errors that meaning is undetectable	
Required Elements	The poster infographic includes all required elements as well as additional information	All required elements are included on the poster infographic	Some required eleemts are included on the infographic	Very few required elements are included on the infographic	No required elements are included on the infographic	
					TOTAL	

Stage 7. Socialization and Completion Decision Stage

The students' will have an exhibit of their output. They will provide a poster stating the rationale of creating the prototype, a brief explanation of the process of designing the prototype and the results of the test and evaluation. They will also include how they incorporate the comments and suggestions based from the result of the test and evaluation of the prototype.

3. Conclusion

This paper showed how to provide STEM education through team teaching based on Sutaphan and Yuenyong [11] context based STEM education learning approach. The local issue of the problem on garbage in Iligan City, Mindanao, Philippines could be provided for the Philippines STEM classroom. In the identification of potential solution, it seems to enhance students to develop their possible ideas of designing something as solutions. And, physics, chemistry, and biology class will support students to find the possible designing solutions regarding on the physics, chemistry, or biology content. In physics class may allow students to design about a simple garbage/waste compactor which requires knowledge of force and pressure. In biology class will enhance students to develop the ideas of designing indicators of water quality from organisms in the water. In chemistry class, teacher will enhance students to develop their designing for clean waste water which require knowledge of chemical treatment waste water. Then, classroom will move to the need for knowledge stage where students will investigate scientific knowledge related to their possible designing solutions. After students learn some more scientific knowledge, they could develop their prototypes or products which concurrent through knowledge based in physics, chemistry, and biology classroom. On development prototypes or products, students could not only apply various kinds of knowledge (e.g. science, math, economic, law, values, culture and so on) but also fluid skills (cooperation, thinking skills, leadership, partnership and so on) to optimize their prototypes or products. In test and evaluation of the solution stage, it allows students to apply scientific and other knowledge to develop framework of testing and evaluating their prototypes and products. Here again, students may learn some engineering process. In the socialization and completion decision stage, it enhance students to learn how it works as human need and revising their ideas through learning from people reflections on final prototype final exhibition. This may share some ideas of how to develop STEM education learning activities through team teaching in school setting.

References

[1] Abukari, A. (2005). Conceptualising Lifelong Learning: A Reflection on Lifelong Learning at Lund University (Sweden) and Middlesex University (UK). *European Journal of Education, 40*(2), 143-154. Retrieved from http://www.jstor.org/stable/3496962

[2] Bryce, J., & Withers, G. (2003). Engaging secondary school students in lifelong learning. *Indigenous Education*, 14.

[3] Cage, B. N. (2004). The Effects of an Integrated, Activity-Based Science Curriculum on Student Achievement, Science Process Skills, and Science Attitudes Tammye Turpin, Louisiana Department of Education. *Electronic journal of literacy through science*, *3*, 1.

[4] Colley, K. E. (2006). Understanding ecology content knowledge and acquiring science process skills through project-based science instruction. *Science activities*, *43*(1), 26-33.

[5] Demirel, M. (2009). Lifelong learning and schools in the twenty-first century. *Procedia-Social and Behavioral Sciences*, *1*(1), 1709-1716.

[6] Echevarria, M. (2003). Anomalies as a catalyst for middle school students' knowledge construction and scientific reasoning during science inquiry. *Journal of Educational Psychology*, *95*(2), 357.

[7] Hager, P., & Holland, S. (Eds.). (2007). *Graduate attributes, learning and employability* (Vol. 6). Springer Science & Business Media.

[8] Netwong, Titiya (2018). Development of Problem Solving Skills by IntegrationLearningFollowing STEM Education for Higher Education. International Journal of Information and Education Technology, Vol. 8, No. 9, doi: 10.18178/ijiet.2018.8.9.1114.

[9] SEI-DOST & UP NISMED, (2011). Science framework for philippine basic education. Manila: SEI-DOST & UP NISMED.

[10] Schunk, D. H., & Zimmerman, B. J. (Eds.). (2012). *Motivation and self-regulated learning: Theory, research, and applications*. Routledge.

[11] Sutaphan, S. and Yuenyong, C. (2019, in press). STEM Education Teaching approach: Inquiry from the Context Based. Journal of Physics: Conference Series.

Students' Performance in Context – Based Lessons in Mathematics Classroom

Potchong M. Jackaria[1], Amelia Buan[2*], and Chokchai Yuenyong[3]
[1] Corazon Abubakar Iran Memorial Central Elementary School, Division of Tawi-Tawi, Philippines
[2] Department of Science and Mathematics Education, College of Education, MSU – Iligan Institute of Technology, Iligan City, 9200 Philippines
[3] Faculty of Education, KhonKhaen University, Thailand
*Corresponding author's e-mail address: amelia.buan@g.msuiit.edu.ph

Abstract. This study aimed at developing contextualized lesson plans integrating writing in mathematics using the pupil's mother tongue. It also aimed at determining pupil's improvement in the writing math activities for a period of one month. A Grade 3 class with 31 pupils was purposively selected because of its heterogeneous composition. Using a comparative-descriptive design, this study utilized both quantitative and qualitative data from pupils' weekly writing outputs, pretest, posttest and transcripts of classroom observations and interviews. Five lesson plans were developed using the backward curriculum design based on the K to 12 curriculum standard. The data revealed that the pupils made significant improvement in their writing math performance irrespective of ability group. However, the high-performing pupils made significant learning gains compared to the developing pupils. In-depth analysis of outputs showed improvements in pupils' writing mathematics skills. Most pupils organized their ideas even after the question prompts have been removed on the last week of implementation. This study claimed that the use of scaffolding, contextualization, and mother tongue helped pupils in their writing and improved mathematical reasoning.
Keywords: Contextualization, Mother-Tongue-Based Education, Integration of Math and writing, Writing in Math

1. Introduction
The future problems students will face require them to draw solutions from an interdisciplinary point of view. Pupils need deeper understanding of issues rather than rote memorizing facts. These principles are embodied in the K-12 mathematics curriculum where emphasis is given to important skill sets such as problem solving and critical thinking [3]

The traditional conception about teaching is to teach mathematics consisting of memorizing facts, formulas, and procedures while reading and writing are part of languages and social science disciplines. However, two of the most pressing academic problems in the primary grades are reading together with writing and numeracy. Pupils who failed to attain fluency in these areas will have hard time learning in the succeeding year[6].

The integration of math with other subject helps pupils appreciate the importance of math in the real world [2]. However, mathematics is oftentimes integrated with sciences [7][10]but not with reading and writing. Finding better ways of successfully integrating both skills is crucial and beneficial to pupils' academic success[11], citing the National Council of Teachers of Mathematics reports, strongly recommended that the process of writing in mathematics become an integral part of mathematics curriculum.

Another emerging issue is the use of pupils' native language in classroom instruction to allow pupils to express ideas and develop in-depth understanding, confidence and self-efficacy in mathematics[5]. In the implementation of the mother-tongue-based education, the Department of Education (DepEd) recognized 19 regional languages which are the bases in the teaching of mother tongue in schools[3]. However, this poses problems to schools and teachers whose learners have a different home language, such as the Sinama speaking learners. In such a case, the schools and teachers should adapt strategies and develop their own contextualized instructional materials based on one of the recognized languages.

For these reasons, the researcher conducted this study with the aim of contributing contextualized lesson plans and activities in Sinama. This study also aimed at determining the effectiveness of integrating writing activities using the Sinama language in a Grade 3 mathematics class. Specifically, this study aimed to:

- Develop context- based math lesson plans
- Determine the improvement of the Grade 3 pupils' performance on the writing math activities

Conceptual Framework. The DepEd's mathematics framework states that the twin goals of mathematics education in basic education are problem solving and critical thinking[4]. Teaching pupils to write about mathematics promote in-depth understanding. Writing Mathematics Activities is grounded on the child-centered learning theories, notably, Vygotsky's constructivism which underscores the importance for the learners to construct their own knowledge instead of being the passive recipient. The use of scaffolding is also anchored on Vygotsky's concept of the Zone of Proximal Development[15]. Another emerging concept is the use of the Mother-Tongue-Based Multilingual Education as a subject and as a medium of instruction. This is also based on Vygotsky's sociocultural theory which posited that conceptual development among children is influenced by cultural practices including pupil's home language.

Figure 1 shows the conceptual framework of the study. The developed lesson plans integrating writing math activities contextualized in Sinama served as inputs which were used throughout the duration of the study. Prior knowledge in mathematics and writing are important variables which may influence the result of the study. The developed lesson plans and activities underwent validation and try-out before the actual implementation. Expected outputs include the developed lesson plans and improved pupils' performance on the writing math activities.

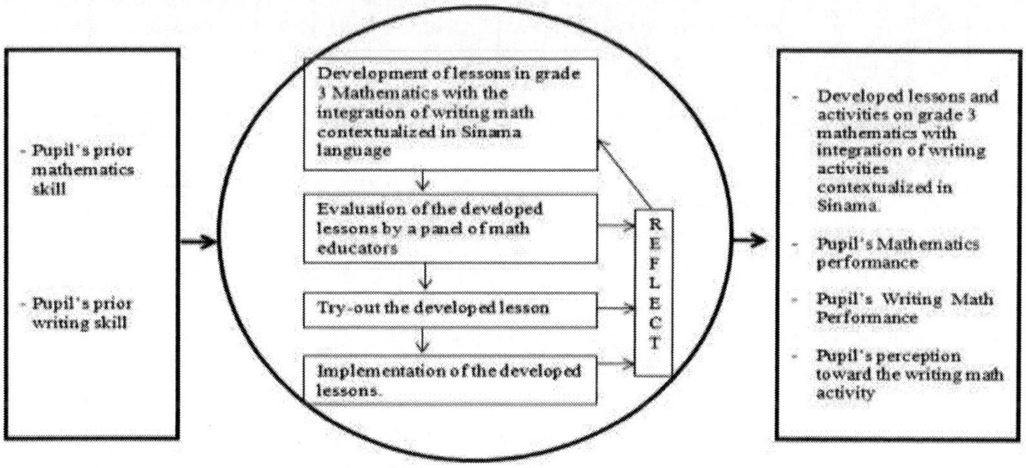

Figure 1. The Conceptual Framework of the Study.

2. Methodology

2.1 Sample of the Study.

One Grade 3 class in the district of Tawi – Tawi during the school-year 2018-2019 was purposively selected as the subjects of the study because of its heterogeneous composition in terms of academic ability and the pupils' use of single mother tongue, i.e., Sinama. The class was composed of 31 Grade 3 pupils who were then classified as high-performing, average-performing, and developing based on their first quarter grades in mathematics.

2.2 Data Collection.

After having done the development process and followed the experts' suggestions, the researcher, with the consent of the thesis adviser asked the permission of the Schools Division Superintendent of the Department of Education (DepEd), Division of Tawi-Tawi for the pilot testing, try-out, and finally the implementation of the lessons to the selected Grade 3 pupils. Similar letters were sent to the district supervisor, school head, and cooperating teachers explaining the purpose of the study. The researcher implemented the developed lessons and activities in the Grade 3 class with the help of the Grade 3 mathematics teacher and a research assistant for the duration of one month for 50 minutes per day. The pretest and posttest were administered before and after the implementation of the study, respectively. Actual classroom observations were done by three trained teachers using the STAR observation tool during the first and the last week of the implementation. Individual interviews were conducted twice involving all pupil respondents. The interviews were done during the first and the last week of the implementation to determine pupils' perceptions and experiences on the writing math activities.

3. Results and Discussions

3.1 Process in the Development of Lesson Plans

The development of lesson plans follows the backward approach to curriculum design. The basic rationale of choosing backward curriculum design is that it ensures the development of the desired output, i.e., the development and integration of writing activities. According to Grant Wiggins and Jay McTighe [17], backward design follows three stages, these are 1.) Identify the desired result 2.) Determine assessment evidence and 3.) Plan learning experiences and instruction.

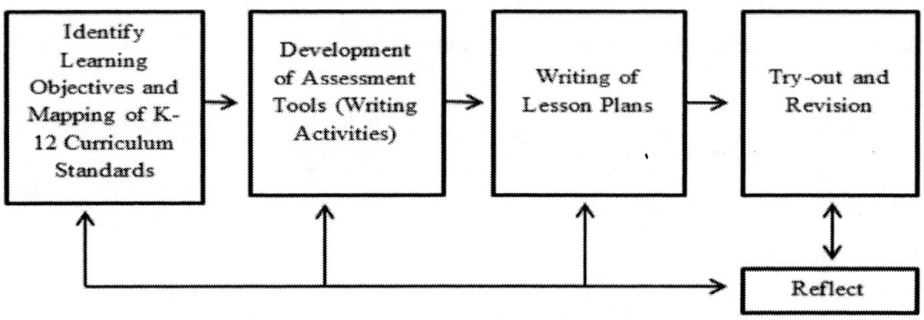

Figure 2. Process of Development of Lesson Plans and Activities

The development started with the researcher's interest in the integration of writing in math class using the pupils' mother tongue and determination of the pupils' progress. The researcher determined how these objectives can be best integrated into the curriculum by mapping of topics based on the K to 12 Curriculum Guide in mathematics. This was followed by the selection of appropriate writing math activities among the many strategies and assessment tools to be used. Then the writing of the lesson plans was based on the above objectives and tools. Afterward, the lesson plans underwent try-outs and revisions. Making reflection is part of every phase of the development.

3.2 Development of Writing Math Activities.

The selection of the appropriate writing math activities was important in this study because these were used to achieve the general objective, i.e., the development of mathematical thinking and problem solving through writing. The developed writing activities were then embedded in the lessons and used to measure pupils' improvement.

The researchers started with researching literature on different writing math activities that can be applied in the context of the study. Considering the grade level of the pupils and their experiences with the writing math activities, the researcher, upon the recommendation of the research adviser, decided to focus on Write-to-Explain math activities.

The researcher recognized the importance of guiding the pupils using prompts during the first few weeks of implementation. It was also decided that the prompts would eventually be removed from the activity as soon as the pupils had developed their writing math skill. This is based on Vygotsky's ideas of Zone of Proximal Development (ZPD) and scaffolding where the quality and quantity of assistance need to change over time with learner assuming more responsibility [15] Finally, there were five question-prompts developed and translated into Sinama. Aside from the use of the pupil's mother tongue, i.e., Sinama, the story problems used in the developed lesson plans were contextualized by relating these to the local setting. This is based on the idea that pupils learn better mathematical concepts by working with things of his culture and the language he is familiar with[9].

3.3 Writing of Lessons Plans.

Five lesson plans were developed by the researcher covering addition and subtraction of three to four digit numbers. These topics were selected from Grade 3 curriculum guides for mathematics and based on the researcher's expected time frame of the actual implementation. The unique features of the developed lesson plans are the utilization of Sinama language and the integration of writing math activities. Each lesson plan has three write-to-explain math activities where pupils are asked not only to compute but to explain and justify their answers in writing. Each lesson plan is expected to be taught in three days. The lesson plan model is based on the DepEd Teacher's Guide (TG) for Grade 3 mathematics consisting of the preliminary activities, developmental activities and assessment.

3.4 Try-out of Lesson Plans and Writing Activities.

The try-out aimed to determine the appropriateness of the developed lesson and writing math activities for possible improvement and the length of time of implementation of each lesson plan was based on the capacity of the pupils faced with the writing math activities. It also looked into difficulties experienced by the pupils with the writing math activities. It was also during this try-out where the pretest was also pilot-tested in the actual classroom.

3.5 Improvement in Pupils' Performance.

To determine the improvement in pupils' performance in the writing math activities, pupils' outputs were rated on a weekly interval. Figure 4.5 shows the pupils' performance in the writing math activities during the four-week implementation. The pretest scores of developing (M=2.73, SD=1.11) and average performing (M=3.96, SD=2.75) were on the 'Beginning level', while those of the high-performing pupils (M=6.58, SD=3.28) were on the 'Developing level.' After a week, pupils' performance improved irrespective of ability groups compared to their pretest. The high-performing group made the largest learning gains (M=8.61, SD=3.42). Significant improvement was also achieved by the average-performing group (M=6.99, SD=3.85) and developing group (M=5.58, SD=2.33).

All groups also improved in their writing math performance in terms of overall mean gains during the second (M = 2.50, SD= 3.56) and the third week (M = .23, SD = 3.07). On the fourth week of implementation, there was a decrease in the pupils' performance in writing math compared to their third-week performance. This decrease in the pupils' scores was probably due to the kind of activity given. It was at this period when the pupils were given Write-to-Explain math activities without the question prompts. However, compared to their pretest scores, the developing and average-performing group are now improving to 'Developing level' (M=7.78, SD=2.15) and 'Proficient level' (M=12.56, SD=2.59), respectively. Similarly, the high-performing group is now in the 'Advanced level' (M=16.29, SD=2.81).

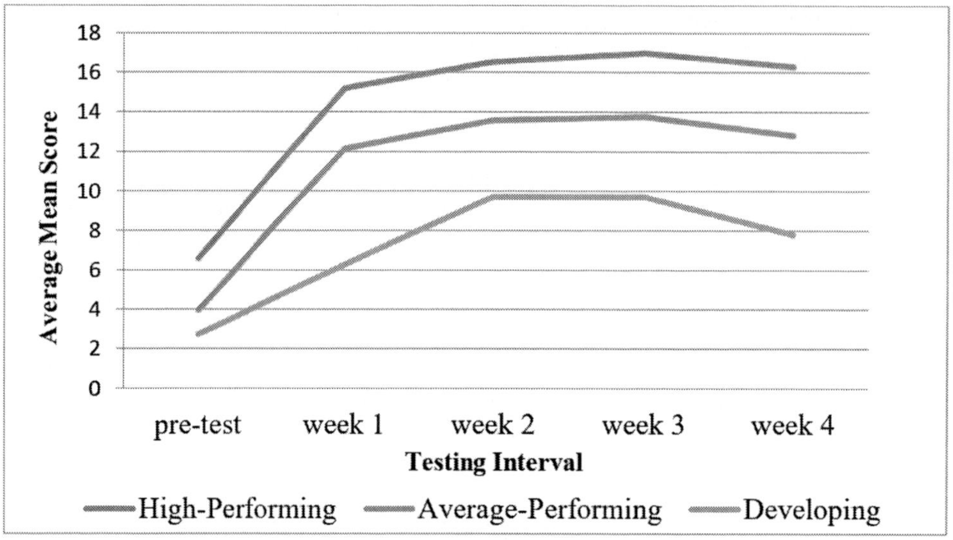

Figure 3. Pupils' Performance in Writing Math Activity during the One-Month Implementation

These findings show that pupils' outputs in writing math improved when pupils were exposed to this strategy which is similar to the findings of Petersen, McAuliffe & Vermeulen (2017) who found improvements in the strategies and explanations learners used after having been taught math writing and to the study of [8]on the use of question prompts that improved pupils' performance in math problem solving. We can also infer that the use of the mother tongue helps pupils in organizing their answers to the math problems. This supports those studies favoring the use of mother tongue in class instruction [13]; [9] [1]. Some pupils made frequent spelling and grammatical mistakes showing that their level of ability in the language used in math classes affects their performance in the writing math activity.

3.6 Pupils' Performance in Pretest and Posttest.

The result of the pretest and posttest shows a highly significant difference in the pupils' performance after one month irrespective of their ability group. The finding is supported by [14] and[12] who found significant improvements in pupils' writing strategies and skills after having been taught writing in mathematics.

Table 1: Pupils' Performance in the Pretest and Posttest

Group	Pretest	Posttest	Standard Deviation	t-value	p-value
High-Performing	6.582	15.519	3.78	6.25	< 0.001*
Average -Performing	3.962	11.294	2.67	8.77	< 0.001*
Developing	2.73	7.356	2.87	6.25	< 0.001*

*Highly Significant at $\alpha<0.05$

3.7 Comparison of the Pupils' Gain Scores by Group.

Table 3 shows the comparison of the mean gain scores by ability groups. The gain scores were computed by subtracting the pupils' pretest scores from their posttest scores. The result shows significant difference in the average gain scores, F (2, 27), p = 0.025 at α = 0.05 level among the three groups of pupils. This indicates that there was an improvement in the pupil's writing math skills after the one- month study.

Table 2: *The Result of the One-way Analysis of Variance (ANOVA) on the Mean Gain Scores Among the Three Groups*

		Sum of Squares	df	Mean Square	F	Sig.
Gain	Between Groups	77.44	2	38.72	4.22	0.025*
	Within Groups	247.83	27	9.18		
	Total	325.26	29			

* Significant at α = 0.05

Table 4 shows the result of the post hoc analysis of the average gain scores among the three groups using the Least Significant Difference (LSD) Test. The result shows no significant difference in the learning gains between the high-performing group (M = 7.33, SD = 2.64) and the average-performing group (M = 8.94, SD = 3.78), p = .292. No significant difference was also found in the mean gain score of the high-performing (M = 7.33, SD = 2.64), and developing group (M = 4.98, SD = 2.87), p = .076 indicating that both groups compared made similar improvements in terms of gain scores. However, comparison in the learning gains between the high–performing group (M = 8.94, SD = 3.78), and the developing group (M = 4.98, SD = 2.87) yields a significant result, p=.010 indicating that high-performing pupils benefited more in the writing math activities compared to the developing group in terms of learning gains. This finding is similar to that of [8] who found that the high-performing pupils benefited more in activities using prompts in mathematical problem-solving.

Table 3: Comparison of Mean Gain Scores Among Groups.

Group		Ave. Gain Score	Mean Difference	Std. Error	p-value
High	Average	8.936	1.60	1.49	.292
	Developing	4.980	3.96	1.27	.010*
Average	Developing	4.980	3.96	1.42	.076

*Significant at α = 0.05

3.8 Pupils' Perceptions on the Writing Math Activities.

The researcher conducted a one-on-one interview to understand pupils' perception about the writing math activities during the first week and after the implementation. During the first week, only 14 or 45% of the pupils said they found the writing math activity easy while 17 or 55% of the pupils found the writing math activity difficult. However, after the one-month implementation, majority or 27 out of 31 pupils described the writing math activity as easy compared to only four or 13% of the pupils who still found writing math activities difficult. Among the top three reasons the pupils found the writing activities easy were: (1) doing the computation, (2) use of question prompts helped them in writing, and (3) looking for what are known and what is asked in the problem.

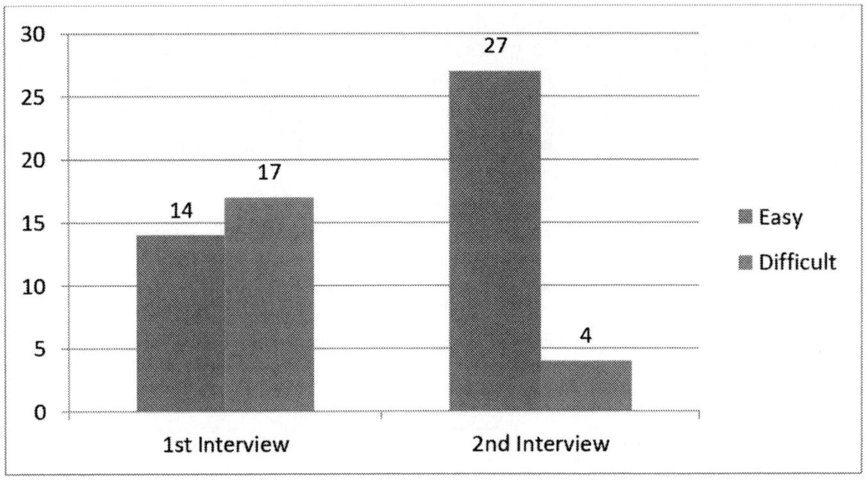

Figure 4. Pupils' Perception of Writing Math Activities Using Sinama

Among the top three reasons mentioned by the pupils as to why they found writing activity hard were: (1) difficulty in identifying the given and questions in the story problems, (2) some pupils had difficulty in making a conclusion or final answer that answers the questions being asked, and (3) difficulty in reading and understanding the story problems. This finding supports [16] which implies that a better understanding of the language subjects used in math class contributes to pupils' performance in the writing math activity.

3.9 Result of Teachers' Observations.

During the first observation, teacher-observers noted that pupils are still new to the activity and many have difficulties. However, marked improvements were noted during the last week observation. The teacher-observers noted that the pupils are finishing the tasks on time and were asking question to clarify doubts about the story problem. On the other hand teachers gave lesser scaffoldings. However, one teacher-observer noted that pupils seemed getting bored after doing the same activity for a long time. This indicates that pupils should be exposed to different writing math activities. Overall, it can be inferred that integrating writing in math class can be a good strategy to be adopted although some pupils will find it challenging at first when they are new at writing in math. Teacher's role and amount of scaffoldings shifted to mostly pupils doing the task when integrating writing in math class.

4. Conclusion and recommendations

The development of lessons integrating writing activities followed a backward curriculum design.. Pupils' performance in writing math improved significantly with the high-performing group benefiting more from the lessons compared to the developing group. The use of scaffoldings, mother tongue and the local context helped the pupils in understanding mathematics problems and in writing math. Pupils' perception on the writing activities in using the Sinama language improves over time as they became familiar with the activities. Similarly, teacher-observers noted changes in the teacher's and the pupils' actions in terms of teacher's scaffoldings, more pupil activity and lesser time in completion of tasks.

This study recommends that: (1) Teachers integrate writing activities into their math lessons; (2) The use of mother tongue helped the pupils in their understanding of math problems and in writing their ideas; therefore, teachers may contextualize math problems and activities to make these more interesting to pupils; (3) The DepEd Tawi-Tawi conduct teachers' trainings on the integration of writing into math lessons and on contextualization using Sinama language; (4) Math teachers use question prompts as scaffolds when introducing the writing in math for the first time; (5) Developing learners be given individualized or special sessions; (6) The write-to-explain math activity be used in combination with other writing activities when integrating writing in math classrooms.

References

[1] Awopetu, Anna V. (2016) Impact of Mother Tongue on Children's Learning Abilities in Early Childhood Classroom. *Procedia - Social and Behavioral Sciences* 233 (2016) 58 – 63, Elsevier Ltd. DOI: 10.1016/j.sbspro.2016.10.131

[2] Chiappetta, C. (2009, February 2) Re: Why it is Important to Integrate Math with Other Subjects? Retrived from https://www.k12academics.com/articles/why-it-important-integrate-math-other-subjects

[3] Department of Education (2014) Grade 3 Mathematics Teacher's Guide. DepEd Complex, Meralco Avenue, Pasig City.

[4] Department of Education, K to 12 Curriculum Guide Mathematics. DepEd Complex, Meralco Avenue, Pasig City.

[5] Department of Education Order No. (74). s. 2009. *Institutionalization of the MTBMLE in Nolasco, RMD, Datar, FA, Azurin, AM (Eds.) Starting where the children are: A collection of essays on Mother Tongue-Based Multilingual Education and language issues in the Philippines, Quezon City, 170,* 70-72.Complex, Meralco Avenue, Pasig City, December 2012.

[6] French, G. (2013). Early Literacy and Numeracy Matters. *Journal of early childhood studies*, OMEP, Vol. 7 April 2013 retrieved from https://arrow.dit.ie/cgi/viewcontent.cgi?article=1065&context=aaschsslarts

[7] Furner, J., and Kumar, D. (2007) The Mathematics and Science Integration Argument: A Stand for Teacher Education. *Eurasia Journal of Mathematics, Science & Technology Education*, 2007, 3(3), 185-189.

[8] Guro, Norolain D. (2016) Enhancing Problem Solving Through Mathematical Problem Solving. Master's Thesis. Mindanao State University-Iligan Institute of Technology. Iligan City.

[9] Hafiz, M., & Farik, M. (2016). Effectiveness Of Teaching & Learning Mathematics Using Children's Home Language And Cultural Tools. *International Journal of Scientific & Technology Research*, 123-127.

[10] Kurt, K. & Pehlivan, M. (2013). *Integrated programs for science and mathematics: review of related literature*. International Journal of Education in Mathematics, Science and Technology, 1(2), 116-121.

[11] Kuzle, A. (2013). Promoting writing in mathematics: Prospective teachers' experiences and perspectives on the process of writing when doing mathematics as problem solving. *CEPS Journal: Center for Educational Policy Studies Journal*, *3*(4), 41.

[12] Petersen, B., McAuliffe, S., & Vermeulen, C. (2017). Writing and mathematical problem solving in grade 3. *South African Journal of Childhood Education (SAJCE), 7*(1) doi:http://dx.doi.org/10.4102/sajce.v7i1.48

[13] Prianes, J. I. (2013). Using Filipino as a Medium of Instruction in Teaching Mathematics: Basis for Policy Formulation. *NEU School of Graduate Studies Research Journal*, *1*(1), 1-1.

[14] Smith, M. J. (2008). Written Solutions of Mathematical Word Problems. Action Research Projects. 31. http://digitalcommons.unl.edu/mathmidactionresearch/31

[15] Thompson, I. (2013). The Mediation of Learning in the Zone of Proximal Development through a Co-constructed Writing Activity. *Research in the Teaching of English* , *47*(3) , 247-276. Retrieved from http://www.jstor.org/stable/24397856

[16] Vilenius-Tuohimaa, P. M., Aunola, K., & Nurmi, J. E. (2008). The Association Between Mathematical Word Problems and Reading Comprehension. *Educational Psychology*, *28*(4), 409-426. DOI: 10.1080/01443410701708228

[17] Wiggins, G., & McTighe, J. (2011). Understanding by Design. Retrieved from http://www.ascd.org/ASCD/pdf/siteASCD/publications/UbD_WhitePaper0312.pdf.

IOP Conf. Series: Journal of Physics: Conf. Series **1340** (2019) 012048 doi:10.1088/1742-6596/1340/1/012048

Linear Equations in Two Variables STEM Education Learning Activities: Developing the Household Power Consumption Calculator App

Crispin D. Ebal Jr.[1*], Mary Joy F. Luga[1], Maria Rebecca O. Flores[1], Dante Joma P. Zabala[1], Amelia T. Buan[2], and Chokchai Yuenyong[3]
[1]Integrated Developmental School, College of Education, MSU – Iligan Institute of Technology, Iligan City, 9200 Philippines
[2] Department of Science and Mathematics Education, College of Education, MSU – Iligan Institute of Technology, Iligan City, 9200 Philippines
[3] Faculty of Education, KhonKhaen University, Thailand
*Corresponding author's e-mail address: crispinjr.ebal@g.msuiit.edu.ph

Abstract. The lesson was developed with the collaboration of experienced teachers through brainstorming and small-group conference. Social issue about electric consumption usage and rate were identified as part of the lesson. The content and skills to be learned in this lesson aligned to the K-12 curriculum. The topics identified were linear equation in two variables for mathematics and renewable and nonrenewable sources of energy for science. In this unit, students develop a prototype of a mobile application that calculates actual household electric consumption. The application aims to help household units to closely monitor their electric consumption in order to lessen the said consumption and to promote household budget savings. The outcome of this lesson contributes to globe's quest to fight against climate change and global warming since electric consumption of households will be lessened. The paper may has implication for STEM education.
Keywords: linear equaltion, K-12 curriculum, STEM education

1. Introduction

Educational institutions in the Philippines are gradually recognising the relevance of science, technology, engineering and mathematics (STEM) Education in supporting innovation and driving future economic growth. In a bid to increase the number of graduates who enroll in science and mathematics-related courses at the tertiary level, the K-12 programme now provides senior high school students with an option to pursue STEM-related fields through the academic track. To this end, significant activity is under way to boost students engagement in these subjects, and teacher quality[7].

Data from World Economic Forum shows that the quality of mathematics and education in the Philippines ranks among the lowest regionally, resulting in relatively low numbers of STEM graduates[7]. In addition, Cordero (2018) cited in his article about TIMSS 2003 results that "The Philippines' scores were below the 400-level. (to pursue opportunities).

STEM education lesson plan may be framed to address a social issue in the community. Krug and Shaw (2016) added that science education reform initiatives have been mobilized as a response to national issue and economic downturns. STEM education prepare students for the professional dispositions required to succeed – and to innovate – in increasingly competitive and globalized 21st century 'knowledge economies'[4].

It is deemed important to develop STEM Lesson plans for a more meaningful classroom experience of both the learners and the teachers. Exposing the learners to the society's issues will help nurture their sense of community involvement as an individual of the society, and this STEM lesson plan can help cater this aspect. This study aimed to contribute to STEM teaching in mathematics classroom. Specifically, this study aimed to present the process of developing STEM lesson learning activity in teaching linear equation in two variables.

2. Developing STEM Education Learning Activities

Iterative process was employed in the designing of the lesson plan. The iterative process in designing the activities was developed regarding on Sutaphan and Yuenyong [11] the context-based STEM education learning approach as shown in Figure 1. the context-based STEM education learning approach included (1) Identification of social issues, (2) Identification of potential solution, (3) Need for knowledge, (4) Decision-making, (5) Development of prototype or product, (6) Test and evaluation of the solution, and (7) Socialization and completion decision stage. As the figure 1, the identification of social issues suggest teachers to bring social or technological issues to motivate students to develop some possible solutions. The 7 stages will allow students using applying scientific and other knowledge for designing the solutions, and provided the context of instruction requires solving a real-world problem or task through teamwork. And, students will have also chance to apply their scientific and other knowledge for problem solving in context of engineers, technology, or entrepreneurship [11].

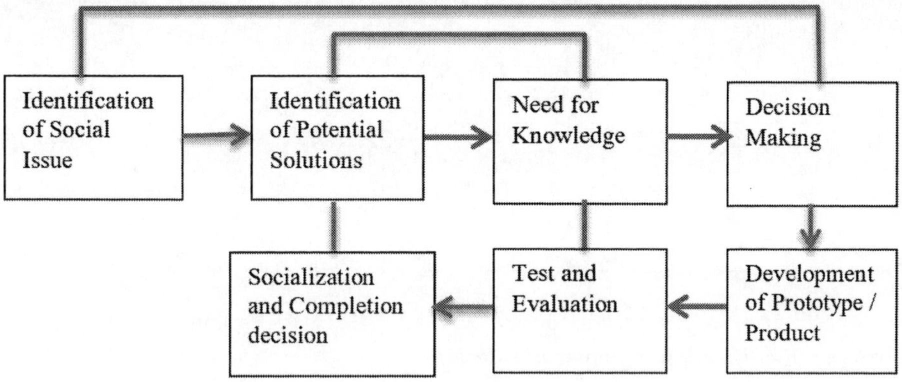

Figure 1. Iterative Process in designing the STEM activity

The design of STEM activities started with the Identification of social issue. The social issue on "electricity consumption" in Mindanao was posted as a community problem to be solved in this activity. The second step is identification of potential solution to the problem. In this step students discuss the possible activities to be done on conservation of energy and identifying possible strategy to help household decrease the energy consumption of electricity at home. To support student's investigation on conservation of energy, a mini lesson was conducted on describing the phenomenon using linear equation on two variables. Science class also discuss about conservation of energy. Brainstorm activity was done to set- up their data collection on the electricity consumption at home. Working in groups, students will develop a prototype on an application in computing the electricity consumption of appliances used in the household. Mathematical equations supported with data will be presented by each group and get feedback from peers. The students will go back to the issue and reflect on their prototype. With this process, STEM experience is beneficial to the students in learning and in the building of confidence in math and science courses[8]. The activities of each stage will be clarified as following.

2.1 Activities in the Identification of social issue

The activities of identification of social issues were developed as the figure 2. In the first stage, the students are assessed of their awareness of the issue on the increase in the usage of the electric consumption as well as its rate. Next, the current situation of the electric supply in the Philippines will be presented in order to have a wide idea of the situation by comparing yearly consumption and analyzing the growth of the rate wherein different factors are considered. After this, contextualization follows where the topic is trimmed down to the local scenario. This gives the students a clear idea of the scenario that they will work with. This stage also provides the expected output of the students which is to make a mobile application that would allow them to compute the amount to be paid based on their consumption. Aside from computation, the students are also given the freedom to include some other portions in the application that they think would make it more interesting and appealing to the consumers (e.g. personal account, brief history, warnings, etc.)

Figure 2: the activities of identification of social issues

Stage	Activity
1. Identification of social issues	1.The teacher will ask the students if they are aware of the electric consumption in their respective household. 2.With their respective current usage, did it somehow affect their way of living? 3.The teacher will present a video on the current situation of the power supply in Philippines and, particularly, in Iligan City such as scheduled power interruption, increase in the rate of per kWh. 4.The teacher will then ask the students "If you have the capacity to create an app to calculate household monthly consumption, how are you going to design your app?".

2.2 Activities in the Identification of potential solutions

The activities of identification of potential solutions were developed as the figure 3. In this part of the lesson, the teacher will divide the class into a group of at least 5 students. Each group will be provided a Household Power Consumption Form for them to complete- by choosing a household to monitor. The students will be tasked to collect data on the amount of power each of the selected-household electrical appliance and/or gadget consumes in kilowatts and record the average number of hours in a day each of the appliance and/or gadget is in use in reference to their monthly usage. They will then compute for the cost each of the appliance or gadgets consume in a month. In this manner, the students will be able to identify which of the appliance and/or gadgets contribute/contributes the most to the cost of the household power consumption.

Figure 3: the activities of identification of potential solutions

STAGE	ACTIVITY
2. Identification of potential solutions	1. The teacher will ask the students to identify the electrical appliances used in their household and collect their electricity bill for the past 3 months. 2. The students will discuss the reason/cause of paying that much for electricity consumption. 3. Take the students to list down estimated kilowatt per hour for each of the electrical appliances. 4. Determine that number of hours per day (in average for a month) each appliance is in use.

2.3 Activities in the Need for knowledge

The activities of the need for knowledge were developed as the figure 4. The topic linear equations in two variables will be discussed and learned by students through inquiry based approach that will make use of a real life situation. A problem about De la Cruz' family electric consumption will be presented. Prior to dividing them by groups, a short exchange of ideas on how the charging of electric consumption is being computed. The students will be given a KWL activity about conversion. Through the KWL activity, the teacher will be aware of their prior knowledge about computing electric consumptions. They will then be divided in groups and will be given time to come up with their solutions on how to compute the electric bill of De la Cruz family for one month. They will be asked to represent their answers algebraically. Through this activity they will be able to realize that algebraic concepts such as linear equations in two variables are present in real life situations. Their prior knowledge of basic calculation will help them arrive with different electric consumption with their respective rate; through these results, they can differentiate the input which serves as the independent variable and output which is the dependent variable. They will be able to read the trend and predict consumptions and rate depending on the given time.

This activity will also give them a chance to play the role as the head of the household and decide what they think is the best budget for their household and to figure out the best plan on how they can make the most out of the electric appliance that they want to operate. That is where they can practice evaluating linear equations in two variables.

Figure 4: the activities of the need for knowledge

STAGE	ACTIVITY
3. Need for knowledge	Mathematics : Mini – lesson on linear equations in two variables Physics : Lesson on Conservation of energy

2.4 Activities in the Decision making stage

The activities of the decision making stage were developed as the figure 5. The students will compare their gathered data to the monthly power consumption based on the 3 months power consumption bills they have gathered. From this, the students will be setting up an equation to represent their power consumption, as well as the power consumption rates, depending on the variables such as time, power rate and appliance power consumed in kilowatts.

Figure 5: the activities of the decision making stage

STAGE	ACTIVITY
4. Decision-making	1. investigate in respective households their electricity consumption. 2. Set-up an equation for their solutions to the issue.

2.5 Activities in the development of prototypes and products

The activities of the development of prototypes and products stage were developed as the figure 6. Each group will study how their formulated equations work by applying them to their respective selected household, and will compare each of their computed result to the three months power consumption bills they have gathered in the first part of the lesson for verification. When the formula is finally tested and proven accurate, the students will then use their equation to design an app that calculates household power consumption and consumption rates. This allow students to provide how the equation can be placed in the design.

Figure 6: the activities of development of prototypes and products stage

STAGE	ACTIVITY
5. Development of prototype or product	1. The students will apply their equation to their household. 2. Each group will develop an app of their own design that calculates household power consumption and consumption rates.

2.6 Activities in the test and evauation stage

The activities of the test and evaluation stage were developed as the figure 7. Each group will present their formulated equation and show in class how their equation works by applying it to their household power consumption and by comparing their computed result to their gathered 3 months power consumption bills of their selected household. They will then present the design of their app ideal for a household power consumption computation for a month.

Figure 7: the activities of the test and evaluation stage

STAGE	ACTIVITY
6. Test and evaluation	* Present their designed Household Power Consumption App

2.7 Activities in the socialization and completion of de stage

The activities of the test and evaluation stage were developed as the figure 7. Through class conference, the whole class will agree on which application will be the best among all the developed applications. This application will then be subjected to further improvement before it will be released to the public for free.

Figure 8: the socialization and completion of de stage

STAGE	ACTIVITY
7. Socialization and completion stage	*Amongst the design presented for the computation of monthly household power consumption rate, the class will decide which design is most comprehensive and convenient to use.

3. Conclusion

Development of STEM Education Lesson plans requires a thorough process of planning. Some aspects to be considered during planning the lessons is the timely relevance of the activities incorporated in it. 21st century skills must also be embedded along with the objectives to ensure learners of a more meaningful experience on the lessons' activities. And the context of activities must adhere to the society's current needs in contrast to the traditional classroom setting. But it must be emphasized that the lessons should still be able to address the basic needs of the learners such as the knowledge and skills that are related to the subject matters of the lessons.

References:

[1]de Castell, S. (2016). A pedagogy of production: An Introduction, New Media Modules, [Video file] Retrieved from https://vimeo.com/181978126.

[2]Cordero, Ted (2018) "Based on 2003 Score PHL education underperforms in East Asia Pacific - WB Report". GMA News Online. Retrieved from

[3]Krug, D., & Shaw, A. (2016). Reconceptualizing ST®E(A)M(S) education for teacher education. Canadian Journal of Science Mathematics and Technology Education, 16(2), 183–200

[4]Olssen, M., & Peters, M. A. (2005). Neoliberalism, higher education and the knowledge economy: From the free market to knowledge capitalism. Journal of Education Policy, 20(3), 313–345.

[5]Talaue, Frederick (2014) Social Equity and Access to a Philippine STEM School. Retrieved from https://www.researchgate.net/publication/266755328_Social_Equity_and_Access_to_a_Philippin e_STEM_School. Accessed on March 11, 2019.

[6]Department of Energy (2018). Philippine Power Statistic. Republic of the Philippines. Retrieved from https://www.doe.gov.ph/philippine-power-statistics. Accessed on March 11, 2019.

IOP Conf. Series: Journal of Physics: Conf. Series **1340** (2019) 012048 doi:10.1088/1742-6596/1340/1/012048

[7]Oxford Business Group(2019) Education research chapter of The Report: Philippines 2018. Retrieved from https://oxfordbusinessgroup.com/analysis/seeds-stem-initiatives-multiple-levels-look-strengthen-tuition-and-outcomes-science-technology. Accesed on March 11, 2019.
[8] Hayword, Jacob (2016) .An Analysis of Secondary Integrated STEM Lesson Plans: Common Characteristics, Learning Expectations and the Impact from the Teacher ' s Definition of I-STEM Retrieved from:
https://scholarworks.uark.edu/cgi/viewcontent.cgi?referer=https://www.google.com/&httpsredir=1&article=3343&context=etd

IOP Conf. Series: Journal of Physics: Conf. Series **1340** (2019) 012049 doi:10.1088/1742-6596/1340/1/012049

Development of an Interactive Module Incorporating Financial Literacy in Teaching Decimals/Fraction

Fhelmar I. Rondillas[1]*, Amelia T. Buan[2]

[1]Basic Education Department, Senior High School Discipline Coordinator, St. Michael's College, Iligan City 9200, Philippines

[2]Department of Science and Mathematics Education, Faculty, Mindanao State University – Iligan Institute of Technology, Iligan City 9200, Philippines

[2]*Corresponding author's e-mail address: amelia.buan@g.msuiit.edu.ph

Abstract. This study sought to develop an interactive module incorporating financial literacy in teaching fraction and decimals for Grade 7 Mathematics. This study used pretest-posttest one-group design with qualitative support to distinguish the effect of the interactive module on students conceptual understanding, financial literacy and motivation to achieve. Based on the quantitative and qualitative analyses of the study, there was a significant improvement in the students' pre-test and post-test scores on conceptual understanding, financial literacy and motivation to achieve. Results also showed that students were able to understanding the concept of fractions and decimals; see application of fraction and decimals into different fields and most specially its connection to financial literacy. In addition, students realized the importance of savings and budgeting, and even became financially wise. The developed interactive modules incorporating financial literacy were integrated with interactive websites, financial literacy videos, and online assessments. The researcher recommends the use of the interactive module incorporating financial literacy in secondary mathematics classes to facilitate the learning process in teaching fractions and decimals. Also, similar interactive modules incorporating any real life skills must be developed.Introduction

1. Introduction

Today's generation of learners is way too different from the past generations. Most students are inert. Schools that were once a sanctuary for young minds to be developed has now become a daily battlefield for teachers trying to inculcate knowledge to these young minds effectively. Thus the need to create effective teaching materials such as interactive module is one strong defense for teachers to win this said battle for inculcation of knowledge to learners.

Learning mathematics is crucial and vital because it develops cognitive thinking in young children and it is a challenging task for parents and teachers to educate them. One of the alternatives is to provide interesting learning environment to the children, by adapting interactive multimedia application. Previous research showed that interactive multimedia influenced the effectiveness of learning process [3].

In traditional method, most of the children use exercise books as their source of information. If this method is being used as a single approach in classrooms, then it will be difficult to build their understanding and imagination towards mathematical subject matter. It believes that this method can cause them to be easily distracted and hard to pay attention, resulting in a dull and boring environment. The lack of attractive approaches in children's learning process could affect their insight.

Tedious learning environment could cause the children to lose interest in understanding mathematical problem solving. Therefore, a Mathematical Learning Courseware is developed to cater to this problem [28].

Teachers need to implement a different learning approach which includes elements of sound, text, graphic and animation in order to provide a better learning approach. These elements have been proven to be more interactive, therefore multimedia learning will improve students' performance and enhance the teaching and learning process [24]. With the pursuit of the Department of Education in the Philippines to adapt the K to 12 Curriculum which aims to upgrade the educational system of the country, it gives a chance to students to be globally competitive and develop the 21st century skills. In K to 12 Curriculum, subjects are taught from the simplest concepts to more complicated concepts through grade levels in spiral progression.

One of the least learned competencies in the national achievement test is fractions and decimals with an above mastery rating. Furthermore, the Percentage of Correct Responses (PCR) per learning competency measured by subject area showed that the pupils' performance in Mathematics is very low. Educators confirmed that a significantly low achievement in Mathematics and a relatively low self-efficacy among students who are impatient in solving mathematical problems posed great challenges to present day Mathematics educators. Proper preparation must be done to address the needs of the students in terms of coping with the new approaches and to improve the performance of students in the national achievement test in Mathematics. As one of the tools to motivate students to react and learn independently with minimal assistance from the teacher and make learning more interesting, learning module is developed. The self-efficacy level of the respondents was also measured to determine the ability to succeed in performing a given task [2].

Fraction and decimal arithmetic are crucial for later mathematics achievement and for ability to succeed in many professions. Unfortunately, these capabilities pose large difficulties for many children and adults, and students' proficiency in them has shown little sign of improvement over the past three decades. To summarize what is known about fraction and decimal arithmetic and to stimulate greater amounts of research in the area, we devoted this review to analyzing why learning fraction and decimal arithmetic is so difficult. We identified and discussed two types of difficulties: (1) inherent difficulties of fraction and decimal arithmetic and (2) culturally contingent difficulties that could be reduced by improved instruction and prior knowledge of learners. We concluded the review by discussing commonalities among three interventions that have helped children overcome the challenges of mastering fraction and decimal arithmetic [15].

In this study the researcher developed an interactive module incorporating financial literacy. Modules, like interactive type, present logical representation of a subject matter using effective learning activities complete with a clear related instruction that could be easily followed by the students in solving his/her own problem. The interactive module that is developed in this study incorporated financial literacy for students to see its application in every topic being inculcated in the module. Knowing the real life application on financial aspects of the topic is one effective technique to catch students' attention and increase their motivation towards learning the subject matter since students will directly see its importance to the society upon using the module. Utilization of an interactive module for a lesson is now considered an effective instructional technique. Hence, interactive modules are packages of learning activities that have to be completed by the learners.

2. Objectives of the Study

The main objective of the study is to develop an interactive module incorporating financial literacy to help enhance the conceptual understanding of the Grade 7 students in fractions and decimals.

Specifically, the study seeks to:

1. describe the process in the development and validation of the interactive module;
2. compare the performance of the students in the pretest and posttest in terms of:
    a. Achievement Test

b. Financial Literacy Test
   i. Management Skills
   ii. Financial Attitude
   iii. Peer Influences
c. Motivation to Achieve
   i. Intrinsic Value
   ii. Test Anxiety
   iii. Self-efficacy

3. determine the students' perception towards on the use of interactive module.

Null Hypothesis

H_o. There is no significant difference between the pretest and posttest performance of students in terms of:
  a. Achievement Test
  b. Financial Literacy Test
     i. Management Skills
     ii. Financial Attitude
     iii. Peer Influences
  c. Motivation to Achieve
     i. Intrinsic Value
     ii. Test Anxiety
     iii. Self–efficacy

3. Conceptual Framework

This study employed the pretest-posttest one group design with qualitative support. The research paradigm in Figure 1 depicts a detailed guide for the progress of this endeavor. It shows that this study was divided into three phases: (1) determination of prior knowledge of students on decimals and fractions, financial literacy and motivation to achieve; (2) development and validation of the interactive module through classroom utilization of the module; and (3) developed interactive module on Decimals/fraction, analysis of the learners' conceptual understanding in terms of knowledge, financial literacy and motivation to achieve.

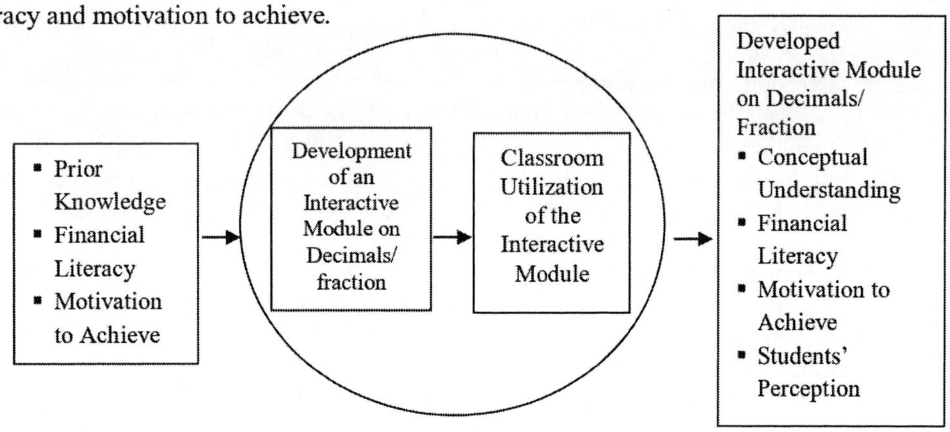

Figure 1. Conceptual Framework

It shows in figure 1 the research paradigm of the study. It was divided into two phases: the implementation of the pre-test in terms of conceptual understanding, financial literacy and Motivation to achieve and the Development of the interactive module.

The interactive module was then evaluated by panel of experts to determine its readiness for utilization. After the ratings, comments and suggestions of the panel, the researcher did the revision of the module. An interactive module was then developed. After its face and content validity were established, it was then implemented in the classroom. The researcher took note of challenges that was encountered and come up with ways to address them. These modifications were then reflected in the revised module.

Conceptual understanding of students was checked by evaluating student performance in the achievement test. Development of financial literacy was measured through assessment methods, such as financial literacy test, student journals and anecdotal notes. Motivation towards the subject prior to the utilization of the module was compared to respondents' responses after the utilization of the module through the use of a motivation scale. Qualitative data derived from student journals and interviews were subjected to in-depth analysis to support quantitative data.

4. Theoretical Background

This study is anchored on Constructivist theory by Jerome Bruner and E-learning theory by Richard E. Mayer and Roxana Moreno.

A major theme in the theoretical framework of Bruner is that learning is an active process in which learners construct new ideas or concepts base upon their current/past knowledge. The learner selects and transforms information, constructs hypotheses, and makes decisions, relying on a cognitive structure to do so [7].

Constructivism is an approach in education which explains how learners construct their own understating and knowledge from their experience so that the role of teaching is not to lecture, explain mathematical concept, but to let students to construct their own understanding that will develop their critical thinking [7].

In constructivist instruction, students are encouraged to use their own methods for solving problems. They are not asked to follow others thinking but encouraged to process their own. Through interaction with mathematical tasks and other students, the student's own mathematical thinking becomes more abstract and powerful. The role of the constructivist teacher is to guide and support students' mathematical concept. Constructivist in teaching approach, teachers believe that students learn improving knowledge that students already know in previous [7].

Meanwhile, e-learning theory established within the scientific literature a set of multimedia instructional design principles that promote effective learning [20].

This theory enables the learners to integrate learning with the help of technologies and provide positive feedback towards achieving common goal and advancement to school implementation [22].

5. Research Design and Methodology

This study presents in detail the methods and procedures in the development and validation of the interactive module incorporating financial literacy in teaching decimals and fractions. Specifically, it comprises the locale of the study, the type of respondents, the research design, the instruments used, the procedure used in gathering data, the statistical tool used in interpreting data and reliable statements concerning the hypothesis, and the coding of qualitative data.

Research Design

This study employed a pretest-posttest one group design with qualitative support. In order to evaluate the effect of the interactive module incorporating financial literacy on students' understanding, skill and motivation to learn, students answered the achievement test and were asked to answer the lesson closure at the end of the module.

Specifically, this study used a nonequivalent dependent variable design (NEDV) wherein there was only one group of subjects but there were other dependent variables, namely:

(1) Achievement Test
(2) Financial Literacy Test
(3) Motivation to Achieve

The primary independent variable in this study was the interactive module that will be implemented in the class for the duration of the experimentation.

Research Setting

This study was conducted in one of the private schools located in Northern Mindanao for S.Y. 2017-2018. The school is a Catholic institution of higher learning; it was administered by the Religious of the Virgin Mary (RVM), the first Filipino religious congregation founded by Mother Ignacia del Espiritu Santo. The school was well known as the oldest school in Lanao area and is a PAASCU Level III grantee.

Research Instrument

Interactive Module

The interactive module was one of the instruments used by the researcher in this study. The researcher made the interactive module based on the learning competencies found in the K to 12 Curriculum guide for Mathematics. The interactive module was then evaluated by panel of experts to determine its readiness for utilization. In gathering the data, a rating scale served as Evaluation Questionnaire by the researcher. This was adapted from the study of Richa [29] and revised by the researcher. This contains the characteristics of the interactive module that was rated by the evaluators. The highest rating is "4" which means "excellent" and the lowest is "1" that means "needs for improvement".

After the ratings, comments and suggestions of the panel, the researcher revised the module and an interactive module was then developed. After its face and content validity were established, it was then implemented in the classroom. The researcher took note of challenges that were encountered and came up with ways to address them. These modifications were then reflected in the revised module.

The interactive module was incorporated with financial literacy where students reflected their knowledge of decimals and fractions through their responses in the lesson closure and process questions after watching the Cha-Ching videos. Each unit contained the objectives, lesson proper, self-assessment questions, and the references. Two (2) modules were prepared by the researcher for the duration of the learning process for decimals and fraction. To make it more interactive, Cha –Ching videos from Youtube were incorporated in the lesson proper wherein students enjoyed watching while learning about financial literacy.

The student will have to correctly answer 75% of the given assessment test before they can proceed to the next module. In case they can't reach 75%, students will have to repeat the activities inculcated in the module.

Achievement Test

The Achievement Test Questionnaire was designed by the researcher to evaluate students' conceptual understanding of decimals and fractions and respondents' knowledge and skill in financial literacy. The achievement test was a 15-item multiple choice test which served as the pre-test and post-test given to the respondents. The test went through face validation, reliability test, item analysis and pilot testing before it was given to the respondents. To ensure content validity, a table of specifications was made prior to test construction.

Achievement Motivation Scale in Secondary Mathematics

The Achievement Motivation Scale used by Villadolid [33] was modified by the researcher to suit Grade 7 students. The motivation scale was administered to the students before and after the module to determine if there had been a change in students' motivation to achieve.

Interview Guide

Interviews were conducted by the researcher on the randomly selected respondents to have a qualitative support to quantitative data obtained from the questionnaire and motivation scale. Questions in the interview were concerning to students' insights about the module, their conceptual understanding of decimals/ fraction and financial literacy. Interviews was recorded and undergone content/ thematic analysis in order to categorize the responses of the students.

Data Gathering

The development of an interactive module incorporating financial literacy underwent some process. First, learning objectives aligned to the competencies from the K to 12 curriculum were identified.

Assessment tools were formulated (Achievement Test, Financial Literacy Test and Motivation to Achieve Test) after which was the development of the interactive module. The module was then evaluated by a panel of experts using a rating form and revision of the module was made based on their comments and suggestions.

The motivational level of the respondents in Mathematics was determined through the modified motivational scale of Villadolid [33]. Before students started using the interactive module, a pretest was administered. Multiple methods of assessment as indicated in the assessment timeline of the interactive module employed throughout the duration of using the module. Data on students' conceptual understanding after the use of the module was derived from the posttest through the achievement test and financial literacy test.

Data Analysis

All quantitative and qualitative data was analyzed to answer the questions raised in the study. Related sample t-test was used by the researcher to compare the scores in the pretest and posttest. Thematic analysis was applied to qualitative data in order to categorize the responses of the respondents. In thematic analysis the researcher identified themes or patterns that have emerged from the qualitative data. The researcher defined each theme sufficiently so that it is clear in categorizing the responses of the students.

6. Results and Discussion

Development of the Interactive Module

The development of an interactive module incorporating financial literacy was anchored on Backward Curriculum Design of Wiggins [34] as cited by Rodriguez [30] and underwent a strategic process as shown in figure 2 below. The procedure included: a.) Identifying of the learning objectives aligning to the competencies from K-12 curriculum; b) Formulating of the assessment tools: Achievement Test, Financial Literacy Test and Motivation to achieve Test; c) Developing of the interactive module; d) Evaluating the interactive module by the panel of experts using a rating form; after which e) revising the module was done based on the comments and suggestions of the panel of experts.

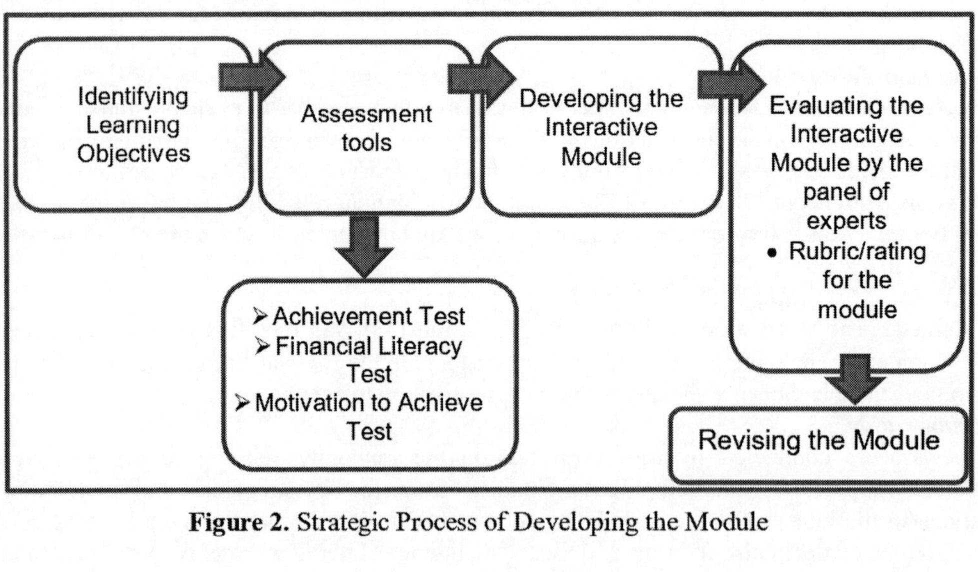

Figure 2. Strategic Process of Developing the Module

Identifying Learning Objectives. From the K to 12 Curriculum guide for Mathematics, the researcher identified specific learning competencies on fractions and decimals. The researcher chose fractions and decimals since it is one of the least learned competencies in the national achievement test with an above mastery rating [2].

Formulation of Assessment Tools. After identifying the learning competencies, the researcher made the assessment tools: Achievement Test, Financial Literacy Test and Motivation to Achieve Test. The 15-item multiple choice achievement test was designed by the researcher to evaluate students' conceptual understanding of decimals and fractions and respondents' knowledge and skill in financial literacy. To ensure content validity, a table of specifications was made prior to test construction. The Financial Literacy Test consists of three components: management skill, financial attitude and peer influence. It was designed by the researcher and underwent validation by the experts and paper adviser. The Achievement Motivation scale used by Villadolid [33] was modified by the researcher to suit Grade 7 students. The motivation scale was administered to the students before and after the module, to determine if there had been a change in students' motivation to achieve. All assessment tests went through face validation, reliability test, item analysis and pilot testing before it was given to the respondents.

Evaluation and Revision of the Module. The interactive module was evaluated by two Mathematics teachers of a private school teaching in the Junior High School as presented in Table 1. A rating or score sheet was used by the evaluators, giving "4" as the highest rating (meaning "excellent") and "1" as the lowest (representing "needs for improvement"). The evaluators also gave their comments and suggestions for further improvement of the module. Changes were then made in the Interactive Module based on the suggestions and comments of the evaluators.

Evaluation of the Developed Module

Table 1. Evaluation of Math Teachers

Content of the Module	Mean for the Faculty Rating	Description
Overview	3.33	Excellent
Lesson Proper	3.59	Excellent
Layout	3.33	Excellent
Over-all Content	3.75	Excellent
Writing Mechanism	3.5	Excellent
Self-Assessment Questions	3.78	Excellent
Citation	3.84	Excellent
Average	3.58	Excellent

Revision of the Module

The module made by the researcher was revised based on the comments and suggestions of the teacher evaluators. The revision was mostly on the physical appearance of the module making the pictures and illustrations more attractive and appealing using appropriate colors aside from black and white. The researcher double-checked and tested the embedded websites in the modules making it easier for the students to open. Lastly, the researcher added self-reflection questions and process questions in the interactive module to make it more engaging, enabling the students to integrate their knowledge to real-life situation.

Students' Pretest and Posttest Results

Table 2 shows the mean difference and t – value of the pretest and posttest scores measuring the students' conceptual understanding of fractions and decimals.

Table 2. Achievement Test Result for Pretest and Posttest

Achievement Test		Mean Difference	t – value	P-value
Pretest Mean	8.4583	-2.6806	-12.1297	.000 < .01
Posttest Mean	11.1389			

Results of the pretest and posttest of the achievement test showed significant difference in students' conceptual understanding (mean difference = -2.8056). A t-value of -12.12976 (p-value =0.000) indicates that there was a significant increase in the posttest scores of the students. This improvement in content knowledge is consistent with the results of Villadolid [33] where the respondents exhibited positive conceptual understanding after research employ project-based unit incorporating entrepreneurial skills in the teaching of percent to elementary pupils.

The financial literacy test is a test that measures the management skills, attitude and level influences of the students. The mean difference and t-values of the pretest and posttest scores are shown in Table 3.

Results of the paired t-test show that there was a significant difference between the pretest and posttest scores on financial literacy in terms of management skills, attitude and influences. In fact, there was an improvement in the skills of the learners in terms of financial matters upon actively performing the activities in the module. This is indicated by the negative t – values for management skills (-0.1707), financial attitude (-0.1874), and peer influences (-1.8348). These results are comparable with those in the studies of Moten [23] which showed that financial literacy should be introduced into the K to 12 curriculum and that there is a need for standardized teacher education in financial literacy.

Table 3. Pretest and Posttest for Students' Financial Literacy Test

		Mean Difference	t – value	P-value
Management Skills				
Pretest Mean	18.0556	- 0.6111	- 0.1707	.000 < .01
Posttest Mean	18.6667			
Financial Attitude				
Pretest Mean	14.8333	- 0.8334	- 0.1874	.000 < .01
Posttest Mean	15.6667			
Peer Influences				
Pretest Mean	14.6161	- 4.5179	- 1.8348	.000 < .01
Posttest Mean	19.1349			

Table 4 shows the mean difference and t – value of the students' pretest and posttest motivation to achieve.

Table 4. Pretest and Posttest for students' motivation to achieve

		Mean Difference	t – value	P-value
Intrinsic Value				
Pretest Mean	14.4000	- 0.1067	- 0.0214	.000 < .01
Posttest Mean	14.5067			
Test Anxiety				
Pretest Mean	15.2250	0.4250	0.7486	.000 < .01
Posttest Mean	14.8000			
Self – efficacy				
Pretest Mean	16.8438	- 0.3437	- 0.2919	.000 < .01
Posttest Mean	17.1875			

Based on the result of the paired t – test, there is a significant difference between the pretest and posttest scores on students' motivation to achieve. The negative t – values for intrinsic value (- 0.0214) and self-efficacy (-0.2919) point to an increase in the importance ascribed by students to Mathematics as well as in their confidence in their own capabilities in the said field. On the other hand, the positive value for test anxiety (0.7486) shows that there was a decrease in the students' discomfort when performing mathematical tasks. Analogous results can be found in the studies of Reibman [27] where a positive relationship between conscientiousness and academic achievement motivation was established. Also, the results are similar with the findings of Villadolid [33] which specified that students become more self-confident and motivated to learn and their test anxiety decreased.

Students' Perceptions on the Interactive Module
The students were asked to write what they have learned upon using the module and how they can apply their learnings in real life situation. Responses of the students commonly revolved around the following themes: (a) understanding the concept of fractions and decimals; (b) application of fraction and decimals; and (c) connection to financial literacy.

Understanding the concept of Fractions and Decimals
Most of the students expressed that they understood the concept of ordering and comparing fractions and decimals, converting fractions and decimals, application of fractions and decimals in real-life and its connection to financial literacy. Students emphasized that they learned how to convert fractions to decimals and vice versa, solve real-world problems involving fractions and decimals and also on how to compare and order a set of fractions or decimals. Students also realized the connection of fractions and decimals to financial matters in which they highlighted in their responses that it made them more knowledgeable about real-life applications of the concept most specially on financial matters.

Application of Fractions and Decimals
Students responded on how they can apply the concept of fractions and decimals in real life situations through self-reflection and their life experiences. Based on the responses, there are four most cited applications of fraction and decimals; financial problems, the monetary system, life experiences and savings. Respondents stated that they can solve any financial problem applying their knowledge in fractions and decimals. They cited a clear application of fractions and decimals in the monetary system as it helped them identify the differences in bills and coins. Also, students used their knowledge in fractions and decimals to compute prices while buying goods in the store. In daily life experiences the concept of fractions and decimals is very useful such as in budgeting time and money, measuring any quantity (for example in cooking), and buying goods for daily living. Saving money is one of the primary skills that a child must learn while growing up. Respondents gave strong statements on how they learned more about saving while working with the modules. One respondent stated, "I can use the concepts in real life for depositing money in my bank account or withdrawing salary or allowance."

Connection to Financial Literacy
Through the different financial literacy videos students responded on how fractions and decimals can be used in financial aspects. Respondents also shared on how fractions and decimals as applied in the videos aided them to realize the importance of savings, budgeting and becoming wiser in dealing with money. Upon knowing the application of fractions and decimals in financial matters, students learned that saving money could make their future brighter and more successful. Also, saving will help students to become financially secure and will provide safety and security in case of emergency. Aside from that, students also stated that saving will help them buy things they wanted but their parents can't provide. Respondents gave an idea on how they will apply savings while they are still students. Their daily allowances will be divided into how much they will save and how much they will spend every day.

Random Interviews

The researcher conducted an interview and confirmed that students appreciated the module because it gave them the opportunity to understand the concept of fractions and decimals. Students shared that using the module made them realize the positive effects of learning online. They also expressed that after using the module they are now mindful of how they are dealing with money. Most of the students responded that using the module made them grasp the importance of savings and budgeting, and on how to divide or balance things specially about money.

However, students also encountered some challenges upon using the module. Some answered that they found it hard to open the links in the module due to compatibility with computer they are using. Others also stated that sometimes they found it hard to express their ideas since they do all tasks alone.

In general, the students had positive responses toward the module and they are looking forward for other learning opportunities applying different ways of learning a topic specially using computer or online activities.

7. Summary of Findings, Conclusions and Recommendations

Findings of the Study

Based on the quantitative and qualitative analyses of the study, the following findings were highlighted:

1. The interactive module incorporating financial literacy was developed through a systematized process involving Identifying the learning objectives aligning to the competencies from K-12 curriculum; formulating the assessment tools, namely, Achievement Test, Financial Literacy Test and Motivation to Achieve Test; developing of the interactive module; evaluating the interactive module by the panel of experts using a rating form and; lastly, revising the module based on the comments and suggestions of the panel of experts. Changes were made of the Interactive Module incorporating the given suggestions/comments of the evaluators. The developed interactive modules incorporating financial literacy were integrated with interactive websites, financial literacy videos, and online assessments.
2. There is a significant difference in the students' pretest and posttest results in terms of conceptual understanding, financial literacy and motivation to achieve in mathematics.
   a. Results show that students scored higher in the posttest on the conceptual understanding of fractions and decimals.
   b. Peer influence is a big impact on students' financial literacy test. Hence, there was an improvement in students' management skills, financial attitude and peer influence based on the result.
   c. The intrinsic value students gave to mathematics and their discernment of their own efficacy was increased while their test anxiety decreased.
3. Students' perceptions of the interactive module revealed that it helped them understand fractions and decimals. In the real-life application of the concept, responses of the students commonly revolved around the following themes: (a) understanding the concept of fractions and decimals; (b) application of fraction and decimals; and (c) connection to financial literacy. Students also learned to solve problems involving fractions and decimals and apply their knowledge to financial matters. Through the different financial literacy videos, respondents express how fractions and decimals aided them to realize the importance of savings, budgeting and economically wise.
4. In interviews, students shared on how the module gave them the opportunity for a new learning and strategy most specially in understanding fractions and decimals. While completing the module, students faced various challenges such as opening the links, checking compatibility of the interactive module in their own computer, answering the questions alone and expressing one's ideas. Students gave their diverse reactions on how the module made them realize the application of fractions and decimals to financial aspects. Respondents grasped the importance of savings, budgeting and dividing or balancing things specially about money.

Conclusions

Based on the findings of the study, the following conclusions are drawn:

1. The interactive module incorporating financial literacy enhanced students conceptual understanding of fractions and decimals, increased financial knowledge and improved motivation to achieve in mathematics.
2. Through the interactive module, students realized the importance of savings, budgeting and becoming financially wise.

Recommendations

The researcher, therefore, recommends the following based on the findings of the study:

1. Since the interactive module incorporating financial literacy enhanced students conceptual understanding of fractions and decimals, increased financial knowledge and improved motivation to achieve in mathematics, it is highly recommended for use in Grade 7 Mathematics classes.
2. From the difficulties encountered by the students while complying the interactive module, clearer instruction and proper citation of the links must be considered. Additional interactive links incorporating financial literacy aside from videos to show more application of fraction and decimals to real life scenario be incorporated.
3. Since there was a progressive effect of the interactive module on students' management skills, financial attitude and peer influence, a continued study be conducted to measure how the students apply their financial knowledge in action.
4. Having known the positive effects of the interactive module on students, similar interactive modules incorporating real life skills be developed.

References

[1] Acelajado, Maxima J. (2005). ''The Modular Teaching Approach in College: An Alternative to Improving the Learner's Achievement, Persistence and Confidence in Mathematics.'' Experimental study at De La Salle University, Manila.

[2] Adora, Nelia M. (2017). Mathematics Performance of Grade Six Pupils in the National Achievement Test, Division of Northern Samar: Basis in Developing a Workbook in Elementary Mathematics VI. International Journal of Science and Research (IJSR) ISSN (Online) : 2319-7064 Index Copernicus Value (2015): 78.96 | Impact Factor (2015): 6.391

[3] Ahmad, S. ; Rosmani, A. ; Ismail, M. and Shakeri, S. ''An Introductory of Mental Arithmetic Using Interactive Multimedia for Pre-School Children'' 2010. Online May 4, 2015. http://www.ccsenet.org/journal/index.php/cis/article/view/6553/6115

[4] Alsmadi, I. ; Nusir, S. ; Al-kabi, M. ; Sharadgah, F. ; Acta Didactica Napocencia ; Cluj-Napoca. ''Studying the Impact of Using Multimedia Interactive Programs at Children Ability to Learn Basic Math Skills" 2012. Online: May 4, 2015. http://search.proquest.com/pqcentral/docview/839765673/5C4E035D67174A73PQ/4?accountid=141440#

[5] Ball, D. L., & Cohen, D. K. (1996). Reform by the book: What is - or might be - the role of curriculum materials in teacher learning and instructional reform? Educational Researcher, 25(9), 6–8, 14. doi:10.3102/0013189X025009006

[6] Bicoy and Teopez (2013). Enhancement and Implementation of Cosinometer on Laws of Sine and Cosine. CED, MSU – IIT. Tibanga, Iligan City.

[7] Bruner J. (1986). Actual minds, possible worlds. Cambridge, MA: Harvard University Press.

[8] Buan, Amelia T. (2010). Utilization of Web 2.0 Tools for Learning. The Mindanao Forum. vol. 23 no. 1.

[9] Cagas and Padua (2002). Development of Instructional Module on Human Nervous System and Sense Organs. CED. MSU-IIT. Tibanga, Iligan City.

[10] Cales and Velasco (2013). Development and Validation of a Laboratory Activity and Teacher's Guide in Teaching Radiant Energy Using Modified (DIY) Differential Radioscope. CED. MSU-IIT. Tibanga, Iligan City.

[11] Capilitan, Diamer B. (2010). Development of an Interactive Instruction System in Science and Health VI. CED-SGS, MSU – IIT. Tibanga, Iligan City.

[12] Crisostomo, Ricardo M., et. al. (2013). Our World of Math 7. Vibal Publishing House, Inc.

[13] Daguan, Jeanie O. et.al. (2000). Development of Modular Instruction Materials for Basic Architectural Drafting and Design, Iligan City; CED, MSU-IIT.

[14] Fonseca, Jaime R. S., and Margarida G.M.S. Cardoso. (2007). Mixture-Model Cluster Analysis using Information Theoretical Criteria, Intelligent Data Analysis.

[15] Forgues, H., Tian, J., and Siegler, R. (2015). Why is learning fraction and decimal arithmetic so difficult? Online: April 21, 2018. http://www.psy.cmu.edu/~siegler/2015-LF-etal.pdf

[16] Germata, Jeanylyn R., Rusam Brechie M. Ongue, Junry U. Narbasa (2008) Development and Validation of Module in Wood Finishing. Unpublished Undergraduate Thesis. Bachelor in Industrial Arts Technology, CED, MSU-IIT, Tibanga, Iligan City.

[17] Kell, Peter . ''MoneySmart Teaching: bringing financial literacy lessons to life in the classroom.'' 2012. Online: April 20, 2015. file:///C:/Users/eljane/Documents/Fhelmar/Curriculum%20&%20Leadership%20Journal%20_%20MoneySmart%20Teaching_%20bringing%20financial%20literacy%20lessons%20to%20life%20in%20the%20classroom.html

[18] Lardizabal, Amparo S. (1990). Principles and methods of teaching - 3rd ed. - Quezon City Phoenix Publishing House, Inc. - xiv, 307 p. col. ill.; 25 cm.

[19] Lusardi, Annamaria and Mitchell, Olivia S. "Financial Literacy around the World: An Overview." 2011. Online: April 21, 2015. https://www.cambridge.org/core/journals/journal-of-pension-economics-and-finance/article/financial-literacy-around-the-world-an-overview/0488F901318E0FBC4C92DC6E964AB89C

[20] Mayer, Richard E. (2014). Principles for MultiMedia Learning. https://hilt.harvard.edu/blog/principles-multimedia-learning-richard-e-mayer.

[21] Melosantos, Luis Allan B., et al. (2015). Math Connections in the Digital World. Sibs Publishing House.

[22] Moreno, Roxana (2006). Learning in High-Tech and Multimedia Environments. http://journals.sagepub.com/doi/abs/10.1111/j.0963-7214.2006.00408.x

[23] Moten, James M., Jr. (2011). Examining the Disparity in Financial Literacy Between High School Seniors of Different Ethnicities and Income Level. Northcentral University, ProQuest Dissertations Publishing. https://search.proquest.com/central/docview/912191994/abstract/A6EB7DD0A305416DPQ/1?accountid=167112

[24] Nott, C., Nott, G. & Lee, C.C. (2000). Simply simulation: an interactive cd-rom-based approach for learning simulation concepts. Proceedings of the 2000 Winter Simulation Conference, Society for Computer Simulation International.

[25] Pascasio, Arlene A., et al.(2013). Math Ideas and Life Applications. Abiva Publishing House, Inc.

[26] Paschke J. & Paschke C. (2005). Flashing into a child's mind: The educational impact of children's multimedia on learner satisfaction. Unpublished dissertation, Mercer University.

[27] Reibman, Shayna Elizabeth (2002). Achievement motivation in Latino college students: The role of personality characteristics, acculturation, and familism. Alliant International University, Fresno, ProQuest Dissertations Publishing. https://search.proquest.com/central/docview/305484554/abstract/CBFE85994B394432PQ/15?accountid=167112

[28] Resnick, Mitchel (2007). All I Really Need to Know (About Creative Thinking) I Learned (By Studying How Children Learn) in Kindergarten. ACM Creativity & Cognition conference. Washington DC.

[29] Richa and et. al. (2010). Development and Validation of Self Learning Kit for Similarity. CED. MSU-IIT. Tibanga, Iligan City.
[30] Rodriguez, C. A. and Buan, A. T. (2018). Development of A Microlesson in Teaching Energy Levels of Atoms.
[31] Smith, E. et. al.(1968). Discoveries in Modern Mathematics Course 2. Charles E. Merril Publishing Co.pp. 126-129.
[32] Ulpina, Jisela N. and Tizon, Lydia T. (2014). Math Builders 7. Jo-es Publishing House, Inc.
[33] Villadolid, Zarina and Buan, Amelia (2012). Development of a Project – Based Unit Incorporating Entrepreneurial Skills in the Teaching of Percent to Elementary Pupils. SGS. MSU-IIT. Tibanga, Iligan City.
[34] Wiggins, G., & McTighe, J. (2011). Understanding by Design. http://www.ascd.org/ASCD/pdf/siteASCD/publications/UbD_WhitePaper0312.pdf.
[35] Ye, Stuart S. (2010). Understanding and Addressing the Achievement Gap Through Individualized Instruction and Formative Assessment. Assessment in Education: Principles, Policy & Practice, 17:2, 169-182, DOI: 10.1080/09695941003694466
[36] Zabala, Dante Joma P. (2011). Development of a Project – Based Unit Incorporating Entrepreneurial Skills in the Teaching of Percent to Elementary Pupils. SGS. MSU-IIT. Tibanga, Iligan City.

IOP Conf. Series: Journal of Physics: Conf. Series **1340** (2019) 012050 doi:10.1088/1742-6596/1340/1/012050

STEM education program: manufacturing mixture of phosphate and potash fertilizer straws and waste of animal bones

Nguyen Mau Duc [1*], Tran Trung Ninh[2]; Ngo Thi Toan[3], Kieu Thi Hai [4], and Chokchai Yuenyong[5]

[1] Thai Nguyen University of Education, Thai Nguyen, Vietnam
[2] Faculty of Chemistry – Hanoi National University of Education, Vietnam
[3] Ly Nhan high school – Ha Nam, Vietnam
[4] Phuc Tho high school – Ha Noi, Vietnam
[5] Faculty of Education, Khon Kaen University, Thailand
[*]Corresponding author email: nguyenmauduc@dhsptn.edu.vn

Abstract. STEM educational orientation program is one interdisciplinary approaching way during study which consists of academic terms with principle characteristics combines with lectures in real life. Joining in STEM educational orientation program, students will have practical experience, thus they have opportunities to develop their capabilities including ability to apply knowledge and skills (AAK&S), help students to improve study skills and methods such as planning, analyzing, summarizing information, suggest creative solutions; encourage motivation, benefits and study activities. Based on the foundation of utilizing STEM educational orientation program in teaching, it can be a direction to develop student's capabilities. In this study, the author instructed students to research "Manufacturing mixture of phosphate and potash fertilizer from straws and waste of animal bones" following STEM educational orientation program in improving abilities and applying Chemistry knowledge to life for the student. The result shows STEM system could handle the theoretical weight situation, help students shape and develop core competencies, specialized Chemistry and especially, the ability to apply knowledge and skills in a systematic and solid manner.

Keywords: STEM, mixture of phosphate and potash fertilizer, straws, waste of animal bones

1. Introduction

There are many concerning with the STEM education in shaping innovation and development. This push the policy of many countries continues under the orientation of STEM education [8]. However, there are three consensus issues of discussion about STEM education including (a) instructionally (b) as a set of integrated or interconnected disciplines, or (c) as more dependent on the stakeholders or context in which it is viewed or conceptualized. For this study, we will hold Moore et.al. [7] definition of STEM education for enhancing teachers. They define it as ''the teaching and learning of the content and practices of disciplinary knowledge which include science and/or mathematics through the integration of the practices of engineering and engineering design of relevant technologies.'' To them, five characteristics distinguish integrated STEM instruction from other teacher pedagogy: (a) the content and practices of one or more anchor science and mathematics disciplines define some of the primary learning goals; (b) the integrator is the engineering practices and engineering design of technologies as the context; (c) the engineering design or engineering practices related to relevant technologies requires the use of scientific and mathematical concepts through design justification; (d) the development of 21st century skills is emphasized; and (e) the context of instruction requires solving a real-world problem or task through teamwork. This conceptualization of STEM is grounded in learning research [12].

STEM education appeared in Vietnam a few years ago. It has just been the stage of media announcement and experimental. It has not been an official education activity in high school. However, STEM education with the tasks of providing crucial knowledge and skills for students in 21st Century will become a widespread education model in the future [6]. Therefore, STEM education is highly needed the concern and awareness of the whole society.

STEM is the integration of Science, Technology, Engineering and Mathematics with real life situations, close and attractive to the learner and they need to explain, analyze, argue or conduct experiments, build models to solve problems. On the other hand, the current demand for jobs related in the STEM fields is growing and the skills of the 21st century workers also require much of the related skills that STEM education can provide. Therefore, the use of STEM educational orientation program in teaching at high schools could bring good effects and meet the requirements of society [8]; [1]. Thus, how to organize STEM activities so that students can develop the most effective their ability to apply knowledge and skills (AAK&S)?

Considering the issues of environmental pollution and human being, the applying chemical knowledge and others in everyday life is needed for people. In the recent years, a large of straws after harvesting is burned indiscriminately causing environmental pollution and many potential risks for human being. Moreover, a number of animal bones from restaurants are dropped to environment as solid waste causing polluted environment. However, if the waste is treated thoroughly, it could bring high economic benefits and also reduce environmental pollution. With "Manufacturing mixture of phosphate and potash fertilizer from straws and waste of animal bones", it could develop AAK&S for students, create confidence, excitement for them and improve effectiveness in studying, meet the target of innovating teaching methods and improving the quality of education. As suggestions of literature review [6], [9], [10] the manifestations of AAK&S were mentioned about (1) be creative in dealing with practical issues, (2) understanding of practical problems and applying chemical knowledge to solve practical problems, (3) finding out the content of chemical knowledge is applied in real life, (4) analyzing and summarizing chemical knowledge, and (5) systematizing knowledge, classifying chemical knowledge.

In addition, this helps students familiarize themselves with the research planning, process design. Students may have chance to design the process with a lot of advantages. These included, for examples; (1) the method of using straws to break bones that saves time, electricity and cost, (2) manufacturing mixture of phosphate and potash fertilizer with %P2O5 = 33,62% and %K2O = 36,8%, and (3) low cost of fertilizer. On the other hand, the greatest significances of compound fertilizer made from straws and waste of animal bones included (1) providing nutrition for plants to achieve high productivity and good quality, (2) adding nutrients to the soil, improving soil fertility, contributing to soil improvement, and (3) reducing the risk of environmental pollution (http://binhdien.com 2018) [3]' [4].

Regarding on manufacturing Mixture of Phosphate and potash fertilizer straws and waste of animal bones, it could provide students chance to apply STEM knowledge for their developing AAK&S. Therefore, STEM education intervention of manufacturing mixture of phosphate and potash fertilizer straws and waste of animal bones may develop AAK&S of students as well as other abilities such as solving problem skill, cooperation skill, creativeness, and experiment skill. Students may analyze the current situation and environmental impacts of burning straws and directly leaving unprocessed animal bones into environment from restaurants. From that, it could build the idea to utilize those wastes because straws contain potassium, animal bones contain calcium and phosphorus, and also suggest using straws and wastes of animal bones to produce mixture of phosphate and potash fertilizer [5], [11].

2. Methodology

Methodology regarded experimental design in order to evaluate ability to apply knowledge and skills (AAK&S) of high school students through STEM activities.

2.1. Target group

In order to evaluate ability to apply knowledge and skills (AAK&S) of high school students through STEM activities, the authors carried out "the STEM education intervention of manufacturing mixture of phosphate and potash fertilizer straws and waste of animal bones" on two groups of student, each group has 10 people at Ly Nhan High School- Ha Nam Province. The experiment time took more than 3 months from March 10th, 2018 to June 28th, 2018.

2.2 Methods of Inquiry

According to development objectives for students (including ability to apply knowledge and skills), the author made initial observations: STEM activities could develop ability to apply knowledge and skills of readers. From that, the author built the survey to find out difficulties and also actual demands of students about STEM educational activities.

The author developed a system of STEM activities and proposed the process of organizing these activities with the target of developing abilities to apply knowledge and skills for students. The effect of the impact will be measured on the measurement scale of ability to apply knowledge and skills in solving problems of students with assigned tasks. Analyzing the impact will help the author answer the question: Is it possible to develop ability to apply knowledge and skills of students through STEM activities? How should STEM activities be organized to the best development of students' ability to apply knowledge and skills?

With the number of students taking part in at the same time, the author also found it difficult to monitor and evaluate the development of each student's ability to apply knowledge and skills. Therefore, the author conducted pedagogical experiments with the students volunteering to participate under the agreement of the school and parents. For the most accurate evaluation ability to apply knowledge and skills of each student, the author combined the combined the assessment through the survey, peer review and self-assessment. To evaluate the effectiveness of each group, the authors based on the results of analysis of fertilizer samples, experimental results on plants and product cost. Especially, in order to check the survey, the author recorded the whole process they worked and discussed.

2.3 STEM education intervention of manufacturing mixture of phosphate and potash fertilizer straws and waste of animal bones

Topic of "Manufacturing mixture of phosphate and potash fertilizer from straws and waste of animal bones" is one expended theme of STEM. After giving the production process, students have to proceed the fertilizer and then analyze the content of nutritional elements N, P in fertilizer, evaluate the quality of fertilizer, count the price and do the experiment on plants. Analysis of fertilizer samples is expensive and experiments on plants take time and effort, thus, the author found it very difficult to organize the whole class to participate in activities that the group had designed. This intervention aimed to develop AAK&S of students as well as other abilities such as solving problem skill, cooperation skill, creativeness, and experiment skill. To explain the figure of intervention; the manifestations of AAK&S of students, STEM knowledge in the intervention,

2.3.1 Manifestations of AAK&S of students in STEM activities

Manifestations of AAK&S of students who are participating in STEM activities of manufacturing mixture of phosphate and potash fertilizer straws and waste of animal bones could be clarified as follows:

- Applying chemical knowledge to explain, prove the practical problems.
- Understanding and explaining chemical applications with different practical problems and areas.
- Understanding and explaining practical issues related to chemistry.
- Applying chemical knowledge and interdisciplinary knowledge (kiến thức liên môn) to explain some natural phenomena, applications of chemistry in life.
- Ability to analyze and summarize chemical knowledge in order to review and evaluate the impact of a practical problem.
- Evaluation: Applying the knowledge to debate, inspect the impact of a practical problem.
- Creation: Applying knowledge to propose some new methods, model designing, problem solving plan.
- Appropriate behavior in situations related to the health of themselves, their families and society. Appropriate behavior follows the natural requirements of sustainable development of society and environmental protection.

2.3.2 STEM knowledge in the intervention

- Science (S): Fertilizer knowledge about nutritious elements, roles, ingredients, methods in order to suggest ideas about resources and production process from straws and animal bones. Designing experiments in order to inspect the quality of fertilizer for plants.

- Technology (T): Using straws and animal bones could reduce environmental pollution as well as make use of raw materials to produce fertilizer. Fertilizer sample not only contains high levels of phosphorus, potassium, calcium but also improve the soil.

- Engineering (E): Fetilizer production process, machines which measure content of nutritious elements in fertilizer sample.

- Math (M): Method to build a baseline measuring P (phosphorus), K (potassium- kalium) and Ca (calcium) in fertilizer sample.

2.3.3 Process

STEM education intervention of manufacturing mixture of phosphate and potash fertilizer straws and waste of animal bones was developed regarding on the process in the figure 1.

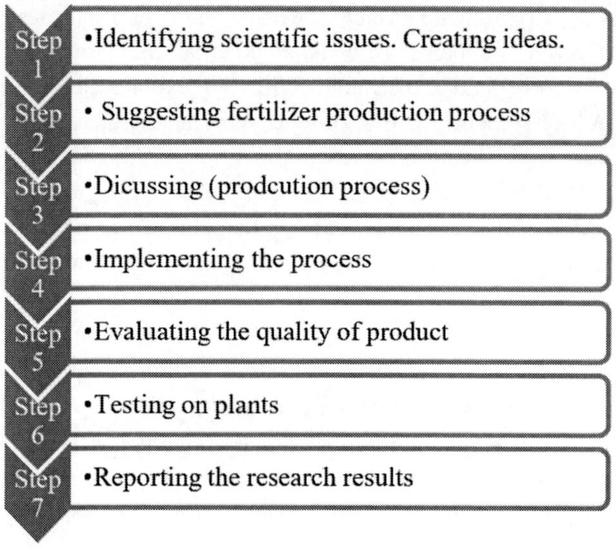

Figure 1: process of developing STEM education learning activities [6]

Step 1: Identifying scientific issues. Creating ideas, teacher prepares activity-oriented questions as following:

- Searching science information through documents such as Chemistry textbook, Technical Chemistry textbook, and Analytical Chemistry. Teachers provided these questions:
  - What are the chemical elements of phosphate and potash fertilizer?
  - What is the role of phosphate and potash fertilizer for plants?
  - In industry, what are raw materials using to produce phosphate and potash fertilizer, and how to produce this kind of fertilizer?
  - If these raw materials are increasingly exhausted, what alternative sources of raw materials can be used?

- Organization for student to do research, analyze about economic and social situation in Ha Nam and identify the scientific issues. Teachers provided these questions:
  - Ha Nam is an agricultural province; the area of rice cultivation is large so there is the amount of straws after harvesting. To prepare for the next crop, farmers usually burn straws on the fields and it creates smoke that causes the environmental pollution. How to reduce this problem?
  - A number of animal bones from restaurants are dropped out the environment as solid wastes making the environment polluted. However, if the waste is treated thoroughly, it could bring high economic benefits, not only reduce environmental pollution. In animal bones contain mainly compounds of P are calcium phosphate Ca3(PO4)2 and hydroxyl calcium apatiteCa5(OH)(PO4)3) which could be made use of nutritious elements providing for plants (Dumitrescu, Manciulea, Sauciuc, Zaha 2009).

Figure 2. Burning straws indiscriminately

Figure 3. Waste of animal bones

Step2: Suggesting fertilizer production process. There are three phases including processed materials, phosphate fertilizer production, and phosphate and potash fertilizer production as following details.

Phase 1: Processed materials: Breaking and crushing bones into powder, laying straws in ashes. Students were provided two options including option 1: Using electricity to burn bones and crush into powder, burning straw on the fields to collect ashes; and option 2: Using straws to burn the bones in the incinerator, then collecting the ashes, and crushing the bones into powder.

Phase 2: Phosphate fertilizer production. There are two stages. The stage 1 is bone meal hydrolysis which provided two options including (1) option 3: using H_2SO_4 solution and (2) option 4: using H_3PO_4 solution. The stage 2 is neutralizing residual acid, rotting after neutralization to phosphate fertilizer.

Phase 3: Phosphate and potash fertilizer production which provided 2 options. These included (1) option 5: mixing ashes and phosphate fertilizerat a random proportion, and (2) option 6: mixing ashes and phosphate fertilizerat a specified proportion

Step 3: Discussing (Manufacturing process). The direction of the activities in the step 2 should be discussing in the following aspects.

In phase 1 processed material: every group follows option 1 and 2 respectively. In option of using electricity to burn bones and burning straw on the fields, students use an oven to break bones around 24 hours, and it does not take advantage of the heat generated by the straws burning process. In option of using straws to burn the bones in the incinerator, students do experiments using straws to break bones in 1 hour so that it saves time, electricity and production costs. And, straw ashes for planting also helps making soil easier, providing some minerals such as Ca, Silica, Potassium, Phosphorus and some other trace elements. Besides, these ashes contain alkaliso it could neutralize acidity in soil. Based on students doing two options, they may draw conclusion to choose the option of using straws to burn the bones in the incinerator.

Phase 2: Phosphate fertilizer production: Every group follows option 3 and 4 respectively.

- Option 3: Using H_2SO_4 solution
    + Stage 1: Hydrolyzing by H2SO4 solution with concentration about 57% (some documents applying concentration from 62% to 68%).
    + Stage 2: Neutralizing residual acid in fertilizer (If having) by NH3solution or Ca(OH)2 solution or bone meal.
- Option 4: Using H3PO4 solution

+ Stage 1: Hydrolyzing by H3PO4 solution diluted to concentration with the appropriate value.
+ Stage 2: Neutralizing residual acid in fertilizer (If having) by NH3 solution or Ca(OH)2 solution or bone meal.

Discussion in phase 2, students should discuss between 2 options above, if Option 3 is chosen, H_2SO_4 solution with concentration about 57% is a strong acid with high concentration, residual acid could create high acid environment that makes plants die. If Option 4 is chosen, H_3PO_4 solution is an average acid with low concentration, it does not contain residual acid but it provides phosphorus for fertilizer. Then, they may draw conclusion to choose the option 4.

Phase 3: Phosphate and potash fertilizer production: Every group follows option 5 and 6 respectively. Different crops have different demand for nutrient elements. Therefore, each crop has a different rate of mixing of phosphate fertilizer made from bones and potash fertilizer extracted from straw ashes as provided in the table 1 and 2. Based on these experiments, they may draw conclusion to choose the option 5.

Table 1: Dividing the proportion of phosphate to potash fertilizer outside the market which should be applied to some agricultural plants.

Crops	Productivity (Quintal/Ha)	The amount of nutrients used to produce 1 quintal of harvested product (kg/Quintal)	
		P_2O_5	**K_2O**
Rice	30	0.4	2.7
Corn	20	0.6	3.0
Sweet potato	200	0.1	0.7
Cassava	100	0.1	0.6
Peanut	20	0.7	2.5
Soybean	10	0.7	2.2

Table 2: Dividing the proportion of phosphate fertilizer to straw ashes (%P2O5 = 33,62%, %K2O = 36,8%) which should be applied to some agricultural plants.

Crops	Productivity (Quintal/Ha)	The amount of nutrients used to produce 1 quintal of harvested product (kg/Quintal)	
		P_2O_5	**K_2O**
Rice	30	0.33	1.57
Corn	20	0.3	1.06
Sweet potato	200	0.28	1.5
Cassava	100	0.42	1.75
Peanut	20	0.32	0.8
Soybean	10	0.4	0.78

IOP Conf. Series: Journal of Physics: Conf. Series **1340** (2019) 012050 doi:10.1088/1742-6596/1340/1/012050

Step 4: Implementing the process. These included preparing and implementing.

Preparing:

- Ingredients: Wastes of animal bones: 10 kg; Straws: 40 kg
- Chemical substance: H3PO4solution (concentration 80%); NH3solution, Ca(OH)2 solution
- Tools:Glasses with different capacity 50ml,100ml,200ml; grinder, hopper; measuring tube, glass chopstick; scale; oven,glass funnel, triangular vase, blue filter paper; beam bottle.

Implementing: Teachers divide the group and ask the students to follow the designed process as the figure 4 and experiment as the figure 5.

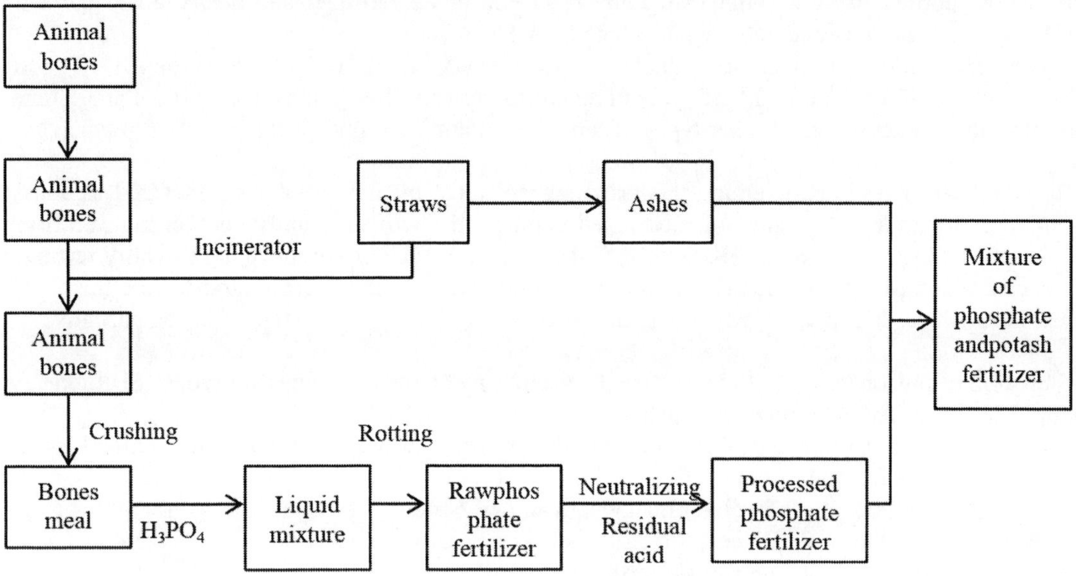

Figure 4: the designed process of implementing

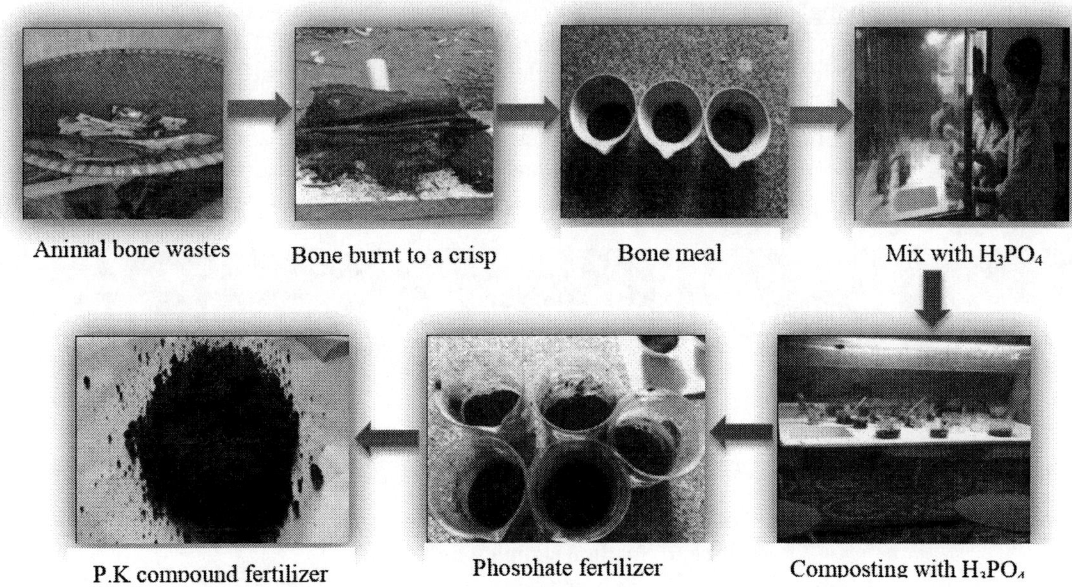

Figure 5: The diagram of P.K. compound fertilizer modulation

Step 5: Evaluating the quality and calculating the price of the final product. Students had to calculate average price of manufacturing cost for making 1 kg. For example, they may calculate as following.

- Making the bone wastes into bone meal: 0.043USD
- Dust disposal (activated charcoal, calcium hydroxide solution): 0.043USD.
- H_3PO_4 (80%) solution (industrial type): 0.54 USD.

From 1kg of bone with chemicals and costs above, the researchers were able to modulate 1,8kg of M2 fertilizer. Hence, approximately 1kg of M2 fertilizer (%P_2O_5 = 33.62% ,%K_2O = 36.8%) will be modulated at the cost of 0.35 USD (this is the modulation cost in the laboratory) while having the nutritious proportion twice as high than Lam Thao phosphate fertilizer and nearly up to par with potash fertilizer manufactured at Supe phosphate Lam Thao factory.

Comparing with Lam Thao phosphate fertilizer (%P_2O_5 = 16÷16,5%) costing 0.17 USD and(%K_2O = 50%- 60%) at 0.43 USD, modulated fertilizer can offer much cheaper price,at 3.5 times lower than the market prices of other types of fertilizers (Lam Thao phosphate, potash) (binhdien.com 2018).

Besides, it is possible to reduce the price significantly by having it manufactured as mass-production. However, as samples of modulated compound fertilizer usually contain high nutrition (%P_2O_5 from 27,82% ÷ 33,62%,%K_2O = 36,8%)), it could be used to produce high-quality fertilizer for special decorative plants/flowers. Also, compound fertilizer made from straws, rice husk, and animal bone wastes has significance to the environment as it could reduce the pollution partially.

Step 6: Testing on crops. Student started to experiment on two types of crops: Scallion and Kohlrabi with 5 different fertilizers, which are:

Sample 1: With Compound fertilizer modulated from straws, rice husk and animal bones (buffalo/etc.)
Sample 2: With Phosphate fertilizer made from animal bones.
Sample 3: With Potash fertilizer.
Sample 4: With Lam Thao phosphate fertilizer.
Sample 5: No fertilizer.

Results on scallion sample

After one month of experiments (From 21/5/2018 to 21/6/2018) in Crops at Nhân Chính commune - Lý Nhân District - Hà Nam Province, students measured the height and number of shoots in each scallion sample and got the final results as follow:

Final resuls from scallion (Density: 100 scallions/m^2), it showed that the sorting of quickest developed sample to sloweset is sample 1, 2, 3, 4, and 5; respectively as showed height of scallion shoots in the table 2. And, pictures of samples testing were provided in the figure 6.

Figure 6: Scallion samples for testing

IOP Conf. Series: Journal of Physics: Conf. Series **1340** (2019) 012050 doi:10.1088/1742-6596/1340/1/012050

Table 3: Average height of scallion samples

	Average height (cm)								
Date	**21/5**	**25/5**	**29/5**	**3/6**	**7/6**	**11/6**	**15/6**	**19/6**	**21/6**
Sample 1	9.1	10.2	11.5	12.5	13.5	15.3	17.2	19.75	22.2
Sample 2	7.5	8.3	9.4	10.3	11.2	13.2	15.4	17.5	19.4
Sample 3	6.6	7.2	8.1	8.5	9.5	10.5	11.5	12.75	14.6
Sample 4	5.8	6.5	7.4	8.2	9.2	10.2	12.5	14.25	15.8
Sample 5	2.4	3.2	4.5	5.8	6.7	7.3	8.5	10.3	12.5

Hence, the modulated compound fertilizer has positive effect on scallion testing samples as it made the green shoots developed quicker and higher than other samples (which were manured with other types of fertilizer, such as Lam Thao phosphate fertilizer or no fertilizer at all).

To increase the practical value, the researcher conducted the experiment on several crops of local farmers:

+ Size: Height 5m; Width 1.2m

+ Density: 100 scallions/m^2

+ Time: From 21/5/2018 to 21/6/2018

+ Place: Crops at Nhân Chính commune - Lý Nhân District - Hà Nam Province

Table 4: effect of the modulated compound fertilizer on scallion testing samples

SAMPLE / Result	S1	S2	S3	S4	S5
Average height (cm)/ scallion	14.6	12.5	9.9	10.0	6.8
Weight (kg)/6m^2	2.5×6=15	2.30×6=13.8	2.15×6=12.9	2.1×6=12.6	1.7×6=10.2
Profit/1 scallion furrow (USD)	13	12	11.2	11	8.9

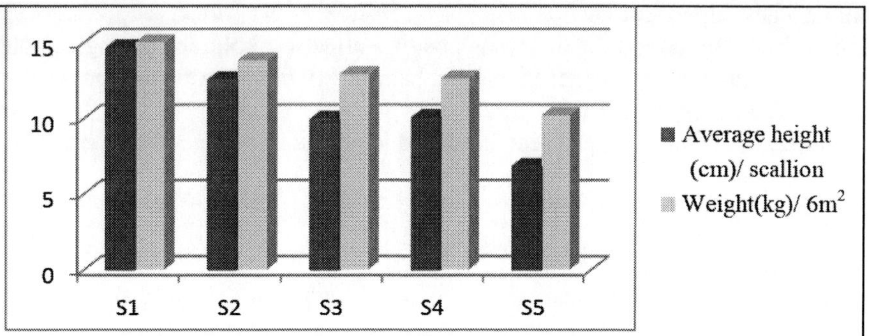

Figure 7. The graph show the testing results of scallion

Final results on kohlrabi sample

S1 S2 S3 S4 S5

Figure 8: *Kohlrabi samples for testing*

Table 5: Average size of kohlrabi bulbs with 5 different samples

	Average size per bulb (cm)								
(Date)	21/5	25/5	29/5	31/5	1/6	3/6	5/6	7/6	9/6
Sample 1	3.55	4.88	6.975	9.52	13.035	14.02	15.04	17.88	19.045
Sample 2	3.2	4.28	6.34	9.345	11.53	12.68	13.99	16.33	18.535
Sample 3	2.95	4.235	6.185	8.155	9.99	10.735	12.785	15.845	18.465
Sample 4	2.65	4.035	5.835	8.035	9.485	10.48	12.485	14.99	17.11
Sample 5	1.85	2.985	3.81	5.13	6.375	7.97	9.285	10.28	11.54

From the result above, shows that:

Sample 1 developed quickest
Sample 2 developed slower
Sample 3 developed slower than sample 2
Sample 4 developed the same with sample 3
Sample 5 developed slowest

Hence, the modulated compound fertilizer has positive effect on kohlrabi testing samples as it made the bulbs developed quicker and bigger than other samples (which were manured with other types of fertilizer, such as Lam Thao phosphate fertilizer or no fertilizer at all).

To increase the practical value, the researcher conducted the experiment on several crops of local farmers:

+ Size: Height 5m; Width 1,2m
+ Density: 25 kohlrabi bulbs/m^2
+ Time: From 21/5/2018 to 21/6/2018
+ Place: Crops at Nhân Chính commune - Lý Nhân District - Hà Nam Province

Table 6: effect of the modulated compound fertilizer on s kohlrabi testing samples

SAMPLE / Result	S1	S2	S3	S4	S5
Diameter/ 1 bulb (cm)	10.5	10.1	9.2	9.1	7.3
Average weight (kg)/6m^2	0.35×150= 52.5	0.3×150 = 45	0.26×150= 39	0.25×150 = 37.5	0.15×150 = 22.5

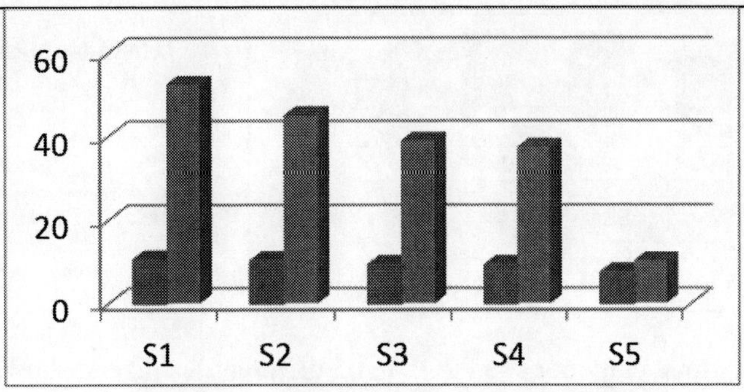

Figure 9. The graph show the testing results of kohlrabi

Step 7: Reporting the final results

1.Students started to test and found the optimal conditions for the procedure of modulating compound fertilizer from animal bones, which are:

- Mixing proportion: Weight of bone meal/ volume of H_3PO_4solution = 50g/150ml.
- Acid H_3PO_4 solution proportion: H_2O (v:v) appropriate proportions are (75:75); (60:90); (50:100) and (37,5:112,5). Especially, (60:90) is most suitable.
- Time for liquid acidic composting is 14 days.

2. Result for finding P content in phosphate fertilizer

The results of finding total content of % P2O5 in animal bone meal in 5 different fertilizers are presented in the Table 7 below

Table 7: Results of finding P content in modulated fertilizer samples

Sample	Animal bone wastes	Straws burning time (minutes)	Composting time (days)	P content in sample (μg/g)	P (%)	P_2O_5 (%)
S1	Pig	40	14	266194,60	13,31	30,56
S2	Mixed	45	14	231759,17	11,69	26,59
S3	Mixed	30	14	241228,65	12,06	27,63
S4	Buffalo/Cow	60	14	231532,46	11,58	26,57
S5	Buffalo/Cow	50	14	224335,32	11,22	25,8

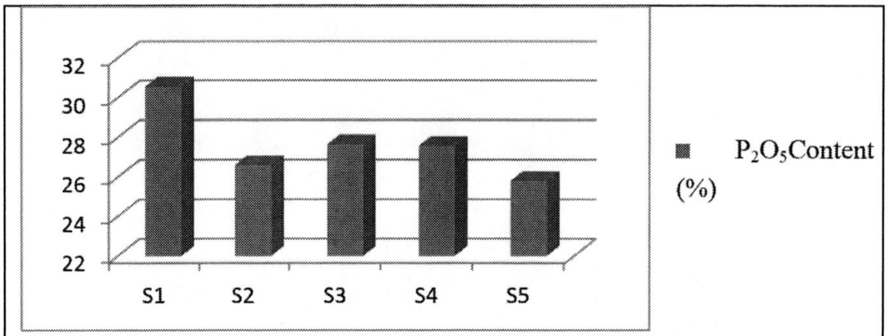

Figure 10. The graph shows P_2O_5content in fertilizer samples

- From the result of finding total content of % P2O5 in 05 different fertilizers, shows that

+ Ability of P in animal bones changing from indissoluble to dissoluble is relatively good.

+ % P2O5 in buffalo/cow is higher than in pig bones. Also, the required straw burning time for buffalo/cow and pig are 60 minutes and 40 minutes, respectively. Thus, it is less time-consuming to burn pig bones than buffalo/cow bones.

3. After actual test on 2 samples, which are Scallion and Kohlrabi, the results showed that modulated compound fertilizer had high economy efficiency and affected positively towards crops:

- The kohlrabi sample manured by compound fertilizer brought better results with bigger bulbs and quicker development, than other samples with different fertilizers: phosphate fertilizer from buffalo/cow bones, Lam Thao fertilizer, potash fertilizer, and no fertilizer.

- The scallion sample manured by compound fertilizer brought better results as it grew taller with more green shoots and quick development.

3. Findings

The process of the pedagogical experiment, provided in the section of 2.3.3, was implemented to students at Grade 11 chosen randomly, in some class 11A4 – Ly Nhan high school, Ly Nhan district, Ha Nam province. The findings clarified the results of implementating in Ly Nhan high school. The findings revealed the following aspects:

3.1 Students and teachers' evaluating on the designed STEM activities provided chance for AAK&S

3.2 Comparing the AAK&S between male and female students

3.3 The development of students' AAK&S during their STEM activities

3.1 Students and teachers' evaluating on the designed STEM activities provided chance for AAK&S

For teacers, in order to evaluate whether STEM activities can be applied as above and create opportunities for students to practice their AAK&S, the researchers has gathered teachers' opinions from 25 different teacher from: Lý Nhân high school; Bắc Lý high school; Nam Lý high school; Nam Cao high school; Lý Nhân commune - Hà Nam province. The result (figure 11) shows that nearly all teachers agreedthat the STEM theme in the unit of "Manufacturing P.K. modulated compound fertilizer from straws and animal bone wastes" provided students chances to practice AAK&S.

IOP Conf. Series: Journal of Physics: Conf. Series **1340** (2019) 012050 doi:10.1088/1742-6596/1340/1/012050

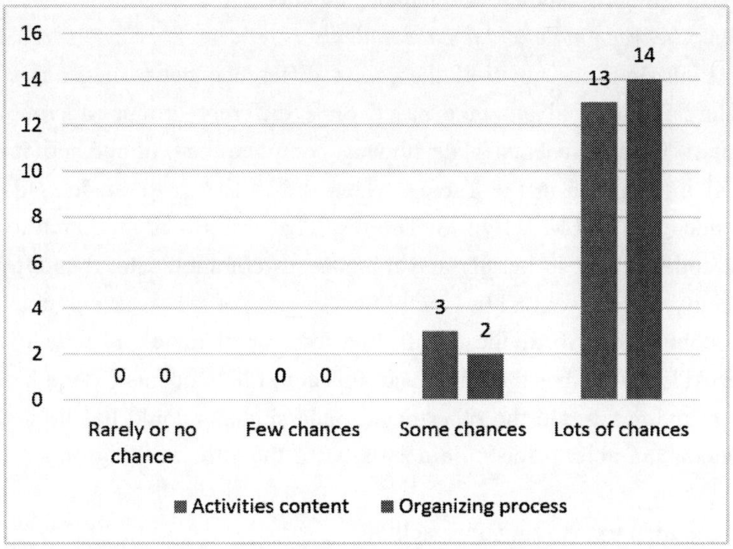

Figure 11. Results showing chances of students in practicing AAK&S

For students: based on the above process of AAK&S, the researchers listed out how much AAK&S students showed during their scientific research period. From data in observation assessment of teachers, the researchers collected initial results. It indicated that the STEM activity could provide good chance for students to practice their AAK&S. Especially, step of actual situations (Step 1) and step of discussing fertilizer manufacturing process (Step 3) could provide the most chances of students showing their AAK&S. This could be seen as the Figure 12.

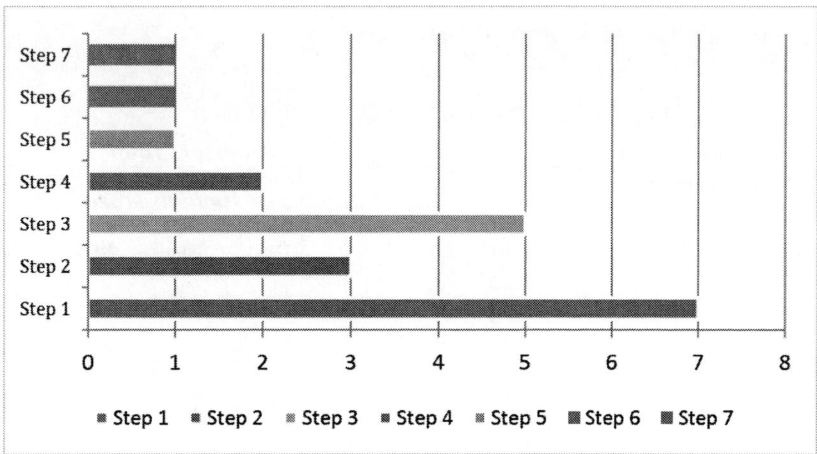

Figure 12: Chances of showing AAK&S in each stage in the organizing process of the STEM activity (approx. 20 students)

IOP Conf. Series: Journal of Physics: Conf. Series **1340** (2019) 012050 doi:10.1088/1742-6596/1340/1/012050

3.2 Comparing the AAK&S between male and female students

3.2 Comparing the AAK&S between male and female students

The researchers were also curious about if there is a difference between genders in terms of developing AAK&S. Here is the final outcome based on 2 different groups: a group of 10 male students and the other of 10 female students. Male students complete the required activities before the female group 10 days as information in the Table 8. This shows that male students do have better AAK&S than female students in STEM activities. The results also allow the research to come to the conclusion that female students tend to be shy and reluctant to communicate, thus affect their own AAK&S in STEM activity. Moreover, STEM activities might not be a strong points of female students. Hence, it is recommended to divide equally the number of male and female students in a group while doing other STEM activities. On one hand, this could let students having higher AAK&S to help other members in the group. On the other hand, male students could handle heavy, physical work, while female students are able to focus on details when they are working on a STEM activity together.

Table 8. Targets and required time to finish the activity

Target	Number	Required time to finish the activity
Male student	10	21/09/2018- 21/10/2018
Female student	10	21/09/2018- 31/10/2018

3.3 The development of students' AAK&S during their STEM activities

After completing the STEM activities systematically, in order to evaluate the development of students' AAK&S during their STEM activities, the researchers propose an observation point assessment. This allows teacher or student to check and give points based on 5 criteria mentioned earlier. Table 9 is the Observation points and will be given to each student by teachers. There in each criterion will be rated with scores range from 1 to 4. 1 means one point, 2 means two points, 3 means 3 points, and 4 means 4 points. The result is described as follow:

Table 9: The result of evaluating AAK&S based on the Observation point assessment

Criteria No.	***Experiment group***					***Control group***				
	The number of students score				***Criteria average point***	***The number of students score***				***Criteria average point***
	1,0	***2,0***	***3,0***	***4,0***		***1,0***	***2,0***	***3,0***	***4,0***	
1	1	2	8	9	3.25	3	7	7	3	2.50
2	1	2	9	8	3.20	4	6	7	3	2.45
3	1	1	9	9	3.30	3	8	5	4	2.50
4	1	2	9	8	3.20	4	6	5	5	2.55
5	2	2	7	9	3.15	4	5	6	5	2.60
Problem-solving capacity average point of experiment group = 3.22						Problem-solving capacity average point of control group = 2.52				
The difference of average point = 0.74										
The S.D. of experiment group = 0.94						The S.D. of control group = 1.10				
Allow independent verification test p= $1,28.10^{-6}$										
ES level = 0,64										

According to the table 9, the number of criteria could be labelled as following. Criteria number 1 is Systematize and categorize chemistry knowledge. Criteria number 2 is Collect and classify knowledge. Criteria number 3 is Recognize chemistry knowledge based on actual application. Criteria number 4 is Recognize problem during actual application and use learned knowledge to solve. And, criteria number 5 is Independent and creative when handling actual problems. The development of students for each criteria could be represented as mean score in the figure 13.

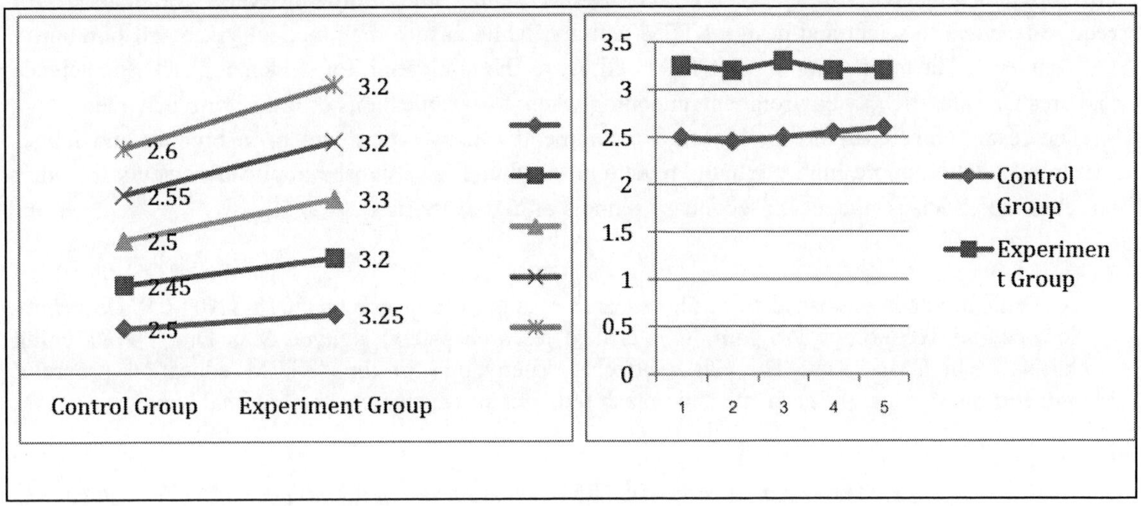

Figure 13. The graphs describe the ability to apply knowledge of students

According to the Figure 13, each criterion of AAK&S that the researchers evaluated during the experimental process of the chosen class, has all increased (as described in the left figure, each line of ability increases; in the right figure, the experiment group's line is higher than the control group).

According to the Table 9, the results in the table prove that the chosen experimental class for STEM model had much higher observation points than a class receiving traditional teaching (The different between Experiment group and Control group is 0,74). That proves STEM in education affect students greatly in the ability of apply knowledge and skills in studying chemistry.

Effect size (ES): ES =0,64. This ES value shows that using STEM model has affected students greatly in the development of their AAK&S. The p value for experiment group and control group is always <0,05, hence proved there is no randomness in the STEM experiment "Manufacturing compound fertilizer from straws, rice husk and animal bones". STEM applied in this activity has brought a great result.

During the process of designing STEM activities, the researcher decided that the first step (1st activity) and the last step (7th activity) would share the same level of difficulty. This was to let each student to show their ability equally. The final data analysis shows, through STEM activities, most students' scores went up (comparison between the 1st activity and the 7th activity), however, the results did vary among students. Some students had significant increase in their scores, while some others remained stable. The results also had no record of students whose AAK&S score decreased. Again, this reaffirms that STEM activity can increase students' AAK&S

4. Conclusion

With the objectives of building and developing students' ability, the discovering and using STEM advantages are totally beneficial and necessary towards Vietnamese education. Applying STEM in teaching science brings great effect in students' study. Student learns about how to collect needed information, show their creativeness, develop and improve their team-work skills or presentation skills. Also, increase their awareness in protecting the environment as well as their passion in pursuing science careers in the future.

From the actual experiment results, the researchers can conclude that it is possible to help student develop their AAK&S through STEM activities. Teaching and following STEM orientation will require teachers to select and design STEM activity, while ensure the practical value and familiarity with students. The teacher could use STEM activity as the motivation for students in studying science and create a safe, friendly environment, encourage them to ask questions or brainstorm new ideas.

The results concluded in this research recommend a new experiment on a broader and longer research to deliver more highly reliable results in the future, and will be applicable widely for other target groups, such as students in secondary schools or in universities.

Acknowledgement

This article is published from the research results of the topic ĐH2018-TN04-05" University of Science and Technology Program "in a critical research paper" Nguyen Mau Duc*; Tran Trung Ninh; Ngo Thi Toan, Kieu Thi Hai, Chokchai Yuenyong– in the "STEM education program: Manufacturing Mixture of Phosphate and potash fertilizer straws and waste of animal bones

References

[1] Brown J. (2012). The current status of STEM education research. *Journal of STEM Education: Innovations and Research*, 13(5), pp. 7-11.

[2] *Dumitrescu L., Manciulea I., Sauciuc A., Zaha C. (2009), "Obtaining biofertilizer by composting vegetable waste, sewage sludge and sawdust", Bulletin of the TransilvaniaUniversity of Braşov, 2 (51) Series I.*

[3] *Hei, L. Jin, P. Zhu, X. Ye, W. Yang, Y. (2016). "Characteristics of speciation of heavy metals in municipal sewage sludge of Guangzhou as fertilizer", Procedia Environmental Sciences, 31, pp. 232 – 240*

[4] *Igbokwe, P. K., Asadu, C. O., Okpe, E. C., and Okoro, S. E. (2015), "Manufacture of Bio Fertilizer by Composting Sawdust and Other Organic Waste", International Journal of Novel Research in Physics Chemistry & Mathematics Vol. 2, Issue 3, pp: (6-15).*

[5] *Lee E. J., Song U. (2010), "Environmental and economical assessment of sewage sludge compost application on soil and plants in a landfill", Resources, Conservation and Recycling, 54, pp. 1109-1116.*

[6] *Ministry of Education and Training - Vietnam Belgium Project (2010), Pedagogical application scientific research, Publisher of Pedagogical University.*

[7] Moore T.J., Johnson C.C., Peters-Burton E.E., Guzey, S.S., (2015). The need for a STEM road map. In: Johnson CC, Peters-Burton EE, Moore TJ (eds) *STEM road map: a framework for integrated STEM education*. Routledge, p 1.

[8] National Science and Technology Council. (2013). *A report from the committee on STEM education*. Washington, D.C: National Science and Technology Council.

[9] *Quang, L. X. (2016), "Problems in teaching Technology based on STEM education orientation", Journal of Science Ha Noi University of Education, 61 (6B), p.211-218.*

[10] *Quang, L. X. (2015), "STEM education – A solution in bulding and developing general education program after 2015", Journal of Education and Society, 6/2015, p. 37-39.*

[11] *Schuchardt F. (2000), "Composting of plant resi-dues and waste plant materials", In: Biotechnology, Volume 11c: Environmental Processes III, Weinheim 2000: Wiley-VCH, pp. 101-125.*
[12] Yuenyong, C. (2019). Lesson learned of building up community of practice for STEM education in Thailand. *AIP Proceeding*, 2081: 020002-1- 020001-6.

IOP Conf. Series: Journal of Physics: Conf. Series **1340** (2019) 012051 doi:10.1088/1742-6596/1340/1/012051

Developing Ellipse Properties Learning Unit with Computer Assisted Learning through GeoGebra software for Thai High School Students

Suphanat Setthaphongsakorn[1] and Chokchai Yuenyong[2]

[1]Khon Kaen University Demonstration Secondary School, Khon Kaen, Thailand,

[2]Science Education Program, Faculty of Education, Khon Kaen University,Thailand,

corresponding author's e-mail address: [1] ksupha@kku.ac.th , [2] ychok@kku.ac.th

Abstract. This study aimed to clarify the developing ellipse properties learning with computer assisted learning through GeoGebra software for Thai high school students. Participants include 40 Grade 10 students who were studying in Khon Kaen University Demonstration secondary school, academic year of 2016. Methodology regarded qualitative research. The intervention of Geometry with GeoGebra was provided for semester. The paper will explain how the GeoGebra will be provided for the opportunity students to construct and dynamically visualize geometric figures. GeoGebra builds a dynamic ellipse to show the characteristic of the points that form the ellipse. And, students may find the relation between two points on the ellipse by using GeoGebra. The unit provide students to generate their representation to explain ellipse definition and properties when they on tasks with the GeoGebra.

1. Introduction

Geometry is not only part of ancient cultures such as Indian, Babylonian, Egyptian, Chinese and Greek but also one of the core topics included in the modern mathematics syllabus that was approved by National Council for Teachers of Mathematics (NCTM). Geometry consists of theorem and many applied problem to solve. Geometry is an integral part of life as it has a bearing on aspects of life such as architecture design, art, building, construction, interior design and so on. It is connected to the visual, aesthetic and intuitive human facilities [6].

Ellipses find wide applications in mathematics, physics, and engineering. Elliptic shapes are used in architectural and design forms. It is difficult to build the ellipses form by hands because they have a continuously varying curvature. If an elliptic arch is to be built, each piece of the structure needs to be of a different shape [2]. Students can make sense of ellipses through a clear geometric derivation of the mathematical theory ($\frac{x^2}{a^2}+\frac{y^2}{b^2}=1$) that governs all four arc constructions of the ellipse. Based on this formula, students could plot graph to display the ellipses. And, also simplified computations using a computer could build the ellipses. It seems that students need to think about moving the continuous points to build the ellipses.

Transformation geometry may comprise of motions such as translation, reflection and rotation and also with dilations or shrinking and enlargement [1,10].explained transformation geometry as the study of figures and properties of figures conserved by reflections, translations, rotations, and their groupings. It provides the opportunity for learners to recognize and perform changes in the coordinates of the image of objects regarding their position, orientation, direction and size within a coordinate plane. In another word, Transformation geometry is defined as "a subset of geometry in which students learn to identify and illustrate movement of shapes in two and three dimensions. Transformation geometry, which can be characterized as the study of geometric objects in the plane, links the properties of transformations to the properties of objects. Geometric transformations should provide discovering and/or proving characteristics of geometric objects; forming patterns like friezes, rosettes, wallpapers; classifying geometric objects; perceiving the chirality of an object. The three types of movement are slides (translations, as when a figure is moved on a page), flips (reflections, that is, when a figure is turned over in three dimensions), and turns (rotations, when a figure is rotated 90° without being flipped)." [7]. In the present study, transformation geometry includes the patterns such as fractals, and the movement of the figures such as reflection, translation, rotation and combination of these.

GeoGebra, an open source dynamic geometry software, provides the opportunity to construct and dynamically visualize geometric figures [4,5], Fahlberg-Stojanovska and Trifunov [4] investigated a study to show how GeoGebra improved students' understanding of construction and geometric proof. They conducted a qualitative exploratory study by using tasks that include construction and proof problems for the relations on the triangles. The results showed that using GeoGebra in these tasks improves the percentage of students that are able to solve the triangle construction and proof problems [4]. This result is consistent with that of Christou et al [3].and Pandiscio [9] in terms of DGS's effectiveness in justification and verification of both geometric and algebraic problems' solutions.

Hohenwarter and Fuchs [5] argued some advantage of the GeoGebra. It promotes guided discovery, cooperative and experimental learning, multiple presentations and students` products in mathematics. Students can manipulate objects by dragging around the plane of drawing or using sliders to test mathematical ideas and see how these movements affect other variables. In this way, students have the opportunity to solve problems by investigating mathematical relations dynamically. Users can personalize their own creations through the adaptation of interface (e.g. font size, language, quality of graphics, color, coordinates, line thickness, line style and other features). The algebra input allows the user to construct new objects or to modify the existing ones by the command line. The worksheet files can easily be published as Web pages.

In order to provide transformation geometry, the GeoGebra need to be introduced to the geometry classroom. GeoGebra has the potential to encourage the student-centered learning, active student participation, collaborative learning, and discovery learning by experimenting mathematical ideas, theorems and using interactive explorations (Preiner, 2008). This study tried to develop the intervention of ellipse learning activities to enhance students to make sense of ellipses as transformation geometry through Geogebra.

2. Intervention of GeoGebra Ellipse Learning activities

Students access the GeoGebra software via http://www.geogebra.org. GeoGebra, an open source dynamic geometry software, provides the opportunity to construct and dynamically visualize geometric figures. The intervention of GeoGebra ellipse learning activities in this study may starting from drawing for understanding of definition of ellipse, and then using GeoGebra for construction and geometric proof, and constructing the ellipse and solving problems of ellipse.

2.1. Drawing on GeoGebra for understanding of definition of ellipse

The GeoGebra will be used to challenge students to construct meaning of ellipse definition. Ask students to draw the ellipse by following the definition of ellipse. The definition of ellipse is a regular oval shape, traced by a point moving in a plane so that the sum of its distances from two other points (the foci) is constant, or resulting when a cone is cut by an oblique plane that does not intersect the base. Students may draw the ellipse as below. This drawing may enhance students to understand the relation of PF + PF'equal a constant which also equal double times of vv'. Students may draw this relation as the figure 1.

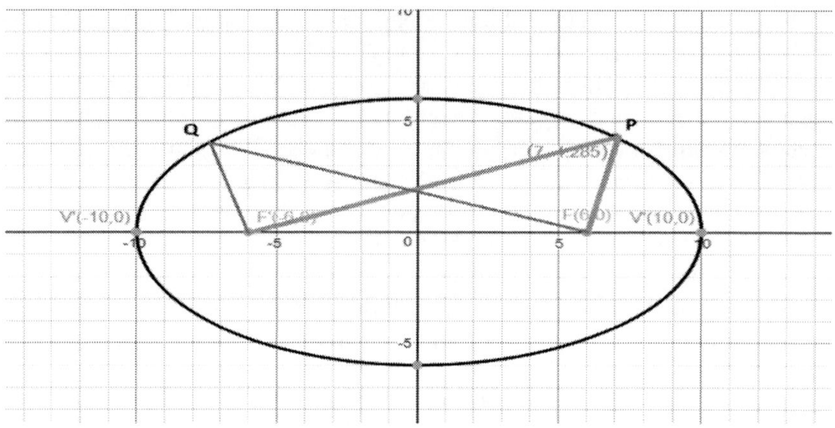

Figure 1: Drawing the relation of PF + PF' equal a constant

2.2. Movement of shapes to proof the representation about ellipse

Constructing meaning of ellipse through some external representations, such as formula and graph, will be provided through GeoGebra transformation geometry which students to identify and illustrate movement of ellipse shapes. The GeoGebra enhances students to recognize and perform changes in the coordinates of the image of objects regarding their position, orientation, direction and size within a coordinate plane. Students will be asked to rotate and move points for shapes to proof the representation about ellipse. Students make sense of ellipses through a clear geometric derivation of the mathematical theory *($\frac{x^2}{a^2}+\frac{y^2}{b^2}=1$)*. They tried to explain the relation the points on the ellipse by rotate the points on the GeoGebra as the figure 2.

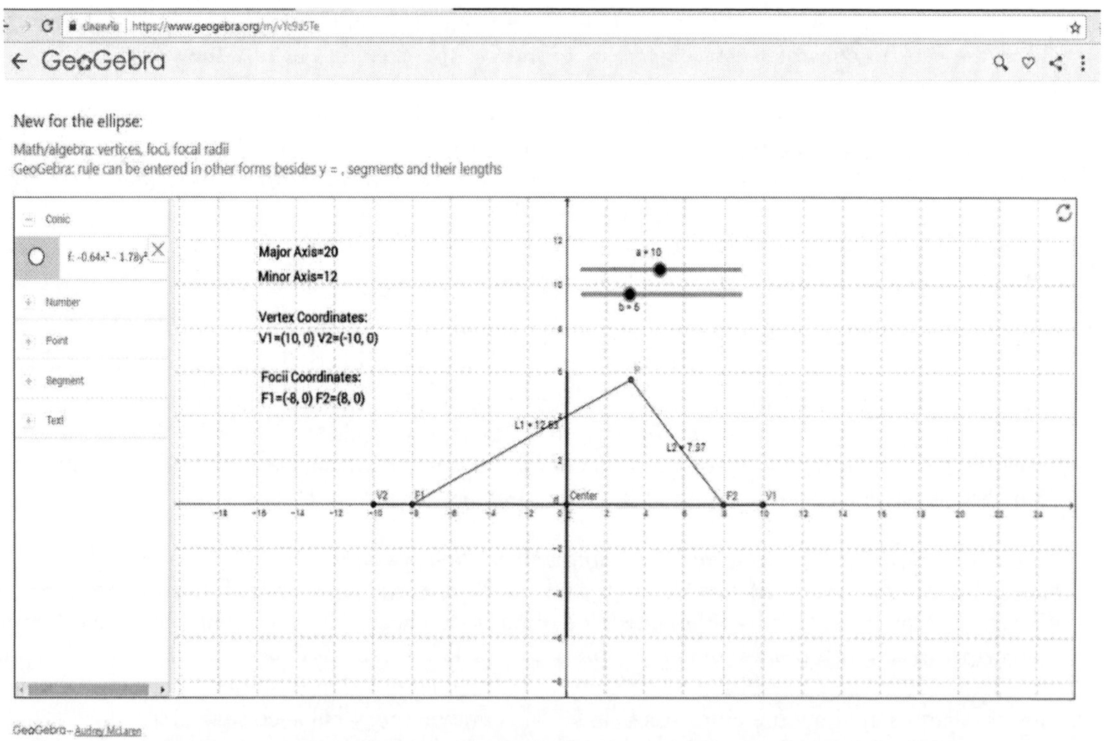

Figure 2. Rotate to explain the relation the points on the ellipse

2.3. Constructing the ellipse and solving problems of ellipse.

Students were enhanced to construct meaning of ellipse through dynamically visualize geometric figures. It indicates that GeoGebra improved students' understanding of construction and geometric proof. They conducted a qualitative exploratory study by using tasks that include construction and proof problems for the relations on the ellipse. The examples of GeoGebra learning activities were provided as following. According to the figure 3, student pointed the position P on the ellipse and then he or she dragged the position P to another point as showed in the figure 4. To do so, student may learn the relation between PF1 and PF2 where sum of PF1 and PF2 is constant. It is a property of ellipse. This could mention that the GeoGebra representationally connected students to construct meaning of ellipse.

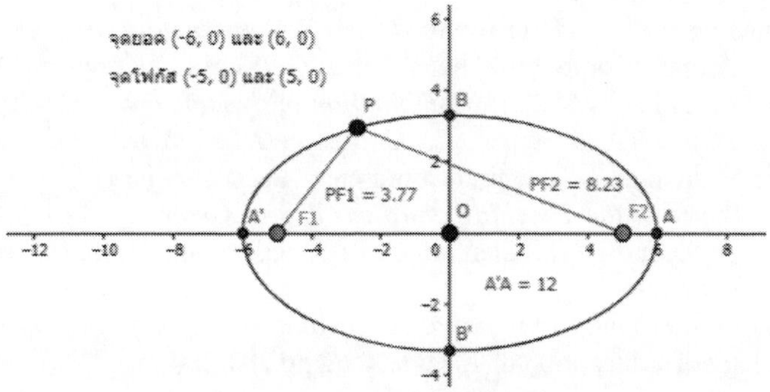

Figure 3. Constructing the ellipse and solving problem

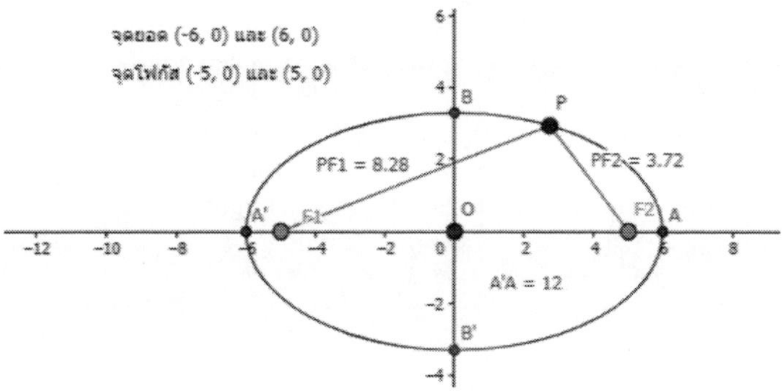

Figure 4. Constructing the ellipse and solving problem

3. Conclusion

The paper could show a method of utilization of computers by the GeoGebra as computer assisted learning about geometry. The GeoGebra could provide students chance to learn geometry as transformation geometry. It provides some ideas to enhance student active learning by experiment mathematical ideas about geometry. The drawing of ellipse is easily done which students could individually make sense definition of ellipse. The activities of rotate and move the coordination on the ellipse could provide the dynamics learning about ellipse. The GeoGebra also could employ to provide activities of proof and solving problems about the geometry.

Acknowledgements

The research was financially supported by Khon Kaen University Secondary Demonstration School.

IOP Conf. Series: Journal of Physics: Conf. Series **1340** (2019) 012051 doi:10.1088/1742-6596/1340/1/012051

References

[1] Akay G 2011 *The effect of Peer Instruction Method on 8th Grade Students' Mathematics Achievement in Transformation Geometry and Attitudes Towards Mathem atics* (Middle East University)

[2] Banchoff T and Giblin P 1994 *On the geometry of piecewise circular curves Amer* (Math. Monthly vol 101) pp 403-416

[3] Christou C, Mousoulides N, Pittalis M and Pitta-Pantazi D 2004 *Problem solving and problem posing in a dynamic geometry environment* (The Montana Mathematics Enthusiast vol 2) chapter 2 pp 125-143

[4] Fahlberg-Stojanovska L and Trifunov Z 2010 *Constructing and exploring triangles with GeoGebra* (Annals. Computer Science Series vol 8) chapter 1 pp 45-54

[5] Hohenwarter M and Fuchs K 2004 *Combination of dynamic geometry, algebra and calculus in the software system GeoGebra* (In Computer Algebra Systems and Dynamic Geometry Systems in Mathematics Teaching Conference, Pecs, Hungary)

[6] Jones K 2002 *Issues in the Teaching and Learning of Geometry.* In L. Haggarty (Ed) (Aspects of Teaching Secondary Mathematics: Perspectives on Practice) pp 121–139 London, Routledger Falmer

[7] Kirby J R and Boulter D R 1999 *Spatial ability and transformation geometry* (European Journal of Psychology of Education vol 14) chapter 2 pp 283-294

[8] Kongpa M, Jantaburom P, Byne D, Obmasuy N and Yuenyong C 2014 *Kindergarten's Scientific Concepts and Skills in the Tree Unit* (Procedia - Social and Behavioral Sciences vol 116) pp 2120 – 2124

[9] Pandiscio A E 2010 *Exploring the link between pre-service teachers' conception of proof and the use of dynamic geometry software* (School Science and Mathematics vol 102) chapter 5 pp 216-221

[10] Pleet L J 1990 *The effects of computer graphics and mira on acquisition of transformation geometry concepts and development of mental rotation skills in grade eight* (Unpublished Doctoral Dissertation) Oregon State University, Oregon, USA

Integrating Nature Conservation and Biodiversity Education in Teaching in General schools in Vietnam

Le Thanh Oai[1], Phan Thi Thanh Hoi[2]

[1]Ministry of Education and Training, Vietnam

[2]Hanoi National University of Education, Vietnam

Abstract. Conservation of nature and biodiversity is the responsibility of the whole society, of all management agencies and of individuals, each of whom needs to be aware of their roles and actions that contribute to nature and biodiversity conservation. The education of nature and biodiversity conservation for school students contributes to creating responsible citizens in the future. On the basis of analyzing a number of key concepts such as nature and biodiversity conservation; general analysis of the situation of biodiversity in Vietnam and degradation, in this article, we introduce some integrated methods, teaching methods and techniques to integrate nature and biodiversity conservation education into teaching in secondary schools in Vietnam which contributes to the development of knowledge and skills on nature and biodiversity conservation for secondary school students.

1. Introduction

Biodiversity is an abundance of genes, species and ecosystems in nature [2]. According to BPV (Biodiversity Philately Vietnam), Vietnam is located in Southeast Asia with an area of about 330,541 km2 ranked 16th in terms of biological resources diversity and is one of the 10 centers with the highest level of biodiversity in the world [4]. Vietnam is highly rich in biodiversity with more than 11,500 animals, more than 21,000 plant species, and more than 3,000 species of microorganisms. Over the past two decades, more than 1,000 new species have been discovered in Vietnam, including discoveries that have surprised the global scientific community [20; p 8].

With the characteristics of geography, climate, soil, ... Vietnam has many types of forests, swamps, rivers and streams, coral reefs ... creating habitat for about 10% of wild bird and animals species all around the world. Vietnam is recognized by the Wildlife Conservation Fund (WWF) (2015) to have three of more than 200 global ecological regions [20; p 5]; International Bird Conservation Organization (BIRDLIFE) recognizes Vietnam as one of five endemic bird regions; IUCN recognized Vietnam has 6 centers of plant diversity [19; p 125].

However, the current status of biodiversity in Vietnam is rapidly declining in recent years. Many species in the Mekong Subregion are at risk of extinction in the next few years if there are no urgent measures to protect their populations and habitats [20; p 6]. The biggest threat now is that the habitat of the species is increasingly lost. Agricultural production, cattle grazing, urban development have been fragmenting and destroying the habitat of terrestrial plants and animals. The situation of over-fishing and exploitation to serve the needs of food and medicines ... also helped to undermine biodiversity. Fishing has destroyed 80% of natural aquatic stocks while deforestation and wildlife hunting in the tropics are pushing many species in the forest to the brink of extinction.

In addition, the various pollution hazards from pesticides and industrial waste are polluting rivers and accumulating in the ecosystem, thousands of dead birds and turtles die each year from eating plastic wrap. Chemical fertilizers and domestic wastewater penetrate water sources and create conditions for algae to proliferate and dead sea areas (due to the absence of oxygen). CO2 causes global warming, which is also a pollutant, causing acidification of the oceans and threatening to eradicate rich biodiversity coral reefs.

Education on nature and biodiversity conservation as well as environmental protection (EP) is an important content in schools in Vietnam, which has been implemented at all levels of education and has been considered as the educational mission of schools. However, the quality and effectiveness of education in nature conservation and biodiversity are not as expected, many issues of nature and biodiversity conservation are still being set; Environmental pollution and biodiversity degradation are happening rapidly. This result has many causes, of which the basic reason is that the competency of educating nature and biodiversity conservation of teachers is still limited, not meeting the requirements of environmental and biodiversity conservation education and sustainable development.

2. Research content and results

2.1. Definition of nature and biodiversity conservation

2.1.1. Conservation and nature Conservation. Conservation means to keep, not to be lost [4; p 39]. Nature conservation can be understood as conservation of living resources including maintaining basic ecological processes, preserving genetic diversity, sustainable use of species and ecosystems. Nature conservation is a prudent management and wisely use of natural resources, aimed at improving the quality of current life while meeting the needs of future generations of natural resources.

2.1.2. Biodiversity and biodiversity conservation. According to the Convention on Biology Diversity, introduced in 1992 at the United Nations Conference on Environment and Development in Rio de Janeiro (Brazil), biodiversity means the diversity among living organisms of all sources include contiguous, terrestrial, marine, other aquatic ecosystems and ecological assemblies that are part of them. This diversity occurs in each species, between species and biological systems [1].

According to the World Wide Fund for Nature (WWF), biodiversity is the prosperity of life on earth, which is the millions of animals, plants, microorganisms, are genes that contain in species and are extremely complex ecosystems that coexist in the environment.

The researchers study biodiversity at three aspects: genetic diversity; species diversity and ecosystem diversity.

Biodiversity conservation

According to the Biodiversity Law (2008), biodiversity conservation is the protection of the abundance of important, specific or representative natural ecosystems; Protecting regular or seasonal natural habitats of wild species, environmental landscapes, unique beauty of nature; rearing, planting and tending species on the list of endangered, precious and rare species prioritized for protection; long-term storage and preservation of genetic specimens [2].

Biodiversity conservation includes zoning for management of high biodiversity areas, developing biodiversity conservation facilities to conserve important natural, specific or representative ecosystems, environmental landscape, nature's beauty, unique, conservation of valuable species threatened with extinction, long-term preservation and preservation of genetic specimens.

2.2. Biodiversity and biodiversity conservation

Ecosystems in Vietnam are diverse, with 95 types of ecosystems belonging to 7 main types of terrestrial ecosystems, 39 types of wetland ecosystems, 20 different types of marine ecosystems. The composition of the communities in the ecosystem is also very diverse, the community structure in the ecosystems is complex, multi-storey and there are many branches. This feature makes the diversity of ecosystems in Vietnam different from other countries in the world [19; p 125].

Species diversity includes: plants with 13,766 species, terrestrial animals with 10,300 species, microorganisms with 7,500 species, freshwater organisms with 1,438 species of microalgae, 800 species of invertebrates; 1,028 species of freshwater fish and sea creatures with 11,000 species. Genetic sources of plants and animals with 14,000 gene sources are preserved and stored [19; p 10].

However, in recent years, biodiversity conservation in Vietnam is facing huge challenges, with the number of endangered, precious and rare species being increased. The main causes of biodiversity degradation include: illegal exploitation and over-exploitation of biological resources (illegal exploitation of timber, hunting and illegal trade of wild animals, fishing by unsustainable methods), changing the purpose of using land and water surface without scientific basis (changing the method of using land, water surface, developing infrastructure); the import of new varieties and alien species; environmental pollution and climate change.

2.3. Education of nature and biodiversity conservation in Vietnam

In the National Strategy on Biodiversity, there are directing points [19]:

- Biodiversity is the foundation of green economy; Biodiversity conservation is one of the key solutions to adapting and mitigating the impacts of climate change.
- Conservation of biodiversity associated with sustainable use of biodiversity, contributing to poverty reduction and improving the quality of people's life.
- Conservation of biodiversity is the responsibility of the whole society, of management agencies, all organizations, businesses and individuals.
- Promote socialization and strengthen international cooperation on biodiversity conservation.
- Integrating biodiversity conservation in national development policies, strategies, planning, in sectors and localities.

Thus, every citizen needs to identify nature and biodiversity conservation as his own responsibility. To do this, education and training play an important role in raising awareness and knowledge of nature and biodiversity conservation for school students. In general schools, the contents of environmental education, nature and biodiversity conservation must be considered orthodox, systematic, quality and effective. The method of inclusion in the general curriculum and training methods can be flexible, but the evaluation of the results must be set in a way that corresponds to the importance of the problem. It is necessary to help students to gain the knowledge, skills by themselves and students themselves have to determine their attitude of having to treat nature as their own home.

In Vietnam, in recent years, the Ministry of Education and Training has introduced the contents of environmental protection education into the education system by integrating into a number of subjects and some content-related lessons in the educational curriculum of different educational levels.

Early Childhood Education: Content of environmental education is implemented through specific activities of children such as play, learning, labor and daily activities.

Primary Education: The study of integrating environmental protection education content into primary school subjects has been conducted within the framework of Project VIE / 98/018. This project has developed a number of environmental protection education modules developed from Primary textbooks. The new Primary Curriculum has been designed and built in line with the contents of environmental protection education in the subjects: Mathematics, Vietnamese, Ethics, Nature and Society, Art, and Exercise....

Secondary and high school education: Although it does not organize a separate subject, the content of environmental protection education has been implemented and is integrated into a number of subjects such as Biology, Chemistry, Geography, and Education of citizens ...

In some professional schools, colleges and universities: Environmental protection education has been put into teaching in the direction of integrating into intensive modules such as: Ecotourism, Sustainable Development Education... In particular, at the departments of training Bachelor of Education in such fields as Geography, Biology, Citizenship, Chemistry,... are developed modules close to environmental education or sustainable development education which is compulsory or elective in teacher training.

In addition, the issue of environmental protection education has been concerned by many researchers in Vietnam and implemented in a number of topics and dissertations such as: Ministry of Education and Training-level project "Environmental education through teaching subjects in Faculty of of Biology-Agricultural Engineering" by Dinh Quang Bao and Duong Tien Sy in 1997 [2]; doctoral thesis "Environmental education through teaching Ecology in grade 11" by Duong Tien Sy (1998) [3]; Nguyen Thi Van Huong with a doctoral thesis "Some measures to improve the quality of environmental education for elementary students" [16]; or Nguyen Ki Loan with a doctoral thesis "Environmental education in teaching Biology 6" and many other studies on environmental education in subjects such as Biology, Geography, Chemistry, Citizen education, ... [6], [12], [14], [16], [17].

Referring to the studies on nature and biodiversity conservation, some scientists have studied in teaching in subjects such as Biology, Geography, Citizen Education [18] and within the framework of the projects, many important documents on the education of nature and biodiversity conservation have also been focused on research, typical documents by the authors: Nguyen Sy Duc (editor) and colleagues [13] that focused on education of nature and biodiversity conservation in secondary schools. In this document, the authors have studied the issues of nature and biodiversity conservation; determine objectives, content and design some lessons to integrate nature and biodiversity conservation in Biology and Geography subjects and extracurricular activities as well as guide the use of some teaching equipment. However, studies on integrating nature and biodiversity conservation in Vietnam are still not many and lack generalization.

2.4. Education of nature and biodiversity conservation in Vietnam

Objectives and contents of education in nature and biodiversity conservation in general schools

2.4.1. Some principles to select the content of education on nature and biodiversity conservation in general schools. The selection of education contents of nature and biodiversity conservation into school subjects should follow some general principles as follows:

- The selected content should be consistent with the characteristics of the mind, the physiology and the development of the students of the grade level;
- The selected content must be associated with the curriculum, textbooks of the school level, not adding new content, overloading the learning process of students;
- On the basis of common goals and contents, it is necessary to define specific objectives and contents on biodiversity for each educational level, class and subject; At the same time, it is necessary to ensure inheritance between levels of education, classes and subjects;
- The selected content must be practical and close in life and production.
- The selected content must be consistent with the socio-economic characteristics and cultural practices of regions.

2.4.2. Basic topics and goals for education in nature and biodiversity conservation in general schools. There are many contents of natural and biodiversity conservation needed to equip school students. However, due to limitations of teaching time, it is necessary to focus on some teaching topics and objectives as follows:

Table 1. Topics and objectives of education in nature and biodiversity conservation at general schools.

Chủ đề	Mục tiêu giáo dục bảo tồn thiên nhiên và ĐDSH
1. Conservation of nature and biodiversity - Concept of nature and biodiversity conservation - The role of nature and biodiversity conservation - Situation of nature and biodiversity conservation in the world and in Vietnam	Knowledge: - Stating some basic concepts such as nature conservation and biodiversity; biodiversity conservation. - Analyzing the role of biodiversity in the environment and human life. - Presenting the situation of natural and biodiversity conservation in the world and in Vietnam. Skills - Behavior: Friendly living, in harmony with natural environmental conditions. Attitude: Respect, love nature. Conscious of preserving and protecting ecosystems.
2. Relationship between people and nature and biodiversity conservation - People are a part of the living environment - The role of people in nature and biodiversity conservation - Human impact on natural and biodiversity conservation	Knowledge: - Identify that people are part of the living environment and people have a huge role in nature and biodiversity conservation. - Present the socio-economic activities of people greatly affecting ecosystems and biodiversity. Skills - Behavior: Detecting and fighting actions of exploiting and using natural resources, especially unreasonable biological resources. Attitude: Critique activities and behaviors that adversely affect ecosystems and biodiversity.
3. The decline in biological diversity - - Decline in biodiversity - Species at risk of disappearance and extinction	Knowledge: - Presenting the manifestations of biodiversity degradation, especially those at risk of losing and extinction. - Analyzing the causes of loss of biodiversity in the process of production development and resources exploitation of human. - Presenting an unreasonable population increase is one of the causes of loss of biodiversity. Skills - Behavior: detecting biodiversity problems in the locality and causes. Attitude: A sense of protection of biodiversity in the process of production and exploitation of natural resources.
4. Measures for nature and biodiversity conservation - The provisions of the law on nature and biodiversity conservation - Measures for nature and biodiversity conservation - National strategy on nature and biodiversity conservation - Students' tasks in preserving nature and biodiversity	Knowledge: - Stating the legal provisions on nature and biodiversity conservation. - Analyzing a number of natural and biodiversity conservation measures. - Describe the Biodiversity Action Plan and the National Biodiversity Strategy to 2020 and vision to 2030 Skills - Behavior: Take measures, take active actions to contribute to solving problems of nature and biodiversity conservation. Attitude: Support and implement activities and policies on natural and biodiversity conservation.

2.5. Methods of integrating the contents of education of nature conservation and biodiversity in general schools

The methods of integrating subjects or integrated teaching has been used relatively popular in many countries.

Khabele (1975) defined scientific integration as an approach to teaching science, in which it presents concepts and principles that express the fundamental unity of scientific thought and avoid

IOP Conf. Series: Journal of Physics: Conf. Series **1340** (2019) 012052 doi:10.1088/1742-6596/1340/1/012052

emphasizing too early or excessive discrepancies between different scientific fields, not recognizing the traditional boundaries of subjects when presenting topics [10].

According to Xavier Roegiers, "Integrated pedagogy is a concept of learning process in which all learning processes contribute to the formation of students with clear, predictable competencies for students, to serve future learning processes, or to integrate students into working life. Thus, integrated pedagogy seeks to make the learning process meaningful "[21; p 73].

In Vietnam, there have been many subjects and educational levels who interest in applying the integrated pedagogical thought into the teaching process to improve the quality of student education (such as Biology, Geography and Citizenship education, Literature,...).

Integrated teaching aims to organize learning activities, in which students learn to use a combination of knowledge and skills in meaningful and close-to-life situations. In the process of such learning, the knowledge of students from different subjects are mobilized and coordinated with each other, forming a unified content, based on the theoretical and practical relationships mentioned in those subjects.

Currently, in Vietnam, integrated teaching has been used to bring the contents of education to protect the environment, nature and biodiversity conservation into subjects in schools. Due to the current structure of the curriculum and textbooks of school subjects, it is aimed at the rigorous system of content, the science of the subject is relatively deep, so the inclusion of the contents of education in nature and biodiversity conservation into subjects must also be carried out by integrated teaching, similar to the introduction of environmental protection education content into the subjects.

The contents of education on nature and biodiversity conservation, as well as the content of environmental protection education, can be integrated into subjects at different levels, namely:

Full integration: Full integration is done when most of the lesson's knowledge is also the knowledge of nature and biodiversity conservation. Then the goal of the lesson is also the educational goal of nature and biodiversity conservation. For example, lesson "Protection of plant diversity" (Biology grade 6); lessons "Diversity of Coelenterata", "Biodiversity" (Biology grade 7); lesson "Ecosystem" and "Protect diversity of ecosystems" (Biology grade 12); lesson "Characteristics of Vietnamese organisms" and "Protection of Vietnamese biological resources", (Geography grade 8),... In these lessons, teachers teach normally like other lessons, just collect additional supporting materials to make the lesson more lively and inspiring for students.

Partial integration: Partial integration is done when a part of knowledge of the lesson has contents about education on nature and biodiversity conservation. For example, lessons "populations of organisms", "communities of organisms" (Biology grade 12); lesson "Hot Zone. Humid equatorial environment" (Geography grade 7); lesson "Rivers and Asian landscapes" (Geography grade 8) ... In these lessons, teachers can integrate the contents of education of nature and biodiversity conservation in some parts of the lessons. However, teachers need to distribute time so that it is reasonable with the general structure of the lesson. Avoid overestimating the content of education of nature conservation and biodiversity without paying attention to other contents of the lessons.

Connected: When some content of the lesson is related to the issue of education of nature and biodiversity conservation, but it is not clear in the content of the lesson. In this case, teachers must exploit subject knowledge and relate them to the contents of nature conservation and biodiversity. For example, the lesson "Tropical Environment", "Population and pressure of population on natural resources, hot zone environment" (Geography 7), "Climate and landscape on Earth" articles, "Vietnam Sea", (Geography 8) ... In these lesson, teachers need to integrate smoothly, attached to the lesson content, avoiding forced connected.

The inclusion of the contents of education in nature and biodiversity conservation in subjects can be carried out in two types of learning organization as follows:

- Type 1: Through lessons in class. In this case, the teacher implements methods that integrate with the levels mentioned above. Teacher activities in this case may include:

1. Study subject curriculum and textbooks to build teaching goals, including objectives of nature and biodiversity conservation education.

2. Determine the specific contents of nature and biodiversity conservation education. Based on the relationship between the subject knowledge and the content of education in nature and biodiversity conservation, teachers select materials and integrated plans, specifically to answer the questions: Which content should be integrated? How to link knowledge of nature and biodiversity conservation with subject knowledge? Duration is how much?

3. Select appropriate teaching methods and means, consider using positive teaching methods, highly effective teaching methods to enhance the visual and excitement of learning of students (such as using pictures, videos, computers,...).

4. Develop a specific teaching process. Here, teachers need to specify students' activities, teachers' support activities.

Type 2: Education of nature and biodiversity conservation can also be implemented as an independent activity but still associated with the application of knowledge of subjects. Activities can be carried out such as: sightseeing, extracurricular, organizing thematic extracurricular groups, project lessons, researching a topic (suitable for students),... With these activities, the level of integration of knowledge and skills of subjects with the contents of nature and biodiversity conservation will be higher. Because in these activities, students learn how to apply knowledge of subjects in situations closer to life.

2.5.1. Forms of teaching in internal curricula to integrate education in nature and biodiversity conservation at schools

2.5.1.1. Teaching in class: usually conducted in three forms of teaching: individual teaching, group teaching, classroom teaching, close coordination in a lesson.

- Individual teaching: to promote the individualization of students' learning, respect each student's individual qualities, create equal opportunities for all students in the class to develop depending on one's forte, ability. On the other hand, individual teaching also trains students to self-study. In personal teaching, teachers organize for each student to work with learning objects (pictures, diagrams, schemas, maps, ...). Students have to answer questions, do exercises,...

- Group teaching is a form of teaching, in which members participate in activities and study together in small groups to develop understanding and awareness of certain learningcontent. In group teaching, students are also trained in cooperation, presentation, rehearsal of organization and control, and a record of promoting the activeness of learners, increasing the interest in learning. In group teaching, teachers divide students into groups depending on the content of the lesson, the number of teaching facilities and the location of group activities, then assign tasks and guide them to work (exchange, discuss, perform a common task, assignment, job, ...). Group teaching is often associated with individual teaching. As a result of working in a productive group, each individual student must have preparation, inquiry, contact or thought about the learning content that will be shared by the whole group.

- Class-based teaching is a form of basic, popular teaching, it is used appropriately in the following cases: checking the old lesson, announcing the objectives of the new lesson, setting the problem, assigning the task to individual or group of students. The main method used in the form of organization of class teaching is presentation. Teachers should pay attention to speaking clearly, concisely, fully and reasonably, in combination with appropriate teaching facilities.

In classroom teaching, teachers always use a combination of these three forms of teaching to accomplish the lesson goal.

2.5.1.2. Outdoor teaching: such as sightseeing, local surveys, interviews, etc. When teaching lessons with practical knowledge, or practical experiences, teachers can organize students to study outdoors. This is a convenient opportunity for students to learn the content of the lesson in a lively way, and have the conditions to practice observation and analysis skills in the field. In order to teach this form effectively, it is necessary for teachers to arrange time and prepare carefully in the field. With these activities, the level of integration of knowledge and skills of the subjects with the contents of education and nature conservation and biodiversity will reach the highest level.

2.5.2. Organization of extracurricular teaching in nature and biodiversity conservation at schools. In the process of teaching in high school, teachers can organize a variety of extracurricular activities. It is a form of voluntary activities in the school, gathering good students who are interested in the subject to expand and supplement the knowledge prescribed in the main curriculum. For the content of education on nature and biodiversity conservation, teachers can organize some extracurricular activities within the framework of subjects on nature conservation and biodiversity associated with students' lives.

In the actual conditions and circumstances of Vietnamese students, it is possible to organize a number of extracurricular activities closely related to the task of educating nature and biodiversity conservation as follows:

2.5.2.1. Organizing club for ones who want to learn about nature and biodiversity conservation, learn how to live environmentally friendly. It is recommended to organize a club according to grade level, schedule activities periodically, once a month with the content around the issues of nature and biodiversity conservation and close to the subject. For example, exhibiting activities on biodiversity, types of plants and animals are in danger of losing and extinction; Discussing on nature and biodiversity conservation: invite experts to talk and exchange with students about nature and biodiversity conservation;

Teachers help students build clubs, then assign students to organize and operate themselves. Teachers play a role of advising, supporting and supervising activities so that students are free and creative in forming ideas and in carrying out activities.

2.5.2.2. Organizing cultural festivals, performing skits with contents related to nature and biodiversity conservation.

2.5.2.3. Organize exhibitions on nature and biodiversity conservation with materials and samples collected by students according to specific topics.

2.5.2.4. Organizing tours and camping to explore the biodiversity of an ecological area; learn about the impact of industrial activities on the local biodiversity,... In particular, according to the school's education plan, every year the school organizes students to camp and visit at a certain location. Camping activities are always interesting for students and it will be more meaningful if teachers stimulate students 'attitude to respect and protect biodiversity, build students' sense of willingness to live environmentally friendly. Therefore, teachers should take this opportunity to carry out the task of educating nature conservation and biodiversity for students.

2.6. Methods and techniques for organizing education in nature and biodiversity conservation in schools

2.6.1. Apply question and answer techniques. Question and answer is a method for teachers to organize lessons through asking questions and asking students to answer, creating an exciting lesson time, everyone participating in a positive exchange, gaining knowledge, at the same time training communication skills. In teaching, the method of questioning is often combined with other methods such as using visual media, practice, experiment, ...

In order to use effective questioning methods, teachers need to clearly define the objectives of the lesson, prepare the system of appropriate questions. In the process of using the question and answer method, teachers should pay attention to exploiting students' knowledge and experience, giving them the habit of thinking, actively learning, applying what is known to new learning situations.

To encourage students to actively participate in learning activities that learn about nature conservation and biodiversity, teachers need to strengthen the use of the "open-ended question" - a type of question that will be answered by many ways.

- Open ended question for information / ideas

This type of question often uses words to ask such as: What motivation ?, What makes...? What do you think about this?; What is your opinion on that issue? Are you most interested in ...?

Example:

- It is said that: "Environmental ecological imbalance is the reason for natural disasters to occur more often." What is your opinion on this issue?

- Please state your point of view: "Protecting biodiversity is protecting our lives".

- In your opinion, what kind of global problems are humanity facing?

• Assume question type

This type of question helps students think beyond the context of the current situation.

This question often uses phrases to ask like: What if ...?; what will happen if ...?; imagine ...; If ... then...?

Example:

- What will happen if we are not interested in the issue of greening the bare hills.

- Imagine, what would happen if we lost the biodiversity?

- What will happen if in an ecosystem completely lost some species?

• Question type of action:

This type of question helps students come up with solutions, ideas, ... to effectively use resources; Sustainable Development; adapting to the environment ... to conserve nature and biodiversity.

Example:

- What do you need to do to protect biodiversity?

- If you are Minister of Agriculture and Rural Development, what measures will you take to protect our country's forest resources?

2.6.2. Applying methods using visual means. In teaching, to integrate education in nature conservation and biodiversity, it is often used with typical visual teaching facilities to express such as maps, tables, charts, diagrams, pictures, video clips,... These types of teaching facilities serve as tools or conditions for teachers to guide students to easily discover knowledge, deepen lesson content, train and develop thinking and manipulating skills. creating theoretical knowledge into real life.

Pictures and videos are also a source of knowledge for students. When using pictures, videos, teachers pay attention to students to observe, describe things, the phenomena expressed in them. Next, find the causes and consequences of phenomena and things; state students' thoughts and feelings about pictures and videos.

Example:

To organize for students to learn about local biodiversity degradation, teachers can use images and combine with suggestive brainstorming or conversational techniques:

- What image does your image relate to about biodiversity issues? Why? What causes that?

- In order to minimize this problem, which problems should localities pay attention to?

2.6.3. Applying problem solving teaching. To conduct problem-solving teaching, the following options can be selected:

- Giving a paradoxical situation: requires explanation. The paradoxical situation is something unreasonable, contrary to common sense recognized by everyone, something unusual for a student's understanding, and sometimes initially heard, seemingly absurd. For example: "Why do many species of organisms be harmful to humans but we still have to protect?".

- Giving a difficult situation, a deadlock: This is the situation between what is known and the unknown and what needs to be discovered and realized; There is still a gap between the amount of scientific knowledge available and the practical knowledge of biodiversity. For example: "Southeast Asia is located in the mineral belt so there are many kinds of minerals, why is the problem of rational exploitation and use of natural resources still an important task for each country?".

- Situation to choose: The teacher offers a problem with many different choices, requiring students to find the most appropriate choice. For example: In the two orientations below, which option is more appropriate? Please state your choice.

+ Economic development with a high growth rate, high profit but environmental pollution;

+ Economic development with slow growth, but continuous and stable, no pollution and destructive environment.

IOP Conf. Series: Journal of Physics: Conf. Series **1340** (2019) 012052 doi:10.1088/1742-6596/1340/1/012052

- Causal situation: This is the case to find the cause of a result, to find the nature of a phenomenon, the deep motivation of a behavior. For example: "Why does globalization put pressure on nature?"

For example: To organize students to learn about the topic "Environment and sustainable development", teachers can create problematic situations to organize for students to solve such as:

- For sustainable development, appropriate use of natural resources and environmental protection;
- Industrial development, urban problems are the fundamental causes of the current global environmental problems such as ozone layer, greenhouse effect, acid rain phenomenon, ...
- Developing countries are places where many environmental issues are concentrated;
- Environmental problems rarely occur in developed countries.

2.6.4. Applying project-based teaching. Example: To organize students to learn about the proper use of natural resources and protect the environment, teachers can use project methods to organize students to design communication posters with topics "We are with natural resources - the environment". The product of this project is the poster showing the problems that are posing to our country's natural resources and environment (depletion of forest resources, loss of biodiversity, degradation of land resources, and use of resources). unreasonable water, environmental pollution, environmental ecological imbalance, ..).

2.6.5. Applying group discussion method

- Applying Jigsaw technique

Example: To organize students to learn about the issues that are posing to our country's natural resources and environment, teachers can use this technique with specific steps such as:

Round 1 (expert group): The whole class is divided into 6 groups (group of 6 people), each group performs a task. Team members work independently and then talk to other members to learn about the issue.

- Group 1: Find out about the problem posed to land resources
- Group 2: Find out about the problem posed to forest resources
- Group 3: Find out about the problem posed to water resources
- Group 4: Learn about the problem posed to biodiversity
- Group 5: Learn about the problem posed to mineral resources
- Group 6: Learn about the problem posed to the environment

Round 2 (group of puzzle pieces):

- Form a new group from the above groups
- Tasks: Drawing a mind map showing the problems that are posing for the natural resources and environment of our country.
- The teacher asked the groups to follow these steps: 1) Member in the first round, exchanging content with in-depth discussions with new team members; 2) New team members discuss the issues that are being raised for our country's natural resources; 3) Drawing a mind map showing the problems that are posing to our natural resources and the environment.

Once the teams have completed their tasks, teachers can use gallery techniques to organize student presentations, exchange and comment on your team's products. Finally, teachers summarize and correct the lesson content.

Using the technique of "Supporting - opposing"

Example: To learn about human impacts on biodiversity, teachers can use this technique to organize cognitive activities for students.

Step 1: Divide the class into 2 groups, each group can have 2-3 small groups. Each group performs a task:

- "Support" group: Prove that people have many positive impacts on the natural environment.
- Group of "Objection": Prove that people have many negative impacts on the natural environment.

Groups can collect members' opinions by taking comments by word or individuals will write their opinions on paper, then discuss and agree on their opinions.

Step 2: The group representative presented and reported their group's arguments.

Step 3: General discussion after the two groups have made all their points, teachers evaluate, summarize, and correct the cognitive content.

- Applying brainstorming technique

Example: To understand the issue of biodiversity degradation, teachers can use brainstorming techniques to ask students to give their opinions about why biodiversity degradation is one of the global issues or what causes loss of biodiversity?

2.6.6. Applying role play method

Example: When organizing students to learn about measures to protect the coastal mangrove ecosystem in the Mekong Delta, teachers can organize students to play the following roles:

- Forest bandits: Deforestation for shrimp farming.

- Farmers living in the buffer zone: Demonstrate the act of cutting down trees for firewood, hunting animals for food or selling.

- People of good faith: Demonstrating sickness due to the consequences of indiscriminate forest exploitation and environmental pollution (natural disasters, epidemics, etc.)

- Forest rangers: Prevent loggers and farmers from clearing forests indiscriminately.

- Local officials represent the law: All unlicensed forest exploitation activities are illegal acts, must be strictly dealt with.

- Local leaders: Solving and arranging jobs for loggers, allocating land to farmers to plant forests to live.

In the course of the characters playing, the whole class pays attention to monitoring situations, gathering information. After that, make additional comments to complete the script content, close to the intended target. On that basis, each student should draw on the role of forests in the environment and human life, the need and the importance of protecting mangrove ecosystems in the Mekong Delta region in particular and our country in general.

2.6.7. Applying sightseeing and survey methods

Example: To organize students to learn about local environmental issues, teachers can use project based teaching and combine them with methods of visiting and surveying to organize for students. Or learn about biodiversity of local fish species that can be organized for students to investigate and survey.

2.7. Testing and evaluating the results of education on nature and biodiversity conservation

2.7.1. Examination and assessment of periodic learning results in integrated education of nature and biodiversity conservation. The content of the assessment is closely followed by the objectives and contents of education on nature and biodiversity conservation at schools. In particular, to achieve the goal of teaching integrated education of nature and biodiversity conservation in the process of assessment, teachers can use the following measures:

- Observe, comment and evaluate nature and biodiversity conservation through pictures, drawings, local reality.

- Collect, process information, write reports and present information on nature and biodiversity conservation.

- Analyzing the relationship between human activities and nature conservation and biodiversity.

Teachers should pay attention to assessing the skill of applying the knowledge and skills learned about nature and biodiversity conservation into situations outside the classroom, right where they live.

Example: Please describe and analyze a specific example in the locality where you live (province / city, district, commune / ward ...) on the issue of declining biological resources? What causes and propose solutions to resolve this situation?

Teachers also need to focus on checking and evaluating the expression of attitudes and behaviors of students in front of problems of losing biodiversity. In particular, we can use the rating scale for attitudes towards natural conservation and biodiversity in 5 levels of R.R.likert: Fully agree; Agree; Hesitate; Disagree; Totally disagree.

IOP Conf. Series: Journal of Physics: Conf. Series **1340** (2019) 012052 doi:10.1088/1742-6596/1340/1/012052

This scale can also be reduced to 3 levels: Agree; hesitated (confused); disagree. Or it can use behavioral scales for natural and biodiversity conservation at the following levels: O (Regularly); R (Rarely); N (Never).

3. Conclusion

With the methods and measures proposed above, we have applied to teaching in some general schools in Vietnam to integrate nature and biodiversity conservation into the content of subjects such as Biology, Geography, and Education of citizens. We have also organized interdisciplinary teaching lessons, or organized extracurricular activities for students. The result is that most students are excited about learning activities, students not only gain knowledge of the subject, but also develop knowledge about nature and biodiversity conservation. At the same time, students develop the skills of observation, data collection, cooperation, presentation in the learning process.

References

[1] Convention on biodiversity 1992
[2] Dinh Quang Bao–Duong Tien Sy 1997 Environmental education through teaching subjects in the Department of Biology *Project of Ministry of Education and Training*
[3] Duong Tien Sy 1998 *Environmental protection education through teaching Ecology in grade 11 in Vietnamese high school* Doctoral thesis Hanoi National University of Education
[4] Hoang Phe 2003 *Vietnamese Dictionary* (Da Nang Publishing House)
[5] http://www.biophilavn.com/
[6] Le Thanh Oai–Duong and Thu Huong–Nguyen Thi Quyen 2015 Develop topics of convergence to integrate environmental education and climate change in teaching Biology at high school level in Vietnam *Journal of Education* pp 178–179
[7] Ministry of Natural Resources and Environment 2010 *National Environmental Report 2010* (Hanoi)
[8] Ministry of Natural Resources and Environment 2011 National Report on Biodiversity.
[9] National Assembly of the Socialist Republic of Vietnam 2008 Biodiversity Law
[10] New trends in integrated science teaching 1979 Volume IV, V, UNESCO, published by the United Nations
[11] Nguyen Ki Loan 2016 *Environmental education in Biology teaching 6* Doctoral dissertation, Hanoi Pedagogical University
[12] Nguyen Phi Hanh–Nguyen Thi Thu Hang 2002 *Environmental education through Geography* Publishing House of Teachers University
[13] Nguyen Sy Duc (editor) - Mai Sy Tuan - Tran Ngoc Hai - Nguyen Thi Minh Phuong - Duong Quang Ngoc - Ngo Van Hung - Nguyen Van Hien - Bui Ngoc Diep - Do Thi Thanh Huyen 2015 E*ducation of nature and biodiversity conservation in secondary schools* Ministry of Education and Training, International Fund for Nature Protection
[14] Nguyen Thi Suu - Pham Hong Bac 2013 Integrating environmental education through project teaching methods in non-metallic chemistry in high school *Journal of Education* **315** pp 45–47
[15] Nguyen Thi Van Huong 2002 *Some measures to improve the quality of environmental education for elementary students* Doctoral thesis in education, Hanoi University of Education
[16] Phan Thi Thanh Hoi - Le Thanh Oai 2015 Building a theme of teaching Biology grade 7 to integrate environmental protection education *Journal of Education* **370** pp 48–52

[17] Phan Thi Thanh Hoi - Pham Thi Nga 2016 Develop themes of integrating education on environmental protection and climate change in teaching Biology and Environment (Biology 9) *Journal of Education* pp 175–180
[18] Phan Thi Thanh Hoi 2017 Applying the method of playing the integrated teaching of nature conservation and biodiversity at junior high school level *Journal of Education* **404** pp 50-53
[19] Prime Minister 2013 National strategy on biodiversity to 2020, vision to 2030
[20] WWW Vietnam 2015 *Twenty years and outstanding achievements* Report twenty years.
[21] Xavier Roegiers 1996 *Pedagogical faculty integrates or how to develop competencies at school (translator: Dao Trong Quang, Nguyen Ngoc Nhi)* Education Publishing House

IOP Conf. Series: Journal of Physics: Conf. Series **1340** (2019) 012053 doi:10.1088/1742-6596/1340/1/012053

The Interpretation of Thai Kindergarten Students' Making Sense of Speed and Velocity

Introng Nitayarote[1*] and Chokchai Yuenyong[2]

[1]Khon Kaen University Kindergarten Demonstration School, Khon Kaen, Thailand

[2]Science Education Program, Faculty of Education, Khon Kaen University, Thailand

* Corresponding auther's email address: intrni@kku.ac.th

Abstract. This paper aimed to examine some issues of kindergarten students' learning about Physics. Students' making sense of speed and velocity will be interpreted. Participants included twenty kindergarten students (4-5 years old) who were studying in Khon Kaen University Kindergarten Demonstration School, Khon Kaen, Thailand. Methodology regarded interpretive paradigm. The goal of kindergarten schools includes development of motor, emotion and mind, social, and cognitive skills. On the way of development of these skills, they may also make sense of science concepts. The students' making sense of speed and velocity will be interpreted during students' on activities for development of gross motor skills. Tools of interpretation included participant observation and informal interview when students were participating on the activities for development of gross motor skills as playing with the ball. The activity was carried out for two months. Students' activities of playing with the ball indicated that some of students made sense of the speed and velocity concept when they showed that they play well on the ball as their perception of the relation of hand, eye, and moving of the ball. It could be mentioned that they not only developed on gross motor skills but also concept of ball moving as speed and velocity. This study may have implications for enhancing kindergarten students to learn science.

1. Introduction

The goal of kindergarten education in Thailand aims to support children's development of body, mind, emotion, and intellectual skills. These aims included having a healthy body; healthy mind and being happy, being ethical with high standards of moral behaviour and also appreciating art, music, movement, and exercise. Further it involves having the ability to work independently; love and preserve nature, the environment, and culture and also to be respectful and promote Thai identity. The ability to communicate; thinking skills and solving problems; imagination and creativity; and positive attitude toward learning and inquiring are also very important goals [4].

The goal of science is to understand the natural world through a process known as scientific inquiry. Scientific knowledge helps us explain the world around us, such as why water evaporates and plants grow in particular locations, what causes disease, and how electricity works. Many jobs involve science, such as electrician, horticulturalist, architect, and car mechanic. And people of all ages learn about the world through actions that begin to approximate scientific practice. Many activities, by scientists and nonscientists, whether happening in the laboratory, in the field, or at home, have in common the active use of the basic tools of inquiry in the service of understanding how the world works. Children and adults, experts and beginners, all share the need to have these tools at hand as they build their understanding of the world [5].

Enhancing early childhood under the natural world as way of knowing could support them develop scientific habit of mind in future [6]. [5] suggested some ways of science learning for young children as engaging in the practice of science. These included (1) doing science is a natural and critical part of children's early learning; (2) children's curiosity about the natural world is a powerful catalyst for their work and play; (3) with the appropriate guidance, this natural curiosity and need to make sense of the world become the foundation for beginning to use skills of inquiry to explore basic phenomena and materials of the world surrounding children; and (4) this early science exploration can be a rich context in which children can use and develop other important skills, including working with one another, basic large- and small-motor control, language, and early mathematical understanding.

The content of science for young children is a sophisticated interplay among concepts, scientific reasoning, the nature of science, and doing science. It is not primarily a science of information. While facts are important, children need to begin to build an understanding of basic concepts and how they connect and apply to the world in which they live [1]. Some study provided kindergarten students to construct meaning of the world from explaining the nature around students. For example, [3] organized kindergarten school student learning unit about the trees around us. Knowledge about the tree could be constructed meaning by young children. The unit provided students chance to clarify the part of the tree, classify of the tree, and find advantage of the tree. Students communicated their information of these activities through drawing. These activities allowed children learning scientific concepts and skills. There are many kinds of activities in early childhood which may allow kindergarten students to build up their understanding of basics concepts of science such as drawing, playing the ball, slider, and so on. This study aimed to examine kindergarten students' learning about Physics when they playing with the ball. To playing the ball, students need to practice their muscle for playing and perceive the space and time of moving ball. This activity may allow them to make sense of speed and velocity.

Speed is defined as the rate of change in distance with respect to time. Velocity is defined as the rate of change in displacement with respect to time [2]. For the world of language using in everyday life, we understand motion as a change in position in the space over a certain period of time. Lengths and time periods are conditions for the quantification of such motions. On this basis, the (average) velocity is defined as the quotient of the distance travelled over the time required. However, time depends on a movement and vice versa, because time is measured using motion. In the hourglass sand runs through, in an analogous clock the pointer moves, and for the period of a year, we follow the cycle of the earth around the sun. This could be mentioned that the idea of space is connected to motion, for only through movement we perceive the space (Möller, 2015).

2. Methodology

The paper regarded the interpretive paradigm to understand kindergarten constructing meaning about the speed and velocity through students' playing on the ball. Interpretation of students' making sense about speed and velocity could be observed through students' developing skills of knocking the ball.

2.1. Participants and setting

Participants were 20 kindergarten students (4–5 years old students). All participants' family stays in down town Khon Kaen. Sixty percents of their parents work in Khon Kaen University. The others work for companies and their own business in down town of Khon Kaen.

2.2. Method of Inquiry

Normally, Khon Kaen University Kindergarten School organized many kinds of activities in order to develop motor, emotion and mind, social, and cognitive skills. Playing with the ball is one of activities which allow students to develop motor skills. This activity has enhanced students to be fun and develop skills of knocking the ball for 4 weeks. The activity aimed to gain students' gross motor development and relationship between eyes and hands. However, students' developing skills of knocking the ball will be also interpreted how students' making sense about speed and velocity. The improving of controlling the ball could be view as gross motor development and relationship between eyes and hands which could be interpreted that students perceived the sense of speed and velocity. To examine students' making sense of speed and velocity, students' skills of knocking the ball will be categorized.

3. Findings and discussion

The first week of students' controlling the ball revealed that many of them could not control the ball. However, students' skills have been improved for the following. This indicated that they make sense of speed and velocity. Students' developing skills of knocking the ball could be seen as their improving on controlling the ball during four weeks as the Table 1.

3.1. Controlling the ball with no confidences

There are some students who control the ball with no confidences. This could be interpreted that they may weak on gross motor development. Their skill of relation between eyes and hand is low which reflect that they do not make sense of speed and velocity.

Table 1. Frequency of students on each category of skills of controlling the ball (N = 20 students)

Categories of skills of controlling the ball	1st week	2nd week	3rd week	4th week
Controlling the ball with no confidences	10	4	3	0
Well controlling the ball by moving their hands under level they are.	8	7	3	3
Well controlling the ball with their hand on level they are.	1	4	10	14
Well controlling the ball by moving their hands higher level they are.	1	5	4	3

3.2. Well controlling the ball by moving their hands under level they are

Students could control the ball which represents their good gross motor development. However, students control the ball by moving their hands under level they are. This indicates that they could not perceive the distance of moving the ball depending on the time going by. It could be mentioned that this category of students could not well make sense of speed and velocity.

3.3. Well controlling the ball by moving their hands higher level they are

Students could control the ball which represents their good gross motor development. However, students control the ball by moving their hands higher level they are. This indicates that they could not perceive the distance of moving the ball depending on the time going by. It could be mentioned that this category of students could not well make sense of speed and velocity.

3.4. Well controlling the ball with their hand on level they are

Students could control the ball which represents their good gross motor development. And, students control the ball by moving their hands on level they are. This indicates that they could perceive the distance of moving the ball depending on the time going by. It could be mentioned that this category of students could well make sense of speed and velocity.

4. Conclusion

Students' playing with the ball showed that they understand the space and time when they could wait the ball to move to their hands. This indicated that they make sense speed and velocity.

5. Acknowledgements
The research was financially supported by Khon Kaen University Kindergarten School.

References

[1] Duschl Richard A, Schweingruber Heidi A and Shouse Andrew W 2007 *Taking science to school: Learning and teaching science in grades K-8* (Washington, DC: National Academies Press)

[2] Hewitt P G, Suchocki J and Hewitt L A 2008 *Conceptual Physical Science, Fourth Edition* (USA: Addison-Wesley)

[3] Kongpa M, Jantaburom P, Byne D, Obmasuy N and Yuenyong C 2014 Kindergarten's Scientific Concepts and Skills in the Tree Unit *Procedia - Social and Behavioral Sciences* **116** 2120–2124

[4] Ministry of Education (MOE) 2003 *Thailand Kindergarten Curriculum* (Bangkok, Thailand: External Relations Division, Office of the Performance Secretary)

[5] Worth K 2010 Science in Early Childhood Classrooms: Content and Process SEED Papers Available: http://ecrp.uiuc.edu/beyond/seed/worth.html

[6] Yuenyong C and Narjaikaew P 2009 Scientific Literacy and Thailand Science Education *International Journal of Environmental and Science Education* **4** 335–349

Students' Applying STEM Knowledge in Learning on the STS-STEM Education Wave Learning Unit

Jiraporn Tupsai[1*], Somsak Bunprom[1], Janjira Saysang[1] and Chokchai Yuenyong[2]

[1]Khon Kaen University (KKU) Demonstration School (Suksasart), Khon Kaen, Thailand

[2]Science Education Program, Faculty of Education, Khon Kaen University, Thailand

* Corresponding auther's email address: tjirap@kku.ac.th

Abstract. This study aims to explore Grade 11 activity of student on Science Technology Engineering Mathematics (STEM) with the integration of learning about wave based on Science Technology Society (STS) approach [8]. The participants were 50 students in Demonstration Secondary School, Khon Kaen University, Khon Kaen Province, Thailand. The methodology is in the respect of interpretive paradigm. The teaching and learning about wave through STS approach carried out over 5 weeks. The Wave unit through STS approach was developed based on framework that consists of five stages including (1) identification of social issues, (2) identification of potential solutions, (3) need for knowledge, (4) decision making, and (5) socialization stage. To start with, the question "How to change wave energy to electricity" was challenged in the class in order to improve students to find the problem of how to design wave energy to electricity from Clean Energy form force wave. Students were accepted to apply scientific and other knowledge to design of wave energy to electricity. Activity of students on STEM were collected during their learning by participant' observation and students' tasks. Them did on STEM were classified when they applied their knowledge for designing the wave energy to electricity. The findings indicated that students cooperatively work to solve the problem when applying knowledge about the content of Science and Mathematics and processing skill of Technology and Engineering. It showed activity that students held the integration of science, technology, engineering and mathematics to design their possible solutions in learning about wave energy to electricity. The paper also discusses implications for science teaching and learning through STS in Thailand.

1. Introduction

Researcher believe that if Physics is integrated into STEM education lessons, then students will become scientifically competent taught within well-designed. The STEM education lessons provide students opportunities to learn science in contexts that they understand and that can lead to cross-discipline connections to well skill. We plan many activities commonly found in technology classrooms have the potential to develop students' scientific literacy [9] We also believe there are activity pass of important benefits to within instruction. When students define a technological design problem scenically they develop more cultured and understandings of those solutions. Finally, in this era of high stakes responsibility, contributing to Physics instruction helps convince school and district administration that help should continue to be supported.

There, research conducted by our team suggests that, just because the physics is present in an activity. Our team has conducted research in science classrooms to teach science, technology, engineering, and mathematics (STEM) education—our focus has been on physics. We have found that in the design and setup of the lesson make a dishing in what students learn. In this article, we share our experiences in redesigning a lesson that uses Wave unit through STS approach in order to generate some general principles.

Knowledge in the 21st century has focused on globalization, information society, the expansion of the technology services industry, competition in the economy, and workforce needs to be creative. To compete in the future world, the government plan sets out to utilize STEM education in training future Thai citizen and Thai STEM workforce to compete in the ASEAN economic community (AEC). The plan outlines the goals, the processes and the participation from the governmental agencies, private enterprises and society as well as international and foreign agencies [3]. The communities are becoming more diverse socially, racially, ethnically, even politically. Changing technologies are disrupting traditional lifestyles and ways of doing business. Consequently, the individuals need to develop their understandings of the world. New technologies and new media are radically reshaping not only when, but how and why people learn. The boundaries of when, where, why, how and with whom people can and do engage in science learning experiences are increasingly blurring. Many research suggested that the 21st century science learning should provide integration of STEM content and pedagogy in ways that more accurately reflect learning as a lifelong process that occurs across settings situations. [2][3][4],[5], [6][10].

To provide for the future, the STEM education in school setting, therefore, becomes the key success. The literatures suggested some ideas related to the STEM curriculum design which required combining theories and practices in real situations, designing process, problem-solving ability and other aspects related to finding the possible solutions. STEM learning could be provided through context of, Science Technology and Society (STS) education [13].

Design-based learning, STEM activity requires students who could integrate and support creativity within activity. As creativity is considered to be an essential element of problem solving, critical thinking, science, and engineering design. The STEM education supports students to on tasks which are considered be ill-structured problem-solving tasks, which are complex, open ended, and more in line with real-world problems. Solving ill-structured tasks places a high level of cognitive demand on students [1] [6][8].

As the literature in this field [4], pondering the bipolarity of some cognitive dimensions is a suitable option in advancing toward a construct definition of creativity. Yuenyong [13] present some topics in order reach a better definition of creativity as a cognitive characteristic. For example, The teaching and learning about wave through STS approach carried out over 5 weeks. The Wave unit through STS approach was developed based on framework that consists of five stages including (1) identification of social issues, (2) identification of potential solutions, (3) need for knowledge, (4) decision making, and (5) socialization stage. To start with, the question "How to change wave energy to electricity" was challenged in the class in order to improve students to find the problem of how to design wave energy to electricity from Clean Energy form force wave. Students were accepted to apply scientific and other knowledge to design of wave energy to electricity [12].

This paper aimed to clarify the developing STEM education learning activities to enhance students' creative thinking. The learning activities were developed for Grade 11 students who will study in the subject of Physics of Demonstration School, Khon Kaen University, Thailand.

2. Participants

Participants include 50 Grade 11 students who enrolled to study in the subject of Physics Demonstration School, Khon Kaen University, Thailand.

3. Methodology

The subject of subject of Physics provided the learning activities to enhance students to design something to solve problems in the real world. The question "How to change wave energy to electricity" was challenged in the class in order to improve students to find the problem of how to design wave energy to electricity their creativity for an essential element of problem solving, critical thinking, science, and engineering design. Learning STEM education to develop the creativity of students. The intervention of STEM was provided for one 5 weeks as the below.

3.1. The Wave EPD-STS-STEM Education Learning Unit

The Intervention of wave EPD-STS-STEM education learning unit was designed based theoretical framework of Engineering Process Design and Yuenyong (2006) STS approach [13]. The ideas of learning activities, firstly, were provided regarding on five stages of Yuenyong (2006) STS approach [13]; then the ideas of the Engineering Design Process (EPD) was provided along each stage of STS approach.

3.2. Five stages of Yuenyong (2006) STS approach

In this research, participants developed the STS unit regarding Yuenyong(2006)'s STS approach. Teaching and learning are started from society realm and moved to attaining technology, science concepts and skills. In conclusion, students have chance to take action in society. Yuenyong (2006) [13] developed science unit through STS approach that involved of five stages including identification of social issues, identification of potential solutions, need for knowledge, decision-making, and socialization stage.

1. Identification of social issues stage. The issue of question asking student about wave phenomenon Tsunami occurred in Japan, Clip scenes, taken from the documentary called as "Japan Tsunami - How It Happened," sinami in Thailand on December 26, 2004, 10:00 am, occurred in the andaman sea. Issue form this clip, what do you think that this huge energy is caused by something? and What is the source of energy that can be devastated? ". This video was projected to encourage students to identify some issues and then find imaginable solutions by applying knowledge, mainly knowledge about wave energy.

2. Identification of potential solutions stage. Students plan to solve the social problem related to wave energy, we used clip Kinetic Wave Power Station power station. It can harvest kinetic energy and turn it into electricity. We challenge them to see the benefits of wave energy. This stage supports students to concern with technological aspects for find the possible solutions. Technological aspects are skills to support student decision making. In the STS wave energy unit, students need to think of what, where, and how ideas also design, systems, volition of application scientific knowledge work for designing. The question for challenge "How can scientists bring the waves to electricity?" Students do brainstorm to list what knowledge could be applied and what they need to know in order to develop the possible solutions.

3. Need for knowledge stage. This stage involves developing scientific knowledge. Social questions and technological knowledge can create science content. Student explorer wave properties and composition of the wave tray simulation and creates concepts wave. These activities may provide chance to develop knowledge based for designing safety Electrical and Energy.

4. Decision-making stage. This stage with student involves in making a decision on how to use wave energy knowledge and technology. This aspect public rhetoric about Wave related technological and social issues. It's becomes dominates would like to project and communicate wave energy. Teaching strategies may be used discussion among students' group, and brainstorming to allow students designing the wave energy. On the process of students' designing, they need to be enhanced to develop their possible solutions, select the best solution, and construct a prototype for Electrical and Energy.

5. Socialization stage. Students have to share their prototype for their best solution to project wave energy They may exhibit their prototype in public and show new product in science project, or sharing

in exhibition. The reflection or comments from public may allow students to communicate the solution, and also test and evaluate solution. Then, students need to be asked to write what they learn from those reflections which they may get some more ideas to redesign and completion decision. According to above activities of the STS wave unit, students probably learn science from social issues and then apply other knowledge on the way of their finding the best solution for the social issues.

3.3. STEM Education and Engineering Design

The pattern of the concept of engineering design and learning of Science, Technology and Mathematics is unique to STEM learning organization. As students attempt to learn, understand and practice skills in Science, Technology and Mathematics, they have an opportunity to apply the knowledge gained to design a product or a method to meet their daily life problem solving needs [7].

The Engineering Design Process (EDP) refers to organizing ideas to improve decision making in order to develop high quality solutions and/or products to problems. The main objectives EDP are 1) students become engineers, meaning that teachers need to listen to students, and 2) classroom environments need to change properly to enable learning through the EDP.

Specifically, skills and abilities associated with engineering design for high school student consist of nine stages according to including (1) identify need or problem, (2) research the need or problem, (3) develop possible solution(s), (4) select the best possible solution, (5) construct a prototype, (6) test and evaluate the solution(s), (7) communicate the solution(s), (8) redesign, (9) completion decision (see Figure 1) [3].

Frequently, the imaginative ideas emerged from overlooking the flaws, or a different approach may become apparent through work on the challenge. Throughout this process, students repetitively assess and challenge their ideas by repeating steps every now and then, even restarting from the beginning. [7],[12] ,[13].

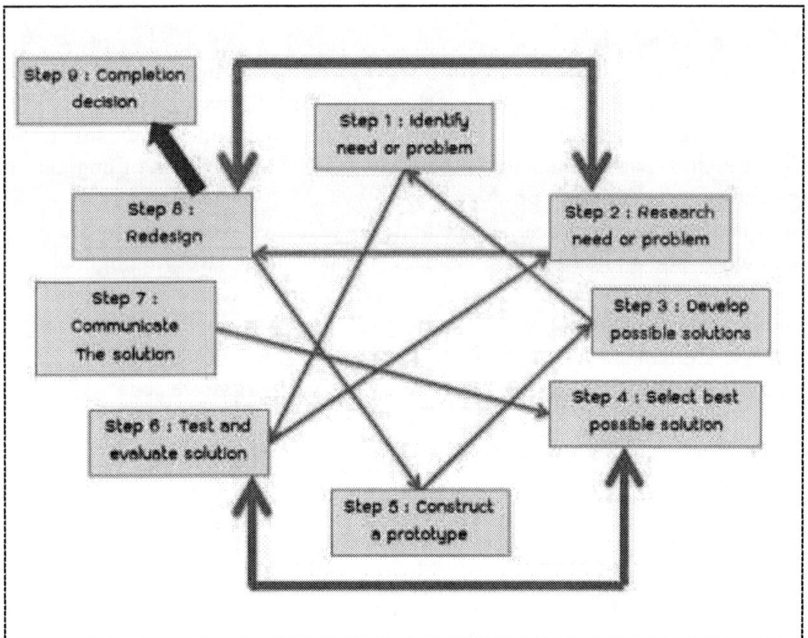

Figure 1. Engineering Design Process [7].

Table 1. Intervention of STEM education learning activities via STS.

Week	Learning activities
1	Course orientation was explained STS stage also, identification of social issues and identification of potential solutions stage and STEM stage identify need or problem, and research the need or problem. Researcher stimulates the student's interest by using a clip to show students about the wave and the phenomenon of the wave. Clip 1 is a wave phenomenon Tsunami occurred in Japan. Clip 2 is a clip from a documentary that describes natural phenomena using simulations the tsunami in Japan in 2011. And clip 3 is a tsunami that took place in Thailand. The news was born in Thailand on December 26, 2004 at 10:00 am in the Andaman Sea. We challenged student by video powered by Hashem Al-Ghaili _Kinetic Wave Power Station. This power station can harvest kinetic wave energy and turn it into electricity (Correction: 10 m = 33 ft).
2-3	Students search the meaning of single of following word – science, technology, engineering, and mathematics. And, sharing their result of searching. Course orientation was explained STS stage also, need for knowledge stage and STEM stage. Students study waves form simulate the waves, creating the concept for some properties of waves because wave is abstracting to create understanding for students.
4	Working in group, students search for innovation which was designed for everyday life. And, each group of students, then, clarifies what aspects of STEM knowledge was applied for designed those innovations. Course orientation was explained STS stage also, Decision-making stage and STEM stage: select the best possible solution, construct a prototype, test and evaluate the solution, communicate the solution, redesign.
5	Course stop students present innovation which was designed final examination design wave energy to electricity from Clean Energy form force wave. Course orientation was explained STS stage also, Socialization stage and STEM stage: completion decision.

4. Data Collection and Analysis

In this study, the STS electrical current unit for STEM learning has been taught by the first author for 16 periods in a month. The participant observation, students' discussion and questions, worksheets, and unstructured interview were interpreted to examine what and how students applied their knowledge of science, and mathematics and others as well as their process of engineer process design to develop best solution as technology products.

Students' ideas in applying these knowledge and processes were clarified in each stage of the STS electrical current unit. The dialogues and students' tasks were also highlighted to represent the interpretation. Then, the interpretation was rechecked as peer debriefing in order to ensure the credibility since the interpretation could be subjective, varied from the particular researchers. The following the story of students on tasks of designing, making, and presenting their innovations will be clarified for representation of what and "How to change wave energy to electricity".

4.1. Design activity Physics of wave motion using STS and STEM process

4.1.1. Improvement emphasis was explained STS stage also, identification of social issues and STEM stage identify need or problem. Researchers challenge the interest of students see the importance of wave power. Example: wave phenomenon Tsunami occurred in Japan. This is the country where this phenomenon occurs most, it is the most affected country by sea waves. The phenomenon was born in 2011. A clip from a camera with a personal view of the waves moving near themselves. And the waves can destroy the walls that scientists created to prevent tsunami destruction, but the walls are unprotected. The wave destroys building blocks from the water pressure, which can't destroy 100%. And Clip scenes, taken from the documentary called as "Japan Tsunami - How It Happened," with focus on only a part with 3D-animated presentation of geological process causing Tsunami that swept Japan on March 11, 2011. That describes natural phenomena using simulations. The tsunami in Japan in 2011, the movement of the underwater crust is beginning in the tsunami. The enormous force of the sea around sea as then energized see the phenomenon. They can saw wave composition move to the shore by simulate wave motion from deep water to shallow water. Notice the amplitude of the wave with the forward motion, the wavelength when it reaches the shore, and the impact on the building and car. And sinami in Thailand on December 26, 2004, 10:00 am, occurred in the andaman sea. A phenomenon was a damaging event the death of about 10000 people in the news was born in Thailand December 26, 2004 at 10:00 am.

Figure 2. https://www.youtube.com/watch?v=UvYmZFPRzT4 [22-7-2017].

Figure 3. https://www.youtube.com/watch?v=JFU9hqAN0t0 [21-5-2017].

Figure 4. https://www.youtube.com/watch?v=cAsbmYRGvFk [22-7-2560]

IOP Conf. Series: Journal of Physics: Conf. Series **1340** (2019) 012054 doi:10.1088/1742-6596/1340/1/012054

There clip reflected critical issues as perceived by students below.

Teacher: From the clip, what did the students notice?

Stedent1: I saw the movement of wave moving and the boat was in the sea, and it was moving around.

Student 2: Sinami, there is enormous energy that destroys houses, which makes the house completely destroyed.

"Issue form this clip, what do you think that this huge energy is caused by something? and What is the source of energy that can be devastated? ".

Form clip help student saw phenomena of wave which allows students to see the energy of the waves from nature. The destruction can't protect in front of them to destroy and humanity. This occurrence was in the human hand, it is only to escape or accept the occurrence of nature.

4.1.2. Improvement emphasis was explained STS stage also, identification of social issues and and STEM stage identify need or problem. We used clip Kinetic Wave Power Station power station. It can harvest kinetic energy and turn it into electricity. We challenge them to see the benefits of wave energy.

Figure 5. https://web.facebook.com/ScienceNaturePage/videos/818130261652567/?pnref=story [21-5-2017]

They learned energy of wave by observe wave to related to various fields. Which is punishable and useful in life, they analysed why tsunamis are so powerful in their destruction and what the characteristics of the waves are. How can scientists bring the waves to electricity? Answer of student interpretive clip was solved situation to finding data, example in below.

Student 1: See the difference that one clip creates huge damage and another human clip can bring energy from the waves.

Student 2: Seeing the concept and ability of scientists who can take the dam from the dam to create a benefit to humans.

Also, the challenge clip and stimulate student interest about wave. The part of the wave with wave components and wave properties Influence on humans. So, from the clip on the wave will notice the wave motion and the power of waves. Where energy is hugely tries to find energy to replace it? If students are engineers from this VOD study, students can design power generation. What's around the students? They could to emphasis explained social issues and identify need or problem for thinking in next station.

4.1.3. Improvement emphasis was explained STS stage also, need for knowledge stage and STEM stage develop possible solution. Student explorer wave properties and composition of the wave tray simulation and creates concepts wave. They studied properties of waves to create understanding to a conceptual element of wave. They studied two main types of waves: mechanical and electromagnetic.

Mechanical waves propagate through a physical matter, whose substance is being deformed. Restoring forces then reverse the deformation. For example, sound waves propagate via air molecules colliding with their neighbours.

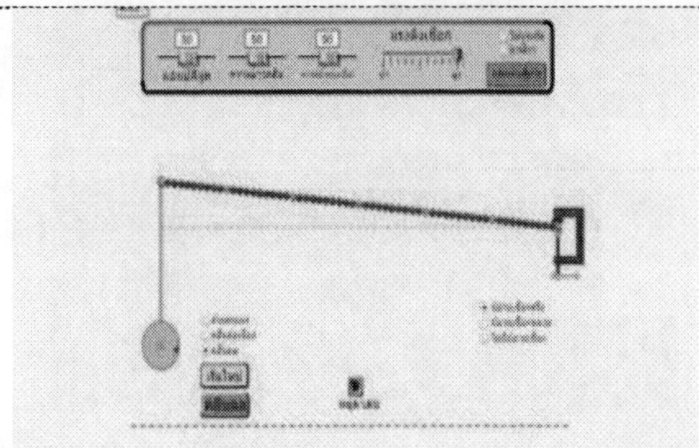

Figure 6. https://phet.colorado.edu/sims/wave-on-a-string/wave-on-a-string_th.html [21-5-2017]

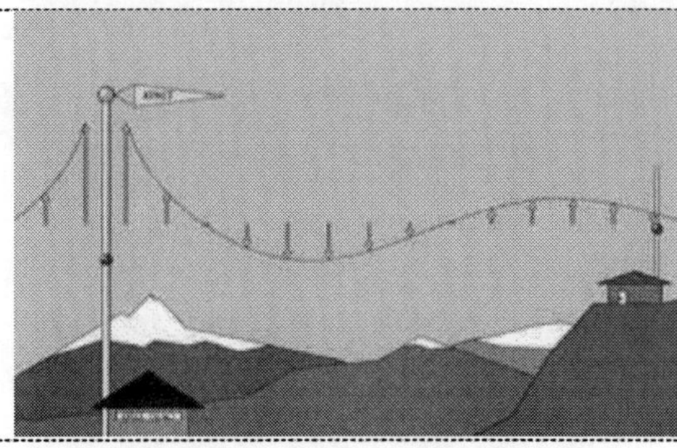

Figure 7. https://phet.colorado.edu/th/simulation/radio-waves [21-5-2017]

A standing wave, also known as a stationary wave, is a wave that remains in a constant position. This phenomenon can occur because the medium is moving.

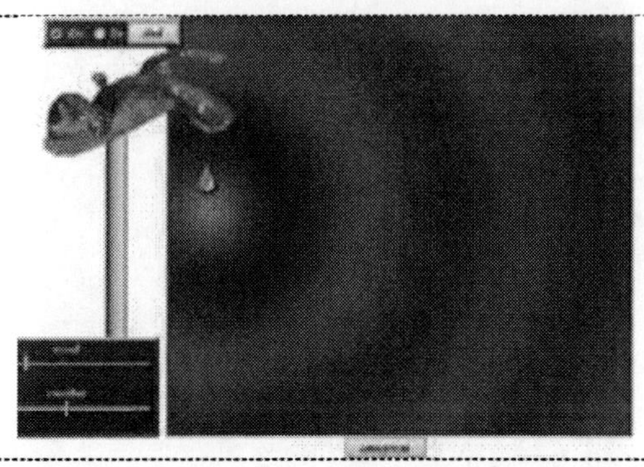

Figure 8. https://web.facebook.com/ScienceNaturePage/videos/818130261652567/?pnref=story [21-5-2017]

Student studied waves that encounter each other combine through superposition to create a new wave called an interference patter. Important interference patterns occur for waves that are in phase.

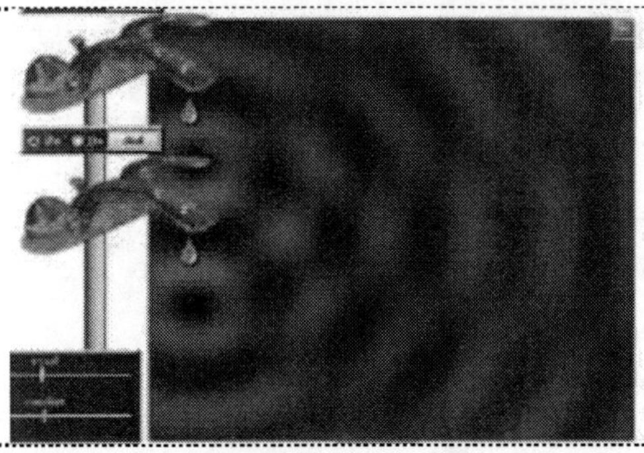

Figure 9. https://web.facebook.com/ScienceNaturePage/videos/818130261652567/?pnref=story [21-5-2017]

The diffraction occurs whenever propagating waves encounter such changes, its effects are generally most pronounced for waves whose wavelength is roughly comparable to the dimensions of the diffracting object or slit.

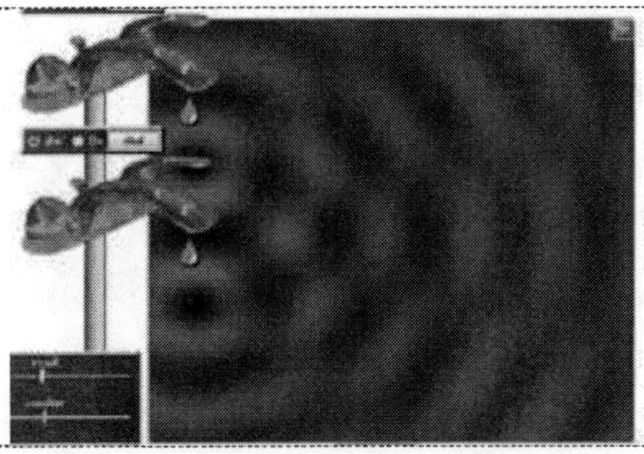

Figure 10. https://web.facebook.com/ScienceNaturePage/videos/818130261652567/?pnref=story [21-5-2017]

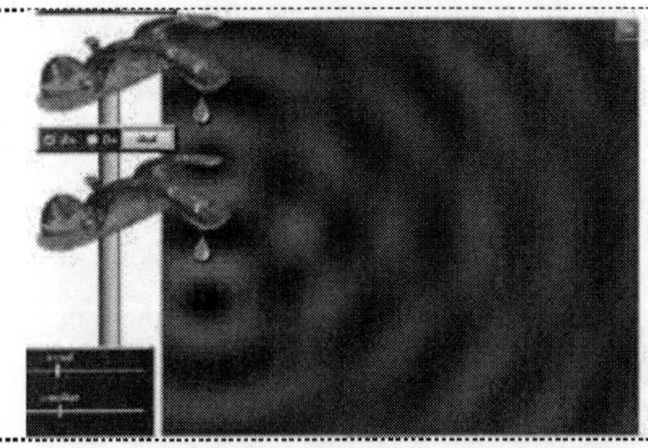

Figure 11. https://web.facebook.com/ScienceNaturePage/videos/818130261652567/?pnref=story [21-5-2017]

IOP Conf. Series: Journal of Physics: Conf. Series **1340** (2019) 012054 doi:10.1088/1742-6596/1340/1/012054

The diffraction occurs whenever propagating waves encounter such changes, its effects are generally most pronounced for waves whose wavelength is roughly comparable to the dimensions of the diffracting object or slit.

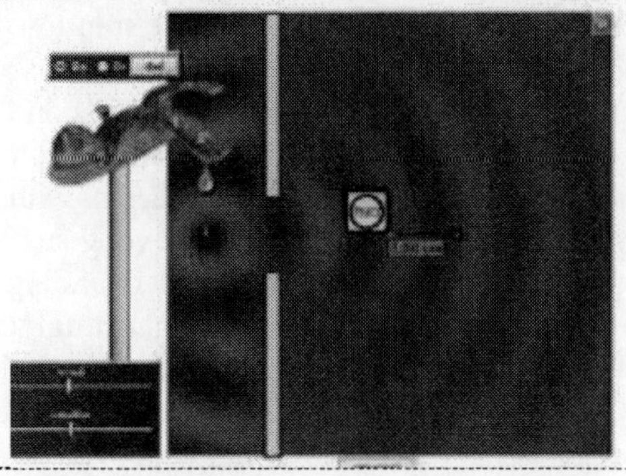

Figure 12. https://phet.colorado.edu/th/simulation/legacy/wave-interference [21-5-2017]

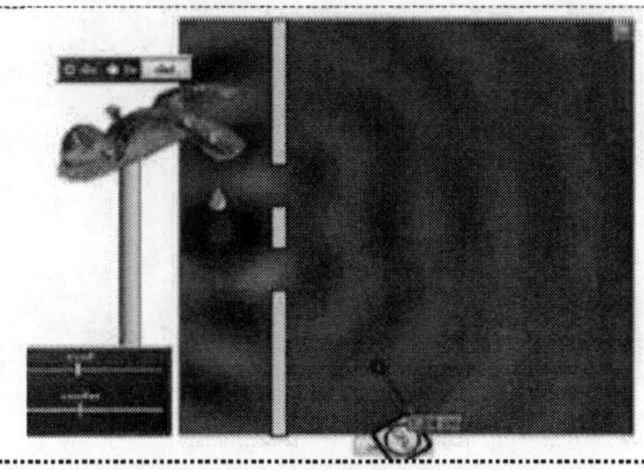

Figure 13. https://phet.colorado.edu/th/simulation/legacy/wave-interference [21-5-2017]

We had question enhancing understanding of student about simulation wave.

Teacher: What does wave occurs?

Student 1: Wave had a wave power station, the waves arriving because the water in the chamber to rise and fall, which means that air is forced in and out of the hole in the top of the chamber.

Teacher: How was wave move?

Student 2: Wave can move energy forward to around but wave make water rise and fall in chamber.

Teacher: What were character of wave?

Student3: Wave have 4 specifics basic: Reflection, Diffraction, Refraction, and Interference. And the action of the waves moves the device, pumping hydraulic fluid to a shore station to drive a generator in my project. Because wave water was transverse wave so were contrast force and moving to diverge.

Teacher: How were they wanted to solve problem wave energy to use to people?

Student could inquiry content wave form activity and simulation about character of wave: Reflection, Diffraction, Refraction, and Interference. They were finding as ocean waves are caused by the wind as it blows across the sea. Waves are a powerful source of energy. Once they've thinking built a wave power station, the energy is free, needs no fuel and produces no waste or pollution. More ideas about how to extract energy from waves are being proposed all the time.

4.1.4. Improvement emphasis was explained STS stage also, Decision-making stage and STEM stage develop possible solution, select the best possible solution, construct a prototype, test and evaluate the solutions, communicate the solutions, redesign. The ability of student to think of decision making in plan for develop possible solution. They could select best possible for construct project wave power. Then, they would to test project evaluate and communicate the solutions with friend, if it projects not possible, they could be redesign new project. They studied force of wave simulation and activity, they were decision-making develop possible their ideas, generate and construct prototypes, and balance competing constraints and criteria in order to create an optimal solution and communicate their work to stakeholders. The fundamental goals for providing students with opportunities to engage in engineering design in class physics. Research had cognitive goals include improving students' problem-solving abilities; system thinking abilities; and understanding of science, technology, and mathematics concepts, while the affective goals include generating interest in engineering and enhancing self-efficacy to solve engineering problems. Research would to create experiences that allow students to draw connections to real engineering projects such that students are motivated to engage in the design problem for reasons other than evaluation. We can also make real world explicit about design wave force and energy to electric. Example student design in below:

Figure 14. Student design wave force and energy to electric.

Models and modeling are widely used in engineering education. Engineers construct models that are based on physical laws and mathematical descriptions, but these models are simplifications of the world around us. For example, an engineer might construct a model of wave force and energy to electric. The model can provide the student who is designing bridge trusses useful guidance on the types of structures that will be most efficient in their design. As students engage in an authentic engineering design task in engineering design may shift, allowing them to embrace the opportunity to learn why something does not work well and to redesign it for optimal functionality. students may engage in iterative engineering design, not because it is what their teacher requires, but because they are invested in the design process and motivated to solve the problem. Moreover, students engaged in the search for solutions to design challenges recognize the need to collaborate and communicate effectively with each other learn to contribute systematically to the design process. Having opportunities to engage in processes that afford such tinkering may also help students expand their repertoire of resources situated within the context of classroom-based of The Wave EPD-STS-STEM Education Learning Unit[1][2][5]).

4.1.5. Improvement emphasis was explained STS stage also; Socialization stage and STEM stage develop completion decision. Students were designing recycling devices, they developed more creative ideas in contexts relation of content physics. They were that solved ill-structured problems those who had initially encountered failure in attempting to solve ill-structured problems outperformed their counterparts who had initially confronted and solved well-structured problems [8]. Researcher would employ different levels of scaffolding throughout an engineering design process depending on students' experiences with open-ended problems, strengths in content areas, and the degree of support they require to overcome obstacles. Accordingly, we can deliberately design learning experiences that activation and differentiation of prior knowledge in relation to the targeted concepts. They could to discussion around specific features or unexpected occurrences organization, assembly, and synthesis of the critical conceptual features into the target concepts; and reflection throughout the iterative design process. Because there is no single correct way to design and implement an engineering design experience, we would vary levels of complexity, dimensions of the problem, or applications. we would choose to introduce students to engineering through well-structured problems, providing explicit criteria. Constructing a prototype is usually viewed as a "hands-on" activity. The term, hands-on, implies an active learning process. Product life cycle management also poses a range of choices, such as the need to provide for efficient disposal or recycling. Example student design in below:

Figure 15. Model wave force and energy to electric.

Students have encouraged on theme's judgment, even when that judgment is based on the teacher's own informed understanding of the design requirements. They were come to see their positive evaluations as the primary goal in and of itself. An alternative situation in which students are able to assess and evaluate their designs according to the requirements of the design problem may help the students to become better design decision makers. Having the requirements drive the students' own assessments and evaluations helps to monitor progress in the middle steps of the design process and determine conditions that will meet the given requirements. The engineering design process must be iterative so that the quality or the functionality of the design can be improved. In order to bring some sense of closure to the engineering design process. When documentation has been completed and edited, and the design team has made oral and written presentations describing the final design solution and demonstrating the prototype to our or stakeholders[6][8].

5. Conclusion

This product is not simply the result of passing a set of predefined tests, but is based on whether or not students believe they have sufficiently optimized their product to the selected constraints. In this step, students make a decision that they have sufficiently met the design requirements and are ready to implement their prototype as a final product. These facets of understanding provide a framework when identifying the desired results of instruction or specific learning outcomes. Within the context of engineering design challenges, these outcomes include systems thinking, the engineering design process. An important consideration in assessing student learning outcomes in the engineering design challenge paradigm is the use of multiple, layered assessment strategies that provide a holistic picture of student learning. Engineering design challenges that target several domains of learning in a complex environment typically require multiple strategies for collecting evidence.

In research to start with, the question "How to change wave energy to electricity" was challenged in the class in order to improve students to find the problem of how to design wave energy to electricity from Clean Energy form force wave. Students were accepted to apply scientific and other knowledge to design of wave energy to electricity. Activity of students on STEM were collected during their learning by participant' observation and students' tasks. Them did on STEM were classified when they applied their knowledge for designing the wave energy to electricity. This study uses of inquiry for 5 stages STEM education including step (1) set long term and short-term goals of fundamental STEM education concept, (2) know how they are spending their time generating creative thinking from prototype, (3) arrange typical tasks according to priorities revised ideas, (4) engineering ability Distinguish between 'urgent' and 'important' tasks recognize the need to schedule blocks of time monitor and evaluate their time usage , and (5) presentation and discussion Apply these skills to typical student time demands. The use of inquiry for STEM education support students to structure the problems to get start designing and making their innovations or products. This research indicated that a design-based STEM education activity could enhance students to creatively design products. Clarifying and focusing questions on students' designing and products make students' thoughts more explicit on originality and elaboration creative thinking. Sharing of applying knowledge and designing on their products could enhance students' fluency and flexibility creative thinking. This study may have implications for providing of design-based The Wave EPD-STS-STEM learning activities. Project management in ways that can be implemented in the Physics subject setting.

References

[1] Apedoe X S Reynolds B Ellefson M R and Schunn C D 2008 Bringing engineering design into high school science classrooms: The heating/cooling unit *Journal of Science Education and Technology* **17** pp 454–465

[2] Atman C J Kilgore D and McKenna A 2008 Characterizing design learning: A mixed-methods study of engineering designers' use of language *Journal of Engineering Education* **97** pp 309–326

[3] Chulavatnatol M 2015 Preparing Thai STEM Workforce with 21st Century. Skills by STEM Education *Paper presented in the 2nd International Conference on Innovation in Education (ICIE 2015)*

[4] Fairweather J 2008 Linking evidence and promising practices in science, technology, engineering, and mathematics (STEM) undergraduate education *Board of Science Education, National Research Council* (The National Academies, Washington, DC)

[5] Hammer D, Elby A, Scherr R E and Redish E F 2005 Resources, framing, and transfer. In J. Mestr (Ed.), Transfer of learning from a modern multidisciplinary perspective (Greenwich, CT: Information Age) pp 89–120

[6] Householder D L and Hailey C E 2012 Incorporating engineering design challenges into STEM courses

[7] Hynes M, Portsmore M, Dare E, Milto E, Rogers C, Hammer D and Carberry A 2011 Infusing Engineering Design into High School STEM Courses *National Center for Engineering and Technology Education* **165** pp 8–13

[8] Kapur M 2011 A further study of productive failure in mathematical problem solving: Unpacking the design components *Instructional Science* **39** 561–579

[9] Litowitz L S 2009 Addressing mathematics literacy through technology, innovation, design, and engineering. *e Technology Teacher* **69** 19–22
[10] Ministry of Education 2014 Basic Education Curriculum (Bangkok: Kurusapa Ladprao Publishing)
[11] Reeve E M 2013 What is STEM Education?[Online]. (Bangkok: The Institute for the Promotion of Teaching Science and Technology (IPST): Special Initiative Project Division)
[12] Tupsai J and Yuenyong C 2018 Investigating students' view on STEM in learning about electrical current through STS approach *In AIP Conference Proceedings* **1923** p 030054
[13] Yuenyong C 2006 Teaching and learning about energy: Using STS approach (Unpublished PhD thesis, Kasetsart University)

IOP Conf. Series: Journal of Physics: Conf. Series **1340** (2019) 012055 doi:10.1088/1742-6596/1340/1/012055

Current Situation of Problem Solving Ability Development for Chemistry Teaching at High Schools of Lao PDR

Khongvilay VOLAYUTH[1], Tran Trung Ninh[2], and Chokchai Yuenyong[3]

[1]Savannakhet University - Lao PDR

[2]Chemistry Faculty-Hanoi National University of Education, Vietnam

[3]Faculty of Education, Khon Kaen University, Thailand

Corresponding author's e-mail address: [1]vorayuth@gmail.com: [2]ninhtt@hnue.edu.vn: [3]ychok@kku.ac.th

Abstract. Education in the Lao People's Democratic Republic (Lao PDR) plays an important role in the development of national construction and defense. Among subjects at High schools, Chemistry is one in the field of natural science with many applications. However, nowadays, despite many achievements, the reality of teaching chemistry at High school has showed a lot of inadequacies. This article addresses some of the problems in the development of problem solving ability in chemistry-teaching at high schools in the Lao People's Democratic Republic.

1. Introduction

Ability development-oriented education is a modern pedagogy trend [1-5, 8,9]. The Lao People's Democratic Republic (Lao PDR) is also in the same general trend. In the national education system of Lao PDR, high school education is the middle level of education, followed by secondaryeducation. Article 18 of the Education Law of the Lao PDR has stated"General education is a regular educational activity. High school education lasts for 3 years and is responsible for upgrading and expanding the content learned from secondary school. In addition, high school education has to go into some subjects to develop the students' awareness, skills, abilities and talents." [5]. High school education plays a very important role in the national education system. The high school's education goal is defined in Article 18, Chapter III of the Education Law of Lao PDR as follows:"High school education must consolidate and develop the content of primary and secondary education, ensure that students have a common understanding of culture, mathematics, history, natural science, social science, law and vocational guidance, etc."[7].

Therefore, the Lao government has set a vision of socio cultural development by 2030 as follows:"Developing human resources is to ensure a competitive economy like other countries in the region and in the world, is a strong production capacity to meet the demand of socio economic development. The people of Lao PDR can therefore receive education, health care, quality, efficiency and increase their average life expectancy up to 75 years old"

Basing on these vision and strategy, the Lao Ministry of Education and Sports has set a vision by 2030 which is "By 2030, all Lao people can receive a quality education that is fair to provide them with the opportunity to develop themselves to become good citizens of the country, having good morals, good health, high level of knowledge for the nation's sustainable development and can be associated with other countries in the region and the world"[7], [9].

2. Content of research

2.1 Research methods

2.1.1. Theoretical research methods. Study documents on policies and educational development of Lao PDR. Study theoretical chemistry teaching on problem solving ability development through chemistry teaching integrated with environmental education. Study the methods and techniques of active chemistry teaching.

2.1.2. Research methodology. The survey content was developed in the form of questionnaires in teacher poll and sent to upper secondary schools in some provinces and cities in the country with 226 teachers including 79 teachers who directly lectured Chemistry.

- Send directly to teachers and students, collect questionnaires and comments.
- Use information technology to get feedback: To facilitate the investigation, processing, and analysis of data, Google Drive's "Forms" tool was created. After designing the contents of the questionnaire, we sent the link to teachers and students to get feedback on the design contents, the results obtained will be processed by the functions available in google combining the part Microsoft Excel to handle the slip and send directly.

Findings and discussions

2.1.3. Current status of the use of active teaching techniques

- Onthelevel of use of teaching methods and techniques

Table 1. Level of use of some methods and techniques of teaching in high school.

No.	Methods and techniques of teaching	Frequent	Sometimes	Rarely	Not used
1	Use visual aids	28,0%	35,1%	27,6%	9,3%
2	Use chemical exercises	14,8%	61,4%	21,1%	2,7%
3	Presentation of the problem	26,8%	48,2%	21,4%	3,6%
4	Inquiry Based learning	23,1%	51,2%	20,3%	5,4%
5	Problem Based learning	37,8%	48,9%	11,6%	1,8%
6	Project Based Learning	33,5%	48,2%	16,1%	2,2%
7	Cooperative learning	28,7%	52,0%	15,7%	3,6%
8	Work in corner	16,5%	56,7%	20,5%	6,3%
9	Mind Mapping Technique	21,8%	51,1%	21,8%	5,3%
10	KWL technique	15,6%	51,3%	21,9%	11,2%
11	Brainstorming Technique	18,4%	48,2%	21,2%	12,2%
12	Technique JIGSAW	15,3%	50,9%	21,6%	12,2%

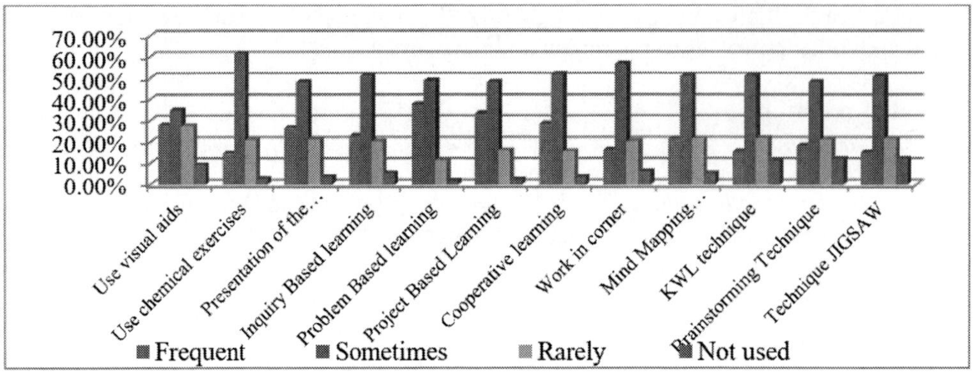

Figure 1. Level of use of some methods and techniques of teaching in high school

Based on Figure 1., teachers have applied the methods and techniques of active teaching in the teaching process. However, the use of active teaching methods and techniques Frequency is low. Some active teaching techniques, such as KWL, brainstorming, JIGSAW have over 10% of total teachers never used in teaching.

- On the level of difficulty using problem based learning method

Table 2. Difficulties in using problem based learning method

Difficulties	Teachers' comments	
	Agree	Disagree
1. Design a problematic situation in teaching chemistry	83,2%	23,7%
2. Chemistry curriculum is not flexible	76,3%	23,7%
3. Lack of equipment for learning and teaching	77,7%	22,7%
4. It takes a long time for design lesson plan	79,1%	20,9%
5. Student's capacity is limited	80,2%	19,8%
6. Teachers do not understand the problem based learning method	73,3%	26,7%

Through the analysis of Table 1.2, teachers face many difficulties in using teaching methods to solve problems. Most teachers said that the development of problem solving capacity for students is difficult; most teachers said that 83.2% of the teachers found it difficult to design the situation problem. As many as 73% of teachers said that they did not know the content of using Problem Based learning methods. Thus, although the survey data in Table 1.1 has 37.8% of teachers regularly use Problem Based learning method; however, they also realize that they have not mastered the Problem Based learning method.

- On the difficulty level of using Project based learning

Table 3. The difficulty level of using Project based learning

Difficulties	Agree	Disagree
1. It takes a long time	79,4%	20,6%
2. Student's capacity is limited	70,8%	29,2%
3. Chemistry curriculum is not flexible	67,2%	32,8%
4. Lack of equipment for learning and teaching	74,0%	26,0%
5. Teachers do not understand the project based learning method	69,5%	30,5%

IOP Conf. Series: Journal of Physics: Conf. Series **1340** (2019) 012055 doi:10.1088/1742-6596/1340/1/012055

Based on the analysis of Table 1.3, teachers face many difficulties when using Project based learning. When being asked about the difficulties of using project based learning, 79.4% of teachers said that it took a lot of time to prepare and finish the project. As many as 70.8% of teachers said that difficulty due to student's capacity is limited. Most of the total teachers, 74% of the teachers said that the school did not meet the necessary facilities such as internet computer system and other machines when using Project based learning. There are 69.5% of teachers said they are not mastering the Project based learning.

2.1.4. The status of environmental education in Lao PDR

- The use of integrated teaching of environmental protection through the teaching of chemistry.

Table 4. Level of use of integrated teaching of environmental protection through the teaching of chemistry

No.	Content	Level			
		Totally Agree	Agree	Disagree	Totally Disagree
1	Integrated in mainstreaming form	30,6%	47,7%	18,0%	3,6%
2	Internal subjects integration	21,6%	57,7%	18,9%	1,8%
3	Related subjects integration	22,9%	50,2%	20,2%	6,7%
4	Cross subjects integration	24,1%	48,2%	16,4%	11,4%

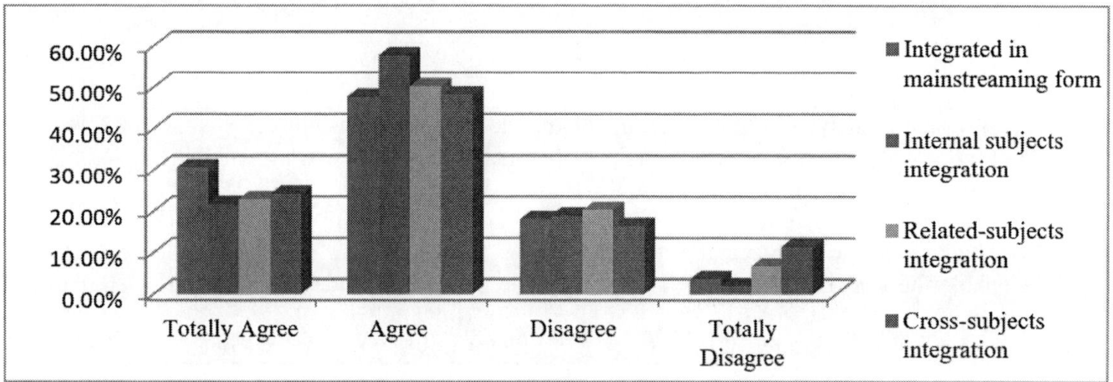

Figure 2. Level of use of some methods and techniques of teaching in high school

As a result, the majority (> 70%) of the teachers in the surveyed schools agree with the levels of environmental education integration. In particular, most teachers agree to use the Internal subjects integration (79.3%) and Integrated in mainstreaming form (78.3%), the teachers assume that Internal subject's integration and Integrated in mainstreaming form are the integrated way which is easy to access and the content is also more diverse compared to other levels such as Related subjects integration or Cross subjects integration.

- On the importance of environmental education to the sustainable development of Laos.

Table 5. On the importance of environmental education to the sustainable development of Laos

No.	Content	Level			
		Totally Agree	Agree	Disagree	Totally Disagree
1	Sustainable development is the harmonious development of both socio-economic and environmental	45,5%	46,0%	7,7%	0,9%
2	Environmental education includes relevant atmospheric, hydrological, and local knowledge	22,5%	62,2%	14,9%	0,5%
3	Chemistry is a science that has a strength in environmental education for students	29,9%	53,4%	14,0%	2,7%

When talking about the sustainable development of Laos, the majority of teachers are very enthusiastic and want to come up with solutions such as 81.5% of teachers agree and fully agree on the concept of sustainable development. Up to 83.2% of teachers identified that chemistry is a subject that can promote the development of environmental education for students.

- On the importance of environmental education to the sustainable development of Laos.

Table 6. Using the level of environmental education integrated exercises in chemistry teaching in high school

Purpose	Agree	Do not agree
1.To consolidate and systematize the chemical knowledge for the students	87,0%	13,0%
2.To practice learning skills (such as using Chemistry language, writing, solving problems in chemistry ...)	74,0%	26,0%
3. To form and improve ability (problem solving, awareness, thinking)	73,6%	26,4%
4. To test and evaluate study result of the students	82,3%	17,7%
5.To enable the students to enjoy learning, have the positive attitude, be active in learning	80,0%	20,0%
6.To use exercises as a source of knowledge for students to study new knowledge.	82,9%	17,1%

On the basis of consultations with teachers who are teaching directly in high schools, the majority of teachers totally agree with testing, consolidating and systematizing chemical knowledge for students; 73.6% of teachers vote for the examination and evaluation organized in parallel with ability development

2.1.5. The current situation of chemistry teaching with the development of problem solving ability in Lao People's Democratic Republic

- Teaching objective

Table 7. The difficulty level of using Project based learning

Survey content	Chosen solution	
1. How often do you care about ability development for the students?	A. Usually	15,5%
	B. Sometimes	42,5%
	C. Rarely	33,0%
	D. Never	9,0%
2. During the process of chemistry teaching, how often do you care about the development of problem solving ability for the students?	A. Usually	53,2%
	B. Sometimes	24,0%
	C. Rarely	12,5%
	D. Never	10,3%
3. In chemistry teaching, which teaching objective do you particularly concentrate on?	A. To convey theoretical content to the students	6,5%
	B. To instruct the students to do types of exercises in the high school chemistry program	57,5%
	C. To develop specific ability of the subject for the students	37,0%

Regarding teaching objective, 57.5% of teachers supposed to mainly instruct the students to do types of exercises in the high school chemistry program according to the textbook of the Ministry. In the teaching process, 53.2% of teachers considered development of problem solving ability for the students as the main ability in comparison to other abilities.

- On assessing the importance of developing the capacity of solving problems through teaching the topic of integrated environmental education.

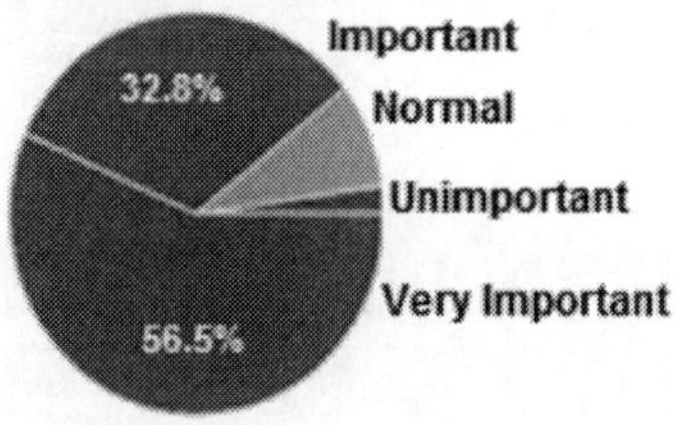

Figure 3. The important level of developing problems solving capacity through teaching the topic of integrated environmental education.

89,3% of teachers said that the development of problems solving capacity of students is important and very important in teaching Chemistry through integrated environmental education topics.

2.1.6. Current status of facilities for chemistry teaching in Lao People's Democratic Republic.

- On facilities for Chemistry teaching. Statistical tables of facilities conditions for Chemistry teaching of high schools in Lao People's Democratic Republic.

Table 8 Facilities conditions of surveyed schools

No.	Contents	Own	Doesn't own
1	Projector or TV connected to computer	48,0%	52,0%
2	Chemistry classrooms	36,0%	64,0%
3	Sufficient chemicals and equipment according to the minimum equipment list	37,8%	62,2%

According to the survey data, there are many high schools do not have sufficient chemistry classrooms, chemicals and laboratory instruments that in the list of school equipment. Some schools still lack of projector or TV connected to computer which is the necessary conditions for favorable chemistry teaching and learning, meet the demand for innovation in teaching methods.

2.2. Results of student surveys

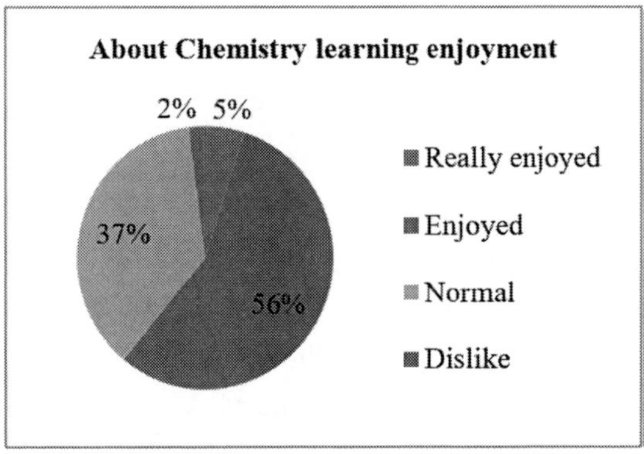

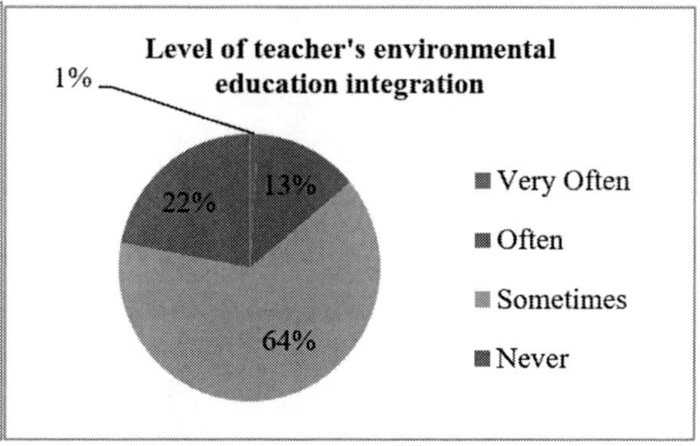

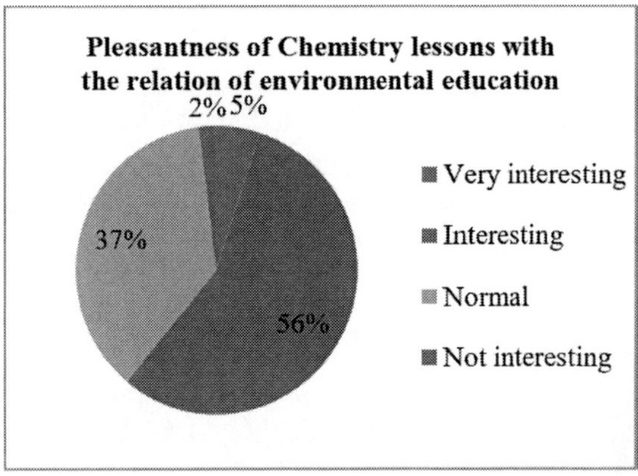

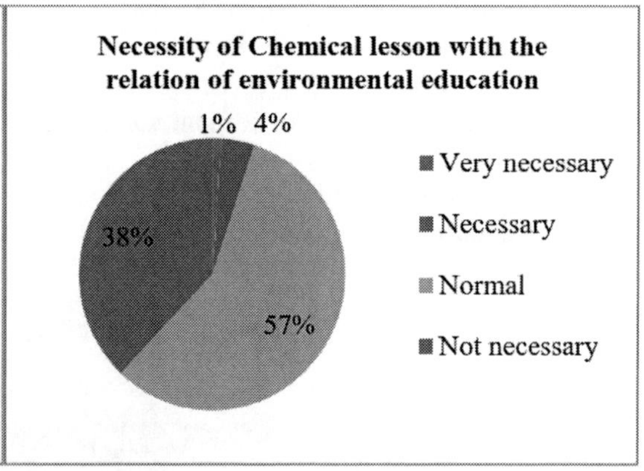

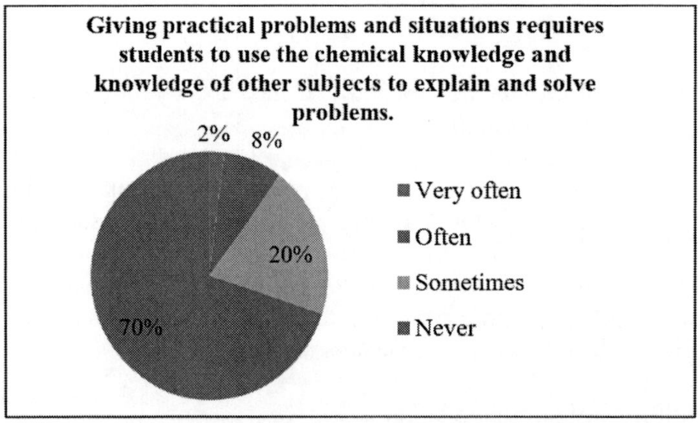

56% of students are interested in chemistry and 37.59% of students find it normal and are not interested.

Level of lecturer's environmental education integration: 64, 40% of students said that the integrated environment education had sometimes been used, 22% of students said that it had never been used.

About giving practical problems and situations requires students to use the chemical knowledge and knowledge of other subjects to explain and solve problems: 70% of students sometimes used it and 2, 14% of students never used it.

Pleasantness, necessity of Chemistry lessons with relation of environmental education: 56,03% of students found it interested, 56,62% of students found it necessary ; 1,93% of students found it not interested and 1,23% of students found it not necessary .

Through the surveys of teachers and students of some high schools in Laos, some comments can be made as follows:

- Integrated teaching, capacity development, especially problem solving capacity is a very new concept for teachers and students of some high schools in Laos.
- There are still many high school students don't know how to integrate Chemistry with other subjects.

3. Conclusion

From the results of survey teachers and students in the provinces, we can see the positive situation of using positive teaching methodologies and techniques, difficulties in applying the Teaching methodologies of Solving Problems and Teaching project. The survey results show that the development of the topic of Integrated Teaching and organizing Integrated Teaching contributing to development of capability of solving problems for high school pupils through teaching Chemical is very necessary in order to reach the target of high school education of the Lao People's Democratic Republic in present.

In order to successfully implement the basic and comprehensive reform of education and sports of the Lao People's Democratic Republic, it is necessary to carry out synchronously the renewal of teaching and learning contents and methods with a view to the stressed importance of training for students' comprehensive capability, especially problem solving ability at all levels. Fostering and development of capacity for students, especially, a key issue is extremely important in the educational and sports development strategy to provide indigenous and high quality human resources for the country. I investigated and studied the situation of teaching chemistry and investigated through teachers and students to assess the current status of solving problem ability of high school students in the Lao People's Democratic Republic. Measures for solving problem capacity development for the students are being researched and will be published in next papers. The current situation of problem solving ability development for students through chemistry teaching at high schools of lao people's democratic republic.

4. Summary

Education in the People's Republic of Laos (Lao PDR) plays an important role in the cause of national construction and defense. In general school subjects, chemistry is a subject in the field of natural science with many applications. However, despite the achievements, the reality of teaching chemistry at the upper secondary level is still inadequate. This article addresses some of the problems in the development of problem solving capacity in high school chemistry teaching in Lao PDR.

References

[1] Bernd Meier-Nguyen Van Cuong 2014 *Modern teaching theories* (Publishing house of University of Education)

[2] DinhQuangBao 2014 *Approach to capacity in the general education program* Proceedings of the workshop (The general education program in the new general education program Hue)

[3] DinhQuangBao (2014) *Integrative teaching - Methods of developing pupils' capacity* Workshop on general education program in the new general education program, HCMC, November 2014 (Ministry of Education and Training) pp 38-44

[4] DinhQuangBao 2013 *Textbooks of Vietnam in the orientation of capacity development after 2015* Proceedings of the International Conference on Innovation and Modernization of

IOP Conf. Series: Journal of Physics: Conf. Series **1340** (2019) 012055 doi:10.1088/1742-6596/1340/1/012055

Programs and Textbooks in the direction of sustainable development Workshop organized by Vietnam Education (Publishing House in Hanoi)

[5] Laos Ministry of Education and Sports 2010 *General Strategy and Plan for Education Development 2010-2020* (Vientiane)

[6] Ministry of Education and Training 2014 *Scientific workshop: "Improving the capacity of training teachers in teaching integrating Natural Science in University of Education"* (Publishing house of Hanoi University of Technology)

[7] Ministry of Education and Sports 2012 *Final report and capacity development plan for Laos secondary school pupils* (Publishing house of Vientiane)

[8] Nguyen Anh Dung 2014 *Integrative Teaching in General Education*, Workshop on Comprehensive General Education Program in New General Education Program, Ho Chi Minh City, November 2014 (Education and Training) pp 25-37

[9] Yuenyong, C. (2017). Enhancing Thai Students' Thinking Skills about Energy issues: Influence of Local Values. *Chemistry: Bulgarian Journal of Science Education*, 26 (3): 363 – 376

Developing the innovative inquiry-based lesson plan through lesson study

Everlita E. Canalita[1], Amelia T. Buan[1], Nikki B. Amboayan[1], and Jabber I. Mindalano[1]

[1]Department Science & Mathematics Education, College of Education, Mindanao State University-Iligan Institute of Technology, Iligan City Philippines
*Corresponding author's email address: everlita.canalita@g.msuiit.edu.ph

Abstract. This study aims to develop an innovative inquiry-based lesson plan through lesson study (LS) [2]. LS is considered as an updated professional development which aims at improving the teaching process and students' learning [16]. The lesson plan makes use of 7E learning cycle [5] embedded with inquiry activities that transform an ordinary classroom setting into an active student-centered learning environment [8]. The lesson model includes the following stages (1) Elicit, (2) Engage, (3) Explore, (4) Explain, (5) Elaborate, (6) Evaluate and (7) Extend. The research lesson made by the lesson study (LS) team was innovated gradually every after each implementation cycle. Revisions were done based on observations such as students' responses, teacher-student interaction or student-student interaction depending on how students view the lesson and how the group activity work smoothly. The innovative lesson plan has refined three times, based on the needs of the students. It is also evident in the implementation that students' learning and motivation are dependent on the scaffolding questions and the ability of the teacher to connect relationships that would help students contextualized the concepts in real life experiences.

1. Introduction

One of the major problems prevailing the fabrics of Philippine education today is the low achievement scores of Filipino students. Department of Education revealed in 2012 that the mean percentile score of Science-NAT in the country was 66.11% for the elementary and 40.53% for the secondary level [4]. Research studies revealed that teacher quality has a significant effect on students' achievement [21]. Thus enhancing teachers' quality is vital in improving the students' learning outcome. Also, to enhance the ability and quality of teachers and the effectiveness of their lessons, teachers may engage in research activities like lesson study [14] The study pointed out that high-quality professional development focusing on higher order thinking skills and diversity issues do not appear to strongly influence classroom practice. Further studies showed that students' success to understand the lesson depends on the ability of the teacher to facilitate the lesson properly [1]. This is the promising goal of inquiry-based learning and lesson study cycle which is intended to improve instruction given the realities in different classrooms.

Literature cited that Lesson Study (LS) in Japan is a fundamental driving force for mathematics teacher development. LS has been the primary mechanism of professional development for both prospective teachers and practicing teachers since the Japanese public education system started [14, 17]. It is a school-based collaborative activity for teachers mainly characterized by a continuous cycle of meticulous planning, prudent and mindful demonstrating, and perceptive improving a lesson. It is a wide scope, the on-site professional development process that involves a small group of teachers. The teachers are of different levels of ability but with interest in working collaboratively, with specific objectives for lesson planning, to carry out the planned and researched lesson [7].

Through lesson study (LS), teachers can work together to develop an effective lesson for the students to maximize their learning experience [6]. Research studies also showed that the use of lesson study leads to instructional development since teachers become more aware of how students learn and think and how instruction affects student thinking. LS is a process in which teachers progressively strive to improve their teaching methods by working with other teachers to examine and critique one another's teaching technique [10]. So, LS is not only for the good of students but for the teachers as well.

With this aforementioned views, this study elucidates how an innovative lesson is developed through lesson study cycle composting of two (2) in-service teachers and two (2) pre-service teachers. The product of LS is a refined lesson plan that make use of 7E Model [5]. This model or learning cycle consists of seven stages: elicit, engage, explore, explain, elaborate, evaluate, and extend, which allows teachers to present lessons that incorporate inquiry activities and promotes a student-centered classroom.

2. Theoretical Background

This study is anchored on the Constructivism Theory of Jean Piaget. Piaget suggested that students construct new knowledge from their experiences through "accommodation and assimilation"[18,19]. The study also pointed out that constructivism as a learning theory views learning as a process in which "students actively construct or build new ideas and concepts based upon prior knowledge and new information" [19] Moreover, Constructivist teaching is based on the belief that learning occurs when learners are actively involved in a process of meaning and knowledge construction as opposed to passively receiving information [17]. Lastly, constructivism activates the student's inborn curiosity about the real world to observe how things work and assume that all knowledge is constructed from the learner's previous knowledge, regardless of how one is taught. Hence, Constructivism modifies the role of teacher that he/she facilitates and help students to construct knowledge rather than to reproduce a series of facts [11].

A study claimed that there was an improvement in the competencies in the cognitive and manipulative skills of students, credibility, and commitment after the implementation of the constructivist teaching-learning process in chemistry laboratory subject [19]. In addition, students accept the constructivist teaching-learning process in their chemistry laboratory positively. Using a constructivist approach significantly improved how motivated are the students and the conceptual understanding of the students increased from low mastery to average mastery level [17].

This is the primary goal of Inquiry-based Approach. The inquiry-based instruction model stresses the learning and thinking process rather than just the acquisition of specific skills. It promotes the thinking process and teaches students how to process information in addition to skill and knowledge development. The focus of inquiry-based instruction is more on the actual process of learning than the understanding of specific concepts [3]. A further study claimed that inquiry-based teaching is a pedagogical approach as well as a learning strategy. Through the use of questioning, the core value of inquiry-based pedagogy puts stress on discovery learning and the development of learners' cognitive skills and metacognitive strategies [12].

Hence, Huber et.al (2000) as cited by Easterly III (2011) taught that students using inquiry-based instruction have higher perceptions of science and have an increased achievement in science. The findings of Von Secker (2002) report that students being taught using inquiry-based instruction had higher content knowledge achievement than did students taught through other means [3]. Indeed, the inquiry-based approach is the heart of constructivism. Although it is not easy to be a constructivist teacher, this can be achieved through reforms in teachers' training like engagement in a collaborative learning community such as lesson study.

3. Methods

This study is a kind of phenomenological inquiry that seeks to develop an innovative inquiry-based lesson plan through a community of learning - Lesson study. LS is a process that follows a cycle in which a group of science teachers works together, identify goals for the lesson, collaboratively developing a lesson plan, implementing the lesson (also called the *Research Lesson*) with observation by colleagues and other experts, analytically reflecting on the teaching and learning that happened, and revising the lesson. The lesson plan made by the team was revised along the three (3) implementation cycles. Briefly, the lesson study team entails the following: setting long-term goals for student development, planning a lesson together n an agreed topic, implementation of the lesson by a teacher while others observe and collect data on the lesson and students' responses, post-lesson reflection and discussion and final revision of the lesson.

3.1. Planning of a Research Lesson

Prior to the lesson development, the LS study team started with a planning session that discusses and agree on how the lesson will flow from beginning to end. The team also decided to follow the 7E learning cycle of Eisenkraft, 2003 which involves inquiry-based activities that would stimulate students to think critically. Basically, the group does the following: share how to teach the lesson and how the students learn the topic; study available teaching and learning resources on the lesson; organize questions to be used in the processing of the students' responses; developing lesson's main ideas and assign the specific task of each member.

What roles do members of the lesson study team play?

3.1.1. Facilitator. The member who lead the meetings and be responsible for coordinating the activities of the group. The facilitator also moderates the post-lesson discussion and synthesizes the discussion made by the group and help integrate revisions into the lesson plan. 2) Documenter. The member of the group who what goes on during meetings, as well as agreements and decisions that the group will make. 3) Implementer. The member who is responsible for carrying out the lesson plan which the group has developed. 4) Observer. The main task is to observe and collect data about the lesson (student-teacher interaction) and students' responses and reactions. The end product of the planning is a lesson plan embedded with inquiry-based activities using the 7E model.

3.2. Implementing the Research lesson

The research lesson was implemented to the 3 sections of grade 7 in Iligan City National High School namely: Grade 7-loyalty, grade 7-courage, and grade 7-prudence. During the implementation, a team member acts as a classroom teacher and carry out the lesson as planned while the other members just observed. The observers take note of what actually happened during the implementation such as modifications or deviations made; problems encountered by the teacher, difficulties experienced by the students and how these things were addressed; how the students responded to the lesson and what it revealed about their thinking and most importantly whether the learning competencies or objectives of the lesson were achieved.

3.3. Post-Lesson Discussion

After each lesson implementation, the LS group met for a post-lesson reflection and discussion. During this session, the members discuss what went well, how are the responses of the students, what problems or troubles encountered by the implementer and what things could be done to improve the lesson. The team also reflected on how students think and learn as indicated by evidence gathered during implementation. The focus of post-lesson discussion is not the teacher but on the lesson and the responses of the students. The research lesson was then revised based on the observations made and insights of the group and other ideas that resulted in the group discussion.

4. Results and Discussion

The figure 1 below is the process of implementing the lesson study. It shows how the (LS) Lesson study team work together to develop an innovative inquiry-based lesson plan. Moreover, an inquiry-based lesson plan had adopted the 7E learning cycle model of Eisenkraft (2003), developed and implemented to grade 7 students in junior high school. The teachers then revised the lesson plan during the post-lesson discussion based on the learners' difficulties and learning.

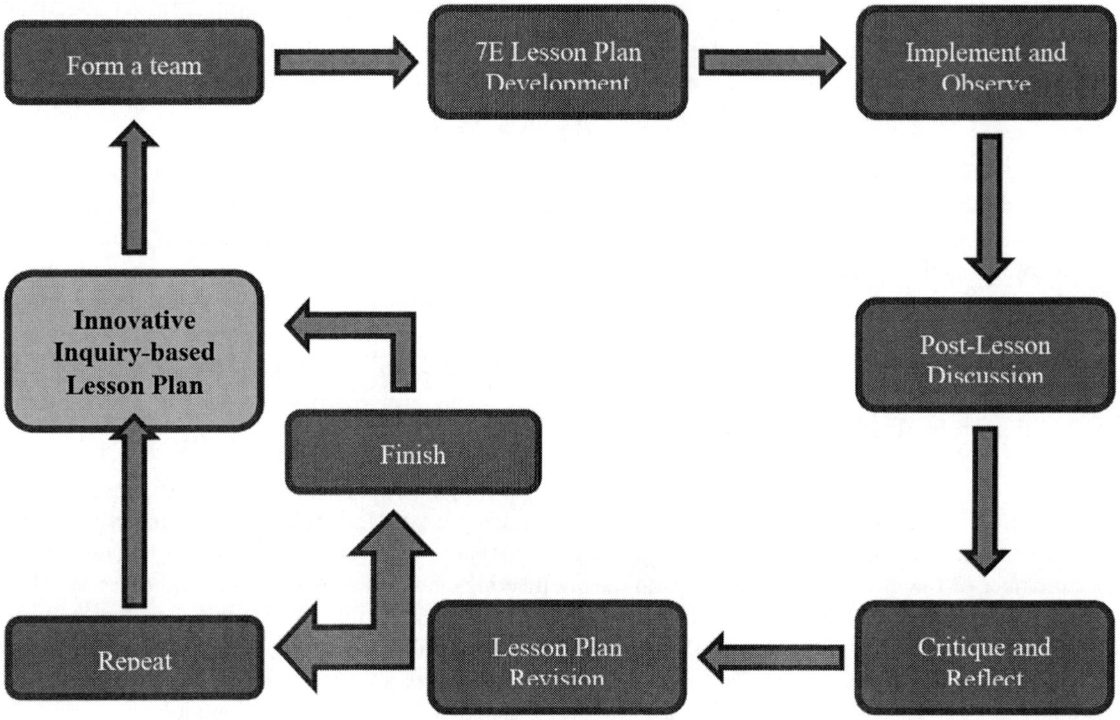

Figure 1 Lesson Study Process

Throughout the three implementation cycles, several changes in the inquiry-based activities, questioning technique and in sequencing details were made. These changes were based on the responses of the students during the student-teacher interaction and student-student interaction towards the lesson. Post lesson discussion and lesson revisions were done every after implementation cycle.

Elicit Phase. In the elicit phase, a game called "4 pics one word" were given to the students in order to access the students' prior knowledge. The implementer showed four pictures on the board, with blank spaces that correspond to the number of letters of the word. The students would come up with a word that best describes the four pictures. They would raise their hand if they know the answer. After the game, an image of an identical twin; which is Raymond and Richard, is shown to the students and questions about the image would be thrown to them. The questions were: "Why are identical twins look similar?", "What makes them similar?" and "Do they have the same genetic material?"

ELICIT (Access prior knowledge) (5 mins.)

To access and check the prior knowledge of the students towards the topic, the teacher will let the class play a "4 pics 1-word" game/ activity in which they will guess the hidden word behind the pictures that will be shown. The teacher will list down the words given by the students such as organisms, twin, identical, two, similar, reproduce, replicate, etc. Let the students construct a sentence out of the given words.

The teacher will then connect their answers with the idea of producing identical twins.
After the game/ activity, another picture of an identical twin will be shown to the students and some questions will be raised by the teacher to support the picture.

The questions will be the following;

1. Do you know them?
2. How do they look? Similar or different?
3. Why do identical twins look similar?
4. Do they have the same genetic material? (If students would answer YES,
5. How are identical twins produced? Allow the students to explain on their own.

Answer: To form an identical twin one fertilized egg (ovum) splits and develops into two babies with exactly the same genetic information.

The concept behind identical twins is the cell division known as the mitosis. Mitosis is one of the two types of cell division that produced two identical cells. Another type of cell division is the meiosis that has the same process and its only difference is that the process in meiosis happens twice and its product is 4 unique haploid cells.

The game catches the attention of the students. They participated excitedly and were always eager to be the first one to guess the right words.

Picture 1 responses:
S1: "divide"
S2: "division"

Picture 2 responses:
S1: "twin"
S2: "identical"
S3: "the same"

Picture 3 responses:
S1: "growth"
S2: "develop"

Picture 4 responses:
S1: "wound"
S2: "heal"

The students answer the questions actively.
Question: do you know them?
Response: "yes"
Question: how do they look? Similar or different?
Response: "they look similar"
Question: do they have the same genetic materials?
Response: "yes they have"
Question: how are identical twins produced?
Responses:
S1: "they have the same parents"
S2: "they have the same father"
S3: "they came from the same sperm"

During the first implementation cycle, the observers and the knowledgeable others observed that the students had some difficulties in getting the correct word that sums up the 4 pictures. The pictures were confusing for them because it does not clearly describe the words and others were images that they were not familiar with. On the other part, students were not able to answer the questions, it was so hard for them since they were not familiar with the deeper concepts of identical twins. In the whole phase, the students were not able to participate actively. So in the second implementation cycle, in order to catch the attention of the students and they would be able to give the right word, the researchers decided to change the pictures that were confusing for the students. In the word "twin or identical", one of the pictures was changed into an image of Richard and Raymond which the students were familiar with, while in the word "growth" other pictures were changed into much more obvious images like a growth of a human being and a plant. For the second part, scaffolding questions were added for the students' aid. The questions were; "Do you know them?", "How do they look?". However, in this implementation cycle, another problem arises; the students were so noisy and the class was a mess because they all want to be called by the implementer to give the right word. The implementer was not able to control the eagerness of the students to answer. Others were even angry for they were not called by the implementer when they say they were the first one to race their hand. So for the third implementation cycle, to lessen the noise and to make the class organized, the researchers decided to make the activity by the group in order to avoid such a scenario. The class was group into 7 groups and each group was given a tag board and chalk where they would write their answer and whoever raises their tagboard first that had the correct answer gained the points. To attract more participation from the students, the researchers were also giving prizes to the groups or students that was the winner or answers the questions correctly.

Engage Phase: In this phase, in order to continue to excite and engage students mind in the lesson, the implementer showed an image of a fraternal twin; which the students were familiar with, and that was a picture of Mavy and Cassy Legaspi. The implementer then threw some interesting questions to the students and discuss the relation of fraternal twins and identical twins. The questions given to the students were; "Why are fraternal twins look different?", "Do they have the same parents?"

ENGAGE (Get the students' mind focused on the topic.) (7 mins.)	
This time, a picture of a fraternal twin will be shown so that the students will be curious about what are the differences between the two types of a twin. It will help the students to focus their mind on the topic.	The students listen attentively to the teacher and answers actively to the questions.
What do you think is this kind of twins? Answer: Fraternal twins The teacher might ask some follow up questions to the students to support the picture above. 1. Why do fraternal twins look different? 2. Do they have different parents? So how come that they look different? Fraternal twins look different even they have the same parents. This is an application or a product of a type of cell division called meiosis.	**Question**: What do you think is this kind of twins? **Response**: that twin is called the fraternal twin. **Question**: why do fraternal twins look different? **Responses**: **S1**: "they do not have the same father." **S2**: "they came from a different cell." **S3**: "they do not have the same parents."

Unfortunately, in the first implementation cycle, the observer noticed that the students had no prior knowledge about the questions asked and they were not also familiar to the explanation to why fraternal twins look different. Because of that, their answers to the questions were mostly guessed. So, for the second implementation cycle, in order to avoid confusion and guessing, the researchers gave scaffolding questions to the students, these questions were: "how do they look?" and "are fraternal twins look different?". Other questions that may be hard for the students to answer were diminished and were placed in optional. As a result, in this implementation cycle, students were more responsive and showed interest in the discussion about the fraternal twins. So the researchers agreed not to revise it for the third implementation cycle.

Explore Phase: In order for students to explore new knowledge, the implementer gave a group activity that made the students solve problems on their own with the aid of instructions and given materials. In this activity, the students were given one set of drinking straw and four plastic cellophanes in a paper box. Then the implementer would instruct the students to divide the straws equally into four. In connection to what they just did, the implementer would ask questions to the students. After the activity, a short overview of meiosis would be discussed by the implementer.

In the first implementation cycle, the researchers saw that not all of the groups were able to follow the instructions well. Other groups divided the straws into 2 or 3. The instruction was not clearly stated and was not repeated which made it confusing to the students on what they really need to do. Observers also stated that the questions were difficult for the students since it was too confusing. Lastly, students also find it difficult to imagine the discussion of the overview of meiosis without any instructional materials. Hence, for the second implementation cycle, in order for the students to follow the instructions correctly, the researchers emphasized that the instructions must be clearly stated and should be repeated many times so that all of the groups would be able to follow correctly. Secondly, unnecessary questions were removed to avoid confusion, from 9, the researchers reduced it into 4 questions which were: "what are your bases in dividing it that way?", "what does the plastic container represent? Pair of straws?", "do you have similar combinations of straws per container?", and lastly "what are the combinations of straws do you have?". In the discussion of the overview of meiosis, since most of the students were visual learners, and so that students wouldn't have a hard time imagining the discussion, the researchers decided to make a 3d model that shows the division of cells and its uniqueness. However, during this implementation cycle, the observers saw that students had some confusions in the questions. This is because the questions were shuffled and the students were having a hard time connecting the dots. So, another revision was made for the third implementation cycle, this was to make the questions much easier to answer and to avoid confusions. The questions were rearranged from simple to complex. As a result, students were able to answer the questions without difficulty and confusion.

Explain Phase: This phase is where the implementer discussed the lesson, which is the meiosis. The implementer would explain thoroughly and clearly each phase of meiosis which were; Prophase, Metaphase, Anaphase, Telophase. Interaction and active participation from the students were encouraged here. During the discussion, simple questions were thrown to the students as an opportunity for them to participate.

Unfortunately, in the first implementation cycle, students were shy to raise their hands and others don't even bother to participate. The observers could observe that the students were a little intimidated because it was their first time meeting the implementer. So, they were not yet comfortable to answer questions. So, for the second and third implementation cycle, an alternative way to get students attention and participation was made. This was giving food or candy prizes to the students who choose to participate. Each student that raises his/her hand and give the right answer shall have a prize. It went well and made the students more eager to answer simple questions.

EXPLAIN (Teach the concept. Should include interaction between teacher and students). (22 mins.)

In this phase, the students will discuss their idea of the 8 stages of the meiosis. The implementer will guide them to the right track and give them clues.	
Meiosis is a type of cell division that results in four daughter cells each with half the number of chromosomes of the parent cell. It has eight stages namely: prophase 1, metaphase 1, anaphase 1, telophase 1, prophase 2, metaphase 2, anaphase 2, and telophase 2. This is divided into two parts which are the meiosis 1 and meiosis 2.	The students participate in the discussion and give their ideas on the 8 stages of meiosis.
Prior to meiosis, preparation of the cell takes place, chromosomes and centrosome will duplicate.	
Question: do you have any idea what phase is this?	**Response**: "just like in mitosis, this phase is called the *interphase*"
After that, the first phase of meiosis follows.	
Question: what is the first phase of meiosis 1?	**Response:** "the first phase if meiosis is prophase 1"
Question: what will happen in prophase 1?	**Response:** "In prophase 1, chromosomes will coil and become shorter and thicker. The nucleolus will disappear, spindle fibers will start to form and centrosome migrate towards the opposite poles."
Prophase 1 is the longest and the most important part of meiosis. Aside from the activities that is already mention. synapsis which is the pairing of homologous chromosomes will occur and crossing over will take place. Crossing over is the process that can give rise to genetic recombination.	
Question: how about the second phase of meiosis 1?	**Response:** "the second phase of meiosis is the metaphase 1"
Question: what happened during metaphase 1?	**Response:** "In metaphase just like in mitosis, centrioles are at the opposite poles of the cell. The chromosomes will align at the middle of the cell which is called the metaphase plate and spindle fibers from one pole of the cell will attach to one chromosome of each pair."
The third phase of meiosis is anaphase 1.	
Question: what is the activity during anaphase 1?	**Response:** "During anaphase 1, chromosomes separate and start moving toward opposite poles of the cell as a result of the action of the spindle fibers."

Elaborate Phase: This phase provides an opportunity for students to apply their knowledge to new areas. So the implementer provides a group activity that will assess the students learning or knowledge about the lesson and enhanced the students' collaboration. In this group activity, the implementer would assign one phase for each group. A strip of paper that has a certain phase and it's meaning was given to each group. They would draw and explain the phase that was assigned to them.

ELABORATE (Students apply the information learned in explain.)

In this phase, the implementer will show a picture of prophase that is ready for recombination and let the students predict the possible color combination if it will undergo the first meiotic division.

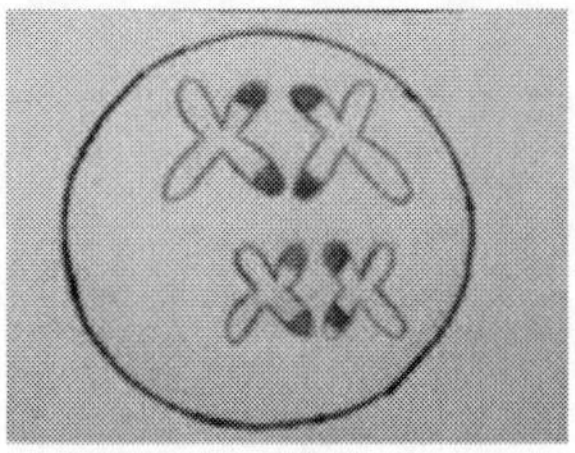

The picture above is an example. The following question will be thrown to the students.
Questions:

1. What stage of meiosis is shown in the drawing?
2. What if this prophase will undergo first meiotic division?
3. What are the possible color combinations in each cell?
4. Can you think of other combination aside from these two?

After the questions, the implementer will then discuss the activity. The implementer will let the students realized that one pair of chromosomes could have thousands of possible combinations. This is why siblings or even fraternal twins may have their similarities but they will not be exactly alike.

During this activity, the students were eager to give the possible color combinations. They were so excited that even other students went to the front to give their answer.

Question: what stage of meiosis is shown in the drawing?
Response: the stage shown in the picture is prophase 1.
Question: what if this prophase will undergo first meiotic division?
Response: "homologous chromosomes will separate"
Question: what are the possible color combination in each cell?
Response: "big green and small red or big red and small green"
Question: can you think of other combination aside from these two?
Responses:
S1: "big green and small green"
S2: "big red and small red"

In the first implementation cycle, the group activity did not go well. Majority of the group outputs did not match to the phases that were given to them and they cannot explain the phase that they drew. The students were not able to familiarize the pictures of the phases from the discussion. In connection with this result, the activity was too difficult for the students. So, the researchers decided to change the whole activity for the next implementation cycle into something that students can enjoy and at the same time learn. In the second implementation cycle, a picture of one of the stages of meiosis is shown to the students. The implementer then asked a series of questions which were; "What stage of meiosis is shown in the drawing?", "What if this prophase will undergo first meiotic division?", "What are the possible color combinations in each cell?" and "Can you think of other combination aside from these two?". This activity drew students attention that made them participate actively and attentively. Despite being one of the last stages, students were excited to answer what stage is shown and eager to predict the color combinations of the chromosomes. Without even noticing, while having fun, the students were able to learn and develop their predicting skills. Given the result, the researchers decided not to revise this phase for the next implementation cycle.

Extend Phase: In order for the students to apply their knowledge to real life situations the implementer showed an image of the Barreto sisters and asked two questions which were; "Why do sibling who have the same parents look so different?" and "What specific activity that results to the diversity of individual?". Throughout the three implementation cycle, most of the students were able to answer the questions confidently. They were able to relate the topic to why siblings with the same parents look so different. This implies that the students learned from the lesson and were able to apply their new knowledge to real life setting.

EXTEND(Deepen conceptual understanding through the use of new context)	
In this phase, the teacher will show the image above (Barretto sisters) and asked the following questions. Questions: 1. Why do siblings who have the same parents look so different? 2. What specific activity that results in the diversity of the individual? This context is in connection with the concept of meiosis, which is the cell division that results in four unique haploid cells. Each cell is different that is why even siblings that have the same parents look different. During the process of meiosis in the prophase 1, crossing over takes place. This is an activity where genes from the mother and the father recombine resulting in the diversity of each individual.	In this phase, the students will apply their new knowledge to real life setting. **Question:** why do siblings who have the same parents look so different? **Response:** "siblings who have the same parents look different because of the cell division called meiosis." **Question:** what specific activity that results in the diversity of the individual? **Response:** "the activity that results in the diversity of individual takes place during prophase 1 and that is the *crossing over*" The students were amazed by the application of the lesson and they were so thankful that they learn something meaningful.

5. Conclusion

Based on the findings of the three implementation cycles, LS provides a good opportunity for the team to develop and innovate a good lesson plan based on the responses of the students during the student-teacher interaction and student-student interactions. It also promotes cooperation and collaboration among members of the LS team to create a meaningful learning environment. Despite the limited resources, teacher implementer was able to show better ways to teach the particular topic because of the critical reflection during the post-lesson discussion. LS is a good practice in the education industry, to come up, not with the best lesson but the most appropriate lesson for the learners.

6. Acknowledgments

The authors would like to thank Philippine DOST-SEI project STAR for sharing with us the idea of lesson study, 7E Model and inquiry-based learning approach. We also greatly appreciate the involvement of the respondent school (Iligan City National High School) for giving the researchers enough time and classes for implementation.

References

[1] Beitenbeck J 2011 Teaching Practices and Student Achievement: Evidence from TIMSS (Masteral thesis) Retrieved from https://www.cemfi.es/ftp/mt/t1104.pdf

[2] Cerbin B and Kopp B 2011 Retrieved June 11, 2017, from https://www.uwlax.edu/sotl/lsp/overview.htm

[3] Easterly R G 2011 Inquiry–Based Instruction for Students with Special Needs in School Based Agricultural Education *Journal of Agricultural Education* **52** pp 36–46 Retrieved March 18, 2019

[4] Department of Education 2010 Discussion paper on the Enhanced K+12 basic Education Program Retrieved from http://ceap.org.ph//upload/download/201210/17115829500c1.pdf

[5] Eisenkraft A 2003 Expanding the 5E model, a proposed 7E model emphasizes 'transfer of learning' and the importance of eliciting prior understanding *Science Education* **5** pp 57–59

[6] Elipane L E 2012 Towards the Embodiment of Competency Standards: Incorporating the Elements of Lesson Study in the Pre-Service Mathematics Teacher Education in the Philippines *The Asia-Pacific Researcher* **21** pp 365–375

[7] Fong M W 2015 *Effects of Lesson Study incorporating phase-based instruction on grade seventh students' geometry achievement* In: 7th ICMI-East Asia regional conference on mathematics education Cebu City, Philippines

[8] Hernadez T and Roleda L 2015 *The Effectiveness of Using 7E Learning Cycle Model in the Learning Achievement of Grade 8 Students* Science Education Department, De La Salle University March 2015

[9] Huber R A, Smith R W and Shotsberger P G 2000 *The impact of a standards guided equity and problem solving institute on participating science teachers and their students* ERIC publication ED 443621

[10] Isoda M, Stephens M, Ohara Y and Miyakawa T 2007 *Japanese Lesson Study in mathematics: its impact, diversity and potential for educational improvement* World Scientific Publishing, Singapore

[11] Khalid A and Azeem M 2012 Constructivist Vs Traditional: Effective Instructional Approach in Teacher Education *International Journal of Humanities and Social Science* **2** pp 170–177 Retrieved March 18, 2019

[12] Lee H 2014 Inquiry-based Teaching in Second and Foreign Language Pedagogy *Journal of Language Teaching and Research* **5** pp 1236–1244. Retrieved March 18, 2019

[13] Lewis C 2000 *Lesson Study: the core of Japanese professional development* Paper presented at the AERA annual meeting

[14] Lomibao L S 2016 Enhancing mathematics teachers' quality through Lesson Study *Springer Plus* **5** pp 1–13 doi:10.1186/s40064-016-3215-0

[15] Makinae N 2010 *The origin of Lesson Study in Japan. Paper presented at the 5th East Asia regional conference on mathematics education* In: Search of excellence in mathematics education, Tokyo

[16] Marzban A and Haghighifard M 2016 Teachers' Viewpoint on the Practical Implementation of Lesson Study in EFL Classrooms *International Journal of Humanities and Cultural Studies* **3** ISSN 2356-5926

[17] Mirana V P 2016 Effects of Computer Simulations and Constructivist Approach on Students' Epistemological Beliefs, Motivation and Conceptual Understanding in Physics. *International Conference on Research in Social Sciences, Humanities and Education* pp 89–93 Retrieved March 18, 2019

[18] Muijs D and Reynolds D 2011 *Effective Teaching – Evidence and Practice* 3rdedition. London Sage Publications

[19] Reyes P B 2013 Implementation of a Proposed Model of a Constructivist Teaching-Learning Process – A Step Towards an Outcome Based Education in Chemistry Laboratory Instruction *Asia Pacific Journal of Multidisciplinary Research* pp 174–187 Retrieved March 18, 2019
[20] Secker C V 2002 Effects of inquiry–based teacher practices on science excellence and equity *The Journal of Educational Research* **95** pp 151–160 doi: 10.1080/002206702 09596585
[21] Wenglinsky H 2001 https://www.ets.org/Media/Research/pdf/RR-01-19-Wenglinsky.pdf. Downloaded 20 Apr 2016

Enhance elementary school students to learn mathematics: Case of the use of mother tongue-base language (MTBL) students

L G Maminta[1*] , J V Mendoza[2*], and C Yuenyong[3]
[1]Department of Education, Iligan City East Central School, Iligan City, Lanao Del Norte 9200, Philippines
[2]Integrated Developmental School, Mindanao State University-Iligan Institute of Technology, Iligan City 9200, Philippines
[3]Faculty of Education, Khon Kaen University, Khon Kaen, Thailand
*Corresponding author's e-mail address:buding062000@gmail.com; josievic.mendoza@g.msuiit.edu.ph

Abstract. The study of using Mother Tongue-Based Language (MTBL) in remediation on Least Learned Competencies in Elementary Mathematics to two sections in Grade VI at Tambo Central School was carried out. Both groups were exposed to team-assisted individualized remediation.The data needed for this study were obtained through 30-item achievement test and seventeen (17) Learning Materials constructed by the researcher which were construct-validated and evaluated by mathematics experts. Identified competencies of the materials were based on the National Achievement Test result and in the K to 12 Curriculum on Rational Numbers. The findings of the study revealed that the use of MTBL in team-assisted individualized remediation evidently improved the performance of the respondents in the three competencies on fractions namely comprehension of fractions; multiplication of fractions and its applications; and division of fractions and its applications.

1. Introduction

Mathematics is essential in many fields of endeavor and is one of the keys to the advancement of modern world. It is observed that in the elementary level, it appears as the least appreciated and liked subject. Many teachers have observed that majority of the pupils found difficulties in learning this subject. The pupils' difficulties may be caused by lack of interest and adverse attitudes in mathematics [1], [2].

One of the factors identified by some researchers is the teaching methodology or approach being implemented by the teachers. It is in this context, that remediation with the employment of MTBL is seen to make mathematics more meaningful and enriching to the pupil [1].

This study determined the effect of Using Mother Tongue-Based Language in Team-Assisted Individualized Remediation on Least Learned Competencies in Elementary Mathematics among Grade VI pupils of a public elementary school situated east of Iligan City. More specifically, this study sought to answer the following questions: (1.) What is the performance of the respondents during the remediation?; (2.) What is the performance of the respondents on Comprehension of Fractions, Operation on Fractions and its applications before and after remediation?; (3.) Is there a significant difference in the performance of the respondents in the pretest and post test scores?; (4.) Does MTBL enhance the use of TAIR in improving the respondents' competencies on fractions?; (5) What are the

IOP Conf. Series: Journal of Physics: Conf. Series **1340** (2019) 012057 doi:10.1088/1742-6596/1340/1/012057

least learned competencies after remediation for the: (a.) MTBL Group and (b.) Non-MTBL Group [2], [3], [4], [5].

2. Methodology

2.1 Design

The paradigm of the study is presented in Figure 1. The Use of Mother Tongue-Based Language in Team-Assisted Individualized Remediation is based on the pupils' knowledge on how to comprehend fractions, do the operations on fractions and its application. After the pretest, the Team-Assisted Individualized Remediation was implemented.

This kind of remediation served as an intervention to improve the pupils' achievement specifically in comprehension of fractions, operations on fractions and its application and hoped to help establish the high performance of the pupils in the National Achievement Test (NAT).

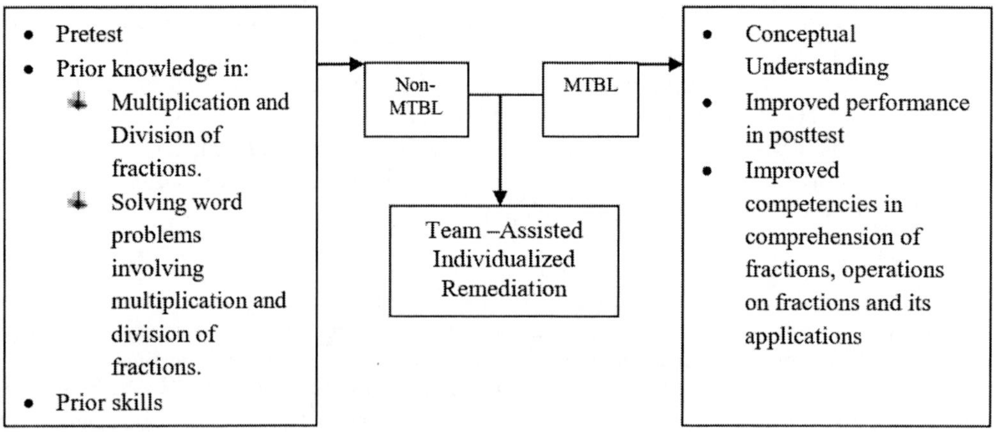

Figure 1. Research Paradigm

2.2. Procedure

The respondents in both sections were exposed to team-assisted individualized remediation (TAIR). However, only section Loyalty as the experimental group used the Mother Tongue-Based Language (MTBL). The control group section (Sincerity) was closely supervised by the researcher to ensure that only the English language is used during the group discussions.

Seventy-six (76) pupils with a grade below 80% in mathematics comprising the middle sections of the grade six level, who are officially enrolled at Tambo Central School for the academic year 2016-2017, were the respondents of the research study. The distributions of the respondents were classified by sections with 38 pupils, respectively. The respondents in each section were grouped into five (5) considering its proximity. The grouping was done after the pretest.

This study is a quasi- experimental research investigating the effect of Mother Tongue- Based Language (MTBL) in teaching concepts and skills on rational numbers using the same teaching technique in remediation. The remediation was done after the regular class hours. The experimental group used the mother tongue language which is the Visayan to discuss the missed items in the answered learning materials while the control group used the English language. Before the start of every remediation, each respondent was given an activity sheet to be answered individually for fifteen (15) minutes. After fifteen minutes, the pupils formed into groups upon the instruction of the researcher and checking of answers were done. Afterwards, the activity sheets are returned to the owner and group discussion follows facilitated by the teacher and assisted by a group leader which encourages a member to ask questions. If the group disagrees on a certain concept, the group consults

IOP Conf. Series: Journal of Physics: Conf. Series **1340** (2019) 012057 doi:10.1088/1742-6596/1340/1/012057

the teacher to settle the conflicting interpretation. Also the group reviewed together to prepare themselves for board work. The board work was done before giving the next activity.

The K to 12 Curriculum became the guide of the researcher in determining the specific topics to be considered in the construction of the learning materials and in the preparation of the Table of Specifications (TOS) required before the achievement test was constructed.

The achievement test is a 30-item multiple choice questions on rational numbers, specifically on comprehension of fractions, multiplication and division of fractions and its application. The collection of data started with the giving of a pretest to the respondents using the researcher–made achievement test a week before the remediation started. The posttest was administered after the three weeks remediation. Both the pretest and the posttest were answered by the respondents for only an hour. The checking and scoring of the achievement test was done by the researcher. Each correct for each item received one point.

The learning material which consists of seventeen (17) activity sheets prepared by the researcher is the other instrument used in this study. These activity sheets were face validated by some elementary mathematics teachers within Iligan City. It was evaluated with an excellent rating using a rubric adapted and revised by the researcher.

3. Findings/Analysis

To determine the difference in the effects of using MTBL in implementing Team-Assisted Individualization Remediation on Least Learned Competencies in Mathematics, the two-tailed t- test is performed so that the results could be seen in either direction making use of degrees of freedom (df) to identify the number of pieces of useful information within the data set at 5% level of significance.

The performance of the respondents after remediation on competencies classified as Comprehension on Fractions was determined after giving the posttest. It revealed that the MTBL group is performing better than the non-MTBL group in all items in the achievement test except on two items. These items are competencies on identifying the reciprocal of a given number and reducing fractions to lowest terms.

The performance of respondents on multiplication of fractions after remediation showed that the number of respondents at MTBL group who answered correctly the items classified as Multiplication of Fractions is more than the number of respondents who answered correctly in the non-MTBL group.

The performance of the respondents after remediation on the items classified as Division of Fractions revealed that there is an equal number of respondents (10) who answered correctly on dividing a whole number by a mixed number. It was observed that pupils have misconceived the term reciprocal which results to incorrect procedure in performing the operation. The pupils have difficulty in identifying the reciprocal of a number. Maybe, the pupils were not able to unlock the meaning of the word reciprocal.

The mean difference, as shown in Table 1, in the pretest and posttest scores of this experimental group (MTBL) is higher by 5.434 as compared to the control group (non-MTBL) with t-value equal to 7.749 and standard error difference of 0.78449 is found to be highly significant. This means that the respondents in the experimental group have better performance during the team-assisted individualized remediation as compared to the control group. In general, the mean difference of the two tests showed that the respondents performed better in the posttest. The result also shows a two-tailed significance of 0.000 which is less than 0.05. This suggests to reject the null hypothesis "there is no significant difference in the performance of the respondents in the pretest and posttest scores".

The posttest mean score (14.71) of the respondents in the achievement test is greater than the pretest mean score (9.57) by 53.81%.

The pretest mean score (8.37) of the non-MTBL group in the achievement test is less than the posttest score by 2.10.

The posttest mean score (18.95) of the MTBL group in the achievement test is greater than the pretest score (10.76).

The mean difference (1.58) of the pretest and posttest scores of the MTBL group in the competency on Comprehension of Fractions is higher to the mean difference (1.55) of the non-MTBL group by 0.03.

The mean difference (4.66) of the pretest and posttest scores of the MTBL group in the competency on Multiplication of Fractions is much higher to the mean difference (0.66) of the non-MTBL group by 4.00.

The mean difference (1.95) of the pretest and posttest scores of the MTBL group in the competency on Comprehension of Fractions is higher to the mean difference (0.11) of the non-MTBL group by 1.84.

Table1. Performance of the Respondents in the Achievement Test

Control Group (Non-MTBL Respondents)				**Experimental Group (MTBL Respondents)**			
Section-Sincerity	Pre-test	Post-test	Difference	Section-Loyalty	Pre-test	Post-test	Difference
Mean (X)	8.368	10.47	2.11	Mean (X)	10.76	18.95	8.18
X (Mean Pretest-Post)	9.421			X (Mean Pretest-Post)	14.855		
Mean Difference (Experimental vs. Control)				6.07895			
t-value				7.749			
Standard Error Difference				0.78449			
Significance				0.00000			
Remarks				Highly Significant			

The computed t-value of the pretest and posttest mean scores of the respondents in the achievement test is 9.81 with 75 degrees of freedom. Hence, there is a significant difference between the pretest and posttest scores of the respondents in the achievement test.

The computed t-value of the pretest and posttest mean scores of the non-MTBL respondents in the achievement test is 4.12 with 37 degrees of freedom. Therefore, there is a significant difference between the pretest and posttest scores of the non-MTBL group in the achievement test.

The computed t-value of the pretest and posttest mean scores of the MTBL respondents in the achievement test is 13.76 with 37 degrees of freedom. Therefore, there is a significant difference between the pretest and posttest scores of the MTBL group respondents in the achievement Posttest.

The computed t-value of the mean difference scores of the non-MTBL and MTBL group in Comprehension of Fractions is 0.06 with 74 degrees of freedom. Therefore, there is a significant difference between mean scores of the MTBL and the non-MTBL groups with respect to Comprehension of Fractions.

The computed t-value of the mean difference scores of the non-MTBL and MTBL group in Multiplication of Fractions is 9.91 with 74 degrees of freedom. Therefore, there is a significant difference between mean scores of the MTBL and the non-MTBL groups with respect to Multiplication of Fractions.

The computed t-value of the mean difference scores of the non-MTBL and MTBL group in Division of Fractions is 5.27 with 74 degrees of freedom. Therefore, there is a significant difference between mean scores of the MTBL and the non-MTBL groups with respect to Division of Fractions.

The five least learned competencies of the non- MTBL and the MTBL groups after remediation are: (a.) Dividing fractions by other fractions, (b) Dividing a whole number by a mixed form, (c) Dividing a mixed number by a mixed number, (d) Dividing a whole number by a fraction and (e) Solving problem involving division of fraction

Based on the findings of this study, the following conclusions have been drawn:

The employment of the team-assisted individualized remediation improved the performance of both the MTBL and non-MTBL groups.

The MTBL group performed better than the non-MTBL group in the three identified mathematics competencies namely: Comprehension of Fractions, Multiplication of Fractions and Division of Fractions.

The use of the mother tongue-based language in remediation enhances the performance of the MTBL group.

Most of the least learned competencies after remediation are topics which involved Division of Fractions.

4. Recommendation

Develop and evaluate instructional and learning materials in elementary mathematics using team-assisted individualized remediation and mother tongue-based language in least learned competencies like division of fractions.

Conduct further studies on Team-Assisted Individualized Remediation versus MTBL using different respondents for two quarters.

Develop a training program to equip teachers with knowledge and skills on how to facilitate the learning process like team-assisted individualized teaching or cooperative learning.

References

[1]DepED Order No. 74,s.2009. Institutionalizing Mother Tongue-Based Multilingual Education (MLE).pp.1-5.

[2]NETRC.2011National Achievement Test for Grade Six. *Certificate of Rating*.Institutional Performance Profile.

[3]NETRC.National Achievement Test. Key Programs. July 05,2011. http://www.deped.gov.ph/about_deped/organization.links.asp?id=16

[4] Slavin, Robert E.(1984a). Team-Assisted Individualization. Cooperative Learning and individualized Instruction in the Mainstreamed Classroom. Research Project. John Hopkins University.

[5] Slavin, R. E. ,(1984b) *Cooperative Learning: Theory, Research, and Practice*.2nd edition. Englewood Cliffs, NJ:Prentice Hall

[6]Zulaut, Madelere (1994). Three-Year Experiment on Extended Must Teaching in Switzerland: The Different effects Observed in a Group of French Speaking Pupils. Research Seminar Journal.

IOP Conf. Series: Journal of Physics: Conf. Series **1340** (2019) 012058 doi:10.1088/1742-6596/1340/1/012058

Development of Mathematics Collaborative Problem-Solving Skills Scale

Joneil B. Medina[1]*, Amelia T. Buan[1], Josie Vic D. Mendoza[2] and Grace P. Liwanag[2]

[1] Department of Science and Mathematics Education, College of Education, Mindanao State University – Iligan Institute of Technology, Iligan City 9200, Philippines
[2] Integrated Developmental School, College of Education, Mindanao State University – Iligan Institute of Technology, Iligan City 9200, Philippines
*Corresponding author's e-mail address: joneil.medina@g.msuiit.edu.ph

Abstract. Collaborative Problem Solving (CPS) is one of the most critical and necessary skills in education and workplace; hence, its assessment is a necessary subject of research. CPS assessment tools must be crafted to establish the effectiveness of the pedagogies to CPS development. Therefore, to address these gaps, this study developed a Likert scale called Mathematics Collaborative Problem Solving Skills (MaCoPSS) Scale. The scale development followed the process: (1) item generation, (2) item evaluation, (3) try-outs, (4) Exploratory Analysis, (5) final evaluation. Open-ended questionnaires were administered to 273 respondents. After analysis, 551 statements were produced as initial item pool. The items are evaluated and only 59 of them are used for the draft scale for the try-outs. The scale was administered to 648 students for the try-outs and responses were cleaned then subjected to Exploratory Factor Analysis (EFA). Two factors, namely social skills (SS) and problem-solving skills (PSS) were extracted. To establish internal consistency, the factors are correlated to one another ($r=0.708$), and factor to total raw score (TRS) correlations are highly correlated: SS-TRS ($r=0.955$) and PSS-TRS ($r=0.894$). The final scale consists of 27 SS and 17 PSS items. The overall reliability coefficient of the scale is $\alpha=0.941$. The resulting scale is considered valid and reliable.

1. Introduction

Lev Vygotsky in his Mind in Society, notes that social interaction is essential to learning and that learning implies development [24]. It is in this sociocultural perspective of learning that collaborative learning is seen as an effective means of facilitating learning even in the context of developing problem-solving skills. The idea of collaboratively solving problems is not new in literature [17]. However, it is only recently that collaborative problem solving is extensively studied in terms of teaching, learning and assessing.

Collaborative problem-solving (CPS), as a construct, brings together individual problem-solving and the collaborative social process of learners working together [17]. In the 21st century, CPS is one of the most critical and necessary skills both in education and the workplace [12]. Project Based Learning (PBL) approach to Science Technology Engineering and Mathematics (STEM) education [1]; National Council of Teachers of English [26] recognizes the importance of CPS. Furthermore, it is included as one of the dimensions in work readiness assessment [20]. These only show the importance of understanding the dynamics of CPS. Hence, many researchers find it necessary to find a framework of developing and assessing the said skill [10, 11, 12, 13, 19, 20, 25].

With several frameworks for CPS emerging from different studies, methods of assessing CPS is not far from being realized. For example, Organization for Economic Cooperation and Development's (OECD) Programme for International Student Assessment (PISA) and Assessing and Teaching 21st Century Skills (ATC21S) projects conceptualized frameworks for assessing CPS in large assessment contexts. PISA first introduced CPS as one of the innovative domains to assess during the 2015 PISA. PISA assessment items are referred to as CPS units where the students solve a series of problems. The students will select predetermined responses in the "chat space" provided [12]. On the other hand, while the ATC21S project used a similar system, the main difference lies in the Human-Human interaction between students. Notice that both assessment systems have cognitive skills and social skills dimensions but inevitably utilize an online platform for assessing. With the emergence of such assessment systems integrating Information and Communication Technologies (ICT), the availability of easily accessed methods of assessment is meager. It is in this context that the problem arises.

In the Philippines, Teaching Mathematics through Problem Solving (TMPS) as an innovative approach utilizing collaborative problem solving [2, 5] is being promoted by the University of the Philippines' National Institute for Science and Mathematics Education Development (UP-NISMED) since 2006 [15] and the Department of Science and Technology, Science Education Institute's (DOST-SEI) Project Science Teachers Academy for the Regions (STAR) since 2014 [6]. The approach's effects on student achievement have been fairly documented in the literature. However, the approach's effects on the development of CPS skills is not yet established.

Therefore, to establish the effect of an approach to the development of CPS skills means that a tool for measuring the level of CPS skills must be devised. With the limitations of the existing CPS assessment systems, this study addresses the gap of CPS assessment in the classroom setting by providing an alternative method of measuring CPS skills through developing a Mathematics Collaborative Problem Solving Skills (MaCoPSS) Scale. To develop a scale for collaborative problem solving as a tool for measuring collaborative problem-solving skills of the students. Specifically, to attain this general objective, this study aims to (1) Describe the process of the development of the MaCoPSS Scale; (2) Determine the dimensions of collaborative problem solving; (3) Determine the construct validity of the Mathematics Collaborative Problem Solving Skills (MaCoPSS) Scale; (4) Establish the reliability of the Mathematics Collaborative Problem Solving Skills (MaCoPSS) Scale.

2. Theoretical Background

The theoretical consideration for the scale development of the mathematics collaborative problem-solving instrument is primarily anchored upon the Collaborative Problem Solving framework of [13]. The same framework was used by the ATC21S project of the University of Melbourne undersigned by CISCO, Intel®, and Microsoft®. The framework is selected as it illustrates some essential skill of the highest cognitive domain for Mathematics [16]. Furthermore, the framework provides analytic criteria for the specified skill as it utilizes a developmental approach to learning and assessment [10].

The framework is composed of two domains: cognitive and social domains. Each domain has specified skills and behavioral indicators (See Table 1). Such dimensions are also seen across different known frameworks. The ATC21S assesses collaborative problem solving in a manner that social skills and cognitive skills are assessed parallel-wise, meaning social skills are measured differently from the cognitive skills [25]. The advantage of this framework is that it provides a clear picture of what to improve concerning CPS skills.

Table 1. Collaborative Problem Solving Skills and Indicator Design

Element	Behavioral Indicators Based on
Social skills	
Participation	
Action	Activity within environment
Interaction	Interacting with, prompting, and responding to contributions of others
Task completion/perseverance	Undertaking and completing a task or part of a task individually
Perspective taking	
Adaptive responsiveness	Ignoring, accepting, or adapting contributions of others
Audience awareness	Awareness of how to adapt behavior to increase suitability for others
Social Regulation	
Negotiation	Achieving a resolution or reaching compromise
Self-evaluation	Recognizing own strengths and weaknesses
Transactive memory	Recognizing the strengths and weaknesses of others
Responsibility initiative	Assuming responsibility for ensuring parts of the task are completed by the group
Cognitive Skills	
Task regulation	
Organizes	Analyzing and describing a problem in familiar language
Sets goals	Setting a clear goal for the task
Resource management	Managing resources or people to complete a task
Flexibility and ambiguity	Accepting ambiguous situations
Collects information	Exploring and understanding elements of the task
Systematicity	Trying possible solutions to a problem and monitoring progress
Learning and knowledge building	
Relationships	Identifying connections and patterns between and among elements of knowledge
Contingencies/rules	Using understanding of cause and effect to develop a plan
Hypothesis	Adapting reasoning or course of action as information or circumstances change

Source: [11, 13]

3. Methods

This study employs the sequential exploratory mixed methods design [4]. The study started with gathering of qualitative data from an open ended questionnaire answered by junior high school students and pre-service teachers. The data were analyzed to be the input for drafting the scale. Quantitative data were then gathered from the drafted scale. Psychometric analysis methods were employed to extract latent factors, establish validity and reliability.

3.1. Research Subjects

The open-ended questionnaires were administered to 237 students and pre-service teachers. The developed Mathematics Collaborative Problem Solving Skill (MaCoPSS) Scale was tried-out to 648 Grade 8 to Grade 10 junior high school students in Iligan City for the S.Y. 2018-2019. The respondents were from six (6) schools from the ten (10) randomly selected public high school. The selected classes will have to pass the basic criterion of having the students undergone collaborative activities during the current grading period during the scale administration.

3.2. Data Collection

The data gathering procedure was divided into the development phase and the try-outs phase; which is also the validation phase. The development phase was also divided into sub-phases: item generation phase, drafting phase and item evaluation phases. Try-outs phase was divided into two sub-phases: massive try-outs phase, factor analysis phase, and the final evaluation phase. After the series of tests, the developed scale was revised accordingly. Then, the panel of specialists evaluated the final scale.

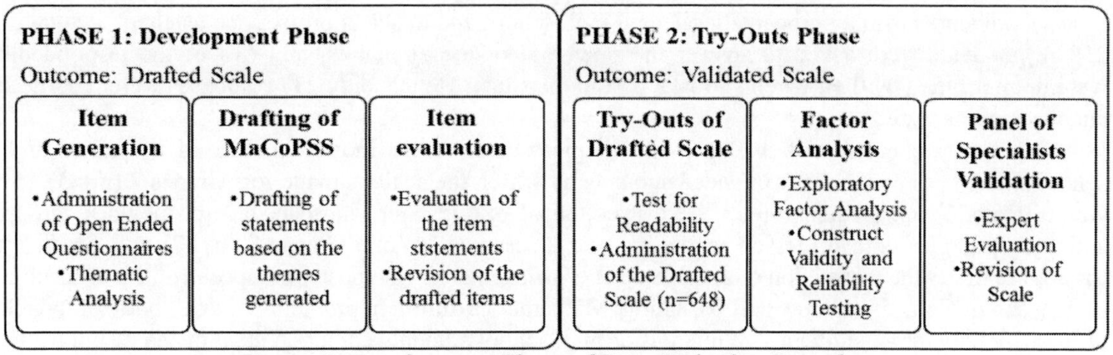

Figure 1. Development Phase of Data Gathering Procedure

In the Development Phase, Item Generation started with the administration of the open-ended questionnaires based on the CPS framework. The resulting pool of items was thematically analyzed by identifying keywords. Such keywords form the basis for the derived statements. The statements are then categorized to the framework of the study. Statements 'fitting' the framework were then be subjected to evaluation and were revised. The outcome of this phase is the "drafted MaCoPSS Scale."

For the large-scale try-outs of the drafted MaCoPSS Scale, all 26 public high schools in the division of Iligan City were considered. Random sampling was used to select ten (10) public high schools. Out of the ten (10) selected schools, only six (6) schools responded favorably. Factors like proximity limited the administration of the scale to these six (6) schools until the quota of 590 was achieved. This quota is based on [3] noting a commonly acceptable ratio of 1:10 for exploratory studies in factor analysis. This ratio indicates that there should be at least ten (10) observations per generated item. After the massive try-outs, the factor analyses were conducted.

Validity and reliability tests were also computed. At this point, a set of identified evaluators evaluated the MaCoPSS Scale. After their evaluation, the resulting outcome was termed the "validated MaCoPSS Scale."

3.3. Research Instrument

3.3.1. Open Ended Questionnaire. This questionnaire is a researcher made open-ended questionnaire, based on the collaborative problem-solving framework adapted for this study, administered to generate the statements for the scale development. The questions are as follows: (1) 1. How do you solve a math problem? (2) How do you solve a mathematics problem if you are working in a group?

3.3.2. Item Statement validation guide. This guide is taken from the guidelines on scale development by [7]. This guide was used in the evaluation of the generated statements before try-outs.

3.3.3. Scale Evaluation Rubric. This rubric is an adopted rubric from an unpublished graduate thesis and guidelines of [7], which was used by the panel of evaluators to evaluate the developed Mathematics Collaborative Problem Solving Skill (MaCoPSS) Scale.

IOP Conf. Series: Journal of Physics: Conf. Series **1340** (2019) 012058 doi:10.1088/1742-6596/1340/1/012058

4. Results and Discussion

This chapter presents the discussion on the development of the MaCoPSS Scale, from the item generation down to the exploration of its dimensions.

4.1. Development Phase

4.1.1. Item generation. The validated open-ended questionnaire was then administered to select junior high school students in a purposively selected high school and to select pre-service teachers. A total of 239 respondents were asked to answer the open-ended questionnaires, and two of the respondents were not permitted by their parents to take part in the study. Hence, only 237 responses were recorded, encoded and analyzed.

The respondents of the open-ended questionnaires are mostly comprised by junior high school students as they are the intended subjects by which the scale is made for: Grades 7 (n=43) and Grade 9 (n=56) are students under the instruction of one teacher whose pedagogy includes 'group activity' strategies both on lesson modelling and assessment. While Grades 8 (n=57) and Grade 10 (n=58) are under the instruction of another teacher whose pedagogy mostly composes of the utilization of the 5E model of learning and Teaching Mathematics through problem-solving, both of which utilizing collaborative strategies. While junior high school students comprise most of the respondents, taking the pre-service perspective is also important. Pre-service teachers have been asked to answer the open-ended questionnaires to provide a perspective of an instructor, as they are trained to facilitate different teaching pedagogies that employ both collaborative and individual teaching methods. This set of pre-service teachers are selected on the basis that they have been sufficiently exposed to a strategy, teaching mathematics through problem-solving, which collaborative problem solving is at its core [2, 5]. As teachers on training, they provided an additional perspective not observed by the junior high students.

The first question in the open-ended question: "How do you solve a math problem?" was intended to explore students' behavior, strategies and skills in solving problems individually, while the question "How do you solve a mathematics problem if you are working in a group?" was intended to explore students' behavior, strategies and skills when solving problems in a group. The first question is necessary as it provides a baseline as to how students individually manage tasks. This is also one dimension of CPS, while the second question primarily focuses on the social aspect of problem solving. The figure below shows the process of item generation forms the open-ended questionnaire responses to the derivation of the declarative statements for the drafted MaCoPSS Scale.

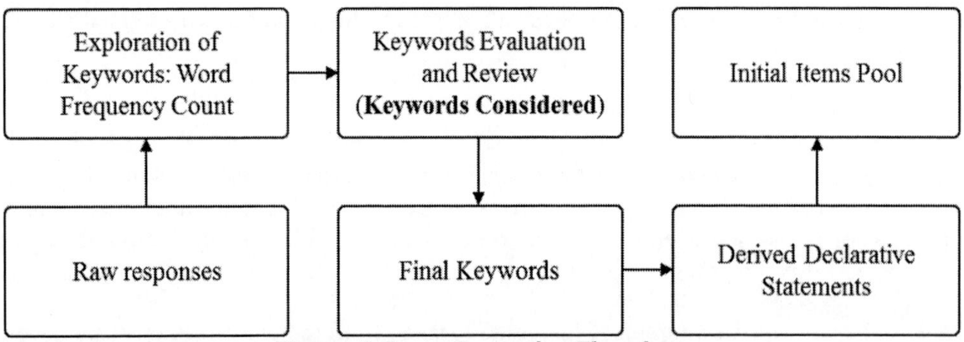

Figure 2. Items Generation Flowchart

The raw responses were encoded and translated to English, for the analysis. The thematic analysis utilizes the strategy of word frequency count. Frequently occurring words describe the concepts the students answer in the open-ended questions. Similar words are counted as one, e.g., solving has 450 occurrences which also counts solve, solved and solving. Each of the words counted as 'frequently occurring' is all reviewed and evaluated. The evaluation is focused on whether a word signifies a course of action, that is, an action word; and a word that would signify a concept relevant to CPS. From the generated pool of considered keywords, declarative statements are then derived from them. Each declarative statement has at least one keyword from which it is derived. A sample is shown below how a declarative sentence is derived.

Raw Response:
S3: [1] I always try my best not to be seen as useless and unhelpful. [2] And so I end up trying to solve all of the problems by myself first and then ask questions if there are some things that I don't quite understand or needs to be clarified. [3] I don't usually share what I know unless they ask me because I don't want to be seen as a "know-it-all." [4] However, I do like it when I am able to solve a problem, and then they ask me questions regarding it, I feel like I'm being helpful for the first time. C33

Keyword: ASK
Breakdown of sentence [2]
- *I end up trying to solve all of the problems by myself first. S3, 2a, C33*
- *I **ask** questions if there are some things that I don't quite understand or needs to be clarified. S3, 2b, C33*

Statements derived [2b]
I ask questions if there are things that I don't understand. C33
I ask questions if there are things that need clarification. C33

Figure 3. Sample derivation of statements

Statement three (S3) is a sample raw response of the respondent C33 for Q2. The keyword under consideration is 'ask,' and the sentence by which it appears is sentence 2. Sentence 2 is then divided into two declarative statements. Considering the keyword 'ask,' then, only from S3-2b will the declarative statement be derived. The researcher then simplified this raw response to its simplest form possible without losing the sense of meaning from the raw response. Then, two final statements are derived from keyword 'ask'. Reiterating the process of deriving declarative statements from the raw responses with the observed keywords, several declarative statements are derived from each keyword. This process resulted in producing 551 declarative statements. These statements were cross referred to the adapted framework by [13] and identified 119 statements that is closely related to the behavioral indicator defined by the framework.

4.1.2. Drafting of MaCoPSS Scale. A framework of Collaborative Problem Solving (CPS) was adopted in this study after a review of literature. This framework is from the work of [13] for the ATC21 Project. The framework identifies skills and indicators of the two major dimensions of CPS: Social Skills and Cognitive skills. The derived declarative statements are screened using this framework.

These statements were subjected to a simultaneous item evaluation and content validation conducted through a group discussion of teachers to maximize the efficiency of the evaluation process. The evaluation is guided by the validation guide adapted from a chapter in the book of [7] on scale development. It comprises of characteristics encompassing (1) item clarity and (2) presence of negatively stated and positively stated items. The evaluators selected have at least ten (10) years of experience in the field as a math teacher and are implementors of collaborative strategies of teaching mathematics. Table 2 shows some sample statements evaluated.

With the researcher facilitating the item evaluation, the teacher evaluators evaluated the items and content validated them based on their actual classroom experience. Out of the 119 drafted statements, 60 statements were rejected with the following reasons: statement length, clarity, and inapplicability. Notice sample items No.1 and No.4 in Table 2. Sample statement 1 is observed to have contained two concepts and was broken down into two separate statements. Whereas, sample

IOP Conf. Series: Journal of Physics: Conf. Series **1340** (2019) 012058 doi:10.1088/1742-6596/1340/1/012058

statement 4 was rejected due to clarity of statement. After the evaluation and content validation of the statements, the drafted MaCoPSS Scale was produced having 59 statements.

Table 2. Sample Statements Evaluation and Validation

No.	Statement	Q No.	Accept	Reject	Remarks
1.	I will allow my group mates to correct me, I will not assert that my answer is correct (67)	2	X		Double Barrel: Break down into two.
4.	I analyze the clue words that can help me solve the problem. (7)	1		X	Analyze: too broad and is not very clear.

4.2. Try-outs Phase

4.2.1. Administration of Drafted Scale. The respondents were distributed across the grade levels 8-10 for the try-outs phase of the scale development. Of each of the school randomly selected, three classes were selected one from each grade level from grade 8 – 10. Due to schedule conflicts, some classes which were selected for the try-outs were not able to answer the try-outs survey. Although there was a total of 235 Grade 8 students from 5 schools, 211 Grade 9 students from 5 schools and 202 Grade 10 students still, from 5 schools, only 565 of the responses were considered valid.

4.2.2. Exploratory Factor Analysis. An exploration of the collected data was conducted, and several low extremes and outliers were identified and removed to prevent misleading interpretation of the results. These outliers were based on the average means of each of the respondents' answers. The kurtosis (-0.279) and skewness (-0.131) values of the means revealed that the data collected is relatively not normally distributed. Considering the normality of the data set, [3] noting that there is an advantage when using Principal Axis Factoring in a relatively non-normal data. Hence, the extraction method used is principal axis factoring or PAF.

4.2.3. Factors Extracted. Before the conduct of the factor analysis, Kaiser-Meyer-Olkin Measure of Sampling Adequacy (KMO-MSA) and Bartlett's Test of Sphericity was tested with the given data [8]. Results showed that the data sample is adequate with KMO value of 0.925. The KMO-MSA measures whether the data is suited for factor analysis. [14] noted that values 0.9-1.00 as 'marvelous,' meaning that the data is well suited for factor analysis. Bartlett's Test of Sphericity tests the hypothesis that the correlation matrix is an identity matrix. Thus, if it is an identity matrix, there is no relationship between variables. This correlation matrix must not be an identity matrix to proceed to factor analysis. The results (sig. $p=0.000<0.05$) indicated that the correlation matrix for the data is not an identity matrix as there are significant relationships between variables. Therefore, the given data is suitable for factor analysis.

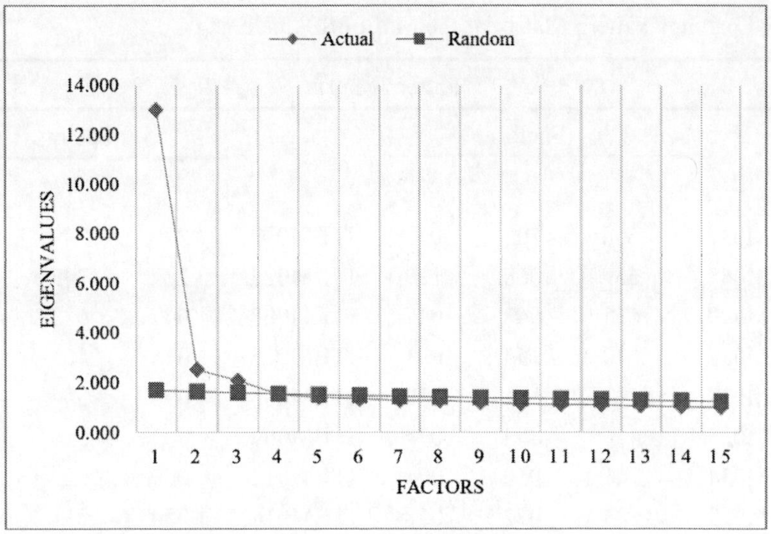

Figure 4. Parallel analysis scree plot

There are two major dimensions of CPS considered in this study, that is, social and cognitive. But to verify and ascertain the number of factors to be retained, a parallel analysis was conducted (Figure 4). The parallel analysis makes use of the actual eigenvalues and a randomly generated eigenvalues given the number of variables and the sample size. The scree plot of the actual and the randomly generated Eigenvalues intersects at factor 4. With the actual Eigenvalue of factor 4 lesser than the randomly generated eigenvalue, suggests that the scree plot "flattens" from factor 4, indicating further that only the three (3) factors are to be retained as suggested by [21, 22, 23].

4.2.4. Explored Dimensions of MaCoPSS Scale. Table 3 shows the rotated factor loadings of the conducted exploratory factor analysis. As a rule of thumb, only variables with a minimum factor loading of 0.32 are interpreted [3]. Factor loading of <0.32 that the variable does not correlate well with the factor extracted. Therefore, based on the factor loadings, 9 statements or variables were discarded. Table 6 shows the factor loadings from the EFA.

Most of the items loaded to factor 1 categorized as Social Skills (SS) in the framework of [13]. Essentially, this factor is the social skill factor. The social skill factor contains statements indicating skills in areas of participation, perspective taking, and social regulation. This confirms the existence of a social dimension in the CPS construct.

Statements loading under factor 2 are categorized as Problem Solving Skills (PSS). The statements loading on factor 2 indicates skills in task regulation and learning and knowledge building. This coincides with the CPS construct of bringing together social dimension and cognitive dimensions of learning.

Finally, factor 3 is categorized as Individualistic Tendencies (IndT). While these items load significantly on a factor, they do not necessarily contribute to successful collaboration. [18] suggested that one of the mechanisms that contribute to poor performance in groups is called social loafing where members of the group do not engage believing that someone else will accomplish the task at hand. Items such as "I usually find it better to work alone" or "If I perceive that my group mates want to work alone, I will leave them be" indicates such cases. Reliability tests also indicates that this factor is not reliable. Hence, this factor was not considered.

Table 3. Factor Loadings Pattern Matrix of the MaCoPSS Scale

Pattern Matrix							
	Factor				Factor		
	1	2	3		1	2	3
ITEM037	.668	-.102	.035	ITEM019	.270	.249	.012
ITEM048	.665	-.068	-.139	ITEM026	.262	.247	-.255
ITEM049	.635	-.042	.017	ITEM008	-.047	.633	-.070
ITEM032	.610	-.154	.091	ITEM015	-.018	.563	-.031
ITEM039	.602	-.141	.019	ITEM005	-.098	.554	-.045
ITEM029	.560	-.093	.074	ITEM009	-.001	.550	-.122
ITEM054	.541	.050	-.004	ITEM012	.009	.529	.079
ITEM050	.533	.072	-.058	ITEM018	.054	.511	.043
ITEM058	.514	.115	-.010	ITEM007	-.103	.509	.193
ITEM040	.496	.013	-.102	ITEM024	-.063	.479	.311
ITEM036	.478	.157	.032	ITEM025	-.108	.467	.195
ITEM047	.477	.154	-.134	ITEM022	.040	.446	.062
ITEM059	.475	.100	-.196	ITEM023	.056	.445	-.023
ITEM052	.474	.049	-.003	ITEM006	.064	.445	.038
ITEM044	.473	.095	-.056	ITEM003	.006	.437	-.110
ITEM030	.472	.138	.044	ITEM004	.055	.431	-.024
ITEM035	.471	-.088	.300	ITEM021	.189	.409	-.094
ITEM046	.458	.042	-.022	ITEM017	.213	.382	.027
ITEM011	.458	.074	-.025	ITEM002	.131	.361	-.175
ITEM057	.449	.066	.020	ITEM016	.235	.308	-.010
ITEM053	.445	-.099	.170	ITEM013	.071	.293	.175
ITEM045	.441	.151	.007	ITEM001	.100	.227	.103
ITEM055	.412	.163	-.001	ITEM014	.169	.172	.092
ITEM056	.407	.091	.088	ITEM010	.147	.162	.084
ITEM033	.402	.247	-.011	ITEM028	-.020	.128	-.489
ITEM031	.401	.192	.031	ITEM020	-.135	.266	.466
ITEM034	.367	.052	.148	ITEM043	.125	-.064	.462
ITEM027	.349	.024	.219	ITEM051	.003	-.023	.364
ITEM041	.310	-.051	.181	ITEM038	.127	.197	.330
ITEM042	.287	.215	-.003				

Note:
Extraction Method: Principal Axis Factoring. Rotation Method: Promax with Kaiser Normalization.
Rotation converged in 6 iterations.

4.2.5. Establishing Construct Validity. The construct validity is established by determining the correlation of the retained extracted factors, SS and PSS. Total raw scores (TRS) for the retained factors are also correlated to the scale TRS. This is to establish that each of the factors contributes significantly to the total raw score. Lastly, a panel of evaluators evaluated the scale developed from the retained items.

Table 4. Correlations of Factors

Factor-Factor	Correlation
Social Skills – Problem Solving Skills	0.708
Social Skills – Individualistic Tendencies	-0.068
Problem Solving Skills – Individualistic Tendencies	0.028

A correlation value of r=0.781 indicates that there is a high correlation between factors SS and PSS, while IndT does not significantly correlated to any of the other factors. The removal of the third factor is also supported by literature [18]. Hence, only SS and PSS factors are retained to achieve a constructively valid instrument.

Table 5. Correlations of Factor Scores and Total Raw Scores

Factor-TRS	Correlation
Social Skills Scores – TRS	0.955**
Problem Solving Skills – TRS	0.864**

** Correlation is significant at the 0.01 level (2-tailed)

A correlation value of r=0.955 and r=0.864 for SS-TRS and PSS-TRS, respectively, means that This means that both the social skills factor and the problem-solving skills factor scores significantly contribute to the total raw scores. With both coefficients being significant at 0.01 level of confidence, we conclude that the scores obtained in the identified dimensions of the scale for collaborative problem-solving skills have a significant contribution to the total raw score.

4.2.6. Final Evaluation. Using a modified rubric, each dimension can be rated with the highest rating of twenty (20) and the lowest of four (4). Two Mathematics Education evaluators and a Registered Psychometrician and Guidance Counselor evaluated the scale. The social skills dimension was rated with an average of 16.5, interpreted as 'Very Good' scale while the Problem-Solving Skills Dimension was rated with an average of 18.33, interpreted as 'Excellent' scale. Overall, the whole scale is evaluated with the average rating of 17.42 out of 20, interpreted as 'Excellent'. One item was removed from the SS as it was redundant with another item.

4.2.7. Establishing Reliability. The reliability of the scale and its explored factors are established by computing Cronbach's alpha. These tests would indicate the reliability and internal consistency of the scale. An acceptable rule of thumb for reliability is that it must have a Cronbach alpha value of not lesser than α=0.70 [9]. The developed scale has a total reliability value of α=0.931 with the third factor removed. The problem-solving skills factor has a reliability of α=0.846 while the social skills have reliability of α=0.912. These indicate that the scale is highly reliable with the given sample (See Table 8).

The final validated instrument contains 44 items 27 of which are under social skills and 17 items under the problem-solving skills (See Appendix). After establishing the construct validity and the reliability of the scale, the scale can be used to measure collaborative problem-solving skills in two dimensions, namely, social skills and problem-solving skills. The items under social skills dimension measures skills based on indicators of participation, perspective taking and social regulation skills. Participation indicators have seven (7) items, Perspective taking have four (4) items, and Social regulation has 16 items. The items under problem-solving skills dimension measure the skills based on indicators of task regulation and Learning and Knowledge building. Task regulation has 13 items, and learning and knowledge building has four (4) items. The two dimensions identified highly correlate to one another, and each of the dimensions' scores significantly contributes to the total raw score of the scale.

Table 6. Reliability Coefficient of the Developed Mathematics Collaborative Problem Solving Skills Scale

Collaborative Problem Solving Dimensions in Mathematics	No. of Items	Cronbach's α
Problem Solving Skills	17	0.846
Social Skills	27	0.920
Developed MaCoPSS Scale	44	0.941

5. Conclusion and Recommendation

Collaborative problem-solving skills have two dimensions namely, the social skills and problems solving skills. Social skills include participation skills, perspective taking skills and social regulation skills, while Problem Solving skills include task regulation and learning and knowledge building [13]. The scale developed is theoretically sound, valid and reliable as the tests indicate. While the developed scale is found to be theoretically sound, valid and reliable, this is only true for the given sample. Furthermore, the scale format may affect the responses of the respondents. Hence, the following are recommended: (1) The scale can be administered to another group of respondents to strengthen the construct validity of the scale. (2) Additional scale items may be considered as one of the mathematics education specialists suggested. (3) A 5-choice format may be adapted to verify whether the number of scale choices affects the student responses.

Since the developed scale was found to be theoretically valid and reliable, hence the following are also recommended: (1) The scale shall be used in actual classroom setting to verify its applicability and usability (2) A factor indicating individualistic tendencies indicates that there is a need to develop CPS through relevant pedagogies in the classroom, especially in the mathematics classroom.

Acknowledgements

The corresponding author would like to thank the Department of Science and Technology, Science Education Institue (DOST-SEI) for the scholarship grant under the project Science and Technology Regional Alliance of Universities for National Development (STRAND); The Department of Education Region Office 10, Division of Iligan City and the Mindanao State University – Iligan Institute of Technology, College of Education, Department of Science and Mathematics Education, Iligan City.

References

[1] Asghar, A., Ellington, R., Rice, E., Johnson, F., & Prime, G. M. (2012). Supporting STEM education in secondary science contexts. Interdisciplinary Journal of Problem-Based Learning, 6(2), 4.
[2] Cai, J., & Lester, F. (2014). Why Is Teaching With Problem Solving Important to Student Learning? (J. Quander, Ed.). National Council of Teachers of Mathematics. Retrieved from

http://www.nctm.org/uploadedFiles/Research_News_andAdvocacy/Research/Clips_and_Briefs/Research_brief_14_-Problem_Solving.pdf

[3] Costello, A. B., & Osborne, J. W. (2005). Best practices in exploratory factor analysis: Four recommendations for getting the most from your analysis. Practical assessment, research & evaluation, 10(7), 1-9.

[4] Creswell, J. W., Klassen, A. C., Plano Clark, V. L., & Smith, K. C. (2011). Best practices for mixed methods research in the health sciences. Bethesda (Maryland): National Institutes of Health, 2013, 541-545.

[5] Degner, K. & Fi, C., & (2012). Teaching through Problem Solving. MATHEMATICS TEACHER,105(6), 455-459. Retrieved November 20, 2014, from http://edcg669-f12-gilbert.wikispaces.umb.edu/file/view/8.4TeachingthroughProblemSolving.pdf

[6] Department of Science and Technology-Science Education Institute (2014). Annual Report. Retrieved from http://www.sei.dost.gov.ph/index.php/annual-reports

[7] DeVellis, R. (1991). Scale development: Theory and applications. In Bickman, L., and Rog, D. (S. Eds.). Applied Social Research Methods Series Vol. 36. Sage Publications Inc. Newberry Park, California.

[8] Factor Analysis: SPSS Annotated Output (n.d.) From: https://stats.idre.ucla.edu/spss/output/factor-analysis/ Retrieved September 27, 2017

[9] Gliem, J. A., & Gliem, R. R. (2003). Calculating, interpreting, and reporting Cronbach's alpha reliability coefficient for Likert-type scales. Midwest Research-to-Practice Conference in Adult, Continuing, and Community Education.

[10] Griffin, P., Woods, K., Mountain, R. and Scoular, C. (2013). ATC21s Module 1: Using a developmental model to assess student learning. Victoria Australia, Assessment Research Centre, Melbourne Graduate School of Education, University of Melbourne. Retrieved from: http://www.atc21s.org/uploads/3/7/0/0/37007163/pd_module_1_for_web_2014.pdf

[11] Harding, S. M. E., Griffin, P. E., Awwal, N., Alom, B. M., & Scoular, C. (2017). Measuring Collaborative Problem Solving Using Mathematics-Based Tasks. AERA Open, 3(3), 2332858417728046.

[12] He, Q., von Davier, M., Greiff, S., Steinhauer, E. W., & Borysewicz, P. B. (2017). Collaborative Problem Solving Measures in the Programme for International Student Assessment (PISA). In VonDavier, AA and Zhu, M and Kyllonen, PC (Ed.), INNOVATIVE ASSESSMENT OF COLLABORATION (pp. 95–111). https://doi.org/10.1007/978-3-319-33261-1_7

[13] Hesse, F., Care, E., Buder, J., Sassenberg, K., & Griffin, P. (2015). A framework for teachable collaborative problem solving skills. In Assessment and teaching of 21st century skills (pp. 37-56). Springer, Dordrecht.

[14] Kaiser, H. F. (1974). An index of factorial simplicity. Psychometrika, 39(1), 31-36.

[15] Lomibao, L. S. (2016). Enhancing mathematics teachers' quality through Lesson Study. SpringerPlus (2016) 5:1590. DOI 10.1186/s40064-016-3215-0

[16] Lindquist, M., Philpot, R., Mullis, I. V. S., & Cotter, K. E. (2017). TIMSS 2019 Mathematics Framework. In I. V. S. Mullis & M. O. Martin (Eds.), TIMSS 2019 Assessment Frameworks (pp. 11-25). Retrieved from Boston College, TIMSS & PIRLS International Study Center website: http://timssandpirls.bc.edu/timss2019/frameworks/

[17] Luckin, R., Baines, E., Cukurova, M., Holmes, W., & Mann, M. (2017). Solved! Making the case for collaborative problem. https://www.nesta.org.uk/sites/default/files/solved-making-case-collaborative-problem-solving.pdf

[18] Nokes-Malach, T. J., Richey, J. E., & Gadgil, S. (2015). When is it better to learn together? Insights from research on collaborative learning. Educational Psychology Review, 27(4), 645-656.

[19] OECD. (2015). Pisa 2015 Draft Collaborative Problem Solving Framework March 2013. Oecd, (March 2013), 89. https://doi.org/http://dx.doi.org/10.1787/9789264266490-en

[20] Oliveri, M. E., Lawless, R., Molloy, H. (2017). A literature review on collaborative problem solving for college and workforce readiness (GRE Board Report No. 17-03). Princeton, NJ: Educational Testing Service. http://dx.doi.org/10.1002/ets2.12133

[21] Parallel Analysis. (n.d.). Retrieved from https://analytics.gonzaga.edu/parallelengine/

[22] Patil, V. H., Singh, S. N., Mishra, S. and Donavan, T. (2007). Parallel Analysis Engine to Aid Determining Number of Factors to Retain. [Computer software]. Available from http://smishra.faculty.ku.edu/parallelengine.htm

[23] Patil, V. H., Singh, S. N., Mishra, S. and Donavan, T. (2008), "Efficient Theory Development and Factor Retention Criteria: A Case for Abandoning the 'Eigenvalue Greater Than One' Criterion," Journal of Business Research, 61 (2), 162-170.

[24] Vygotsky, L. S. (1978). Mind in Society: The Development of Higher Psychological Processes. Human Development. Cole, John-Steiner, Scribner & Souberman, (Eds.) Harvard University Press. https://doi.org/10.1007/978-3-540-92784-6

[25] Woods, K., Mountain, R., Griffin, P. and Scoular, C. (2013). ACT21S Module 4: Use of a developmental framework for teaching. Victoria Australia, Assessment Research Centre, Melbourne Graduate School of Education, University of Melbourne. Retrieved from: http://www.atc21s.org/uploads/3/7/0/0/37007163/pd_module_4_for_web_v2.pdf

[26] Zollman, A. (2012). Learning for STEM literacy: STEM literacy for learning. School Science and Mathematics, 112(1), 12-19.

IOP Conf. Series: Journal of Physics: Conf. Series **1340** (2019) 012058 doi:10.1088/1742-6596/1340/1/012058

APPENDIX A

VALIDATED MATHEMATICAL COLLABORATIVE PROBLEM-SOLVING SKILLS (MACOPSS) SCALE

Mathematics Collaborative Problem-Solving Skills (MaCoPSS) Scale

Last Name: (Optional)____________ First Name: ______________ MI: _____ Grade Level*:____

School*:__________________ Gender*: [ ] Male [ ] Female

**Please provide necessary information*

Instructions: Read each statement carefully. Place an X mark on the choice corresponding the degree of your agreement to the statement. Answer all items with honesty. There are four choices for each statement which corresponds to the following:

1 – Strongly Disagree 2 – Disagree 3 – Agree 4 – Strongly Agree

The example below indicates that you **AGREE** with the statement: 'I ask my teacher if I don't understand.'

Statement	**1**	**2**	**3**	**4**
1. I ask my teacher if I don't understand.			**X**	

STATEMENT	1	2	3	4
Problem Solving Skills				
1. I repeatedly read the problem if it is difficult.				
2. I take note of important details in the problem.				
3. I mark the important data or words to guide me in my solving.				
4. I make sure that I know the meaning of the words in the problem.				
5. I try to illustrate the problem so that I could understand it better.				
6. I use variables to represent quantities.				
7. I always make sure that I have the goal in mind when solving.				
8. After understanding the problem, I will identify what is unknown to solve the problem.				
9. If I can't find a suitable answer, I seek help from knowledgeable people on how to solve it.				
10. After finding an answer using one method, I use another method to confirm my answer.				
11. I will identify the given quantities first.				
12. I try to figure out if the problem is already discussed in class or not.				
13. After obtaining necessary information from the problem, I determine what series of operations should I use.				
14. I solve step-by-step so that I can look back and know what I am doing.				
15. I look for patterns in the problem for me to solve it.				
16. I recall of my past lessons that I can apply on the given problem.				
17. I associate the given problem to the real world to come up with practical ways to solve the problem.				
18. I manipulate the formula after determining what variables are missing.				

APPENDIX E (Contd.)

Social Skills				
19. I can communicate well my solution to my classmates.				
20. I ask my group mates how do they understand the problem.				
21. If somebody explains his answer to the group, I listen to him or her and react if necessary.				
22. We talk as a group on all possible solutions that will help us come up with an answer.				
23. I compare my answers to my group mates' after solving the problem on my own.				
24. I read and analyze the problem myself before sharing anything to my group mates.				
25. I must consider the ideas of my group mates whether their ideas are applicable or not.				
26. While listening to my group mates' understanding of the problem, I also compare it to my understanding and determine which is more reasonable.				
27. If we arrive at different answers, then I will ask my groupmates why their answers are like that.				
28. If I don't have ideas on my own, I will listen to the ideas of my group mates.				
29. I will allow my group mates to correct me.				
30. When working as a team, the team members must have a common goal.				
31. We agree which method to use by comparing and contrasting ideas given by the group.				
32. Everyone in the group must agree with the final solution.				
33. We make sure that all of us understands the process of solving the problem.				
34. When I can't solve the problem on my own, I ask my group mates how to solve the problem.				
35. I share my thoughts if I think my solution is correct.				
36. If I know the answer, I would help my group mates so that they would get the correct answer.				
37. If my answer differs from my group mates, I will try to find my mistake.				
38. Even if my answer is not correct, at least I am able to give an idea that might guide us to solve the problem				
39. I pay attention if others will not understand the problem to assist them				
40. I first ask my group mates if they can properly comprehend the situation, if there are questions, then clarify				
41. I would assign problems/tasks for each one, so no one will be doing nothing				
42. I try to keep everyone moving, not relying on one person				
43. I try to get everyone's attention before asking questions how they would solve the problem				
44. I engage my group mates to share their ideas on how they solve the problem to make better solution for the problem				

IOP Conf. Series: Journal of Physics: Conf. Series **1340** (2019) 012059 doi:10.1088/1742-6596/1340/1/012059

Assessment of Science Teacher Competence in Teaching Secondary Science with Spiral Progression Approach

Emily M. Maing[1], Almaira T. Mangadang[1], Monera A. Salic-Hairulla[1], Everlita E. Canalita[1], Fernando R. Sequente Jr.[1] and Chokchai Yuenyong[2]
[1] Department of Science and Mathematics, College of Education, MSU-Iligan Institute of Technology, Andres Bonifacio Iligan City 9200, Philippines
[2] Faculty of Education, Khon Kaen University, Khon Kaen, Thailand
*Corresponding author's e-mail address: monera.salic@g.msuiit.edu.ph

Abstract. This study aimed to assess the competence of science teacher in teaching secondary science with spiral progression approach in terms of their knowledge on subject matter, teaching strategies, personality traits and teaching approach. Science teachers in both public and private schools within Lanao del Sur and Lanao del Norte served as the respondents of this study. The study used questionnaire as methodology and utilized the frequency distribution, correlation coefficient and regression analysis method. The result showed that the profile of the respondents specifically age, tenure and year in teaching science, and salary grade are correlated with science teacher's competence in knowledge of subject matter. The result also showed that aside from age and tenure rank or position of the respondents is also correlated with the teaching strategies. Length of service correlates only with the personality traits of the respondents. Civil status, years in teaching, highest educational attainment correlates with the respondents teaching approach.

1. Introduction

Over a century, many policy makers around the world have tried their best to revise and improve the science education by altering the curriculum based on the perception that creating a new curriculum that will influence teachers to shift from their traditional routines in the classrooms, making students learn far better in science [8]. In order to cope with global challenges in science education experienced by many countries around the world especially Philippines, reforms in education is continually occurring [5]. Philippines' new educational system was implemented starting in 2011. This is the Republic Act 10533 also known as the Enhanced Basic Education Act of 2013 enabled the implementation of the K-12 in the country [13].The features of the K-12 Enhanced Basic Education Program include the strengthened Science and Math education which follows a spiral progression. The use of spiral progression avoids disjunctions between stages of schooling and allows learners to learn topics and skills appropriate to their developmental and/or cognitive stages [4].On that teacher are required to develop the essential knowledge skill, attitudes, and values of the student to enable the cope with these changes. It is known that there has been a focus upon the student's learning performance bur there has also been an acknowledgement that students's learning was not liely improve markedly until teacher were given the opportunity to further support and develop as well ti increase their teaching skills [2]. Teacher competencies is defined as "set of knowledge, skills, attitudes, values, beliefs and experience necessary for future, which manifests in the teachers' activities [12] Teachers need to improve knowledge and skills to enhance and explore their teaching practices. In fact, teacher's professional development should be defined for sustainability. Sililarly, teacher are responsible for operating educational system and they need strong and effecient professional competence [6].

Data analysis showed that the profile of the respondents were correlated with there teaching competencie.Therefore, in this work, we first intended to contribute to the search for good and effective curriculum of science subject. Specifically, it seeks to explore that may beneficial to both public and private schools, which all seek to assess science teachers' competencies teaching science in the spiral progression approach mandated in the K-12 program. Findings in this research may also give rise to the role of progressive schools in educational research and development of alternative, pedagogical systems.

The purposes of this study were to to assess the competence of secondary science teacher teaching science in spiral progression approach Specifically, the study sought to answer the following:

1. What are the competencies of a Science teachers in the spiral progression approach in terms of:
   1.1. Knowledge of Subject Matter
   1.2. Teaching Strategies
   1.3. Personality Traits
   1.4. Teaching Approaches
2. What are problems encountered by the science teacher using spiral progression approach?
3. Is there a significant correlation between the competencies of science teachers and their profile?

Null Hypothesis (H0): There is no significant relationship between the competences of science teachers and their profile.

Alternative Hypothesis (Ha): There is a significant relationship between the competencies of science teachers and their profile.

2. Theoretical Background

According to the principles of Jerome Bruner's Theory [4] Instruction comes in three ways these are the: readiness which the Instruction must be concerned with the experiences and contexts that make the student willing and able to learn, spiral organization which the Instruction must be structured so that it can be easily grasped by the student and last but not the least is the principle that going beyond the information given in which Instruction should be designed to facilitate extrapolation and or fill in the gaps. The Spiral progression approach has main philosophies behind these are the Constructivism, Progressivism and Behaviorism. A major theme in the theoretical framework of Bruner [4] is that learning is an active process in which learners construct new ideas or concepts based upon their current/past knowledge. As far as instruction is concerned, the instructor should try and encourage students to discover principles by themselves and they should engage in an active dialog. Curriculum should be organized in a spiral manner so that the student continually builds upon what they have already learned. Teachers' competency in teaching secondary science with spiral progression approach can be achieve through assessment of science teachers' competence in teaching Secondary Science with Spiral Progression Approach and comparing the relationship of the profile of the teacher to the teachers' competency by the profile of the teachers and teachers' pedagogy in teaching science with Spiral Progression Approach.

3. Methods

The study was conducted in selected public and private high schools in Lanao del Sur namely the MSU-Saguiaran, Saguiaran National High School, MSU – University Training Center – Main and Experimental Campus, MSU – Integrated Laboratory School. MSU – Institute of Science Education, Marawi National High School, Ibn Siena Integrated School and one (1) school in Lanao del Norte namely Pantar National High School. It obtained 55 Science teachers with their profile terms of sex, age, civil status, educational attainment, length of service, rank or position, years in teaching science, tenure, salary grade and which are officially teaching Science subject during the academic year of 2017-2018. It used correlational research design with purposive sampling method to provide description of the correlation between the competencies of science teachers and their profile. used questionnaire which underwent several revision and face validation by the researchers' panel members and experts in education before it was used as a tool for collecting data. Used survey questionnaire contained the items that referred to assess the competencies of Science teacher in teaching secondary Science with spiral progression approach. The respondents in this study completed three sections of questionnaire, the profile section, the competencies towards knowledge to subject matter, teaching strategies, personality traits and teaching approach section and problems encountered by the Science teachers in teaching secondary Science with spiral progression approach. The data gathered were immediately tabulated, analyzed and interpreted. The collected data were entered in the Statistical package. The results were generated according to the objectives of the study using frequency distribution table, correlation coefficient and regression analysis as a statistical tools.

4. Results and Discussion

Here is the presentation of the various data according to their respective arrangement as what have been mentioned in the statement of the problem. Significant implications then reinformed each tabular finding.

4.1 Profile of the Respondents in Term of:

Sex from the gathered information, female has the highest frequency of 84.78% while male has 15.22%. This implies that there were more number of women gets in educational courses. Age the data tells us none of them has an age of 59 and above since teachers in that age were in their retirement period. Moreover, the data also implies that there are more fresh graduates ranging from age 23-28. In addition, the result of this study is in line with older workers participating in work beyond the statutory retirement age indicated several preconditions and motives for staying in the work force. Civil Status as for the result that shows, out of 55 respondents 35 of them were single and 20 of them were married. There were no respondents answered widow nor separated. This data shows that most of the respondents are single and only few are married. This is because most of the respondents were fresh graduates. Educational Attainment there are more teachers who are currently enrolled in their Master's Degree than Doctoral Degree. This is because Master's Degrees are much easier to accomplish within 2 years. However, in PhD this could take longer years such as 4 years. Length of Service this implies that most of the teachers were on their first to fifth year of service with a frequency of 23 compared to 16 years and above length of service. This is because most of the respondents are fresh graduates and least of them approaching on their retirement. Rank or Position there were most number of Teacher I with 30 respondents compared to Master Teacher I. This is because most of the teachers are fresh graduates and they first land on teacher I before they get promoted. Years in Teaching Science please take note that the respondents of this study were all Science teacher and handling Science subject presently this only means that the length of service and the year in teaching science are mostly likely the same. Most of the teachers are in their first to fifth year in teaching science compared to 16 years and above. Tenure stability in teaching profession tells us most of the respondents are permanent in comparison to substitute teachers. This means that the newly hired fresh graduate teachers landed permanent. And this also implies that there were many vacant slots for new teachers. Teacher Salary Grade out of 55 respondents 32 of the have the salary grade of 11 compared to salary grade of 20. This is because most of the respondents are newly hired teachers.

4.2 Competencies of Science teacher in the spiral progression in term

4.2.1 Subject Matter

It described the distributions of respondent's level of competence about the subject matter. It showed that respondents preferred to teach the Subject matter if made applicable to the real world situation and strongly agreed with 75% and had a frequency of 41 out of 55. This implies that most of the teachers agreed and strongly agreed the level of competence of the subject matter. This was supported by the study of Cochran and Jones [9] in which they defined SMK rather than content knowledge as the overarching category which included content knowledge considered as substantive knowledge only together with syntactic knowledge and beliefs about subject matter.

4.2.2 Teaching Strategies

This showed that 76% of them used the teaching strategies that respondents preferred to relate the subject matter to real-life situations; give tests, quizzes, projects, etc. that cover the most important points of the topic or chapter. This implied that most of the teachers' response on competencies in teaching strategies were mostly agreed and strongly agreed. The result was in line with the statements teaching is the opportunity to help others to live their lives fully, which means we help to give to our learners' lives through their physical, emotional, intellectual and social growth [14].

4.2.3 Personality Traits

It shows about personality traits were possessed by the Science teachers and some of the respondents strongly agreed on the following statements that 57% or 29 of them were flexible, and likes new and different ideas. This implied that distributions of competencies on respondents' personality traits were mostly agreed. This result was in line with the Palardy and Rumberger [15] study, in which teacher aspects such as beliefs, attitudes, practices, and personality were the most relevant variables to teacher effectiveness.

4.2.4 Teaching Approaches

It indicates that the teachers used various approaches which they strongly agreed on the statements that the students should work in a group or teams to help each other as well as students who wants only to work themselves. This was supported by Dyer and Osborne [15]in their study entitied "Effects of Teaching Approach on Achievement of Agricultural Education Students with Varying Learning Styles" proposed that "the selection of an appropriate teaching approach is one of the most important processes to have teaching success and student achievement" (p. 260). [15] further stated that "students react differently to different teaching methods, and that the selection of the proper method is critical to the learning style of those being served by the instruction" (p. 260).

1. **Problem Encountered by the Science Teachers**

As for the problem encountered by the Science teachers, the most perceived problems encountered by the Science teachers were lack of teaching equipment, lack of instructional materials and in-sufficient in-service training. This because of the shifting of curriculum. Teachers are adjusting to the new curriculum hence they need more equipment, sufficient instructional materials and in-service training in order for them to function effectively in the new curriculum. This is supported by GMA News that the unavailability of learning materials is just one of the problems still hounding the country's new basic education program, K to 12, in the three years of its implementation as well as in-service trainings for teachers.

2. Significant correlation between the competencies of science teachers and their profile

a. Profile of Science Teachers and their Teacher- Knowledge to Subject Matter

There is a significant relationship between the age, years in teaching science, tenure and salary grade of the respondents and their knowledge of the subject matter. This is because the neophyte teachers' knowledge varies from the veteran teachers. According to Meister and Melnick [16], over the past thirty years, numerous studies confirmed that new teachers do not have the requisite knowledge to understand the complex interrelationships among management, behavior, and academic tasks. This lack of knowledge prevents new teachers from focusing on student learning; instead, they are preoccupied with their own behavior as they try different workable procedures. An important missing piece in the literature is how the concerns of experienced teachers differ from those cited by beginning teachers. Moreover, teachers may vary based on their tenure because the higher the tenure and salary grade the higher the significance of the teachers' knowledge. It is in lined in utdallas.edu that It is important to recognize that the changes in salaries and student demographics reflect both teacher preferences and district hiring and retention decisions. Because some teacher transitions result from involuntary job separations, these figures likely understate teacher preferences for both higher salaries and specific student characteristics.

b. Profile of Science Teachers and their Teacher- Teaching Strategies.

There is a significant relationship between the age, rank or position and tenure of the respondents in terms of teaching strategies. First, this implies that the age of the teachers affects their strategies. For instance, a fresh graduate teacher versus a veteran teacher. Their strategies really differ because fresh graduate teachers are more inclined to what he/she learned in his/her bachelor's degree while a veteran teacher can become flexible since the number of teaching experience are longer as well as its tenure. According to oecd.org, teachers' beliefs, practices and attitudes are important for understanding and improving educational processes. They are closely linked to teachers' strategies for coping with challenges in their daily professional life and to their general well-being, and they shape students' learning environment and influence student motivation and achievement. Teachers are more inclined to regard students as active participants in the process of acquiring knowledge than to see the teacher's main role as the transmission of information and demonstration of correct solutions. This is most true in northwest Europe, Scandinavia, Australia and Korea and least true in southern Europe, Brazil and Malaysia where teachers fall between the two views.

c. Profile of Science Teachers and their Teacher- Personality Traits.

Based on the results there is a significant relationship between the length of service and personality traits. This is because as a teacher gain experience through the years, his/her personality also improves. According to Azer [1]. Excellent teachers serve as role models, influence career choices and enable students to reach their potential. Some of the necessary qualities are inherent, others can be acquired through the years of teaching.

d. Profile of Science Teachers and their Teacher- Teaching Approaches

Based on the results there is a significant relationship between the civil status, educational attainment and the number of years in teaching of the respondents in terms of teaching approaches. This implies that teacher's civil status has an effect on what his/her teaching approaches are. According to Gill [11] every teacher has her or his own style of teaching and as traditional teaching styles evolve with the advent of differentiated instruction, more and more teachers are adjusting their approach depending on their students' learning needs. Moreover, educational attainment also could influence on how the teacher teach his/her students. As what Gill [11] said that teaching approach could also be improved by continuing study in every aspect. Although it is not the teacher's job to entertain students, it is vital to engage them in the learning process. Selecting a style that addresses the needs of diverse students at different learning levels begins with a personal inventory—a self-evaluation—of the teacher's strengths and weaknesses. As they develop their teaching styles and integrate them with effective classroom management skills, teachers will learn what works best for their personalities and curriculum.

References

[1] Azer, S. (2005). The qualities of a good teacher: how can they be acquired and sustained? Journal of the Royal Society of Medicine. Retrieved from: https://www.ncbi.nlm.nih.gov/pmc/articles/PMC1079387/

[2]Azuelo A. et al (2014) Science Teachers' Teaching Competencies and Their Training Experience to Classroom Pedagogical Approaches, 2014.

[3] Barlongo, C.J. (2015 May 26) . Reforms in the Philippine education system: The K to 12 Program(Business Mirror).Retrieved from http://www.businessmirror.com.ph/reforms-in-the-philippine-education-system- the-k-to-12-program/

[4] Bruner, J. (1977). The process of education. Cambridge, MA: Harvard University Press

[5] Canakan,M. (2017). Spiral Progression Dilemna. Retrieved from:www.Academia.edu/29017585/Spiral_Progression_Delimma

[6] Carlgrens I. (1999).Professionalism and Teachers as Designers. Journal Curriculum Studies.43-56.

[7] Clark, S. (2010). Jerome Bruner: Teaching and Learning and Spiral Curriculum. Community and Thought in Education. Retrieved from: http://Spiral%20Progression%20Approach/jerome-bruner-teaching-learning-and- the-spiral-curriculum2.pdf

[8] Cuban, L. (2012). Some Thoughts on Science Education Reforms in the Past Century. Retrieved from: http://eric.ed.gov/?id=ED538282

[9] Cochran, K. F., &L. L. (1998). The Subject Matter knowledge of Pre-service Science Teachers. In B. J. fraser & K. G. Tobin (Eds), International handbook of Science Education (Vol. 2,pp. 707-718). Dordrecht: Kluwer

[10] Dyer, J. E., & Osborne, E. W. (1995). Effecats of teaching Approach on Achievement of Agricultural Education Students with varying Learning Styles. Proceedings of the 22nd Annual National Agricultural EducationResourch Meeting, 22,260-271

[11] Gill, E. (2013), What is your teaching style? 5 effective methods for Your classroom. Retrieved from

[12] Katane I. (2006) Teacher competence and further education as priorities for sustainable development of rural school in Latvia.Journal of Teacher Education and Training.Vol.6, 4 -59.

[13] Resurreccion, J. A. et al Adanza, J. (2015) Spiral Progression Approach in Teaching Science in Selected Private and Public Schools in Cavite. Retrieved from http://www.dlsu.edu.ph/conferences/dlsu_research_congress/2015/proceedings/LLI/017LLI_Resurrecion_GF.pdf

[14] Carkhuff, R. R. (1981). The Skilled Teacher. Human Resource Development Pr.

[15] Palardy, G. J., & Rumberger, R. W. (2008). Teacher Effectiveness in First Grade: The Importance of Background Qualifications, Attitudes, Instructional Practices for Student Learning. Educational Evaluation and Policy Analysis, 30, 111-140. http://dx.doi.org/10.3102/0162373708317680

[16] Meister, D. G., & Melnick, S. A. (2003). National new teacher study: Beginning teachers' concerns. Action in Teacher Education, 24(4), 87–94.

Acknowledgements

The authors would like to thank their family for unending support. Also, we greatly appreciate MSU-Iligan Institute of Technology, Iligan City, Philippines for providing research space, classroom, and meeting rooms to us

Development of webquest using Google Site in teaching Circulatory System

Lea May L. Agad[1*], Daryl Jean A. Pitonang[1], Tyra Faye Terrado[1], Monera A. Salic-Hairulla[1], Rizalina G. Gomez[1], Edna B. Nabua[1] and Chokchai Yuenyong[2]

[1] College of Education, Mindanao State University – Iligan Institute of Technology, Iligan City, Philippines
[2]Faculty of Education, Khon Kaen University, Khon Kaen, Thailand
*Corresponding author email: leamay.agad@g.msuiit.edu.ph

Abstract. This study aimed to develop a webquest in teaching circulatory system using Google site for Grade VI pupils. The developed webquest was an instructional material designed to enhance pupil's performance in engaging learning activities especially in science lessons based from the K-12 Science Curriculum. The developed webquest was evaluated and rated Excellent by the ICT Experts, In-Service Teachers and Pre-Service Teachers. The comments and suggestions were used to improve the developed webquest. The researchers also made use of the Flesh-Kencaid Readablity Test to determine the readability of the developed webquest and the results revealed that it is appropriate for Grade VI pupils.

1. Introduction

Social change has a great impact on education. The society is changing and advancing. Education as well is evolving. As teachers being the builders of the future, they must be competent and advanced in all aspects especially to create instructional activities based on the new concepts corresponding to characteristic development of learners in the 21st century, which enables the learners to have creativity and ability to generate new innovation [1]. With the rapid development of the technology, the passion of internet is boosting among learners. The World-Wide Web (WWW) is becoming the most important media for collecting, sharing and distributing information. Similarly, in the case of education, it can be a mean of collecting the relevant information about the concerned subject. This important media can serve both teachers and learners in teaching and learning activities. In fact, the number of internet users in the Philippines increased by 13 million or 27 percent from January 2016 to January 2017, according to a report by social media agency "We Are Social". Both teachers and students have benefited from various educational technologies, teachers have learned how to integrate technology in their classrooms and students are getting more interested in learning with technology [2].

To make teaching more effective and innovative, teachers should select and make instructional materials that meet the needs of their pupils. With today's generation, the researchers seek to develop webquest as an instructional tool in teaching sixth graders about Circulatory System. The researchers have chosen to use the webquest about Circulatory System because the researchers believed that it was the one of the least learned topic in science. When educational technology tools are used appropriately and effectively in science classrooms, students actively engage in their knowledge construction and improve their thinking and problem solving skills [3]. Research findings also support in bridging the gap between technology and education. WebQuest encourages inquiry based learning which allows students to research a chosen topic while effectively using the internet and create a final product. It is designed to use learners' time well, to focus on using information rather than looking for it, and to support learners' thinking at the levels of analysis, synthesis and evaluation [4].

2. Objectives of the Study

This study hopes to achieve the following objectives:

1. To develop webquest that will help the learners understand better the topic about circulatory system and kindle their interest in science
2. To evaluate..
3. To help science teachers have an instructional material to further leverage the learners inquisitive minds

3. Method

This study used both quantitative and qualitative methods of research. In the quantitative research method, the data were obtained from the ratings of the evaluators on the developed webquest. In the qualitative research method, the data were obtained from the comments and suggestions of the evaluators. The researchers followed steps in developing WebQuest; a.) Selection of the topic on Circulatory System, b.) The identification of standards that was based on the K to 12 Science Learning Competencies 2016 for teaching circulatory system c.) Making of lesson plan d.) Developing Webquest e.) Testing the Readability f.) Validation and g.) Revision of the developed webquest. The WebQuest was rated by the ICT Experts, In-Service teachers, and Pre-Service Teachers through a rubric. The figure below shows the process for the development of webquest.

4. Results and Discussions

The development of the Webquest was organized based on the process of method.

4.1. Selection of Topic

The researchers preferred the topic on Circulatory System because the topic is one of the least learned topics in science for the elementary learners. The researchers want to create a new set of instructional material that will mold and enhance the students' interest and love towards science and will give ease on the teacher's side through the use of webquest.

4.2. Identification of Standards

The developed webquest was designed based on the standards and learning competencies provided by the K-12 Basic Education Curriculum. The learning objective of the learners is to be able to explain how the organs of circulatory system work together.

4.3. Making of Lesson Plan

The researchers made a lesson plan based from the K-12 Curriculum Guide of 2016. The learning objective formulated is given from the K-12 Curriculum Guide of 2016. The lesson plan was face validated by sceince teachers experts.

4.4. Developing Webquest

After selecting and assuring, the researchers conducted library research, interview and interenet surfing about the subject and also studied the different designs of webquest. The researchers developed webquest in the Google Site. The content in the Google site was based on the outline as arranged in order as listed, to wit: (1) Home (2) Output Timeline (3) Webquest (4) Teacher Support (5) Sample Output (6) References

Under the developed Webquest are listed, to wit: (1) Introduction (2) Task (3) Process (4) Materials (5) Procedures (6) Resources (7) Evaluation

Figure 1. Home Page of the Entire Webquest

The developed webquest is entitled "A Ride through your Heart: A Webquest on the Circulatory System". An awesome adventure on wheels is going to start! This Webquest will utilize the student's knowledge about the circulatory system. They will be grouped by seven and will be asked to make an alternative heart model (which will be their output in this Webquest) that will show the parts and functions of the circulatory system and the flow of blood throughout the body . The theme of the model will be like a road trip. The students will be using a toy car that will represent as the blood. This toy car will run throughout the alternative heart model. The students will use their understanding about the Circulatory System to be able to finish their output. After the production of the output, presentation will follow. The students will demonstrate the use of the alternative heart model. This means, the ride through the heart is going to start.

WebQuests are not just activities that utilize the Internet [5]. Students can experience both individual and team learning. This can be done by encouraging them to produce an authentic end product that is creative and applicable to real life. Webquests should be real, rich, and relevant. Students work in teams, learn to cooperate and collaborate because they not only access the Internet for information, but they also apply the information in a new way. This leads students to use their higher thinking skills for a deeper understanding and more independent learning. As a result, they become responsible for their own learning [5].

4.5. Readability

The researchers made use of the Flesh-Kencaid Readablity Test to determine the readability of the developed webquest and the results revealed that it is appropriate for Grade VI pupils. The developed webquest has low percentage of passive sentences. The total number of words is 673. The total number of unique words is 307 which is 46% of the total text. Total number of repeat words is 366 which is 54% of the total text. Average number of words per sentence is 10. Total number of characters is 3, 136 and average number of characters per words is 4.7. Average number of syllables per word is 1 and total number of syllables is 984. Readability formulas are used widely in education and has demonstrated moderate to strong predictive correlations with reading comprehension [6]. This means that the developed webquest is easy to read and understand. It has a result of 72.6 of Flesch Reading Ease which means that the developed WebQuest uses shorter sentences and easy to understand. It showed that the webquest has a Grade Level of 5.7 which indicates that it is expected to be understood by an average student in the 6th grade.

4.6. Validation of the Developed Webquest by the ICT Experts, In-Service Teachers and Pre-Service Teachers

After testing the readability of the developed webquest, the researchers met some experts regarding this matter. Three (3) ICT experts, five (5) In-Service teachers and ten (10) pre-service teachers have been given a copy of the WebQuest for them to check and give their comments and recommendations. The ICT experts evaluated the technical aspect of the WebQuest using rubric while the in-service teachers checked the content and the strategies using the rubric.

Table 1. Summary of Ratings of the developed Webquest by the experts through Rubric

Criteria	Average Mean Rating	Description
Content	1.11	Excellent
Layout	1.44	Excellent
Content Accuracy	1.17	Excellent
Work Ethic	1.22	Excellent
Overall Quality	1.17	Excellent

The table shows that the ICT Experts, In-service Teachers and Pre-service Teachers rated excellent developed Webquest. In content, 88.9% rated the site has a well-stated clear purpose and theme that is carried out throughout the site. It indicates that 16 out of 18 evaluators rated excellent. In layout, 55.6% rated the site has an attractive and usable layout. It indicates that 10 out of 18 evaluators rated excellent. It is easy to locate all important elements. In content accuracy 15 out of 18, 83.3% rated all information provided by the researchers on the site are accurate and all of the requirements of the assignment have been met. In work ethic 14 out of 18, 77.8% rated that researchers always use classroom project time well. And in over-all quality 15 out of 18, 83.3% rated that the site is of high quality. Teachers have access to more sophisticated tools that can break the monotony of everyday lectures and facilitate the learning experience in different ways. In the modern age, teaching is no longer limited to the physical classroom [7]. Any experienced teacher can tell you that no two people learn the same way. There are visual learners, auditory learners, read-write learners, and kinesthetic learners. And because all these types of learners are mixed in one classroom, teachers need to be able to find ways to accommodate these disparities. Thus, Webquest can be a tool in helping learners.

4.7. Revision

With the ratings, the researchers used the comments and suggestions of the ICT Experts, In-service teachers and pre-service teachers to improve and revise the presentation, layout, graphics, organization of themes and pages of the Webquest so that it will be an effective instructional tool in teaching circulatory system. Lecturers' awareness of 21st century skills, especially the knowledge and skills in the integration of ICT, is a constraint in the training process and therefore, requires special attention especially to per-service teachers [8].

5. Conclusion

The developed webquest was rated Excellent from the ICT Experts, In-service Teachers, and Pre-service Teachers. It was evaluated based on its content, layout, content accuracy, work ethic and overall quality. It is readable and appropriate for the Grade VI Pupils. From the comments, the developed webquest was engaging and the site was exemplary. For science teachers, the webquest was helpful in teaching circulatory system.

Acknowledgement
Sincere gratitude is due to the following: To our dear parents for the unconditional love and gave us words of encouragement and to Mindanao State University – Iligan Institute of Technology for providing the researchers' an avenue for learning and growth. This accomplishment would not have been possible without them. Finally, to the Almighty God for the guidance and protection.

References

[1] Chatwattana, P., & Nilsook, P. (2017, May). A web-based learning system using project-based learning and imageneering. *International Journal of Emerging Technologies in Learning (iJET)*
[2] Ramey, K. (2013, March 14). The use of technology - in education and teaching process. *Use of Technology*
[3] Guzey, S. S., & Roehrig, G. H. (2009). Teaching science with technology: Case studies of science teachers' development of technology, pedagogy, and content knowledge. *Contemporary Issues in Technology and Teacher Education*
[4] Azzopardi, A. (2012). Webquest. A Primary School Teacher
[5] Deutsch N. (2014). Why Use WebQuests for Today's Learners. *The WizIQ Blog*
[6] Humphreys, A. H., & Humphreys, J. T. (2013). Reading difficulty levels of selected articles in the journal of research in music education and journal of historical research in music education. *Music Education Research International, Volume 6*
[7] Natividad, N. (2016, September 9). Technology and teaching: how the internet opened opportunities for education. *Rappler*
[8] Manik, M. M., Qasim M., & Shareef, A. F. (2014). Embedding 21st century skills in pre-service teacher training: a case study from the Maldives. *Teaching and learning in the 21st Century: Challenges for Lecturers and Teachers*

Implementation of Webquest in teaching Circulatory System

Monera A. Salic-Hairulla[1*], Lea May L. Agad[1*], Daryl Jean A. Pitonang[1], Tyra Faye Terrado[1] and Chokchai Yuenyong[2]

[1] College of Education, Mindanao State University – Iligan Institute of Technology, Iligan City, Philippines
[2]Faculty of Education, Khon Kaen University, KhonKaen, Thailand
*Corresponding author's e-mail address: leamay.agad@g.msuiit.edu.ph

Abstract. This study aimed to determine the performance of the respondents in using the developed Webquest in the Google site. The study used the quasi-experimental design method making use of pretest-posttest design. The result of the study showed that there was a significant improvement in the respondents' conceptual understanding in the concept of Circulatory System. The respondents exposed to the developed webquest had significant high test scores in the post-test as compared to the pre-test. The developed instruction was implemented to 30 Grade 6 Elementary pupils and another 30 pupils for traditional lecture method from selected public schools who were enrolled during the SY 2016 - 2017. Moreover, the result also showed that there is a significant difference between the post test scores of control group and experimental group. Result implies that the developed Webquest has a positive impact in teaching Circulatory System.

1. Introduction

The impact that technology has had on today's schools has been quite significant [1]. This widespread adoption of technology has completely changed how teachers teach and students learn. Teachers are learning how to teach with emerging technologies (tablets, iPads, Smart Boards, digital cameras, computers), while students are using advanced technology to shape how they learn. By embracing and integrating technology in the classroom, teachers are setting their students up for a successful life outside of school. Moreover, textbooks are a great source of reliable information and ready-made activities, but the content they provide can be generic and not particularly engaging for students [2]. By leveraging the instructional potential of web-based resources, learners can increase student engagement, expose them to authentic content, and engage them in collaborative activities that trigger critical thinking and creativity. It is fact that it can offer students a "virtual teacher" because students can access the instructional materials anytime, anywhere. This allows students who were absent from the opportunity to access instructional materials away from school, and even the possibility to accommodate students in a course when their schedule is full [3].

Recent advancements in educational technologies have yielded positive results in our education sector [4]. The idea of webquest, a web-based teaching material, has become increasingly popular. A Webquest engages students in an inquiry, problem-based learning activity that integrates teamwork, higher order thinking, and access to information on the Internet [5] The use of web in learning by students is increasing as it enriches the experience in learning, making the students to be responsive of the existing development. The role of educators has become more challenging as the education sector to date requires the teachers to have knowledge and skills in ICT. Significantly, it is pertinent because of the technological advancement in developing ideas and creativity among teachers and students [6].

This paper will examine the effects on the implementation of the developed webquest in teaching circulatory system among Grade Six pupils.

2. Statement of the Problem

This research aimed to utilize Webquest in teaching Circulatory System among Grade Six pupils in Sgt. Miguel Canoy Memorial Central School. Specifically, this study seeks to find answers to the following:

1. What is the profile of the respondents according to electronic devices
2. What is the performance of the respondents in the developed Webquest in?
    a. Pre-test
    b. Post-test
3. Is there a significant difference on the performance of the experimental between pre-test and post-test?
4. Is there a significant difference of the control group's performance and experimental group's performance?

3. Methods

This chapter presents in details the methods and procedures of this study. It includes subjects of the study, research design, instruments of the study, and data gathering procedure.

3.1. Subjects of the study

The respondents were 30 Grade 6 Elementary pupils for the experimental group and 30 pupils for the traditional group from selected public schools who were enrolled during the SY 2016 - 2017.

3.2. Research Design

This study used the quasi-experimental design method making use of pretest-posttest design. The developed instruction was utilized to the experimental group and the traditional lecture method was implemented to the control group. In the selection of the respondents, a purposive sampling was used in choosing the subjects of the study. Alternatively, purposive sampling method may prove to be effective when only limited number of people can serve as primary data sources due to the nature of research design, aims and objectives. Correspondingly, the researchers chose Grade VI pupils who belong to the first section both in experimental and control group.

3.3. Instruments of the Study

The researchers formulated one set of questionnaire each consisting 30 items that was pilot-tested and validated by the thesis adviser and science teachers; pre-test and post-test, with exactly the same content to be handed before and after the implementation of the class discussion. This was face validated by the Science teacher experts, panel members and thesis adviser. And learner's perception checklists were also provided. The questionnaire was used to determine the learners' motivation through the Webquest in learning the circulatory system. It was composed of 5 items that corresponds whether they agree or disagree to the perception on the developed Webquest as an instructional material for teaching circulatory system.

3.4. Data Gathering Procedure

The researchers submitted a letter-request to conduct a study on the said schools from the Schools Division Superintendent. After the permit was approved, the researchers secured a permit from the School Principal of Sgt. Miguel Canoy Memorial Central School and Hinaplanon Elementary School to push through with the study. The first group, the control group, will be exposed on the traditional method which is the Grade IV pupils from Hinaplanon Elementary School. The second group, the experimental group will be exposed on the use of Webquest which is the Grade VI pupils from Sgt. Miguel Canoy Memorial Elementary School. Internet connection must be considered for the experimental group. Prior to the administration of the two groups with the varied method mentioned, they answered a pre-test to ensure a justifiable correlation to the post test results, which was given after administering the two methods.

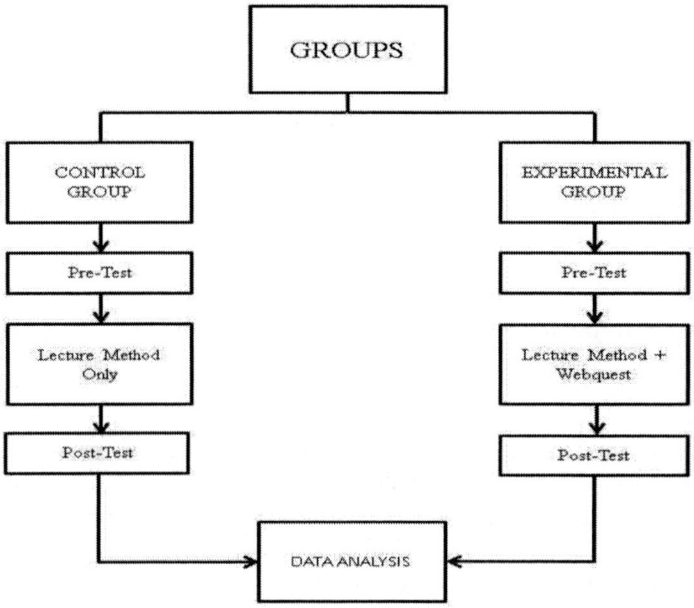

Figure 1. Flow Chart of the Data Gathering Procedure

4. Results and discussion

The researchers found the profile of the respondents who owned electronic devices to determine their accessibility to internet. With technology playing a bigger role in the classroom and children utilizing computers, tablets, and apps to learn, making sure children know how to use these different types of media is becoming more important [7]. 32% of the respondents use mobile phone followed by tablet which is 24%. On the other hand, only 6% use Ipad. This means that the biggest slice of the pie which is mobile phone is used as the means to connect on the internet and the least is Ipad. It indicates that 10 out of 30 respondents in Sgt. Miguel Canoy Memorial Central School has mobile phone. Mobile phones are a popular form of communication in today's society, and respondents of all ages own cell phones [8].

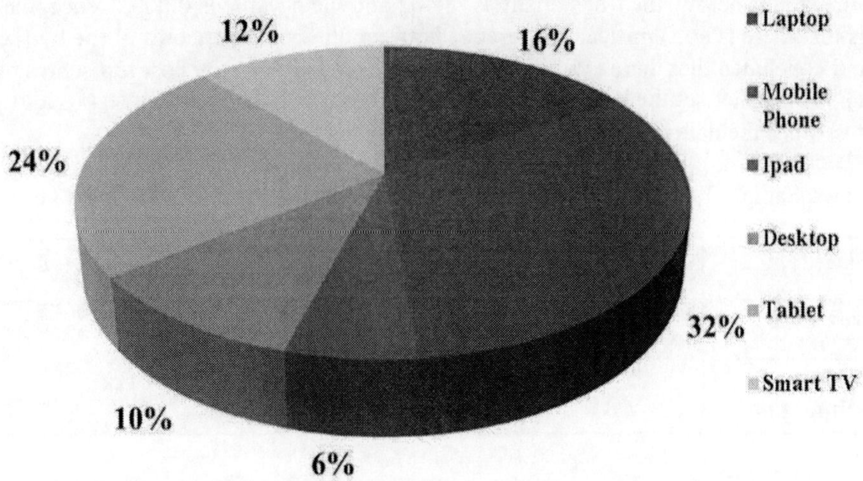

Figure 2. Profile of the Respondents According To Electronic Gadgets

The researchers used 30-item test for the pretest and posttest. The table below shows the descriptive statistics of the performance of the respondents in using Webquest. Their pre-test scores have an average of 14.43 with minimum score of 7 and maximum of 26. The total score of the respondents is 433. The standard deviation is 3.91 which means that the scores are dispersed from the mean. For the post test, scores are way better than the pre-test scores. The average score is 21.13 and the total score is 634. The minimum score is 13 while the maximum is 27. The standard deviation of 3.58 is also good because it indicates that scores are closer to the mean. It shows that the respondents performed well during the discussion in association with the implementation of Webquest.

Table 1. Performance of the Respondents in Using Webquest

Variable	Mean	Standard Deviation	Minimum	Maximum	Sum
Pretest	14.43	3.91	7	26	433
Posttest	21.13	3.58	13	27	634

The paired t-test result shows that there is a difference between the pre-test and post test scores of the respondent's performance. The null hypothesis of the test is that, there is no significant difference between the two test scores is rejected since there is a significant difference between the two test scores. The researcher set the level of significance to be 0.05. The statistical test used is paired t-test since the data are results of before-and-after situation. Based on the results below, the difference between the means of the test scores is -6.7 and the p-value is 0.000. Hence, the null hypothesis is rejected and the researchers conclude that there is a significant difference between the test scores.

Table 2. Paired t-test Result

Variable	Mean Difference	t-statistic	p-value
Pretest-Posttest	-6.7	-10.841	0.000

Based on the results below, the test statistic is -8.462 and the p-value is 0.000. Hence, the null hypothesis that there is no significant difference between the post test scores of the two schools is rejected and concluded that there is a significant difference between the post test scores of the two schools. The researcher set the level of significance to be 0.05. The statistical test used is independent sample t-test since the data came from two independent samples.

Table 3. Two Sample T-Test For The Post-Test Of The Control And Experimental Group.

Variable	Mean	t-statistic	p-value
Posttest Control	11.07	-8.462	0.000
Posttest Experimental	21.13		

The table below shows the distributions and percentage of learners' perception on the developed Webquest as an instructional material for teaching the circulatory system. All of the respondents agreed that they enjoyed the topics of the circulatory system by using the Webquest and the Webquest activity is interactive. This means that the introduced Webquest can be enjoyed in an interactive way. There were 93.30% of the respondents agreed that they can easily understand the topics on the circulatory system by using the Webquest and it motivates them to learn more, and 6.70% of them disagreed to the statements. This means that the introduced Webquest is simple and instructional. There were 86.70% of the respondents agreed that the Webquest develops their critical thinking skills and 13.3% of them disagreed. This means that the Webquest aims to develop the Higher Order Thinking Skills (HOTS) of the learners. There were 80% of the respondents agreed that the Webquest can be easily manipulated without the supervision of the teacher and 20% of them disagreed. This means that without the aid of the teacher the learners can learn through their own. Overall, the table says that the learners have learned and enjoyed the Webquest. From the student's perception questionnaire, the respondents found that the Webquest was enjoyable, motivational, entertaining, interactive, easy to learn, and helped them develop their critical thinking skills.

Table 4. Percentile distributions of learners' perception on the Webquest

Learner's Perception	**Agree**		**Disagree**	
	f	%	F	%
1. I can easily understand the topics on circulatory system by using the Webquest.	28	93.3	2	6.7
2. The Webquest develops my critical thinking skills.	26	86.7	4	13.3
3. I have enjoyed the topics on the circulatory system by using the Webquest.	30	100		
4. The Webquest motivates me to learn more.	28	93.3	2	6.7
5. The Webquest activity is interactive.	30	100		

5. Conclusion

The profile of the students showed that they were able to use the developed Webquest because of its gadget and internet ownership. There is highly significant difference between the pre-test and post test scores of the respondents based on their performance. There is a difference between the post test scores of control group and experimental group. This also concludes that the Webquest is helpful and effective in teaching Circulatory system among Grade 6 pupils. The Webquest was really enjoyable, entertaining, interactive, easy to learn and easy to use. It really helped the learners to develop their collaboration skills and creativity especially in making their outputs.

Acknowledgement

Sincere gratitude is due to the following: To our dear parents for the unconditional love and gave us words of encouragement and to Mindanao State University – Iligan Institute of Technology for providing the researchers' an avenue for learning and growth. To the DEPED office of the division of Iligan City, the Schools Division Superintendent, School principals and the Teachers of the schools of Sgt. Miguel Canoy Memorial Elementary Central School and Hinaplanon Elementary School, for allowing us to conduct this research. This accomplishment would not have been possible without them. Finally, to the Almighty God for the guidance and protection.

References

[1] Cox, J. (2015). Benefits of technology in the classroom. *K-12 Teachers Alliance*. Retrieved May 21, 2017
[2] Simon, E. (2015, April 28). Teaching With Web-Based Resources. *George Lucas Educational Foundation*
[3] Mathew, N., & Dohery-Poirier, M. (2000, March 6). Using the world wide web to enhance classroom instruction. *First Monday*
[4] Ramey, K. (2013, March 14). The use of technology - in education and teaching process. *Use of Technology*
[5] Deutsch N. (2014). Why Use WebQuests for Today's Learners. *The WizIQ Blog*
[6] Nordin N. et.al (2012). Development and Evaluation of Webquest for Science Subject. *The International Journal of Learning*
[7] Locklear, M. (2017, May 12). Even very young children require media literacy to learn from technology. *Blog on Learning and Development.*
[8] Watson, E. (2010). Advantages of cellphones for students. *Leaf Group Ltd. Leaf Group Media.*

IOP Conf. Series: Journal of Physics: Conf. Series **1340** (2019) 012062 doi:10.1088/1742-6596/1340/1/012062

Experienced difficulties of BSE-Chemistry students in physical chemistry and suggested enhancement

Mc Rollyn D. Vallespin[1*], Angelie Aliman[2], Esmar N. Sedurifa[3], Monera Salic-Hairulla[4], Joel H. Jorolan[5]

[1,2,3,4] Department of Science and Mathematics Education, College of Education, Mindanao State University – Iligan Institute of Technology
[5] Department of Chemistry, College of Science and Mathematics, Mindanao State University – Iligan Institute of Technology

*Corresponding author's e-mail address: mcrollyn.vallespin@g.msuiit.edu.ph

Abstract. This study aimed to determine the experienced difficulties in Chemistry 70 (Introduction to Physical Chemistry) among ten (10) selected BSE-Chemistry students who had undergone Chemistry 70 and two professors of Physical Chemistry. Qualitative research design with purposive sampling was used in this study. It was found that out of the total 10 BSE-Chemistry respondents, majority got a passing grade of 1.50 to 3.00 in Chemistry 70, representing their academic performance for this particular study. They considered "Chemical Thermodynamics and Equilibrium" as "much difficult" and classified "teacher factor" as the primary reason for the difficulty in understanding the subject. Teachers play a great role in the learning of the students in Physical Chemistry. They must be equipped effectively with the concepts needed for the efficient delivery of lessons in the classroom. Moreover, teachers must be able to relate both theories and concepts to real-life situations for the students to have a better understanding of the subject matter. It was suggested that students should be able to master the basic skills and concepts in Math 61 to improve their mathematical skills which are very much needed in the study of Physical Chemistry.

1. Introduction

The wide expanse of the universe poses great challenge for humanity. A human being strives to exist with the environment, thus, every individual needs to find ways to survive and co-exist with the environment.

In order to endure life, each one has to learn more, hence the need to pursue education, that is, to acquire knowledge, skills, values, beliefs, and habits. Education takes place with the guidance of educators, but learners may educate themselves not only in the four corners of the classroom but also from the environment where they live.

Each learner may be able to perceive things and evaluate them as good or bad as he or she enters school seeking to acquire knowledge to continually grow and prosper. It is the teacher in school who facilitates the learning process using different techniques and strategies to ensure holistic development of the learner [1].

In the Philippines, the curricular offerings demand so much effort from each learner if he or she really needs to succeed in the academe. One of the curricular offerings of the Mindanao State University – Iligan Institute of Technology in Iligan City is the Bachelor of Secondary Education

Major in Chemistry or BSE-Chemistry offered by the Department of Science and Mathematics Education (DSME) which admits enrollees with SASE score of not less than ninety (90).

This four (4) year education program is among the difficult courses not only in College of Education (CED) but in the whole institute due to the fact that Chemistry is a highly specialized subject and loaded with other courses from other fields of natural sciences like Physics and Mathematics. Hence, a number of BSE-Chemistry students shift to other programs after finding so much difficulty in the program. This made BSE-Chemistry fall short in number in terms of producing graduates every year.

In this degree course, one of the most challenging subjects is Physical Chemistry which students generally consider as a difficult course because it deals with specific mathematical calculations together with concepts which is needed to be fully understood, otherwise they would not learn anything at all. Thus, the subject was deleted from the curriculum in the past years, but later restored in 2010 because of the recommendation coming from the Commission on Higher Education (CHED) – a development regarded as a return to the old curriculum. Subsequently, because many students found Physical Chemistry difficult, many would flunk and/or retake the subject several times causing them to be delayed in graduating.

Notwithstanding the complaints of students, Physical Chemistry remains an essential course, specifically for those who want to pursue their master's degree and to those who wish to be future Chemistry teachers.

As a branch of Chemistry, Physical Chemistry demands not only knowledge of theories but more importantly, on having practical skills especially in solving problems in Mathematics. Many students find the subject challenging because it does not only require the ability to identify given facts but also necessitates comprehension and analysis of the problems solution.

However, this requires a thorough background and strong foundation of basic chemistry and mathematics courses to support a more intellectual and broader scope and more advanced study of physical chemistry. It was on this premise that this research undertaking was conceptualized and developed to determine the difficulties of BSE-Chemistry students in the subject Physical Chemistry. In the learning process, it is important to determine the students" difficulties and learning styles for the teachers to help them look for specific strategies to handle such challenges, to decrease the number of failing grades in this particular subject, and increase the number of BSE-Chemistry graduates in the Department of Science and Mathematics Education, College of Education, Mindanao State University – Iligan Institute of Technology, Iligan City.

2. Statement of the problem

This study aimed to determine the experienced difficulties in Chemistry 70: Introduction to Physical Chemistry of the BSE-Chemistry students during the school years 2014-16. Specifically, it sought to answer the following questions:

1. What is the academic performance of the BSE-Chemistry students in Chemistry 70: Introduction to Physical Chemistry?
2. What topics in Chemistry 70 are experienced as great difficulties by the BSE-Chemistry students?
3. Is there a sufficient evidence to conclude that the students find difficulties in the covered topics in Chemistry 70?
4. What are the possible reasons or causes for such difficulties?
5. Is there a significant relationship between the possible reasons or causes and their academic performance in Chemistry 70: Introduction to Physical Chemistry, course?
6. What are the suggested enhancements to address such difficulties in Chemistry 70?

3. Research Methodology

This chapter presents in details the methods and procedures of this study. It includes subjects of the study, research design, instruments of the study, and data gathering procedure.

3.1. Subjects of the study

The subjects of the study were ten (10) BSE-Chemistry students of the Mindanao State University – Iligan Institute of Technology who were officially enrolled in Chemistry 70 'Introduction to Physical Chemistry' during the 1st semester, school year 2014-2015. All 10 respondents were females; 5 of them took the subject during the 1st semester, SY 2014-2015 while the rest took the subject during the 1st semester of school year 2015-2016. Some of the respondents were BSE-Chemistry graduates while others were still taking up their remaining minor courses.

3.2. Research design

The researchers used the Qualitative Survey (in-depth interview) and Purposive-Sampling method of research. The data were obtained through a One-on-One in-depth interview with ten (10) students who had undergone Chemistry 70.

A Physical Chemistry teacher was needed in the study to provide feedback on experienced difficulties of the students and the possible causes or reasons that made Physical Chemistry difficult from their own perspective. Through the said interviews and the gathered experienced difficulties of both students and teachers when they took up and taught Physical Chemistry, revealed the deepest reasons of difficulties.

3.3. Instruments of the study

The instrument utilized in the conduct of the study was the questionnaire that had been face validated by expert educators. The instrument was divided into three (3) parts: 1) Part A was the ranking of the topics according to their level of difficulty; 2) Part B solicited their reasons or causes for considering the subject in Chemistry 70 as difficult; and 3) Part C, an in-depth questionnaire was used during one-on-one interview with respondents.

In Part A, respondents were asked about their experienced difficulties in Introduction to Physical Chemistry with the use of the face validated questionnaire. The respondents were going to rank the topics according to their level of difficulty: 1 as the 'least difficulty', 2 'less difficulty', 3 'average difficulty', 4 'much difficulty', and 5 as 'extreme difficulty'. In Part B, respondents had to put a check mark on the four (4) classifications (Student factors, Teacher factors, and Personal factors and Resources factors) of possible reasons or causes that the respondents experienced when they took Physical Chemistry. Part C was made up of open ended questions which the respondents had to answer accordingly and honestly.

Content validity of the questionnaire had been established by acquiring the Chemistry 70 topic outline from the College of Science and Mathematics (CSM), Chemistry Department for the formulation of the questions and by submitting the original form of the questionnaire to the adviser for comments and suggestions. The questionnaire was composed of questions regarding the experiences encountered at the five (5) chapters: 1) The Ideal and Real Gases, The First Law of Thermodynamics; 2) The Second and Third Laws of Thermodynamics; 3) The Chemical Thermodynamics; 4) Equilibrium; and 5) Phase Equilibrium.

3.4. Data gathering procedure

As soon as the questionnaire had been face-validated by the chemistry experts and Physical Chemistry educator, respondents were requested to answer the formulated questionnaire and checklist. For those respondents who were not in the vicinity of Iligan City, they were contacted via cell phone, Facebook chat or video call, asking them regarding their difficulties on the five (5) chapters in Chemistry 70 by ranking the topics based on the levels of difficulty from the least to extreme difficulty.

In Part B, the respondents were to put a check mark on the reasons or causes for difficulties they have personally encountered in Introduction to Physical Chemistry. There were 4 classifications of possible reasons or causes that the respondents experienced while taking Physical Chemistry: 1) student factors; 2) teacher factors; 3) personal factors; and 4) resource factors.

Student factor is about how the student performed and behaved themselves during the subject. Criteria that comprised student factors were the readiness and preparedness of the student on the basic concepts, their understanding of the lesson during class discussion, how they cooperated and participated in class discussion, how they aimed at getting good grades and how they reacted if the class was cancelled/interrupted or the teacher was not present.

Teacher factor is how instructors present the lesson to their students effectively and keep the positive relationship between them. Under this factor were: relationship established by teacher with his/her students, how firm the teacher was in making decisions, the openness of the teacher to suggestions and opinions raised by the students, the usefulness of the various tools, strategies and techniques used in presenting the lesson, and how the subject matter was systematically presented.

Personal factor describes students' issues regarding the relationship within their families, financially stability, as well as the spiritual, physical and emotional aspects in their personal lives. Criteria that belonged to personal factors were how the parents were concerned with their studies, an established relationship with their classmates, handling time management, taking other responsibilities to support their study, and being bothered from the other things aside from studying.

Lastly, resources factor mentioned the learning materials that the student needed for the subject as guide or source of information. Resources factors involved: books being sufficient to cater the needs for learning, giving handouts to the learner, using PowerPoint presentations as they presented the lesson, incorporating technology as one of their resources in solving problem sets and establishing peer tutoring with their classmates in comprehending the concepts and even in doing calculations.

In part C, the students underwent an in-depth interview and were asked various questions regarding their experiences when they took Physical Chemistry. The researchers also asked them on how they handle the difficulties they experienced in those times. After knowing the difficulties that students experienced, the researchers proceeded to interview the Physical Chemistry educators to know also their insights on teaching Chemistry 70. The researchers asked the educators regarding the students' performance and what were their responses to the difficulties experienced by their students. In-depth interviews between the students and Physical Chemistry educators were conducted to find the difficulties experienced by each individual, as a student or teacher in Chemistry 70, and how Physical Chemistry educators could help their students to overcome such difficulty.

After the interview with the students through Facebook and via call, the researchers gathered all the data they obtained from the interview. The data were assessed and evaluated through interviewing the educators. The interview conducted on the teachers included their experienced difficulties in handling the students and the course in general. It was followed by the gathering of answers from the respondents and teachers, compared and listed the congruent responses or significant findings using the triangulation method to successfully assess the responses by both the student-respondents and the Physical Chemistry educators, also known as "mix method" (also known as triangulation) research. The purpose of the triangulation was to give more detailed information in facilitating deeper understanding of the situations between the teachers and students. After analysis, formulation of learning enhancement on Physical Chemistry was done.

The answers given by both student-respondents and Physical Chemistry teachers were recorded including the conversation at Facebook or voice recorder and were then evaluated.

4. Results and discussion

From the grades being collected, it shows that the respondents "academic performance wherein four (4) or 40% of the respondents were very good performers based on their grades in Chemistry 70; one (1) or 10% was a good performer; three (3) or 30% performed satisfactorily; while two (2) or 20% got a passing grade of 3.00. It could be implied that none of these respondents excelled in Chemistry 70. The students' anticipation of physical chemistry as a difficult course can greatly affect their academic performance as well as how the students perceive and deal with the topics; resulted to lower students" success rate of getting an excellent performance [2].

Using the triangulation method with a survey questionnaire, interview conducted to both student-respondents and Physical Chemistry teachers showed that the topic "Chemical Thermodynamics and Equilibrium" (rank 1) and "Phase Equilibrium" (rank 2) were considered as the most difficult topic in Chemistry 70 since it dealt with theories and laws which were abstract in nature. Dealing with problem solving made the respondents felt uneasy – the very reason why they were not able to grasp the topic well. This may also root from their lack of mathematical skills and basic chemistry concepts.

This contention was supported [3] that many of both high school and university students experienced difficulty in chemistry, specifically on thermodynamic concepts since they never mastered the fundamental concepts in Chemistry [4].

A similar observation was cited [5], this time with Mathematics. The structured course identified certain issues which limit the students to develop their mathematical abilities, thus, sufficient support from the instructor must be available. However, the student themselves still had to do extra effort to master the subject in preparation for higher lessons [6].

On the other hand, cited the following as possible sources of learning difficulties in Chemical Thermodynamics and Equilibrium [7]: 1) lack of knowledge on the fundamental thermodynamic concepts and application of algorithms due to lack of conceptual understanding; 2) memorization of scientific laws and statements without deep understanding thus resulting to inappropriate over-generalization; and 3) confusion of fundamental ideas and socioeconomic and methodological aspects of the teaching-learning environment.

The checklist revealed that teacher and student factors greatly affected students' performance on the said subject specifically on the relationship between the teacher and students, the teacher's portrayal of their role during classroom discussion, the strategies used in line with the topic outline, the openness of the teacher to suggestions and opinions raised by the students, and how firm the teacher was in making decisions. Even themselves, the students have great effect in their performance specifically on how they performed and behaved during the discussion of the lesson, their preparedness on basic concepts, their understanding on the lesson, how they reacted if they have a class interruption, aiming for getting good and high grades, and lastly, on how they cooperate and participate to their discussion. Based on the triangulation process, it was found out that the teachers are competent when it comes to teaching Physical Chemistry concepts. However, some of the educators were trying to adjust their teaching strategies employed based on the capabilities of their students.

Both chemistry lecturers and learners had experienced difficulties in the subject Physical Chemistry as evidenced by the results of the survey where the lecturers' perceptions on the students' learning difficulties were quite different from that of the students [8]. Students were worried about the course content, availability of the resources, the lecturers and the way their teachers taught Physical Chemistry. Meanwhile, the lecturers generally focused on the course related factors such as overcrowded classes, lack of staff and resources as well as the academic background of the students and socioeconomic conditions.

Row mean was the calculated mean between the academic performance (grade) and the four (4) classifications of possible reasons and causes (factors) which made Physical Chemistry a difficult subject. According to the results, the computed f_1 (academic performance) of 2.857 indicates that the students" academic performance did not significantly affect the factors which caused them difficulties in Chemistry 70. Column mean was the calculated mean between four (4) classifications of possible reason and causes (factors) and the academic performance (grade) of the students. The computed f_2 of the column mean (factors) was 9.566 which would indicate that the factors causing difficulties did significantly affect the students' academic performance on the subject. Interaction mean was the calculated mean of the academic performance and the factors having the computed f_3 resulted to 4.950 showing that there was a significant relationship between the factors which caused difficulties and the students" academic performance on Introduction to Physical Chemistry.

On the other hand, Chemical Thermodynamics and Equilibrium were considered the most difficult topics among the 5 topics in Physical Chemistry as experienced by the respondents. Results could be parallel to the findings [9] that the advanced concepts of "Thermodynamics and Kinetics" was perceived by many students as the most difficult topics in Physical Chemistry. Thermodynamics discusses bulk properties in exploring the behaviour of individual atoms and molecules in the use of spectroscopy, rate and mechanisms of chemical change analysis [10], with the following main divisions: 1) equilibrium, 2) structure, and 3) change.

Chemical Thermodynamics and Equilibrium have several concepts that need internalization by both learners and mentors especially that problem solving in these fields greatly need mathematical skills. As these concepts are not only significant in the academe but also in our daily lives, it is of importance that learners must be exposed to practical or hands-on experiences for better understanding of the lessons. The most important subsets of variables were found to be Chemistry and Mathematics [11]. This contention had been supported [5] stated that, "[an] important skill that is needed to master is mathematics; it is required and recommended by the chemistry department in studying higher mathematics, but the relevance of Mathematics was not explained to the students." This may be one of the reasons why many students fall short in their academic performance in the subject as the practical significance of the topics isn't clear to them.

Table 1. *Suggested Enhancement to the Experienced Difficulties in Chemistry 70 Topics*

Difficulties in Chemistry 70	Possible reasons or causes for such difficulties	Suggested Enhancement
1. Some parts of Gas Law (Virial coefficient and Van der Waals' Equation)	The problem sets in Virial Coefficient & Van der Waals" Equation needed a thorough analysis to distinguish whether the given problem is an ideal gas or real gases. Unclear classification between ideal and real gases. Most of the problem sets dealt on quadratic equation.	Review in Chem 15 is needed in recalling the basic concepts of Gas Laws. Constant practice and solving problems is needed to master quadratic equations. Looking for more resources in the internet and scanning previous notes in Math 2 is highly recommended.
2. Phase Diagram	The dependence of stability on the conditions of the pure substance was too broad and was not taught into simpler concepts.	Go to the library or look for other reliable references in the internet and read more about on the aspects of phase transitions of the pure substances.
3. First Law of Thermodynamics	There were too many problem sets involved that dealt with derivation. Faster pace in dealing with problem sets may cause confusion to the students.	Analyzing and solving problem sets are needed to recall or master the specific mathematical skill in deriving equations. Reviewing Math 51 is highly recommended especially on Derivatives.
4. Chemical Equilibrium and Kinetics	Problem sets were difficult to understand and there were many problem sets that involved integration and derivation.	Constant practice and solving related problem sets is needed to master the derivation or integration part of the problem. A review on Chem 16 is needed in recalling the basic concepts of Chemical Equilibrium and Kinetics. Review on Math 61 is highly recommended also especially the techniques on how to derive and integrate equations.

5. Engines and Thermodynamics	The applications on thermodynamics in machinery were not explained well by the teacher. The techniques in solving fundamental equations were not introduced properly.	Watching some video clips in the internet related to machinery and Read more About thermodynamics to understand the lesson well. Constant practice in solving and deriving fundamental equations is highly recommended. Techniques in solving math related problems can be found in the previous Math 61 topics.
6. Chemistry 70 (as a subject)	Concepts were difficult to understand, many problem sets and equations involved, and the teaching method was not being utilized well by the Chemistry 70 teacher.	Look for more resources in the library or internet Mastery on mathematical skills is essential in answering problem sets. Solve problem sets regularly to practice the techniques needed in the problem Peer tutoring to help each other in understanding the concepts and in solving problem sets. Incorporate technology in teaching by utilizing multimedia facilities/ resources.

Table 1 shows the suggested enhancement to the difficulties encountered by the students in Chemistry 70 topics. Enhancement was based on the possible reasons or causes that were encountered by the students and how they will respond in dealing with such difficulties. Here are some recommendations also [12] when seeking to support chemistry students with strong foundations in mathematics: 1) the importance of mathematics to any chemistry subject should not be hidden; 2) tutorials and peer-assisted learning sessions; 3) there should be continuum in developing their mathematical skills through all aspects of their major; 4) Universities should consider the appropriate time frame to introduce mathematical concepts; 5) key mathematical ideas, concepts and results must not be hidden from the students rather, presented in an easy manner for them to understand the lessons easily.

5. Conclusions

Majority of respondents got only a passing grade that served as evidence of their academic performance in the subject Chemistry 70 "Physical Chemistry", with grades ranging from 1.50 to 3.00. Students had difficulty in the dealing with the subject matter: Introduction of Physical Chemistry especially on the topic: Chemical Thermodynamics and Equilibrium, which they considered most difficult. Regarding classifications of possible reasons for causes of difficulties, the following ranked classifications were discovered: 1) teacher factors; 2) student factors; 3) personal factors; and 4) resource factors.

6. Recommendations

1. Teachers must help students in the improvement of the latter's academic performance in Chemistry 70: Introduction of Physical Chemistry by not depending on only one book but also look on other references;
2. The Physical Chemistry teacher/s must not focus on only one method of teaching but find other ways to present the lesson more effectively by utilizing multimedia facilities/resources;
3. Advise to shift will be given or provide enhancement if the student gets a failing grade in Math 2 and Chemistry 15, as these are the most basic subjects of this course;
4. Chemistry 70 should be taught not only as requirement for passing the subject but also to appreciate its very vitality to actual life situations;
5. There should be a continuum in developing students' mathematical skills in all aspects (laboratory and lecture sessions) as strong support to understand Chemistry subjects. With this, the teachers need to clarify the significance of the mathematical techniques in imparting lessons directly related to Chemistry;

6. Teachers in mathematics should introduce the essence of having such calculations in the students' respective courses especially for the BSE-Chemistry students for them to be able to integrate it in their careers as future educators; and lastly,
7. It is recommended that more studies be done in the future with more sets of BSE-Chemistry students in Physical Chemistry as respondents so as to support and enhance further the results of this study.

Acknowledgements
Sincere gratitude is due to the following: To our dear parents who supported us financially and gave us words of encouragement. DR. ELLEN V. INUTAN, for allowing us to interview about her experiences in teaching Physical Chemistry. The respondents of this study, for patiently answering our calls messages for the in-depth interview as well as for answering the questionnaires provided. DR. CHOKCHAI YUENYONG of Khon Kaen University for his expertise in making this research paper more appropriate and publishable. And Mindanao State University – Iligan Institute of Technology for providing the researchers' an avenue for learning and growth. This accomplishment would not have been possible without them.

References
[1] Logan 2015 Education [Abstract] Retrieved October 20, 2016, from http://www.edubilla.com/articles/education-trends/education-4912/
[2] Shadreck M 2013 University students' performance in physical chemistry at undergraduate level: Perceptions of zimbabwean Chemistry undergraduates and lecturers. *International Journal of educational science and Research* (IJESR) ISSN 2249-6947 Vol. 3, Issue 2. TJPRC Pvt. Ltd.
[3] Sozbilir M & Pynarbapy 2010 Prospective chemistry teachers conceptions of chemical thermodynamics and kinetics. *Eurasia journal of mathematics, Science and Technology education*,6(2),111-120,5
[4] Holme T A, Luxford C J, & Brandriet A 2015 Defining Conceptual Understanding in General Chemistry
[5] Shallcross D E, & Yates P C 2014 Skills in mathematics and statistics in chemistry and tackling transition. Higher Education Academy
[6] Gagan M 2008 Review of the student learning experience in chemistry. Higher education academy physical sciences subject centre
[7] Woldamanuel M 2014 What makes chemistry difficult. AJCE, 2014, 4(2), Special issue (Part 1)
[8] Sozbilir M 2004 What makes physical chemistry difficult? Perception of turkish chemistry undergraduates and lecturers, *Journal of chemical education*, 81(4).
[9] Sozbilir M, Pinarbasi T, & Canpolat N 2009 Prospective chemistry teachers' conceptions of chemical thermodynamics and kinetics. *Eurasia Journal of Mathematics, Science & Technology Education*. 2010, 6(2), 111-120.
[10] Atkins P & de Paula J 2010 Physical chemistry (9th edition). New York: W.H Freeman and Company
[11] Derrick M Elizabeth 2002 Predictors of success in physical chemistry. *Journal of chemical education*. 2002, 79(8)
[12] Grove M & Pugh S 2015 Is a conceptual understanding of math vital for chemistry? Retrieved November 16, 2016, from http://www.rsc.org/eic/2014/12/mathematics-problem-chemistry-university

IOP Conf. Series: Journal of Physics: Conf. Series **1340** (2019) 012063 doi:10.1088/1742-6596/1340/1/012063

Thai students' existing ideas about plant biological energy

Nannabhat Surat[1], Siripohn Charuchun[1], and Chokchai Yuenyong[1*]

[1]Science Education Program, Faculty of Education, Khon Kaen University, Thailand

* Corresponding auther's email address: ychok@kku.ac.th

Abstract. This study explored students' existing ideas about biological energy. The study employed the Interview semi-structured interviews. Findings indicated that students explained that the plants are not hot or have high temperature, they think that things which have energy are always hot, have high temperature, can explode or can destroy anything. Moreover, Students called plants energy as 2 frameworks, Include Plants energy including to the plant processes and Physics energy framework. And they instanced plants processes that use energy, include Photosynthesis, Respiration, Water-Nutrients Transportation process, Transformation energy process, Combustion process, Osmosis, Nitrogen fixation process, Stimulus responding process, and Reproductive process. Their explanations are based on their understanding. Regarding constructivist perspective, Teachers need to know students' prior knowledge in order to take into consideration these viewpoints for designing and implementing teaching interventions (Trumper, 1990; Duit & Haeussler, 1994)

1. Introduction

Energy is one of the science concepts that cuts across all science disciplines and is experienced in our everyday life situations (Saglam-Arslan & Kurnaz, 2009). Teaching and learning is so difficult because energy is not realized by seeing that is realized from Degradation energy. and There is difficulty to choose suitable pedagogical for the Energy concept instruction. We can describe about that "Teaching energy concept is mystery" This mystery of the energy concept is related with its usability in different disciplines, as well as its abstract and theoretic identity (Diakidoy et al., 2003; Lemmer & Lemmer, 2006).

Diagnosing students' understanding of energy and its related concepts in biological context explored that student understand that Biological Systems Have a Unique Kind of Energy. Many students used processes to make up a name for a type of energy, such as respiration energy, photosynthetic energy and nutritional energy. (e.g. Kruger et al., 1992; Mann, 2003; Trumper, 1997).

In fact, Photosynthesis is the chemical process by which green plants convert sunlight into sugar. In essence, this process transforms a wave of light energy into chemical potential energy, which the plant then stores in the molecular bonds of sugar molecules. A sequence of chemical reactions transfers the sun's light energy into the chemical bonds that hold together special, energy-carrying molecules (the most common of which are called ATP). The energy originating from the sunlight is being stored in the ATP molecules as chemical potential energy.

Many people do not know that plant have energy to live example there need energy to manufacture food, transporting water, soluting across cell membranes, using during reproduction, germinating seeds, and processing all of metabolic in plant cells. [4]

In Thailand, Science is explicitly separated. Physics rarely link with others science. Energy is mostly known that about Physics, Few Thai people realize that Energy connects all of science. Student who study about Science is the best of representative and bring reliably results of the research. Therefore, the aim of this research "Student's understanding of plant biological energy concepts.' is studying understanding of plant biological energy concepts of 4 major from Science education student Khon Kean University.

In Science, Plant have two principal forms of energy, there are kinetic energy and potential energy. (Kinetic energy is the energy of movement such as the motion of molecules. The second form of energy, potential energy is stored energy in cell.). Plants kinetic energy such as Radiant energy of Photon from Photosynthesis, the moving electrons energy, and other energy from charged in atom. In addition, Plants potential energy is chemical potential type, most of them is made by diffusing across cell wall.

One statement tested students on whether living matter (biological organisms) is governed or has a unique kind of energy different from non-living matter. This kind of thinking is referred to as vitalism (Barak et al., 1997), and it was tested using the statement living organisms have a kind of energy which is different from the energy we learn about in physical science. (Kruger et al., 1992). Barak et al. explained that this erroneous thinking has been exacerbated by the idea that living matter is made of special constituent materials which differed from non-living matter and requires a 'vital force' (i.e. a unique form of energy) different from other forms of matter. The current scientifically acceptable view of energy is that whilst living systems are unique, biological phenomena can be understood within the framework of concepts applied to physics and chemistry [1].

Faculty of Education, Khon Kean University, is educating student who is going to be teacher. Science education is one of popular major that the youth want to admission in. There are four programs educating, include Physics, Chemistry, Biology, and General Science. Each program has different instruction, but the most important thing is all of program must have the right science concept, especially Energy concept that is related every science branch.

Thus, the research is going to record and analyze data to conclude what students' plant energy ideas are, how students understand the different between process and energy, and How different the plant energy ideas concepts between 4 majors of science.

2. Methodology

2.1. Method
The present study was a naturalistic inquiry conducted within a constructivist paradigm. Students' existing ideas about energy were elicited by dialogue between researchers and participants (Johnson & Gott, 1996).

2.2. Participants
The sample consisted of 29 students from the science education, including 6 chemistry students, 6 physics students, 9 biology students, and 3 general scientists from Khon Kaen University.

2.3. Data collection
Data collection identifies students' existing ideas about energy using semi-structured interviews. All interviews were recorded and recorded in full. Interviews started with researchers asking questions about whether plants have energy. And ask the next question. To test the understanding of energy in plants. If students say that the plant does not have energy Researchers will ask questions to guide students but without any effort to modify their initial ideas.

3. Results
All interviews were audio taped and fully transcribed. Students' responses relating to each questions were analyzed and then categorized as students' framework as plants energy. Each category was

IOP Conf. Series: Journal of Physics: Conf. Series **1340** (2019) 012063 doi:10.1088/1742-6596/1340/1/012063

explained with students' answers of 6 question. These include Do the plants have energy?, Why do you think plants have energy?, Which energy do plants have?, What are type of each energy in plants?, What do processes inside plants?, How different between energy plants and process plants?. How related between energy plants and processes plants? And the result of this study are:

- Do the plants have energy?

Students referred that the plants do have energy. Students' explanations from year 2nd and year 3rd based on their prior knowledge. The researcher interviewed the 29 participants from 117 students of Faculty of Education, Science at Khon Kean University. The data were analyzed by the statistical include of frequency, percentage are as follow 20 females (69%) and 9 men (31%), they are separated as 6 Physics major students (20.7%), 6 Chemistry major students (20.7%), 9 Biology major students (31%), and 8 General Science major students (27.6%). The results of this study are 27 students who answered plants have energy (93.1%), 2 students who answered plants do not have energy (6.9%)

- Why do you think plants have or have not energy?

Students explained their reason about plants have energy. Researcher separated their answers as 4 framework, and the answers of this question are as follow:

Category	Students' responses
Plants have processes that produce and use energy	9
Plants is nature, All of nature must have energy no matter what they are	2
Plant are producer who need energy for works.	2
The law of Thermodynamics	1
Plant do not have high temperature.	1

60% Of students replied that The Plants have energy because they have processes that produce and use energy e.g. Respiration, Photosynthesis, Reproduction, and Growth. 13.34% of students think that plants is nature, all of nature must have energy no matter what they are. 13.34% of students told researcher that Plant are producer who need energy for works, and 6.66% of students replied that They have energy is as a results of the law of Thermodynamics that explain transformation of energy. and 6.66% replied that plants do not have energy because They are not hot or have high temperature, they think that things which have energy are always hot, have high temperature, can explode or can destroy anything.

Students know what Photosynthesis is and what this process produce that bring many benefit for others organisms, Moreover, They know that Plant use chemical energy for photosynthesis and respiration the most common of which are called ATP. But a few participants have misunderstanding and use prior knowledge about fuel and bomb that can see better that plant energy in their real life, So they answered plant do not have energy as result of They do not look like fuel.

- Which energy do plants have?

Students called plants energy as 2 frameworks, Include:

Category	Students' responses
Call as Plant processes	25
Called as Physics energy	21

Plants energy including to the plant processes e.g. Growth energy, Water Absorption or dehydration energy, Diffussion energy, Photosynthesys enrgy, Respiration energy like ATP ADP NADPH+. According to the plant processes are known by Science student. On the other hand They called plant energy that Physics energy framework e.g. Heat, Chemical potential energy, Light energy (Photon), Machanical energy, Kinetic energy.

- What are type of each energy in plants?

Category	Students' responses
Physics energy framework	32
Nutrients	1
Unknown	1

Most of students have basics knowledge that plants energy can be grouping in Physics energy framework about 94.10%, i.e. Plants energy are Kinetic energy like transportation energy of water and nutrients because there are movement, moreover they were been grouping Absorption and Dehydration energy Photosynthesis energy and Respiration energy as Potential energy because they do not have movement but work. There are 2.95 % of student who replied that plants energy can be grouping in nutrients like Starch, and 2.95% of students do not know how can answer this question.

- What do processes inside plants?
  Student answered as

Category	Students' responses
Photosynthesis	21
Respiration	12
Water-Nutrients Transportation process	11
Transformation energy process	3
Combustion process	1
Osmosis	1
Nitrogen fixation process	1
Reproductive process.	2
Stimulus responding process	1
Total	53

There are known that Plant have Photosynthesis, Respiration, Water-Nutrients Transportation process, Transformation energy process, Combustion process, Nitrogen fixation process, Stimulus responding process, and Reproductive process.

- How different between energy plants and process plants?
  All of student responded Energy is absolutely as different as processes.

Student instanced 9 plants processes that use energy, include Photosynthesis, Respiration, Water-Nutrients Transportation process, Transformation energy process, Combustion process, Osmosis, Nitrogen fixation process, Stimulus responding process, and Reproductive process.

4. Conclusions and Suggestions

The Participants who accounted for 24.8% of Science Education Students at Khon Kean University, include Physics major, Chemistry major, Biology major, and General Science major are interviewed about learning student's understanding of plant biological energy concepts. The data were analyzed by the statistical include of frequency, percentage, and grouping to frameworks in each issues. Interestingly, Students' explanations from year 2nd and year 3rd based on their prior knowledge. The results of this study are as follow 27 students answered plants have energy (93.1%), 2 students answered plants do not have energy (6.9%) and they explained that The plants are not hot or have high temperature, they think that things which have energy are always hot, have high temperature, can explode or can destroy anything. Moreover, Students called plants energy as 2 frameworks, Include Plants energy including to the plant processes and Physics energy framework. And they instanced plants processes that use energy, include Photosynthesis, Respiration, Water-Nutrients Transportation process, Transformation energy process, Combustion process, Osmosis, Nitrogen fixation process,

IOP Conf. Series: Journal of Physics: Conf. Series **1340** (2019) 012063 doi:10.1088/1742-6596/1340/1/012063

Stimulus responding process, and Reproductive process. Their explanations is based on their understanding. Regarding constructivist perspective, Teachers need to know students' prior knowledge in order to take into consideration these viewpoints for designing and implementing teaching interventions (Trumper, 1990; Duit & Haeussler, 1994). Additionally, Science teacher students who will be Science teacher in the future is important person who should be correct their misconception about all of Science knowledge for the better educational in Thailand.

References

[1] Boundless biology Energy and Metabolism [website] Retrieve from URL: https://courses.lumenlearning.com/boundless-biology/chapter/energy-and-metabolism/.

[2] Chokchai Yuenyong, and Jirakarn Yuenyong 2007 Grade 1 to 6 Thai Students' Existing Ideas about Energy Science education international **2007** pp 289–298

[3] Oundless biology. Transport of Water and Solutes in Plants [website] Retrieve from URL: https://courses.lumenlearning.com/boundless-biology/chapter/transport-of-water-and-solutes-in-plants/

[4] Vivien Mweene Chabalengula, Martie Sanders and Frackson Mumba 2011 Diagnosing students' understanding of energy and its related concepts in biological context Retrieve from URL: https://link.springer.com/article/10.1007/s10763-011-9291-2

Develop tools and softwares for the assessment of ethnic minority Vietnamese primary school students

Ngo Thi Thanh Quy1*, Nguyen Thi Thu Thuy[1], Tran Thi Ngoc Anh[1], and Dao Kien Quoc[2]

[1]University of Education - Thai Nguyen University, Vietnam

[2]Institute of Information Technology – Vietnam National University, Hanoi Vietnam

* Corresponding auther's email address: ngothanhquy@dhsptn.edu.vn

Abstract. Language education has been considered a crucial issue by many countries and there have been studies on this area. By applying research achievements in the modern language of the world, this article focuses on the issue of Vietnamese language education for ethnic minority students at primary schools in Northwest Vietnam by a set of standards, tools and assessment software. In order to handle the problem, we use a mixed method in the study of the social sciences and humanities together with quantitative and qualitative research methods. We survey some typical primary schools in which ethnic minorities are living and studying, such as Thai, Muong, Dao, Mong, and Tay ethnic groups. After that, we develop a set of standards, tools and Vietnamese language proficiency assessment software for ethnic minority students in the Northwest of Vietnam. The data, which is composed of four skills including reading-writing-speaking-listening with a total score of 100, is designed by a software that has both online and offline features. Our team made an experimental research with 5000 test times. The research results will contribute significantly to the enhancement of Vietnamese language proficiency for ethnic minority students in the Northwest region of Vietnam in particular and those who learn Vietnamese as a second language in general.

1. Introduction

The Northwest of Vietnam is a key mountainous region with thousands of kilometers bordered by China and Laos. This is home to many ethnic minorities whose educational standard is still limited. In order to improve the quality of education and training as well as develop human resource of this area, it is necessary to enhance Vietnamese language proficiency of the students, especially in primary schools. Educational experts have also spent a great deal of effort developing a program of local integration to improve the quality of education in this multicultural area. These efforts, however, have not yet made a real difference in Vietnamese language proficiency of the students. Attaching to the Northwest region for a long time and deriving from the real teaching situation, the research team proposed a set of tools for assessing Vietnamese language proficiency in reading, writing, speaking and listening skills. We hope to make considerable contribution to the improvement of Vietnamese language skills for ethnic minority students in the Northwest Vietnam. In order to address the issue, we use the interdisciplinary approach of the social sciences and

IOP Conf. Series: Journal of Physics: Conf. Series **1340** (2019) 012064 doi:10.1088/1742-6596/1340/1/012064

humanities with the combination of the qualitative and quantitative research method. In addition, the quantitative research method is employed to process and analyze data that has or can be quantified, such as the processing of test scores and assessments of Vietnamese language proficiency of primary students. Assessment tools are designed and analyzed according to the modern measurement theory. Additionally, other specific methods are employed, such as *Sampling method*. Oriented sampling was carried out in the provinces of Dien Bien, Son La, HoaBinh, Yen Bai and Lao Cai, etc. These selected provinces are representative of multi-cultural regions in the Northwest Vietnam. Based on the approach and specific methods, we use the comparative method to design a set of tools and Vietnamese language proficiency assessment software for ethnic minority students in the Northwest of Vietnam.

2. Methodology

2.1. Target group

In recent decades, the world has revolutionized the field of language competence assessment and measurement of learners with typical examples of the introduction of the Common European Framework of Reference for Languages (CEFR) in 2001 and the adaptation of the TOEFL test (Test of English as Foreign Language). Language educators hope to assess learners' ability from the focus on grammatical knowledge to communication competence in a variety of academic situations. In the field of language measurement and assessment in Vietnam, there has been a shift from knowledge testing to competence assessment of the learners. The theoretical basis for these changes is derived from the development of models of competence and language proficiency in the field of linguistics and language assessment. Researchers are interested in Lado's component mode; Communicative competence model of Canale and Swain [9], Farhad's functional competence model; Bachman's language competence model; Celce Murcia's model of interactive elements of the communication competence. Language competence models have become the basis of language theory for us to develop a set of standards, tools and Vietnamese language proficiency assessment software for ethnic minority students in the Northwest of Vietnam. We have agreed to use six criteria to design Vietnamese language proficiency standards for ethnic minority students in the Northwest region.(1) Reliability for all stakeholders, especially for teachers, parents and students. Reliability depends on: The authenticity of the product and the rationality of the measurements. The criteria are rational and valid, which means that teachers' measurements and standards must be parallel to the values established by the regulators or recognized on a large scale. (2) Usefulness. Assessments must be useful to make a positive contribution to students and teachers. (3) Method balancing, which is linked to practical tasks/ assignments/ tests. (4) Honesty and fairness are required to ensure that each student follows the important criteria. (5) Intellectual and emotional Seriousness - focus on ideas, questions, issues, texts and core knowledge; tests are designed so that the questions/ tasks/ exercises/ exams stimulate students' learning pleasure. (6) Feasibility in terms of resources, preparation, politics, and time to design, adjust, use, evaluate and report students' results / products / assignments.

2.2. Methods of inquiry

The core content of the Vietnamese language proficiency standards for ethnic minority students focuses on the following issues:

1. Communication: Standard 1.1: Students participate in conversation, provide and collect information, express feelings, emotions, and exchange ideas in Vietnamese; Standard 1.2: Students understand, explain in writing and spoken language about a range of topics. Standard 1.3: Students present information, concepts, and ideas about a variety of topics to the listeners or readers.

2. Culture: Students gain knowledge and understanding of the Northwest culture; Standard 2.1: Students demonstrate their understanding of the relationship between the Northwest culture and reality; Standard 2.2: Students express their understanding of the Northwest culture.

3. Connect: Students connect to other fields to get information; Standard 3.1: Students consolidate and acquire knowledge of other subjects in Vietnamese; Standard 3.2: Students are informed and aware of different perspectives through Vietnamese language and regional culture.

4. Comparison: Students develop their understanding of the nature of Vietnamese Language and Culture; Standard 4.1: Students demonstrate their understanding of the core characteristics of Vietnamese through comparison with their mother tongue; Standard 4.2: Students represent their knowledge of Vietnamese culture in general through comparison with their ethnic culture.

5. Community: Students participate in multilingual communities; Standard 5.1: Students use Vietnamese both in and out of school; Standard 5.2: Students show that they are lifelong learners by using Vietnamese for their personal interests and profession.

From the standards and criteria above, we have developed a set of tools for assessing ethnic minority students at primary level with 5 parts, total = 100 points/ 25 questions.

1. Knowledge (10 points/ 3 questions); 2 questions with 2 points each and 1 question with 4 points;

2. Reading comprehension (30 points/ 8 questions): 4 questions with 3 points each, 2 questions with 4 points each and 2 questions with 5 points each;

3. Listening comprehension (17 points/ 5 questions): 3 questions with 4 points each, 2 questions with 4 points each and 2 questions with 5 points each;

4. Writing (28 points/ 3 questions): 2 questions with 3 points each, 1 question with 8 points and 1 question with 10 points;

5. Speaking (15 points/ 4 questions): 2 questions with 3 points each, 1 question with 4 points and 1 question with 5 points.

3. Research Findings

To assess Vietnamese language proficiency by testing method, the quality of the question bank is of crucial importance. Therefore, it is necessary to evaluate whether the set of questions is relevant to learners' competence. There are some important indicators in the assessment of learner competency such as the level of difficulty and distinction. The level of difficulty is a factor that needs to be taken into consideration because it reflects the general level of students. Too easy or too difficult tests may not bring about the desirable results. The level of difficulty is shown on the percentage of students who have a correct answer. To be more detailed, the high percentage of correct answers means that questions are too easy and vice versa, if the percentage of correct answers is too low, questions are difficult. After evaluating the results, it is possible to statistically measure the percentage of correct answers to the questions. The level of distinction indicates that the test results have a reasonable similarity to the student's academic performance. Therefore, the software should provide tools to calculate the correct answering frequency of the students and, once provided with the students' classification of academic performance, can also calculate the correlation between the test and the students' academic performance to determine whether the test is relevant or not. In order to improve the effectiveness of the set of standards, criteria and assessment tools to improve the Vietnamese language proficiency of ethnic minority students in the Northwest, we believe that it is necessary to design and develop a software system to support the assessment of Vietnamese language proficiency of ethnic minority students at the primary level when they learn the second language. The software system should provide complete functions that enable the user to perform all assessment tasks, such as test formation, attendance list, enrolment number, online examinations, automatic evaluation of results, making reports for online exams. During the research, we found that most ethnic minority students in the Northwest are new to computers. Therefore, in order to meet the requirements of the teaching and learning process as well as the assessment of Vietnamese language proficiency of primary students, the software should also consider some criteria, namely easy to understand and easy to use to meet the requirements of teaching Vietnamese. In terms of testing methods, although there are many variations such as choosing right -wrong, matching, arranging and gap filling, etc. choosing the right answer in multiple choices is the simplest. Students are required to select the appropriate option rather than enter data or perform complex tasks. The software should support the ability to display multimedia data such as audio, video, and photo to assist students take the test conveniently and smoothly. In terms of features, the software should have a user-friendly interface which is suitable for the skills of teachers and students. The online software must run on the WEB, can be installed on the LAN of the school and the WAN on the Internet. The assessment of Vietnamese language proficiency through the software system will be implemented comprehensively in terms of: (1) Knowledge of students; (2) Reading comprehension skill; (3) Listening comprehension skill; (4) Writing skill; (5) Speaking skill. With such testing content, the software is designed to be able to create tests with structure and marking distribution as specified, which can be customized according to specific requirements. For the test, the software is able to create an interface for students to do online tests, record the results reliably and accurately, provide teachers with assessment of Vietnamese language proficiency.

With the current infrastructure of the Northwest primary schools, it is included that online tests cannot be conducted on a large scale. For this reason, it is necessary to add a paper-based test. However, instead of being manually marked, it will be using the image-processing method which will scan and transfer the students' answer into the computer. In this way, it is possible to evaluate a large number of students in schools with a DR scanner that can scan a variety of pages at a high speed. This should also ensure the link between the online assessment system and the paper-based exam software. For direct paper examination, the software must have the function of printing answer sheets, importing candidate lists (directly or from the online export software), entering answers, configuring the scanner, scanning tests, identifying tests, correcting students' errors (related to the form of the answer sheet) and reporting results (direct use or transfer to online software for analysis). Along with it is the question banks, test items, list of candidates, answer sheets created by the online system. If the paper-based test is required, the software will mark and then transfer the results back to the online system for storage. For this reason, there is data available to assess the suitability of the question banks.

Thus, the development of software for assessing the Vietnamese language proficiency of ethnic minority students in the Northwest of Vietnam follows the general principles of software engineering. The process of software development includes the following stages: Requirement survey; Overall design; Detailed design; Programming/ testing (module test, integration test, system test and operational test); Transfer/ Maintenance. The detected defects will be corrected and upgraded. Online software is installed on Windows via ISS with SQL/ Server database. Client software runs on popular browsers such as IE, FireFox, Chrome, Safari, Opera, etc. The software is built on the Web with MVC, Unicode support and ASP.NET programming language.

The functions and features of the software meet the requirements of supporting the testing and assessment of Vietnamese language proficiency for ethnic minority students as follow: Online test software. The online test software is designed with the following functional groups: Functional group of examinations: a) Managing examination: Add, Delete, Edit exams; b) Managing students of an exam using various methods such as importing students from excel spreadsheets or directly updating the student list; c) Numbering registration, including complete numbering, partial numbering, complementary numbering or existing registration re-using; d) Dividing examination rooms and printing students list; e) Issuing accounts for students to take the test; f) Assigning sets of tests to an exam and activating the official test set. Functional group of managing question banks, tests and sets of tests; a) Questions management: Questions are in the form of multiple choice so that they can be used consistently in the form of online or paper-based test forms. Other types of multiple choice questions, such as matching, arranging and gap filling, etc. are all about choosing the best option. This ensures that students only need to choose rather than perform complex keyboard typing. The functions include: Adding a single question (direct editing)/ Adding a series in the format of the table set in a word file; Deleting a single question/ Editing a single question; Adding a new group question/ Deleting a group question; Updating component questions (single questions of a group); Editing context; Seeing details. b) Question bank management. Questions bank management not only aims at categorizing content, skills, knowledge, but also automatically generates tests by selecting the questions from the repositories and re-coordinating them in a particular structure which is consistent with the method of assessment. Each test has 25 questions with a 100-point scale. There is a difference in knowledge among classes 1, 2, 3 4, 5 with 5 parts: knowledge, reading comprehension skill, listening comprehension skill, writing skill and speaking skill. Each of 25 questions is to evaluate a skill and cannot be able to coordinate arbitrarily. For this reason, each set of interchangeable questions must be placed in a repository. When a test is formed, only one question of each repository can be selected. For reading and listening comprehension skills, questions have a specific context. Therefore, it is necessary to select a group of questions with the same context. In other words, there needs to be a repository of questions that stores both the context and the component questions in a unified whole. Based on the assessment method, the software must create 65 repositories and then lock the update function to avoid modification that may affect the creation of tests. Management functions include: adding, removing, editing, viewing questions, adding questions to the repositories by either direct editing or importing from a word file. c) Test structure management; The software must have a function to update the structure of the test to determine the number of parts, the name of each part, the number of questions, expected time of allowance; Each section is created by selecting a specific number of questions from one repository. Design functions include: creating, deleting, editing, updating the list of the repositories from which the questions are selected for the structure (also known as the test matrix) locking the structure so that it cannot be modified. d) Test matrix management: Once the structure has been set, it is possible to automatically generate tests from the repositories by stating the number of questions selected from each bank. The test matrix can be defined as a table to indicate the number of questions selected from the repositories to form a specific part of the test. The test matrix is the basis for producing the chosen structure. Functions related to the management of the test matrix include: Creating/ Deleting; Modifying/ Printing. In this research, the test matrix requires that every single question be withdrawn from one repository. The structure and test matrix after being created will be locked to avoid modifications. e) Test management: Tests are generated in each set, with relevant functions including: Adding a new set of tests with a specific test matrix/ Deleting/ Editing/ Creating/ Printing/ Viewing: Exporting a multiple-choice test for students (only 19 questions). For each class, if there are 16 questions in each bank and one question is selected from 13 banks, there will be 16 13 (about 256 trillion) different sets of tests; Producing a full test (25 questions) with key for the teacher. There are three types of generating tests, including independence, permutation and mixed type. For independence type, the system creates enough "n" tests which are named Test 1, Test 2 .... For permutation type, the system produces only one test which is named Test 0 (the original test) and then it is permuted to create Test 1, Test 2 ... For mixed type, the process is the same as in independence type, but when editing, it is possible to delete or select one test and permute from it.

In addition, the research team built an interaction between the online test and offline test software. a) Exporting information. When performing this function, the information about the exam code, exam name, number of tests, test code, number of questions, answer key, list of candidates and registration number is exported into a file to be transferred to the offline test software so that the software can identify information about the examination to export test sheets, making candidate lists, scanning, identifying, marking and sending the results back to the online software. b) Entering score: The score chart of offline test software with the exam code is returned and automatically entered into the results of the online software.c) Functions related to taking online tests of students. Log in: Students sign in with their registration number and a provided password. If students enter correct information, the system will display their full name. The system must take control over students so that they cannot log in more than once or from one or more than one computers. Interaction with the system: Because there are few questions, there is no need to show up in part to get students to gradually download as in the large-scale exams. Layout: On the left side is the question and the options to choose, on the right is the list of questions and test status. If students choose a question, it will be displayed on the screen for students to do or re-do. In the case of a question from a group of questions, the left part is divided so that the context will be displayed on the left. It can be a text, video or audio. At the top of the screen, there is information about the test, test part, connection status and countdown time of the timer. Students can log in again when problems occur. While students are taking the test, the software periodically saves the results. The software not only saves the results but also saves the time remaining. Therefore, students can log back in after problems occur. After the system recognizes students who have been granted the right to revisit the unfinished part, it will restore the clock, test status, and let students do the next task. The software for assessing the proficiency of primary school students in the Northwest is also developed with the function of controlling the exam: monitoring the list of candidates and the status of the test; monitoring technical status (such as network connection); closing the examination and stopping all test activities. The software is capable of analyzing results, calculating indicators of difficulty and distinction. Regarding the feedback on the test status, the number of people answering each question, the number of people choosing options A, B, C, D are updated to the question bank (sum, accumulative) and locked to return only one time. For directly answered questions, the correct answer is recorded; Regarding statistics, grading and point scales, the number of candidates with 0 points and the number of candidates with 1 to the maximum point are also calculated. However, point scales are only meaningful when doing the permuted test. Statistics on the percentage of correct answer for each question are used to assess the level of difficulty; the distinction of the question can be calculated when it comes with students' academic performance. Besides, the software also includes system administration functions, including user management functions such as: adding, deleting, adjusting users, granting permissions; data backup and data recovery. For the paper-based test software with marking by image processing, functional groups include: managing test terms, automatically updating the test data or receiving test data from the online software; monitoring students directly or by data exported from online test software; Printing/ scanning answer sheets; pre-processing photos and correcting students' answers; Identifying/ Marking; Extracting the results to transfer to the software online; Managing the system.

In spite of predicting situations which may occur in reality, the research team encountered a lot of difficulties during the study since the facilities for teaching - learning process in remote areas and ethnic minorities in the Northwest are still limited. In the best schools, only about 10 computers can operate stably. In many places, computers have not yet been connected to the network. The bandwidth for Internet connection is still low. Most computers are not equipped with headphones, which makes it difficult to test listening comprehension skill. For these reasons, the online test software must ensure to be able to run on low-configuration computers. Furthermore, it should be very light in order to be used on a normal computer as a server. In the future, when the Internet bandwidth is improved and servers are equipped on the Internet, the system can be installed in one place and communicate with users over the Internet. The software has been tested twice. The first time was in April, 2017 on a small scale in Yen Mong Primary School, HoaBinh City, HoaBinh Province, Vietnam. The second time was in May 2018 in Cao Phong district, HoaBinh province, and Sa Pa district, Lao Cai province, Vietnam with 5000 test times. Due to the limited number of computers in these schools, only a small number of students can participate in the online assessment. The rest is tested by answer sheets which are then scanned and marked by the software. The results are quite good and the spectrum is quite wide.

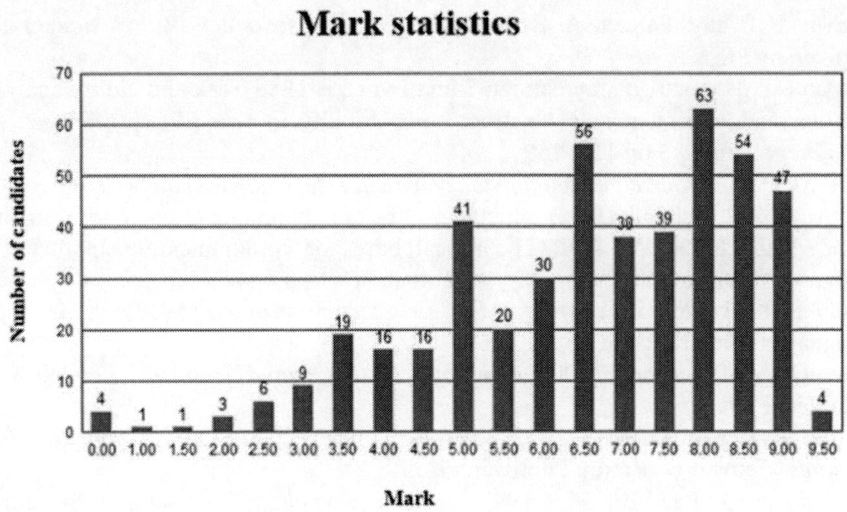

Average mark:	**6.63**	
Number of candidates with excellent mark	168	Account for 35.97%
Number of candidates with good mark	133	Account for 28.48%
Number of candidates with average mark	91	Account for 19.49%
Number of candidates with below average mark	51	Account for 10.92%
Number of candidates with poor mark	24	Account for 5.14%

4. Conclusion and Recommendation

Thus, the development of a set of standards, tools and assessment software for Vietnamese language for ethnic minority students in the Northwest with specific standards and criteria will help teachers and learners to calculate the amount of knowledge and Vietnamese language skill. This makes considerable contribution to the improvement of Vietnamese language skills for ethnic minority students at primary level and the foundation and basic conditions for them to study and integrate into general education. The development of a set of standards, tools and assessment software helps us have orientations in changing curricula, textbooks and methods of teaching Vietnamese at the primary level to improve the proficiency of ethnic minority students. It is also a useful reference guide for teachers to develop exercises, tests, and activities for Vietnamese students. This is also an effective tool for teachers to teach Vietnamese to ethnic minority students in the Northwest./.

References

[1] Bachman L F and Palmer A S 2010 *Language Assessment in Practice* Oxford: Oxford University Press
[2] Bachman L F, Nunnan A, Vanniarajan S and Lynch B 1988 Task and ability analysis as a basis for examining content and construct comparability in two EFL proficiency test batteries *Language Testing* **5** pp 128–159
[3] Canale M 1983 *From communicative competence to communicative language pedagogy* J. Richards & R. Schmidt (Eds.) Language and Communication (New York: Longman)
[4] Canale M and Swain M 1980 Theoretical bases of communicative approaches to second language teaching and testing *Applied Linguistics* **1** pp 1–47
[5] Farhady H 1983 *New directions for ESL proficiency testing* J. W. Oller, Jr. (Ed.) Issues in language testing research pp 253–269
[6] Fulcher G and Davidson F 2007 *Language Testing and Assessment* (London & New York: Routledge)
[7] Hoang H B & Nguyen M T 2012 *Method of teaching Vietnamese language from primary schools* Vietnam Education Publishing House.
[8] Le A, Nguyen Q N and Bui M T 1997 *Methods of teaching Vietnamese* (Vietnam Educational Publishing House, Hanoi)
[9] Le P 2004 "Standard language" and "Speech culture" for primary school students *Education Journal* No 1/2004
[10] Ministry of Education and Training 2006 *Methods of teaching Vietnamese to ethnic students* Teacher training materials (Vietnam Education Publishing House)
[11] Ministry of Education and Training 2012 *Training materials for teaching model of new school in Vietnam* (Hanoi)
[12] Nguyen C H and Vu D N 2015 *The set of criteria for evaluating the Vietnamese language proficiency of international students* (VNU Publishing House)
[13] Oller J W (1979) *Language Tests at School* (London: Longman)
[14] Swain M 1985 *Communicative competence: Some roles of comprehensible input and comprehensible output in its development* S. Gass& G. Madden (Eds.) Input in Second Language Acquisition (Rowley, Massachusetts: Newbury House) pp 235–253

Enhancing Grade 10 Students' Achievement and the 21st Century Learning Skills by Using Information Based on STEM Education

Nittayaporn KINBOON

Science and Technology Teacher, Phuwiangwittayakhom School, Khon Kaen, The Office of the Basic Education Commission (OBEC), THAILAND

*Corresponding author's e-mail address: dr.nittakinboon@gmail.com

Abstract. The purposes of this study are to (1) develop students' achievement of learning technology focusing on solving current environmental and social issues in daily life, and (2) examine the results of using information technology media based on STEM education to enhance the 21st century learning skills of grade 10 students. The action research method was used in this study by dividing into 3 action parts. The instrument used to collect data of students' learning activities was observation form to assess teachers' teaching strategies and abilities to integrate information and communication technology (ICT) knowledge into the contents based on Technological Pedagogical Content Knowledge (TPCK) approach. After the 3 action parts completed, data were analyzed, aiming to develop students' learning achievement. The samples were a total of 40 students in grade 10 in the semester 1 of academic year 2017. The findings showed that using information technology media in teaching based on STEM education in line with TPCK approach developed the 21st century learning skills of students in grade 10 effectively. The approach consists of 5 steps: 1) introduction; 2) cognitive conflict; 3) reflective activity; 4) cognitive restructuring; and 5) summary and evaluation. The 5 representations reflecting students' 21st century learning skills were found at a high level including academic excellence, bilingual communication skills, creativity and innovation skills, problem-solving skills and responsibility for the social issues, attitude towards global awareness. The overall development of the 21st century learning skills of grade 10 students was at the highest level (= 4.57, S.D.= 0.72). The result of students' ONET assessments with specific aptitude skills at a higher level reflected that their learning achievement developed. Moreover, students were able to communicate in Thai and English. Students had competence in designing learning activities to share ideas and projects reflecting their responsibility to solve environmental and social issues. Students also demonstrated their creativity and innovation skills in designing visual webpage and provided presentation showing their global awareness by understanding the diverse culture and religion across the world.

1. Introduction

The demand for personnel in the field of science, mathematics, technology has been rapidly growing in the past two decades while the need in engineering tends to decrease in the 21st century. However, the reports show the decline in students entering to higher education in science, mathematics, technology. This results from the poor educational management of teaching and learning in science and mathematics in schools, resulting in the lack of motivation in learning [3]. In addition, the importance of acquiring knowledge linked to students' everyday life and future career path has not been highlighted. Although developing education for the 21st century requirements aims to provide the meaningful knowledge for the learners to construct necessary and innovative skills in their future profession, students' perception towards science and mathematics learning is that they cannot relate themselves to the content [2].

The development of students' learning achievement and 21st century learning skills using information technology media in activity-based teaching in accordance with the Science Technology Engineering and Mathematics (STEM) Education was implemented with grade 10 students at Phuwiang wittayakhom school in the academic year 2018. The STEM education mainly focuses on providing opportunities to students to construct self-learning in science, mathematics, and technology to apply knowledge in their daily life and future career [2].

Moreover, the STEM education approach addresses on developing students' systematical thinking, collaborative working skills, communication skill, and creativity through variety of learning activities and projects. The activity-based teaching and learning combines the knowledge of science, mathematics, technology, and foundation of engineering design. Therefore, students' completed project would reflect their understanding towards the concepts of the four subjects and be able to optimize the knowledge to solve problems in the future [1].

In the perspective of Thailand's Basic Education Core Curriculum B.E. 2551 (A.D. 2008), the STEM education relates to three teaching and learning departments: science, mathematics, and technology and occupational and technology [2]. Based on STEM education, the teaching and learning management in these three modules aims to develop the students' literacy in science, mathematics and technology which is in line with the objectives of core curriculum of in enhancing the students' achievement and the 21st century learning skills.

2. Objectives

To investigate the learning achievement of grade 10 students in the semester 1 of academic year 2017 after using information technology media in activity-based teaching according to STEM education approach by comparing between pre- and post-test scores.

To enhance the 21st century learning skills of grade 10 students in Phuwiang wittayakhom school in academic year 2017 by using information technology media in activity-based teaching according to STEM education approach.

3. Population and samples

The population used in this study was a total of 365 students in grade 10 in Phuwiang wittayakhom school in academic year 2017. The 40 students in grade 10 class 2 were selected by using cluster purposive sampling technique.

3.1. Research tools

The research tools used in this study are as follows:

1. Teacher's information technology media in 8 contents in the subject of technology and communication and 8 learning plans of STEM education,

2. The assessment used in this study consists of 40 questions with 4-multiple choices to measure students' learning achievement. The reliability of the assessment was 0.83,

3. The observation form is used to observe students' learning structure, and the 5-scale measurement is used to assess students' 21st century learning skills. The reliability of the measurement was 0.84.

IOP Conf. Series: Journal of Physics: Conf. Series **1340** (2019) 012065 doi:10.1088/1742-6596/1340/1/012065

4. Research methodology and data collection

This research was an action research which was divided into 3 action parts. Research tools were teachers' lesson plan, and information technology media based on STEM education. The 1st-3th learning plans were used in the first action part, followed by the 4th-5th learning plans in the second action part, and the 6th-8th learning plans in the final action part, respectively. To collect data of teacher's teaching competence, the observation form was used to assess teacher's teaching abilities in integrating the Technological Pedagogical Content Knowledge (TPCK) approach and ICT knowledge into the content by using TPCK as assessment. The information technology media used in this study were learning resources e.g. applications, webpages, internet, multimedia, and e-books. The students' learning activities are also assessed by using observation form, interview, and quizzes.

5. Data analysis

The data were analyzed by using descriptive statistics, mean, standard derivation, percentage, dependent t-test to identify the difference between pre- and post-test scores. The analysis of content including students' interpretation and the logical reasoning is used to conclusion.

6. Results

6.1. Students can analyze problems and present logical reasoning with mathematic principles to describe phenomena in different contexts. From the teaching and learning activities based on STEM education, the project of packaging design for containing fermented fish was selected by students as they aimed to solve problems in their community. Students reflected their abilities in applying the knowledge of technology and providing presentation by using information technology media effectively.

Furthermore, students sought additional information to develop their design and innovative skills in the project. By implementing the activities, students linked the knowledge of three subjects: science, mathematics, and technology to solve the problems appropriately. Consequently, the post-test scores of grade 10 students were higher than the pre-test score at significant value of 0.01. These results showed that students' achievement developed after the implementation of teaching and learning activities based on STEM education.

6.2 The overall development of the 21st century learning skills of grade 10 students was at the highest level (= 4.57, S.D.= 0.72). The results of 5 representations in TPCK assessment are described as follows:

The representation of students' technology skills was at a high level *(*= 4.*76*, S.D.= 0.*79)*. There was evidence that students used technology to seek more information and link it with the content efficiently; consequently, they interpreted the content, made accurate conclusion and provided presentation properly. This reflected that students can apply knowledge to solve problems in their daily life effectively.

The representation of students' academic excellence was found at the highest level (= 4.*66*, S.D.= 0.*64*) from the empirical findings reflecting students' specific aptitudes (e.g. projects to solve community's problems of packaging designs of fermented fish).

The representation of students' creativity and innovation skills was at the highest level (= 4.*55*, S.D.= 0.*65)* from the evidence that students planned and designed learning activities and projects of packaging designs. The students were creative and innovative to design the webpage and share ideas among groups to seek solutions for the community's problems.

The representation of students' attitude towards global awareness was at the highest level (= 4.*53*, S.D.= 0.*69)*, meaning that students were able to construct self-learning and manage complexity under the difficult circumstances of multi religious and cultural community. Students demonstrated their positive attitudes towards responsibility for the social issues and firmly held global awareness.

The representation of students' bilingual communication skills was at the high level (= 4.*34*, S.D.= 0.*84)*.

IOP Conf. Series: Journal of Physics: Conf. Series **1340** (2019) 012065 doi:10.1088/1742-6596/1340/1/012065

7. Acknowledgement
Sincere to The Office of the Basic Education Commission (OBEC), Thailand that supporting The scholarship for published and oral presentation in I AM STEM 2018.

References
[1] Koehler M and Mishra P 2009 What is technological pedagogical content knowledge? **Contemporary Issues in Technology and Teacher Education 9** pp 60–70
[2] Office of the Basic Education Commission Ministry of Education 2008 Indicators and Core Learning Objectives Occupation and Technology Learning Department Core Curriculum Basic Education 2008 (Bangkok: Printing Agriculture Cooperatives of Thailand)
[3] National Research Council 2012 A Framework for K-12 Science Education: Practices, Crosscutting Concept, and Core Ideas. Committee on New Science Education Standards, Board on Science Education, Division of Behavioral and Social Science and Education (Washington, DC: National Academy Press)

IOP Conf. Series: Journal of Physics: Conf. Series **1340** (2019) 012066 doi:10.1088/1742-6596/1340/1/012066

Explicit Nature of Science in the STS Contact Lens "Big Eyes" Unit

Nonglak Attapan[1] and Chokchai Yuenyong[2]

Science Education Program, Faculty of Education, Khon Kaen University 40002, Thailand

Corresponding author's e-mail address: [1]oilandbee@hotmail.com, [2]ychok@kku.ac.th

Abstract. This paper aimed to explain Grade 11 students' view of nature of science (NOS) in the STS contact lens "big eyes" unit and explicit NOS. The STS contact lens "big eyes" provided students to learn about refraction, thin lens, eyes and vision through science technology and society (STS) approach and learning activities of explicit NOS. Participants included 15 Grade 11 students who study in Kaewsadet school, Kalasin, 2nd semester, academic year of 2011. Students' views of NOS were interpreted through students' responses in learning activities of explicit NOS, participant observation, journal writing, and interviewing. Students' responses were categorized regarding 8 aspects of NOS. These 8 aspects included (1) scientific knowledge demands evidences; (2) tentativeness of scientific knowledge; (3) laws and theories; (4) scientific methods (no universal scientific method); (5) subjective in that science is a human endeavor and investigations are conducted within the context of particular theoretical frameworks; (6) partly the product of inference, imagination, and creativity; (7) socially and culturally embedded (can be influenced by contextual factors outside of the scientific community); and (8) developed from a combination of observation and inferences. The paper will discuss students' views of NOS that were showed in explicit NOS activities of the STS contact lens "big eyes unit. This paper has implications for enhancing students' understanding of NOS in Thailand

1. Introduction

Present science education around the world have made significant contributions to the understanding of the nature of science. And present in the core curriculum of basic education in 2008 AD, the nature of science and technology learning at 8 by the standards of learning that Students will be able to use the scientific and spiritual science in the quest for knowledge to solve problems as a natural phenomenon that occurs mainly in the form that it can describe and verify the information and tools that are available in during that time. Understand that science, technology, society and environment. With their relationships. The Institute for the Promotion of Teaching Science and Technology [8] Science and Technology has set standards that teachers must first understand the nature of science and technology. The process of inquiry and problem solving. They can apply their knowledge to create learning experiences that make subject matter meaningful to students. Therefore, many countries have focused on identifying the nature of science in education. Due to the nature of science, which is a key component of learning science. It allows individuals to seek self-knowledge. Learning throughout life. Can think of. Think creatively. Solve problems in a systematic manner. And led to the creation of new knowledge that is grounded in social development. Especially in the economic competition between the countries. Manufacturers must have the ability in science. Employees must be skilled. Have a good understanding and ability in science that meet the needs of the country. (Yuenyong & Narjaikaew, 2009) To understand the nature of the science of the people, everyone is important. And to increase the competitiveness of the global economy.

Researchers have commented that the teaching of science to promote an understanding of the nature of the science curriculum is a curriculum based on science, technology and society (STS). The teaching of science by Science. Technology and society. Is the integration of teaching science and related science context. Technology and society. Promote the development of scientific knowledge. Due to the nature of science is developed through a process of social processes and the findings of scientists and scientific knowledge is public knowledge, reliable and have references. But the quest to develop the scientific knowledge shows that science has changed. Of scientific knowledge is reliable knowledge over time. This is not a real-time. But that has changed scientists. Examine and develop additional knowledge and credibility. The phenomenon can be explained with the changes [12]. The concept of learning by using technology and social sciences to the curriculum that emphasizes real-world experience for students [4]. Focusing on problems and issues in the main. Encourage students to apply scientific knowledge and technological applications in accordance with their context. The initial implementation issues related to science and technology, the actual. Instead of teaching the concepts and processes. Encourage students to think critically and apply the concepts in real situations. Students can link classroom learning with real-life situations. The local context of the learner. (Wilson & Livingston, 1996, referred to in [7]) The teaching of science concepts. Technology and society. This approach encourages students to use inquiry process. Using science and technology issues related to the actual context of the society. A catalyst for students interested in learning content. And encourage students to share learning activities and learning about the inquiry process. e process of learning by discovery of knowledge related to social issues. The understanding of the nature of scientific knowledge through the process of the quest to understand the nature of the social world, which makes the students understand the nature of science as well.

Learning based on science, technology and society, the concept of Yuenyong (2006) [9], with social issues. Technological and social context. To encourage students to learn. The students have the opportunity to use their own ideas and creative concepts to explore the knowledge associated with prior knowledge. Learning theory of Constructivism. Of development. Or change existing ideas of the students. The conversion concept is to create and accept new ideas. Or the structure of the original idea that is new. Students to generate ideas, rather than absorb new ideas. To fill the empty brains of students to the full. Or not the acquisition of the new ideas of the students. But who is the creator of your own knowledge [6]. Gives students the knowledge and skill quest information. Students can study the self. The process can be applied in everyday life. Learning based on science, technology and society, the concept of Yuenyong (2006) is the one that will help the students understand the nature of science. The learning efficiency. I can bring that knowledge to solving problems is to live appropriately in line with economic conditions. Society and culture.

For learning themes. In physics. The Grade 11. For learning about contact lens "big eyes" unit in physics. The Grade 5. The purpose of learning so that students understand and can explain the refractive lens of the eye and some vision. The contact lens is critical to human life. To be utilized in the visible. Contact lenses is associated with the daily lives of people from past to present. The use of contact lenses is what is happening in society. The science and technology and other facilities. And meet human needs. But teaching is the knowledge of the refractive lens of the eye and some vision. In human terms that do not involve the determination of the latent image object distance and focus distance learning students do not understand why they need to learn about some of the refractive lens of the eye and vision. In addition to learning to solve problems that can score. Score, which is due to a lack of awareness of the relationship between science and technology influence each other. Teachers and students were not aware of the importance of learning the nature of science. Students lack an understanding of the nature of science. Learning and teaching in the refractive lens of the eye and some vision. The goal should be to help the students understand about refractive lens of the eye and some vision. Associated with the daily lives of students, Promote and develop an understanding of the nature of the science of learning. The students in Science, Technology and Society. This is another way to help students enhance their knowledge on their own. Presented their ideas properly. The knowledge used in the appropriate context. And thinking skills. Classified as critical thinking, reasoning with a difference. The conclusions and decisions. The relationship between science and recognizing. Technology and society. Environment in terms of influence and impact each other. Science is a spiritual virtue. Good ethics [10]. Understanding of all aspects of Scientific Literacy, This includes understanding the nature of science and Science Knowledge and Habits of Mind. They can bring knowledge to solve the problem. Application of scientific knowledge about the social, technological, economic and cultural environment. The aim of science education in Thailand. Lead to the development of national social and economic development to expand the group next.

In the past decade, scientists have studied the nature of scientific research and understanding of teachers and students, the teachers and students understand the nature of science. In particular, the lack of scientific understanding of the nature of science teachers, science and the resulting impact on student learning is the students' understanding of the nature of scientific error. Teaching science is often little connection with reality. And they are of the view that scientists must strictly adhere to the scientific approach to the knowledge of the truth does not change. Knowledge and understanding that the theory is not proven and that the rule is exact and accurate. Many students do not see the connection between what they learn in science class and what students know about the real world. (Bell, n.d.) For a study on the nature of science is the study and understanding of the nature of science teachers only. A study to develop in students the nature of science. In order to promote understanding of the nature of scientific knowledge that can allow people to develop an understanding of the nature of science to students. According to the study of the core curriculum of basic education, BE 2008. The scientific study of the proposed solutions. And the nature of scientific knowledge in a variety of methods such as case studies. Role-plays. The inquiry. Field of the debate, etc. These are the basic principles of the technique. "The contextual and pointed out" the nature of scientific knowledge should not be separated. It should be integrated with scientific knowledge, in different and emphasize with the students the opportunity to comment on the nature of science. (McComas, 2004 referred to in [5] The nature of science learning that, Learning the nature of science explicit. The learning activities and experiences with respect to the nature of science and allows learners to understand. Discussion by asking about the nature or the nature of science is clearly in the lesson. Therefore, the nature of science teaching and learning about science, to develop a better understanding of the nature and the scientific method. And the relationship between science and society. Should be learning by pointing out the importance of the issues about the nature of science and learning that emphasizes explicit [2].

This study is intended to be Grade 11 students' view of nature of science (NOS) in the STS contact lens "big eyes" unit and explicit NOS. To guide the development of teaching natural science, especially physics and the next.

2. Methodology

This research was conducted as a Mixed Method for the Interpretive Paradigm for the purpose Grade 11 students' view of nature of science (NOS) in the STS contact lens "big eyes" unit and explicit NOS.

2.1. Participants

Participants included 15 Grade 11 students who study in Kaewsadet school, Kalasin, 2nd semester, academic year of 2011.

2.2. Methods

Grade 11 students' view of nature of science (NOS) in the STS contact lens "big eyes" unit and explicit NOS. Students will experience learning about the nature of science that allows learners to understand the nature of science by asking questions and discussions about the nature or presentations

IOP Conf. Series: Journal of Physics: Conf. Series **1340** (2019) 012066 doi:10.1088/1742-6596/1340/1/012066

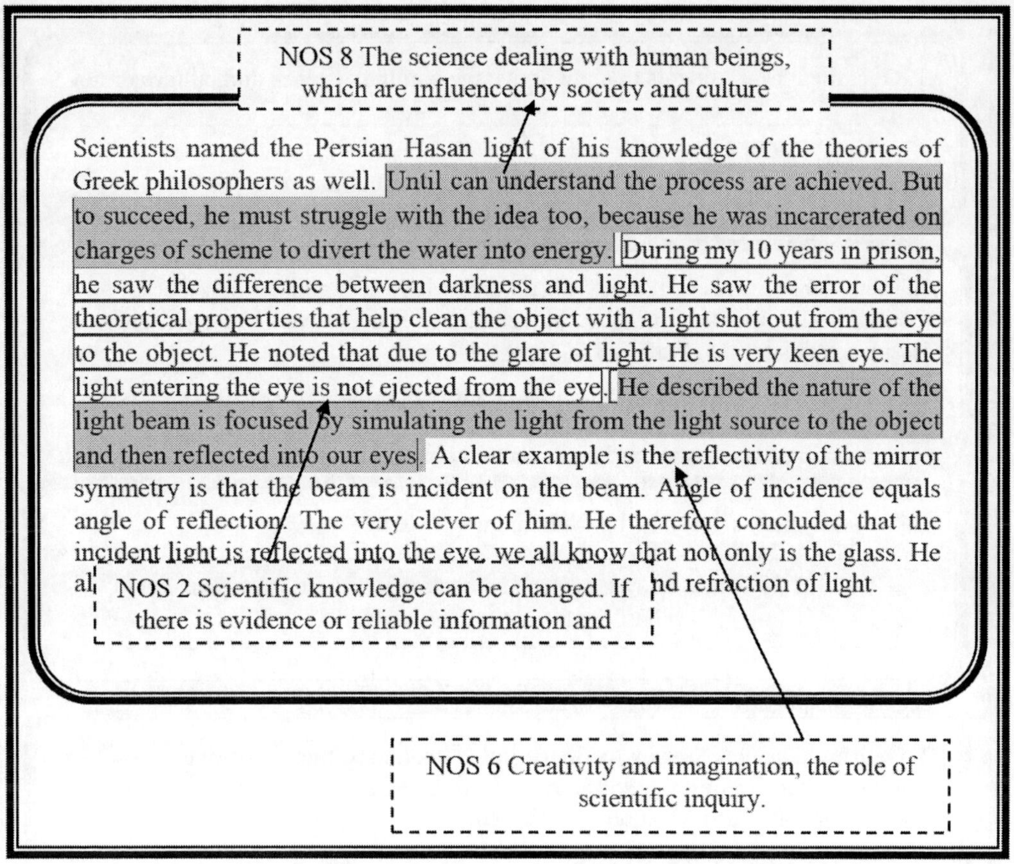

Figure 2a. indicates the nature of scientific activity, the 2, 3, 6 and 8.

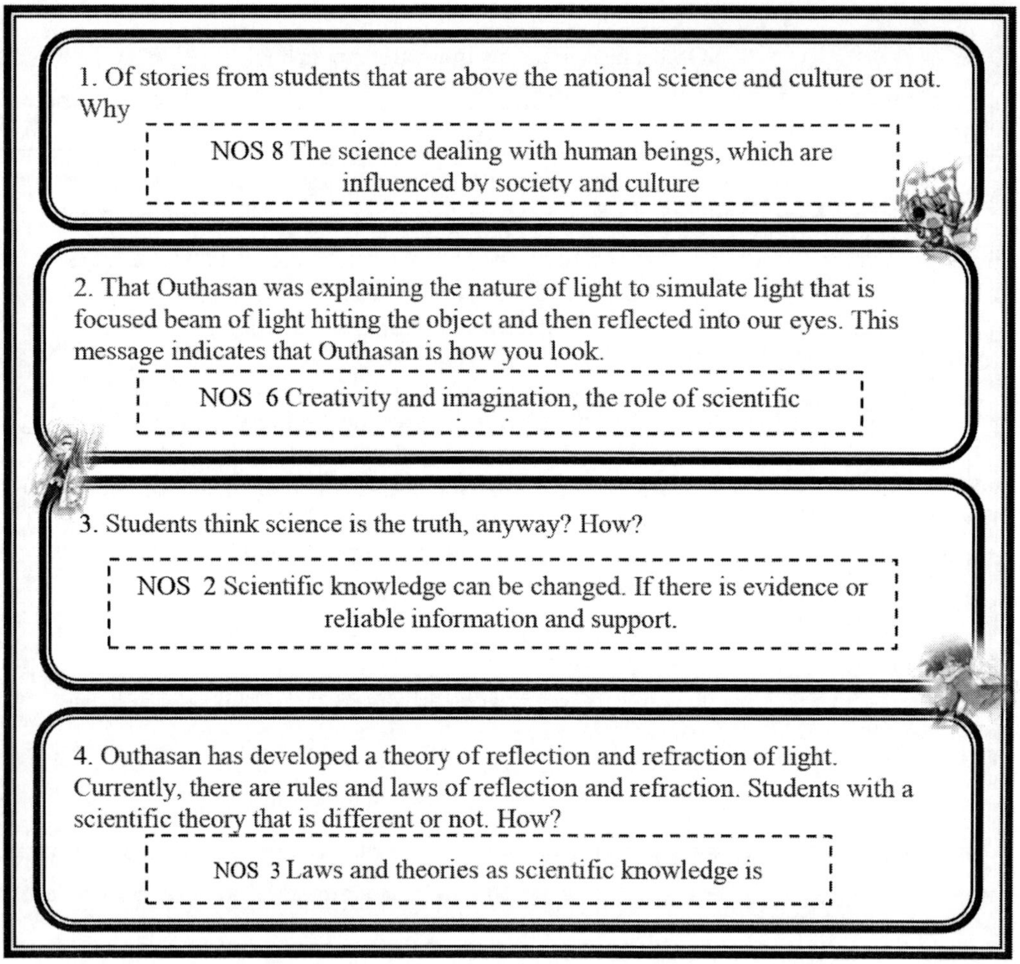

Figure 2b. indicates the nature of scientific activity, the 2, 3, 6 and 8.

In addition, after teaching a lesson plan, students must reflect a student's notes about the incident to express their own feelings and experiences. Or the normal operation of the daily diary The students analyze the text of the interpretive understanding of the nature of scientific theory as a framework for the analysis of students' understanding of the nature of science.

After learning of the concept of science technology and society Yuenyong (2006) and indicate the nature of science, the Contact Lens "Big Eyes" Unit. Focus Group Interviews Academy students will be working together. To better understand the issues that students are not as detailed. Information is not clear. The conclusion is not. After removing the tape recordings and images from a video camera during the course. The text of the Journal and to the interpretation of the data analysis to understand the nature of science as a theoretical framework to analyze and understand the nature of science students.

3. Findings and discussion

The research this time is Grade 11 students' view of nature of science (NOS) in the STS contact lens "big eyes" unit and explicit NOS, 8 issues. On the Nature of Science 1 (NOS 1) The scientific study of natural phenomena. This requires evidence through critical thinking and rationality. On the Nature of Science 2 (NOS 2) scientific knowledge is subject to change. If there is evidence or reliable information and support. On the Nature of Science 3 (NOS 3) laws and theories as scientific knowledge is different. On the Nature of Science 4 (NOS 4) for a variety of methods of scientific inquiry. On the Nature of Science 5 (NOS 5) the science by inquiry, observation and inference are different. Observations provide evidence for the inference. Scientific knowledge and the inference from the evidence obtained by observation. On the Nature of Science 6 (NOS 6) creativity and imagination, the role of scientific inquiry. On the Nature of Science 7 (NOS 7) Science is a human activity that is directed or guided by theory, practice and experience the emotions of people. And the nature of science, the 8 (NOS 8) The science dealing with human beings, which are influenced by society and culture. Based on an analysis of the nature of science students. Analysis of the data can be displayed as shown in Table 1.

Table 1. shows the frequency and percentage of students' understanding of the nature of scientific knowledge in the contact lens "big eyes"

STS approach	nature of science	The behavior of the students' understanding of NOS.				
		Frequency			And frequency	Percent
		The activity	Diary	Observation		
Identification of Social Issues Stage	NOS 1	8	-	-	8	23.53
	NOS 8	15	11	-	26	76.47
Identification of Potential Solutions Stage	NOS 1	7	-	-	7	6.14
	NOS 4	15	1	15	31	27.19
	NOS 6	15	1	15	31	27.19
	NOS 7	4	-	15	19	16.67
	NOS 8	15	11	-	26	22.81
Need for Knowledge Stage	NOS 1	45	2	18	65	33.85
	NOS 2	13	1	-	14	7.29
	NOS 3	15	-	-	14	7.29
	NOS 5	15	-	-	15	7.81
	NOS6	15	2	3	20	10.42
	NOS7	15	-	4	19	9.90
	NOS 8	30	12	3	45	23.44
Decision-Making Stage	NOS 6	15	-	-	15	16.67
	NOS 7	15	-	-	15	16.67
	NOS 8	60	-	-	60	66.67
Socialization Stage	NOS 8	15	-	-	15	100.00

Table 1 shows that learning about themes. The teaching model based on science. Technology and Society (STS) and indicate the nature of science. To promote understanding of the nature of science students in the eight issues, which they have expressed opinions on the nature of science as well as 8 points to the following description.

Students commented on NOS 1 that scientific research is looking at reliability. Because every time they try to take notes to use as a reference a proof. The students' answers to the following.

"All the scientific knowledge acquired in the experiment, the experimental device. The experiment. The results of the experiment. And the time is recorded as a reference" (S8)

"Why is the scientific study of the experimental observation that the test to find out" (S1)

Students commented on NOS 2 is that the scientific literature on this subject to change at any time. Some other scientists are discovering new knowledge to prior knowledge or refute the inquiry was to

IOP Conf. Series: Journal of Physics: Conf. Series 1340 (2019) 012066 doi:10.1088/1742-6596/1340/1/012066

modify the original knowledge. If there is evidence or reliable information and support. The students' answers to the following.

"The science has changed. They are subject to change at all. Some other scientists have discovered that the new knowledge or new knowledge or the knowledge that has ever been discovered. I could not find any that may be more realistic "(S3)

"The science has changed. Because scientific knowledge is subject to change if new evidence or reliable information to support "(S9)

Students commented on NOS 3 that the rules are more reliable than theory. The theory can be changed rather than the rule. But they still can not explain what happened to the laws and theories. Related or not. And equally important, or not. The analysis of students' answers to the quoted text of statement. Substantive understanding of the nature of science. The orientation of the student teachers. In preparation for data collection, the "laws and theories are different. This rule describes the phenomenon that occurs as a formal condition for any theory or explanation of the reasons for the phenomenon that occurs with conventional "(S2)

Students commented on NOS 4 into a scientific inquiry should investigate a variety of methods. In order to get reliable information on various surveys, such as searching for observation and experimentation. This is the quest that will take any appropriate way according to each one. The students' answers to the following.

"Everyone has a different opinion to help make it more intimate. And thus we see the intent of the design specs of the invention "(S14)

"There are many ways to find answers. You can find answers to many of the experiments and observations "(S7)

Students commented on the observation that NOS 5 and the sequence or to a different conclusion. This observation is to monitor and record what happens. The data obtained from observation into a conclusion that is different so that the observations and conclusions are related to each other. The students' answers to the following.

"The observations and conclusions differ. It is a way of looking at the object it is. The conclusion section summarizes the results we obtained from the various trials "(S2)

"The observations and conclusions differ. The observation Is to look at things around us as a concept. The conclusion is to collect all the information to be conclusive "(S9)

Students commented on the NOS 6 that creativity and imagination are essential to the design work of students. Similarly, scientists at the creativity and imagination are essential to the design method. The students' answers to the following.

"Scientists must design the experiment. The need for creativity. Because it may lead to new knowledge of how many scientists, since each person may experience the same "(S8)

"Outhasan is a wise observation, thinking I do not have the imagination" (S3)

Students commented on the inquiry as to NOS 7. Design work. And the interpretation of experimental results with a variety of ways. Due to the interest of each individual is different. And when the scientists who have studied in the same conclusions. Conclusions from the research may be different. The scientists have their own ideas. There are different methods. Are different, it is subjective in that the scientists have to work in science. The students' answers to the following.

"The water is predicted to be light because of a prophecy in mind, each with a different answer is to get out" (S6)

"Prediction of mechanical science. Water and light are the only people that it is not the same "(S9)

Students commented on the NOS 8 that science and technology are related by a relationship as a link. The technology must have the knowledge to know the scientific and social values. To develop technologies that benefit the people. Do not make trouble for the society. Because scientific knowledge is not power that can do just about anything. Not above the society and culture. Have not taken advantage of society and humanity. Scientific knowledge and scientific inventions. Scientists have researched it a society. The students' answers to the following.

IOP Conf. Series: Journal of Physics: Conf. Series **1340** (2019) 012066 doi:10.1088/1742-6596/1340/1/012066

"Science is a part of society and culture. The science and technology. Facilities. The penalty is both useful and interesting. The people are so used various technologies to facilitate their own. But some of them may be used in the production of scientific technology in the wrong way. So science is a part of society and social values "(S3).

"Science does not transcend national and cultural. The story of how scientists make discoveries. I need to get from the national. Religion and culture before. If not, they will be punished by a prison "(S9)

So from the above analysis to understand the nature of science. Students can express their views about the nature of science appropriately. The activities of Learning contact lens "big eyes" By Science Technology and Society (STS) and indicate the nature of science. To promote development of natural science to the students as well.

4. Conclusion

Form grade 11 students' view of nature of science (NOS) in the STS contact lens "big eyes" unit and explicit NOS. The results showed that Students understand the principles and concepts underlying scientific evidence for a reasonable principle. And scientific knowledge can change as a reliable supporting evidence. But understanding the nature of scientific laws and theories of the students are not able to explain the importance of the opinion that the rule, but they are more reliable than theory. The theory is likely to change than laws. For scientific inquiry, students can propose a variety of research knowledge. The scientific quest to understand a variety of ways, such as the search for treatment or observation. The data from the observed data to infer the evidence. The scientific experiments. Scientists have designed an experiment with their creativity and imagination in the design plan. The creativity and imagination are vital to scientific inquiry. The students also have an understanding of the learning experience of science, belief, or social and cultural influence on science. The learning experience ideas or beliefs of scientists. As a result, the conclusions of scientific knowledge is subjective. The social and cultural. Students understand that technology is designed to use scientific knowledge and social needs into the design. These technologies can be developed which scientific knowledge has progressed further. However, science or technology. Must be accepted by society to be reliable and do not affect the hardship to society. Thus, society and culture is essential to the development of knowledge in science and technology.

The curriculum is grade 11 students 'view of nature of science (NOS) in the STS contact lens "big eyes" unit and explicit NOS. Can promote development of students' understanding of the nature of science as well.

5. Acknowledgement

Thank you, the Institute for the Promotion of Teaching Science and Technology (IPST.) that provides scholarships to graduate students in the country and to fund the study. And also thank you for Khon Kaen University Graduate School for the financial supporting of the study.

References

[1] Attapan N 2012 *The Study of Grade 11 Students' Understanding in Learning about Lights through Science Technology and Society (STS) Approach and Explicit NOS* (Master of Education Thesis in Science Education, Graduate School, Khon Kaen University)

[2] Kaewnan C 2009 *Effects of Science Teachers' Understanding and Learning Management of the Nature of Science on Understanding of the Nature of Science of Mathayomsuksa One Students in Basic Educational Extending Opportunity School* (Master of Education Thesis in Science Education, Graduate School, Prince of Songkla University)

[3] Meesri S 2007 *"A Development of Professional Development Program for Enhancing Teachers' Understanding of the Nature of Science and Its Implementation in the Classroom"* (Doctor of Education Thesis in Science Education, Srinakharinwirot University)

[4] National Science Teacher Association [NSTA] 1993 *Science/Technology/Society As Reform in*

IOP Conf. Series: Journal of Physics: Conf. Series **1340** (2019) 012066 doi:10.1088/1742-6596/1340/1/012066

Science Education. (New York: State University of New York Press) pp 3-13

[5] Pongsophon P 2009 *Science teaching* As science (As a science vol 63) chapter 1 pp 84-89

[6] Randy L Bell PH D (n d) *Teaching the Nature of Science* Three Critical Questions (Best Practices in Science Education)

[7] Sonsanam O 2010 *Enhancing Grade 12 students' Everyday Life Problem Solving Ability and Learning Achievement on Human and Sustainable Environment through Science Technology and Society (STS) Theme* (Master of Education Thesis in Science Education, Graduate School, Khon Kaen University)

[8] The Institute for the Promotion of Teaching Science and Technology [PIST] 2002 *Science and technology standards* (Bangkok: Printing krurusapa Ladprao)

[9] Yuenyong C 2006 *Teaching and Learning about Energy : Using STS approach* (Doctor of Education Thesis in Science Education, Graduate School, Kasetsart University)

[10] Yuenyong C 2007 *Since the concept of human life in science, technology and social learning in science can find* (Journal of Basic Education Commission. Ministry of Education vol 10) chapter 2 pp 29-34

[11] Yuenyong C & Narjaikaew P 2009 *Science Literacy and Thailand Science Education* (International Journal of Environmental and Science Education, July) pp 335-349

[12] Yuenyong C & Sumranwanich W 2009 *Teaching of Science, Technology and Society* Khon Kaen : Department of Science Education (Faculty of Education Khon Kaen University)

Developing Competences of English Teachers in Secondary STEM Education Schools Track in Thailand

Nujaree Sukkasame[1*], Watcharaporn Buanil[1], Prapaporn Udomsiang[1], and Metta Marwiang[1]

Department of English, Faculty of Education, Khon Kaen University, Thailand

* Corresponding auther's email address: nujasu@kku.ac.th

Abstract. The purposes of this study are to investigate the competences of English teachers in Thailand and propose the teaching competence enhancement approach to meet the needs of 21st century English teachers for STEM track school. The documentary study was used as methodology in this study. The findings of this study demonstrated that the activity-based teaching allows students to practice communication skills effectively. Furthermore, English teacher's competences to develop students' English learning can be implemented by using five-step communicative teaching approach, including warm up, presentation, practice, production, and wrap up.

1. Introduction

Due to the rapid change of today's globalization, English has become increasingly essential for communication. English is one of the most core subjects that students must acquire for greater social understanding. The language is also massively used in technology-based scheme, particularly in the direction that country is moving towards ASEAN Economic Community [17]. Therefore, incompetence of language is a huge barrier of country's development, especially in terms of human resources development. A key success to students' English competence development is to equip them with the meaningful knowledge of native speaker' language and culture, so that they are able to use English to communicate fluently, accurately and confidently. Thus, students should be provided with opportunities to practise their language skills in and out of the classroom [2, 8].

In Thailand, the Ministry of Education has recently placed emphasis on the English teaching and learning management to meet the challenges of the 21st century. Accordingly, the students are highly anticipated to gain experience and develop English competence in line with the social context [1]. Hence, communicative language teaching (CLT) has increased the importance in educational institutes. In addition, students are encouraged to become active learner, instead of being passive learner in the traditional English teaching approach. Students should be provided with opportunities to demonstrate their English skills with confidence and apply them in students' everyday lives [6, 21].

However, the traditional English teaching approach secondary schools remains in Thailand. Teacher-centered approach in most English classes focusing on reading and writing; as a result, students mainly learn from the textbooks with passive classroom learning environment. The interactive English teaching and learning for communication are limitedly implemented in classroom. Furthermore, it is clearly seen that exam-oriented education in Thailand aims to boost students' college readiness, instead of teaching to enhance students' skill and experience. The report showed that 90 percent of Thai students learn to memorize vocabularies, sentences structures by heart, regardless communication skills, resulting in the students' poor communicative English learning outcomes [3].

Due to the growing needs of 21st century knowledge and skills in terms of communication and technology literacy, the Ministry of Education has highlighted the importance of English teaching and learning as a key subject to access other knowledge to pace up with the global competitiveness. Recently, the leading schools in Thailand have conducted a European educational framework, called the Common European Framework of Reference for Languages (CEFR) the English teaching and learning. The framework is designed to improve the students' English ability to reach the level of B1, or Independent User, meaning that the students are able to communicate, write, and comprehend the general contexts. The framework also guides English teachers to implement their teaching in accordance with the framework's objectives.

On the other hand, English teaching and learning in the secondary and undergraduate level fails to meet as learners cannot communicate effectively. One of the unsuccessful factors that is highly claimed is the poor English teaching approach. In the previous study [4], it was noted that the four widely seen hindrances in English teaching were that English teachers have limited understanding of the alternative teaching method that focuses on practicing English for communication, rather than memorizing sentences structures. Teachers are lack of confidence to teach English speaking and listening skills since they are unfamiliar with the new teaching approach. As a consequence, teachers mainly conduct the traditional teaching approach that learning refers to memorizing repeatedly. This causes the students' lack of opportunities to develop the communicative skills.

The second failure in English teaching and learning is that teachers are concerned about time limitation. Teachers perceive that learning activities require much time that they may not be able to complete teaching session which is massively based on textbooks. The third hindrance in English teaching and learning is that incompetence of teachers to design the creative and innovative teaching materials that would inspire learners' motivation. Another obstacle that causing the ineffective English teaching and learning is that teachers have misunderstanding of the evaluation and assessment of the students' learning outcomes.

The results of the study are consistent with the report of Department of Academic Affairs [14] stating that challenges in English teaching and learning in Thailand have long rooted in teachers' teaching quality. The report concluded that teachers are incompetent to teach English for communication purposes from the evidence that they mainly teach reading, writing, and word-by-word translation.

The issues have existed in schools for decades and never been adequately resolved although the continuous teachers' professional development in line with the country's educational development has been implemented over time. The author, as an English teacher with 20 year teaching experiences, believe that it is important to take these issues into action seriously. The problems found in English teachers' training-based professional development are summarized as follows: 1) inadequacy of training, 2) inability to meet teachers' requirements, 3) theory-based training, and 4) lack of follow-up evaluation and practical implications as further development guidance for teachers.

Moreover, it is indicated that the teachers in Thailand should have bilingual competences; Thai as the first mother-tongue language and another as the second language [18]. Particularly, 21st century teachers must have good command in English. The training programs offered for professional development should be on-the-job-training based on practice.

Therefore, this study aims to investigate the competences of English teachers in Thailand and propose guidance for further English teachers' competences development.

2. The theoretical frameworks of English teachers' teaching competences

Previous studies investigated the teachers' competence and suggested several development methods as presented below.

Teacher competence is an observable behavior. Proficient teachers must be able to transfer knowledge and urge students to demonstrate what they have learned in class. In other words, competent teachers must lead students to reflect learning behavior in consistency with the teachers' teaching objectives [11].

The competence of teacher refers to "the ability to address the problems." What extent teachers are able to solve problems in particular is considered as competence. Thus, teachers with high competence are those who are successful to eliminate problems in various areas [7].

The competency based education emphasizes on the following five key qualities of teachers that have an influence on students' learning outcomes [24].

1. Teachers have strong competency, knowledge, skills, and attitudes
2. Teachers utilize the potentials wisely.
3. Teachers assess and evaluate students accurately.
4. Students must be motivated to be greater and progressive, and
5. Teacher must improve the content continuously.

From the related studies, it can be concluded that teachers' competencies refers to the ability or behavior of the teacher that reflects the knowledge, understanding, skills, attitudes, and attributes which lead teachers to perform teaching effectively. The teachers' strong competencies have a positive learning outcomes of students in the way that they can sustain their physical, mental, emotional and social well-being development. As a consequence, students can apply taught knowledge to their daily lives.

The teachers' competencies development can be implemented by using guidelines for activities-based teaching and learning as presented below.

2.1. Guidelines for activities-based teaching and learning

1. Develop curriculum at classroom level by designing teaching structure in sequential order, starting from easy-to-difficult content, or from the natural understanding to theories.
2. Transfer knowledge and contents that must be applied in students' everyday lives.
3. Provide students with exciting learning activities to learn and to meet students' needs
4. Design teaching and learning activities that promote critical-thinking and problem-solving processes
5. Direct students to do good deeds and sustain life-long learning by building active classroom environment, using psychological tests.
6. Evaluate the student's overall development, rather than the result of academic tests.
7. Develop curriculum at local level by designing compulsory and elective curriculum to allow students to select the subjects that interest them.

According to the guidelines for teaching and learning activities, certain techniques are suggested to be used for activity-based teaching effectiveness as further explained.

2.2. The activity-based teaching techniques

Teaching technique refers to the specific methods chosen by the instructor to be implemented in the classroom activities. The teaching technique must be developed in accordance with the content, the objectives of learning, and the ability of learners.

1. The teaching technique used in the group activities should be designed for more than two students so that they interact with each other. The common activities to motivate activities include games, role-play conversation, simulation, short play, case study, group discussion, brainstorming, field trip, etc.
2. Teaching techniques should be consistent with the collaborative learning that creates active classroom learning environment. Students collaboratively brainstorm in small groups by sharing ideas through activities e.g. storytelling, group discussion, etc.
3. Teaching technique must emphasize students' creativity and drive students to seek more information. Certain activities can arouse students' critical thinking process including metaphors, interpretations, criticism, etc.

Teaching English at the elementary level should be in accordance with the learners' nature of understanding. That is, elementary school students in grade 1-6, aged between 6-12 years should be provided with direct experience by using self-learning activities since the common personality of young-aged children is curiosity to learn surrounding environment. The activities and lesson should be structured in an easy-to-difficult order. Moreover, the activities should offer opportunities to learners to express themselves; for example, games, video presentation, using visual aids e.g. cartoons along with teaching. The technique is to divide students, mingling between female and male students, with high, moderate, and low GPA, into small groups or learning center groups. The teacher act as facilitator encouraging students to participate in group activities or learning centers.

Activities-based teaching and learning methods and techniques should focus on providing learners opportunities to develop communicative skills; listening speaking, reading, and writing so that they can practice and eventually communicate in English effectively. Therefore, it is important to understand the techniques and methods selected in teaching and learning to prompt students with communicative skills in a particular context.

2.3. Competencies in teaching and learning management

Teaching activities is the most important element of teaching in this approach. Thus, the instructional activities means 'all activities performed in class must be time effective and bring about students' learning process" with the collaboration between the instructor and learners. The students' experience of self-learning from activities can be developed to become a life-long learning behavior [20, 23].

It is suggested that teaching activities are key element of teaching when they are aligned with the content of the lesson effectively, the students can learn from both theories and practice [5]. Teaching activities should meet the following requirements.

1. Teaching activities are consistent with the purposes of curriculum that is to develop learners' problem-solving and analytical thinking skills.
2. Teaching activities are consistent with the teaching objectives that focus on enhancing certain learning behaviors in particular.
3. Teacher must organize activities in accordance with learners' age and their ability to learn.
4. Teacher must design activities in accordance with the nature of the subject; for example, types of facts, types of thinking, and level of creativity.
5. Teacher must manage activities in sequence to allow the students to develop their learning so that they are not confused and seek the linkage to the content of the study.
6. Teacher must create interesting activities by using the appropriate teaching materials.

7. Teacher must provide students with opportunity to do activities in order to achieve self-directed learning. Teacher acts a facilitator, encouraging students to acquire knowledge by themselves.

8. Teacher must ensure that activities used in class positively challenge students' ways of thinking development so that they learn to seek more information and to solve problems.

9. Teacher must perform variety of activities to suit the current situations, so that they are more related to students' every lives and increase their attention.

10. Teacher must control the class with positive characteristics (e.g. fun and friendly), making the students feel comfortable. Teacher's emotions while teaching also have an influence on students' attitude towards the subject. Therefore, the effectiveness of the classroom learning environment depends on the personality of the instructor.

11. he activities must be assessed to determine the strengths and weakness for future improvement. The assessment should be during and post-activities implementation by using observations, questionnaire, examinations, or tests.

From the suggestions in the previous studies, it can be concluded that teaching activities refer to teaching element that direct students to develop learning accurately and effectively. Moreover, the teaching activities must meet the objectives of curriculum and suit for the age of students to develop learning appropriately.

This statement is consistent with the study of Millman [12] in ways that teaching behavior in the classroom plays a vital role. When the teacher pays attention to the students' learning and controls his or her teaching behavior accordingly, learning goals are likely to be achieved. Despite various factors have an influence on students' learning outcomes, teacher's teaching attitude and behavior remain important in teaching.

However, apart from the content of subject, teachers are expected to have other basic knowledge. To combine the holistic knowledge, English teachers need to have strong understanding of English linguistics and teaching English [15]. The following four topics are critical for teaching English.

1. Nature of language
2. Psychology of Learning
3. Uniqueness of English
4. Teaching foreign language

Hence, whether the English teacher is a native speaker or not, it is important to recognize the difference between formal English (e.g. used in the group discussion and presentation) and informal English (e.g. used in conversations for everyday lives at work, home, schools, shops, and among friends). The listening and speaking skills must be highlighted as means of communication.

From the scholars' theoretical work, it can be summarized that the qualities of competent English teacher are the abilities to lead students to accomplish the learning goals by acting as facilitator.

3. Concepts of English teachers' competencies development in teaching and learning for communication

English is key command required in all today's careers because it is a universal language used to communicate among people worldwide. In terms of education, it is such a communicative tool that makes it possible to reach out wider knowledge that is available endlessly throughout the world. Due to the benefits and importance of English, Thailand's government has placed an increasing effort in English teaching and learning methods focusing on communication purposes. The objectives of English teaching and learning using communicative approach are to prompt learners to utilize English knowledge and skills for the effective communication in their future career [22].

According to previous studies, the widely seen competencies required for English teachers addressing on communication are listed as below.

3.1. Teaching English as a second language

The purpose of today's teaching focuses on the development of communicative competence that means the appropriateness and acceptance of language use in the social situations. Teaching English for communication is a teaching approach that includes theories of communication, structure of language, situations, and human psychology. In addition, Finocchiaro [9] proposes the following guidelines for teaching and learning communication.

1. Teacher must plan to provide students with experience carefully.
2. Teacher must design the simulation of situations and teaching materials to help students understand the meaning and present them clearly to students.
3. Each lesson should include language training with communicative activities to allow students to practise.
4. The existing language of students should be monitored. Teacher should suggest higher level of language use. However, when teachers explain important information, the easy-to-understand language should be used.
5. Teacher must motivate students to practice together as a group, or even as an individual.
6. Teacher must have the expertise in teaching that allows students to practice in pairs.
7. Teacher must ensure that time is adequately offered to students to practice communicative conversations as well as articles. This helps students understand the content of the lesson and develop their communication skills.
8. Teacher must be flexible to prepare the manuscript for audio recording and communicative activities that can be practiced in the classroom appropriately.
9. Teacher must analyze how the students develop their reading and writing an early stage.
10. Teacher must lesson plans to develop students' reading to help them increase both knowledge and experience.
11. Teacher must design learning activities that increase the creativity and ability to write freely.
12. Teacher must encourage students to understand the culture of native speakers from the teachers' designed activities e.g. formal versus informal dialogues.
13. Teacher must shape students to be proud of their country and themselves despite the differences of native speakers' culture.
14. Teacher must select audio and visual aids to help students develop their language learning and understand native speakers' culture effectively.
15. Teacher must conduct student's learning outcomes assessment and evaluation to ensure that the student's learning achievement as well as teachers' teaching effectiveness are satisfactory.

From the concepts of English teaching and learning for communication. It can be concluded that teaching should emphasize the opportunities of students to practice the use of language in order to develop communication skills. The teachers act as facilitator organizing the activities for students to experience both simulation of situations (e.g. role-play conversation) and actual situations outside classroom. The objective of teaching English for communication is to allow the students to apply knowledge in their everyday lives effectively. Teaching should not focus massively on structure, or widely called 'grammar'. Thus, the communicative teaching approach based on activities can enhance the ability of students to use language to communicate properly with greater confidence.

3.2. Teaching procedures

In the communicative teaching approach, 5-step teaching procedure is recommended [19].

1. Warm up: The first step is to provide activities to prompt students prior learning any new content. It can be a review of the previous lessons, or the importance of the new content. The suggested activities used in this stage can be music, games, tales, dialogues or performances.
2. Presentation: The second step is a language input to the students. The teachers' language must be used commonly in everyday life and in the classroom. Teachers present information while the students observe teachers' speaking. This step is to develop students' listening skills.

3. Practice: This step focuses on students' practice of language. This step is to demonstrate how students should perform in the actual situations. Teachers play a role as a conductor, allowing students to practice and participate in the learning activities.

4. Production: This step is to allow students to apply taught knowledge of language for communication like actual situations in their every lives. The new situations can be similar to the situations that the teachers have presented. The production step also encourages learners to utilize the language they have learned in the previous lessons. The teachers plays a role as an advisor to guide and assist learners in applying the language. Students are responsible for production of language use.

5. Wrap up: The final step is an activity to provide students with a summary of the lessons that can be music, games, exercises, or assignments.

In order to achieve in teaching and learning activities management, the teachers must understand and realize the teachers' role and students' role in each step. The teaching procedures help teachers design teaching activities more effectively [16].

4. Evaluation of the communicative teaching approach

According to the Ministry of Education [13], the assessment of teaching foreign language for general use can be conducted by using the questionnaire to measure and evaluate students' ability. The language proficiency test can be divided into two types as follows.

1. Discrete Point Tests: are used to measure students' specific knowledge, including phonetics, spelling, grammatical structure, and vocabulary.

2. Integrative Skills Tests: are used to measure students' overall skills, including the fluency in general language [10]. Students' communication skills tests are generally measured by each skill at a time; listening, speaking, reading and writing. By conducting the test, teachers should consider the following items:

- The ability to communicate, unlike typical language knowledge should be tested from listening and speaking competence.
- There should be a clue (e.g. keyword) in the text that students have experienced in lessons.
- Sentences or passage should be included in the reading tests, so that students can gain knowledge from reading.

5. Conclusion

The today's concept of teaching English to Thai students should focus on promoting learning by using the communicative teaching approach. The ultimate goal of teaching English is to develop learners to apply the language in everyday lives for communication effectively. It is necessary to modify the traditional teaching methods that massively address the understanding of grammatical principles and vocabularies by memorizing. English teachers should provide instructions of language use for communication based on learning activities that encourage learners to practise the communicative language.

The communicative activities must meet the interests of the students in line with the recommendation of student-centered teaching method. Teachers' task-based instruction helps learners improve their English skills, critical-thinking skills, problem-solving skills, and collaboration skills. Also, the activity-based teaching encourages students' creativity, innovation and actively seeking more information; as a result, learners can build new knowledge upon the existing knowledge. In addition, the suggested productive learning is in line with the social needs that students should be prepared for future career in the 21st century.

From the previous studies, two main problems in competencies development of English teachers are that (1) teachers remain mainly implement traditional teaching method, and (2) teachers are lack of spirituality as teachers. Moreover, in the 21st century, teachers must have seven competencies: 1) student-centered teaching competencies, 2) assessment and evaluation of ability development with strong understanding of background knowledge differences among individuals, 3) computer and technology competencies, 4) competence in pursuing ethics and code of conducts, 5) teamwork and sharing skills, 6) cross-cultural competencies, and 7) competency in facilitating and guiding students.

Recommendations from this study will be beneficial to the school management and education-related agencies for teachers' competencies development planning to meet the needs in the 21st century.

References

[1] Batang B 2014 Communicative Competence and Language Learning Styles of Perspective Teachers of English *International Referred Research Journal Issue*

[2] Cao Y 2011 *Investigation Situational Willingness to Communities within Second Language: Learning Teaching Assessment* (Hong Kong: Cambridge University Press)

[3] Chaisaeng J 2014 Launch 6 policies for upcoming semester 2014 Retrieved 8 July 2018 from https://www.thairath.co.th/content/404166

[4] Chaitiamwong P 1998 *Teaching Techniques for Elementary Students* (Bangkok: Chulalongkorn University Printing)

[5] Chaitiang A 1997 Principles of teaching (Bangkok: OS Printing)

[6] Chuanchayasit C, Prabparn K 2009 *Strategic study of communication: English of graduate students in Thailand* (Bangkok: Language Institute Chulalongkorn University)

[7] Coker H 1976 Indemnifying and Measuring Teacher Competencies the carrel county Project *Journal of Teacher Education* **47** pp 168–171

[8] Edwards E 2015 Seeking out challenges: Developing Self-Confidence: A Language Journey to Proficiency *Journal of English as A Second Language 8* **4** pp 1–24

[9] Finocchiaro M 1982 Reflections on the Past, Present and Future *Journal of Education* 20 vol **7** pp 192–200

[10] Harris D P 1969 Testing of English as a Second Language (New York: MacGraw Hill Book Company)

[11] Lindsey M 1973 Performance-Based Teacher Education *Journal of Teacher Education* **44** pp 180–182

[12] Millman J 1987 *Handbook of Teacher Education* (Condon: Sage Publications)

[13] Ministry of Education 2016 Techniques for organizing learning activities to drive development focus: Quality of learners in foreign languages (English) (Bangkok: Chamchuri Product)

[14] Ministry of Education, Department of Academic Affairs 1998 *Summary of Current Teaching English Based on the English Curriculum 1996 for the Elementary Level* (Bangkok: the Office of Educational Research, Department of Academic Affairs)

[15] Noisri P 2017 Challenges in teaching reading in high schools Bangkok: Publishing of Ramkhamhaeng University

[16] Office of the National Primary Education Commission 1999 *Education in Thailand* 1999 (Bangkok: Kurusapa Ladprao Publishing)

[17] Onwimon S 2014 *English for becoming ASEAN community* Retrieved 8 July 2018 from http://www.dla.go.th/upload/ebook/column/2013/5/2060_5264.pdf

[18] Patanaphumi S 1999 Concepts of Quality Assurance *Journal of Teacher* **19** pp 9–16

[19] Silapatham S 2007 *Research and Development Report: The development of knowledge and skills in practice for English teachers in secondary school* (Bangkok: Good Print)

[20] Tasanee S 2010 *Elementary school curriculum and management* (Bangkok: Odean Store)

[21] Tassanakornkul M 2014 *English Ability Test* (Bangkok: the Office of Elementary School)

[22] Thongchai P 2017 The Role of Language Teachers in the Thailand Policy 4.0 Journal of Humanities 1 pp 99-106

[23] Valaiporn K 2012 Principles of teaching (Bangkok: Rajabhat Thonburi University Publishing)

[24] Vogier D E 1972 Identifying and Validation Professional Occupational Teacher Education Competencies *Journal of Competencies for Teacher* pp 2–3

Cognitive attributes, physical and psychosocial aspects of learning environment: Its relationship to learners' chemistry achievement

Maris Jade Q. Orongan[1], Edna B. Nabua[2]*, Manuel B. Barquilla[2], Amelia T. Buan[2], Ellen N. Inutan[2], and Chokchai Yuenyong[3]

[1]Central Mindanao University, Bukidnon, Philippines

[2]MSU-Iligan Institute of Technology Tibanga, Iligan City, Lanao del Norte, Philippines

[3]Khon Kaen University, Thailand

*corresponding author's e-mail: edna.nabua@g.msuiit.edu.ph

Abstract. The objective of the study was to determine the cognitive attributes, physical aspect and psychosocial aspects of students' learning environment on their science performance. A total of one thousand one hundred twenty-three Grade 10 science students in six divisions of Region X, Philippines participated in the study. A triangulation technique was applied in which data were obtained from the survey instruments, semi-structured interview, class observation and literature review to ensure the validity and accuracy of the results of the study. This study utilized descriptive correlational and causal comparative research design. The results revealed that students performance in science was of low mastery level with a very satisfactory average grade during elementary, having a satisfactory average of all the science subjects taken from Grade 7 to Grade 9 and with a description of moving towards mastery in their scientific ability of the National Career Assessment Examination (NCAE) result. In the psychosocial aspects, the participants generally practiced their science activities in both laboratory and classroom environments. They possessed moderate confidence in their emotion and self-efficacy in chemistry, having average science process skills and with very satisfactory science teachers. Cognitive attributes such as Junior high school science general weighted average, elementary general weighted average, and National Career Assessment Examination result and psychosocial aspects like classroom learning environment and teacher quality were significantly correlated with science performance of which the Junior high school science average ($\beta = 0.194$), elementary average ($\beta = 0.096$) and classroom learning environment ($\beta = 0.132$) were the predictors of science performance. The study puts forth the notion that the most fitting structural model of students' performance in science is when it is anchored on the cognitive ability of the students and supported by their physical aspect of science leaning environment. Students with high cognitive ability supported by conducive, well-equipped physical facilities in science, expectedly perform better. It is recommended that the stakeholders may anchor their learning modules and activities on the different levels of learners' cognitive abilities and needs. Likewise, they may provide public secondary schools with adequate learning resources to have an optimal learning in science to cope with the demands of 21st Century towards quality science education for all.

1. Introduction

Competencies can escalate from multiple areas of education- in natural sciences or social sciences, for instance. However, many literature sources have cited that students find areas most especially science-related subjects, and in this case, chemistry, as a difficult and irrelevant subject because of its abstract nature (Cardellini, 2013; Sirhan, 2007; Treagust, Duit and Nreswandt, 2016; Woldeamanuel and Engida, 2014). This notion results to students failing in their chemistry subjects (Arbutante, 2010).

Moreover, the globally poor performance of students in science was also reported in the International Science Study and International Assessment of Educational Achievement (Imam, Mastura, Jamil and Ishmael, 2014), and Trends in International Mathematics and Science Study (TIMSS, 1997 and 2003). These assessment studies conclude that Filipino students may still lack the scientific knowledge and skills to meet the globalization demands of the 21st Century education. For almost three decades, there is a significant growing literature of students' perception of the learning environment which was measured using Science Laboratory Environment Theory (SLEI) (Che Ahmad et al., 2010, 2013, and 2014; *Osman, Che Ahmad, and Halim, 2011).* What Is Happening In This Class (WIHIC) (Aldridge, Fraser and Huang, 2010; Iram and Ambreen, n.d.; Murugan and Rajoo, 2013 & Treagust, 2003) and self-efficacy (Dorman and Fraser, 2009; Kolo, Garba, Jaafar, and Binti, 2017). Many literatures cited have pronounced that conducive and favorable learning environment can contribute to the enhancement of students' academic achievement. In addition, psychosocial variables such as emotional intelligence (EI or EQ), high school chemistry self-efficacy, science process skills and teacher characteristics are found to have a positive and significant influence towards students' academic performance (Abungu, Okere, and Wachanga, 2014; Kolo et al., 2017; Oloyede, 2012 & Preeti, 2013).

Hence, this study was conducted to ascertain the factors that could significantly contribute to junior high school (JHS) students' academic performance. Results of this study are expected to serve as imperative for government stakeholders such as DepEd officials, school administrators, curriculum makers and teachers on the areas where improvements can be made to promote the quality of teaching and learning in science. Likewise, the study sought to provide effective remedies to the problems faced by the young students who are performing low in science. Specifically, the study geared to:

1. find out the level of performance of students in Grade 10 science:
2. determine the level of cognitive attributes of Grade 10 students with respect to each student's:
   - General Weighted Average (GWA) in Grade 6
   - General Weighted Average (GWA) in science in the Grade 7 to Grade 9 and
   - scientific ability derived from National Career Assessment Examination (NCAE)
3. ascertain the level of the physical aspect of the science laboratory environment of Grade 10 students in terms of:
   - furniture and equipment/learning space
   - technology
   - lighting
   - air quality
   - safety items
4. correlate academic performance to the cognitive attributes, and physical aspects of science learning environment of the Grade 10 science students

2. Methods

This research used a mixed-method approach involving quantitative and qualitative data. This method applied the triangulation technique in which data obtained from the survey instruments, semi-structured /one-on-one interview, class observation and literature review. This was done to strengthen and ensure the validity and accuracy of the results of the study.

In the quantitative approach, descriptive-correlational and causal-comparative designs were employed. The descriptive-correlational design was utilized to determine the results of students' performance of their science learning environment- cognitive factors, physical and psychological aspects in relation to academic performance in science. The causal-comparative research design was employed to determine the influence of the cognitive attributes and physical aspects of the students' classroom/laboratory environment on their science academic performance. Structural equation modeling (SEM) was used to find out what model best links to the academic performance of Grade 10 students in science. This would serve as basis of improving the science learning environment of the students towards effectiveness in learning.

In the qualitative approach, a semi-structured interview was employed to validate the students' answer on their performance with respect to science learning environment, emotional quotient, high school chemistry self efficacy, science process skills and teacher quality. A class observation was also conducted to some science teachers to further validate the students' responses on the Teacher Quality instrument. An intensive literature review and studies were gathered for the different variables employed in the study.

Moreover, the data collected from the respondents for the three methods (in yellow, Figure 1) were compared and then were triangulated in the interpretation phase of the study.

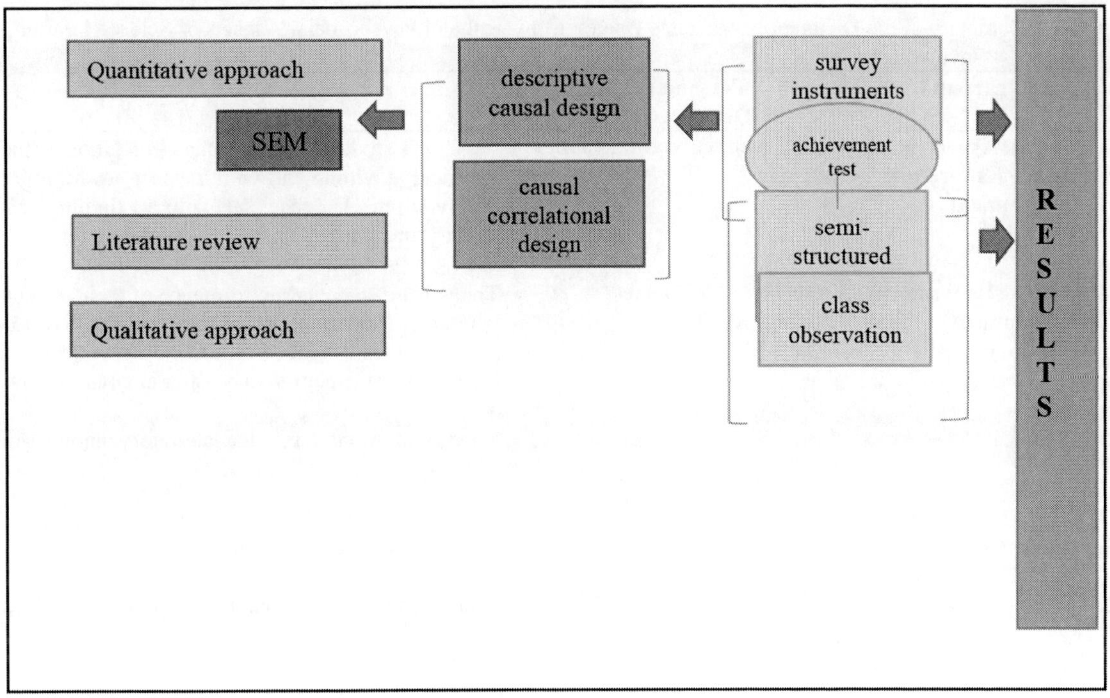

Figure 1 Research Paradigm

2.1. Research Setting

This study covered nine (9) public secondary schools in randomly selected school divisions of Region 10, Philippines. It comprises five provinces: Camiguin, Misamis Oriental, Lanao del Norte, Bukidnon and Misamis Occidental and two cities classified as highly urbanized, all occupying the north-central part of Mindanao island, and the island-province of Camaguin. The regional center is Cagayan de Oro (https://en.wikipedia.org/wiki/Northern_Mindanao).

The province of Bukidnon is comprised of three divisions, namely; Bukidnon, Malaybalay City and Valencia City. However, DepEd- Division of Bukidnon is the biggest bureaucracy of DepEd in Region X (http://www.depedbukidnon.net.ph/). Cagayan de Oro City, Iligan City and Misamis Oriental, the other divisions were included in the study. These public schools have a large population of high school students. The researcher strongly believes that they can provide reliable information for the study.

2.2. Research Instruments

There were eight instruments employed during the collection of data. This was to determine the level of performance of Grade 10 science students in terms of physical and psychosocial aspects of the science learning, emotional quotient, high school chemistry self-efficacy, science process skills, teacher quality and achievement test in science. After the content validation, the instruments were pretested and piloted to Grade 10 students for reliability to ensure internal consistency using Cronbach's alpha. The content validity and reliability of these instruments (ranging from 0.78 to 0.96) were carefully done for temporal, spatial and ethical considerations. These instruments were pilot-tested to more than 100 students from the three sections in Grade 10 of Central Mindanao University-Laboratory High School (CMU-LHS) last May 2016.

Experts on content and pedagogy validated the content of the survey instruments and science achievement test. Three of them hold a degree of Doctor of Philosophy (PhD) in Science Education (Major in Chemistry) from the University of the Philippines (Diliman) (1) and De La Salle University (2), Manila respectively. They are professors of University of San Carlos (1), Cebu City and Notre Dame University, Cotabato City. The fourth one earned her degree of Doctor of Philosophy in Chemistry at the University of Adelaide, Australia. She is a full professor, Chemistry Department, College of Arts and Sciences of Mindanao State- Institute of Technology. The latter local expert only validated the content of the achievement test in science.

Table 1 Summary of the Different Instruments Used for Physical and Psychosocial Aspects of Science Learning Environment.

Instrument	Person(s) Developed	Use(s)
1.Physical Aspect of Science Laboratory Environment (PASLEI)	Che Ahmad, Osman & Halim (2014)	• To assess students' performance in relation to the physical environment with respect to furniture and equipment, space, technology, lighting, air quality and safety items.
2.Science Laboratory Environment Inventory	Fraser, McRobbie & Giddings, 1993	• To evaluate students' performance of their science laboratory environment in terms of the five (5) scales: student cohesiveness, open-endedness, integration, rule clarity, and material environment, and • To examine what makes the laboratory unique.
3.What is Happening In This Class (WIHIC)	Fraser, McRobbie & Fischer, 1999	• To measure the students' performance in their learning environment in terms of student cohesiveness, teacher support, involvement, investigation, task orientation cooperation, and equity.

2.3. Research Subjects

The participants were Grade 10 students of the public secondary schools in six divisions of Region 10. The names of the schools' respondent are listed in Table 2 with the corresponding number of Grade 10 enrollees (SY 2017-2018) and the desired sample size per stratum. For Structural Equation Modelling, the more samples, the more fitting the model is, however, a good rule of thumb is >15 cases per predictor/indicator was considered (Stevens, 2012).

Table 2 Respondents from Different Secondary School Divisions in Region 10.

DepEd Divisions	Code	No. Grade 10 enrollees SY 2017-2018	Sample size
Cagayan de Oro City	C1	205	37
Bukidnon	B1	96	86
	B2	205	132
	B3	529	112
Iligan City	I1	558	145
Malaybalay City	M1	1309	271
	M2	108	78
Misamis Oriental	MO1	403	85
Valencia City	V1	940	177
TOTAL	**9**	**4353**	**1123**

The researcher humbly requested permission from the Regional Director of Department of Education (DepEd). Another letter of request was forwarded to the School Division Superintendents of the six divisions in Region X -Northern Mindanao in their respective school divisions. A letter of request was then forwarded to the School Principals asking permission to conduct the study involving their Grade 10 high school students as research participants. Second, through the students' class advisers and subject teachers. A Consent Form with attached cover letter was provided to the students.

In this research study, Central Mindanao University-Laboratory High School was chosen school for the pilot testing of the seven instruments and the science achievement test. The school has shifted its academic calendar this SY 2016-2017 and has a different science curriculum compared to DepEd secondary schools.

Stratified random sampling using proportional allocation was employed to obtain the students who served as participants. There were nine (9) national high schools. However, a total of 1000 respondents or more was set. To obtain the percentage proportional allocation of each division, the researcher simply computed the ratio of the desired sample size and the total number of participants and multiplied by 100.

The data necessary for the study were collected through the seven survey questionnaires for cognitive characteristics, physical aspects of science laboratory environment, psychosocial aspects of science learning environment, and the achievement test in science. Before the survey questionnaires was administered to the participants, the researcher provided them a preliminary orientation about the study. They were informed that the complete results of the survey and their performance in the achievement test will be kept confidential. The data collected were then tallied, tabulated, and analyzed.

On the qualitative aspect of the study, a semi-structured interview was conducted. Five (5) students for each school respondent were interviewed, however, the same ethical process for the survey questionnaire were undertaken by the researcher for this aspect. Students' responses were patiently recorded, transcribed and translated the vernacular responses to English language for thematic analysis purposes that is in terms of frequency. Data that were obtained from this interview would further deepen the results of the study and provide tangible, compelling.

2.4. *Cognitive Attributes of Grade 10 Students*

This study used the students' general weighted average in elementary (Grade 6) and in science grades from Grade 7 to Grade 10. This is to ascertain the relationship with respect to their level of academic performance. The following grading scale and descriptive rating were employed in the study based on DepEd Order #8, s.2015:

Grading scale	Descriptive Rating
90-100	Outstanding
85-89	Very satisfactory
80-84	Satisfactory
75-79	Fairly satisfactory
74 and below	Did not meet the expectations

Their NCAE results taken in Grade 9 level (i.e. SY 2015-2016) were formally requested from their School Guidance Counsellor in their respective school. This served as secondary data.

2.5. Qualitative Data

The results of the quantitative data of the study were underpinned by the gathered review of literature and studies of the different variables, students' oral interview and class observation of the science subject teachers handling the chemistry lessons for the Fourth Grading, SY 2017 to 2018.

Before the one-on-one interview, permission from the section adviser/science subject teacher was humbly asked to conduct the interview for five Grade 10 students only per school for validation purposes of the instrument.

A brief orientation was given to the students on the purpose of the interview, and requested to fill up the consent form for the research participants voluntarily. The researcher provided a simple introduction about herself and greeted the students warmly to establish a positive atmosphere before the interview, Guide questions (validated by three local experts) were also provided to students to internalize the questions by encouraging them to use the papers to organize their answers. A formal one-on-one interview was followed to have an in-depth

IOP Conf. Series: Journal of Physics: Conf. Series **1340** (2019) 012068 doi:10.1088/1742-6596/1340/1/012068

response of the student participant from the given questions. All students' response were recorded, transcribed, consolidated and subjected to content analysis in terms of frequency.

Validation of the instrument on teacher quality was done through class observation of the subject science teacher, either announced or unannounced. Permission from them was sought before the conduct of the class observation. At the end of the observation, the researcher showed the filled up form. There was an exchange of ideas, suggestions on the current status of science education, especially in DepEd system. The subject teacher signed the form. Results of the teacher's actual observation were summarized, and consolidated to determine the frequency of the items in the Observation Checklist.

3. Results and Discussion

Out of one thousand one-hundred twenty-three (1123) student participants, only three of 0.27% are on the range of 66 to 85% rating described as moving towards mastery (Table 3). It can be noted that almost 40% of the students were on the average mastery (38.02%). Further, more than 60% of the students obtained a percentage rating score ranging from low mastery to no mastery at all. The overall performance of students for the six divisions of Region X is 32.42%, indicating low mastery. This implies that majority of the respondents had difficulty to grasp the scientific concepts introduced to them.

Table 3. Students' Level of Performance on the Science Achievement Test.

RANGE (%)	FREQUENCY	PERCENTAGE (%)	DESCRIPTIVE EQUIVALENT
96 - 100	0	0	Mastered
86 - 95	0	0	Closely approximately mastery
66 - 85	3	0.27	Moving towards mastery
35 - 65	427	38.02	Average mastery
16 - 34	659	58.68	Low mastery
5 - 15	33	2.94	Very low mastery
0 - 4	4	0.89	Absolutely no mastery
TOTAL	**1123**	**100%**	
OVERALL MEAN	**32.46%**	**LOW MASTERY**	

The results further implied that the students obtained low performance towards mastery of the learning competencies in understanding the concepts, theories and principles in science. Thus, students still lack the scientific knowledge and skills to meet the globalization demands of the 21st Century Science Education. York (2015) found that students' academic performance in the form of grades and GWA can assess the cognitive skills to measure students' academic success.

Banerjie (2016), King 'Aru (2014) & Mulela (2015), report that shortage of learning resources such as laboratory rooms, facilities, materials, and unfavorable learning environment are the perennial reasons for the poor performance of students in their science subjects. Other factors that possibly account for such unsatisfactory performance of students include teachers' heavy teaching assignments, traditional teaching strategies used in imparting the subject matter, limited students' learning experiences, students self-efficacy, less perseverance in understanding the science concepts as the topics become complicated, students' low confidence with the science knowledge and skills, lack of parents' and teachers' support and low level of emotional quotient (Oyinloye, cited in Nwadinigwe & Azuka-Obike, 2012, p. 395).

The results of students' achievement test agree with the literature of Atagana & Engida, (2014), Cardellini, (2013), Sirhan, (2007) & Woldeamanuel et al. (2016), which put forward the notion that many students find science subjects (such as chemistry) very difficult to comprehend because of its abstract nature. Likewise, the reports of Benito (2014) and Briones (2014) on the low performance of Filipino students in the two national examinations, NCAE and NAT (in science area) also conform to the research outcomes of this study.

Knowingly, science teachers play a crucial role in the holistic development of student's personality. To meet this challenge, they may allot more time for students to achieve mastery of the subject matter, acquisition of 21st century skills (e.g. critical thinking, creativity) and values to ensure that the topic has been understood by heart. According to Newmann and Wehlage, as cited by Doll (1996, p.89), science teachers perhaps create a classroom environment that is well-organized to facilitate students' learning.

Therefore, all stakeholders responsible to alleviate students' poor performance may work together to improve the science curriculum that will expectedly yield positive impact towards the improvement of students' academic achievement and transform them into life-long learners.

3.1. General Weight Average (GWA) in Elementary Level

Table 4 shows the students' general weighted average in their elementary level. The overall mean percentage score (MPS) is 85.37, which denotes very satisfactory. This implies that students had a very satisfactory grade performance in all their subjects during their primary level.

Generally, teachers provide summative evaluation in terms of general weighted average (GWA), or subject grade to students to determine their level of content and procedural knowledge and what they have learned about their subject. According to Airasian, cited by Lipnevich & Smith (2008, p. 4), grades can serve as motivation to encourage students to try harder. GWA and grades are the most appropriate variables to assess students' academic success (York, 2015).

The findings conform to the improved results of the National Achievement Test (NAT) in the elementary level (68.88%), however it was not able to reach the 75% MPS target for national standards. Inadequacy of school facilities, weak reading comprehension, lack of science literacy and computational skills may be attributed to satisfactory performance of students in their GWA.

Table 4. Distribution of Students' General Weighted Average in Elementary

RANGE (%)	FREQUENCY	PERCENTAGE (%)	DESCRIPTIVE EQUIVALENT
90-100	138	12.29	Outstanding
85-89	532	47.37	Very satisfactory
80-84	364	32.41	Satisfactory
75-79	89	7.93	Fair satisfactory
74 and below	0	0	Did not meet the expectation
TOTAL	**1123**	**100%**	
OVERALL MEAN	**85.37**		**VERY SATISFACTORY**

Thus, elementary teachers may be innovative and creative in their learning materials and activities that will effectively enhance students' reading comprehension, and hone their skills in science and mathematics as well. As a result, students will yield a remarkable performance in all subject areas, especially in science.

3.2. General weighted average (GWA) in Science Grade (Grades 7 - 9)

The overall MPS is 84.42%, which equates to very satisfactory. This implies that Grade 10 students in the six division of Region X (Northern Mindanao) still needs to improve their conceptual and procedural knowledge in science to master the learning competencies set by their science teachers. In effect, this will expectedly produce good results in students' learning performance.

Table 5. Distribution of Students' General Weighted Average in Science Grades.

RANGE (%)	FREQUENCY	PERCENTAGE (%)	DESCRIPTIVE EQUIVALENT
90-100	198	17.63	Outstanding
85-89	517	46.04	Very satisfactory
80-84	277	24.67	Satisfactory
75-79	130	11.58	Fair satisfactory
74 and below	1	0.089	Did not meet the expectation
TOTAL	**1123**	**100%**	
OVERALL MEAN	**84.42%**		**SATISFACTORY**

The results of this study supports the findings of Bernardo et al. (2008) that assessment procedures of students performance and the use of science grade in different higher levels (Grade 7 to Grade 10) did not reflect their actual learning. Students may be given credit for the successful attainment of the science curriculum goals without actually doing it. According to Page (quoted by Lipnevich & Smith, 2008, p.7), grades could be effective for the promotion of students' learning when accompanied by a comment. Therefore, science teachers may provide assessment tools that would align with the learning goals in order to measure the learned knowledge and acquired skills of the students.

3.3. Students' Performance on the National Career Assessment Examination (NCAE)
The overall NCAE performance of the students has 66.47% MPS, best described as moving towards mastery. The result implies that students' performance in terms of scientific ability in NCAE is moving towards mastery of the scientific concepts studied in their science class, in fact, one step higher than average mastery.

Table 6. Distribution of Students' General Weighted Average in NCAE

RANGE (%)	FREQUENCY	PERCENTAGE (%)	QUALITATIVE DESCRIPTION
96 - 100	146	13.00	Mastered
86 - 95	166	14.78	Closely approximately mastery
66 - 85	402	35.80	Moving towards mastery
35 - 65	261	23.24	Average mastery
16 - 34	104	9.26	Low mastery
5 - 15	40	3.56	Very low mastery
0 - 4	4	0.36	Absolutely no mastery
TOTAL	**1123**	**100%**	
OVERALL MEAN	**66.47%**		**MOVING TOWARDS MASTERY**

The performance of students may be attributed to students' personal experiences in their environment by giving them opportunity to construct new knowledge, and ideas. This is in accordance to Piaget's Cognitive Development. He believes that a constructivist classroom must provide students multiple activities that will challenge them to accept individual differences, discover new ideas, and construct their own knowledge (Bation & Sabaldana, 2018).

However, result of this study does not conform to the findings of Bernardo et al. (2008), Imam et al. (2014), and the reports of Benito (2014) & Briones (2014) for the low performance of Filipino learners in the NCAE and NAT in the area of science respectively. Thus, teachers may possibly apply effective teaching strategies to cultivate students' interest and radiate the positive scientific attitudes towards an improved educational performance in science.

3.4. Physical Aspect of Science Laboratory Environment
The physical aspects of the science laboratory environment are composed of (a) furniture and equipment, space, (b) lighting, (c) technology, (d) air quality, and (e) safety items are germane to effective students' learning and academic performance. Learning is the core of Department of Education for students' holistic development. Physical facilities play an important role in providing quality performance of students in conducting laboratory activities.

3.4.1. Furniture, Equipment/Learning Space Scale. The mean scores of the furniture, equipment and learning space of physical aspects of science learning environment inventory were computed and assessed. This ascertain the strengths and weaknesses of the science laboratory environment of the student participants, as basis for its enhancement towards effective learning in science. The study found that the layout of furniture (3.02) and chairs (3.00) in students' laboratory environment make them comfortable while doing their laboratory work; the space is enough for them to move and their laboratory is also equipped with adequate materials (2.98). The lowest mean (2.68) accounted to the statement that said the number of students in a science laboratory is not more than 50. The overall mean (2.92) indicates that the level of furniture, equipment and learning space is moderate.

This results find support to the findings of Che Ahmad et al. (2010, 2011, 2014) and Puteh et al. (2015), which revealed that furniture and learning space is in moderate level. The design of tables and chairs make students feel uncomfortable to be used for an extended period of time. The learning space seems not suitable for the class size where the students cannot freely interact with others for discussion and sharing of insights about the activity. This will create an effect on the learning outcomes of the students.

Below are some of the verbatim answers of the students during interview:

Yes, our laboratory have enough space for the students to work their experiments, but sadly, we don't have the equipment/materials for the laboratory and for the students. (B1.5)

No, because it is too small ... for us.(B2.2)

Our classroom as laboratory ...enough... enough cia mam...dili ka maka ingon na gu-ot cia mam... (*Our classroom serves as our laboratory room. It is enough and you cannot say it is congested.*) (B3.3)

No ma'am, we can't move easily because of the occupied space and some of the students are just standing. (I1.5)

Gamay... gamay ra...(*The laboratory working area is only small.*) (M2.5)

Sometimes, kung daghan tao, sikit kayo pero kung gamay ra luag kaayo...(*Sometimes if there are many students in the laboratory, we are very close to each other. The area will be spacious if there are few students in the room.*) (MO1.3)

No ma'am, the space is not enough to conduct the experiment because the tables ma'am... are huge and they are so close to each other... so we cannot work properly. (V1.3)

The responses of the students expose the limited space of their laboratory room in carrying out an activity, which may affect the accuracy of their resulting data, and performance as well. This will also affect their interest in doing science activities.

Hence, this physical feature of the laboratory be given attention by the School Heads and other government officials to formulate provision for a comfortable movement of students around the laboratory. To the teachers, they may create a safe working science laboratory environment that will allow students to work comfortably and can sustain their interest in performing their tasks completely.

3.4.2. Lighting Scale. The following statements under the lighting scale include brightness of light in the laboratory that can illuminate the science laboratory across the room (2.94), and the lighting that can be adjusted according to the needs of their learning activities (2.80). The science laboratory can be darkened when the lesson calls for it (2.65). The overall mean (2.80), which is described as moderate. This implies that the lighting condition in the laboratory is poor. It does not fully performed its function.

The finding of Puteh et al. (2015) agrees with the result of this study, but it negates to the findings of Che Ahmad et al. (2010, 2011, & 2014), who found that there is high level of fitness for lighting scale. This study suggests that subject teachers may also conduct a regular check of the lighting condition in the science laboratory to motivate students to focus on their laboratory work as to enhance their understanding of the science concepts.

3.4.3. Technology Scale. The students responded positively to the statements that their laboratory is equipped with functioning television and LCD (2.83) and are in good condition (2.80). This produced a mean score of 2.81, which shows that the level of fitness for technology is moderate. This implies that the Information and Communication Technologies (ICT) in the science laboratory room for the nine school respondents still do not have the complete ICT equipment. The findings of this study conform to the results of Che Ahmad et al. (2010, 2011 & 2014) & Puteh et al. (2015).

3.4.4. Air Quality Scale. With respect to the air quality scale of PASLEI, it has been found that the science laboratory has a suitable temperature to allow the learning process to take place comfortably (2.88), and has adequate number of electric fans to ensure good airflow in the laboratory (2.72). Also, the exhaust fan installed in the laboratory increases the efficiency of airflow (2.58). The students' positive responses gave an overall mean of 2.74, which can be described quantitatively as moderate level of fitness. This may be attributed to unavailability or limited number of exhaust fans, and poor ventilation in the laboratory room.

In addition, the study also found that the safety items aspect scale had a mean score of 3.09, which indicates high. Specifically, the science laboratory has two doors that function properly (3.23), equipped with a first-aid kit (3.12) and fire extinguisher (3.03). Overall mean (3.09), denotes high level of fitness. The findings of this study conform to the results of Puteh et al. (2015), but it does not agree with previous literature of Che Ahmad et al. (2010, 2011 & 2014) who reported that there is a moderate level of fitness for safety aspects.

Based on the transcription of the oral interview, majority of the participants mentioned that there is a unit of fire extinguisher available in their science laboratory. However, out of 45 students, 15 (33%) claimed that an exhaust fan is available, and 22 (49%) claimed to have located an emergency kit found in their science laboratory room. Here are some of students' responses which evidently support the above statements, coming from selected school divisions of Region X-Northern Mindanao: Divisions of Bukidnon (B1.2, B2.2, & B3.2), Cagayan de Oro City (C1.2), and Valencia City (V1.4).

> Naa mi laboratory glasswares, naa man emergency kit, naa man fire extinguisher, exhaust fan, naa. (*We have laboratory glasswares, emergency kit, fire extinguisher, and exhaust fan.*) (B1.2)
>
> Our science laboratory is not complete with equipments like exhaust fan, some emergency kit and others. (B2.2)
>
> In our laboratory apparatus and equipment.... Science laboratory is kuan mam kuan kanang tarung pagka-plastar...emergency kit –naay daghan ug sulod na tambal, fire extinguisher gibutang sa butanganan, walay exhaust fan. (*In our science laboratory, the laboratory apparatus and equipment are orderly arranged. There is an emergency kit with a lot of medicines inside. The fire extinguisher is placed on a receptacle; no exhaust fan.*) (B3.2)
>
> Ahm... murag para sa akoa mam kay okey...okey ra cia murag ... naay ...naay...murag dili kumpleto naa mi laboratory apparatus and equipment... ang kulang sa amoa ang exhaust fan. (*For me, our science laboratory is okay, but it is not complete with the laboratory apparatus and equipment. There is no exhaust fan.*) (C1.2)
>
> By... kanang... kanang naka butang sa order ug kanang ug pud mga mga... mahitabo cia dali ra kuha-on. Fire extinguisher- naa mam, emergency kit – naa mam, exhaust fan – naa mam. (*The laboratory apparatus and equipment are placed in order and are accessible to the students. There are fire extinguisher, emergency kit and exhaust fan in the laboratory room.*) (V1.4)

The narratives of students imply that there is a dire need for DepEd officials such as the School Principals/Division Superintendents to provide the public secondary schools with sufficient learning resources to effectively educate the students to facilitate quality learning and teaching in science. They can possibly look into providing the schools adequate facilities and equipment needed in the realization of laboratory experiments. This may widen students' knowledge and develop the 21st Century skills necessary in coping with the globalization demands towards quality science education.

4. Conclusions

The level of Grade 10 students' academic performance measured using the standardized teacher prepared achievement test in science exhibited low in terms of mastery of the scientific concepts. This implies that students lack the necessary scientific knowledge, skills and learning experiences to unleash their difficulties in understanding concepts in science and appreciate the importance of science in their real life situations.

The cognitive attributes of the students obtained very satisfactory and satisfactory ratings for their elementary general weighted average and general weighted average in science grade, respectively. Further, their performance in NCAE scientific ability displayed moving towards mastery of the science concepts taught.

Moreover, physical aspects of science laboratory environment of the students showed moderate level of fitness. Thus, providing a well-equipped laboratory with advanced materials and equipment to cultivate students' interest to exert effort in understanding the complexity of science.

The Grade 10 students had demonstrated positive responses in various dimensions of psychosocial aspects towards their science learning environment. Laboratory, and classroom environment were moderately practiced. Cognitive attributes such as elementary GWA, Junior high school GWA of all the science subject taken and the NCAE scientific ability found to have a significant correlation to students' academic performance. Thus, science teachers may use the science grades as feedback to their performance.

Three variables that came out to be the predictors of science performance, namely: JHS science GWA, Elementary GWA and Classroom learning environment. Students' grades and conducive classroom environment would serve as motivation to enhance their science learning. The latter variable may also the students with stimulating learning experiences that will help them augment the conceptual knowledge and understanding of their science lessons. This kind of environment will likewise provide students' opportunities to develop the necessary 21st Century in depth knowledge, skills and values towards learning competence.

Generally, students with high cognitive ability supported by conducive, well-equipped physical facilities in science, the better is the performance. The cognitive attributes were best measured in terms of students' grades in elementary and junior high school science subjects together with their NCAE performance (NCAEPER) in scientific ability. Thus, theoretically, the best fitting structural model of science performance is anchored on students' cognitive ability coupled with the physical learning environment aspects in science.

It is recommended that Science teachers may find ways in updating or innovating strategies, methods, approaches, and assessment tools to support instructional materials that cater to the current needs, skills and abilities of secondary students for an effective and meaningful science learning. Hence, this will improve the students' understanding about the science lesson and will eventually lead to better academic performance. Pre-service teachers may continue to have a one-week enhancement training to develop the necessary scientific knowledge, and skills to become competent teachers in their own field of specialization.

School Administrators may ensure and encourage science teachers for continual professional growth through attending seminars, symposia, short term courses that enhance the current trends and approaches in the teaching of science. They may inspire teachers for collaborative researches in the area of science education, as well.

Government offices (like DepEd, DOST, CHED and others) most likely will continuously sponsor seminars, trainings, conferences and scholarships for public science secondary teachers to provide and strengthen their knowledge and experiences in their field of specialization in local and international arena. The Scholarship program on Capacity Building in Science and Mathematics Education (CB in SME) may be revived in the current DepEd administration. This will probably improve the quality of science education in the country and develop more competent public secondary teachers that will motivate students' interest towards their attitudes in science. Hence, this may yield good results in the performance of students academically. Government officials, especially at the Department of Education, may find ways of improving and maintaining physical plan in uplifting and updating state of the art science facilities for actual and experiential learning of students considering the nature of the science subject. Further, a laboratory technician or personnel be employed and properly trained to handle such equipment and facilities to maximize science activities especially in the laboratory. Further studies may be conducted using the results of this study as benchmark in improving performance of students not only in science but also to other fields of specialization.

5. Acknowledgement

The authors would like to thanks different entities and personalities who contributed to make this study a success. The Regional Director of DEPED in particular as well as the Schools Division Superintends of the nine schools division. The students also who participated as the respondents were highly acknowledged

References

[1] Abd Rauf R A, Rasul M S, Mansor A N, Othman Z and Lyndon N 2013 *Canadian Center of Science Education* **9** doi: 10.5539/ass.v9n8p47

[2] Abdullah M C, Elias H, Mahyuddin R and Uli J 2004 Emotional intelligence and academic achievement among Malaysian secondary students *Pakistan Journal of Psychological Research* **19** pp 105–121 Retrieved from: http:// pjprnip.edu.pk/pjpr/index.php/pjpr/article/download/108/104

[3] Abrigo J 2002 *Factor Correlates and Science Education Literacy* (Master's thesis), Bukidnon State University

[4] Baanu T F, Oyelekan O S and Olorundare S A 2016 Self-efficacy and chemistry students'academic achievement in senior secondary schools in North-Central, Nigeria *Journal of Educational Science* **4**

[5] Baharon S, Khoiry M A, Hamid R, Mutalib A and Hamzah N 2015 Assessment of psychomotor domain in a problem-based concrete laboratory *Journal of Engineering Science & Technology* pp 1–10 Retrieved from: https://ukm.pure.elsevier.com/.../assessment-of-psychomotor-domain-in-a-problem-ba

[6] Caballo A K 2009 Education:Analyzing the status of math & science education Retrieved from: https://growthrevolutionmag.wordpress.com/tag/trends-in-mathematics-and-science-study-timss/

[7] Cabansag M G 2014 Impact on the K-12 science program in the enhanced basic education curriculum in provincial school *Journal of Arts, Science & Commerce* **5** pp 29–38

[8] Cakiroglu J, Capa-Aydin Y and Woolfolk Hoy A n.d. Chapter 3 *science teaching efficacy beliefs*. Retrieved from: http://www. link.springer.com/chapter/10.1007%2F978-1-4020-9041-7_31

[9] Dagondon A A 2008 *Chemistry laboratories of selected public secondary schools division of Cagayan de Oro City SY 2006-2007* (Master's thesis), Xavier University

[10] Dehyadegary E, Cheragi S, Ebrahimi N, Bagheri M and Nakhaei N 2013 Relationship between emotional intelligence self-efficacy, academic involvement and academic performance among adolescents in Kerman-Iran *Journal of American Science* **9** pp 130–137 Retrieved from: http://www.jofamericanscience.org

[11] Dehyadegary E, Cheragi S, Ebrahimi N, Bagheri M and Nakhaei N 2013 Relationship between emotional intelligence self-efficacy, academic involvement and academic performance among adolescents in Kerman-Iran *Journal of American Science* **9** pp 130–137 Retrieved from: http://www.jofamericanscience.org

[12] Dorance R L and Dietrich M C 1980 Psychosocial abilities of science and nonscience high school students *Journal of Research in Science Teaching* **17** pp 495–502 doi:10.1002/ten.3660170602

[13] Dorman J P and Fraser B J 2009 psychosocial environment and affective outcomes in technology-rich classrooms: testing a causal model *Soc Psychol Educ* **12** pp 77–99 doi: 10.1007/s11218-008-9069-8

[14] Ferrer F P, Geiser and Dela Cruz R. J 2017 Correlation of stem students' performance in the National Career Assessment Examination and academic subjects *International Journal of Social Sciences* **3** Retrieved from http://dx.doiorg/10.203191/pijss.2017.s31.532541

[15] Ferido and Marlene B The Spiral Progression Approach in Science. Retrieved April 19, 2018 from http://ceap.org.ph/upload/download/20136/3144824758_1.pdf

[16] Fernandez G R 2010 Secondary science and mathematics gaps Retrieved February 21, 2017, from http://mis.wvsu.edu.ph/librarium/pub/results/311

[17] General Chemistry PHS 1015 Practice Exam 4 ...Accessed February 13, 2017 ,from http://www.teachersites.schoolworld.com/.../general%20chemistry%20phs%201015%20practice%..

[18] Hannah R 2013 The effects of classroom environment on students' learning *Honors Theses* pp 6–12 Retrieved from: http://scholarworks.wmich.edu/honors_theses

[19] Hashemi S, Ahamad K, Ali H, Seyed M, Shirpoor A and delaviz A 2014 The relationship between emotional intelligence and self-efficacy and academic performance of students *World Essays Journal* pp 65–70 Retrieved from: http://www.worldessaysj.com

[20] Higher Chemistry Unit 2 - Section 1 Fuels Multiple Choice Questions.Accessed February 13, 2007 from: http://www.calderglen.s-lanark.sch.uk/.../Multiple%20choice%20PPs/.../Higher%20Chemistry

[21] Ibe E, Nworgu L N and Anyaegbunan N J 2016 Influence of Tecahers' characteristics on academic achievement of secondary school biology student *British Journal of Science* **13** pp 33–41 Retrieved from www.ajournal.co.uk/pdfs/BSvolume13(2)/BSVol.13%20(2)%20Article%203.pdf

[22] Imam O A, Mastura M, Jamil H and Ismael Z 2014 Reading comprehension skills and performance in science among high school students in the Philippines *Asia Pacific Journal and Education* **29** 81–89 Retrieved from: http://www. apjee.usm.my/APJEE_29_2014/Art%205(81-94).pdf

[23] Jackson J W 2002 Enhancing self-efficacy and learning performance *The Journal of Experimental Education* **70** pp 243–254 Retrieved from: https://www.jstor.org/stable/20152681

[24] Jenaabadi H, Shahidi R, Elhamifar A and Khademi H 2015 Examine the relationship of emotional intelligence and creativity with academic achievement of second period high school students *World Journal of Neuroscience* **5** pp 275–281 doi: 10.4236/wins.2015.54025

[25] King'Aru J M 2014 Factors contributing to poor performance of science subjects: a case of secondary schools in Kawe division, Kinondoni municipalit (Doctoral Thesis) Retrieved from http://repository.out.ac.tz/598/1/JAMES_MUCHWE_KING%E2%80%9DARU_Final_Thesis.pdf

[26] Lansangan M J G, Quiambao D, Baking E G, Nicdao R C, Nuqui A V and Cruz R C 2015 *Journal of Business & Management Studies* pp 1–7

[27] Lipnevich A A and Smith J K 2008 Response to assessment feedback: the effects of grades, praise, and source of information research report Retrieved from https://www.ets.org/Media/Research/pdf/RR-08-30.pdf

[28] Mulela M M 2015 Effects of availability and use of laboratories on students performance in science subjects in community secondary schools in Kinondoni municipality (Doctoral thesis) Retrieved from http://www. repository.out.ac.tz/1155/1/Mercy.doc

[29] Murugan A and Rajoo L 2013 Students' perceptions of mathematics classroom environment and mathematics achievement: a study in Sipitang, Sabah, Malaysia *Journal of Social Science Research*

pp 325–343

[30] Oginni A M, Awobodu V Y, Alaka M and Saibu S O 2013 School factors as correlates of students' achievement in chemistry *International Journal for Cross-Disciplinary Subjects in Education* **3** pp 1515–1522 Retrieved from: http://www.nfonomics-society.org/.../School-Factors-as-Correlates-of-Students-Achievement-in-d.oi:10.5901/jesr.2014.v4n3p397

[31] Slavin R E 2003.*Educational Psychology*.Seventh edition Boston: Pearson Education, Inc

[32] Stein C M, Morris N J and Nock N L 2012 Structural equation modeling In *Methods in Molecular Biology* **850** pp 495–512 (Methods in Molecular Biology; Vol. 850) doi: 10.1007/978-1-61779-555-8_27

[33] Stevens J P 2012 Applied Multivariate Statistics for the Social Sciences: Analyses with SAS and IBM's SPSS, Sixth Edition Retrieved from: http://www.amazon.com/Applied-Multivariate-Statistics-Social-Sciences/dp/0805859039

IOP Conf. Series: Journal of Physics: Conf. Series **1340** (2019) 012069 doi:10.1088/1742-6596/1340/1/012069

Trends and Research Issues of STEM Education: A Review of Academic Publications from 2007 to 2017

Paranee Chomphuphra, Pawat Chaipidech and Chokchai Yuenyong*

Science Education Program, Faculty of Education, Khon Kaen University, Khon Kean 40002, Thailand

*Corresponding author's e-mail address: ychok@kku.ac.th

Abstract. STEM education (Science, Technology, Engineering, and Mathematic) research has been become more attractive area in science education for a decade. This study was aim to analyze research articles in a SCOPUS database and two journals which were not indexed in SCOPUS including Journal of STEM Teacher Education, and International Journal of STEM Education. The research articles during 2007 to 2017 were reviewed and analyzed according to the authors' nationality, journals, STEM research topics. The research findings indicated that there were 56 published papers related to providing STEM learning activities in school setting, top three countries which published STEM papers over the decades were United States (46), Australia (2), Canada (2). Besides, the journal with the greatest number of published papers wasJournal of STEM Teacher Education, with a total of 16 papers, and the second is International Journal of STEM education. The three popular topics which published in STEM papers were innovation for STEM learning, professional development and gender gap and Career in STEM, respectively.

1. Introduction

STEM is an acronym which stands for Science, Technology, Engineering, and Mathematics. At the beginning, STEM education has often been called a Meta discipline, the creation of a discipline based on the integration of other disciplinary knowledge into a new whole [4]. Now, many educators and policy makers have concerned for emphasis on science, technology, engineering and mathematics (STEM) disciplines integration in our schools. The challenges the world faces today call for a global society that is multidisciplinary and may "require the integration of multiple STEM concepts to solve them" [8]. Therefore, it is important to train and prepare a diverse STEM-literate workforce with the capability to understand and comprehend the technological world [5].

Over the past decade, numerous researchers have been conducted STEM education in many contexts and subject areas. For instance, in a high school level, [7] studied an engineering curriculum in school, identify teaching strategies used to increase math and science literacy, and discover challenges and constraints that occur during its development and delivery, as well as what strategies are used to overcome these obstacles. Besides, in a university level, [2] conducted the research to provide a model, an example, and suggestions for establishing and fostering meaningful partnerships to construct authentic and relevant STEAM learning experiences for pre-service teachers.Obviously, the given example of the studies about STEM education were varies used.Therefore,to understand the trends of STEM education research in science education, literature on SCOPUS database and two journals; Journal of STEM teacher education, and International Journal of STEM education from 2007 to 2017 were analyzed in this study.The research questions of this study were:

1. How did the authors' nationality and journals of STEM education article publish across these decades (2007-2017)?
2. How about the topics of the published articles in journals vary across these ten years (2007-2017)?

2. Research Methodology

We use the content analysis method which helps us find and identify some of the features that appear in the research paper. This method is provides a simple, easy and understandable and is the most recent one used among the research trends in science education.

2.1 Selection Research Papers Related to STEM Education

The criteria by which studies are to be included in, or excluded from, the review were determined. The studies which appear to meet these criteria were listed by means of electronic atabase searching, and then the abstracts of the studies were screened to see if they meet the inclusion criteria

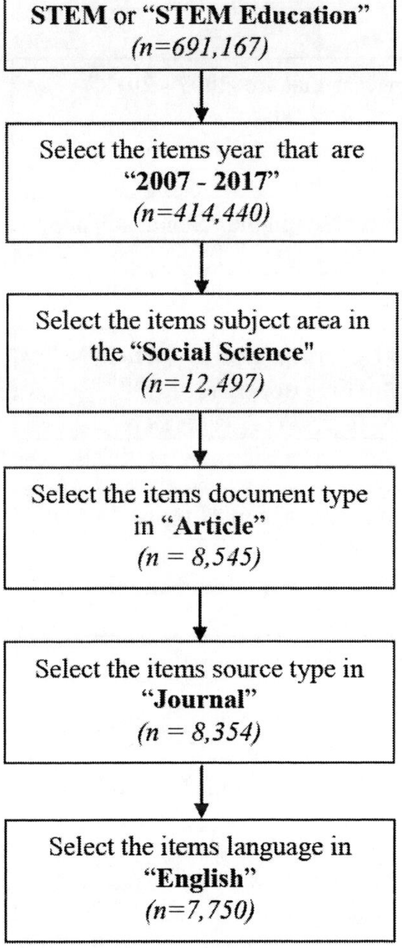

Figure 1. Scopus database searching steps on October 19st, 2017.

2.1.1 Criteria for selection research papers Related to STEM Education:

The research articles of each journal related to STEM Education between 2007 and 2017 were searched in the Scopus database on October 19st, 2017, where the criterion was to include the keyword 'STEM' or 'STEM Education' to detect all words related to STEM sources, STEM Education contexts and STEM learning. The result from searched in Scopus database found that, there were 691,167 papers were related to STEM or STEM Education, among them 414,440 papers were published in 2007 - 2017 and 12,497 papers were related to social science. In addition, we selected document type in 'Article', source type in 'Journal' and language in 'English'. Finally we found a total of 7,750 were selected in the final list for analysis, as shown in Figure 1. However, the result from searching by used the keywords 'STEM' or 'STEM Education' were too large and wide range that was difficult to analyze. So, the next step we focus on keyword that we interested in. That was providing STEM learning activities in school setting.

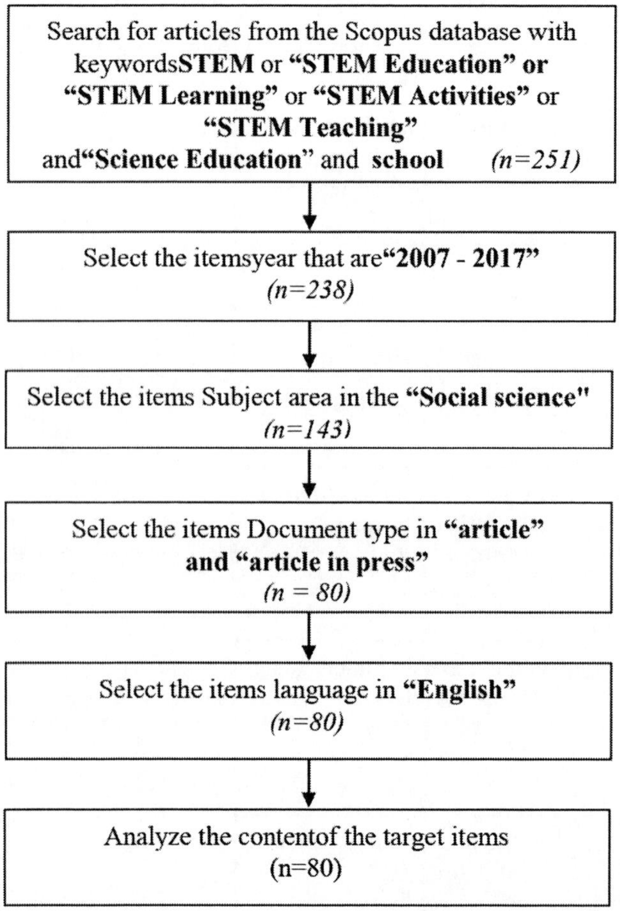

Figure 2. Scopus database searching steps on October 20st, 2017.

2.1.2 Criteria for Document Selection for Data Analysis:

In addition, the study based on the review of publishing article in Journal of STEM Teacher Education (JSTE) from 2011 to 2017, and International Journal of STEM Education (IJSE) from 2016 to 2017. The rationale for the choice of these two journals is that they are the major journals that focused in STEM education research. The number of total access was 53 articles.

A second-level in-depth screening by reading the abstracts resulted in the exclusion of some of the articles since the keyword STEM or "STEM Education" or "STEM Learning" or "STEM Activities" or "STEM Teaching" and "Science Education" and school was relevant to another context in STEM Education. This step eliminated irrelevant 80 articles. The resulting number of research articles is provided in Table 2.

2.2 Identifying and generating systematic categories

2.2.1 The categorization of coding system:

Considering STEM education research topic, the categories were created as a coding system to use for a structured observation. Research articles could be identified and categorized by a coding system that describe the characteristic of the article, allow the researcher to literate and categorize clearly. The Table 1 showed the categories to identify STEM education articles.

Table 1 Summarization of the categories in a coding system.

No.	Categorize	Description
1.	Professional Development	Pre-service teachers in math and science experiencing with integrated STEM task, Technological pedagogical content knowledge (TPACK) for science teacher, Promote Science Education and Provide Hands-On Experiences
2.	Gender gap and Career in STEM	Gender disparities in STEM, Immigrant–native educational attainment gap in STEM, Students' motivation and promote students' STEM career choice
3.	Integrated Engineering into STEM	Engineering-based STEM integration , apply scientific knowledge and skill to help solve engineering challenges in science class
4.	Innovation for STEM learning	Developed program, promote STEM competency through the game, Innovative Technologies
5.	STEM learning/activity	Metasynthesis focused on STEM teaching and learning practices in middle and high school classrooms
6.	Integration of STEAM/STEAM learning	Development measurement for researchers and educators to use to assess student collaboration in STEAM activities.
7.	STEM School Framework/Model	Critical components of STEM schools and derived a theoretical framework
8.	Other	Paper suggestion/review/perspectives

IOP Conf. Series: Journal of Physics: Conf. Series **1340** (2019) 012069 doi:10.1088/1742-6596/1340/1/012069

Table 2 The distribution of STEM education research articles across Journals.

Journals	Number of STEM education articles
International Journal of STEM Education	12
Journal of STEM Teacher Education	16
Research in Science Education	2
Journal of Research in Science Teaching	3
Cultural Studies of Science Education	3
Journal of Science Education and Technology	4
Journal of Pre-College Engineering Education Research	2
International Journal of Science and Mathematics Education	2
ACM Transactions on Computing Education	1
Computers & Education	1
Educational Researcher	1
International Journal of Educational Research	1
International Journal of Technology and Design Education	1
Journal of Chemical Education	1
Pacific Economic Review	1
Social Psychology of Education	1
Social Science	1
Sustainability	1
Technology, Knowledge, and Learning	1
The Curriculum Journal	1

IOP Conf. Series: Journal of Physics: Conf. Series **1340** (2019) 012069 doi:10.1088/1742-6596/1340/1/012069

3. Research Finding and Discussion

The 56 articles reviewed in this research included all aspects of STEM Education published in SCOPUS database as well as in Journal of STEM Teacher Education (JSTE) and International Journal of STEM Education (IJSE) from 2007– 2017. The results and discussion are given in relation to the research questions.

1. How did the authors' nationality and journals of STEM education article publish across these decades (2007-2017)?

Figure 3 showed the publication situation of providing STEM learning papers from 2007 to 2017. The earliest paper was written byYu, Iskander, Kapila, and Kriftcher (2007),and promoted engineering careers using sensor in high school science labs. Fewer than five papers were published each year from 2007 to 2011.It was not until 2012 that it started to receive more attention from researchers. In Figure 2, it could be seen that 30 papers were published in the STEM education domain in 2017. Such a finding is reasonable since, due to the policy have focused on development knowledge and skills in area of STEM. In particular, in recent years, the rapid growth of the STEM career policies around the globe has encouraged in schools which increases science literacy, and enables the next generation of innovators.

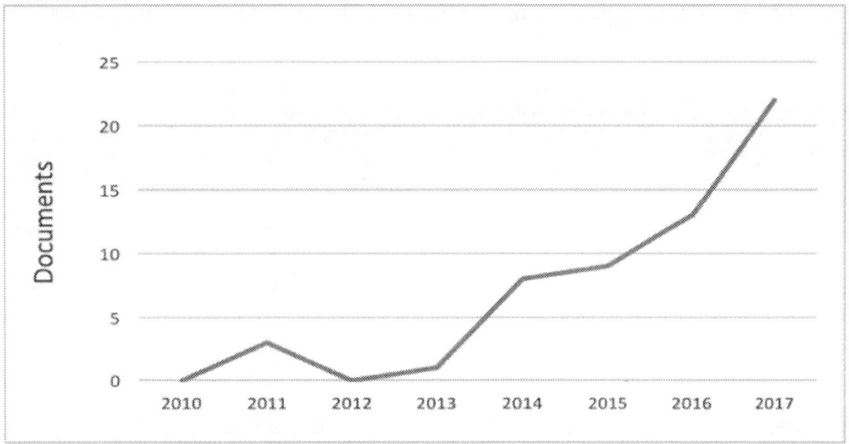

Figure 3. Published papers providing STEM learning from 2007 to 2017.

Only the nationalities of the first authors of the published papers in STEM education were counted in this study. From the results, it can be found that there were many researchers from different countries attempting to conduct STEM education. Fig. 4 shows the distribution of the top 10 countries which published STEM papers over the decades. The top three countries were the United States (46), Australia (2), and Canada (2).

The Figure 5 presented eight journals which published about STEM education paper during year 2007 to 2017. These journals included International Journal of STEM education, Journal of STEM Teacher Education, Research in Science Education, Journal of Research in Science Teaching, Cultural Studies in Science Education, Journal of Science Education and Technology, International Journal of Science and Mathematics Education, and Journal of Pre-college Engineering Education Research. The journal with the greatest number of published papers is Journal of STEM Teacher Education, with a total of 16 papers, and the second is International Journal of STEM education.

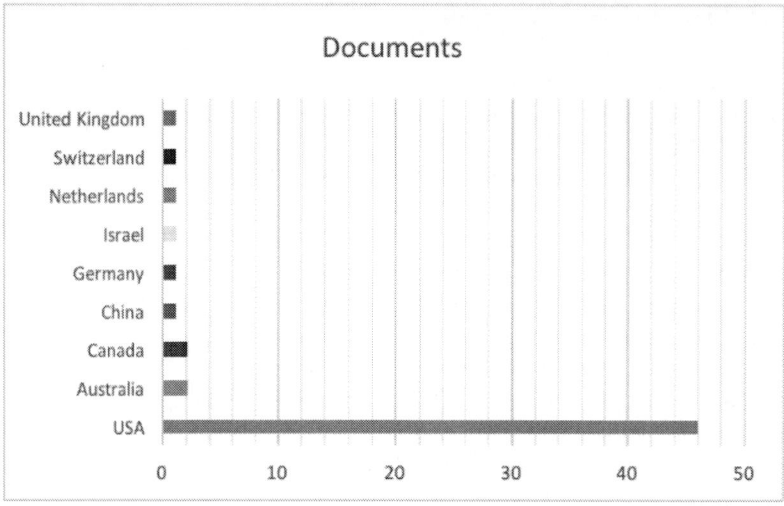

Figure 4. Top ten countries publishing papers on providing STEM learning during 2007 to 2017.

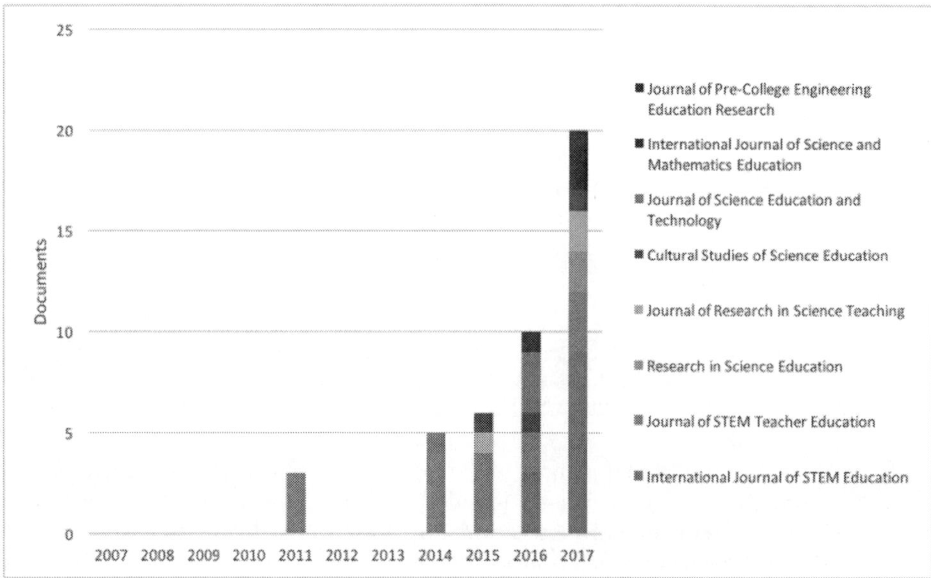

Figure 5. Journals publishing papers on STEM learning from 2007 to 2017.

2. How about the topics of the published articles in journals vary across these ten years (2007-2017)?

The selected 58 STEM education articles were searched from SCOPUS database which keywords related to STEM education and were selected from the obviously STEM education journals including the Journal of STEM Education Teacher and International Journal of STEM Education. A coding system could categorize the aims of research into 8 categories. These included Professional Development (C1), Gender gap and Career in STEM (C2), Integrated Engineering into STEM (C3), Innovation for STEM learning (C4), STEM learning/activity (C5), Integration of STEAM/STEAM learning (C6), STEM School Framework/Model (C7), and Other: paper suggestion /review/perspectives (C8). The frequency of papers in each category could be provided as shown in the Table 3. It found that the most popular STEM education research category (from 2007 to 2017)was

Innovation for STEM learning (C4) (20 papers). The top three ranking on STEM education research categoriesincluded Innovation for STEM learning (C4), Professional Development (C1), and Gender gap and Career in STEM (C2); respectively.

C1: Professional Development

Professional developmant is the second popular topic which concept of pre-service teachers in math and science experiencing with integrated STEM task, technological pedagogical content knowledge (TPACK) for science teacher, promoting science education and provide hands-on experiences.

C2: Gender gap and Career in STEM

Gender gap and career in STEM was categorized to research that focused on the concepts of gender disparities in STEM, immigrant–native educational attainment gap in STEM, students' motivation, promote students' STEM career choice and the integration of STEM practices into the curriculum.

C3: Integrated Engineering into STEM

Integrated Engineering into STEM was described to research that focused on engineering-based STEM integration , apply scientific knowledge and skill to help solve engineering challenges in science class.

C4: Innovation for STEM learning

The innovation for STEM learning is the most clarification on the STEM education research as shown in the Table 3. The innovation learning for STEM learning was labeled to represent to the research that focused on the concept of developed program, promote STEM competency through the game, innovative technologies and technology and design activity.

C5: STEM learning/activity

STEM learning and STEM activities was focused on STEM teaching and learning practices in middle and high school classrooms.

C6: Integration of STEAM/STEAM learning

Integration of STEAM and STEAM learning was concentrated on learning development measurement for researchers and educators to use to assess student collaboration in STEAM activities

C7: STEM School Framework/Model

STEM School Framework and Model was represent to the research that focused on critical components of STEM schools and derived a theoretical framework.

C8: Review STEM perspectives and suggestions

Review STEM perspectives and suggestion on STEM and STEM integration.

Table 3 Frequency of STEM education papers in each category.

Journals	Frequency of STEM education papers							
	C1	C2	C3	C4	C5	C6	C7	C8
International Journal of STEM Education	2	1	-	-	-	1	1	3
Journal of STEM Teacher Education	4	2	2	2	2	1	-	3
Research in Science Education	-	1	-	1	-	-	-	-
Journal of Research in Science Teaching	1	1	-	1	-	-	-	-
Cultural Studies of Science Education	-	1	-	2	-	-	-	-
Journal of Science Education and Technology	-	-	1	3	-	-	-	-
Journal of Pre-College Engineering Education Research	-	-	1	1	-	-	-	-
International Journal of Science and Mathematics Education	-	-	1		-	-	-	-
ACM Transactions on Computing Education	-	-	-	1	-	-	-	-

Journals	Frequency of STEM education papers							
	C1	C2	C3	C4	C5	C6	C7	C8
Computers & Education	-	-	-	1	-	-	-	-
Educational Researcher	-	1	-	-	-	-	-	-
International Journal of Educational Research	-	1	-	-	-	-	-	-
International Journal of Technology and Design Education	-	-	-	1	-	-	-	-
Journal of Chemical Education	1	-	-	-	-	-	-	-
Pacific Economic Review	-	-	-	1	-	-	-	-
Social Psychology of Education	-	-	-	1	-	-	-	-
Social Science	-	1	-	-	-	-	-	-
Sustainability	1	-	-	-	-	-	-	-
Technology, Knowledge, and Learning	-	-	-	1	-	-	-	-
The Curriculum Journal	-	1	-	-	-	-	-	-
Total	**10**	**10**	**5**	**20**	**2**	**2**	**1**	**6**

The most popular topics from 2007 to 2017 was 'innovation for STEM learning' (20 papers), which the concept of developed program, promote STEM competency through the game, innovative technologies and technology and design activity. The methodology of these topics was mixed method, quantitative and qualitative research. Among the participants including: pre-service teachers, students and students of STEM Education course at graduate level. The second popular topics was 'Professional Development' (10 papers), which the concept of pre-service teachers in math and science experiencing with integrated STEM task, technological pedagogical content knowledge (TPACK) for science teacher, promoting science education and provide hands-on experiences. In this concept, the researcher use quantitative, qualitative and mixed method, the qualitative data including: focus group interview transcripts, responses from open-ended survey questions and all researcher notes (including classroom observations and notes from teacher workshops). The participants including: pre-service teachers, science teachers, scientists and educational technology. For the third popular topic was 'Gender gap and Career in STEM'(10 papers), which the concepts of gender disparities in STEM, immigrant–native educational attainment gap in STEM, students' motivation, promote students' STEM career choice and the integration of STEM practices into the curriculum. In 'Gender gap and Career in STEM' topic, the researcher use case study, qualitative, quantitative and mixed method and the participants were students: postsecondary students,school-aged girls.

As discussed by Christine V. McDonald in 2016, the integration of technology and engineering into school education has been proposed as an effective means to enhance student learning and raise student achievement in STEM disciplines [1]. Technology and engineering activities have been shown to develop STEM literacy and increase motivation, in addition to providing real world contexts for learning scientific and mathematical concepts [6]. The majority of research on professional development in STEM disciplines has been conducted in science and mathematics, with findings indicating that engaging in professional development has been shown to be beneficial to teachers [3].

4. Conclusion

A systematic review was used in this research to analyze STEM education in science education published in SCOPUS database and two journals included Journal of STEM teacher education, and International Journal of STEM education from 2007 to 2017. It was found that the number of publish paper rapidly increasing over the decades. Moreover, we found that top three countries which published STEM papers over the decades were United States (46), Australia (2), and Canada (2). Besides, the journal with the greatest number of published papers wasJournal of STEM Teacher Education, with a total of 16 papers, and the second is International Journal of STEM education. The three popular topic which published in STEM papers were innovation for STEM learning, professional development and gender gap and Career in STEM, respectively. Although we reviewed only some of English communicated journal, this report may suggest some trends of research in STEM education.

References

[1] Brophy, S., Klein, S., Portsmore, M., & Rogers, C. (2008). Advancing engineering education in P-12 classrooms. Journal of Engineering Education, 97(3), 369-387. IN Christine V. McDonald. (2016). STEM Education: A review of the contribution of the disciplines of science, technology, engineering and mathematics. Science Education International. 27(4), 530-569.

[2] Bush, S. B. & Cook, K. L. (2016). Constructing Authentic and Meaningful STEAM Experiences Through University, School, and Community Partnerships. *Journal of STEM Teacher Education,* 51(1).Retrieved from https://ir.library.illinoisstate .edu/jste/vol51/iss1/7.

[3] Christine V. McDonald. (2016). STEM Education: A review of the contribution of the disciplines of science, technology, engineering and mathematics. Science Education International. 27(4), 530-569.

[4] Lantz, Jr., B. H. (2009). *Science, technology, engineering, and mathematics (stem) education what form? what function?* Retrieved from http://www.currtechintegrations.com/pdf/ STEME ducation Article.pdf

[5] Merchant, , S., &Khanbilvardi, R. (2011, April).A national framework to integrate remote sensing science in STEM education and training.*Proceedings of the 1st Integrated STEM Education Conference (ISEC),* Ewing, NJ, 6B-1–6B-4. doi:10.1109 /ISECon.2011.6229633..

[6] NRC (2012).National Research Council. A Framework for K-12 Science Education: Practices, Crosscutting Concepts, and Core Ideas. Washington, DC: The National Academies Press. IN Christine V. McDonald. (2016). STEM Education: A review of the contribution of the disciplines of science, technology, engineering and mathematics. Science Education International. 27(4), 530-569.

[7] Sticker, D.R. (2011). A Case Study: Teaching Engineering Concepts in Science. *Journal of STEM Teacher Education*, 48(2). Retrieved from https://ir.library.illinoisstate.edu/jste/vol48/iss2/6

[8] Wang, H.-H., Moore, T. J., Roehrig, G. H., & Park, M. S. (2011). STEM integration: Teacher perceptions and practice. Journal of Pre-College Engineering Education Research, 1(2), 1–13. http://doi.org/10.5703/ 1288284314636

IOP Conf. Series: Journal of Physics: Conf. Series **1340** (2019) 012070 doi:10.1088/1742-6596/1340/1/012070

Magnet and Pin kit: Connection Symbolic and Submicroscopic Representations of Lewis dot structure and Molecular geometry

Phalakorn Kamkhou[1,a] and Chokchai Yuenyoung[2,b]

[1]Khon Kaen University Demonstration Secondary School (Suksasart), Khon Kaen, Thailand,

[2]Science Education Program, Faculty of Education, Khon Kaen University, Thailand,

Corresponding author's e-mail address: [a]phalka@kku.ac.th [b]ychok@kku.ac.th

Abstract. This work aims to connect the symbolic and submicroscopic representations of covalent compound's Lewis dot structure and molecular geometry by using the Magnetic and Pin kit (MP-kit), which was constructed from magnets and pins. The MP-kit was created to facilitate the students. The kit has two sets, the first set, called Magnet set, consists of a 30x30 cm iron plate, an alphabet A with magnet (refer to central atom), 6 alphabets B with magnet (refer to surrounding atom), 18 white circular magnets (refer to valence electron of A), 42 black circular magnets (refer to valence electron of B) and 3 yellow circular magnets (refer to extra electron of molecule, from ion generation). Which was used to create the Lewis dot structure of covalent molecules. The second set, called Pin set, consists of a spherical eraser (refer to central atom), 6 pins (refer to bond and surrounding atom) and 3 droplet pins (refer to lone pair electron of central atom). Which was used to make the molecular geometry of covalent compounds, that predicted by using VSEPR theory. The users can use the Magnet set to create the Lewis dot structure without frequent drawing and rubbing on the papers. The magnets on the plate are moved for anytime when the users know it's wrong. Moreover, this set can be used to show the number of bond pair electrons, types of bond and number of lone pair electrons on central atom. The Pin set is used to build the 3D molecular structure and specify the bond angles. The kit was used to link the symbolic and submicroscopic representations and encourage the student's understanding. Finally, the teacher can use the MP-kit to check the student's misconception, leading to solve problems immediately.

1. Introduction

Gilbert and Treagust presented the types of representation in chemistry for three levels, there are macroscopic, submicroscopic and symbolic. The macroscopic level (or phenomenological type) consists of representations of the empirical properties of matters. These properties are perceptible in chemistry laboratories and in everyday life and are therefore able to be measured. Examples of such properties are mass, density, concentration, pH, temperature and osmotic pressure. The submicroscopic level (or model type), it is usual to produce models built from entities such as atoms, ions, molecules and free radicals, for phenomena described with the macroscopic representation. For examples, in terms of electron density, shape of atomic and molecular orbitals (including the use of valence electron repulsion theory) and diagrams or graphs. The symbolic level involves the allocation of symbols to represent atoms, whether of one element or of linked groups of several elements; of signs to represent electrical charge; of subscripts to indicate the number of atoms in an individual ion or molecule; of letters to indicate the physical state of the entity (e.g. solid (s), liquid (l), gas (g), aqueous (aq) or other solution). This depiction is then followed by the inclusion of these representations as appropriate within all conventions of chemical and ionic equations, with the use of prefixed coefficients to show the conservation of matter during a reaction. This level of representation also can be used both in respect of the first, the phenomenological representational type, when dealing with bulk quantities of reactants and products in stoichiometric computations, and with a wide range of models of the second type of representation when describing physical changes (e.g. changes of state and dissolution of solutes) and the chemical changes taking place during reactions [1].

Gilbert Lewis formulated that atoms combine in order to achieve a more stable electron configuration. Maximum stability results when an atom is isoelectronic with a noble gas. When atoms interact to form a chemical bond, only their outer regions are in contact. A chemical bond in which two electrons are shared by two non-metal atoms, called covalent bond. The compounds that contain only covalent bonds are called covalent compound. The structures are used to represent covalent compounds are called Lewis dot structures. A Lewis dot structure is a representation of covalent bonding in which shared electron pairs are shown as pairs of dots between two atoms, and lone pairs are shown as pairs of dots on individual atoms. Only valence electrons are shown [2].

Molecular geometry is the three-dimensional arrangement of atoms in a molecule. However, there is a simple procedure that enables us to predict with considerable success the overall geometry of a molecule or ion if we know the number of electrons surrounding a central atom in its Lewis structure. The basis of this approach is the assumption that electron pairs in the valence shell of an atom repel one another. The valence shell is the outermost electron- occupied shell of an atom; it holds the electrons that are usually involved in bonding. In a covalent bond, a pair of electrons, often called the bonding pair, is responsible for holding two atoms together. However, in a polyatomic molecule, where there are two or more bonds between the central atom and the surrounding atoms, the repulsion between electrons in different bonding pairs causes them to remain as far apart as possible. The geometry that the molecule ultimately assumes (as defined by the positions of all the atoms) minimizes the repulsion. This approach to the study of molecular geometry is called the valence-shell electron-pair repulsion (VSEPR) model, because it accounts for the geometric arrangements of electron pairs around a central atom in terms of the electrostatic repulsion between electron pairs. We can predict the geometry of molecules (and ions) in a systematic way. For this purpose, it is convenient to divide molecules into two categories, according to whether or not the central atom has lone pairs. A molecule's geometry affects its physical and chemical properties, such as melting point, boiling point, density, and the types of reactions it undergoes [2]. If we predict a mistake on a molecular geometry, we will describe their properties error.

For molecules in which the central atom has no lone pairs. If we consider molecules that contain atoms of only two elements, A and B, of which A is the central atom and B is a surrounding atom. These molecules have the general formula ABx, where x is between 2 and 6. Determining the geometry of a molecule is more complicated if the central atom has both lone pairs and bonding pairs. In such molecules there are three types of repulsive forces—those between bonding pairs, those between lone pairs, and those between a bonding pair and a lone pair. We designate molecules with lone pairs as ABxEy, where A is the central atom, B is a surrounding atom, and E is a lone pair on A. Both x and y are integers; x = 2, 3, …, and y = 1, 2, … Thus, the values of x and y indicate the number of surrounding atoms and number of lone pairs on the central atom, respectively [2]. This work aims to construct the Magnet and Pin kit for using as the tool, leading to connect the symbolic and submicroscopic representations of covalent compound's Lewis dot structure and molecular geometry [2].

IOP Conf. Series: Journal of Physics: Conf. Series **1340** (2019) 012070 doi:10.1088/1742-6596/1340/1/012070

2. Method

The construction of Magnet and Pin kit and their components

The Magnet and Pin kit has two sets, the first set, called Magnet set, Figure 1. (a), consists of a 30 x 30 cm iron plate, an alphabet A with magnet (refer to central atom), 6 alphabets B with magnet (refer to surrounding atom), 18 white circular magnets (refer to valence electron of A), 42 black circular magnets (refer to valence electron of B) and 3 yellow circular magnets (refer to extra electron of molecule, from ion generation). The second set, called Pin set, Figure 1. (b), consists of a spherical eraser (refer to central atom), 6 pins (refer to bond and surrounding atom) and 3 droplet pins (refer to lone pair electron of central atom).

The Magnet set was used to guide the Lewis structure writing for describe the bonding pair electron between atoms and lone pair electron of central atom in covalent molecules. After that, the Pin set was used to construct the molecular geometry, which was predicted by Valence Shell Electron Pair Repulsion theory (VSEPR).

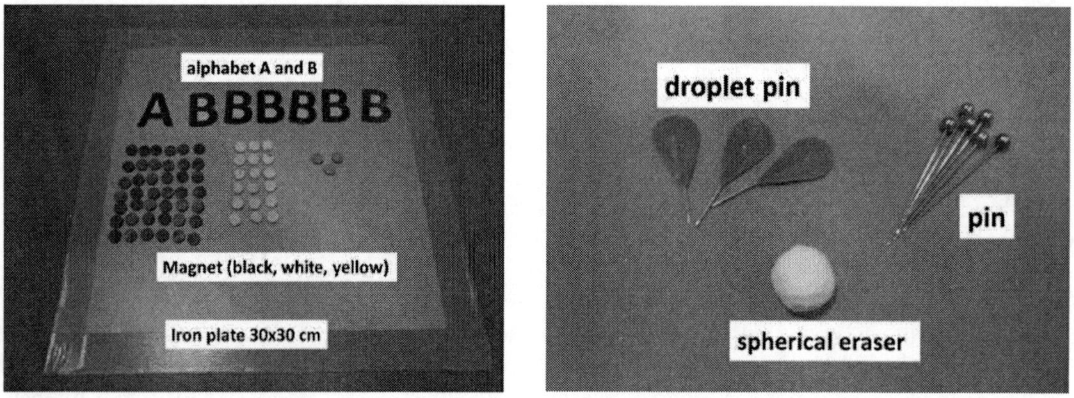

(a) (b)

Figure 1. The components of (a) Magnet and (b) Pin kit

3. Results and Discussion

The Magnet and Pin kit was used to guide the Lewis structure writing, which leads to predict the molecular geometry of covalent compounds and it was used to connect the submicroscopic and symbolic representation.

3.1. The Magnet and Pin kit was used to guide the Lewis structure writing, which leads to predict the molecular geometry of covalent compounds and it was used to connect the submicroscopic and symbolic representation.

3.1.1. Symbolic Representation: Lewis structure and molecular geometry of CO_3^{2-}. The molecule of CO_3^{2-} ion consisted of a carbon atom as a central atom, which is rounded by three oxygen atom. It is the covalent compound, resulting from electron sharing between non-metal atoms. Two carbon-to-oxygen bonds are single bonds, involving one electron pair and the third is a double bond, involving two electron pair. Moreover, two extra electrons are accepted into molecule in order to have the electronic structure of a corresponding noble-gas atom. The central Carbon there are no lone pairs of electron. The Lewis dot structure of CO_3^{2-} are shown in Figure 2. (a). Based on Valence Shell Electron Pair Repulsion Theory (VSEPR), the electron clouds of the atoms around the carbon atom will repel each other. As a result, they will be pushed apart giving the CO_3^{2-} molecule a trigonal planar molecular geometry. The CO_3^{2-} bond angle will be about 120 degrees, as shown in Figure 2. (b).

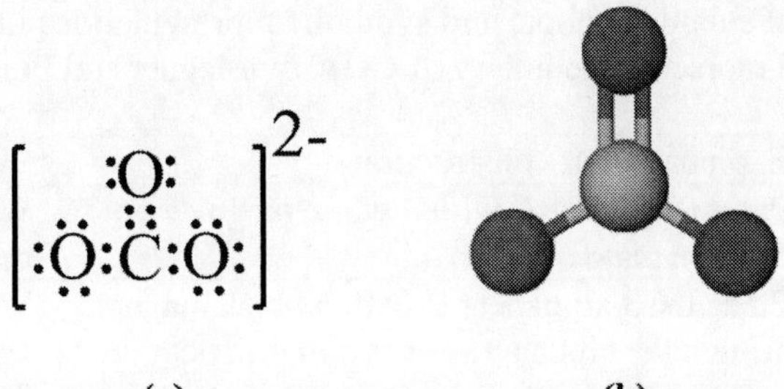

(a) **(b)**

Figure 2. (a) Lewis dot structure and (b) molecular structure of CO_3^{2-}

3.1.2. Submicroscopic Representation: The construction of CO_3^{2-} Lewis structure and molecular structure. The CO_3^{2-} molecule consists of a C-atom (4 valence electrons), as a central atom and three O-atom (6 valence electrons), as surrounding atoms. Alphabet A and B refer to C and O-atom, respectively. The Lewis structure was constructed by following step; (1) put an alphabet A with 4 white magnets on the center of iron plate, (2) place the 3 alphabets B with 6 black magnets around the alphabet A at symmetric position. (3) move a black magnet of each B to pair with white magnet around alphabet A, this step, they are equal to 3 pairs of magnets around A. (4) count the total magnet (black and white) around each alphabet B, they are equal to 7. (5) move a black magnet of B to pair a white magnet of A. (6) count the total magnet around B again, an alphabet B has completely 8 magnets but the other is not. (7) add a yellow magnet for each alphabet B. (8) count the total magnet around B again, an alphabet B has completely 8 magnets. The Lewis structure construction is finished in this step. After that, (9) the molecular structure was constructed by impaling 3 pins on spherical eraser with the widest angle. The angles between each pins are about 120°. This model refers to the geometry of CO_3^{2-} molecule as trigonal planar. The Steps of construction the CO_3^{2-} Lewis dot structure and its molecular geometry are shown in Figure 3.

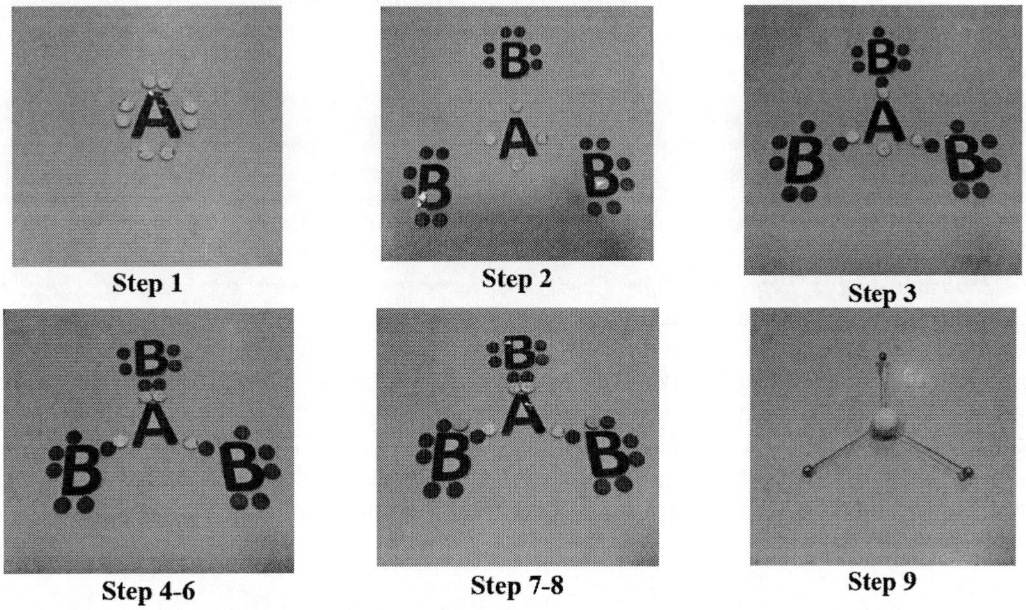

Figure 3. The Steps of construction the CO_3^{2-} Lewis dot structure and its molecular geometry

The Conversion of submicroscopic and symbolic representations in construction of Lewis structure and molecular geometry of CO_3^{2-} are presented in Table 1.

IOP Conf. Series: Journal of Physics: Conf. Series **1340** (2019) 012070 doi:10.1088/1742-6596/1340/1/012070

Table 1. Conversion of submicroscopic and symbolic representations in construction of Lewis structure and molecular geometry of CO_3^{2-} by Magnet and Pin kit

Encoding Steps	Explanation of construction	Symbolic Representation
Step 1	Put an alphabet A with 4 white magnets on the iron plate	Writing the Lewis structure of a central atom
Step 2	Place the 3 alphabets B with 6 black magnets around the alphabet A at symmetric positon	Writing the Lewis structure of surrounding atom and preparation to form bonds
Step 3	Move a black magnet of each B to pair with white magnet around alphabet A, this step, they are equal to 3 pairs of magnets around A	Sharing of the electron to form covalent bonds
Step 4	Count the total magnet (black and white) around each alphabet B, they are equal to 7	Checking the number of electron of surrounding atom following Octet rule
Step 5	Move a black magnet of B to pair a white magnet of A	
Step 6	Count the total magnet around B again, an alphabet B has completely 8 magnets but the other is not	
Step 7	Add a yellow magnet for each alphabet B	Receiving extra electron into molecule for following Octet rule and charge begetting
Step 8	Count the total magnet around B again, an alphabet B has completely 8 magnets	Checking the number of electron of surrounding atom following Octet rule
Step 9	Construct the molecular structure by impaling 3 pins on spherical eraser with the widest angle. The angles between each pins are about 120°. This model refers to the geometry of CO_3^{2-} molecule as trigonal planar.	Prediction the molecular geometry following VSEPR

3.2. The Magnet and Submicroscopic and Symbolic Representation of XeF_4 molecule.

3.2.1. Submicroscopic Representation: Lewis structure and molecular geometry of XeF_4. The molecule of XeF_4 consisted of a xenon atom as a central atom, which is rounded by four fluorine atom. It is the covalent compound, resulting from electron sharing between nonmetal atoms. Four xenon-to-fluorine bonds are single bonds, involving one electron pair. The central xenon there are two lone pairs of electron. The Lewis dot structure of XeF_4 are shown in Figure 4. (a). Based on Valence Shell Electron Pair Repulsion Theory (VSEPR), two lone pair electron clouds of the Xe-atom will repulse bond pair electrons, giving the XeF_4 molecule a square planar molecular geometry. The angles between Xe and F atoms are 90°. Each lone pair electrons are in position of top and bottom of the molecular plane, as shown in Figure 4. (b).

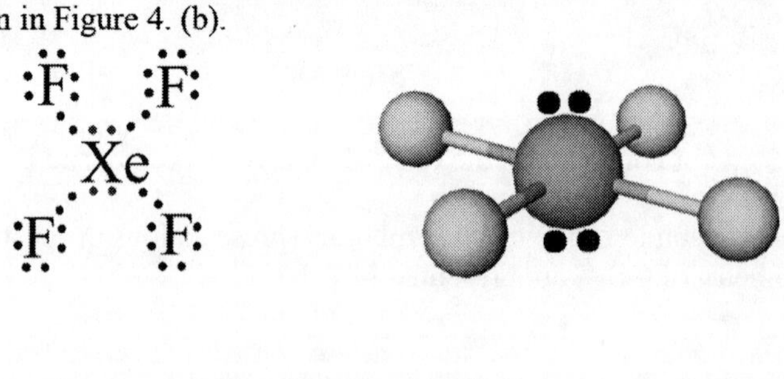

Figure 4. (a) Lewis dot structure and (b) molecular structure of XeF_4

3.2.2. Submicroscopic Representation: The construction of XeF4 molecular structure. The XeF_4 molecule consists of a Xe-atom (8 valence electrons), as a central atom and four F-atom (7 valence electrons), as surrounding atoms. Alphabet A and B refer to Xe and F-atom, respectively. The Lewis structure was constructed by following step; (1) put an alphabet A with 8 white magnets on the center of iron plate, (2) place the 4 alphabets B with 7 black magnets around the alphabet A at symmetric positon. (3) move a black magnet of each B to pair with white magnet around alphabet A, this step, they are equal to 4 pairs of magnets around A. (4) Count the total magnet (black and white) around each alphabet B, they are equal to 8. This step is finished. Then, we have to count the black magnets around A that are not paired with white magnet of B. There are 2 pairs. (5) The molecular structure was constructed by impaling 4 pins and 2 droplet pins on spherical eraser with the widest angle. The angles between each pins are 90°, the angle between two droplet pins is 180° and the angles between pins and droplet pins are 90°. This model refers to the geometry of XeF_4 molecule as square planar. The Steps of construction the XeF_4 Lewis dot structure and its molecular geometry are shown in Figure 5.

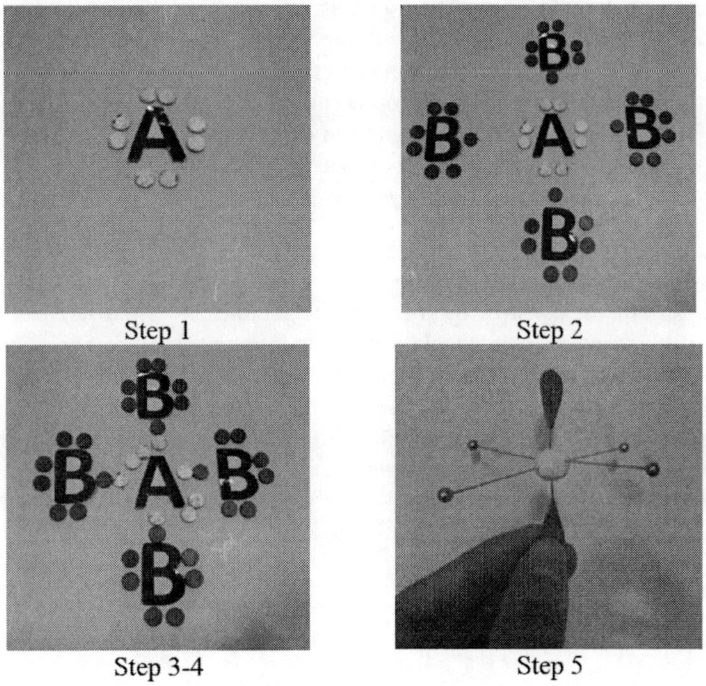

Figure 5. The Steps of construction the XeF_4 Lewis dot structure and its molecular geometry

The Conversion of submicroscopic and symbolic representations in construction of Lewis structure and molecular geometry of XeF_4 are presented in Table 2.

Table 2. Conversion of submicroscopic and symbolic representations in construction of Lewis structure and molecular geometry of XeF_4 by Magnet and Pin kit

Encoding Steps	Explanation of construction	Symbolic Representation
Step 1	put an alphabet A with 8 white magnets on	Writing the Lewis structure of a

	the center of iron plate	central atom
Step 2	place the 4 alphabets B with 7 black magnets around the alphabet A at symmetric positon	Writing the Lewis structure of surrounding atom and preparation to form bonds
Step 3	move a black magnet of each B to pair with white magnet around alphabet A, this step, they are equal to 4 pairs of magnets around A	Sharing of the electron to form covalent bonds
Step 4	Count the total magnet (black and white) around each alphabet B, they are equal to 8. This step is finished. Then, we have to count the black magnets around A that are not paired with white magnet of B. There are 2 pairs	Checking the number of electron of surrounding atom following Octet rule
Step 5	Constructed by impaling 4 pins and 2 droplet pins on spherical eraser with the widest angle. The angles between each pins are 90°, the angle between two droplet pins is 180° and the angles between pins and droplet pins are 90°. This model refers to the geometry of XeF_4 molecule as square planar	Prediction the molecular geometry following VSEPR

4. Conclusion

The Magnetic and Pin kit was constructed from magnets and pins. The users can use the ***Magnet set*** to create the Lewis dot structure on iron plate without frequent drawing and rubbing on the papers. The magnets on the plate are moved for anytime when the users know it's wrong. Moreover, this set can be used to show the number of bond pair electrons, types of bond and number of lone pair electrons on central atom. ***The Pin set***, constructed from pin and eraser, is used to build the 3D molecular structure and specify the bond angles. This kit was used to connect the symbolic and submicroscopic representations and encourage the student's understanding. Finally, the teacher can use it to check the student's misconception, leading to solve problems immediately.

References

[1] John K Gilbert and David Treagust 2009 *Multiple representations in chemical education* (Springer)
[2] Raymond Chang 2010 *Chemistry* (The McGraw-Hill Companies, Inc)

IOP Conf. Series: Journal of Physics: Conf. Series **1340** (2019) 012071 doi:10.1088/1742-6596/1340/1/012071

Challenges of implementing Lesson Study in Cambodia: Mathematics and Science Teaching by using Lesson Study at Happy Chandara School

Visa Kim[1*], Makara Douch[2], Savrin Thy[3], Chokchai Yuenyong[4], Sampan Thinwiangthong[5]

[1] Mathematics Teacher, Happy Chandara School, Phnom Penh, Cambodia
[2] Head of Mathematics Subject, Happy Chandara School, Phnom Penh, Cambodia
[3] Principal of Happy Chandara School, Phnom Penh, Cambodia
[4, 5] Lecturer of Faculty of Education, Khon Kaen University, Thailand Type the author addresses here

*corresponding author's e-mail address: visa.kim@happychandara.org

Abstract. This study diagnoses the challenges of implementing Lesson Study (LS) in teaching mathematics and science at the junior high school and the senior high school of Happy Chandara School (HCS), Cambodia. This action research was selected during July 2017 to July 2018. The study seeks to answer two main questions: 1) what are the advantages of implementing LS? 2) What are the challenges of implementing LS? To collect the data, we carried out 5 steps: classroom observation, assessments of pedagogical content knowledge (PCK), surveys of teachers and school directors, introducing LS, and implementing LS. As the result, it found that LS helped teachers: to clarify the level of learning outcomes to which the instruction should be directed; to make a qualified lesson plan (it is not individual work); to become more cooperative to improve their teaching and students' learning; and to build up the professional capacity among the group work of their colleagues with the same subject. Furthermore it sheds light on content and instruction, instructional support systems, the implementation of teacher standards, and time management as the challenges.

1. Introduction

Science, technology, engineering and mathematics (STEM) collectively known as one of the richest sources of employment and economic growth will be jobs that require skills [2]. In Cambodia, we need STEM-related talent to compete globally, and we will need even more in the future. STEM has become prioritized development agendas to promote economic and societal development of the country as stipulated in the Industrial Development Policy of Cambodia 2015-2025 [5] - science and mathematics are part of the STEM multi-dimensional strategy and play a crucial role. In preparing the students of today to become successful individuals of tomorrow, we need to ensure that our science and mathematics education is effective. According to the Global Competitiveness Report 2017–2018 presents that the quality of math and science education in Cambodia is at the ranking 111 among 124 countries. To be able to ensure and increase the quality of the science and mathematics education in

Cambodia, the researchers cited an action research at Happy Chandara School (HCS) – Junior High School and Senior High School, which located in Prek Thmey Commune (Outskirt of Phnom Penh city), one of the country's poorest areas. Students in Prek Thmey perform worse in regional mathematics and science tests (HCS students' score record book) and they hate these subjects. According to teachers' interview, the teachers blamed that "this poor performance is because their family is poor and lack of socioeconomic." This problem was discussed by the researchers with the school directors focusing on the classroom observation and the assessment of teachers' pedagogical content knowledge (PCK). We found a common point is we cannot be blamed on students, should look at the teachers. The ways of teaching the subject was not make it comprehensible to others. It is not another attempt to bash teachers or blame them but we want to know how their teaching is. The teachers was unable to help students understand specific subject matter and how particular subject matter topics, problems, and issues can be organized, represented, and adapted to the diverse interests and abilities of learners, and then presented for instruction. School learning will not improve markedly unless we give teachers the opportunity and the support they need to advance their craft by increasing the effectiveness of the methods they use [7]. By understanding why students' performance is poor, the researchers and school directors used Lesson Study which has been successful implementing in many countries as the professional development for the teachers and making classroom have meaning for students. A substantial amount of research studies have noted that LS improves teachers' learning and supports teachers to grow professionally [cited from 9]. LS embodies many features effective in changing teacher practice, such as using concrete practical materials to focus on meaningful problems, taking explicit account of the contexts of teaching and experiences of teachers, and providing on-site teacher support within a collegial network [cited from 8]. Furthermore, teachers in LS groups are not only improving their own knowledge and skills but are also contributing to a knowledge base that may, potentially, inform more permanent improvements over time [7]. LS has the potential to radically transform schools in Cambodia into learning environments in which teachers working collaboratively, can investigate, share and verify what works well for their students [7].

2. Research Methodology

The present study adopts of LS consists of three phases: collaboratively design a research lesson; collaboratively observe the research lesson; and collaboratively discuss and reflect on the research lesson [3] as a means to improve teachers' profession. The analysis is conducted using theoretical concepts involving PCK [6]. This study was cited at the junior high school and the senior high school

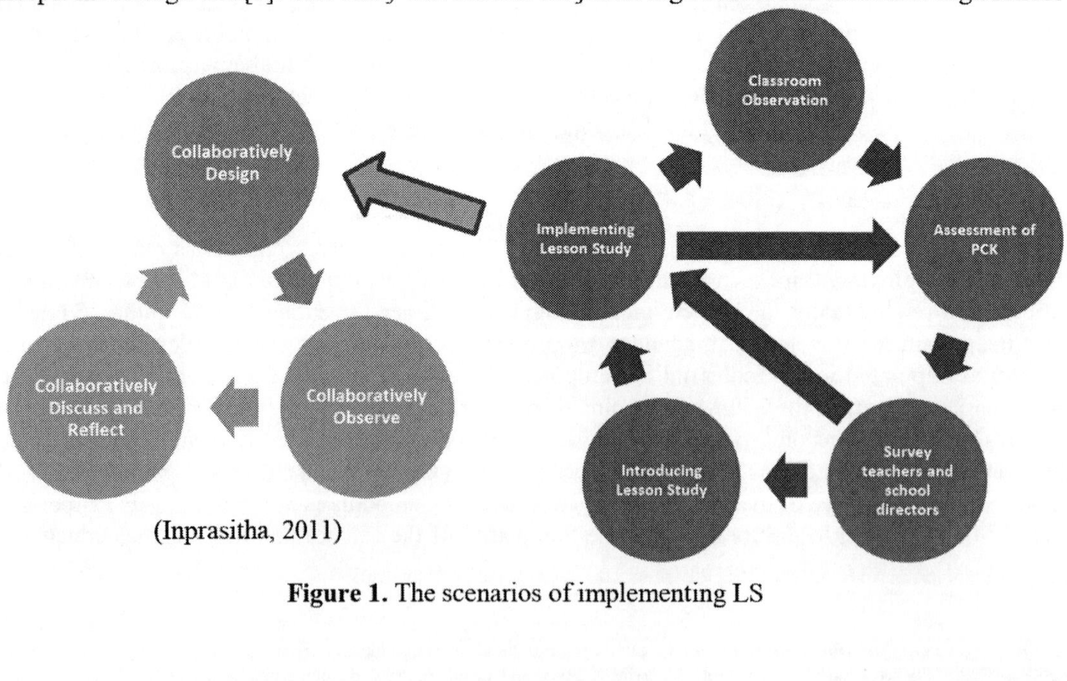

Figure 1. The scenarios of implementing LS

of Happy Chandara School (HCS), Cambodia. This action research was selected during July 2017 to July 2018. The study seeks to answer two main questions: 1) what are the advantages of implementing LS? 2) What are the challenges of implementing LS? To collect the data, we carried out 5 steps: classroom observation, assessments of pedagogical content knowledge (PCK), surveys of teachers and school directors, introducing LS, and implementing LS (Figure 1). The required instruments to collect data are field note, questionnaire, and interview. The field notes for recoding how the students learn during conducting the lesson plan in the classroom and for recording the teacher's PCK in all steps of LS. The questionnaire and interview are used for recording teachers' feedback after they implemented LS to seek for the two questions in the aim.

3. Result

As the result, it found that LS helped teachers: to clarify the level of learning outcomes to which the instruction should be directed; to make a qualified lesson plan (it is not individual work); to become more cooperative to improve their teaching and students' learning; and to build up the professional capacity among the group work of their colleagues with the same subject. Furthermore it sheds light on content and instruction, instructional support systems, the implementation of teacher standards, and time management as the challenges.

4. Conclusion

LS is the key to success of every teacher in teaching mathematics and science. By writing a lesson plan together, the teachers can have the clarity concerning the level of learning outcomes to which the instruction should be directed. To make a qualified lesson plan, It is not individual work. By observing the classroom together, the teachers can learn how to teach. For teachers, administrators, and the school, to serve as vehicles for professional growth rather than performance evaluations have multiple benefits. By reflecting together, the teachers build up the professional capacity among the group work. Furthermore, the teachers understand what makes the learning of specific topics easy or difficult, and how to help students understand specific subject matter.. It identifies the distinctive bodies of knowledge for teaching, and represents the blending of content and pedagogy into an understanding of how particular topics, problems, or issues are organized, represented, and adapted to the diverse interests and abilities of learners, and presented for instruction.

5. Acknowledge

This research is (partially) supported by the Happy Chandara School, the Faculty of Education Khon Kaen University, and the Ministry of Education Youth and Sport Cambodia

References

[1] Ali F 2017 *Global Competitiveness Report 2017-2018*
[2] Engler J 2012 *STEM education is the key to the US's economic future* (US News & World Report)
[3] Inprasitha M 2011 *One Feature of Adaptive Lesson Study in Thailand: Designing a Learning Unit* (Journal of Science and Mathematics Education in Southeast Asia vol 34) chapter 1 pp 47-66
[4] Inprasitha M 2015 *Preparing Ground for the Introduction of Lesson Study in Thailand* (Lesson Study: Challenges in Mathematics Education) pp 109-120
[5] Ministry of Education Youth and Sport Cambodia (MoEYS) 2016 *Policy on Sicence, Technology, Engineering and Mathematics (STEM) Education.*
[6] Shulman L S 2015 *PCK: Its genesis and exodus* (In Re-examining pedagogical content knowledge in science education) pp 13-23 Routledge
[7] Stigler J W & Hiebert J 2009 *The teaching gap: Best ideas from the world's teachers for improving education in the classroom* (Simon and Schuster)
[8] Takahashi A 2015 *Lesson study: An essential process for improving mathematics teaching and*

learning (In LESSON STUDY: Challenges in Mathematics Education) pp 51-58
[9] Visa K et al 2016 *Students' Cognitive Processes in Mathematics Classroom Using Lesson Study and Open Approach with GeoGebra* (Proceedings of the 11th APEC-Khon Kaen International Symposium and the 9th International Conference on Educational Research)

What Teachers learned from STEM Education Project: Case Study of High School Teachers

Wisanugorn Nammungkhun[1*], Napaporn Yutthaisong[1], Wanphakorn Jumphonnoi[1]

[1]Banphaisuksa school, Khon Kaen Secondary Educational Service Area Office, Ministry of Education, Thailand.

*Corresponding author: n.wisanugorn@gmail.com

Abstract: The necessity of highly qualified supported scientists, technologists, engineers and mathematicians as STEM perceived by Thai government. STEM, an acronym for Science, Technology, Engineering and Mathematics, it has become a key driver in interdisplinary. This study revealed what and how teachers learned from participated in STEM Education Project (SEP). The SEP is a project which use driven STEM Education in Banphaisuksa school; 1) STEM training for teachers, 2) Hands-on activities for students ,3) Inspiring students to STEM through control robotic system activities. Four teachers who participated in SEP were participant in this study. Methodology regarded ethnographic research Through participant observation, interviewed teachers, and collected their documents and authors as a participant. The data analysis was carried out from a community of practice (CoPs) approach. Teachers as apprenticeship learned through legitimate peripheral participation in SEP community of practice. They developed an understanding of STEM Education from scientist and engineer. They learned what and how to teach STEM from implementation STEM activities in classroom. This study highlights the need of established teachers' collaboration to design STEM lesson with experts and their peers.

1. Introduction

The basic goal of education is to prepare students to able to live, work, and contribute to the society in which the live. A STEM education provides students with knowledge and skills of these areas; Science, Technology, Engineering, and Mathematics, connect them to real world experience and promote student-centered teaching approach [4]. In addition, STEM literacy has increasingly important as a 21 st century competencies, ability is developing a STEM-capable workforce and boosting interest and engagement in STEM [6]. That is, STEM Education was global debate for improving national education system and reach competitive capability in especially economy. Current, Thailand encourages teachers and stakeholders come to learn and develop their teaching approach, promote STEM to students in classroom. STEM Education is new issues for Thailand. [5] explored perception of teachers toward teaching STEM in Thailand, they found that the majority of teachers never heard about STEM Education. While the interesting of STEM Education was increasing in global issues, but the studies of STEM Education or promote STEM to students have insufficient. Moreover especially, the role of teachers for supporting STEM into their classroom.

Community of Practice (CoPs) is an evolutionary process for learning in community which have passion in something together [8]. The key point of CoPs is community knowledge, in sum community knowledge is greater than sum of individual knowledge ([3]. Investigated teachers enhance students' thinking project, such encourages teachers to learn how to do the research and develop students thinking [5]. That study emphasizes an important of CoPs as a crucial tool to understand people learn from new comer to old timer under social perspective. In addition, studied the collaboration in a Community of Practice working to integrate Engineering Design in Elementary Science Education [1],[7]. The studied suggest that the collaboration among community of practice participants are importance to success of school-based STEM Education reform. That is, collaboration of practitioners as community of practice is the key practice to accomplish STEM Education in school. The present study was to investigate what teachers have learn from the community of practice in STEM project in a secondary school.

2. Method.

The study is qualitative inquiry, interpretative approach was used to understanding and make meaning of participant' experiences. The instrument to collect data is semi-structure questions of interview.

3. Participant.

Three teachers were participants of this study, they are member of STEM project. Each participant teaches subject matters in school. The first one, teacher N she teaches biology in high school level and she have experiencing in teaching in over 25 years. The next one, teacher P she teaches chemistry in high school level and general science for junior high school level, she was 5 years in teaching job. The last one, teacher T she is social study teacher, but she interesting in STEM activities. Teacher T experience in teaching about 3 years.

4. School setting and STEM Project.

STEM Project is a school-based project to drive STEM Education in classroom. Such project carries out over two years scholar. The first year, we look for visited and learn with each other which they have experiences in STEM Education. In addition, two STEM camps have launched for encouraging students and also teachers to learn more about STEM. The last years, Teachers have integrated STEM into the classroom. They will work together to design, teach and also reflect the teaching practices. Moreover, they have learned closely with scientist and engineer.

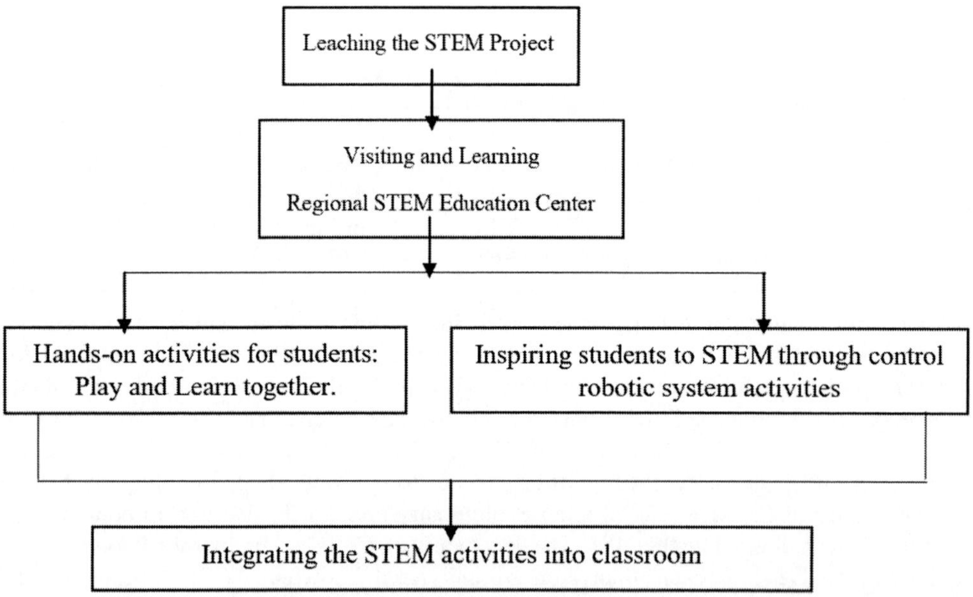

Figure1. Flowchart school-based project to drive STEM Education

5. Data collection

The source of data from semi-structured interviews. The interviews protocol includes of 5 questions to elicit what teachers' understanding and learning of STEM both ideas and practices. These questions show in below;

What do you think about STEM Education?

Do you think teaching STEM differ from traditional teaching? How?

How you participated in STEM Project in your school?

What did you learn of participate in STEM Project?

What do you think about the aspects of STEM should be teach in classroom?

6. Data Analysis

The interviews were transcribed verbatim. Three interview transcripts word by word. All researchers discussed to resolve an inconsistency of transcriptions. This establishment of trustworthiness, credibility of data analysis.

7. Findings and Discussion

This investigation is to examine what the teachers learn from they participate in STEM project of secondary school in Thailand. Three teachers from difference subject matter were source of data. This section will show and interpret what they have learn as well as their experience in STEM Education.

7.1 STEM Education as the new challenging and opportunities

STEM is a new topic for school in Thailand. The rising of the STEM topic come from Ministry of Education by Institute for the Promotion of Teaching and Technology (IPST). MOE propose the policy of STEM is the nation agenda and launch many projects for training the teacher and develop the activities for students over the country. That is, STEM is a new thing for several teachers include the teacher in this project. They perceived STEM Education as new opportunity for students' learning such as

"STEM Education differ from traditional instruction. Such integrate four discipline, Science-Technology-Engineering-Mathematics. The integration has useful to solve the real-life problems. The tradition separates them into independent subject matters, not link them to the classroom."

The participants have reflected their experience that in common classroom teachers take only their subject to students. STEM Education come to full fill the gap, such gap is the separating a problem issues into each subject (science, math, technology and engineering), instead integrate these subjects to operate an issue. They face with the new challenging for integrating many subjects in classroom, because they need to work with other teachers.

7.2 The Collaboration strengthen the learning activities

"For teaching (STEM) we have been collaborate with peers. Its quite difference from previous, Teachers in other subjects come to design, teach, and reflect the activity together. We use the content knowledge in each our subjects for helping students learning. I think it's a good thing for us and especially students"

The key factor driven the STEM project is teachers' collaboration. School administrator provided the time for teachers' professional learning community (PLC). Such community make a strengthening of teachers' professional. The teachers from each subject to support students learning, co-thinking, sharing, decision-making and reflection. The studied shown that professional community of lecturer in university, Supervisor, teacher and pre-service teacher for teaching STEM make a strong community as Community of Practice [2]. Moreover, CoP of teachers have built up the STEM Education in School context for enhancing students' learning performance [1].

Reference.

[1] Means B, Wang H, Young V, Peters V L and Lynch S J 2016 STEM-focused high schools as a strategy for enhancing readiness for postsecondary STEM programs **J Res Sci Teach 53** pp 709–736

[2] Wenger E 1998 Communities of practice: learning, meaning, and identity (New York: Cambridge University)

[3] Nammungkhun N, Satchukorn S, Saenpuk N, Yuenyong C and Chantaranuwong W 2017 Teachers' learning about research for enhancing students' thinking skills in science learning **IP Conference Proceedings1923** https://doi.org/10.1063/1.5019563

[4] National Academy of Engineering and the National Research Council 2014 STEM integration in K-12 education: Status, prospects, and an agenda for research (Washington, DC: National Academies Press)

[5] Chamrat S S 2016 STEM Education on the Road of Socially-Engaged Scholarship: Game Changer for Future Learning **Kasetsart Educational Review 31** pp 34–47

[6] Gherardi S and Nicolini D 2000 The organizational learning of safety in communities of practice **Journal of Management Inquiry 9** pp 7–18

[7] Aduldej T, Kulpatsorn S and Yuenyong C 2018 "Buildinup STEM education professional learning community in school setting: Case of Khon Kaen Wittayayon School" **AIP Conference Proceedings 1923**, 030067 (2018); https://doi.org/10.1063/1.5019558

[8]Srikoom W, Hanuscin D L and Faikhamta C 2017 Perceptions of in-service teachers toward teaching STEM in Thailand **Asia-Pacific Forum on Science Learning and Teaching 18**

Needs Assessment for the 21st Century Teaching and Learning in STEM Education to Promote Students' Problem Solving in Thailand

Wutcharapong Kinboon[1*], **Supphachai Sanghuaypai**[1], **and Artha Nantachukra**[1]
[1] Development, Rajabhat Mahasarakham University, Thailand

*Corresponding author's e-mail address: E-mail: w.kinboon@gmail.com

Abstract. The purposes of this study were 1) to investigate the current issues and the needs for the 21st century teaching and learning management in science and technology based on STEM education to solve issues in schools under the Office of Khon Kaen Primary Education Area 4, 2) to study the human resource development methods to enhance technology knowledge and skills to apply in solving science learning issues in schools by adopting STEM education approach, and 3) to propose the guidelines to support information technology and educational management based on STEM education. The population used in this study was a total of 2,185 teachers and staff in educational institutes under the Office of Khon Kaen Primary Education Area 4 in the academic year 2018. The 445 samples were selected by using cluster purposing sampling technique. The research tools used in this study were semi-structured questionnaire with 5-Likert scale questions divided into 2 parts: general information of respondents in the first part and open-ended questions in the second part asking the respondents' suggestions on future development of educational management. Data were analyzed by using descriptive statistics, percentage, mean, and standard deviation. The interpretation was also used in the content analysis. The findings in the study are as follows: 1) Behavioral abilities related to the technology adoption to solve science learning management issues were found in 5 aspects, including seeking, using, collecting, selecting, and writing information skills. From the overall behavioral assessment, the level of knowledge in technology and technology skills for communication was found at a high level)$\overline{X}$ = 4.52, S.D.= 0.66(.2) The human resources development methods to enhance and apply technology knowledge and skills in solving science learning issues in schools were effectively implemented by adopting STEM education approach. The technology used in science learning issues should be verifiable, adaptable, and up-to-date. 3) The guidelines developed the teachers' and staff's knowledge and skills of information technology, which in turn, supported educational management based on STEM education effectively in practice.

1. Introduction

The 20-year national strategy B.E. 2060 – 2579 (A.D. 2017 - 2036) has emerged in the relation to the 12th National Development Plan aiming to solve the country's sluggish development. The several plans and strategies for the future of the nation are directly manipulated by the government as well as the budget allocation is strategically arranged in proportion under the vision: [1] "the country will become a developed country with prosperity and sustainability under the philosophy of sufficiency economy to bring happiness to Thai people."

The schools under the Office of Khon Kaen Primary Education Area 4 are regarded as qualified educational institutes providing basic education in accordance with the nation's educational standards. Schools pursue strong administration and management of education, abiding the principles of good governance and philosophy. The professional development of teachers are constantly highlighted to sustain professional standards, ethics, and skills, especially in effectively adopting technology. In general, students in the educational service area are provided with basic education equally.

Although the use of science and technology in solving learning issues is prevalent, teachers and staff have limited skills in seeking information to design science teaching media for students. Moreover, the educational personnel's skills in collecting, selecting, using, and writing information need to be developed due to the rising demand for information technology adoption in students' science learning. In addition, teachers and staff were found with poor ability to write newsletters on website for communication purposes. Teaching is mainly based on theoretical method, instead of activity-based teaching approach. This results in the students' lack of interest and motivation to learn. Students have different levels of learning abilities, consequently, they perceive and interpret the knowledge differently [3].

Thus, the objectives in this study aim to understand the current issues and needs for using technology in science education based on STEM education in schools under the Office of Primary Education Area Khon Kaen Area 4 to develop students' learning skills with the 21st century requirements. The STEM education is further implemented to enhance knowledge and skills of educational human resources to adopt technology in solving students' learning problems effectively.

2. Objectives

To investigate the current issues and the needs for the 21st century teaching and learning management in science and technology based on STEM education to solve issues in schools under the Office of Khon Kaen Primary Education Area 4,

To study the human resource development methods to enhance technology knowledge and skills to apply in solving science learning issues in schools by adopting STEM education approach, and

To propose the guidelines to support information technology, leading to the effectiveness of educational management based on STEM education.

3. Population and samples

The population used in this study was a total of 2,185 teachers and staff in educational institutes under the Office of Khon Kaen Primary Education Area 4 in the academic year 2018.

The 445 samples were teachers and staff in educational institutes under the Office of Khon Kaen Primary Education Area 4 in the academic year 2018. The samples were selected by using cluster purposing sampling technique [2].

4. Research tools

The research tools used in this study are list below.

4.1 Semi-structured questionnaire consisted of questions asking about 1) knowledge of science and technology, 2) techniques of presenting information and activities, 3) channels and methods of publishing online information, 4) science and technology skills for communication purposes on websites, and 5) technology skills for distance learning information technology (DLIT). The questions were designed with 5-Likert scales, ranging from 5 as highly effective performance, 4 as effective performance, 3 as moderate performance, 2 limited performance, and 1 very limited performance. The final part of the questionnaire was the open-ended questions asking the respondents' suggestions on future development of educational management. The research tools were tested with the reliability of 0.87.

5. Data collection

The triangulation approach, survey, and feedbacks of the samples divided into groups of educational institutes were used to collect data from various resources. The respondents were those who had experience working in schools, attending the operational management meetings, brainstorming in discussion and sharing ideas for educational management.

6. Data analysis

Data were analyzed by using descriptive statistics, percentage, mean, and standard deviation. The interpretation was also used in the content analysis.

7. Results

7.1 the current issues and the needs for the 21st century teaching and learning management in science and technology based on STEM education to solve issues in schools under the Office of Khon Kaen Primary Education Area 4

The assessment of the use of technology to solve issues in teaching and learning science based on STEM education in school and the effectiveness of technology adoption are described below.

1. The findings showed the behavioral abilities related to the technology adoption to solve science learning management issues including seeking, using, collecting, selecting, and writing information skills. The overall assessment of teachers' and staff' abilities in 5 aspects was found at a high level *)* $\overline{X}$ = 4.52, S.D.= 0.66); consisting of knowledge of science and technology, science and technology skills for communication purposes on website, technology skills for distance learning information technology (DLIT), techniques of presenting information and activities, and channels and methods of publishing online information. Among the 5 aspects of teachers' and staff' abilities in adopting technology to solve issues in teaching and learning science, the first three aspects with highest level are presented as follows:

1) The science and technology skills for communication purposes on website was found at the highest level ($\overline{X}$ =4.67, S.D. = 0.63). The use of communication technology on website for seeking information among relevant agencies and the use of communication technology on website for operating information were significantly implemented.

2) The techniques of presenting information and activities was found at high level *)* $\overline{X}$ = 4.58, S.D.= 0.6*(*7. The techniques of writing contents, newsletters with headlines, introduction, and content bodies with detailed description were clearly presented.

3) The technology skills for distance learning information technology (DLIT) was found at high level *)* $\overline{X}$ = 4.54, SD = 0.72)). The evidence showed the presentation of activities on the website, presentation of information on educational management, and news for public relations in social media. Moreover, the knowledge of science and technology was found at a high level ($\overline{X}$ = 4.45, S.D. = 0.67). The channels and methods of publishing online information was also found at a high level ($\overline{X}$ = 4.35, S.D. = 0.74), respectively.

7.2 Guidelines for the human resource development methods to enhance and apply technology knowledge and skills in solving science learning issues in schools based on STEM education approach

1. Readiness at all time (Timeliness): the information must be available and accessible at all time; whenever and wherever.
2. Accuracy: the information must be correct, precise to the reality, and up-to-date.
3. Speediness: the information must be convenient and timely to transfer.
4. Completeness: the information must be complete and most currently updated.
5. Relevance: the information must be relevant to the users' need as much as possible.
6. Verifiability: the information must be certified to ensure its accuracy and reliability.
7. Adaptability: the information must be adaptable to be up-to-date.

8. Discussion

The guidelines to promote science education based on STEM education in relation to information technology are effective in all aspects including behavior, skills, and practices of technology adoption in solving learning management issues. Teachers and staff were able to seek, collect, use, select, and write information effectively. They also had strong technology knowledge and skills to publish information online and use of web-based communication technology to acquire information among agencies. Moreover, they had skills in presenting information, activities and image and writing information e.g. contents, news, and articles that were accurate, fast, convenient to access, and beneficial for the users, which in turn, develop the effectiveness of science education management.

References

[1] Chester I B 1968 The Function of Executive (Cambridge, Harvard University Press)
[2] Lee J C 1977 Educational psychology (New York: Harcourt Brace Jevanoich. Inc)
[3] Phillip B and Susan G 1992 Electronic Book for Early Learners **Educational and Training Technology International** pp 281–290

IOP Conf. Series: Journal of Physics: Conf. Series **1340** (2019) 012074 doi:10.1088/1742-6596/1340/1/012074

Investigate Grade 7 Students' Knowledge Construction About Diffusion and Osmosis

Phattaraporn Pondee and Chokchai Yuenyong

Department of Science Education, Faculty of Education, KhonKaen University, 40002, Thailand

Corresponding author's e-mail address: phattaraporn.sp@gmail.com, ychok@kku.ac.th

Abstract. This research aimed to investigate the students' constructing scientific knowledge about diffusion and osmosis with analogy approach. The subjects were selected by purposive sampling method including 30 Grade 7 students in Yasothonpittayakom School, Muang District, Yasothon Province, Thailand, in second semester of 2014 academic year. Gathering data by semi-structured interviews and open-ended questionnaire. The data analyses were qualitative analytical techniques. The results of study revealed that almost students held the concept of dissolve to mention diffusion and held the concept of the particulate and the movement of particulate. Their explanations of diffusion and osmosis mentioned various phenomena.

1. Introduction

Science does not exist as a body of knowledge separate from knower. On contrary, science is viewed as a set of socially negotiated understandings of the events and phenomena that comprise experienced universe [11]. The phenomena of science classroom of Thailand in the junior high school that will improve because the teachers emphasize rote learning, the Thai curriculum load overabundant content, and teachers lack the building motivation to students and the examination focus on the memorization [10]. So the teacher will relate the students to learn by themselves. The dominant theory of learning today is called constructivism. This theory claims that knowledge is actively constructed by the student, not passively absorbed from textbooks and lectures. The knowledge is constructed from experience [4]. The constructivist view comes in many variants in science education literature on students' learning [7]. Learners construct their own meaning, using their own ideas, beliefs, experiences, etc., to interpret the message conveyed by teachers [8]. Thereby the teacher will chose the appropriate approach to teach specific science content to promote meaning learning. There are several reasons why we should focus on diffusion and osmosis concepts in biology. Diffusion and osmosis are the key to understanding many important life processes [9]. Diffusion is the primary method if short distance transport in a cell and cellular systems. An understanding of osmosis concepts is the key to understanding water intake by plants, water balance in land and aquatic creatures, turgor pressure in plants, and transport in living organisms. In addition, diffusion and osmosis are closely related to key concepts in physics and chemistry such as permeability, solutions, and the particulate nature of matter [6]. [1] report about misconceptions were detected in six of seven conceptual areas: the particulate and random nature of matter, concentration and tonicity, the influences of life forces on diffusion and osmosis, the process of diffusion, and the process of osmosis. [3] reported that college students in a "constructivist" course learned significantly more traditional biology course. It is motivation of learning. Various instructional methods can be used for meaningful learning. Using analogies in science classroom help students make connections between everyday life and the concepts. Humans use analogical reasoning naturally, especially when trying to explain something to others (S. Rena Smith and Sandra K. Abell). Analogy is attractive because it is a simple way to explain abstract ideas in familiar terms [2]. Analogies are considered because of their potential to compare one object or situation to another [5]. The participants are in the enrichment science classroom which this curriculum develop and promote students who talent in science, held integrating education; child center. To develop potential of individual student and emphasize higher-order thinking skill, problem solving skill, critical thinking skill and creative thinking skill. Students must to inquire knowledge by themselves. Accordingly this study presents to investigate students' knowledge construction about diffusion and osmosis. The teachers can use the findings to develop students' understand and chose appropriately approach to teach diffusion and osmosis.

2. Methodology

The study involved investigating Grade 7 students' knowledge construction of diffusion and osmosis. This study was designed for collecting and analyzing data in order to answer.

2.1. The qualitative research method was applied as an interpretive strategy

2.2. Participants

The target group, selected by purposive sampling, was composed of 30 Grade 7students (From 1/14) at Yasothonpittayakom School, who were in enrichment science classroom.

2.2.1 School background. Yasothonpittayakom School located at 34 Muang District, Yasothon province, Thailand which is part of the secondary educational service area office 28 and taught from grade 7-12. There are 3,291 students and 165 teachers.

2.2.2. Target students background. Target students included 30 Grade 7 students of from 1/14 of Yasothonpittayakom School who were in enrichment science classroom. There were 14 males and 16 females.

2.3. Research instruments

Research instruments included the diffusion and osmosis understanding test (DOUT) that is open-ended question. There were 2 items constructed and developed by myself and peer.

The first item is "After the boy finished watercolor brush, the boy took the brush dipped in water containers. The next time the boy seen the movement of colors spread over an area of water. Why the colors are so dispersed?"

The second item is "When you squash the orange, the orange juice move to lower part but the crude fiber cannot move. Why is it that?"

These questions ask the students to check their existing idea and construct knowledge about diffusion and osmosis. The question determine situation that can show concept of diffusion and osmosis.

The second instrument was the semi-structured interview form for the students' understanding of the diffusion and the osmosis. This interview was conducted in order to confirm the results obtained from the questionnaire.

2.4. Data collection

2.4.1. School background, including student, teacher, and general data of target students were collected.

2.4.2. The researcher administered the questionnaire on students' understanding of the diffusion and the osmosis

2.4.3. Some students were selected for an interview who answered questions interestingly based on the semi-structured interview from the diffusion and the osmosis that confirmed the result of the questionnaire.

2.5. Data analyses
The data obtained from the questionnaire administrated with the students, and the semi-structured interview conducted with the students was analyzed as follows:

2.5.1. 1) The school baseline data was written in a descriptive style to reflect the holistic view of the school.

2.5.2. The practicum students' understanding of the diffusion and the osmosiswere analyzed. Each students' answer was read to find key words, encoded to represent the constructing the diffusion and osmosis from analogy situation. The answers were read again to classify ideas.

2.5.3. Students' understanding of the diffusion and the osmosis from the questionnaire was analyzed into good understanding, unclear understanding, and understanding with discrepancies.

2.5.4. Students' interview on their understanding of the diffusion and the osmosis were transcribed to confirm the conclusion drawn from the questionnaire.

2.5.5. The analyzed results were interpreted thestudents' understanding and constructing of diffusion and osmosis to categorize group of the answer with the supervisor and research assistant.

3. Results
The results of this study are presented in second sections. First, the alternative concept of students' understanding of diffusion and osmosis from questionnaire are reportedin Table 1. Second, the students' knowledge construction of diffusion and osmosis that resulted from the questionnaire and semi-structured interviews are reported.

3.1. Alternative concept

3.1.1. Alternative concept about diffusion. The students' answers from the questionnaire consist of two main concepts such as dissolve and reproduction that illustrate frequency of each concept in table 1.

Table 1. Alternative concept about diffusion of grade 7 students' understanding of the questionnaire

Topic	Concept	Frequency)people(	Percentage
Diffusion	Dissolve	29	96.67
	Reproduction	1	3.33
	Total	30	100.00

- Example of students' alternative concept about diffusion. Question: After the boy finished watercolor brush, the boy took the brush dipped in water containers. The next time the boy seen the movement of colors spread over an area of water. Why the colors are so dispersed?

- Description of diffusion concept. Diffusion is the net movement of a substance from a region of high concentration to a region of low concentration. The dipping a paintbrush is the diffusion and the route of the color is the movement of substance. The movement of the color liken substance that move from a region of high concentration to a region of low concentration.

- Example the answer that students held the concept of dissolve

"The color contact with the water, it spread around the area. (S01)" This answer show the concept about solvent and solute. The water is solvent but the color is solute.

"Because the color mixes the waterto occur diffusion. It similar the combination between solution and solvent (S05)." This answer illustrates the explanation about the combination between colors and water which as dissolve of solution.

"The color molecule insert in the water molecule (S06)." This answer shows that the explanation about insertion of molecule.

"Because the color has less volume than the water. The color can dissolve in the water (S10)." This answer shows the concept about volume of solvent (water) and solute (color) that volume means size of something which as 3D shape. The principle of dissolve uses the quantity of substance.

"Because the water breaks up bonding force of the color which make to occur the distribution of color (S21)." The answer shows explanation the breaking up of bonding force of color. The substance in nature is huddle. The energy is used to separate substance.

- Example the answer that students held the concept of reproduction

"The color seeks a mate that the distribution of color makes to occur new color (S04)." The student understands that the color breed between two color to occur new color. The principle of reproduction means the addition descendants that same their parent. This example shows misconception about diffusion.

3.1.2. Alternative concept of osmosis. The students' answers from the questionnaire consist of two main concepts such as movement of particulate and size of particulate that illustrate frequency of each concept in table 2.

Table 2. Alternative concept about osmosis of grade 7 students' understanding of the questionnaire

Topic	Concept	Frequency (people)	Percentage
Osmosis	movement of particulate	7	23.33
	Size of particulate	23	76.67
	Total	30	100.00

- Example of students' understanding about osmosis. Question: When you squash the orange, the orange juice move to lower part but the crude fiber cannot move. Why is it that?

- Description of osmosis concept. Osmosis is the spontaneous net movement of solvent molecules through a semi-permeable membrane into a region of higher solute concentration. When you squash the orange, the orange juice move to lower part but the fiber cannot move. It can explain that a semi-permeable allow some substance through a region of low concentration to a region of high concentration.

- Example the answer that students held the concept of the movement of particulate

"The color cannot dissolve in the water and molecule move to around the water. (S08)"This answer show explanation concept the movement of molecule. But the student misunderstands about dissolving of color in the water.

"The water move from a region of high pureness to a region of low pureness (S29)."This answer uses the concept of the feature of water to explain water movement.

- Example the answer that students held the concept of the size of particulate

"The fiber has bigger size than filter so the fiber cannot through the filter (S02)."This answer shows the explanation size of particulate between particulate and container.

"The fiber is fluid. It can through the filter." (S14) This answer use concept of state of matter to explain situation.

"Because the fiber has bigger than the filter. The orange juice is colloid and mixture. It is osmosis. The roots of the plant absorbed the water and mineral that must be small size enough to pass (S30)."This answer use the kind of matter to refer osmosis concept and illustrate example about osmosis phenomena.

3.2. Students' knowledge construction

This section presents the students' knowledge construction from the semi-structure interviews that use analogy approach to ask participants.

3.2.1. Example of students' knowledge construction about diffusion.

Interviewer: Can you explain situation that it similar with the brush dipped in water?

S30: Many events that can be easily seen.

Interviewer: Can you instance situation?

S30: My friend broke wind in the classroom. Then smell spread around the classroom.

Interviewer: Does it relate to diffusion?

S30: Yes, it is the diffusion of smell into the air.

3.2.2. Example of students' knowledge construction about osmosis.

Interviewer: Can you explain easy situation that it similar with the squashing the orange?

S05: I think...if the fat man and the thin man walk through the narrow channel. The fat man cannot through the narrow channel but the thin man can through this channel.

Interviewer: Why does it similar with the squashing the orange?

S05: Because the thin man has smaller than the channel. So the thin man can through the narrow channel. Thus the thin man likens the orange juice.

Interviewer: Does it relate to osmosis?

S05: Yes, the men like substances that can movement through channel by the size of body. It is special characteristic that select substances to pass into cell of organism.

The both of constructing concept illustrate that the students can construct situation to relate each concept.The student can explain situation to relate their concept which similar situation in the test.

4. Conclusion and discussion

These research findings indicated that grade 7 students' knowledge construction about diffusion and osmosis were unclear because the most students' can explainpart of concept. The situation may not clear to relate to the concept of diffusion and osmosis. Alternative concepts illustrate students' understanding about their concept which according to major misconception about diffusion and osmosis. Similarly, Arthur (1995) report major misconception. The particulate and random nature of matter look alike alternative concept in the research.From the result, students held concept of dissolve to explain about diffusion that students do not understand concept of concentration of solution. The learning activity will engage about the concept of dissolve to conduct about the concept of concentration of solution. The students' constructing concept of diffusion and osmosis mentioned various phenomena. Good constructing of knowledge that students develop their existing knowledge to change conceptual. Students will construct the knowledge from prior experience by themselves. Christianson and Fisher (1994) reported about constructivist course learned significantly more diffusion and osmosis concepts. It was suggested that motivation and learning could be enhanced by: Allowing teacher-student and student-student discussion, allowing time for prediction, using concept mapping to anchor concepts and construct meaning, using the a variety of teaching method. Different students will respond to different techniques.In conclusion, outcome of this research study can improve learning activity to appropriate to construct diffusion and osmosis.

References

[1] Authur Louis Odom 1995 *Secondary & College Biology Students' Misconceptions about Diffusion & Osmosis* (The American Biology Teacher vol 57) chapter 7 pp 409-415

[2] Aubusson P J 2006 *Metaphor and Analogy in Science Education* (New York: springer)

[3] Christainson R G & Fisher K M 1994 *Analysis of student learning about diffusion and osmosis in three different biology classes using Odom and Barrow's Diagnostic Test* (Unpublished manuscript)

[4] Christie A 2005 *Constructivism and its implications for educator*s Retrieved February 20, 2015, (from http://alicechristie.com/edtech/learning/constructivism/index.htm)

[5] Duit R 1991 *The role of analogies and metaphors in learning science* (Science Education vol 75) chapter 6 pp 649-672

[6] Friedley Y Amir R Tamir P 1987 *High school students' difficulties in understanding osmosis* (International Journal of Science Education vol 9) chapter 5 pp 541-551

[7] Good R Wandersee J & Julien J St 1993 *Cautionary notes on the appeal of the new "Isms" (constructivisms) in science education* In K Tobin Ed The practice of constructivism in science education (Washington, DC: AAAS Press) pp 71-87

[8] Gunstone R 1990 *"Children's science": A decade of developments in constructivist views of science teaching and learning* (Australian Science Teachers Journal vol 36) chapter 4 pp 9-19

[9] Louis Odom and Lloyd H Barrow 1993 *Freshman biology majors misconceptions about diffusion and osmosis* Retrieved March 17, 2015, (from https://archive.org/stream/ERIC_ED362399/ ERIC_ED362399_djvu.txt.)

[10] Thomas 2012 *Critique the phenomenon of Thai classroom* Retrieved February 20, 2015, (from http://www.qlf.or.th/Home/Contents/499)

[11] Tobin K Ed 1993 *The practice of constructivism in science education* (Washington, DC: AAAS Press)

Applying the philosophy of sufficiency economy and STEAM knowledge of Grade 11 Students in the STS biodiversity Unit

Phimnaphat Pagsangkanae and Chokchai Yuenyong*
Faculty of Education, Khon Kaen University, Khon Kaen 40002, Thailand

*corresponding author: ychok@kku.ac.th

Abstract. Learning biology through social context local wisdom helps to encourage students to appreciate the value of what is available in the community. This research has purposed to develop Science Technology and Society (STS) based on Yuenyong, 2006 according the Philosophy of Sufficiency Economy (PSE) with science on biodiversity learning activities consist of 5 steps: (1) Identification of social issues, (2) Identification of potential solutions, (3) Need for knowledge, (4) Decision-making, and (5) Socialization stage. In addition, learning activity was used according to PSE that includes three elements: moderation, reasonableness and self-immunity, as well as two conditions which are knowledge and moral. Eleventh grade student from high school in Khon Kaen, Thailand were selected for the study. However, this research is a qualitative research. Focus on interpretive paradigms of phenomena occur in the classroom. It is a way to study and analyze the Philosophy of Sufficiency Economy in Biology classes about biodiversity of the tadpoles that local name is Huanh. There are tools to collect data. Record of observation of class notes after teaching and student work. As a result, learning of STS approach enables stimulation students to apply Philosophy of Sufficiency Economy and STEAM knowledge in the biology class through social cohesion related to the local wisdom of the local community.

1. Introduction

Education in the 21st Century focuses on learning for life. Its main purpose is study for education, society, economy, politics and culture. It is well known that a good education is the driving force that keeps society has evolved and progressed to the good side. The changes in education over the past century have resulted in science becoming a central part of learning. Science is also an important intermediary in advancing technology because we bring science to new technologies. As a result, science and technology are inextricably linked. However, this century has transformed education into a special aspect that is important for future-oriented education. The first thing to focus on is the integration of knowledge and application to create more knowledge into the application. This has resulted in a major turning point in education not just the use of knowledge. It is the knowledge to innovate to apply in the daily life reasonably creative and virtuous [3]. Especially, linking science to the well-being of local people and local wisdom helps students to develop and promote the creation of new technologies that consistent with the need and demand for people in the community. Moreover, the science associated with the local help students realize the value and importance of the resources that are around the students. They can create new things from the wisdom that they may have forgotten to return to the new value again. To see that science is not just a question of what we suspect. They can create new things from the wisdom that they may have forgotten to return to the new value again. It is the knowledge and understanding that is used to develop the environment and resources around us to be stable, balanced and sustainable [4].

However, to create a balanced and sustainable community economy, it is necessary to have quality human resources development. Create intensity Which is an important learning resource for quality human resource development, which is school by developing knowledge in science, technology, engineering and mathematics (STEM) [7]. STEM education guidelines that have been integrated knowledge between subjects such as scientific knowledge, technological knowledge, engineering knowledge and mathematical knowledge combined [5]. In 2016, Georgette Yakman [8] and her team put forward STEAM education (science, technology, engineering, art and mathematics) on the basis on STEM education. From the perspective of science and technology, interdisciplinary concept of different subjects will be integrated for the development of modern society to provide excellent human resources support. Schools are learning STEAM learning programs to provide students with the skills and knowledge necessary to develop in the 21st century. But will produce designers and engineers tomorrow only to develop innovative ideas and problem-solving abilities to ensure that our students become technology builders, not just consumers. Student should study ensures contextualized and meaningful learning.

In Thailand, His Majesty King Bhumibol Adulyadej has brought the philosophy of sufficiency economy (PSE). This is an approach that will encourage people to become stable, balanced and sustainable. It can be applied to any situation, not just science education or country. The philosophy of sufficiency economy includes three elements: moderation, reasonableness, and self-sufficiency, and requires two conditions in order to work: knowledge and moral [6]. The application of the philosophy of sufficiency economy of science class will encourage students to learn science that is close to real conditions. Yuenyong and Narjaikaew [10] reported learning education science technology and society (STS) approach in conjunction with the application of the philosophy of sufficiency economy support encourage students to learn throughout life.

In 2004, UNESCO [1] reported that an important tool in teaching students to understand the issue of sustainability in schools is good. Teachers should teach about biodiversity. Because diversity affects all living things in the community and affects the ecosystem. In addition, biodiversity can directly affect people in the community from ecological dependence. Thus, learning about biodiversity does not only encourage students to learn about local organisms. It also aims to develop social, political, economic and cultural life of people in the community. Encourage students conserve resources with the community [2].

For this reason, this study focuses on the application of the philosophy of Sufficiency Economy and STEAM knowledge in biodiversity by learning concepts through Yuenyong (2006) [9] science technology society (STS) class of biology students in grand 11th.

2. Research Methodology

This research is a qualitative research focusing on the paradigm of interpretation to study the application of the Sufficiency Economy Philosophy of students in grand 11th in During the first semester of 2018, Khon Kaen University Demonstration School a high school in Khon Kaen, Thailand. Focus on education on the application of sufficiency economy in biology classes. During STS learning activities, the application of the philosophy of economic sufficiency and the study of how students apply the philosophy of sufficiency economy observed of planning behavior among groups, work piece and journal writing.

2.1. Participants

The participants consisted of 70 students that were divided into groups of 5-6 persons from a high school in Khon Kaen, Thailand who entered in science program in grand 11th about biology class. Most students in the classroom have a different way of lifestyle than a rural one. It is a family town. Therefore, students do not understand the way of people in local the community and do not know the local biodiversity as they should.

IOP Conf. Series: Journal of Physics: Conf. Series **1340** (2019) 012075 doi:10.1088/1742-6596/1340/1/012075

2.2. Lesson plan
Lesson plans, the concepts based on biodiversity, science, technology and society of Yuenyong (2006). The purpose is to study the application of the philosophy of Sufficiency Economy and can link the scientific knowledge and local wisdom in the biology classes. The lesson plan was developed by the following steps:

- Study the curriculum content and structure Standard measure and indicator in Teaching Plan based on Science Curriculum in Basic Education Core 2008 on biodiversity of students in grand 11th.
- Create a Lesson plan on concept of Science, Technology and Society (STS) approach of Yuenyong (2006). The concepts based on biodiversity for student can the application of the philosophy of Sufficiency Economy. The teaching of STS consists of 5 steps: including step 1 identification of social issues stage, step 2 Identification of potential solutions stage, step 3 Need for knowledge stage: step 5 Decision-making stage and step 5 Socialization stage.
- Expert panel for reflections on the STS Biodiversity lesson plan.
- Improve learning management plan based on the guidance of Experts.
- Use learning plans with target students.

2.3. Intervention
Activities learning in biology class does not have a strategy to entice students to learn in a context that can be linked to local communities. Therefore, activities to define the intervention strategy by using the STS method. It encourages and encourages students to learn through social context and to know what is available in the community. The pedagogical intervention took place over two weeks. In title biodiversity and applying the philosophy of Sufficiency Economy in biology class, it covers the organisms existing in the Northeastern of Thailand community. Classes were developed to encompass support to love the homeland, watch videos that link to the philosophy of Sufficiency Economy, group activities, discussion and creation the project.

Table 1. Overview of "Huanh-Kob (Tadpoles)" Biodiversity Lesson Plan through STS Yuenyong (2006) and applying of the philosophy of Sufficiency Economy (PSE)

Stage	Learning activity
1. Identification of social issues stage.	1.Teacher ask student about organism in around for into meaning of biodiversity. 2. Teacher opened music VDO " Hor Mok Huak Bpai Faak bpaa " meaning bring steamed of Tadpoles bake to leave offer aunt. With music about eating culture and broad-mindedness of people in the Northeast of Thailand  **Figure 1.** Song " Hor Mok Huanh Bpai Faak bpaa " (Source: https://www.youtube.com/watch?v=FudAnrJkOb4) Questions for students:

	1. How are our daily lives linked to biodiversity? 2. How can people in the community make money from biodiversity? 3. "If students are member in the community who like to eat tadpoles, student will be what are some ways to make biodiversity of tadpoles and the community of students have revenue from tadpoles a stable and sustainable how is the balance caused?"
2. Identification of potential solutions stage.	1. Students identification of potential solutions to the problems of the community as possible about generating revenue and maintain a variety of sustainable tadpoles. 2. Each group of students brainstormed on problem solving about revenue and biodiversity of tadpoles. "what the existing knowledge solve this problem and have any knowledge that students need more used to solving the problem? 3. Student see the video on the application of the philosophy of Sufficiency Economy in daily life.  **Figure 2.** Sufficiency Economy? (Source: https://www.youtube.com/watch?v=2CjE6RWrB5I)
3. Need for knowledge stage	In need for knowledge stage used 5E inquiry for learning activity: 1. Engagement: Teacher opened the VDO about "Father African Bullfrog help tadpoles" for show the relationship of tadpoles with anther organism in ecosystem.  **Figure 3.** Father African Bullfrog help tadpoles (Source: https://www.youtube.com/watch?v=325Xz26Z7Tk&t=4s) Questions for students: How do tadpoles and frogs interact with the ecosystem? 2. Exploration: Teacher gives the question on worksheet about biodiversity and the application of the philosophy of sufficiency economy and allow each group to brainstorm ideas for answers.

	3. Explanation: Each student grout showed the idea the question on worksheet and the information is summarized as their own knowledge. 4. Elaboration: Teacher explain about general knowledge about frogs and tadpoles, habitat, culture and living by applying the philosophy of sufficiency economy in daily life. 5. Evaluation: Student present what they have learned from their classrooms.
4. Decision-making stage.	1. The students in each group to share their opinions and decisions by choosing the best way to how to solve diversity and generates revenue from tadpoles to the community.
5. Socialization stage.	1. The students in each group to share their solutions by infographic through an online community on Facebook for 1 week. 2. The students in each group written report describing the work of the group.

3. Data Collection and Data Analysis

The researcher collected data by using 2 Science Knowledge Management Plans based on Science, Technology and Society (STS) approach and apply to the philosophy of Sufficiency Economy titled "Huanh-Kob (Tadpoles)" with the target group, total duration of 4 hours. The participant's observations were recorded report, worksheet and student performant was taken through a paradigmatic interpretation. To examine the application of philosophy of sufficiency economy of students in each of the STS stages of learning in biodiversity content. The information is subject to quality assurance by experts, depending on the researcher's framework. In designing this learning activity, STS will support students to apply the philosophy of sufficiency economy. It consists of includes three elements: moderation, reasonableness, and self-sufficiency, and requires two conditions in order to work: knowledge and moral.

3.1 Data collection and data analysis of worksheet.

Table 2. Data collection and data analysis of worksheet by discusses and determine in each group.

Science, Technology and Society (STS) approach	The Philosophy of Sufficiency Economy	Example answer
	Three Elements:	
1. Identification of social issues stage: Questions for students in work sheet. - What is biodiversity? - What are the tadpoles? - If we eat too much tadpoles, how will it affect us? - How do tadpoles affect	1. Moderation: Sufficiency, not much and not too little, without causing detriment to others.	Students realize the consumption of tadpoles. Growth of tadpoles in each age range. Recognizing that eating too much tadpoles can affect the ecosystems of the offspring and other organisms. Students are aware of their livelihoods and need to rely on

biodiversity? - How can students make money from tadpoles? - Can the tad be enough for the needs of the community? - What should I do to get rid of		the environment to balance the ecosystem. Students are aware of the consequences after the loss of biodiversity.
the tadpoles from nature? - How does the loss of tadpoles from nature affect the ecosystem? - Do students agree with eating tadpoles? 2. Identification of potential solutions stage. Questions for students in work sheet. - What knowledge do students need to address the social issues associated with tadpoles and the sustainability of their community?	2. Reasonableness: The decision to do something carefully and through contemplation.	Students need to know why these issues are important. Students must know why sufficiency economy is essential to life. Students can use scientific reasoning to answer questions. Students are aware of the importance of losing diversity by bringing the scientific knowledge to the rational explanation. Students plan the feasibility of doing business under the possibilities
3. Need for knowledge stage. Questions for students in work sheet. - General Information about tadpoles. - How to make a tadpole? - The importance of tadpoles to the ecosystem. - The main reason for the number of tadpoles down. - Why study the diversity of tadpoles? - How to business planning? - How to market planning? - How to prepare for business? - How do the tadpoles in the community?	3. Self-sufficiency: To prepare for the impact of the changes that occur	Bringing scientific knowledge to solve problems related to tadpoles. Predict the feasibility of business planning and marketing using tadpoles. Prepare to cope with the problems and consequences of eating tadpoles. Know how to make money and reduce the impact caused by the loss of biodiversity on Tadpole population decline. Indicates possible ways to address diversity and downstream tadpoles in their communities.
- How to promote the income of people in the community? 4. Decision-making stage. Questions for students in work sheet. Questions for students in work sheet. - How does the student solve the problem in order to sustain the	Two conditions: 1. knowledge: Round knowledge in different fields of knowledge that will be instrumental in linking the planning and decision-making.	Bringing scientific knowledge to help solve biodiversity problems without the emotions of the students. Planning guidelines for answers that make sense. It can link knowledge about biodiversity and knowledge in other fields.

community sustainably and by making money from tadpoles to balance the biodiversity?		The knowledge that exists to solve the problem.
5. Socialization stage. Questions for students in work sheet. Questions for students in work sheet. - "If students are member in the community who like to eat tadpoles, student will be what are some ways to make biodiversity of tadpoles and the community of students have revenue from tadpoles a stable and sustainable how is the balance caused?"	2. moral: Being aware of things in an ethical and honest intelligence unit in lifestyle.	Students are aware of the loss of habitat and extinction of tadpoles. Students aware of the consequences that would happen to the environment and recognizing the importance of tadpoles. Students are aware of the impact that will come to the community during do business.

4. Finding

The results showed that the teaching of biodiversity based on Science Technology and Society (STS) approach through Yuenyong (2006) promotes and supports the application the philosophy of Sufficiency Economy (PSE). This shows students make the process revenue from biodiversity in local. The learning process of the students is under the application the philosophy of Sufficiency Economy includes three elements: moderation, reasonableness, and self-sufficiency, and requires two conditions in order to work: knowledge and moral could be analyzed. On the way of students' developing solutions related the tadpole, it indicated that they applied STEAM knowledge.

4.1 Students' generating ideas of application of sufficiency economy philosophy and STEAM knowledge. Students' generating ideas of application of sufficiency economy philosophy and STEAM knowledge could be interpreted through students' completion on the worksheet.

It is a worksheet that occurs during the STS learning activities. The worksheet aimed student can connect biology education style with social issues and application of sufficiency economy philosophy in science classroom. In addition, student should have created new idea or innovative for to solve that the problem. The worksheet provided student can connect biodiversity content with PSE and STEAM knowledge by STS learning activities.

This worksheet may illustrate the students' thinking processes about the application of sufficiency economy philosophy in the biology class. The worksheet is stored procedurally. Identification of the social issues stage, Identification of potential solutions stage, need for knowledge stage and Decision-making stage, as shown in the following table.

Table 3. Result of applying the philosophy of sufficiency economy of worksheet in Identification of social issues stage.

Group	Description	Involved from PSE

IOP Conf. Series: Journal of Physics: Conf. Series **1340** (2019) 012075 doi:10.1088/1742-6596/1340/1/012075

1	Students can link to listen to the song and help each question. The question brings a sense come into the question. Why take such foods made from tadpoles to leave her, what other things to leave? Some questions and represents a realized number of tadpoles in the community, if people eat a lot of tadpoles. By the question "Is there any way to breed tadpoles to meet their needs?" In addition, the students also take into account the ecological system in that case "If control of the tadpoles and release of the tadpoles back to the natural habitat would result in tadpoles' survival and as a result, the environment would not have any impact on the food chain"	From the description of the students found, students can positively address social issues and question their scientific knowledge. And that the tadpoles can significantly affect the balance of ecosystems. It represents an important link with the main philosophy of sufficiency economy are moderation, reasonableness, knowledge and moral.
2	After listening to music, students can identify social issues. Why tadpoles are popular among people in the community. However, the students went on to explain that the fact that people in the community eat tadpoles makes the tadpoles in the community balanced because community members are part of the ecosystem in the community.	Based on the student's suggestions, show that Students can link scientific knowledge to social issues. And it can be said that the eating of people in the community does not encroach on life. But it shows that people and the environment are dependent on each other. It illustrates the linkage the philosophy of sufficiency economy that students can link the philosophy of sufficiency economy are moderation, reasonableness and knowledge.
3	From listening to music to identifying social issues. Students use the feeling to ask questions. Why do I have to leave the tadpoles? Is it possible to leave? And explain that in the community there is enough volume. Because of the source of the farmland, which is the frog's habitat, there are frog spawn.	By explaining that students do not understand the way of life of the rural Northeast. And students do not understand about the environment of the tadpoles and frogs. It is based on the knowledge that students have in predicting the wellbeing of life. Which links the main philosophy are reasonableness knowledge and moral.
4	Students identify social issues. "How to need to eat tadpoles in any season to be appropriate. And to explain that should have control of the population to eat, the survival rates of tadpoles suit the environment.	Indicates the amount of tadpoles that are sufficient for survival in the community to grow into frogs. This can be linked to the philosophy of Sufficiency Economy are moderation, reasonableness and knowledge.
5	The social aspect that students identify. "How do tadpoles' taste? Is there any way to make delicious taste of tadpoles?" And explaining that eating tadpoles represents the local identity of the Northeast.	It shows that people have not eaten tadpoles but know that they are local foods. This can be linked to the philosophy of Sufficiency Economy are reasonableness and knowledge.
6	Social issues identified by students. "Why are people in the community to eat the tadpoles." Eating tadpoles during spawning season. As a result, the tadpoles are reduced	Students' explanations show that students are aware of eating too much natural tadpoles. This can be linked to the philosophy of Sufficiency Economy are moderation, reasonableness and

	in nature. People in the community should keep their natural balance. Although tadpoles and frogs are local food.	knowledge.
7	Identify social questions "How do people stop eating tadpoles?" And explaining that eating too much tadpoles can result in lower tadpoles in the community. As a result, the ecosystem is not fertile. Frogs may be extinct in the future.	Students demonstrate awareness of the nature and loss of ecological balance in the community. This can be linked to the philosophy of Sufficiency Economy are moderation, reasonableness knowledge and moral.
8	The question of social determinants is that. "How delicious, why eat?" The students explained that. In the Northeast there are many interesting creatures. Why do you have to eat tadpoles? Eating tadpoles and frogs, resulting in the balance system. If the tadpoles decrease, the frogs will decrease. And more insects. Cause the effect on the food chain because the frog is a member of the ecosystem.	The explanation shows that students never eat tadpoles and do not know how tastes are. Proposed solutions to the problem of eating other organisms instead of eating tadpoles and linking them to how ecstasy affects the ecosystem. This can be linked to the philosophy of Sufficiency Economy are moderation, reasonableness knowledge and moral.
9	The questions that students make about social issues. "How does the population of tadpoles in the water dive?" Overdose may affect the survival of frogs in the community. Therefore, people in the community do not eat up too necessities.	Show that students are aware of the tadpoles' survival rate and point out that they should be eaten in moderation that will not affect the ecosystem. This link the philosophy of sufficiency economy moderation, reasonableness, knowledge and moral.
10	Social issues What is the survival rate of frogs? People in the community are very tadpoles, resulting in shortages and income in the community. Although tadpoles are the food that people in the community like to eat. If we let it grow into a frog, it will help to generate more income.	The students' explanations show that students can bring social issues to explain about the income generation of people in the community. And point out that students are into the tadpole's way of life towards the community. This can be linked to the philosophy of Sufficiency Economy are reasonableness knowledge and moral.

Table 4. Result of applying the philosophy of sufficiency economy of worksheet in Identification of potential solutions stage.

Group	Description	Involved from PSE

IOP Conf. Series: Journal of Physics: Conf. Series **1340** (2019) 012075 doi:10.1088/1742-6596/1340/1/012075

1	Students explain the two main approaches to finding answers: how to make the right income for the community. Increasing variety of tadpoles and breeding methods. And students link to whether they should educate people in the community to make money by making food from tadpoles. Export of tadpoles and tadpoles are OTOP products.	self-sufficiency by planning to search information the feasibility of business planning and marketing using tadpoles.
2	The student explains that the student's response to the question is to have knowledge of the environment and the life of the tadpoles, Farming methods, Cost, Breed selection, how to allocate space that does not affect the environment and how to make money in the community.	self-sufficiency by students plan their research by taking into account business planning and general knowledge about tadpoles.
3	Students are interested in researching on the philosophy of sufficiency economy and information on raising frogs and behaviors that farmers need to know.	self-sufficiency by recognize that the planning of problem solving requires students to have knowledge.
4	Students plan how to solve problems by farming frogs and tadpoles.	self-sufficiency by planning to search information the feasibility of business planning
5	Students interested in planning a farm system, ecology of the tadpoles, breeding methods, production marketing and advertising.	Students connect by self-sufficiency of business planning and general education about tadpoles.
6	It is planned to study the processing of products for export, general knowledge about tadpoles, seeding and marketing.	Linking the economic philosophy by planning and marketing products related to tadpoles, which corresponds to self-sufficiency.
7	Learn about the tadpoles, allocation of space business, administration tadpole conservation in the community and to generate revenue through online methods.	Students have a sense of community-based knowledge and business planning in line with the philosophy of Sufficiency Economy is self-sufficiency.
8	A study of general knowledge about nature and living tadpoles, how to preserve the tadpoles, how to advertise, how to farm Life style of the community and the season should not eat tadpoles.	From the description corresponds to Can link the philosophy of sufficiency economy by self-sufficiency.
9	The thing to know about tadpoles and economic philosophy is just enough. Study of the life of the tadpoles. Create a new image for the tadpoles. Exports are genetically modified farming product design.	Can link the philosophy of sufficiency economy by Can link the philosophy of self-sufficiency and knowledge.

10	Study the environment suitable for tadpoles, market demand, natural tadpoles and off-season eggs of frogs.	Students use knowledge as a way to find answers and plan marketing. The three links to the philosophy of sufficiency economy as follows self-sufficiency and knowledge.

Table 5. Result of applying the philosophy of sufficiency economy of worksheet in Decision-making stage.

Group	Description	Involved from PSE
1	Group members offer solutions to problems. Whether it's a frog farm to rearing tadpoles. The tadpoles that no longer cause disease. However, students make a decision on how to farm frogs. There is a negligence in educating people in the community. The community members have ideas and different businesses in the community.	Which show moderation, reasonableness, self-sufficiency, knowledge and moral.
2	Students propose ways to solve many problems, such as the use of biotechnology to increase the number of tadpoles to meet the needs of the community. To create a product that is stored for three cars, such as pickling, drying extends the life. Community educators are encouraged to cultivate the tadpoles naturally. Reduces chemical contamination. As a result, people in the community eat non-toxic food.	Can be linked moderation, reasonableness, self-sufficiency, knowledge and moral.
3	Students present frog farming. Fed by natural methods that are readily available in the community. Fundraising from the community. To educate and bring productivity to be built into the product. To make money for the community. People in the community have a professional well-being. And the balance of ecosystems. Reduce pollution in the community due to natural products. The tadpoles that have been processed such as meatballs, Tadpole-fried and tadpoles dry.	Can be linked moderation, reasonableness, self-sufficiency, knowledge and moral.
4	Frog culture is sufficient to meet the needs. When large populations of frogs are found, tadpoles are sufficient for the needs of the community. The frogs must take into account factors related to many frogs. In order to get the right amount of eggs for the needs of the community.	Can be linked moderation, reasonableness, self-sufficiency and knowledge.
5	Set up a community breeding center. To increase the population of tadpoles to balance the needs of the community. Tadpoles can be processed into the OTOP products to generate revenues were adding value to productivity such as steamed tadpoles, tadpoles dry and tadpole's basil. Import duties on processed products exported to other communities. Establishment of ecotourism area to educate tourists and people in the community.	Can be linked moderation, reasonableness, knowledge and moral.

Group	Description	Involved from PSE
6	The tadpoles have been introduced to educate and advertise tadpoles to the public and promote local conservation. Genetically engineered for different and diverse varieties.	Can be linked reasonableness, self-sufficiency and knowledge.
7	Do breeding tadpoles the system closed which makes the tadpoles are clean and safe. This care can be thoroughly. The survival rate is higher than the natural system. The benefits of choosing this method is that many tadpoles. And the impact frogs and tadpoles may lose due to various breeding in consanguinity.	Can be linked moderation, reasonableness, self-sufficiency, knowledge and moral.
8	Creating a breeding ground for frogs and tadpoles in the community by raising funds from people in the community and creating jobs for people in the community. Raising and breeding frogs for sale. The villagers eat and sell throughout the year. And processed foods are to add value.	Can be linked reasonableness, self-sufficiency and knowledge.
9	Farm produce tadpoles and increase biodiversity by genetically modified brush to get the frog and tadpole species. Then bring the community products such as soap-tadpoles, fried-tadpoles, and then send them online, inside and outside the country. Production processes by the people in the community.	Can be linked reasonableness, self-sufficiency and knowledge.
10	Increase the number of tadpoles to meet the needs. And is distributed outside the community. The production and the sale will be assigned to those knowledgeable in the community helping each other accountable. There is growing awareness in the community to conserve natural resources for the tadpoles grow into frogs next.	Can be linked reasonableness, self-sufficiency, knowledge and moral

4.2 Student Product and application of PSE and STEAM knowledge

Students have a way to resolve the social issues by creating products about Tadpole on the biodiversity community to have a stable and sustainable prosperity. The analysis of data from Student performant of students on the application of the philosophy of sufficiency economy and STEAM knowledge. Which can be grouped as follows:

Table 6. Result of applying the philosophy of sufficiency economy of Student performant.

Groups	Description	Involved from PSE and STEAM knowledge
1. Farm for created new product.	Frogs were cultured in a natural way to increase numbers. Bring frogs across the breed to meet the needs of consumers. And to process a variety of products on demand. Evaluate the quality consistently to meet customer needs. 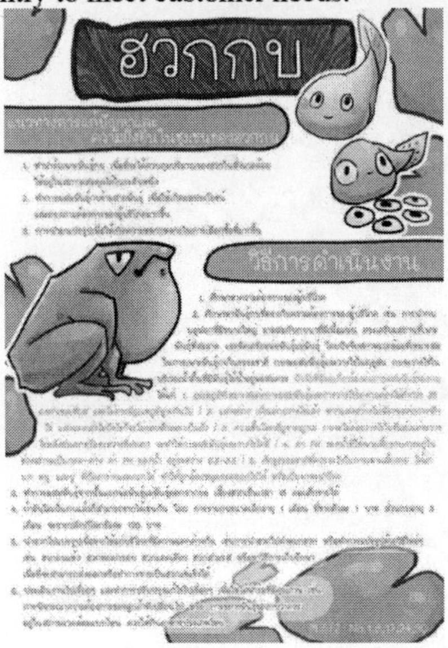 **Figure 4.** Group work of Farm for created new product.	They can connect PSE are reasonableness, self-sufficiency and knowledge. From picture show that student use STEAM knowledge. Students can connect knowledge in science and technology by interbreeding many tadpoles. It is illustrated science, technology, engineering, art and mathematics.
2. Dim Sum Huak (Tadpoles)	Applied as a food of another culture to offer food in a new way. Learn about the tadpole propagation of habitat from natural sources using technology to help in propagation. Plan your marketing by giving customers and give your neighbors tastes to improve the recipes. To be formulated into dumplings stuffed tadpoles by cooperation of local members. 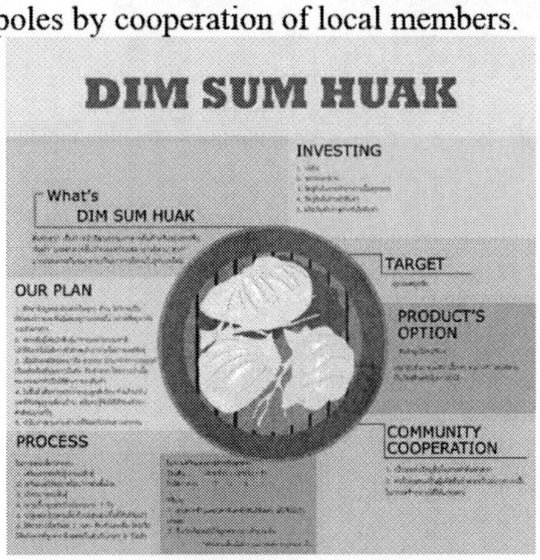 **Figure 5.** Group work of Dim Sum Huak (Tadpoles)	They can connect PSE are reasonableness, self-sufficiency, knowledge and moral. From picture show that student use STEAM knowledge by plan, design, create food and science education. It is illustrated science, technology, engineering, art and mathematics.

Groups	Description	Involved from PSE and STEAM knowledge
3. Frozen Food Mok Huak (tadpoles baked)	Planning the finished product from the tadpoles. Action Plan, Brainstorming, Strategy Development Business analysis, testing and market introduction the goal of the community and people interested. Can eat during hustle. Investment from farming Cooking and advertising materials. 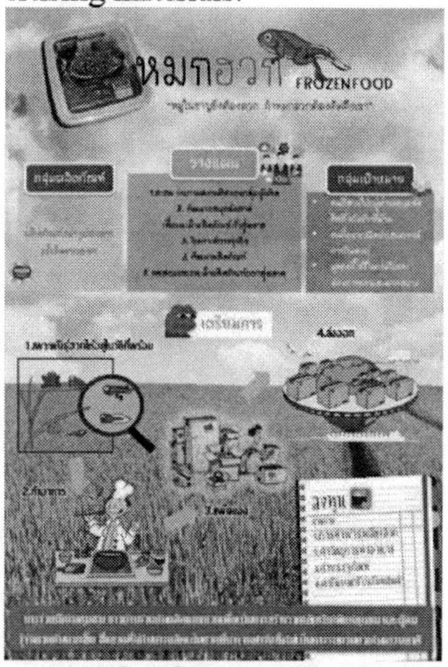 **Figure 6.** Group work of Frozen Food Mok Huak (tadpoles baked).	They can connect PSE are reasonableness, self-sufficiency, knowledge and moral. From the piece showing how students can connect to STEAM knowlage by designing frozen food from local animals to be in line with the present age. It is illustrated science, technology, engineering, art and mathematics.
4. Spray drying of Tadpoles	Start by selecting a target market. Design a memorable product. Thai people in foreign countries Export focus Prepare by finding a marketer who makes space in online advertising. Create jobs and income in the community. 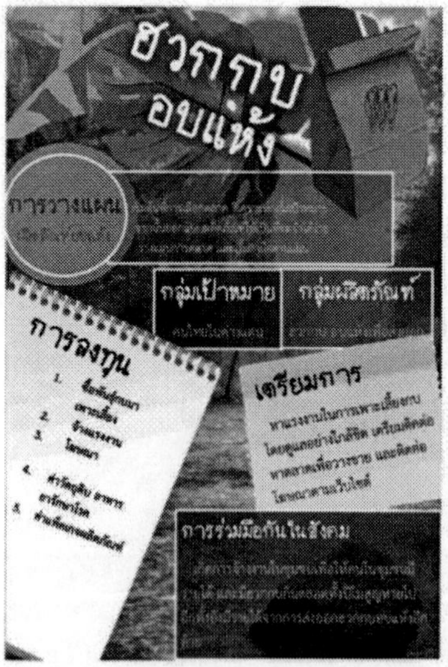 **Figure 7.** Group work of Spray drying of Tadpoles.	They can connect PSE are reasonableness, self-sufficiency and knowledge. The work shows that students can connect STEAM knowledge by processing food to be able to keep for a long time and export to sell abroad. It is illustrated science, technology, engineering, art and mathematics.

Groups	Description	Involved from PSE and STEAM knowledge
5. Chinsuwan farm	It is a tourist attraction. Information on how and how to grow. Farm layout design Breeders Raise funds to people in the village, government and loans. Make a deposit from the tadpole product.  **Figure 8.** Group work of Chinsuwan farm.	They can connect PSE are reasonableness, self-sufficiency and knowledge From the implementation plan, it shows that students can connect STEAM knowledge by planning, operating and planning finances. It is illustrated science, technology, engineering, art and mathematics.
6. Farm and Research Centre.	Founder of the research on tadpoles to reduce the problem of loss of children on the land. By breeding frogs and tadpoles raised area. The frogs and tadpoles were used to generate income for the community by making museums and aquariums. Export products from tadpoles and restaurants.  **Figure 9.** Group work of Farm and Research Centre.	They can connect PSE are reasonableness, self-sufficiency, knowledge and moral. From produce show that students use STEAM knowledge by science knowledge to cultivate, apply engineering knowledge to design farm structures Which is a new technology for breeding tadpoles, art by design project and math.

IOP Conf. Series: Journal of Physics: Conf. Series **1340** (2019) 012075 doi:10.1088/1742-6596/1340/1/012075

Groups	Description	Involved from PSE and STEAM knowledge
7. Chips Tadpoles	Farm to increase the number of tadpoles to meet the needs. The product is exotic. Expand the market from neighboring communities to foreign export.	They can connect PSE are reasonableness, self-sufficiency and knowledge. The work shows that students can link STEAM knowledge by creating a farm. Cooking, processing and making products for export abroad. It is illustrated that they can connect STEAM knowledge (science, technology, engineering, art and mathematics).

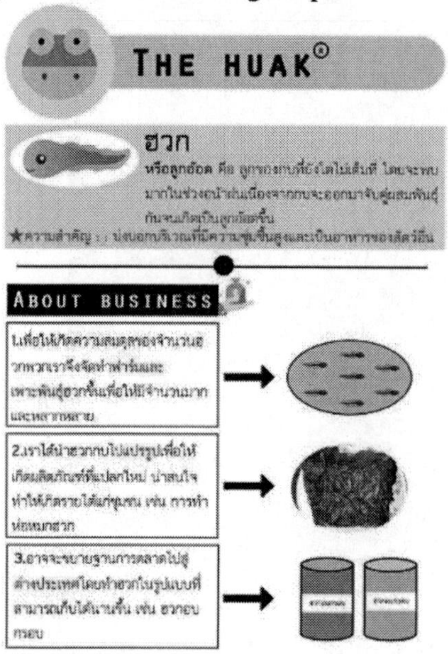

Figure 10. Group work of Chips Tadpoles.

Groups	Description	Involved from PSE and STEAM knowledge
8 Tadpoles Can	Plan to analyze product objectives, find promotional events Quality as required, Ask for product reviews from customers, Ads by people in the community, Investment from various sources.	They can connect PSE are reasonableness, self-sufficiency and knowledge. The work shows that students have designed and developed technologies to preserve food for a longer life. Which can link STEAM knowledge. It is illustrated science, technology, engineering, art and mathematics.

Figure 11. Group work of Tadpoles Can.

Groups	Description	Involved from PSE and STEAM knowledge
9. GMOs tadpoles	Genetically modified tadpoles to a variety of species. Leading breeder and breeder rearing frogs to sell to people in the community. Conservation of local zoning tadpoles. Product Processing Such as smoothies, health tadpoles	They can connect PSE are reasonableness, self-sufficiency and knowledge. From the work piece showing that students use science knowledge Technology and engineering in planning concepts for creating GMOs and art in design work.

Figure 12. Group work of GMOs tadpoles.

10. Chips Tadpoles	Business planning is by providing information about the tadpoles. Design, manufacture, package, trial, sell, refine, formulas and distribute by shop. The appropriate amount of tadpoles was prepared using natural methods. The investment consists of food tadpoles. Frogs and raw materials used in processing. The products are offered for all ages	They can connect PSE are reasonableness, self-sufficiency and knowledge. From the picture shows that students have designed and developed technology to preserve food to be eaten in free time. Students are aware of the income, expenditure in making products linked to mathematics. It can be seemed that they can connect STEAM knowledge (science, technology, engineering, art and mathematics).

Figure 10. Group work of Chips Tadpoles.

5. Conclusion and Discussion

From the learning process, STS, based on through Yuenyong (2006), can promote the application of the philosophy of sufficiency economy and STEAM knowledge as follows:

5.1 Students can apply knowledge in product design. To solve the problem of income to the community.
Students can use the knowledge gained from science in decision making to solve problems, which methods can support the best approach. Students have invented a new concept with local creatures. Which some have not yet found before in the local area of Thailand, showing students being linked to technology. In addition, students also have a plan to solve social problems that occur systematically in accordance with the application of engineering knowledge. Student has used the design knowledge to design the work to increase the value of the product. And some students can use mathematical knowledge to calculate costs and planning on using money.

5.2 Students can link social issues to the events.
Students can relate to what students encounter in everyday life. Although what students have learned from social issues is far away. Such as crispy tadpoles which is similar products that students like to eat and are sold in the market.

5.3. Students make plans to solve the problem in a systematic and teamwork.
However, in the application of the philosophy that students can promote their applications in all stages. STS, but each step may not cover all aspects of the philosophy of sufficiency economy. Moreover, this study allows students to learn in a new social context in which students have never met before in biology class.

6. Acknowledgement

This research was financially supported by Thailand Research Funding (TRF)–Contract Number: MRG5180130

References

[1] Choi S-H 2004 *Encourage Private Sector: Pre-school Education Reform in Morocco* Paris, UNESCO (UNESCO Policy Brief on Early Childhood, 20.)
[2] Jakel L 2013 *Interest and Learning in Plant Biodiversity, as Influenced by Teaching Contexts* (Interest in plant biodiversity) pp 1-22
[3] Singh R and Raja B 1991 *Education for the twenty-first century: Asia-Pacific perspective* Bangkok, UNESCO (Asia and the Pacific Programme of Educational Innovation for Development) pp 93
[4] Parmin et. al. 2016 *Preparing Prospective Teachers in Integrating Science and Local Wisdom through Practicing Open Inquiry* (Journal of Turkish Science Education vol 13) chapter 2 pp 3-14
[5] Siripatharachai P 2013 *STEM Education and 21st Century Skills Development* (Srinakarinwirot Business Journal vol 2) pp 49-56
[6] UNESCO *2014 Learning about Biodiversity: Multiple-Perspective Approaches Education for Sustainable Development in Action Learning & Training Tools* (United Nations Educational, Scientific and Cultural Organization vol 7) pp 1-33
[7] Warde and Sah 2014 *STEM Education Reform Considerations for the Caribbean* (Business Barbados) pp 1-4
[8] Yakman G 2016 *Developing STEAM Education to Improve Students' Innovative Ability* (STEAM Education Business Offices in MA) pp 1-12
[9] Yuenyong C 2006 *Teaching and Learning about Energy: Using STS approach*. (Bangkok, Thailand: Thesis of Doctoral Degree in Science Education, Kasesart University)
[10] Yuenyong C and Narjaikaew P 2009 *Scientific Literacy and Thailand Science Education* (International Journal of Environmental and Science Education vol 4) chapter 3 pp 335-349

Exploration of Students' Metacognitive Experience in Physics Classroom

Jirutthitikan Pimvichai[1], Soonthareeya Sanium[2] and Khajornsak Buaraphan[2]

[1]Faculty of Education, Khon Kaen University 123 Nai-Muang, Muang, Khon Kaen 40002 Thailand

[2]Institute for Innovative Learning, Mahidol University 999 Salaya, Phuttamonthon, Nakhon Pathom 73170 Thailand

E-mail: fern_699@windowslive.com

Abstract. Metacognition has been accepted as one desirable attribute for learners in the 21st century. There are a number of researech studies related to how to promote metacognition in school students. This study aimed to explore students' metacognitive expereince in one physics classroom in the equilibrium of moments topic. The research participants were 35 Grade 9 students from one secondary school located at Nakhon Ratchasima province in the northeastern region of Thailand. After learning physics, the participating students were asked to respond to the Metacognitive Experience Questionnaire (MEQ). The studnets' metacognitive experience was analyzed for mean and standard deviation. The finding revealed that the students derived the neutral to frequently levels of expereience in metacotnition in the physics classroom. This finding suggests to add more metacognitive expereience for Grade 9 students.

Keywords: Metacognitive experience, physics classroom, secondary school, Thailand

1. Introduction

Metacognition generally refers to thinking about thinking or knowing about knowing. It is commonly believed that metacognition can help individuals learn more effectively or reach higher learning achievement. Empirically, several studies affirm that when students develop their own metacognition; from developed metacognition, they subsequently develop other desirable characteristics as learning achievement, critical thinking skill, problem-solving skill, decision making skill, and awareness and environmental conservation behaviors [5,13].

There are several teaching strategies effectively develop students' metacognition as the 5E and 7E learning cycle, open-ended problems, constructionist activities, and Tri-Sik-Kha principle-based teaching [14-20]. The questions generated during teaching influenced how students learned and studied, especially their metacognition [8].

Young children have more trouble than older ones in a comprehension process. Moreover, young children have less ability to develop metacognitive knowledge than older ones. They face the difficulty in feeling, planning, monitoring and evaluating their learning when they face difficult problems or situations to solve [7]. In other words, young children face the difficulty regarding metacognitive awareness; they hardly realize that whether their learning is progressing well or failed. Metacognitive awareness takes part in metacognitive experience in a sense that metacognitive experience prefers awareness linking the present with the past learning experiences in order to facilitate learning in the present and in the future. Therefore, reflection upon learning experience of oneself can benefit individual's metacognition [6]. Up to this, metacognitive experience and an ability to reflect upon it are regarded as the important component of effective learning for students in science classrooms.

The research question of this study is: What are the metacognitive experiences that the Grade 9 students derive during their learning in the physics classroom? Regarding this research question, the research objective is: To explore the characteristics of metacognitive experienc that the Grade 9 students derive during their learning in the physics classroom.

2. Literature Review

This section illustrates the literature review about impact of metacognition, development of metacognition and metacognitive experience.

2.1. Impacts of metacognition

Ruangsri [18] studied about a relationship between metacognition and critical thinking of Grade 6 students. The researcher found that the multiple correlation coefficients between the subcategories of metacognition and the sum of all subcategories of critical thinking were significant at the .01 for all values and the multiple correlation coefficients between the subcategories of metacognition and each subcategory of critical thinking were statistically significant at the .01 level for all values. The beta weight values in each subcategory of metacognition positively contributed to the sum of all subcategories of critical thinking for all values. However, some subcategories of metacognition were negatively contributed to critical thinking.

The effect of a metacognition training with attribution training to effort on problem-solving skills of Grade 7 students was studied by Wongmak [26]. This research conducted with the three groups: the control group, the experimental group 1 received a training on metacognition and attribution to effort and the experimental group 2 receiving a metacognition training only. After training, the experimental groups 1 and 2 received higher problem-solving skill scores at the .05 level of significance; while the control group received lower problem-solving skill scores.

Prasertsook [17] studied Grade 8 students' scientific problem solving ability, metacognition development and scholastic achievement in work and energy topic by using metacognitive strategies. Four metacognitive procedures were 1) analyzing problems, 2) planning to solve problems, 3) managing problems and 4) evaluating problems. Metacognition helped students develop scientific problem-solving ability as students analyzed and planned to solve scientific problems. The students applied metacognitive procedure to solve situation they received. The metacognitive learning affected students' achievement.

Learning management by using metacognition strategy to develop the scientific problem solving skills for Grade 7 students was studied in Peungsaijai [14]. The efficiency of the metacognition learning plan to develop the scientific problem-solving skill was at 85.86/ 81.66. When the students used metacognition strategy, they had higher scientific problem-solving skill and learning achievement at the .05 statistically significant level. The students satisfied with the metacognition learning management at the highest level.

Nonjuy [12] studied the effect of metacognitive strategy on decision making and problem solving abilities in force and motion topic for Grade 9 students. The results showed that, after learning through metacognitive strategy, the student's post-test average score of decision making and problem-solving abilities were significantly higher than pre-test at the .01 statistically significant level.

Decision making and problem solving abilities in science learning through metacognitive strategy was studied by Senruang [22]. The metacognitive strategy increased the students' post-test mean score of decision making and problem-solving abilities to be significantly higher than pre-test at the .05 statistically significant level.

Teprattananan [23] studied the effects of cooperative learning activities management integrated with metacognitive thinking on learning achievement and metacognitive thinking ability of Grade 10 students. The results revealed that the students who learned with cooperative learning integrated with metacognitive thinking has higher learning achievement and metacognitive thinking ability than those who learned with cooperative learning only at the .05 statistically significant level.

Development of knowledge and understanding, critical thinking, awareness, environmental conservation behaviors of Grade 12 students using the good science thinking moves with metacognition techniques was studied in Ladawan [9]. The lesson plans using the good science thinking moves with metacognitive techniques had an effectiveness index of .7290. The students in the experimental group developed higher knowledge and understanding, critical thinking, awareness and environmental conservation behaviors ($p<.001$). In overall, gender did not affect different learning outcomes.

Lertchuwongsa [10] used problem-based learning to promote metacognition and mathematics achievement of Grade 10 students. The problem-based learning increased the students' metacognition mean score from a moderate level to a good level. Also, most of the students (more than 80%) had mathematics achievement mean scores higher than the minimum criteria at 60%. The mathematics achievement mean score, in overall, was 14.9 of 20 points (74.5%).

In summary, many studied confirm that when teachers developed metacognition in their students; the students also developed other desirable attributes as learning achievement, critical thinking skill, problem-solving skill, decision making skill, and awareness and environmental conservation behaviors.

2.2. Development of metacognition

Chansana, Suksringarm and Parakham [3] compared effects of the 7E learning cycle using metacognitive moves on Grade 5 students' alternative conception in biology and basic science process skills. The results showed that the 7E learning cycle using metacognitive moves helped students gain more understanding about biology concepts at the .05 level of significance. The students developed basic sciences process skills higher than the 50% criterion at the .05 level of significance. In addition, the female students significantly evidenced more interning skills than the male students.

The effects on using open-ended problems to develop metacognition for Grade 9 students was studied by Pitchayanurat [16]. Using open-ended problems could help 80.95% of the students passed the metacognition test criterion of 50%. The students also showed a "good" level of metacognition practice with mean score of 2.96.

Thatsamee [24] developed the physics learning activities to develop problem solving ability with metacognition for Grade 10 students. The results showed that 80.95% of the students passed the physics problem solving ability test with criterion of 70%.

The effects of science learning management on Grade 10 students through learning cycle with metacognitive reflection and awareness were studied by Yutuuam [27]. There was no interaction between types of learning management and basic scientific knowledge. The students who learned with the learning cycle with metacognitive reflection and awareness (LCMRA) performed the highest learning outcome, scientific problem solving ability and metacognitive ability, followed by those learned with the learning cycle with metacognitive reflection (LCMR) and learning cycle (LC) only. The degree of student basic scientific knowledge corresponded with the degree of achieved learning outcome, scientific problem-solving ability and metacognitive ability.

Na Phatthalung [11] studied the results of 5E instructional model with metacognitive reflection strategies on science learning achievement, science concepts and metacognition of upper secondary school students. After learned with the 5E instructional model with metacognitive reflection strategies, the students in a bright, moderate and weak groups had higher science learning achievement, science concepts and metacognition than priori at the .01 of significance. Also, the students had science learning achievement higher than the criteria of 60%.

The development of metacognition by using constructionist activities in the topic of evolution for Grade 12 was done in Sangwong [19]. The constructionist activities could develop Grade 12 students' metacognition in a high level. This was occurred from the characteristic of constructionist activities as working with multiple formats, a teacher allowing students to discuss and share ideas both in and outside the classroom.

Phuaphan [15] studied effects of instruction based on Tri-Sik-Kha principle for developing Grade 10 students' metacognition in the Ecosystems and Human and Sustainable Environment topic. The result showed that the instruction based on Tri-Sik-Kha principles could develop the students' metacognition. The student average score of metacognition was very high at 70.35%.

In summary, many researchers tried to apply several teaching strategies, that is, the 5E and 7E learning cycle, open-ended problems, constructionist activities, and Tri-Sik-Kha principle, in order to develop students' metacognition. Importantly, all teaching strategies can develop the participating students' metacognition in a positive way.

2.3. Metacognitive Experience

Metacognitive experience is a part of metacognitive knowledge involved with knowledge about persons, goals, tasks and strategies. Metacognitive experience generally refers to one's experience regarding feelings, estimation or judgments in his/ her learning process during completing the specific task. Metacognitive experience refers to one's awareness linking the present with the past learning experiences that can facilitate learning in the present as well as in the future.

Cognitive and affective factors are two factors that can affect one's metacognitive experience by triggering or impeding his/ her cognition and behavior related to effective learning in both short- and long-terms [5].

Metacognitive experience takes role in self-regulation that can provide an input to activate one's metacognitive skills to control learning action and behavior [4]. For example, learners' feeling of difficulty can be attributed to task complexity, task demands, or a lack of competence to deal with the task. The outcome of cognitive process and detects the discrepancy from the goal set can also affect one's metacognitive experience. We call these as feedback loops [2].

3. Methods

3.1 Research participants

The research participants were 35 Grade 9 students from one classroom in one secondary school located at Nakhon Ratchasima province in the northeastern region of Thailand. The age of students ranged between 15-16 years. The students were in the final year of their lower-secondary school level. They all volunteered to participate in this study.

3.2 Data collection

After learned in one topic of the physics class, the students were asked to respond to the Metacognitive Experience Questionnaire (MEQ), which was consisted of 17 items. The MEQ is divided into three parts. The first part was consisted of nine items (items 1-9) asking for Monitoring, Evaluation and Planning (MEP). The second part was consisted of five items (items 10-14) asking for Learning Risks Awareness (LRA). The third part was consisted of three items (items 15-17) asking for Control of Concentration (CC).

3.3 Data analysis

The students' responses of MEQ were quantitatively analyze for means and standard deviation (S.D.) by using the SPSS statistical package. The mean scores were interpreted according to these criteria. The mean scores between below 1.00-1.50, 1.51-2.50, 2.51-3.50, 3.51-4.50 and over 4.51 was interpreted as never, rarely, neutral, frequently and always, respectively.

4. Results

Mean scores and associated S.D. of each item in the MEQ were calculated and interpreted as table 1.

Table 1. Grade 9 students' metacognitive experience in each item of MEQ (n = 35)

Metacognitive Experience	mean	S.D.	Interpretation
1. I adjust my plan for a learning task if I am not making the progress I think I should.	2.77	1.00	Neutral
2. I plan to check my progress during a learning task.	2.74	.74	Neutral
3. I try to understand clearly the aim of a task before I begin it.	3.94	.90	Frequently
4. I evaluate my learning processes with the aim of improving them.	3.37	.80	Neutral
5. I consider what type of thinking is best to use before I begin a learning task.	3.40	.84	Neutral
6. I consider whether or not a plan is necessary for a learning task before I begin that task.	2.83	.89	Neutral
7. I stop from time to time to check my progress on a learning task.	2.77	.97	Neutral
8. I try to predict possible problems that might occur with my learning.	3.17	1.01	Neutral
9. I assess how much I am learning during a learning task.	3.49	.88	Neutral
10. I am aware of when I am about to have a learning challenge.	3.40	.84	Neutral
11. I am aware of when I am about to loose track of a learning task.	3.54	.86	Frequently
12. I am aware of when I don't understand an idea.	3.54	1.03	Frequently
13. I am aware of when I have learning difficulties.	4.09	.70	Frequently
14. I am aware of when I am not concentrating.	3.54	1.09	Frequently
15. I adjust my level of concentration, depending on the learning	3.46	1.01	Neutral

16. I adjust my level of concentration depending on the difficulty of the task.	4.34	.59	Frequently
17. I adjust my level of concentration to suit different science subjects.	3.69	.93	Frequently
Overall	3.41	.88	Neutral

Table 1 indicated that, in overall, the participating students frequently derive metacognitive experience in the physics classroom (mean = 3.41, S.D. = .88). Of 17 items, there were seven items rated in a frequently level and 10 items rated in a neutral level of experience related to metacognition.

Considering Grade 9 students' metacognitive experience in MEP, the participating students showed the neutral level of experience concerning MEP (see Table 2). They had a frequently level of MEP experience regarding item 03 "I try to understand clearly the aim of a task before I begin it."

Table 2. Grade 9 students' metacognitive experience in MEP (n = 35)

MEP component	mean	S.D.	Interpretation
1. I adjust my plan for a learning task if I am not making the progress I think I should.	2.77	1.00	Neutral
2. I plan to check my progress during a learning task.	2.74	.74	Neutral
3. I try to understand clearly the aim of a task before I begin it.	3.94	.90	Frequently
4. I evaluate my learning processes with the aim of improving them.	3.37	.80	Neutral
5. I consider what type of thinking is best to use before I begin a learning task.	3.40	.84	Neutral
6. I consider whether or not a plan is necessary for a learning task before I begin that task.	2.83	.89	Neutral
7. I stop from time to time to check my progress on a learning task.	2.77	.97	Neutral
8. I try to predict possible problems that might occur with my learning.	3.17	1.01	Neutral
9. I assess how much I am learning during a learning task.	3.49	.88	Neutral
Overall	3.16	.89	Neutral

Regarding Grade 9 students' metacognitive experience in LRA, the participating students showed the frequently level of experience concerning LRA. They had a neutral level of LRA experience regarding item 10 "I am aware of when I am about to have a learning challenge" as Table 3.

Table 3. Grade 9 students' metacognitive experience in LRA (n = 35)

LRA component	mean	S.D.	Interpretation
10. I am aware of when I am about to have a learning challenge.	3.40	.84	Neutral
11. I am aware of when I am about to lose track of a learning task.	3.54	.86	Frequently
12. I am aware of when I don't understand an idea.	3.54	1.03	Frequently
13. I am aware of when I have learning difficulties.	4.09	.70	Frequently
14. I am aware of when I am not concentrating.	3.54	1.09	Frequently
Overall	3.64	.90	Frequently

Table 4 showed that Grade 9 students derived the frequently level of experience concerning CC. They had a neutral level of LRA experience regarding item 10 "I adjust my level of concentration, depending on the learning."

Table 4. Grade 9 students' metacognitive experience in CC (n = 35)

CC component	mean	S.D.	Interpretation
15. I adjust my level of concentration, depending on the learning.	3.46	1.01	Neutral
16. I adjust my level of concentration depending on the difficulty of the task.	4.34	.59	Frequently
17. I adjust my level of concentration to suit different science subjects.	3.69	.93	Frequently
Overall	3.83	.84	Frequently

5. Discussion

In a holistic view, Grade 9 students derive only a neutral level of experience related to metacognition in their physics classroom. Up to this, we as science or physics educators, need to provide more metacognitive-related experience to our students in physics classrooms in order to help students derive higher level of metacognitive experience, for example, from the neutral to the frequently levels.

Comparing between MEP, LRA and CC, Grade 9 students derive MEP-related experience less than LRA- and CC-related experience. So that, the physics teachers should provide more opportunities for students to plan, evaluate and monitor their own learning. This cannot be accomplished with the lecture-based teaching style as being normally found in physics classrooms in Thailand. Physics teachers should change their teaching paradigm to constructivist teaching paradigm. The physics students should have more chance to act as active learners and, in the same sense, the physics teachers should act as facilitators.

Several constructivist teaching strategies should be employed in physics classrooms such as inquiry, learning cycle, problem-based learning, project-based learning, STEM (Science-Technology-Engineering-Mathematics) and so on.

In addition, to derive more complete picture of metacognitive experience in physics classroom, we need more measurement tools. The findings from this study may face its limitation regarding the use of single instrument and method to collect data about metacognitive experience. Schraw [20] refers to the difficulty of measuring metacognition and states that using a single method is not enough to explore all dimensions of such a complex process as metacognitive process. Regarding this, Akturk and Sahin [1] state that the researchers in metacognition filed need the tool to measure metacognition and its related components. The metacognition researchers should utilize more than one tool to measure metacognition in their research sample or participants.

6. Implication

This study urges for the need to develop MEP-related experience for students in physics classrooms. However, the LRA- and CC-related experience should be developed in physics students as well. In addition, the research about metacognition should apply diverse tools, methods and methodologies in studying about metacognition. Regarding this, the qualitative or mixed-method tools should be considered to apply in metacognition-related research in order to capture the complexity of metacognitive process.

Scoot [21] mentions the power of interview in uncovering students' metacognitive process. Interview questions demand students as the interviewees to expand or clarify their answers. In the opposite way, the use of questionnaire in studying about metacognition should not be neglected. The utilization of questionnaire is useful in a sense that it can be applied with a large sample size. It is easier and quicker for the researchers to interpret the data derived from the questionnaire. The questionnaire can be use for quick survey and evaluation of metacognition in respondents [25].

7. Acknowledgment

This research was supported by the Higher Education Research Promotion and National Research University Project of Thailand, Office of the Higher Education Commission, through the Cluster of Research to Enhance the Quality of Basic Education. The authors would like to express sincere gratitude to the Office of the Higher Education Commission.

References

[1] Akturk A O and Sahin I 2011 Literature review on metacognition and its measurement **Procedia Social and Behavioral Sciences 15**
[2] Carver C 2003 Pleasure as a sign you can attend to something else: Placing positive feelings within a general model of affect **Cognition and Emotion 17** 241–261
[3] Chansana P, Suksringarm J and Parakham M 2008 Comparisons of effects of the 7E learning cycle approach using metacognitive moves on alternative conception of some biology and basic science process skills of grade 5 students with different genders **Rajaphat Maha Sarakham University Journal 2**
[4] Efklides A 2002 Feelings as subjective evaluations of cognitive processing: How reliable are they? Psychology: The Journal of the Hellenic **Psychological Society** 9 p 79
[5] Efklides A 2001 Metacognitive experiences in problem solving Trends and Prospects in Motivation Research p 298
[6] Efklides A 2009 The role of metacognitive experiences in the learning process **Psicothema 21** p 76
[7] Flavell J H 1987 Specul ation about the nature and development of metacognition. (Eds.) (WRHK, editor. NJ: Erlbaum: Hillsdale)

[8] García F C, García A, Berbén A B G, Pichardo MC and Justicia F 2014 The effects of question-generation training on metacognitive knowledge, self regulation and learning approaches in Science **Journal of Psicothema 26** pp 385–90
[9] Ladawan C A 2014 development of knowledge and understanding, critical thinking, awareness, environmental conservation behaviors of mathayomsuksa 6students using the good science thinking moves with metacognition technique (Mahasarakham: Mahasarakham University)
[10] Lertchuwongsa J 2016 Using problem-based learning to promote metacognition and mathematics achievement of grade 10 students (Chiang Mai: Chiang Mai University)
[11] Na Phatthalung N 2011 The results of 5E instructional model with metacognitive reflection strategies on science learning achievement, science concepts and metacognition of upper secondary school students (Songkhla: Thaksin University)
[12] Nonjuy D 2013 The effect of metacognitive stratergy on decision making and problem solving abilities in force and motion topic for mathayomsuksa 3 students (Phitsanulok :Naresuan University)
[13] Panprueksa K, Phonphok N, Boonprakob M and Dahsah C 2012 Thai students' conceptual understanding on force and motion **International Conference on Education and Management Innovation** 9-275
[14] Peungsaijai S 2011 The learning management by using meta cognition strategy to develop the scientific problem solving skills for mathayom suksa 1 students (Chiang Rai: Chiang Rai Rajabhat University)
[15] Phuaphan N 2014 The Effects of Instruction Based on Tri-Sik-Kha Principles for Developing Grade 10 Students' Metacognition in the Topic of Ecosystems and Human and Sustainable Environment (Bangkok: Kasetsart University)
[16] Pitchayanurat S 2006 The effects of using open-ended problems to develop metacognition for mathayomsuksa 3 students (Khon Kaen: Khon Kaen University)
[17] Prasertsook N 2011 Grade 8 students' scientific problem solving ability, metacognition development and scholastic achievement in work and energy topic by using metacognitive strategies (KhonKaen KhonKaen University)
[18] Ruangsri U 2001 A study of Relationship between metacognition and critical thinking of prathom six students (Bangkok: Srinakharinwirot University)
[19] Sangwong P 2013 The development of metacognition by using constructionist activities in the topic of evolution for grade 12 students (Bangkok: Kasetsart University)
[20] Schraw G A 2009 Conceptual analysis of five measures of metacognitive monitoring. Metacognition Learning **Metacognition Learning 4** pp 33– 45
[21] Scoot B M 2008 Exploring the effects of student perceptions of metacognition across academic domains In press
[22] Senruang R 2006 Decision making and problem solving abilities in science of grade level 3 students learning through metacognitive strategy (Chiang Mai: Chiang Mai University)
[23] Teprattananan A 2014 The effects of cooperative learning activities management integrated with metacognitive thinking on leaning achievement and metacognitive thinking ability of mathayom suksa iv students of suan si wittaya school in chumphon province **SKRU Academic Journal 7** pp 31-43
[24] Thatsamee S 2009 The development of learning activities in physics subject to develop problem solving ability with metacognition for mathayomsuksa 4 students (Khon Kaen: Khon Kaen University)
[25] Tobias S and Everson H T 1996 Assessing metacognitive knowledge monitoring (New York: The College Board)

IOP Conf. Series: Journal of Physics: Conf. Series **1340** (2019) 012076 doi:10.1088/1742-6596/1340/1/012076

[26] Wongmak J 2010 Effect of a metacognition training together with attribution training to effort on problem - solving skills of seventh grade students (Bangkok: Chulalongkorn University)
[27] Yatuuam D 2005 The Effects of science learning management on mathayom suksa 4 students through learning cycle with metacognitive reflection and learning cycle with metacognitive reflection and awareness (Bangkok: Srinakarinwirot University)

The Participatory Characterization Based on STEM Education in the Restricted Area for Fishing at Tha Song Korn Temple in Khon Kaen Province

Prakrupariyat Sarakran

Cultural Science, Mahasarakham University, Thailand

* Corresponding author's e-mail address: prakrupariyatsarakran@gmail.com

Abstract. This study aims to investigate the participatory characterization based on STEM Education in the restricted area for fishing in front of the temple on the banks of Tha Song Korn River. The qualitative research methodology was used in this study. Data were collected by using field study methods along with in-depth interview and observation. Data were analyzed by using documentary analysis by using triangulation techniques. The results showed that the participatory characterization in the conservation area for fishing in front of the temple on the banks of Tha Song Korn River developed due to the collaboration of local government agencies, community, and villagers. It was found that the temple helped improve the water pollution in front of the temple by promoting the collaboration of all the stakeholders based on STEM Education in four steps; planning, implementation, taking benefits, and evaluation. The abbots and monks transferred the knowledge of water conservation to villagers along with the assistance of government agencies and the governors in the community, the surrounding area can be developed to be tourist attraction of the public. This holistic method derived from combining knowledge of science, mathematics, technology and engineering design based on STEM education leads to sustain management in the community.

1. Introduction

Natural resource plays a key role in everyday human living. In particular, Thailand is an agricultural country that villagers massively depend on water consumption for their farming e.g. rice farming and fish farming. Recently, the community has rapidly expanded while fish farming along the river is growing [9]. The economic-driven expansion of community causes not only the water pollution problems but also the decrease of fish which affects farming along the river. Thus, the effective management of natural resource is vital to ensure its sustainability and quality [2,5].

The local temple, as one of the religious institutions, is a main source of education for village members in Thais' society [12]. Buddhist monks teach village members about ethics and morality connected with Buddhist ways of living. Conducting Buddhist virtue approaches in physical and mental manners to ethics leads village members to live peacefully together. Religion tightly holds villagers' soul and spirit as it located in every village. Therefore, temple as a heritage of the community is required effective collaboration of all stakeholders in the village to respect and foster sustainably [7].

Likewise, controlling the restricted fishing area around temple is an approach that aims to improve fish farming along the river, strengthen the collaboration among villages, temples, and community, and develop community in long term [3]. The restricted fishing area helps increase number of fish, which in turns, helps the villagers to generate more earnings from fish farming sustainably and wisely. Hence, teaching and learning based on STEM Education should be adopted in the community's sustainable management of natural resources.

Teaching and Learning based on STEM Education using activities, for example, projects that combined knowledge of science, mathematics, technology, engineering design allows greater opportunities to villagers to construct their knowledge and enhance skills in four areas of knowledge by themselves. The campaign of controlling the restricted fishing area around temple leads the village members to learn how to conserve the natural resources, starting from understanding its importance and current issues to seek solutions and prevention of the issues [10].

The village members are encouraged to brainstorm and generate ideas to determine objectives, plan, make decisions, and implement the plan to achieve the goals collaboratively with 4-step collaboration; including planning, implementation, taking benefits, and evaluation. The outcomes of the campaign may or may not be as expected due to the downturn of economic circumstances; however, [7] the sustainable management of natural resources must be seriously taken into action to solve the environment of the nation. Therefore, the villager members must be informed about the issues so that they become aware of their participation in natural resources conservation, which means to its sustainability.

2. Objectives

To investigate the participatory characterization of village members in fish farming around WatTha Song Korn Temple located on bank of Nam Phong River in Khon Kaen Province.

3. Population

1. Key Informants are providers of in-depth data related to aquatic animal conservation around Tha Song Korn Temple located on bank of Nam Phong River; including
    - 1.1 3 Monks
    - 1.2 4 Temple committees
    - 1.3 6 Village leaders
2. Casual Informants are 5 people implementing the aquatic animal conservation around Tha Song Korn Temple located on bank of Nam Phong River.
3. General Informants are 10 Villagers, male and female living in Tha Song Korn Sub-district.

4. Research tools

The study was conducted between March 10, 2061 and April 20, 2061. The qualitative research methodology was used as follows:

4.1 Focus on the phenomenon of the overall picture from various aspects. Adhere to the meaningful concepts from several theories, rather than one concept.

4.2 Follow a long-term plan to investigate the changes and dynamic of the society which can be up to a year period.

4.3 Study the phenomena in the natural environment from the field research method with no control in the laboratory.

4.4 Consider the human nature of the participants. The researcher must respect them and their rights to participate in the research to build a good relationship and trust.

4.5 Analyze with descriptive analysis, rather than statistical analysis.

4.6 Focus on factors or variables reflecting meaning in thoughts underlying in human behavior.

5. Objectives

The research tools used in this study to collect the data are as follows:

5.1 Basic Survey is used to explore the primary data in the community area

5.2 Interviews consisted of structured and unstructured types

1.2.1 structured interviews.

1.2.2 unstructured interviews are mainly open-ended questions to interpret the meaningful data by conducting in – depth interviews with key informants, casual informants and general informant.

5.3 Observation to observe collaboration of villages in the community.

1.3.1 Participant observation: The researcher visited the village and used questionnaire, taking note of the participant's fish farming behaviour.

1.3.2 Non-Observation.

5.4 Focus Group: Discussion conducted in group approximately 5 to 10 participants; including monks, monks, conservationists, and villagers.

5.5 Workshop was also held for the meeting among 10 participants who were key informants, casual informants, and general public to provide information about the collaboration for sustainable management of fish farming along the river.

6. Data Analysis

Data were analyzed by using descriptive statistics analysis, including mean, standard deviation. The researcher presented the results of data analysis according to the purpose of the research [8].

7. Results

It was found that two main factors influence the community collaboration of aquatic animal conservation at the bank of the temple as restricted area. First, the government initiative has been paid to develop river conditions and increase the number of fish to solve the water pollution problems. Moreover, the guidance on sustainable fish farming was provided for the villagers. The policies regarding how the river should be conserved, especially around the temple was explained to the monks, village leaders and head villagers so that they can implement the policies appropriately.

The second reason was that the similar pattern requirements of other communities occurred and found to be successful; as a result, the abbots and monks transferred the knowledge of water conservation to other temples. Along with the assistance of government agencies and the governors in the community, the temple along the river and surrounding area can be developed to be tourist attraction to the public. This holistic method derived from combining knowledge of science, mathematics, technology and engineering design based on STEM education leads to sustain management in the community.

The community collaboration in fish and water conservation at Tha Song Korn Temple on Nam Phong River in Khon Kaen Province can be explained into following four main aspects.

1. The process of collaboration in planning of fish conservation management around the temple in allocating conservation zones should be determined by the designated conservation committee.

2. The process of collaboration implementation requires the staff who is responsible for monitoring the river and report when anybody violate the rules of the restricted area or fish conservation zone. Thus, the conservation zone must be declared to the villages to gain social acceptance.

3. The process of collaboration during taking benefits in both concrete and abstract types. People can gain concrete benefits as they have greater water source for fish farming and safer for breeding, which increases the amount of fish in the river. Moreover, the villagers can obtain the abstract benefits of social and community collaboration as a unity. The elected conservation committee also feel honored to participate in community services.

4. The process of collaboration evaluation. The villagers are self-assessors. In practice, the reflection of the opinions of the villagers and those who have contributed to the project is considered as an

indicator in the effectiveness of water and fish conservation. The implications are to be used for further development.

References

[1] Anders V 2004 "The Social Ecology of Rural Violence : Land Scarcity, the Organization of Agricultural Production, and the Presence of the State **American Journal of Sociology 10** pp 313–348
[2] Arturo E 1998 Whose Knowledge? Biodiversity Conservation and S.Boston : Allyn and Bacon
[3] Bennett T 2004 "Introduction : Cultural Capital and Cultural Policy" **Cultural Trends 13** pp 7–14
[4] Brown, Radcliffe A R 1971 Structure and Function in Primitive Society (London: Coher & West)
[5] Chankaew K 2004 Science and Environment. 6th Ed (Bangkok: Kasetsart University)
[6] Dumentt T E 1871 Primitive Culture (London : John Murray)
[7] Kannasutra C 2007 Variety of fish species in rivers in Thailand (Bangkok: Environment Policy and Planning Office)
[8] Kungsirikul N 2004 Action research. Learning from Wang Nam Kiew (Bangkok: Learning Promotion Center for Community)
[9] Lamphun P K N 1972 Thai Society Characteristics (Bangkok: Liangsiangchongcharoen)
[10] Nilkhong W 2002 Local Fish Conservation in Lam Nam Wa Community, Nam Fang Sub-district, Nan (Bangkok: Research Funding Office)
[11] Richard C 2004 "Stedman. Resource Dependence and Community Well - Being in Rural Canada" **Rural Sociology 69** pp 213–234
[12] Tiannoi W 1990 Natural resources conservation (Bangkok: Aksornwattana)

Formulating Refreshment Drink Activity Utilizing STEM Education for Grade 8 Learners

Rica Mae Guarin[1], Amelia T. Buan[1], Elesar Malicoban[1], Manuel B. Barquilla[1] and Chokchai Yuenyong[2]
[1] College of Education, MSU – Iligan Institute of Technology, Iligan City, 9200 Philippines
[2] Faculty of Education, Khon Khaen University, Thailand
*Corresponding author's e-mail address: ricamae.guarin@g.msuiit.edu.ph

Abstract. This article is a qualitative study using literature review to create a lesson plan in STEM education. The lesson plan employs the pedagogy of purposeful design and inquiry. This proposes combining elements of design, art, geography and marketing into the disciplines in scientific inquiry of the STEM education in order to further strengthen it.The embedded lesson plan focused on identifying and solving social issues while hitting as many as possible learning competencies along the way. This lesson plan will serve as a contextualize application of learning competencies in Grade 8 Integrated science: biology, chemistry and physics as well as mathematics' topic on statistics and social sciences' topic on local geography. This study also include innovation and humanities so students learn skills such as communication, marketing and entrepreneurship. This will also be implemented with the goal of steering students' interests towards a career in STEM and developing more 21st century as well as life skills that they can use in the future.The student activity in this lesson plan will be assisted and evaluated by the teacher through questions that are framed to encourage deep thinking about the concepts integrated in the activity and how it would be used to produce quality product that would solve the social issue presented at the beginning of the lesson.

1. Introduction

In the 1990s, STEM education has been born in the US but up until now, misconceptions and misinterpretations are still being dealt with. By definition, STEM education means "pedagogical applications based design and engineering technology for teaching content and practice in science and mathematics education with the content and practice of educational technology and engineering with simultaneous". [1]

STEM education was defined as "the teaching and learning of the content and practices of disciplinary knowledge which include science and/or mathematics through the integration of the practices of engineering and engineering design of relevant technologies. [2]

From the aforementioned definition, STEM education has been defined and redefined according to the context of how it is being used in school, in the classroom setting, in the education system and even in different countries. It has also been variably defined from a person's perspective to another and teachers as well as administrators are not exempted to this. In the Philippines, STEM as a tract in senior high school is on its pioneering years. Since then, greater emphasis on these fields and for improvements in the quality of curricula and instruction efforts is being paved. [4]Numerous new instructional materials, programs, and specialized approaches are emerging. While most of these initiatives addressing one or more of the STEM subjects separately are good, there are also increasing calls for emphasizing connections between and among the subjects.

Advocates of more integrated approaches to K–12 STEM education argue that teaching STEM in a more connected manner, especially in the context of real-world issues, can make the STEM subjects more relevant to students and teachers. This in turn can enhance motivation for learning and improve student interest, achievement, and persistence. And these outcomes, advocates assert, will help address calls for greater workplace and college readiness as well as increase the number of students who consider a career in a STEM-related field.[3]

In a goal to bridge the gap between the specialized and separate subjects in science, technology, engineering and mathematics and provide students learning experience with integrated STEM disciplines, this study proposes the use of the embedded lesson plan.

2. Developing STEM Education Learning Activities

The lesson plan of formulating refreshment drink activity utilizing STEM education will bring students to practice STEM knowledge as clarifying in the Table 1. The STEM education activities were developed on concept of Context-based STEM education learning approach which the STEM education learning activities should provide not only the ways of investigation and solving problem but also a real world problem solving. Regarding on Sutaphan and Yuenyong [5], the context-based STEM education learning approach included (1) Identification of social issues, (2) Identification of potential solution, (3) Need for knowledge, (4) Decision-making, (5) Development of prototype or product, (6) Test and evaluation of the solution, and (7) Socialization and completion decision stage. The lesson plan could be highlighted as the table 2.

Table 1 STEM Education that the students will learn from the STEM education activity

STEM	Contents and skills
Science (S)	Diseases resulting from nutrient deficiency and ingestion of harmful substances (Biology), Elements, compounds, and mixtures (Chemistry), energy gain (Physics), seasons (Earth Science), Geographic location and ethnomedicine (Social Science)
Technology (T)	Methods, techniques, materials
Engineering (E)	Design, control, and process analysis, the use of engineering process
Mathematics (M)	statistics and marketing

Table 2 Lesson Plan on Refreshment- Making Based on STEM Education Learning Approach

STAGE	ACTIVITY
1. Identification of social issues	1. Ask the students to perform a ROLE PLAY on "What they are going to do during hot summer days?" 2. Processing: "What are the common issues/problems that you can see from the role play?" 3. Let the students focus on the social issue on refreshment vs. nutrition. 4. Observe the School Canteen: "What are the common refreshments bought by the students?"

Table 2 (Cont')

<table>
<tr><th>STAGE</th><th>ACTIVITY</th></tr>
<tr><td>1. Identification of social issues</td><td>5. Let the students make a diagram from their gathered observations in the School Canteen:
<table>
<tr><td>REFRESHMENT</td><td>NUTRITIONAL VALUE/RISK (Let the students identify the substances that pose risk to human health)</td><td>QUANTITY x COST</td></tr>
<tr><td></td><td></td><td></td></tr>
<tr><td></td><td></td><td></td></tr>
<tr><td></td><td></td><td></td></tr>
</table>
6. The teacher then states the topic: "If students are FOOD TECHNOLOGISTS, how can they design a product/refreshment with the healthiest benefits using organic ingredients available within the community?"</td></tr>
<tr><td>2. Identification of potential solution</td><td>1. Students and teachers share the cost analysis in the design of their refreshment products.
2. Students may discuss their possible design on the refreshments regarding on the five capitals: physical, financial, social/technology, human and natural capitals.
Physical – Students can interview or ask help/tap people from food manufacturing companies within their area, tap university resources like Fablab, PRISM, IDEYA
Financial – reducing the cost
Social/Technology – Material selection using appropriate materials, find ways to solve the problem, collaboration with local agencies and manufacturers
Human need – The safety of food and its nutritional value, palatability of the product, aesthetic aspect, cheaper cost
Natural – The use of organic ingredients available in the community, utilization of local raw materials</td></tr>
<tr><td>3. Need for knowledge</td><td>1. Based on the plausibility of the manufacturing of a refreshment product with higher nutritional value and cheaper capital and higher palatability to the consumers.
Financial – marketing and accounting strategies,
Physical – design of product packaging, analysis of nutritional value, food preservation, food manufacturer,
Social – sources for the natural products and raw materials within the community
2. Plan the formulation of the refreshment products, including the formula, composition, raw materials, additives
3. Do an activity on the different mixtures and formulas of the refreshment product.
3.1 Pre-activity homework: ask the students to bring the needed ingredients for the refreshment they are to make
3.2 Activity proper – making a sample of the refreshment product
❖ The teacher should provide students with different glasswares and laboratory apparatus for the activity
3.3 Doing a taste test of the product</td></tr>
</table>

Table 2 (Cont')

STAGE	ACTIVITY
3. Need for knowledge	3.4 Revision of the product 4. Teacher and students discuss the content integration of the activity using questions. Examples: Why did you choose to make this refreshment product? Why did you choose to use the raw materials on your refreshment product? Why do you think this refreshment product is a good substitute of the commercial products? Do you think students your age would buy this product? Why? 5. The teacher discusses and processes the activity and the answers of the students using the following concepts: Biology: Diseases resulting from nutrient deficiency and ingestion of harmful substances Chemistry: Elements, compounds, and mixtures Physics: Energy Earth Science: Wet and dry seasons Social Science: Geographic location, ethnomedicine Mathematics: Statistics and probability
4. Decision-making	1. After the students have the knowledge and education activities. The teacher assigns each group to design their final product. 2. Students work on the design, marketing, raw materials, final formula of the product.
5. Development of the product	1. Students formulate the refreshment product by using all the materials needed. 2. Students can bring local /raw material that can be used as refreshment product. 3. Sample refreshment product
6. Test and evaluation of the solution.	1. The students will be asked on the aspects of the possibility of the product. They need to test the following issues: 1. Is the use of local raw materials for the refreshment product comparable to the commercial product its terms of: a. Nutritional Value b. Energy gain c. Material suspension and homogeneity of the mixture d. raw material availability e. possible ingestion of harmful substances 2. Is the product salable? a. probability of buying the product b. appeal to customer c. availability of consumption throughout the year (seasons of the year) d. packaging material and sanitary process e. Cheap 3. How can you promote this product?

IOP Conf. Series: Journal of Physics: Conf. Series **1340** (2019) 012078 doi:10.1088/1742-6596/1340/1/012078

Table 2 (Cont')

STAGE	ACTIVITY
7. Socialization and completion decision stage.	1. Conduct a free taste in school vicinities. Solicit feedbacks. 2. Design a brochure/leaflets to promote the product. 3. Modify the product based on the feedbacks. 4. Conduct a project proposal for the conduct at the school level.

According to the table 1, identification of social issues stage will be started by situation of summer in the Philippines is rather long and hot. Before presenting the lesson on Digestive System, the teacher presents a scenario to the students: A Hot Summer Day. The teacher then asks the students to do a role play on what they would usually do on a hot summer day. The students would probably focus on activities under the sun as well as some refreshments. In processing the role play, the teacher would ask students: What are the social issues that you can see in the role play? The students would probably answer different issues but with the mention of unhealthy refreshments, the teacher would then single-out refreshments and asks: What refreshments would usually sell on a hot summer day? The students' answers may vary but would mostly include soda and sugary drinks. To test the students' hypothesis, the teacher would group the students in 5 and assign them to observe in the school canteen and different stores nearby. This would be their homework for the whole week. Each group would observe the specific refreshment that people buy. In the worksheet given to them, the students are to list down the names of the refreshment, the nutritional value or health risk it contain as well as the cost which will computed by the total number of bottle sold on that drink for the whole week multiplied by the cost of each bottle. The students will then make an analysis on the issues concerning the refreshments. To solve the issue, the students will act as Food Technologist. They will be ask to design/formulate a refreshment drink made from natural and cheaper ingredients that can be found in the community with high palatability and aesthetic appeal to be a substitute to the unhealthy refreshments that most people avail.

Identification of potential solution, the teacher will create a new groupings of 8 members each. The students will be ask to brain storm on the product cost, design, market, packaging and formulation. They have to come up with a proper product proposal containing the five capitals: physical, financial, social/technology, human and natural capitals. To effectively do this, the teacher will hint the students some possible collaboration to help them with. For the Physical capital, the students may partner with local manufacturers, designers and agencies. They may write a letter to the Premier Research Institute of Science and Mathematics (PRISM) of MSU-IIT for their analysis on the nutritional value of the natural formula that they can come up with. They may also tap FABLAB Mindanao of MSU-IIT to help them with the design of the packaging and the prototype of their refreshment product. The IDEYA of MSU-IIT can also give them a hand on how to effectively start-up a business for their product as well as manage and market it. This would be their financial capital. For the social/technology capital, they may walk around the local market or the neighborhood community for raw materials that are readily available. They can also tap local food manufacturer to learn where they can buy cheaper goods and raw materials. The human need capital can be brainstormed using the data on their observation which could be the safety of the refreshment/health benefits, palatability and aesthetics appeal as well as cheaper cost. For the natural capital, the group must focus on keeping the ingredients and packaging as natural as possible. Using mostly organic and local raw materials.

Need for knowledge, based on the plausibility of the manufacturing of a refreshment product with higher nutritional value and cheaper capital and higher palatability to the consumers, the students must have the knowledge on what agencies, manufacturers or offices to tap. They need to gain knowledge on how to manage the production and marketing of the product, where to get the capital and how to promote and sell it. They also need to know the geographic location of the best raw materials that they need. For example, if they need to use lemon grass, lemon and honey, they need to know where they can find fresh produce within Iligan City. They have to gain knowledge design making. Their design

IOP Conf. Series: Journal of Physics: Conf. Series **1340** (2019) 012078 doi:10.1088/1742-6596/1340/1/012078

should have cultural, social and aesthetic appeal. Lastly, they need to have a background idea on the health benefits of the raw materials that they might use as well as the longevity of the ingredients so they can decide on the natural preservation that they should utilize. With all these things in mind, they have to formulate at least 3 natural combination of possible formula for their refreshment product. They have to bring the raw materials as a pre-activity. On the activity proper, the students will try formulating it in the lab. The teacher would provide the instruments and glassware. Trial and error method would be done on the three formula then it would be presented to the teacher. During presentation, the teacher would ask the following questions: Why did you choose to make this refreshment product? Why did you choose to use the raw materials on your refreshment product? Why do you think this refreshment product is a good substitute of the commercial products? Do you think students your age would buy this product? Why? The teacher will then process their answer as an integration to the following topics: Biology: Diseases resulting from nutrient deficiency and ingestion of harmful substances which is a learning competency under the topic on Digestive System. For Chemistry: Elements, compounds, and mixtures, siting the examples on harmful effects of some substances in commercials refreshments and the health benefits of natural substances found in the natural refreshments. In Physics, the integration will be on the Energy gain after being naturally refreshed by their drink. In Earth Science, the effects of the season to the growth of these natural products. In Social Science, the significance of the Geographic location to the growth and health of the natural ingredients as well as the ethno-medicinal effects of these ingredients. In Mathematics, Statistics would be integrated through their survey, their marketing strategies and feasibility study.

Decision-making, after their presentation and processing of the activity, the students would be ready to create their final product. They can now decide which formula would they use, which ingredients is best and etc. They would then work on their final product design and proposal.

Development of prototype or product, this would be the development of the final product stage. Using the final formula, they would develop their final product and try to package it. They would produce 10 bottles at most for their promotion and marketing.

Test and evaluation of the solution, before the actual promotion activity, the students will be asked to have a product pitching. The teacher will ask them on the aspects of the possibility of the product by testing the following issues:

1. Is the use of local raw materials for the refreshment product comparable to the commercial product its terms of:
   a. Nutritional Value
   b. Energy gain
   c. Material suspension and homogeneity of the mixture
   d. raw material availability
   e. possible ingestion of harmful substances
2. Is the product salable?
   a. probability of buying the product
   b. appeal to customer
   c. availability of consumption throughout the year (seasons of the year)
   d. packaging material and sanitary process
   e. Cheap
3. How can you promote this product?

The product pitching would prepare the students to answer prospect costumer's questions as well as to guide them to have the most effective and informative promotional campaign.

Socialization and completion decision stage, after they pass the product pitching, the students will then conduct the promotional campaign by doing a free taste within the school vicinity. They have to prepare a booth where students would have free taste of the refreshments which the ingredients are not yet revealed. The costumers are then asked to give feedback and reactions on the refreshments. The collected feedback will be their basis for the further refinement of the product and final promotion and product proposal to be submitted to agencies that can support such ingenuity.

3. Conclusion

The lesson plan was created considering all the target learning competencies of different field/subjects in Grade 8 to maximize time, minimize workload of students and produce output that are ingenuine, authentic and of higher order thinking skills.

Conceptualizing STEM education inside the classroom in the Philippine setting is already ambitious while implementing it may be laborious and out of the ordinary. Nevertheless, this new definition and application of STEM education is both liberating and beneficial especially to students with wide-range skills and talents but are boxed to the norms of traditional STEM education. The skills that they can develop in these activities are important life skills that would not only equipped them for higher education but also prepare them for better employment. These kind of activities can also encourage them to take up STEM related courses in college and provide the nation with more work force passionate in STEM fields.

References

[1] Sanders, Mark. (2009). *STEM, STEM Education, STEMmania*. The Technology Teacher,December/January, 20-26. URL: https://www.teachmeteamwork.com/files/sanders.istem.ed.ttt.istem.ed.def.pdf. Retrieved: March 12, 2019.

[2] Moore, T. J., & Smith, K. A. (2014). Advancing the state of the art of STEM integration. *Journal of STEM Education, 15*(1), 5–10.

[3] The National Academies Press (NAP). 2014. *STEM Integration in K-12 Education: Status, Prospects, and an Agenda for Research (2014*) URL: https://www.nap.edu/read/18612/chapter/2. Retrieved: March 12, 2019More references

[4] Bahrum, Suraya,Wahid, Norsalawati, and Ibrahim, Nasir. 2017. *Integration of STEM Education in Malaysia and Why to STEAM.* International Journal of Academic Research in Business and Social Sciences 2017, Vol. 7, No. 6 ISSN: 2222-6990 URL: http://dx.doi.org/10.6007/IJARBSS/v7-i6/3027

[5] Sutaphan, S. and Yuenyong, C. (2019, in press). STEM Education Teaching approach: Inquiry from the Context Based. Journal of Physics: Conference Series.

Appendix

WORKSHEET

Activity 1: Formulating Refreshment Drink

Names: ______________________________ Date Started: ____________
______________________________ Date Finished: ____________
______________________________ Rating: ____________
______________________________ Group No. ____________

Objectives:

1. Determine the type of refreshment drink that people buy in stores and school canteen;
2. Compute the cost of refreshment drinks consumed in a day from the stores and school canteen;
3. Investigate the nutritional value and risks of the different substances and ingredients of most refreshments; and
4. Identify solution to the social issue observed.

Introduction:

According to Section 13 of Article 2 of the Healthy Beverage Options Act of 2011 : *"The State recognizes the vital role of the youth in nation-building and shall promote their physical, moral, spiritual, intelectual, and social well-being. It shall inculcate in the youth patriotism and nationalism, and encourage their involvement in public and civic affairs."* Thus, this Act prohibit schools from selling carbonated drinks inside the school premises and the school will be held liable if proven

IOP Conf. Series: Journal of Physics: Conf. Series **1340** (2019) 012078 doi:10.1088/1742-6596/1340/1/012078

guilty of not following the said Act. Nevertheless, carbonated drinks and most commercial refreshments are still present inside school premises.

Social Issue:

Refreshment vs. Nutrition

Instruction:

Your group of 5 members will investigate the type of refreshments that people purchase in the school canteen as well as in stores near your house as well as its nutritional value, the risks it pose to the human health and the total cost spent to this refreshment in the whole week.

Task:

For this activity, to solve the social issue discussed in the past meeting, you will act as Food Technologists but each member of the group will be assigned to the following specific tasks:

1. Field Researcher (3 members) - the three of you will observe in the school canteen during breaks and in the stores within the neighborhood on the refreshment drinks that people usually buy. Write your findings in the table below.
2. Nutritionist (2 member) - the two of you will go to the library while the other group members are observing in the canteen. You may read books or browse the school's internet resources for the common substances and ingredients of commercial refreshments that can be found in the canteen or nearby stores, their health benefits and risks they pose to the human health as well as the energy gain from this refreshments. Write your findings in your notebook.

Results and Findings

Specific Name of the Refreshment	**Nutritional Value**	**Health Risk**	**Cost (Quantity for 2 days x cost per bottle)**

Questions:

1. Which refreshment has the highest sales for 2 days?
2. What do you think are the factors that people consider in buying a refreshment drink?
3. What are the common substances that can be found in those commercial drinks?
4. Why do you think people still buy refreshments that has potential risks to the human health?
5. Do you think these refreshments are worthy of its costs? Why or Why not?

Conclusion:

What solution can you propose to address this social issue?

Indonesian Students' Attitude and Interest in STEM: An Outlook on The Gender Stereotypes in The STEM Field

Hadi Suwono[1]*, Rifka Fachrunnisa[2], Chokchai Yuenyong[3], and Linda Hapsari[4]

[124]Biology Department Universitas Negeri Malang Malang, Indonesia
[3] Science Education Department Khon Kaen University, Khon Kaen, Thailand

*Corresponding author's e-mail address: hadi.suwono.fmipa@um.ac.id

Abstract. STEM education could be reckoned as a noble thing in the Indonesian Educational System in the last few years. This paper is a preliminary attempt to investigate the students' attitude and interest in STEM regarding the gender perspective of Senior High School's student in Malang, Indonesia. To explore the attitude and interest in the gender stereotypes, this quantitative method study has been conducted. 357 students participated in the survey which comprised 126 males and 231 females. The survey contended 40 items of belief's indicators in the STEM field such as science, technology, engineering, mathematics and the STEM. Besides, at the end of the survey's section, there are some options for future study programs and career to be selected by the learner. Unexpectedly, the finding of this project revealed that male has a more positive belief in the most of STEM field, such as technology, engineering, mathematics, and STEM, rather than female. In comparison, the female has a more positive belief than male in the science only. The marked investigation of emerging from the data comparison was both of male and female have a higher interest in the social field and medicine. However, the additional interest of male and female was significantly different. In males, we found that they have more interest in engineering, while females have an interest in biology. This study also disclosed that the factors, namely family support, family financial, friends, teachers, and hobby, determined both attitude and interest in STEM among students in Malang.

1. Introduction

In the current fast changing world, STEM education is considered as a visible elucidation in preparing new generation to deal with the global pace. STEM education has been running for the program that can help student to foster various strategies in tackling interdisciplinary problems and advancing their scientific leadership and economic growth in the future [14]. Besides, there has been numerous reported about innovation and development as the evident of STEM education's outcome in various countries [4]. To engage with STEM education, thus needed the competencies as the fuels to foster stem in the educational system to fulfil the requirement of the current or future STEM workplaces.

This new movement has impacted globally which induced a massive engagement with STEM education in many countries around the world over past 20 years is proving of the crucial STEM education challenges to be addressed [14]. Countries in the Europe such as France, Germany, The Netherlands, Switzerland and Austria has been struggling to enhance their people in operating and creating technologies to boost their domestic production [3]. In the same way, Australia also improve STEM skills in their educational system intensively due to the fact that there is less student in the college who want to pursue STEM field. The main indicator of this issue was showed by the fewer qualified applicant for engineering compared to natural sciences and medical fields. Likewise, in the USA, they have a similar issue to find qualified employer in technical and engineering skills to fulfil more than 500.000 jobs for the manufacturing sector [12].

In the light of Indonesian education system context, STEM education can be an overlapping part among the three pillars of the current Indonesian curriculum . The three pillars included character, literacy and competency. Developing STEM education can mean developing those pillars too. Knowing that Indonesia was considered as the big seven countries which has the most STEM graduates in the world, the fact of the ratio or comparison with the population rate in Indonesia showed that it was not the big seven anymore. There were only seven people who were graduated from STEM among 10.000 populationc[12]. It was an alert that the current system should improve the STEM graduates. To bring it to fruition, there many reported studies were specifically focused on developing approach for students' carrier pathway to engage with STEM. This was presumably adopted from the replication of the previous studies in the decade ago, when the world paradigm set science education as the focal point for development. Varieties branches of education – such as psychology, sociology and philosophy- also supported the discussion on carrier in science's topic, chiefly scrutinizing the attitude and interest in science. Those two facets are always juxtaposed together with the carrier aspirations discourse. The gender stereotypes, in other way, also completed the puzzle due to many studies disclosed that the stereotypes take into account in determining both of attitude and interest among students significantly.

On the one hand, the attitude in science has been manifested as a perplexing combination of the eagerness to know and understand, to examine the method of the statements, to find data and their meaning, to require for verification, to adhere to logic and to envisage the premises and the consequences [4]. Reported that several large quantitative studies found that overall females' attitude towards science are less positive than boys [2]. They also revealed that the attitude declines more significantly with age. In the other replications, there are seven factors significantly correlated with the females' attitudes towards science, namely equity, access, curriculum, pedagogy, reconstruction the nature, culture of science and identity. The interest, on the other hand, has been rationally operated as a multidimensional framework which requires both of cognitive and emotional categories [14], [13] conveyed that this behaviour attribute represents a specific personal orientation, a reference valuation, or a consciousness of feasibilities for an action. Their replication also revealed that to concrete the action, interest acts as an effect provider for emotional reactions, declarative knowledge, and procedural knowledge of the action's engagement. In term of interest in learning, presumed that interest has positive motivational features for long periods of time.

Steering Indonesian students to choose STEM field as their life carrier would require immense endeavour since carrier pathway in Indonesia highly correlated with the social and cultural discourses. Few researchers have addressed the problem of fostering STEM education in Indonesia, yet it has not been established whether there is a distinct between male and female's attitude and interest in STEM. This present study, therefore, sheds new light on the gender stereotypes in the STEM field among Indonesian students.

2. Methodology

To identify students' attitude and interest in STEM based on gender perspective, this study employed a mixed method approach. On the quantitative stage, we administered survey by deploying a questionnaire. The questionnaire was operationalized through the use of a Likert-Scale questionnaire contented 40 indicators of the two aspects – interest and attitude. However, before the survey was carried out, an initial pilot test questionnaire was conducted to test the questionnaire. There are 126 males and 231 females from Malang's senior high school participated in this survey. Because the participants in this project were under 18 years old, a parental consent form was required before undertaking the survey. Probability sampling (or, specifically, random sampling) was employed in selecting the population sample. The population sample is explained as a small population subset which is representative of the whole [7]. Thus the selected sample (357 students) was determined by the total number of participants whose parents consented to completion of the questionnaire. On the qualitative stage, however, we conducted interview to the four lowest and highest score in questionnaire. The interview was aimed to explore the social and cultural background of the research participants. The result of the interview was then analysed using thematic coding analysis.

3. Results

3.1. Quantitative Result

Based on the Table 1, the attitude toward science of male students is lower than female students with a significance value of 0.041 which means that the two of them are differ significantly. The mean value of mathematics in male students was 182.06 higher than female students with a value of 173.38. The significance value is 0.446 which means that the attitude of male and female students to Mathematics is not significantly different. In the field of Engineering, the attitudes of male students are higher than the attitudes of female students with a significance value of 0,000. This means that the attitudes of male and female students in the Engineering field differ significantly. In the field of technology, the attitudes of male students are also higher than female students with a significance value of 0.030, which means that there are significant differences in attitudes of male and female students in the field of technology. In the field of STEM the attitude values of boys are 194.04 and the attitude of female students is 170.80 with a significance value of 0.041 which means that there are significant differences in attitudes between male and female students in the STEM field.

Table 1. Attitude toward STEM

Subject	Male	Female	Asymp. Sig. (2-tailed)	Significance
Science	170.76	194.11	0.041	Significant
Mathematics	182.06	173.38	0.446	Not significant
Engineering	207.44	163.48	0.000	Significant
Technology	194.96	170.29	0.030	Significant
STEM	194.04	170.80	0.041	Significant

The differences in attitudes of male and female students in each STEM field can be seen from the results of factor analysis. Factor analysis is carried out to identify fundamental factors that can explain the correlation of a series of variables, identification of new smaller variables to replace uncorrelated variables from a series of origin variables that are correlated from multivariate analysis, and identification of small variables that stand out from bigger variable. The requirement for factor analysis is the fulfillment of several assumptions, namely; 1) the magnitude of the correlation between the independent variables must be strong enough (above 0.5), 2) partial correlation, the correlation between the two variables must be small by assuming the other variables are fixed from the Anti-Image Correlation, 3) testing the entire correlation matrix measured with the Bartlett Test of Sphericity or Measure Sampling Adequacy (MSA), which requires a significant correlation between at least a few variables. Based on the results of factor analysis, it is known that there are some differences in attitudes between male and female students towards the STEM field. The differences in attitudes of male and female students can be seen in the communality table. Communality aims to see the proportion of certain variable variances explained by factors. The higher the value of communality means that these variables increasingly have the same factor. Communality test results can be seen in Table 2.

Table 2. Rotated Component Matrix Correlation of Attitude towards STEM among Male and Females' students

	Male		Female	
	1	2	1	2
Science	0.853	0.004	0.868	-0.009
Mathematics	0.750	-0.030	0.750	-0.353

Engineering	0.727	0.314	0.551	-0.320
Technology	0.578	0.016	0.499	0.705
STEM	-0.210	0.953	0.562	0.606
Extraction Method: Principal Component Analysis.				
a. 2 components extracted.				

According to Table 2, there has been identified 6 variables, 2 factors were formed. Determination of variables included in the factor is determined by looking at the greatest correlation value. In men, the greatest STEM correlation at factor 1 is 0.853. Likewise science (0.750), Engineering (0.727), and Technology (0.578) have a greater correlation in factor 1. Mathematics is the biggest correlation at factor 2, which is 0.953. In female students, the greatest correlation value of STEM at factor 1 is 0.868, followed by science (0.750) and engineering (0.551). While Technology and Mathematics are the biggest correlations in factor 2 with a value of 0.705 and 0.606. Based on the description above, it can be concluded that members of factor 1 in male students are STEM, science, engineering, piety, and technology. While members of factor 2 are only Mathematics. This is different compared to female students. For female students, member factor 1 is STEM, Science, Engineering, and Faith. While factor 2 members are Technology and Mathematics.

On the other hand, interest in study programs for male and female students was carried out in a descriptive analysis. Table 3 showed that in the male students the highest mean score is in the social field (3.12) followed by the medical field (3.03) and engineering (2.88). This is almost the same for female students (Table 4). The highest mean score is in the social field (3.18), then followed by the medical field (3.06), and biology (2.95). But when compared to each field, the average value in the fields of physics, chemistry, biology and social was higher for female students than for male students. While the average value of mathematics, engineering, and technology is higher for male students than female students.

Table 3. Descriptive analysis of interest in study programs among male students

	N	Minimum	Maximum	Average	Standard Deviation
Physics	126	1.00	4.60	2.59	0.88
Chemistry	126	1.00	5.00	2.50	1.16
Biology	126	1.00	5.00	2.69	0.95
Mathematics	126	1.00	5.00	2.88	1.03
Medicine	126	1.00	5.00	3.03	1.04
Engineering	126	1.00	4.50	2.77	0.82
Technology	126	1.00	5.00	2.76	1.03
Social Work	126	1.00	5.00	3.12	0.86
Valid N (listwise)	126				

Table 4. Descriptive analysis of interest in study programs among female students

	N	Minimum	Maximum	Average	Standard Deviation
Physics	231	1.00	4.60	2.64	0.78
Chemistry	231	1.00	5.00	2.59	1.13
Biology	231	1.00	5.00	2.74	0.88
Mathematics	231	1.00	5.00	2.95	1.06
Medicine	231	1.00	5.00	3.06	1.04
Engineering	231	1.00	4.60	2.47	0.79
Technology	231	1.00	5.00	2.79	0.97
Social Work	231	1.00	5.00	3.18	0.81

3.2. Qualitative Result

The qualitative approach was devised to understand the experiences of the participants who showed the highest and lowest score in interest and attitudes towards STEM based on score calculated in the quantitative stage. In fact, more than one candidate had the same score, but only one was selected for follow-up interviews. This process was complicated by ethical issues. I had of course chosen those who had consented for the interview process. There were four participants who participated in interview section, which had been conducted in accordance with the plan. The remaining prospective candidates had rejected the consent due to the others matters such as desire to go home earlier, or a preference to do extra-curricular school activities. As mentioned in the Methods section, the analysis was conducted using thematic coding. [6] Affirms that the aims of this analysis are suitable for studies that are theoretically conducted based on group comparison (in this study, extremes group) in relation to a specific issue (here, creativity in the science education). He posited that with this analysis technique, different views can be revealed. In addition, he also suggested the following procedure when undertaking thematic coding: provide a brief description of each case (participant background); develop a thematic structure; and refine the interpretation of thematic domains or make key questions. In this analysis process, the four interview transcripts were read several times.

Afterwards, some codes were extracted from the passages, which were separated based on the group themes of participants (attitude and interest in STEM). Each extracted code was then separated on paper. Following this, these codes were grouped together, and the revealed codes from different groups were selected. Then, analysis revealed one code which appeared in both the low and high groups. Thus, this established a communality amongst the participants in different groups [1]. Based on the theme extraction, the choice of study program interests is strongly influenced by the encouragement of family and close relatives, especially supported by the family's economic background which is sufficient. Other factors that can influence the interest in choosing a study program are friends, teachers, and hobbies. At the time of the interview, most students said that the selection of study programs had been prepared by the family, especially parents, for various reasons, such as continuing the family business, coming from families who all had profession as doctors, or indeed because of the wishes of children who had been supported by parents.

4. Discussion and conclusion

Data analysis revealed that there were differences in attitudes between male and female students in STEM. Male students show a more positive attitude towards the fields of Mathematics, Engineering, Technology, and STEM. Whereas female students showed a more positive attitude in the field of science and piety. The results of this study are supported by the 2012 PISA results which suggest that male are superior to female in mathematics performance. Female feel less motivated to learn mathematics and are less confident in their abilities compared to men [12]. For technology and engineering, [3] states that the field of STEM is dominated by male. Women mostly choose the field of biology or science, so they choose less in the fields of engineering, computer science, and mathematics. The results of research conducted which examined the application of STEM-based PBL for students to improve scientific literacy. In addition, he also disclosed that male is superior to indicators of science-technology (S-T), science-mathematics (S-M), and science-technology-engneering (S-T-E). Whereas the female class excels for science indicators only. According to [9] Women tend to prefer hands-on learning experiences, make intuitive judgments based on feelings, always oriented, and feel comfortable with ambiguity. While men tend to do analysis in learning, think logically and rationally, and enjoy working with symbols like structures. So that the STEM field that reflects an analytical approach tends to be more suitable for boys than girls. [15] also revealed the factors that caused females had lower participation in science field: history of science education; the international situation in science for females; sociocultural accounts pertained with the science attitudes and environment of the classroom; and the policies related to the gender issues.

The analysis shows that male and female students share a high interest in social and medical fields. In male students, social fields that have the highest interest are teachers, and management. Whereas for female students, social fields that have the highest interest are management and linguists, especially Arabic. The second highest interest is in the field of medicine. Most students, both male and female, choose medicine (doctors, dentists, nurses, midwives, pharmacists, and nutritionists) as a choice of their future study programs and careers. This is usually due to the child's own wishes and has been supported by parents or parents' wishes even though the child does not agree. The third highest interest in male students is in the field of engineering while the female students are in the field of science. This is in accordance with the opinion which states that female students are more interested in science issues, support science inquiry, and responsibility for environmental resources than male students. This also supported by the replication of [11] which states that female students prefer to study biology or science related to the choice of work in the future [8]. This is due to the high and low attitudes of students towards science are influenced positively, among them are the work desired by students, teaching and learning activities in the classroom, and the amount of time spent studying science. In career aspirations, girls are more likely to choose careers that will not interfere with their future roles as parents or parents. While boys have higher long-term expectations for themselves [3].

Taken together, this study highlights that in introducing STEM education in Indonesia can be embarked with taking awareness of the plot of attitude and interest among students. It is useful to consider a transformation in science curriculum in Indonesia to flourish those traits in supporting the increase economic and technology of Indonesia, thus Indonesia can be juxtaposed with the other advanced countries around the world. This transformation, however, has to be supported by all parties in the education system including the parent. School-parent engagement in the most mainstream school also still should be improved. In this case, this program can depress the discrepancy between female and male interest and attitude towards science. Moreover, by knowing the plot of those two aspect, teachers in the classroom should be aware to design the learning strategy in accommodating the improvement on STEM interest and attitude among both of female or male students. Furthermore, the stakeholder – Indonesian Minister of Education- should take it as a granted by conducting some programmes that can support the advancement of STEM education. It might be concluded from this that by taking more enactment of developing STEM education by considering the two traits – interest and attitude- can decrease the impact of the determinant factors which have been investigated in this study – such as family support, family financial, friends, teacher, and hobby-. Thus, they will not be stumbling blocks to foster STEM education and increase STEM graduate in Indonesia

References

[1] Abbott A 2014 The system of professions: An essay on the division of expert labor (University of ChicagoPress)
[2] Brotman J S and Moore F M 2008 Girls and science: A review of four themes in the science education literature **Journal of Research in Science Teaching: The Official Journal of the National Association for Research in Science Teaching 45** pp 971–1002
[3] Chant S and Sweetman C 2012 Fixing women or fixing the world? 'Smart economics', efficiency approaches, and gender equality in development **Gender & Development 20** pp 517–529
[4] Educational Policies Commission 1962 Education and the disadvantaged American
[5] English L D 2016 STEM education K-12: perspectives on integration **International Journal of STEM Education 3** p 3
[6] Flick U 2014 An introduction to qualitative research (Sage)
[7] Fowler Jr F J 2013 Survey research methods Sage publications
[8] Hango D W 2013 Gender differences in science, technology, engineering, mathematics and computer science (STEM) programs at university (Statistics Canada Statistique Canada)
[9] Kulturel-Konak S, D'Allegro M L and Dickinson S 2011 Review of gender differences in learning styles: Suggestions for stem education **Contemporary Issues in Education Research 4** pp 9–18

[10] Lacey T A and Wright B 2009 Employment outlook: 2008-18-occupational employment projections to 2018 **Monthly Lab. Rev. 132** p 82

[11] Larson L M, Stephen A, Bonitz V S and Wu T F 2014 Predicting science achievement in India: Role of gender, self-efficacy, interests, and effort **Journal of Career Assessment 22** pp 89–101

[12] Organisation for Economic Co-operation and Development 2014 PISA 2012 results: What students know and can do (volume I, revised edition, February 2014): Student performance in mathematics, reading and science (OECD Publishing)

[13] Renninger K A, Hidi S, Krapp S and Renninger A 2014 The role of interest in learning and development Psychology Press.Sala-i-Martin, X., & Schwab, K. (Eds.). (2011) The global competitiveness report: 2011-2012 World Economic Forum

[14] Sanders M 2009 Integrative STEM education: primer **The Technology Teacher 68** pp 20–26

[15] Scantlebury K, Baker D, Sugi A, Yoshida A and Uysal S 2007 Avoiding the issue of gender in Japanese science education **International Journal of Science and Mathematics Education 5** pp 415–438

IOP Conf. Series: Journal of Physics: Conf. Series **1340** (2019) 012080 doi:10.1088/1742-6596/1340/1/012080

Examine Students' Perception on Their Critical Thinking and Problem-Solving Ability through Journal Writing for Learning about Genetics and DNA Technology

Saranya Wongsila[1*] and Chokchai Yuenyong[1]

[1]Science Education Program, Faculty of Education, Khon Kaen University, Khon Kaen 40002, Thailand

* Corresponding auther's email address: saranyawsl@kkumail.com

Abstract. This study examined grade 12 students' perception on their critical thinking and problem-solving ability when they participated in the STS genetics and DNA technology unit. The participants were 43 grade 12 students in Maung, Khon Kaen, Thailand. The STS genetics and DNA technology unit were provided learning activities through STS approach in framework of Yuenyong (2006) and has been taught by the first author for 3 weeks. Methodology regarded interpretive paradigm. Students' perception on their critical thinking and problem-solving ability were interpreted through their reflection as journal writing for their learning about genetics and DNA technology. Students' journal writing was collected after each class of the STS genetics and DNA technology unit was done. The findings revealed that students perceived on their critical thinking and problem-solving ability when they wrote their journals to reflect what they learned. These reflections provided 4 elements of critical thinking and problem-solving ability. These included reason effectively, used systems thinking, make judgment and decisions, and solve problem. These finding would be a good practices for teaching biology through STS approach to who may take further applications for science teaching and learning.

1. Introduction

The major corporations around the world found that the results of the study have defined the 21st-century skills that most valued in today's complex globalized and rapidly changing world are competencies or learning and innovation skills (4Cs) [7]. There are the basic skills that people in 21st-century must learn that allow them to become innovative in what they do. It would be mentioned that learning and innovation skills will prepared students for more complex life and work environments which is the necessary skills for developing students' thinking both inside and outside the classroom. Learning and innovation skills composed of 4 critical skills included: creativity, critical thinking, communication, and collaboration that is essential for students in the future. This paper focuses on critical thinking and problem-solving ability that is the ability to identify, analyze and evaluate situations, ideas and information in order to formulate responses to problems.

According to the vision, there are some ways of developing scientific knowledge that could promoted students' thinking through science, technology, and society that lead to various of students' thinking. Therefore, the researcher believed that the thinking which develop on science classroom is critical thinking and problem-solving ability that involved to social issues in the present. The researcher follows the framework of P21 [5] of critical thinking and problem solving ability in learning and innovation skills that composed of 4 elements:

- Reason Effectively
  - Use various types of reasoning (inductive, deductive, etc.) as appropriate to the situation.
- Use Systems Thinking
  - Analyse how parts of a whole interact with each other to produce overall outcomes in complex systems.
- Make Judgments and Decisions
  - Effectively analyze and evaluate evidence, arguments, claims and beliefs.
  - Analyze and evaluate major alternative points of view.
  - Synthesize and make connections between information and arguments.
  - Interpret information and draw conclusions based on the best analysis.
  - Reflect critically on learning experiences and processes.
- Solve Problems
  - Solve different kinds of non-familiar problems in both conventional and innovative ways.
  - Identify and ask significant questions that clarify various points of view and lead to better solutions.

These learning and innovation skills will examine students' thinking that could be addressed by scientific knowledge related to science, technology, and society issue. In order to enhance teaching biology through STS approach. Yuenyong [9] stated that teaching science through STS developed many skills as well as thinking and decision-making skills, the perception of the relationship between science, technology, and society. Including knowledge, skills and the confidence to express opinions and take responsible action to address issues. Likewise, [6] studied Thai Students' Decision Making in Societal Issue of Surface Area and Concentrated Solutions as a Factor in the Rate of Chemical Reactions found that STS approach could give students a chance to learn not only scientific knowledge but also developing the model of decision making. Moreover, Chantaranima and Yuenyong (2014) also reported that STS approach could support students to gain their capability of analytical thinking such as thinking for classifying, compare and contrast, reasoning, interpreting, collecting data and decision making. So, the researcher use STS approach of Yuenyong (2006) that consisted of five stages: (1) identification of social issues; (2) identification of potential solutions; (3) need for knowledge; (4) decision-making; and (5) socialization stage to enhance students' critical thinking and problem-solving ability.

Some literature points suggested the investigation to examine students' viewpoints or reflection on learning activity. [1] reported on Scientific Literacy of Grade 12 Student through Teaching and Learning about Genetics and DNA Technology on Science Technology and Society Approach Yuenyong (2006). The study revealed that after the class was done we should emphasize students to reflect their perspectives on journal writing because it can reflect the thoughts, feelings and knowledge of students which enable teachers to be used to adjust and develop appropriate learning activities in accordance with the interests and needs of students. Moreover, [4] stated that the overall majority of the students clearly stated potential benefits of journals writhing as a learning tool. In addition to the satisfaction of students towards writing their own journals is consistent with their views on the benefits of writing. The journal writing is a summary of what is learned and helped developing writing skills and identifying problems and solutions [2]. In the same way, journal writing has been the case partly that the researcher try to add the ability that would be worthy for student's learning and also as the reflection of endless learning process to teaching.

According to the literature of learning activities for biology in order to develop students to learn biology in science, technology, and society. The authors were suggested how to provide design learning activities to improve learning and innovation skills. The 4 elements of critical thinking and problem-solving will be taken into account for biology learning by self-directed learning and teamwork.

Therefore, the researcher has an objective to examine the critical thinking and problem-solving ability in learning and innovation skills on genetic and DNA technology unit though students' journal writing by using scientific, technological and social instructional model of Yuenyong (2006).

2. Methodology

This study was a qualitative study focusing on paradigm in interpreting to search for examination grade 12 students' critical thinking and problem-solving ability, during the first semester of 2018 school year, The Modindaeng Demonstration School of Khon Kaen University. The study would emphasize on the importance of the students' critical thinking and problem-solving from reflecting's learning in their journal writing for show their attitude and motivation toward science studying. Data were analyzed, interpreted, and concluded.

2.1. Participants

The participants were 43 grade 12 students in The Modindaeng Demonstration School of Khon Kaen University, Khon Kaen, Thailand.

2.2. Intervention

This research examined students' critical thinking and problem-solving ability on genetics and DNA teachnology unit. The unit was taught by considers the relationship between science, technology and society (STS) approach of Yuenyong (2006). Consequently, the researcher was design learning activities that promote the critical thinking and problem-solving ability on GMOs issue on the table 1. Finally, students will show their different reflections on their journal writing after the class was done.

Table 1. Overview of the STS Genetic and DNA technology Unit. (a)

Stages	Learning activities
Identification of social issues stage	The issue of GMOs in Khon Kaen province was raised by news on "Let's start to know what GMOs is "headline. The news showed about the effects of GMO' papayas. This news was expected to motivate students to identify some social issues from problem that they may involve with.
Identification of potential solutions stage	Students will need to identification of potential solution on identification of social issues stage from the question. - Students on each group brainstorm to review prior knowledge: "What knowledge can be used to solving the problem?" Moreover, Students create an unknown knowledge question and further research to find
Need for knowledge stage	The researcher use The 5Es Instructional Model on this stage. Engagement: Students watch video about GMOs on human's life. Exploration: Teacher show the news "The effects of GMOs on agriculture in Thailand" and students brainstorm to answer the question on worksheet. Explanation: Teacher and student discussion to summarize "Should GMO products continue to be used?" Elaboration: Teachers present pictures on Power point and explain more GMOs. Evaluation: Teachers use the evaluation tools to be used either interview, observation or checklist etc. To evaluate the topic taught based on the learning objectives set at the beginning of the lesson.

Table 1. Overview of the STS Genetic and DNA technology Unit. (b)

Stages	Learning activities
Decision-making stage	Each group works together to use critical thinking and problem-solving ability on GMOs. From Need for knowledge stage to create possible solution. Moreover, they work together again to make decision for select the best possible solution of group. Next, hand on design on worksheets and journal writing.
Socialization stage	Each group presents "The best possible solution". By record it is a Video File or post poster on Facebook. This performance will open for comments and ideas. Comments and ideas will be revise and develop again to completion.

2.3. Data Collection and Data Analysis

For this study, the researcher follows the framework of P21 [5] of critical thinking and problem-solving ability to design the lesson plans through that composed of 4 elements: (1) reasoning effectively; (2) using systems thinking; (3) making judgments and decisions; and (4) solving problems. The genetic and DNA technology unit has been taught by the researcher for 3 weeks. The researcher was interpreted to examine what and how students' critical thinking and problem-solving from students' journal writing. Then, the interpretation will be rechecked as peer debriefing in order to show the credibility which the interpretation would depend on the particular researcher and each researcher.

3. Findings

The finding revealed that the learning activity about genetic and DNA technology unit promoted students' critical thinking and problem-solving ability followed the framework of P21 [5] including: reasoning effectively, using systems thinking, making judgments and decisions, and solving problems. The journal writing could interpret through the critical thinking and problem-solving ability as follows:

3.1. The critical thinking and problem-solving ability from students' journal writing

The students' journal writing was reflected after the class of each plan was done. The examination from journal writing of students 43 members found that students' journal writing organized into 2 groups are 1) students' journal writing about contents of GMOs; 2) students' journal writing which represent their critical thinking and problem-solving ability about GMOs. Consequently, the researcher focus on the results of examination students' journal writing from 10 members (Table 2) that display students' perception on their critical thinking and problem-solving ability in learning and innovation skills in group 2. This is an important data to analysis that students' critical thinking and problem-solving ability on GMOs issue by using significant 4 features on figure 1.

IOP Conf. Series: Journal of Physics: Conf. Series **1340** (2019) 012080 doi:10.1088/1742-6596/1340/1/012080

Table 2. *The critical thinking and problem-solving ability from students' journal writing.* (a)

Number of students	Reason effectively	Use Systems Thinking	Make Judgments and Decisions	Solving Problems
1	Student stated that genetic engineering lead to GMOs and GMOs are products which human needs such as genetic modification of papaya to increase agricultural productivity.	-	-	-
2	Student provided the reason that GMOs can protect plants from insects. Moreover, increasing crop products on agriculture.	Student related that GMOs help farmers increased their expenses such as insecticide.	Student compared between positive and negative ways of the effects of GMOs on economy and healthy.	Student suggested that we might get an effect from GMOs. So, we all have to determine the right solution that can solve by everyone.
3	Student stated that the advantage is GMOs have the resistance to pest and disease. However, the dis- advantage is GMOs also give disease too.	Student stated that GMOs do not accept in aboard because they concern on the effect of GMOs.	Student thought that after the GMOs class, he more carefully on having meal.	-
4	-	-	Student reflected that GMOs are the issue for solving problem about mostly production of agriculture.	-
5	Student provided the causes that we needed GMOs because GMOs give more choices to us.	-	Student gave the advantage and dis advantage from GMOs.	-
6	Student gave the reason about GMOs that it answer many people for protect their plants from weeds.	-	-	-

Table 2. *The critical thinking and problem-solving ability from students' journal writing.* (b)

Number of students	Reason effectively	Use Systems Thinking	Make Judgments and Decisions	Solving Problems
7	GMOs is the innovations from modification the organism to obey human in the present.	-	-	-
8	-	-	Student stated that GMOs is the new innovation for solving many problem on the other hands GMOs cannot be the solution but it is the problem.	-
9	-	-	Student reflected that GMOs have an advantage and disadvantage. The discovery of the way to solving the gap lead to solution of all problem from GMOs.	-
10	Student provided that GMOs solved the problem from pest but it might bring the concerning to consumers.	Student look that GMOs will give an effect from the residue to their health and environment.	-	-

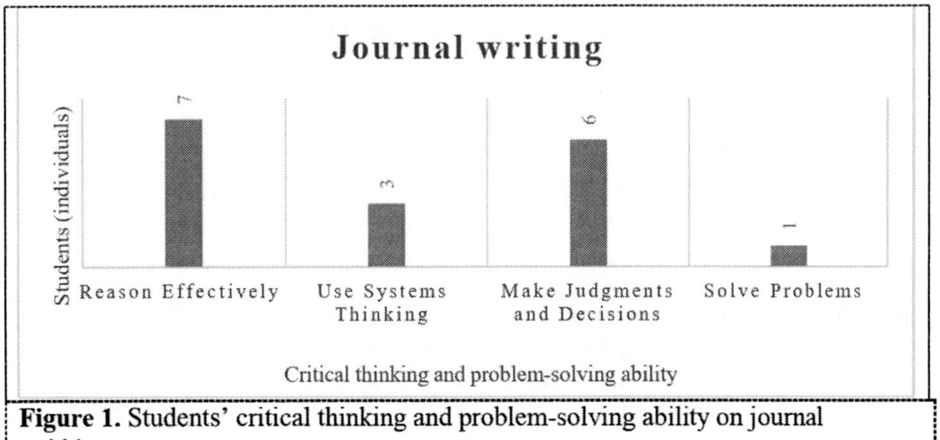

Figure 1. Students' critical thinking and problem-solving ability on journal writhing.

This graph showed the data about the 4 elements which examine by interpreted from students' journal writing from 10 of 43 students in class. The result found that the number of students' reason effectively is 7, used systems thinking is 3, make judgment and decisions is 6, and solve problem is 1, respectively. It can be seen that the examination of journal writing on GMOs STS unit can enhance students' critical thinking and problem-solving ability in learning and innovation skills.

4. Conclusion

The study examined students' perception on their critical thinking and problem-solving ability in learning and innovation skills by journal writing which collected after the class. The learning activity was designed to develop the high school students' critical thinking and problem-solving ability which follows the framework of P21 Partnership for 21st century learning (2015) to learning biology. The focus had a relationship among science, technology and society. This research also reported the process of developing Genetic and DNA technology unit on the STS approach [8]. The finding found that after learning through STS and interpreted from students' journal writing. The number of students' critical thinking and problem-solving ability are 10 members have critical thinking and problem-solving ability. These result show that students' significant 4 features were 1) reasoning effectively: Students gave reasons for GMOs' status in the present, and provided a variety of reasons why GMOs was happened; 2) using systems thinking: Students analyzed GMOs' useful and useless on their health, and gave an implication of GMOs and the direct and indirect effects on environment; 3) making judgments and decisions: Students identified that GMOs is not the best solution for the basic human needs; and 4) solving problems: They reflected an individual's viewpoint and motivation toward possible solution by together with everyone. Consequently, these reflection from journal writing involved with GMOs that promoted students' perception on their critical thinking and problem-solving ability on genetic and DNA technology unit. It indicated that students also applied these results to their experience. Moreover, it could be mentioned that students could develop their ability of critical thinking and problem-solving while they learning STS of GMO issue.

5. Acknowledgements

The research was financially supported by the institute for the promotion of teaching science and technology (IPST) and the promotion of Science and Mathematics Talented Teachers (PSMT) scholarship since my bachelor's degree.

References

[1] Aryowong N, and Sumranwanich W, 2011. Scientific Literacy of Grade 12 Student through Teaching and Learning about Genetics and DNA Technology on Science Technology and Society Approach Yuenyong (2006) *Journal of Education Graduate Studies Research*, **5** pp 21–28

[2] Buaraphan K, 2007 Preservice Science Teachers' Perceptions of Journal Writing. Kasetsart *Soc Sci* **28** 43–57

[3] Chantaranima T and Yuenyong C 2014 The Outcomes of Teaching and Learning About Sound Based on Science Technology and Society (STS) Approach *Procedia-Social and Behavioral Sciences* **116** pp 2286–2292

[4] Good J M and Whang P A 2002 Encouraging Reflection in Preservice Teachers through Response Journals *The Teaching Educator* **37** pp 254–267

[5] P21 Partnership for 21st century learning 2015 P21 Framework definitions *The Partnership for 21st century learning* pp 1–9

[6] Chattabud J, Suwannoi P, Sranamkam T and Yuenyong C 2015 Thai Students' Decision Making in Societal Issue of Surface Area and Concentrated Solutions as a Factor in the Rate of Chemical Reactions *Mediterranean Journal of Social Sciences* **6** pp 56–61

[7] World Economic Forum 2015 The skills needed in the 21st century Retrieved on Febuary 13, 2019 from https://widgets.weforum.org/nve-2015/chapter1 html

[8] Yuenyong C 2006 Teaching and Learning about Energy: Using STS approach *Thesis of Doctoral Degree in Science Education* (Kasetsart University)
[9] Yuenyong C 2013 Enhancing Scientific Literacy in Thailand *Global Studies of Childhood* **3** pp 86–98

Learning Activities to Promote the Concept of Engineering Design Process for Grade 10 Students' Ideas about Force and Motion through Predict-Observe-Explain (POE)

Somsak Bunprom [1*], Jiraporn Tupsai [1], and Chokchai Yuenyong[2]
[1] Khon Kaen University (KKU) Demonstration School (Suksasart), Khon Kaen, Thailand,
[2] Science Education Program, Faculty of Education, Khon Kaen University,Thailand,

*Corresponding author's e-mail address: somsbu@ kku.ac.th,

Abstract. The purpose of this study was to learning activities for promoted the concept of engineering design Grade 10 students' ideas of engineering design process on the problem based about force and motion. The force and motion unit provided the problem based activities through predict-observe-explain (POE). Participants included 60 Grade 10 students who were studying in Khon Kaen University Demonstration Secondary School (suksasart). Methodology regarded interpretive paradigm. The Hynes et.al. (2011) framework of engineering design process (EPD) was applied as reference of students' existing ideas of engineering design process. The Hynes et.al. (2011) EPD consists of 9 steps including (1) identify need or problem, 2) research need or problem, 3) develop possible solutions, 4) select the best possible solution, 5) construct a prototype, 6) test and evaluate solution, 7) communicate the solution, 8) redesign, and 9) completion decision. The force and motion POE problem based was provided in order to allow students to design the possible solutions of inclined plane problem based. The 13 students' tasks of solutions were interpreted to clarify students' existing ideas of EPD. The findings revealed that all group could be interpreted that they held 9 steps of Hynes et.al. (2011) EPD to design their solutions. The paper will discuss the details of students' existing ideas of engineering design process. The paper may have implication for STEM education teaching and learning.

1. Introduction

Education in the 21st century society endures the responsibility for preparing young people to realize their potential amidst rapid social and economic change. Schools are tasked with having to prepare students for jobs that over expectations, technologies that have not yet been invented and problems that we don't yet know will arise [13]. While this is a noble sentiment there are a number of obvious challenges when trying to meet objectives that are shrouded in uncertainty. Globally, the focus of meeting these objectives has led to the promotion of STEM education, where the STEM acronym stands for Science, Technology, Engineering, and Mathematics. The evolution of the STEM acronym can be traced back to the 1990s when the National Science Foundation in the United States invested heavily in science, mathematics, engineering and technology (SMET), stating that the "quality of our future technological achievements depends heavily on the education in science, mathematics, engineering and technology (SMET)" (National Science Foundation, 1998). 'SMET', as an acronym, failed to catch on and was replaced with STEM in year 2001 [11].. STEM became a more widely known acronym because the popularity of books like "The World is Flat" [6] which highlighted how traditional emerging economies like India and China were catching up on the US economy, and could potentially surpass it, due to their investment in STEM education. As a result, the US government invested heavily and the STEM acronym quickly became ubiquitous with western education [12].

The availability of funding for STEM initiatives in the European Union also saw the acronym proliferate among European countries. Horizon 2020 is the research and innovation funding programme of the European Commission and has posed a series of 'Societal Challenges' that will rely on STEM research over the next seven years to help tackle the economic and social challenges facing European society [4]. These challenges range from topics such as health, food, energy, transport, climate, equality and freedom and will receive almost €30 billion in funding — more than 35% of the entire budget for Horizon 2020. This funding is expected to be utilised by researchers in STEM education, particularly in higher education, as noted by Cortese [2] : "If higher education does not lead the sustainability effort in society, who will?". Despite the availability of funding, implementing a coherent approach to supporting STEM education has proven difficult. Attempts to incorporate each of the core subjects into a consistent curriculum have been met with constraints and challenges which has seen the STEM acronym struggle to cement itself in the educational environment. Failures to consider STEM as a true curriculum concept and an overemphasis on science and mathematics to the detriment of technology and engineering are among the chief concerns [7].

What the conceptions of STEM education should be provided for teachers? It found that descriptions of what comprises STEM content and practices and what STEM conceptions look like range in the literature, especially conceptions of STEM education [1]. However, there are three consensus issues of discussion about STEM education including (a) instructionally (b) as a set of integrated or interconnected disciplines, or (c) as more dependent on the stakeholders or context in which it is viewed or conceptualized. For this study, we will hold Moore et.al. [5] definition of STEM education for enhancing teachers. They define it as ''the teaching and learning of the content and practices of disciplinary knowledge which include science and/or mathematics through the integration of the practices of engineering and engineering design of relevant technologies.'' To them, five characteristics distinguish integrated STEM instruction from other teacher pedagogy: (a) the content and practices of one or more anchor science and mathematics disciplines define some of the primary learning goals; (b) the integrator is the engineering practices and engineering design of technologies as the context; (c) the engineering design or engineering practices related to relevant technologies requires the use of scientific and mathematical concepts through design justification; (d) the development of 21st century skills is emphasized; and (e) the context of instruction requires solving a real-world problem or task through teamwork. This conceptualization of STEM is grounded in learning research.

STEM education in Thailand has been strongly driven. The government organizations have pushed the projects to prepare STEM teachers across the nation. The STEM professional development projects have been provided for in-service teachers by many organizations such as the institute for the promotion of teaching science and technology (IPST), the office of basic education commission (OBEC), the office of higher education commission (OHEC), the ministry of energy, the Electricity Generating Authority of Thailand (EGAT), the chevron enjoy science project, and the schools and universities STEM education projects. These projects could provide approximately 5,000 teachers (in the year of 2017) who could generate the learning activities through integrating knowledge, concepts and skills systematically as STEM education.

Thailand Office of Basic Education Commission (OBEC) has pushed in-service teachers to develop his or her competence of organizing STEM education instruction. The OBEC STEM in-service teacher professional development project called STEM smart trainer team organized the workshop across the nation in order to enhance teachers to understand the concept of STEM education and fundamental ideas of utilizing STEM instructional approaches in their classroom. The STEM smart trainer team project enhanced teachers to provide STEM learning activities through engineering design process (EDP).

3.2. Predict–Observe–Explain (POE)

White and Gunstone (1992) have promoted the predict-observe-explain (POE) procedure as an efficient strategy for eliciting and promoting discussion of students' science conceptions. This strategy involves students predicting the outcome of a demonstration, committing themselves to a possible reason for their prediction, observing the demonstration, and finally explaining any discrepancies between their prediction and observation. Whether used individually or in collaboration with other students, POE tasks can help students explore and justify their own individual ideas, especially in the prediction and reasoning stage. If the observation phase of the POE task provides some conflict with the students' earlier prediction, reconstructions and revision of initial ideas are possible (Searle & Gunstone, 1990; Tao & Gunstone, 1997).

3.3. Data Collection and Analysis

3.3.1. Observation of students' behavior in each activity. Observation with the following.

- Predict: What students have done and are consistent with the EDP.
- Observe: What did the students do and was consistent with the EDP.
- Explain: What students do and how they are aligned with the engineering process.

3.3.2. Interviewing students' ideas during activities.

3.3.3. Interpretation of the protocol from the activity sheet.

4. Results

To start, the teacher opens a video about the bowling competition for students to watch. And ask students: Do students know about bowling? How did you play? Then define the problem and conditions for the students.

Teacher had question before activity, How student designed a plate to hit bottles fall as much as they can? By the way we prepared materials for them. The second, They predicted and explained their concepts about solutions and problems when they let bottle fell form inclined floor.

Figure 2. Bowling competition.

Figure 3. materials

The result of learning activities to promote the concept of engineering design process for students' Ideas about force and motion through predict-observe-explain (POE) are as follows.

All student could identify solutions and situations and create a part of the material. So relationship with step 1 The Identify problem of engineer design process (EDP) following figure 3

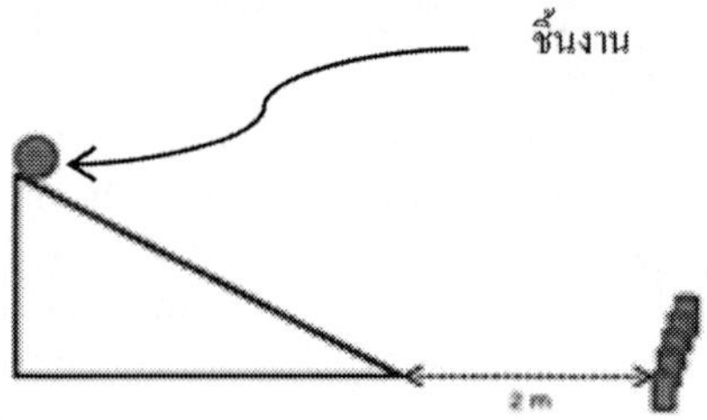

Figure 4. problem and conditions

4.1. Predict

It is a step for students to predict phenomena and the expected results. Students must give reasons for their predictions.

• Students groups have searched scientific concept to design work piece be consistent with step 2 Research of engineer design process (EDP)

• Students have brainstorming and discuss each other in groups to find the best possible materials and have teacher supervision be consistent with step 3 possible solutions of engineer design process (EDP) Example: student interviews

Group 1: They think, should be designed with circular wheels for work to roll away.

Group 2: They think, should be designed to have a cylindrical shape to be able to roll far.

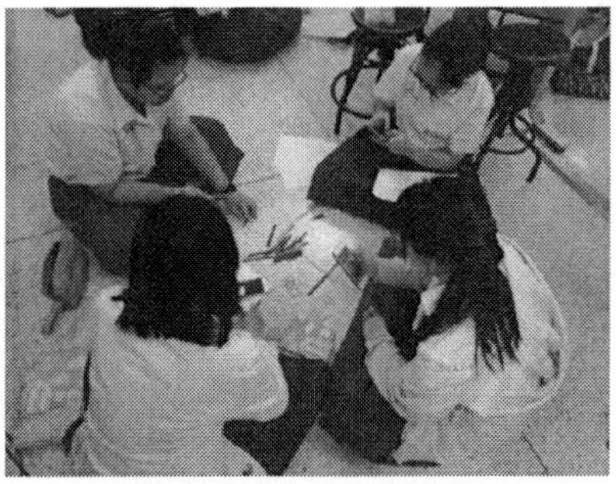

Figure 5. search

Figure 6. discuss to possible solutions

• Students applied scientific principles and analyze to guide the creation of work pieces be consistent with step 4 select the best solution

Example: Interpretation of the protocol

Group 1: They think, should reduce the contact between the material and the floor to reduce friction, causing the work piece to roll further. And they were make roll in circle to make rolling easier and reduce friction.

Group 2: They uses the principle of moment of inertia, which try to make the work piece a cylindrical shape with less radius. For moving parts faster.

4.2. Observing

It is a step that students must take to experiment with what is predicted.

• Each group as created a piece of work From the model designed. Be consistent with step 5 construct a prototype of engineer design process (EDP).

All students will be able to create three main types of work:

Model 1: make the plate a cylindrical shape and use sticks piece to roll.

Model 2: make the work piece by sticks piece a cylindrical shape and use plate to roll.

Model 3: make the plate a cylindrical shape is embedded sticks piece in the core

Example: design and prototype

 group 1	กลุ่มนี้นักเรียนใช้แผ่นฟิวเจอร์บอร์ดม้วนเป็นแกนทรงกระบอก และใช้แท่งไม้ไอติมทำเป็นสามเหลี่ยมเพื่อทำเป็นล้อเพื่อให้กลิ้งไปได้
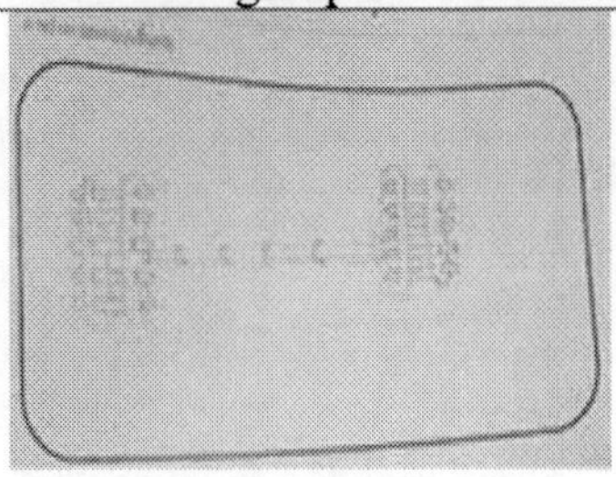 group 2	กลุ่มนี้นักเรียนใช้แผ่นฟิวเจอร์บอร์ดตัดเป็นแผ่นกลม 4 ชิ้น จากนั้นใช้แท่งไม้ไอติมเชื่อมติดระหว่างแผ่นฟิวเจอร์บอร์ดให้มีลักษณะเป็นล้อกลม จากนั้นใช้แท่งไม้ไอติมที่เหลือทำเป็นแกนเพื่อเชื่อมทั้งสองล้อเพื่อให้กลิ้งไปได้
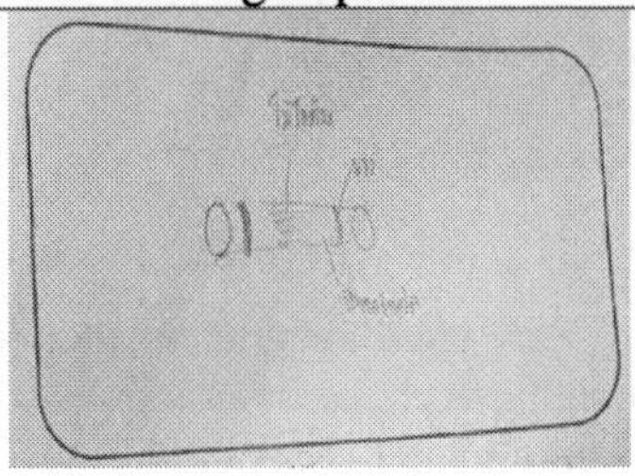 group 3	กลุ่มนี้นักเรียนใช้แท่งไม้ไอติมมัดรวมกันเป็นแกนกลางและใช้แผ่นฟิวเจอร์บอร์ดม้วนให้เป็นเหมือนทรงกระบอกตันเพื่อให้กลิ้งไปได้

Figure 7. design

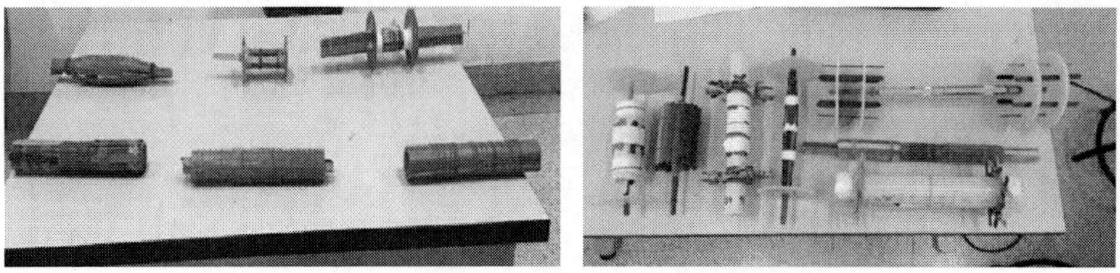

Figure 8. prototype of each group

• Each group created a work piece to experiment on the sloping ground to roll against a bottle of water with step 6 test and evaluate solution of engineer design process (EDP).

Figure 9. test and evaluate solution

4.3. Explain

Students have committed their explanations to paper, discuss their ideas together.

• Students present ideas. And the process of creating a piece of work rational or scientific principles that bring solutions and the integration of STEM, which corresponds to step 7 communicate the solution.

Example: Students present ideas scientific principles and analyze to guide the create of work, have three main types. Following:

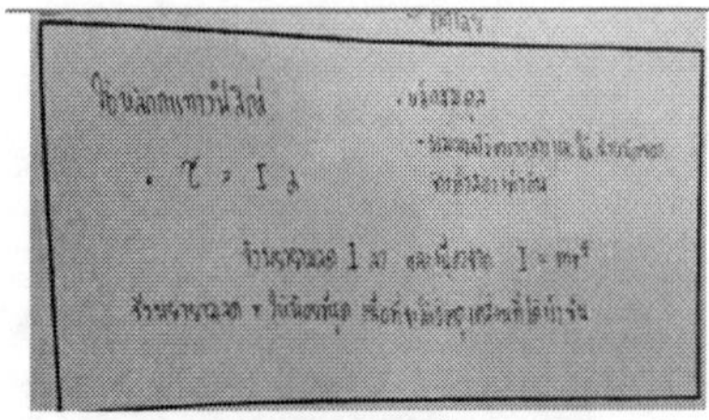 group 1	Main concept of group 1: they were using the principle of moment of inertia to make the workpiece has a cylindrical shape with less radius for moving parts faster.
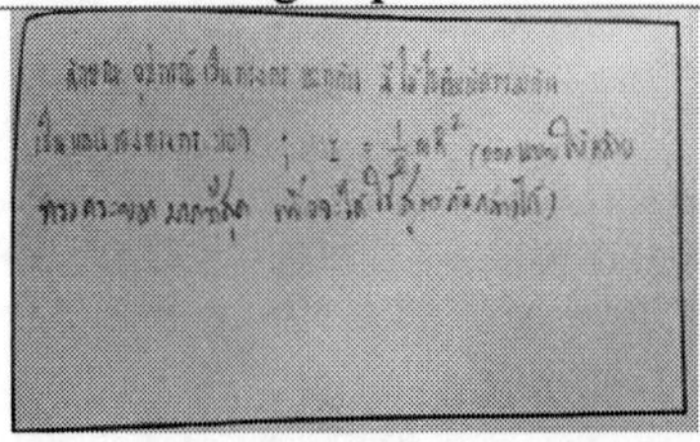 group 2	Main concept of group 2: make the work piece a cylindrical shape.
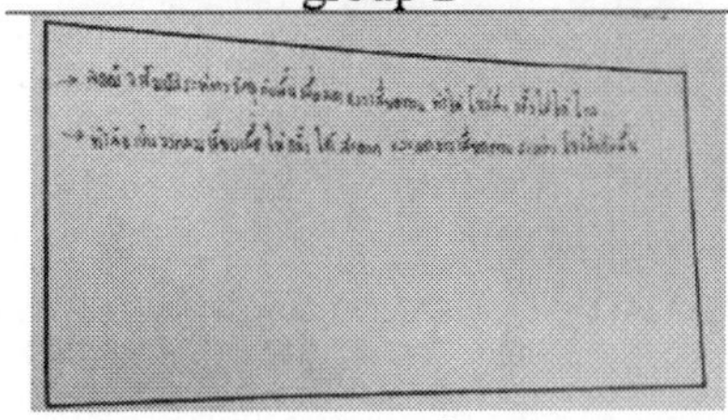 group 3	Main concept of group 3: should reduce the contact between the material and the floor to reduce friction, causing the work piece to roll further. And they were make roll in circle to make rolling easier and reduce friction.

Figure 10. scientific principles of students

• Students found a problem. Then modify the workpiece and then try again, which corresponds to step 8 redesign

Example: how to improve student work are optimize into 4 main ideas:
Method 1: adjusting the wheel size
Method 2: adjusting the shape of the workpiece to the most appropriate target.
Method 3: improve the workpiece to be more balanced.
Method 4: improves wheel alignmen.

• Each group decided and chose the workpiece, which corresponds to step 9 completion decision
It is seen that students have nine engineering processes that appear in this activity.

5. Conclusion

The results show Engineering Design Process Skills that Hynes (2011)'s nine engineering design processes were relation the planning of learning activities, such as predictive observation (POE). It also encourages students to integrate science, technology, engineering and mathematics (STEM) in solving problems or situations under specified conditions.

References

[1] Ananiadou K and Claro M 2009 *21st century skills and competences for new millennium learners in OECD countries*
[2] Cortese A D 2003 *The critical role of higher education in creating a sustainable future* (Planning for higher education vol 31) chapter 3 pp 15- 22
[3] Dede C 2010 *Comparing frameworks for 21st century skills.* 21st century skills (Rethinking how students learn vol 20) pp 51-76
[4] European Commission 2013 *"Horizon 2020 - EU Research and Innovation: Tackling Societal Challenges"* (from http://ec.europa.eu/programmes/horizon2020)
[5] Fairweather J 2008 *Linking evidence and promising practices in science, technology, engineering, and mathematics (STEM) undergraduate education* (Board of Science Education, National Research Council, The National Academies, Washington, DC)
[6] Freidman T 2005 *The world is flat. New York* (Farrar Straus and Giroux)
[7] Herschbach D R 2011 *The STEM Initiative: Constraints and Challenges* (Journal of STEM Teacher Education vol 48) chapter 1 pp 96- 122
[8] Hynes M Portsmore M Dare E Milto E Rogers C Hammer D and Carberry A 2011 *Infusing Engineering Design into High School STEM Courses* (National Center for Engineering and Technology Education vol 165) pp 8-1
[9] Ministry of Education. Basic Education Curriculum 2014 (Bangkok: Kurusapa Ladprao Publishing)
[10] Reeve E M 2013 *What is STEM Education?* [Online] (Bangkok The Institute for the Promotion of Teaching Science and Technology (IPST): Special Initiative Project Division)
[11] Rimes J & de la Barra B A L 2014 *Enhancing science technology engineering and mathematics (STEM) education for girls through a university-school partnership* (Knowledge in Action: University-Community Engagement in Australia) pp 18-34
[12] Sanders M 2008 *STEM, STEM Education, STEM mania* (The Technology Teacher vol 68) chapter 4 pp 20- 27
[13] Schleicher A 2011 *The case for 21st century learning* (Organisation for Economic Cooperation and Development The OECD Observer vol 282) pp 42-43

IOP Conf. Series: Journal of Physics: Conf. Series **1340** (2019) 012081 doi:10.1088/1742-6596/1340/1/012081

[14] Seattha P Tupsai J Sranamkham T and Yuenyong C 2016 *Students' thinking on STEM in learning about Electrical Current through STS approach* In N Rusli W M K A W Zaimi K A M Khazali M J Masnan W S W Daud N Abdullah & Z Yahya (Eds) (AIP Conference Proceedings vol 1775 AIP Publishing) chapter 1 pp 030063

[15] Yuenyong C 2006 *Teaching and learning about energy: Using STS approach* (Unpublished PhD thesis Kasetsart University)

[16] Yuenyong C 2019 *Lesson Learned of Building up Community of Practice for STEM Education in Thailand* (AIP Proceeding)

IOP Conf. Series: Journal of Physics: Conf. Series **1340** (2019) 012082 doi:10.1088/1742-6596/1340/1/012082

Meta-analytic review of studies on the effectiveness of chemistry teaching methods on student transformation

Shalom Grace C. Sugano[1], Edna B. Nabua[2], Manuel B. Barquilla[2], Amelia T. Buan[2], Ellen N. Inutan[2]

[1]Visayas State University, Visca, Baybay City, Leyte, Philippines
[2]MSU-Iligan Institute of Technology Tibanga, Iligan City, Lanao del Norte, Philippine

Abstract. In the hope to provide life-long and quality learning in chemistry education in the Philippines, quality of instruction must be guaranteed first. This calls for exploration and utilization of teaching methods that effectively promote student growth and learning success. In an effort to this, this study utilized the quantitative meta-analysis technique to integrate chemistry research findings that were conducted only in the Philippines covering the period 2005-2016. It aimed to investigate the effectiveness of teaching methodologies in secondary chemistry and their effects on student transformation consisting academic performance and student characteristics. A search procedure and broad literature review were executed in order to identify the potential and qualified chemistry studies from published and unpublished graduate theses, dissertations and journal articles. Meta-analysis was effectively and systematically carried out through study acquisition, study coding, determination of interrater reliability, establishing inclusion criteria, computation of effect size including mean and interpretation of analyses. The meta-analytic review revealed that during the last decade (2005-2016), nine different teaching methods have been implemented and evaluated in chemistry classrooms. It has been determined that incorporation of innovative methods significantly promote positive transformation of students as a whole in secondary chemistry. Moreover, findings provide evidence supporting that academic performance, attitude and motivation can be greatly enhanced if students are placed in an environment in which they can actively connect chemistry instruction to previously learned scientific and chemical concepts and materials through constructivist (inquiry-based and problem-based) and collaborative learning method. The main implication of this research is that it generated empirical evidence supporting the effectiveness of innovative instruction in chemistry.

1. Introduction

Most reforms in chemistry education call for an emphasis on improving teaching quality and effectiveness. Research shows that the role of teachers have a profound effect on student growth and learning success. A chemistry teacher must be a catalyst of change. This change includes one's ability to exceed the typical way of transforming students in the learning environment and to translate new understanding about inquiry and meaningful learning into actual habits of practice [4]. The expected role of a teacher in the traditional manner is a content expert who can impart his or her knowledge to students and shifts to a newer orientation that teachers should act as a facilitator of the learning process. Chemistry teaching is supposed to be result-oriented and student-centered, and this can only be achieved when students are willing and the teachers are favorably prepared to use the appropriate methods and resources in teaching the students [5]. Various teaching methods/strategies have been introduced and adopted by chemistry educators and researchers as considering ways and means of ensuring effective teaching and learning in school. However, questions about the effectiveness of teaching methods on student learning have consistently raised considerable interest in the thematic field of educational research [6]. Chemistry educators and researchers continually made efforts in conducting relevant studies that dealt with teaching-learning strategies, assessment on student outcomes, intellectual and nonintellectual learning skills and others. The accessible large body of literature and periodic outputs of graduate students in the higher educational institution are manifestations of these endeavors. However, observation shows that findings are rarely applied in the actual teaching. These educational researches provide reliable findings that pose an aim to give knowledge on the present status and to improve the chemistry teaching-learning process in school. Therefore, there is a need to collect, integrate, synthesize and repeatedly metaanalyze existing chemistry research findings geared to theory development to better inform teachers and educators. A research that deals with meta-analysis of chemistry education research findings geared to theory development is a pioneer across the available secondary chemistry educational researches in the country. Most of the educational researches conducted do not involve meta- analysis of findings rather empirical studies are abundant. Optimistically, the researcher sees the need to undertake this study and look into the significant contribution to the improvement in the field of educational research and chemistry education. In the hope to provide current status of chemistry education in the Philippines, it is really important to examine the type of methods/strategies used and how it effectively fits into the students' needs and characteristics.

In light of the lack of previous meta-analyses on the issue, this study aims to integrate and meta-analyze chemistry education research findings covering the period 2005- 2016. It also aims to describe and determine whether the use of teaching methodologies contribute significantly to studenttransformations (academic performance and characteristics) in secondary chemistry teaching-learning process based on the chemistry education research findings. This study addressed the following specific objectives:

1. To determine the over-all effect of the chemistry education studies qualified in the meta-analysis with regards to the use of teaching methods on student transformation.

2. To identify the teaching method categories that appear to have the largest effect size in relation to:

- Academic Performance
- Attitude
- Learning Styles, Motivation, and Self-efficacy
- Multiculturality

2. Methods

Research Design Meta-analysis offered a way to broaden the search, examining all available studies (published and unpublished) at once, so that an overall effect size would be calculated and potential moderators would be established. Moreover, this study employed a combination of qualitative and quantitative approaches to examine the relationship between teaching methodologies and the student transformations in chemistry education. Mixed methods research allows for the "opportunity to compensate for inherent method weaknesses, capitalize on inherent method strengths, and offset inevitable method biases" [7].

The qualitative aspect would involve the coding of chemistry education research findings or hyper-analysis mapping of meta-analysis results. On the other hand, studies of impact would be quantitatively synthesized using meta-analysis technique in order to produce a big picture of influence among variables. Research Locale Collection of chemistry education research findings of both published and unpublished studies were done nationwide in the Philippines to avoid publication bias. The search were systematically done in both private and state universities having graduate chemistry education and teaching program. The researcher personally conducted a broad search of possible pool of studies related to chemistry instruction throughout the Philippines and through or by online browsing.

2.1. Research Instruments

2.1.1. Study Qualifying Sheets. These sheets were developed for each cases of chemistry education that served as basis for qualifying the study during the pre-screening process. It contain the set of criteria specified in selecting the appropriate study.

2.1.2. Coding Sheets. The sheets were adapted [2] for each cases of chemistry education research findings. These were modified based on the inclusion criteria being set as well as the classes of variables (i.e. publication year, locale of the study, intervention, etc.). The information taken from these sheets were tabulated and used in the meta-analysis as well as in plotting the mapping matrices in the different categorization.

2.1.3. Data lay-out Matrix. This matrix was used to summarize the information taken from the coding sheets and to facilitate the systematic analysis of science education research findings.

2.1.4. Mapping Matrices. These matrices were developed for independent or dependent variable based on the different sub-categories of such variables. It was simply utilized to plot the different variables based on their sub-categories. The researcher was responsible in facilitating the groupings of variables, in comparing the dataset of each outcome variable and in ranking of the effect size as to determine the categories of teaching methods (independent variables) that demonstrated as the most influencing factor to student transformations (i.e. student motivation, academic performance).

2.2. Data Gathering Procedures

2.2.1. Identification and Collection of Potential Studies. Empirical studies involving student transformation as affected or influenced by different teaching methodologies were assembled. This pool of studiesincludes a particular case of research findings that deals only on the effects of teaching methodologies on student transformations including self-efficacy, learning styles, attitude, motivation, multiculturality and academic performance. A search procedure and broad literature review were executed by the researcher in order to identify the potential and qualified chemistry education studies from the published and unpublished graduate theses and dissertations. Moreover, a computerized search was also conducted from the following; journals of educational organizations for both foreign and local, electronic search through ERIC (Education Resource Information Center) database and search terms using standard search engines Google and Google Scholar.

2.2.2. Pre-Screening. Surface level screening was performed in each of the gathered chemistry education research in order to efficiently exclude a substantial part of studies that did not qualify on the inclusion criteria. In this manner, this was executed on the basis of reading title and abstracts. In the case where the essential study characteristics were not evident from the title and abstract, the full text would be consulted. Several rigorous criteria were employed for the inclusion of studies in this synthesis. The following criteria includes: 1) Studies should focused either on student self-efficacy, learning style, attitude, motivation, multiculturality and academic performance as resulting from the use of chemistry education teaching methods. 2) The study should be completed within 2005-2016. 3) The study should be conducted or carried out in the Philippines. 4) The study should be involved only in secondary chemistry academic domain. 5) The study should utilized the experimental or quasi-experimental research design. 6) The study report effect size (ES) or contain statistical details necessary to calculate effect size (ES) such as mean, standard deviations, sample size, t, f and p values, ANOVA tables, degree of freedom (df), etc. The meta-analysis studied the relationships between the dependent variables and the potentially contributing independent variables [2]. In order to explore some of the heterogeneity of effect sizes among the research findings, identified teaching methods were cast into treatment categories that act as the independent variable of this study. On the other hand, the dependent variable of this meta-analysis is student transformation, presented in six (6) different outcome variable namely; academic performance, self-efficacy, learning style, attitude, motivation and multicultural. Each of these were affected by the teaching method categories.

2.2.3. Coding Process. All studies gathered were coded by the researcher and a PhD level professional in science education. The coders performed the coding process separately using a coding scheme with the study features. The coding sheets were developed to delineate each study features based from the data extracted. Hence, the researcher gave a careful attention to the qualified studies as these served as the framework for analysis. Coding was done for each cases of chemistry education research findings. Each coding sheet was constantly refine as the synthesis proceeded. Both qualitative and quantitative information were coded using the sample coding sheet to permit comparison of findings across studies. The table for quantitative description of the different characteristics included in the case of the study were provided to the coders. Data extracted from each study are coded according to the features of the study.

2.2.4. Data Analysis. Study coding and data were compiled using a Microsoft Excel Spreadsheet 2013 in order to collect and organize the data. An excel form was be created to prompt for all needed features for each study and automatically save the information in a database from which analyses would be conducted. Alternatively, data were analyzed using the Comprehensive Metaanalysis Program/Software (Trial Version License). The following statistical analysis were conducted in this study. Meta-analysis – it provides a systematic method for accumulating and synthesizing statistical evidence from a variety of studies that address a common research question.
The following statistical computation were executed in this meta-analysis:

- Compute Effect Sizes for Study
- Differentiate and determine weighted effect size for each study
- Test of Homogeneity
- Mean Effect Sizes
- Ranking of Mean
- Effect Size and Regression Analysis

3. Results and Discussion

3.1. Meta-analysis for All Studies

One of the goals of the meta-analysis is obtaining a mean effect size of the over-all studies [3]. Using the random-effects model, the computed mean effect size for the 67 studies is 0.982, considered as a large positive result as presented in Table 1. Several factors like sample size and variance can influence the magnitude and direction of the effect size, the estimates of the effect sizes will vary among studies. In such case the effect size of the individual studies may somewhat imprecise and therefore can lead to an unstable finding when multiple small studies are utilized. Thus, weighing of the standard error based on the sample size gives the best precision of the effect size estimates [8]. The standard error of the weighted mean effects is 0.089. Hence, this over-all result can be reflected as a strong high level. At a confidence level of 0.05, the 95% confidence limit was determined. In this manner, the confidence limits were 0.806 and 1.157. Confidence intervals of continuous measure that include zero represent non-significant results [8]. Since the interval never reaches at zero, the null hypothesis that the population effect size is zero can be rejected. The use of chemistry teaching methods is considered effective than the traditional or conventional method.

Table 1. Metaanalysis for all studies.

Over-all Effects	Mean ES	Mean SE	Lower Limit	Upper Limit	Z Value	Over-all P Value
67 Studies	0.982	0.089	0.806	1.157	10.970	0.000

The findings presented agree with [9] who reported that their metaanalysis of the effect of instructional intervention on students' mathematics achievement indicated a statistically significant positive high effect (Cohen's d = 0.58). In the study of [12], it also concluded a significant effect of teaching strategies (d= 0.67) on student achievement in science.

3.2. Meta-analysis for Studies Classified by Treatment Categories

The 67 chemistry education research studies describe the effects of the nine identified teaching method (treatment) categories on student transformation. Student transformation refers to the student characteristics like self-efficacy, attitude, learning style, multiculturality and academic performance in chemistry teaching-learning process. It is a common process of assimilative learning, the type of learning that takes place when students simply acquire new information/knowledge that can easily fit into their pre-existing knowledge structures [11]. With the same meta-analysis procedure, each study was assigned by an effect size, and a mean effect size was determined for each treatment categories. This also permits the ranking of teaching method categories according to their effects on each dependent variable.

3.3. Effects of Treatment Categories on Academic Performance

Table 2 shows the metaanalysis results for the studies investigating the effects of each treatment category tailored to students' academic performance. As seen in the summary table, inquiry-based learning (n = 11), manipulative, models and multiple representations occupy the top teaching methods being studied on academic performance in chemistry. It was followed by cooperative learning and technology-aided instruction with nine (n = 9) and six (n = 6) qualified studies, respectively. There were only five (5) primary studies qualified that fall under individualized instruction and four (n= 4) studies on problem-based learning. The least studied type of teaching methods include project-based learning (n = 3), combination learning (n = 2) and multicultural education (n = 1).

Table 2. Frequency distribution of students on academic performance by treatment categories.

Treatment Categories	N (No. of Studies)	Percentage
Combination of Learning	2	3.92
Cooperative Learning	9	17.65
Individualized Instruction	15	9.80
Inquiry-based Learning	11	21.57
Manipulative, Model and Multiple Representations	10	19.61
Multicultural Education	1	1.96
Problem-Based Learning	4	7.84
Project-Based learning	3	5.88
Technology-Aided Instruction	6	11.7
TOTAL	**51**	**100.00**

The presented results implies that in the field of chemistry, inquiry learning considered a constructivist approach is the commonly studied teaching intervention with regards to academic performance. It can also be accounted that graduate chemistry researchers engaged in the following methods: the use of models, symbols (abstract) and various representations, in small teams (groupings) and integrating technology, as to explore their effectiveness on secondary students. In addition, the very few quantity of multicultural education and combination learning presented in this study would be an enlightening for practitioners and researchers to pursue more of such teaching methods in promoting high quality learning in chemistry. In such case, multicultural education has only one study involved which denotes as weak and unreliable result, which in turn, not included in accounting for the meta-analysis as presented in succeeding part.

Table 3. Meta-analysis of qualified studies on academic performance by treatment categories.

Treatment Categories	**Mean Effect Size**	**Lower Limit**	**Upper Limit**	**Z Value**	**P Value**
Combination of Learning	0.687	0.332	1.041	3.797	0.000
Cooperative Learning	1.268	0.612	1.924	3.79	0.000
Individualized Instruction	0.933	0.118	1.748	2.245	0.025
Inquiry-based Learning	1.783	1.118	2.449	5.253	0.000
Manipulative, Model and Multiple Representations	1.072	0.587	1.557	4.333	0.000
Problem-Based Learning	1.276	0.186	2.365	2.295	0.022
Project-Based learning	1.035	0.377	1.694	3.083	0.002
Technology-aided Instruction	1.017	0.44	1.595	3.451	0.001

In the meta-analysis, all innovative teaching methods presented in Table 3 exhibited a significant positive influence ($p < 0.05$) in terms of their mean effect size on student performance. The weighted mean effect size of the treatment categories ranges from 0.50 to 1.80. Using the random-effect model, the inquiry-based learning ($d = 1.783$) has the largest mean effect size, followed by problem-based learning ($d = 1.276$) and cooperative learning ($d = 1.268$) which are close in effect values. Nearly of the same effect values shared by MMMR (manipulative, model and multiple representation) ($d = 1.072$), project-based learning ($d = 1.035$) and technology-aided instruction (1.017), then individualized instruction that still have positive effect level ($d = 0.933$). Combination learning showed to have moderate positive effect on the performance of students. The 95% confidence intervals for the treatment categories are presented in Table 5. It can be clearly noticed that all the confidence intervals are judged to be significantly different from zero. Confidence interval lend insight into the precision of the treatment estimates included in the studies. Conversely, a wider confidence interval indicates less precise estimates and coupled with a small sample size, can lead to imprecision in measurement [8]. Within the collection of teaching methodology, the inquiry-based learning method such that focus on inquiry questioning, facilitated inquiry activities, guided discoveries and inductive laboratory exercises seem to have the greatest impact on students' academic performance in chemistry. As emphasized by Maxwell, Lambeth and Cox (2015), activities in science classrooms could involve observations, questioning, investigating, predicting, explaining and communicating results. Memorizing facts would not increase skills in students, but on having the freedom to explore and investigate through inquiry learning. Conventional teaching promotes rote memorization that does not develop or supports the development of students' problem solving and critical thinking skills.

3.4. Effects of Treatment Categories on Attitude

A total of twelve (12) studies were reviewed and made it into the meta-analysis in relation with the use of different teaching methodologies in secondary chemistry and as it influenced students' attitude. Table 4 presents the meta-analysis outcome of this study. It can be observed that cooperative learning ($n = 3$) is the top most method being studied on attitude, followed by project-based learning ($n = 2$) and individualized instruction ($n = 2$). The rest of the identified methods except combination learning, only a single study was qualified in the analysis based on the given inclusion criteria. However, no studies were reviewed on combination learning. The findings show that there are still very few available empirical studies conducted in the Philippines that centers the effects on student attitude in the chemistry teaching-learning process.

Table 4. Frequency distribution of studies on attitude by treatment categories.

Treatment Categories	N (No. of Studies)	Percentage
Combination of Learning	0	0
Cooperative Learning	0	27.27
Individualized Instruction	2	18.18
Inquiry-based Learning	1	9.09
Manipulative, Model and Multiple Representations	1	9.09
Multicultural Education	1	9.09
Problem-Based Learning	1	9.09
Project-Based learning	2	18.18
Technology-Aided Instruction	1	9.09
Total Studies	12	100

The main purpose of meta-analysis is to increase precision of a conclusion which the single studies were not able to cater due to imprecision. However, as presented in Table 5, it can be observed that treatment categories in the following; inquiry-based learning, manipulative, models and multiple representations, multicultural education, problem-based learning and technology-aided instruction, possessed of only a single study with no multiple results. Thus, these categories were considered to be imprecise and not interpretable. This implies that more studies are still needed in these teaching method categories in order to derive to a definite conclusion. The findings disclosed that there are still very few available empirical studies conducted in the Philippines that explores the effectiveness of different teaching methods/strategies on student attitude in the chemistry teaching-learning process. Examining the meta-analysis results as shown in Table 4, significant mean effect sizes were observed in the following teaching methods based upon the use of random-effects model; cooperative learning ($d = 0.500$), project – based learning ($d = 0.452$), and individualized instruction ($d = 0.388$). Each of the obtained mean effect sizes corresponds to be small which denotes that these treatment categories to have significant effects toward the given outcome variable. In the same way, it can be deduced that they are effective yet in a minimal manner as compared to traditional teaching.

Table 5. Meta-analysis for qualified studies on attitude by treatment categories.

Treatment Categories	Mean Effect Size	Lower Limit	Upper Limit	Z Value	P Value
Cooperative Learning	0.500	0.311	0.690	5.18	0.000
Individualized Instruction	0.388	0.171	0.606	3.503	0.000
Project-Based Learning	0.452	0.166	0.739	3.091	0.002

Cooperative learning revealed to have largest significant effect size on student attitude followed by project-based learning and individualized instruction. Cooperative learning can be a useful tool to help develop a positive attitude toward learning [10]. Several studies affirm that cooperative method was better than traditional method in terms of stimulating positive attitude [1]. However, the very small effect size observed in project based learning and individualized instruction denotes small effect based on the Cohen's interpretation, presenting a minimal impact on students' attitude. The lower effects of constructivist method such as inquiry-based, problem and project- based learning on attitude outcome is most likely due to the lack of requisite materials needed for carrying out investigations in a chemistry classroom. Since this method mainly focused on questioning, investigation and problem-solving that involved hands-on manipulation and required critical and creative thinking skills, this probably give students the negative impact and perceiving it to be difficult, less enjoyable and inconvenient on their part.

3.5. Effects of Treatment Categories on Learning Styles, Motivation, Self-efficacy

In the efficacy of teaching methods on other dependent variables, learning styles, motivation and self-efficacy, four (4) studies were qualified in the meta-analysis. According to random- effects model as presented in Table 6, only cooperative learning displayed a significant large effect (d = 0.860) on students' motivation. However, inquiry-based learning did not exhibit any significant effect on motivation (d = 0.080). Moreover, no significant effect were identified on learning styles and self-efficacy with respect to the use of inquiry-based learning and cooperative learning. In this meta-analysis, a single study was observed however it has multiple findings presented. These findings indicate that the cooperative learning method was found to have a significant impact than the traditional method in terms of an increasingly positive interest/ motivation towards learning chemistry. It can be addressed based on the findings that cooperative learning method was found to have a significant impact than the traditional method in terms of an increasingly positive interest/ motivation towards learning chemistry. This kind of teaching method is relevant in the chemistry instruction because students learn to work as part of a team to with various tasks and solve problems. In today's world, working in teams is encourage to solve complex problems, so it is imperative that students should be taught and be exposed on how to be an effective member of a team (Quinn, 2006).

Table 6. Meta-analysis for qualified studies on learning styles, motivation and self- efficacy by treatment categories.

Treatment Categories	N (No. of Studies)	Mean Effect Size	Mean Standard Error	Lower Limit	Upper Limit	Z Value
Learning Styles						
Inquiry-based Learning	1	0.199	0.176	-0.146	0.544	1.130
Motivation						
Cooperative Learning	1	0.860	0.241	0.388	1.332	3.571
Inquiry-based Learning	1	0.080	0.176	-0.265	0.425	0.454
Self-efficacy						
Cooperative Learning	1	0.041	0.203	-0.356	0.438	0.202

3.6. Effect of Treatment Categories on Multiculturality

No studies were able to qualify for the meta-analysis on student multiculturality. This study can provide a clear information that there is a need to direct the research focus on student multicultural and to further examine how they can be influenced or changed by the use of a suitable instruction.

4. Conclusion and Recommendations

As a result of this study, it has been determined that teaching methodologies included in the meta-analysis exhibited an over-all large effect size (d = 0.982) on student transformation as a whole. Constructivist teaching approach (inquiry-based and problem-based learning) and cooperative learning were among the studied teaching method categories that exhibited a significant strong impact on students' academic performance as compared to the conventional teaching. In terms of student characteristics, cooperative learning displayed significant effect size on both variables; attitude and motivation. For the purpose of this study, this calls for more empirical research studies particularly those that investigates the effectiveness of an instructional intervention towards student characteristics such as attitude, interest/motivation, learning styles, self-efficacy and multiculturality. It is also important to have more metaanalysis on science education, as in the Philippines there are only a few meta-analysis studies in this field.

It is then recommended based from the findings of this study, chemistry teachers and educators should consider employing one of the various forms of pedagogies and methods that have been shown to be effective in improving student performance and characteristics in chemistry. Also, this study can be a benchmark to venture other meta-analysis studies examining effects of various methods applied in teaching other discipline of science; physics, biology and general science.

5. Acknowledgement
The authors would like to express thanks and gratitude to those in one another have help this paper a success. Specifically, to the different colleges and universities who made a great contributions.

References

[1] Capar G and Tarim K 2015 *Efficacy of the Cooperative Learning Method on mathematics Achievement and Attitude: A Meta-Analysis Research* Educational Sciences: Theory & Practice

[2] Cavanaugh C S 1998 *The Effectiveness of Interactive Distance Education Technologies in K12 Learning: A Meta-Analysis* A Doctoral Dissertation. USA, University of South Florida 26 June 2016 https://www.academia.edu/232521/THE_EFFECTIVENESS_OF_INTERACTIVE_DISTANCE_EDUCATION_TECHNOLOGIES_IN_K12_LEARNING_A_METAANALYSIS

[3] Cavanaugh C S 2001 The Effectiveness of Interactive Distance Education Technologies in K12 Learning: A Meta-Analysis *International Journal of Educational Telecommunications* **7** pp 73–88

[4] Craven J and Penick J 2001 Preparing New Teachers to Teach Science: The Role of the Science Teacher Educator Online 13 January 2013

[5] Edomwonyi-out L and Avaa A 2011 The Challenge of Effective Teaching of Chemistry: A Case Study

[6] Ganyaupfu E M 2013 *Teaching Methods and Students' Academic Performance* International Journal of Humanities and Social Science Invention Online 2 February 2017

[7] Harwell M R n.d. Research Design in Qualitative/Quantitative/Mixed Methods Online 1 March 2017

[8] Israel and Ritcher 2011 A Guide to Understanding Meta- Analysis *Journal of Orthopedic & Sports Physical Therapy* Online September 25 2017

[9] Jacobse A E and Harskamp E n.d. A Meta-Analysis of the Effects of Instructional Intervention on Students' Mathematics Achievement Online 10 January 2017

[10] Johnsen S 2009 Improving Achievement and Attitude Through Cooperative Learning in Math Class *Action Research Projects* **64**

[11] Johnson A 2015 *Tree Views of Teaching: Transmission, Transaction ad Transformation* 21 March 2018 Retrieved from https://www.linkedin.com/pulse/three-views-teachingtransmission-transaction-andrew-johnson

[12] Schroeder C M 2007 A Meta-Analysis of National Research: Effects of Teaching Strategies on Student Achievement in Science in the United States *Journal of Research in Science Teaching* pp 1436–1460

IOP Conf. Series: Journal of Physics: Conf. Series **1340** (2019) 012083 doi:10.1088/1742-6596/1340/1/012083

Learner Characteristics as Consequences of Active Learning

Jiratchaya Phala[a] Suthida Chamrat[b]

a Ph.D student, Doctor of Philosophy Program in Education (Science education), Faculty of Education, Chiang Mai university Email: jiratchaya0942@gmail.com
b Department of Curriculum, Instruction and Learning, Faculty of Education, Chiang Mai University Email: suthida.c@cmu.ac.th (Corresponding Author)

Abstract. This paper aims to explore the consequences of active learning implementation toward students 'characteristics .Literature reviewing, and synthesizing intend to delineate the framework of active learning driven classroom where teachers lessen their own profile from lecturers into facilitators .By deploying active learning, students 'multiple skills are activated and practiced .These skills, gradually, become students 'characteristics .Reviewing and synthesizing from the existing literature and research reports indicate students 'characteristics as possible outcomes of active learning in 12 traits :1) students have a great opportunity to first-hand experience; 2) students create learning process and construct knowledge by themselves; 3) students learn and reflect from their experiences; 4) students develop higher order thinking; 5) students develop their learning skills; 6) students play important roles in learning which are also varied and diverse; 7) students highly participate in learning process; 8) students reach academic achievements with meaningful learning; 9) students build rapport with teachers and their peers; 10) students integrate information with their prior knowledge; 11) students develop positive attitudes and self-esteem; 12) students are enthusiastic, active, happy and joyful.

1. Introduction

Educational development in Thailand mainly focuses on knowledge, morality, learning process, and integration suited to the students. According to Section 22 of the National Education Act B.E.2542, education arrangement shall be consistent with the principles that all students are capable of learning and self-development. Accordingly, students are highly prioritized. In other words, teaching and learning arrangement is based upon the student-centered approach in accordance with active learning. It is, in fact, the learning system which enables students to engage themselves in the learning process, develop their learning to suit their potential, interest, and aptitude. They are also encouraged to actively think and do on their own in order to construct the knowledge through the learning exchanges for their meaningful learning.

2. Definition of active learning

"Active learning" is regarded as one type of teaching and learning arrangements which has been studied by several researchers. It is hand-on learning and experiential learning which is interchangeably used [25] with different names in Thai language such as active learning [39] practical learning [50] holistic learning [68] learning by doing [66], active teaching and learning [69] enthusiastic learning [7]and experiential learning. The word "active learning" is used in this research.

3. Meaning

Active learning refers to real practice and active participation of students while listening to the formal presentation in class [11]. The learning arrangement undertaken by students and based on what they have done [6] is regarded as the learning activity under two assumptions: 1) learning as a natural attempt of human, and 2) distinctive ways of learning among individuals [41].Typically, students are conversed from receivers into co-creators [23] .

Active learning is therefore the learning activity allowing students with an opportunity to learn in a meaningful, valuable, and joyful way in response to their interest. In this manner, students are challenged and encouraged to find the answers through various activities e.g. group working, talking, reading, writing, discussing, questioning, reflecting, and finding answers. These enable students to construct knowledge and comprehension along with enthusiasm for learning, problem-solving skills, data analysis and synthesis systematically.

4. Importance of active learning

The Office of Basic Education Commission (OBEC), Ministry of Education, has developed the educational system in line with Thailand 4.0 framework to forward the country with security, wealth, and sustainability as a developed country under the philosophy of sufficiency economy. Indeed, Thai people of Thailand 4.0 must possess the capability of higher knowledge and skills under the 20-year national strategies so that they can apply the knowledge for innovations, potential developments, and environmental friendliness. In addition, the 12th National Economic and Social Development Plan mentions about the promotion and development of human potential with high quality education based on the international standards and constant self-learning approaches in line with the National Education Plan B.E.2560-2579. As planned, all students shall be developed with the characteristics and learning skills of the 21st century in regard to changes of economy, society, culture, and environment along with rapid changes of advanced science and technology as well as international testing e.g. PISA, which reflects that Thailand is encountering with the educational problems and educational inequality. The educational challenge in the 21st century for students to get prepared for living in the 21st century is essential for current social changes which put an effect on the social living. Teachers must therefore be alert on learning arrangements to get students prepared with skills required for such living. That is, learning skill has influenced the learning arrangement to enable students with knowledge, abilities, and essential skills due to reformation of learning arrangements and other preparations [46]. Active learning is a new concept for traditional education reformation with its focus on the direct transfer of knowledge from teachers to students through varied teaching techniques for designing the learning arrangements and activities. Meanwhile, it can promote class participation among students and improve relations between students and students or between students and teachers. It is therefore another kind of learning arrangement that enhances students' characteristics to keep up with the current trends [45].

5. Definition of students' characteristics of active learning

Students' characteristics of active learning refer to the quality of students brought by the arrangement of active learning through 1) students' characteristics of intellectual ability or intelligence as the ability for solving problems in diverse environments or producing varied works depending on the cultural contexts (Gardner) through eight types of intellectual abilities [30] 2) students' characteristics of physical ability[68] ; students' characteristics of attitudes [3] 4) students' characteristics of society or learning reactions between students and students in three traits [26] through an observation, evaluation, and interrogation.

6. Students' characteristics of active learning

Students' characteristics of active learning synthesized from academic journals can be divided into 12 traits below.

6.1. Students learn through first-hand experiences.

The learning process mainly stresses on students [9] whereby students are deemed to play key roles in their learning[55] . In this regard, students are required to control themselves in doing various activities that will allow them to think and make decisions about speaking, listening, reading, writing and reflecting [36]. Besides, students can learn by doing [56,60] various activities [2] or other actions on their own through an adoption of the framework relating to what they have done

[5,60]. It is regarded as higher order thinking, analytical thinking, synthesizing, and evaluation [69]. Students must have more chances to learn by doing other than passive listening to teachers or only reading [18,37, 45, 63] .Students implement learning-by-doing and construct the knowledge from what is undertaken during lectures [5], mainly prioritizing skill and ability developments in response to their existing knowledge) through an availability of various learning media for an exchange of opinions or experiences [2]. These enable students to learn through first-hand experiences from real practice or activities where perception is drawn through five senses [44], leading to the constructive knowledge from activities during lectures through speaking, listening, reading, writing, and reflecting [48]. Learning-by-doing will build understanding and develop several skills [50] e.g. experiment, survey, practice for intellectual development, problem-solving, analysis or decision making [18] such as application, presentation, and implementation [52]. In addition, students will have an opportunity to make an observation, establish some talks relating to the subject matters with their peers, and communicate for self-reflection [22]. Teachers are to facilitate and encourage students through learning-by-doing [56]. Students may apply the skills of speaking, listening, reading, and reflective writing along with the knowledge obtained from doing other learning activities [9] so that they can successfully implement learning-by-doing [32,56] . Students are able to discover the mathematical concepts from learning-by-doing or observations and scientific experiments for the summaries and rules from those experiments [20]. Their learning also corresponds to their own interest, knowledge presentation, and practice [52]. Students are therefore able to think and do in a meaningful way to apply the knowledge for their daily life [13].

6.2. Students create the learning process and construct knowledge by themselves.

Active learning is the situation-based process that enables reading, speaking, listening, and deep thinking among students. By this process, students manage their self-learning [32,34,56] . Indeed, students are able to learn by themselves [17,22,43] to obtain the direct experience in response to their real problem-solving [35] and create the learning process suited to the natural way of brain functionality [13,70]. Students can also construct knowledge [5,32,35,45,52,56,64,72] and manage their self-learning process [32,34,45,56] .It is required that students themselves search for knowledge, attitudes, values, and experiences [5,22,45]. Accordingly, knowledge is obtained from the experience, knowledge construction, and reviews [32,34,56]. Students are then able to find out the knowledge and apply various skills for their comprehension to create meaningful learning [29,32,34,49,62,67,72] . Students learn by doing and construct knowledge from what is practiced during lectures [5] while determining the concept, plan, acceptance, evaluation, and presentation [35]. In addition, students have an opportunity to learn by themselves to obtain the first-hand experiences to address the problems in authentic situations. They also do and think about what they are doing for construction of new knowledge [6] while choosing and generating their knowledge through numerous activities [65] other than passive listening to teachers [54]. Students must search for the contents to generate the knowledge by talking, writing, reading, and asking questions or the learning process that encourages students' movement [13].Students are viewed to possess the thinking concepts and working systems, have self-discipline, and search for new ideas by themselves through knowledge exchanges [55].

In this process, they are able to apply their knowledge [55] whereas teachers are advisors and facilitators for their learning [5, 22, 32, 45, 72]. It is therefore to reduce the process of content transfer from teachers to students solely whereas it is another way to develop higher order thinking which promotes learning-by-doing rather than passive listening to the lectures and provide feedbacks to students [72].

6.3. Students learn from their experience.

In the arrangement of practical learning, students will have first-hand experience by doing, searching for the data, thinking, and summarizing the results using the media that enable the perception of five senses [8]. In fact, this knowledge is based upon the experience and generated by students [13]. In addition, students will learn experience-based sciences and solve the problems arising from skeptical situations [22, 41] Practical learning enhances self-learning among students to learn by doing [66] and promotes the experience through practices and observations along with talks to themselves and others [22]. In fact, they have more enthusiasm for learning and more activities by exchanging experiences through speaking, writing, and discussion [12, 72]. This helps generate experience-based learning [61] , promotes exchanges of learning experiences [52] ,connect their existing experiences to new ones or explain new problems for further knowledge construction or conclusions [52] . Such encouragement for knowledge and experience exchanges enables students to learn by doing, make an observation, and talk to themselves or others through diverse activities e.g. listening, speaking, reading, writing, and reflecting or responding [16]. These can be done through passing on their thinking and obtained knowledge to their group as the data for solving the problems and improving the learning process of the whole group members [54] . Students will build their understanding and seek the meaning of contents by connecting them with their already existing experiences [65], memorizing the contents and principles through their understanding gained from such experiences. Students can eventually search for knowledge, attitudes, values, and experiences on their own [22].

6.4. Students develop their higher order thinking.

Active learning is the development of thinking ability, problem-solving skill, and knowledge application [32] [34]. Students are then required to read, write, discuss or solve the problems, and implement actions, particularly for the work requiring higher order thinking of analysis, synthesis, and evaluation through an evaluation of higher order thinking, analyzing, and critical thinking [5] . It is believed to develop students with higher order thinking, analysis, synthesis, and evaluation [6,13,22,37,45,73] as well as application [5,7,12,22,45,56,63]. It is also to constitute the skills of analytical thinking, problem-solving, and higher- order thinking. Activities focusing on higher order thinking [56] in which students are required to think and do on their own are essentially arranged [69]. These activities have an influence on the development of thinking skills into higher levels [5,32,45,60,72]. Students are also evaluated for higher order thinking, analysis, and critical thinking [5]. The development of intellectual abilities such as thinking, problem-solving, and knowledge application [17] [56] can promote higher order thinking whereby students must read, write, question, and discuss the subject matters collaboratively in their groups and put it into practice [58] . Hence, it can reduce the process of content transfer to students solely while developing higher order thinking and focusing on practical learning rather than passive listening to the lectures along with the provision of feedbacks to students [72].

6.5. Students develop their learning skills.

Active learning is the learning process aimed to lessen the knowledge transfer from teachers to students and constitute learning skills among students [5,22]. It is also the process which rests on the student-centered approach by allowing students to apply their skills of speaking, listening, reading, thinking, and writing, including expressing their opinions while performing activities [28]. In fact, it is to develop the skills and concepts of students other than those of teachers regarded as the knowledge providers [6] .Learning is a natural attempt of human beings whereas students can develop their skills for knowledge searches independently [13] and learn by themselves to promote the constant out-of-class learning [13,22,45]. Students are then able to express themselves through speaking, listening, reading, writing, and reflecting [41]. Learning activities therefore rest on higher order thinking [56] Activities regarding the development of student skills and concepts [5] typically improve learning skills [5,22,32,45,72] and thinking skills as well as writing skills [13]. Students have an opportunity to develop themselves and their academic performance [15,31], improve skills of knowledge seeking by themselves [12] and promote collaborative learning habits among students by expressing their opinions through listening, speaking, reading, writing, and reflecting [51] . The students' roles are truly conversed from students into co-creators to establish the communicative skills[5]. Indeed, students take roles in their learning through a number of diverse learning activities to help them constructively develop their knowledge and essential skills [39]. Students can implement learning-by-doing and generate knowledge from what is performed during the lectures, with an emphasis on the ability and skill development that matches their existing knowledge. For practical learning, students are able to build their understanding from what they are doing and promote other skills [50] . Meanwhile, students receive the data to form their own concepts or new skills from such learning [41] to help them analyze and solve the problems, synthesize the data systematically [65,75]. Besides, students can develop their thinking ability, problem-solving skill, and knowledge application [17] while transferring the knowledge to their peers or expressing opinions and doing presentations properly. They also determine the concept, plan, acceptance, evaluation, and collaborative presentation. Students therefore must seek the meanings and understanding by themselves or with their peers e.g. finding out the answers, discussion, collaborative presentation, and summary of important concepts [18,73] .

6.6 Students play the key roles in learning and adopt diverse learning approaches.

Learning is a natural attempt of humans whereas individuals have their own distinctive ways of learning [23]. Active learning is regarded as the learning process with its main focus on students [9]. Teachers are therefore viewed as educational facilitators [5,32,45,72] to help students in their learning [13] together with their responsibility for such learning. Regarding a conversion from the lectures provided by teachers into the roles regulated by students [47], students are changed from receivers into co-creators [23,63]. In other words, it is a shift from listeners to participants in learning activities set out by teachers such as reading, writing, corresponding, and problem analyzing [73]. Students are believably benefited from diverse teaching methods whereby teachers are encouraged to adopt different strategies for comprehension of each individual [57]. Learning how to learn is expected among students [65] and they can learn from valid sources through distinctive approaches. Students play the main roles in their learning more actively [39]. In this manner, students must regulate their self-learning at a high level and are independent for their own learning [69]. While performing those activities, they will have an opportunity to think and make decisions about speaking, listening, reading, writing, and reflecting [36,39], corresponding and problem analyzing [37] and forming social relationships [39]. Students additionally engage in learning [32,34] and determining the learning goals as they are encouraged to forward the problems for further analysis [47], pay attention to group discussions, and apply the knowledge for real life [57] . Hence, most students are more satisfied with active learning than passive learning [55,58] .

6.7. Students actively participate in the learning process.

Active learning allows students' participation in the learning process at most [7,12,17,32,45,56,61,63] Students have more participation in learning or doing activities so that meaningful learning is truly established [65]. Students take more roles in their learning rather than gain knowledge or new concepts as passive students [22] . In doing so, students have more engagement in learning [12,61] in addition to just only listening to teachers [5,6,22,23,45,56]. They are conversed from passive students into active ones who engage in learning activities planned by teachers [12]. In fact, there is a shift from receivers into co-creators [5,12,23,63]. It is believed to promote collaborative learning behaviours since students are able to share their opinions and adopt the skills of listening, speaking, reading, writing, and reflecting [51]. Moreover, this typical learning is to promote reactions and collaborations rather than competitions [12,45,56,70].Students must seek the answers and understanding by themselves or with their peers e.g. finding the answers, discussion, presentation, and summary of important concepts [12,73] . Students generate the knowledge and systemize the learning process on their own from collaborative learning [56] whereas they learn together and put the knowledge into practice [61]. Students participate in the teaching and learning activities at the highest level while they take roles of speaking, questioning, discussing, and declaring [54]. Students play the main roles in their learning and a variety of learning activities are arranged for self-development among students to promote more creativity for knowledge and skill enhancements [39]. Students engage themselves in the learning process [5,32,34,45,61,72] which are varied and diverse [32,34]. They are then encouraged to participate in learning activities [12,55] , for example, reading, discussing, and writing [6] and have reactions with learning activities [61]. They may work in groups and collaborate with other students while jointly planning the curriculum for active learning. Students have roles in learning activities and knowledge seeking [4] . They also share their opinions based upon the analysis, synthesis, and application of supportive principles and theories [74] .

6.8. Students attain the academic achievements and meaningful learning.

Active learning enables students to have academic achievements [12,72] and understanding of the main concepts they have learned precisely and correctly which is permanent and well-connected knowledge [45,57]. Students are required to search for the meanings and build comprehension by themselves or with peers [12,73]. Students are then able to express through speaking, listening, reading, writing, and reflecting [41]. Students take part in learning activities [16] and solve or adjust their misconceptions immediately from communications made between peers and teachers [57]. It is viewed as learning by understanding with long-lasting memories and problem-solving skills along with positive attitudes toward the study and motivation [42] . Students have an opportunity to develop themselves and improve their academic performance [15,31]. Students share their participation as a part of learning activities for them to associate their existing knowledge with something new while studying [16]. This process emphasizes students to construct the knowledge and comprehension of contents and principles through experience-based understanding. Students eventually establish more understanding and longer memories [7]. Besides, students can seek the knowledge and apply various skills for their understanding to generate meaningful learning [29,32,34,49,62,67,72]. Meanwhile, students share their roles in learning or doing activities to promote meaningful learning [65]. With cooperation among students, they will have more enthusiasm in doing activities whether it is an exchange of experiences through speaking, writing, or discussion with peers [72].

6.9. Students build rapport between teachers and students.

Active learning is a shift from passive students into co-creators in learning activities planned by teachers [12]. At the same time, it promotes good relations between teachers and classmates [72]. This also makes the lessons more interesting [32,34]. Students build good relations between teachers and classmates [48] and gain the communicative skills [7,72]. As students benefit from reactions in class, they can develop their social experience to work with other people [7,32,34,57]. Students have an opportunity to question, respond, criticize, and appreciate the working system with different methods and perspectives of individuals or groups [57]. Students talk with classmates to build cooperation and rely on each other [61]. Students learn to have joint responsibilities, work discipline, and different duties [17,32,34,45,56]. At the same time, students can learn together [17,22,43]. Typically, students observe and do by themselves and they must communicate with each other about the subject matters and reflect their own thinking [22]. Students share roles in expressing opinions to promote the analysis, synthesis, and application of supportive principles and theories [74] as well as exchanges of learning experiences [35] .

6.10. Students integrate the information with their prior knowledge.

Active learning is an activity that allows students to integrate the information and principles to generate the main concepts [32,34,56,]. Students with enthusiasm and skills are able to select, analyze, and synthesize the information systematically [65,75].Students can integrate the information and technology into analysis, synthesis, and evaluation [17,56] whereas students and teachers receive the feedbacks from reflection quickly [5,45,60] .Students talk about what they have learnt, work on writing related to their prior experience, and apply the knowledge for daily life [42]. Students can integrate the subject matter and relate it to other different subjects. Students' engagement is considered a part of learning activities where they can associate their new and existing knowledge through learning by doing during the lectures [16]. They can also search the meanings of contents and relate them to their prior experience [65] and integrate the new knowledge with their own [5].

6.11. Students develop positive attitudes and self-esteem.

Active learning helps students to learn from what they have practiced through interesting and challenging activities in order to bring them pride and enjoyment [7]. It also develops students in terms of feeling, self-confidence, interest, and self-esteem [22]. Furthermore, students are encouraged to search for the knowledge, attitudes, values and experiences by themselves [22,45]. Active learning is then regarded as valuable and joyful learning in such a way that students can learn what truly interests and challenges them [75]. Meanwhile, it can promote positive attitudes toward the study due to their realization of learning by doing [57] . This process stresses on the importance of surveying attitudes and social values among students [5,22,45]. Active learning develops students in respect of feeling, self-confidence, interest, and self-esteem [22] already existing in students [5,45,60]. Students have enthusiasm for learning e.g. getting actively involved in activities [12,13] while developing positive attitudes toward the study and raise learning motivation [7,32,34,42,72] among students [5].

6.12. Students are enthusiastic, active, happy, and joyful.

Active learning helps students to be cognitively active to achieve more efficient learning [12,55]. They will have more interest and pay more attention to the study in a joyful way [42]. Therefore, active learning is valuable, exciting, and challenging for students [13,75] so that they are proud and joyful [7]. This stimulates their curiosity [66]. Students have more enthusiasm in doing activities whether it is an exchange of experiences through speaking, writing, and peer discussion [12,72]. Students will read, speak, listen, think, and write enthusiastically [13] for their study and problem-solving [75]. Students further develop social intelligence, learn by themselves, work together with other people, and establish good relationships [57].They also possess skills of selecting, analyzing, synthesizing the information systematically [65]. Students learn meaningfully through their collaboration and they have more enthusiasm in doing activities which are varied and diverse in the form of an exchange of experiences through speaking, writing, and peer discussion [72].

7. Application of learning activities for good learner characteristics

Active learning can be used for the learning arrangement in numerous ways e.g. 1) constituting learning among students both in class and out of class, 2) being generated by an individual or group, and 3) managing the learning process with or without technology. Upon an application of active learning, teachers will consider the followings: 1) adopting the proportion suited to the allowable time to help students develop their deep understanding and skills, 2) spending some time providing additional time and advice, 3) applying for and demonstrating what is learnt, and 4) receiving the feedbacks at once from peers and/or teachers [21] .

8. Roles of teachers in active learning for good learner characteristics

Considering the roles of teachers for active learning [33,56] , teachers take the following roles in active learning.

- Adopting the student-centered approach for learning whereas activities must reflect the need for student development and application for real life.
- Creating collaborative learning and responses contributing to good relations with teachers and classmates.
- Arranging the active learning activities to promote students' participation and learning achievements.
- Establishing collaborative learning for cooperation among students.
- Introducing learning activities which are challenging and allow students to learn through diverse teaching methods.
- Planning the time for learning clearly in terms of contents and activities.
- Teachers are open-minded and accept students' ability of expression and thoughts.

Cornell University has applied the teaching techniques based on active learning for effective practice.

1. **Giving an assignment for preparation before the study.**
   1.1 Assigning students to read the contents or subject matters related to their study in advance.
   1.2 Assigning students to study the video and give them the questions so that they can find the answers by themselves.
   1.3 Encouraging students to provide answers to the subject matters of the next lesson.

2. **Adopting the teaching techniques contributing to active learning in class.**
   2.1 Choosing the techniques which are simple and do not require much time for starting the lesson e.g. "one minute paper" and "think-pair-share".
   2.2 Analyzing the learning targets and a few teaching techniques.

3. **Main principles of learning arrangement based on active learning.**
   3.1 Choosing the activities suitable for the contents and what students should practice.
   3.2 Emphasizing the importance of reactions among students.
   3.3 Describing the activities and their expected benefits.
   3.4 Allowing students to bear the responsibility and determining the amount of time and budget.
   3.5 Summarizing the main points and exchanging them before starting the new contents.

4. Applying the technology for effective learning arrangements to promote the activities based on active learning.

9. Conclusions and suggestions

Active learning is another learning process that prioritizes practical learning or learning by doing whereas the knowledge is truly obtained from the experience. In this process, students are required to take actions rather than passively listen to teachers. Hence, activities for students to learn through reading, writing, responding, and problem analyzing must be introduced. Meanwhile, students are encouraged to apply higher order thinking such as analysis, synthesis, and evaluation. Active learning is therefore the process contributing to meaningful learning through collaborations among students. In this manner, teachers must limit their roles in teaching and providing knowledge to students directly, but add the process and activities that engage students in doing more activities. These gradually develop learner characteristics. According to the literature review and knowledge synthesis, learner characteristics generated by active learning are described in 12 traits: 1) students have a great opportunity to first-hand experience; 2) students create learning process and construct knowledge by themselves; 3) students learn and reflect from their experiences; 4) students develop higher order thinking; 5) students develop their learning skills; 6) students play important roles in learning which are also varied and diverse; 7) students highly participate in learning process; 8) students reach academic achievements with meaningful learning; 9) students build rapport with teachers and their peers; 10) students integrate information with their prior knowledge; 11) students develop positive attitudes and self-esteem; 12) students are enthusiastic, active, happy and joyful. For suggestions, learner characteristics from active learning should be acknowledged as they bring positive results to students and can be further applied as the guidelines for designing learning activities in class to benefit Thai children with effective learning in the 21st century. These characteristics also influence the design of learning programs to students in terms of the process and target to promote the full potential development and preparation for the challenging and complicated world at present and in the future.

References

[1] Anunthavorasakul A 2016 Assessment Result of PISA 2015 with The disparity of quality education in Thailand
[2] Attaporn J and Koraneekit P 2014 Development of an Active Online Instructional Model to enhance Learning Behaviors of Undergraduate Chulalongkorn University
[3] Belkin, Gary S and Skydell R H 1979 Foundation of Psychology (Boston : Hougton Mifflin Co.)
[4] Banjongjit S 2008 Active Learning **Journal of Naval Academy 8** 34–42
[5] Bonwell C C Active Learning: Creating Excitement in the Classroom. [Online]. Retrieved October 17, 2017 from https://www.ydae.purdue.edu/lct/HBCU/documents/Active_Learning_Creating_Excitement_in_the_Classroom.pdf
[6] Bonwell C C and Eison J A 1991 Active Learning: Creating Excitement in the Classroom. ASHE-ERIC Higher Education Report, Washington DC: School of Education and Human Development (George Washington University)
[7] Boonprakob M 2001 Research Report No. 80 (Research and Development)
[8] Boonyakanit W 1989 The Result of Implementing Learning Experience Practice and General relation on Young Children's Graduate School (elementary education) (Bangkok: College Srinakharinwirot University. Photocopy)
[9] Brandes D and Ginnis P 1996 A guide to Student Centred Learning (Oxford : Blackwell)
[10] Center for Teaching Excellence, University of Kansas 2000 Teaching Strategies:Active Learning. Retrieved November 27, 2006, from http://www.ku. edu/~etc/resources/ teachingtips/active.htm
[11] Chickering A and Gamson Z 1987 Save principles for good practice **AAHE Bulletin, 39** pp 3–7

IOP Conf. Series: Journal of Physics: Conf. Series **1340** (2019) 012083 doi:10.1088/1742-6596/1340/1/012083

[12] Chaikijpinyo S 2005 How to Teaching Active learning **Journal of Innovative Teaching, 2** pp 12–15
[13] Chamnankit B 2006 Why do you need a knowledge management to learn in higher education knowledge management journal **Nakhon Sawan Rajabhat University 1** pp 1–7
[14] Cornel University Center for Teaching Excellence 2015 Why incorporate active learning Techniques and How can you incorporate active learning into various classroomsettings.Retrieved October 17, 2017 from http://www.cte.cornell.edu/teachingideas/engaging-students/active-learning.html
[15] Colleen J 1996 Designing Web-Based Instruction: Research and Rationale [Online] Retrieved October 17, 2017 from http://www.cc.utexas.edu/~jonesc/research/empaper.html
[16] Dasa J 2009 15 techniques in Active learning **IPST Magazine 36** pp 72–76
[17] Driscoll M 2002 Web-Based Training:Creating e-Learning Experiences. 2 nd.ed (SanFrancisco: Jossey-Bass Pfeiffer)
[18] Detsri P 2002 Active learning **Journal of Science Education Mathematics and Technology 30** pp 53–55
[19] Development Science Instruction techniques based on Science Education Quality (Bangkok: Srinakharinwirot University)
[20] Doctorow M, Wittrock M C and Marks C 1978 Generative processes in reading comprehension **Journal of Educational Psychology 70** pp 109–118
[21] Eison Jim 2010 Using Active Learning Instructional Strategies to Create Excitement and Enhance Learning. University of South Florida. Retrieved October 17, 2017 from http://www.cte.cornell.edu/documents/presentations/EisenHandout.pdf.
[22] Fink L D 2003 Creating Significant Learning Experiences : An Integrated Approach to Designing College Courses (Jossey Bass Higher)
[23] Felder R M and Brent R 1996 Navigating the Bumpy Road to Student-Centered Instruction **Journal of College Teaching 44** 43–47
[24] Good CV 1973 Dictionary of Education Prepared Under the Auspices of PhiDelta Kappa (3rd ed.). New York: McGraw-Hill.McKeachle (1998) Active Learning. Retrieved October 17, 2017 from http://hydro4si.fau.edu/~rjordan/active_learning.html.
[25] Hendrikson L 1984 Active Learning. Retrieved October 17, 2017 from http://www.ed.gov/database/ERIC_Digests/ed253468.html.
[26] JohnVon D W, JohnVon R T and Holubee E J 1994 The Nuts and Bolts ofCooperative Learning (Minnesota : Interaction Book Company)
[27] Kasembundit university n.d. Active Learning. Retrieved October 25,2560, From http://www.archlebu.ac.th/home/research/pdf/academic/academic001pdf
[28] Kamted W 2006 Effects of using Leslie dee Fink's model of active learning in biology instruction on paragraph writing ability and learning achievement of upper secondary school students. Thesis (Science Education) (Bangkok: Graduate Studies, Chulalongkorn University. Take document)
[29] Khammani T 2005 Research-based learning (Bangkok: Chulalongkorn University Printing House)
[30] Khammani T 2012 Science of Teaching: Knowledge to the effective learning process (Bangkok: Chulalongkorn University Printing House)
[31] Khan B H 1997 Web-Based instruction.Englewood Cliffs (New Jersey: Educational Technilogies)
[32] Ketpichainarong W 2014 Introduction Active Learning Sharing Active Learning Experiences. Nakhon Pathom: Institute for Innovative learning, Mahidol University. Retrieved October 25,2560, From http://qa.bu.ac.th/cop/index.php/ component/phocadownload/ category/-km-copteaching?download=18:active-learning
[33] Kaewchaijaroenkit N 2007 The Role of Teachers in the event and how to follow the guidelines of the Active Learning Retrieved October 25,2560, fromhttp://www.itie.org

[34] Kaewsanit M 2015 Active learning through a student-centered approach in the communication arts department **Kasem Bundit Journal 16** pp 113–124

[35] Kruesaeng P 2013 Proactive Technical Education Learning from Shelton College international, Singapore Journal of Humanities and Social Sciences (Nakhon Sawan Rajabhat University)

[36] Kiteak P 2007 The Effect of Using Active learning toward Achievement and Attitude in Chemistry of 12nd Grade Students Pathumthani Province.Master's Thesis (Chandrakasem Rajabhat University)

[37] Kothakhun P 2011 What is Active Learning? Retrieved October 25,2560, from http://prapasara.blogspot.com/2011/09/active-learning.html

[38] Laohapiboon P 2542 Teaching Science (3rd ed.) (Bangkok: Thai Watana Panich Printing)

[39] Manopichetwattana S 2004 The Development of a Science Instruction Model Integrated Management to engage learner participation in active learning about the Anatomy.Ph.D. thesis, Srinakharinwirot University

[40] Meyers Chet and Jones Thomas B 1993 Promoting Active Learning: Strategies for the Collage Classroom (San Francisco: Jossey-Bass)

[41] Meyers C and Jones T B 1993 Promoting Active Learning : Strategies for the Collage Classroom (Oxford: Elmsford)

[42] McKeachie Wilbert J, Paul R Pintrich, Yi-Guang Lin, and David A F Smith 1986 Teaching and Learning in the College Classroom: A Review of the Research Literature **Ann Arbor: Regents of The Univ of Michigan ED 314 999 124** pp MF-01, PC-05

[43] Na Songkhla J 1999 WWW-Based Instruction **Journal of Education Studies 27** pp 18–28

[44] Nateetanon S 2007 The Effect of Active Learning Environment on Creative Thinking of Preschool Children. Master thesis, M.Ed (Early childhood Education) (Bangkok: Graduate School , Srinakharinwirot University Take document)

[45] Pakdeejit Y 2014 Academic Dissemination Documentation: Day promotes academic quality of teaching (Faculty of Education, Nakhon Sawan Rajabhat University)

[46] Panich V 2012 21st century Skills: Rethinking How Students Learn (3rd ed.) (Bangkok: Sodsri-Saridwongso Foundation)

[47] Praprutkit N 2003 Childhood behavior in childhood behavior and teaching Materials (Bangkok: SukhothaiThammathirat Open University)

[48] Pattrakorn S 2009 The Effects of Organizing Active Learning on Problem Solving and Mathematical Communication Abilities of Mathayomsuksa III Students in Probability (Master thesis, M.Ed. (Secondary Education), Srinakharinwirot University)

[49] Pititayanuwat S and Boonterm T 2002 Research methodology : a tool or a way of [40] thinking **Journal of Research Methodology 15**

[50] Phudetch A 1998 Practical learning (Active Learning) **Journal of Private Education, 72** pp 57–58

[51] Prince Michael 2004 Does Active Learning Work? A Review of the Research **Journal of Engineering Education 93** pp 223–231

[52] Paonil P 2003 Training materials: Participatory Learning. Bongkok: Chandrakasem Rajabhat University (Photocopy)

[53] Ramkhamhaeng University 2014 History of the University Retrieved October 25,2560, From http://www.ru.ac.th/th/index.php/site/page?view=history

[54] Rungklab A 2000 Why is that Active Learning? **Journal of Naval Sciences 83**

[55] Roongrangsee C 2006 Active learning management plan in Course 1000101 Human Behavior and Self Development Unit 2 "Self Development" (Nakhon Sawan: Faculty of Education, Nakhon Sawan Rajabhat University)

[56] Ruangsuwan C (n.d.) Active Learning Retrieved October 25,2560, from http://www.drchaiyot.Com

[57] Salemi M K 2001 An lustrated Case for Active Learning. University of North Carolina Retrieved October 17, 2017 from http://www.unc.edu/~salemi/Active_Learning/llustrated_Case.pdf.

[58] Sweller J 2006 The worked example effect and human cognition: Learning and Instruction (New Jersey:EducationalTechnologies)
[59] Sawekngam W 2012 Pipeline Instructional Teaching: How to teach effectively in the context of education that focuses on results **Journal of Education. 40** pp 255–272
[60] Shenker J I, Goss S A and Bernstein B S 1996 Instructor's Resource Manual for Psychology: Implementing Active Learning in the Classroom Retrieved October 17, 2017 from http://s.psych/uiuc.edu/jskenker/active.html
[61] Silberman M 1996 Active Learning. Boston : Allyn & Bacon. Strategies: Active Learning Retrieved October 17, 2017 from http://www.ku.edu/~etc/resources /teachingtips/active.html
[62] Sinlarat P 2002 Research-Based Learning (Bongkok: Chulalongkorn University Printing House)
[63] Sirijanchuen M 2011 The teaching a large group in Gsoc 2101 the development community. Approach to learning by Active learning and using e - leaning to develop a model of teaching in higher education, learning and academic achievement of students: Chiang Mai Rajabhat University
[64] Songkram N 2012 The Development of Blended Learning Model with Active Learning for Knowledge Construction and Creative Problem Solving Ability for Undergraduate Students (Research Grants Chulalongkorn University)
[65] Suwannatthachote P 2008 Active Learning. Retrieved October 17, 2017 from http://www.academic.chula.ac.th/elearning/content/active%20learning_Praweenya.pdf
[66] Tantiphlachiva K 2000 Psycho-Intellectual Model: The line is used to create lesson plans Of Pre-School Children (BangKok: Edison press product.co.,ltd)
[67] Thipakorn B 2007 The Development of tecturer's Potential in higher education institutions. Paitoon Sinlarat (Editor) Professional teacher: Concepts tools and development (Bangkok: Chulalongkorn University Printing House)
[68] Theeradakorn S 1999 Educational Psychology Bangkok: Department of Psychology and Guidance (Faculty of Education Phranakhon Teacher's College)
[69] Thepworachai A 2000 Active Learning (1st ed.) (Bangkok: Academic Welfare Scholarship)
[70] Thomas J 1972 The variation of memory with time for information appearing during a lecture **Studies in Adult Education 4** pp 57–62
[71] Toonthong S 2002 A Development of an Authentic Assessment Model. Dissertation Ed.D.(Testing and Measurement) (Bangkok: Graduate School. Srinakharinwirot University)
[72] Wattana B 2003 Active Learning **Academic Journal 6** 30–34
[73] Walailak University 2012 Active Learning (Copied from Walailak University)
[74] Ward R 1998 Active, Collaborative and case-based learning with Computer based case **Scenarios C 30** pp 103–110
[75] Wuttiwan W 2010 The effect of teaching science by using the active learning method to enhance sciene achievement and problem solving ability for Students in Matthayomsuksa (Master Thesis, Burapa university)

IOP Conf. Series: Journal of Physics: Conf. Series **1340** (2019) 012084 doi:10.1088/1742-6596/1340/1/012084

The Nature of Science Views among Intern General Science Pre-Service Teachers

Somsak Gathong[1] and Suthida Chamrat[2*]

[1] Ph.D student, Doctor of Philosophy Program in Education (Science education), Faculty of Education, Chiang Mai University Email: somsak_gat@hotmail.co.th

[2] Department of Curriculum, Instruction and Learning, Faculty of Education, Chiang Mai University

*Corresponding Author email: suthida.c@cmu.ac.th

Abstract. This research aimed to study the nature of science views of 23 students who are currently intern general science pre-service teachers in one of Rajabhat Universities in the North. The research instrument used in the study was: the nature view of science questionnaire developed by Haidar (1999) and the researcher has improved it in accordance with the curriculum of science and education in Thailand. The nature of science views had three aspects totally 27 items. There were 8 items for scientific world view, 4 items for the scientific knowledge inquiry, 6 items for scientific enterprise and they were totally 18 items and open-ended question on the nature of science views with three aspects of science in the world: 3 items for scientific world view, 3 items for scientific inquiry and 3 items for scientific enterprise. They were totally 9 items. Then the data were analyzed by grouping the answers and finding the percentage. The results of the research revealed that intern general science pre-service teachers had the same view of the nature of science that was consistent with 3 aspects of the traditional view. That were the 2nd aspect; scientific inquiry. The 1st aspect; scientific found that91.30 %better understood consistent with the traditional view that working of a scientist must be based on a scientific method that has a definite sequence of steps. As the 2nd aspect; the scientific knowledge creation found that 78.26% had the consistency of understanding with the traditional concept. It stated that scientific knowledge is created by observation, survey or experiment. The 3rd aspect; evidence and imagination, and creative thinking, it affects the scientific knowledge creation. It showed that 73.91% had an understanding that is consistent with the traditional concept that scientists acquire scientific knowledge using only evidence-based. The findings indicated that intern general science pre-service teachers still have many points of view. The most necessary thing is teachers need to use teaching and learning methods that solve the problem of natural science erroneously in the course of learning science management before the students go for the practicum.

1. Introduction

The nature of science is the heart of the scientific literacy. It identifies the nature of science in the national and international science curriculum [3] because science is a culture of the modern world, a knowledge-based society, Therefore, everyone needs to be developed for science in order to understand nature and the technology that human create and apply it reasonably [19] The main aim of teaching and learning science, which scientists study and organizations around the world have accepted the role of science's nature in teaching science as well as countries that include the nature of science in the curriculum, such as Australia, Canada, South America, the United Kingdom, the United States and New Zealand [12].Thailand raises an awareness of nature of science as importance and it shows on the standards of science learning in the 8th core curriculum of Basic Education 2008 [18].The nature of science enables students to acquire their knowledge by themselves, create new knowledge, understand the nature and limitations of science including the impact of science, and technology to society. These understanding of the nature of science will improve the attributes of science. [20]

Scholars have defined the meaning of nature in many ways [11]. states that the nature of science is directly related to epistemology with values and hypotheses as well as scientific processes. It is also a development of scientific knowledge, where scientific knowledge is independent of thought [14]. states that the nature of science is a harmonious blend of sociological study of science in many aspects, including the history of science, sociology and philosophy of science, as well as the study of scientific knowledge. In addition, [15] argue that the nature of science is a combination of various social sciences, including history, sociology and philosophy of science [7]. also supports that the nature of science refers to the study of scientific knowledge; it is the method by which the scientist derives his knowledge of the work or society of the scientist and the value of science to society. According to this definition shows that the nature of science does not pay only attention to scientific knowledge but also the process of gaining scientific knowledge, skills used in the quest for knowledge and work or society of scientists and the value of science to society associating the relationship between science and technology, and society.

From the results of both local and international research, it has been found that teachers have misunderstandings about the pursuit of scientific knowledge and the influence of society and culture on scientific work [6] Additionally, the study of the nature of science's views among intern pre-service teachers in Promotion Science and Math Teacher Project (PSMT) indicate that the students think creative thinking and imagination have no part in creating science. Moreover, social factors and cultural factors have no impact on records of scientists [10]This is consistent with the study of the students' views among intern pre-service teachers towards Brunei teachers as the scientific knowledge inquiry that imagination and creativity do not contribute to the creation of scientific knowledge [22] Research in the field of science has found that the lack of understanding on the nature of science by science teachers results in the inability to integrate science teaching[17] Moreover, the results showed that the students lack understanding and misunderstand about the nature of science in many aspects, such as the steps and methods of scientific knowledge inquiry, meaning and origin of rules and theories, factors affecting the work of scientists, influence of social and cultural relations on science. As mentioned above, there are research studies in science education to solve the problem of understanding the nature of science about the development of science teachers to develop the curriculum of teaching about the nature of science. The studies review that most of intern general science pre-service teachers have views about the nature of science, but lacking in profound understanding of theology. It requires more effort in teaching and learning.

However, today it has found, students also lack the understanding of the nature of science. The main reason is that teachers have misunderstandings like understanding that science is only scientific knowledge and they do not take into account the process of acquiring scientific knowledge. Organizing science learning activities and the 8th core curriculum will be successful or not, teachers are very important persons to develop students' understanding of the nature of science. Institute for the Promotion of Teaching Science has set the standard of science teachers that "Teachers should understand the nature of science and technology, including the content structure of the curriculum and the knowledge of science and technology, the concept of inquiry and problem solving. It can also be used to create a learning experience that makes the course content meaningful to students, ", which is consistent with teacher standards in the United States that encourage teachers to teach the nature of science to students as well [21].The teachers have accurate idea about the nature of science and have an understanding of the teaching process that corresponds to the nature of science. It improves the efficiency of science teaching and encourages students to become knowledgeable scientists [16]. Learning the nature of science is paramount. Those who do not have the science of nature cannot convey the nature of science to the students [1].Currently, most of educational institutes of science teachers in Thailand have not had any specific courses in the nature of science and what students or students are learning in the course "How to teach science." Of course, if the intern general science pre-service teachers understand the nature of science and realize that the nature of science is important to the students, the teacher training students will be able to integrate the nature of science into teaching. Therefore, the teaching of the nature of science is essential.

In order to be successful in teaching and learning, teacher is one of the key factors contributing to the teaching and learning of the nature of science, as evidenced by the standards of science teachers of the Institute for the Promotion of Teaching Science B.E. 2002 (IPST) as in the 1st standard, teachers must have an understanding of the nature of science and technology in terms of knowledge-based content, consistent with teacher standards in the United States that encourages science teachers [21]. Based on the objectives of the nature of the science teaching management, it emphasizes on the understanding of teacher including problems of teaching and learning science. Therefore, the researchers have the interest to study the nature of science's views among intern general science pre-service teachers to be as the basic information and to develop students to understand the nature of science and to be used in the development and improvement of the curriculum of science teachers in the future.

2. Objective of the research

To study the nature of science among intern general science pre-service teachers practicing professional science teachers.

3. Approach of the study

The research conduction on " The Nature of Science Views among Intern General Science Pre-Service Teachers" is descriptive research, which is an in-depth study with the group with less people to obtain information describing the ideas and insights of intern general science pre-service teachers.

4. Research participants

The target group is 23 of the 5th year's students of Bachelor of Science Program in Science Faculty, majoring in General Science, academic year of B.E. 2560. The samples were selected as the purposive sampling, consisting of 2 males and 21females and they have currently trained as intern general science pre-service teachers were chosen by the purposive sampling.

5. Pre-service teacher professional development

The researchers have developed a questionnaire about the understanding the nature of science as in sequence. The questionnaire is based on the conceptual framework of the question showing the views about the nature of science. Each item contains 2 messages; the left side is the message on the nature of science based on traditional ideas. The message on the right is a natural science based on the theory of creating new knowledge, covering 3 aspects: the scientific world view, the pursuit of scientific knowledge, and scientific enterprise.

5.1 Research Instrument

The questionnaire is about the nature view of science developed by Haidar (1999) and the researcher has improved it in accordance with the curriculum of science and education in Thailand. The nature of science views has three aspects totally 27 items.

5.1.1 Closed-ended questionnaire is the natural view of science. There are 3 aspects: scientific world view, 8 scientific inquiry, 4 science-related questions, and 6 scientific enterprise. They are totally 18 questions

5.1.2 Open-ended questionnaire is on the nature of science views with three aspects of science in the world: 3 items for scientific world view, 3 items for scientific inquiry and 3 items for scientific enterprise. They are totally 9 items

5.2 Data Analysis

Analytical data of the nature of science is in accordance with [8] nalytical criteria, having a philosophical view of the nature of science, are divided into two perspectives: the traditional view of nature and the contemporary science view has the details as follows;

5.2.1 The traditional view of nature is based on two philosophical ideas: the concept of realism: this is a view on scientific theories and explanations. The second one is empiricism: this is a view on scientific knowledge and the development of scientific knowledge.

5.2.2 Nature view of theoretical knowledge as constructivism view is based on two philosophical ideas: instrumentalism s is a view on theories and scientific explanations and constructivism is a view on scientific knowledge and the development of scientific knowledge.

The researcher analyzed the relevance of the data by reading each student's answers to each question, interpreted the answers and grouped their understanding in accordance with the setting criteria. The data was analyzed twice, and the second time was sent to the experts who were scientists to investigate one more time.

6. Research findings

According to the analysis of data, the overall opinions of the intern general science pre-service teachers had the most traditional views on the nature of science, varied from the highest level to the lowest level as steps of scientific method, methods to create scientific knowledge and evidence with imagination and creative thinking that affect scientific knowledge creation. The theory of knowledge creation is in descending order as relationship of scientist, rules and scientific model theories and the collaboration of scientists. It is noteworthy that the second issue, the scientific knowledge inquiry. The intern general science pre-service teachers had the view to be consistent with the three traditional concepts, where each information and sub-topic is shown in Tables 1-3 and the following description.

Table 1. Nature of Science Views as the 1st Aspect; Scientific World View of Intern General Science Pre-Service Teachers

Issues of Nature of Science	Item	No. of Answers (percentage) (N =23)		
		Traditional View	Uncertain	Knowledge Creation Theory
1. Meaning of the nature of science's characteristics.	1	8(34.78)	0(0)	15(65.22)
2. Scientific knowledge as the explanation for understanding nature.	4	15(65.21)	1(4.35)	7(30.43)
3. Development of scientific knowledge.	7	6(26.08)	3(13.04)	14(60.86)
4. The persistence of scientific knowledge.	10	2(8.69)	6(26.08)	15(65.21)
5. Creation of scientific knowledge for nature explanation.	12	8(34.78)	4(17.39)	11(47.82)
6. Characteristics of rules.	13	3(13.04)	6(26.08)	14(60.86)
7. Source of theories,	14	13(56.54)	3(13.04)	12(52.17)
8. Scientific Modelling.	17	3(13.04)	3(13.04)	17(73.91)
		30.44	13.04	56.52

Based on the analysis of the data from the questionnaire on the scientific world view among intern general science pre-service teachers, it was found that 56.52% had a consistent view of the theory of knowledge creation and 30.44 % had a consistent view of traditional idea and 13.04 % of the respondents was uncertain. There are several issues that correspond to the theory of knowledge creation in descending order as scientific modelling 73.91%of the respondents considered it as a "rational or logical simulation in terms of the meaning and nature of science", 65.22 % argued that "scientific knowledge is a human explanation for the understanding of nature, but not everything is real and some scientific knowledge cannot be proved." As the issue of the persistence of scientific knowledge, 65.21% agreed that "knowledge of scientists is acceptable and can be changed". The issue is in line with the traditional view as following order scientific knowledge as the explanation for understanding nature. 65.21% argue that scientific knowledge is a fact that occurs in nature and the source of theories, 56.54 percent had the reason that "scientific knowledge has been accepted because it is proven by the theories to explain what we study."

Table 2. The Nature of Science View: The 2nd Issue: Scientific Knowledge Inquiry of Intern General Science Pre-Service Teachers in each issue.

		No. of Answers (percentage) (N =23)		
Issues of Nature of Science	Item	Traditional View	Uncertain	Knowledge Creation Theory
1. Evidence with imagination and creative thinking affecting scientific knowledge creation.	2	17(73.91)	1(4.43)	5(21.73)
2. Scientific knowledge inquiry approach	5	5(21.73)	5(21.73)	13(56.52)
3. Scientific knowledge creation.	8	17(73.91)	4(17.39)	2(8.69)
4. Steps of scientific methods.	11	21(91.30)	1(4.34)	1(4.34)
		65.22	13.04	21.74

According to the analysis of data from the questionnaire, the views of nature science, among intern general science pre-service teachers as the scientific knowledge inquiry, most of them, *65.22%* had the consistent views with the traditional view and *21.74%* was consistent with knowledge creation and *13.04* % of the respondents was uncertain. There are several issues that are consistent with traditional views in descending order as steps of scientific methods revealed that 91.30% agreed that "Scientists have scientific knowledge by using scientific processes created by the exact sequence of experiments." The next issue, scientific knowledge creation, 73.91% of the respondents argued that "scientific knowledge is based on credible experiments.", then evidence with imagination and creative thinking affecting scientific knowledge creation showed that 73.91% agreed that " Scientists acquire scientific knowledge using scientifically proven empirical evidence derived from experiments. " Moreover, there are also issues that are relevant to the theory of knowledge creation. The issue of scientific knowledge inquiry approach found that 56.52% agreed that "Scientific knowledge could be created by imagination, creative thinking and other things."

IOP Conf. Series: Journal of Physics: Conf. Series **1340** (2019) 012084 doi:10.1088/1742-6596/1340/1/012084

Table 3. The Nature of Science: the 3rd Aspect: The Scientific Enterprise of Intern General Science Pre-Service Teachers in each Aspect.

Issues of Nature of Science	Item	No. of Answers (percentage) (N =23) Traditional View	Uncertain	Knowledge Creation Theory
1. Collaboration of scientists.	3	1(4.34)	6(26.08)	16(69.56)
2. Working for meeting the social need.	6	8(34.78)	8(34.78)	7(30.43)
3. Influence on records of scientists.	9	15(65.21)	4(17.39)	4(17.39)
4. Ethics of scientists towards experimental animals	15	5(21.73)	7(30.43)	11(47.82)
5. Influence on the work of scientists.	16	6(26.08)	2(8.69)	15(65.21)
6. The relationship of science, technology and society.	18	3(13.04)	3(13.04)	17(73.91)
		27.54	21.74	50.72

The analysis of data from the questionnaire indicated on the science world view among intern general science pre-service teachers in the field of science, as the scientific enterprise, 50.72 %had a view consistent with the theory of knowledge creation and ***27.54%*** had the consistency of the original concept and ***21.04%*** was uncertain. There are several issues that are consistent with the theory of knowledge creation varied from the highest level to the lowest level as the issue of the relationship of science, technology and society; 73.91% had the same view that "Science, technology and society are related." Aspect of collaboration of scientist, 69.56% provided the reason that "Scientists work in the scientific community to find the best way to describe nature. They may collaborate with many parties. "The next issue is Influence on the work of scientists, 65.21% agreed that "Scientists report the fact by virtue of prior knowledge and rationality. " In addition, there are issues that are consistent with the original idea, namely, the issue of Influence on records of scientists, 65.21% had the view that "The scientists report the data to match the results without distortions. The former has no effect on the record of the scientist. " The next issue is that working for meeting the social need, ***34.78%*** argued that "Most scientists like to find and pay attention to what they like. "

7. Discussions

Based on the survey of views about the understanding of intern general science pre-service teachers, it revealed that the students had a consistent view of the theory of knowledge creation and some aspects responds to the concept of traditional view classified in each aspect of the following areas. The 1st aspect is in line with the theory of knowledge creation on scientific modelling and the persistence of scientific knowledge. But there is also a view that corresponded to the traditional one as scientific knowledge as the explanation for understanding nature with the reason that scientific knowledge is a fact that occurs in nature, is acceptable since it has been proven by the theory to explain what we have learned because knowledge is true. This view is not consistent with the nature of science view by [1] which state that scientific knowledge is a temporary reality can be changed. However, from the survey, the view of understanding the nature of science as in the 2nd aspect on scientific knowledge inquiry shows that intern general science pre-service teachers with the view that they correspond to the original idea, *91.30%* agreed that the scientists realize the knowledge using the scientific process created by the exact sequence of experiments supported by the research of [10] ;the view of students in the project of promotion and science and math production is that scientific knowledge derived from scientific methods with the accurate steps respectively. Misconceptions about the nature of science [8].

state that science teachers often manage learning sequentially according to scientific methods. It understands that the scientific method is the same as the scientific inquiry and is the best approach, in line with [10] say that the scientific method is to determine the problem, data gathering, hypothesis testing, and conclusions. Furthermore,[5]who studied pre-service teachers, found that most teachers have misunderstandings that scientific method is the only way to investigate scientific knowledge. Based on science textbooks or from the form of a scientific action has a sequence of experimental methods, the steps make teachers understand that it must follow the sequence. However, scientific methods can be varied. The method is not limited to scientific methods only. The issue of scientific knowledge creation and evidence, and imagination and creative thinking that affect the scientific knowledge creation found that most students see that scientists gain scientific knowledge by using scientific processes with empirical evidence from experiments. Referring to all the data, students are given the opinion that scientific knowledge can be obtained from reliable experiments. This is in line with [6] who found that even the students as the samples in the study can be told of the conclusion of the observation. But students also focus on the experiment. This can convey that the students have a view to acquire knowledge, which emphasizes the importance of the experiment. But the truth of the scientific knowledge is not only from the experiments but is also from the observation and other information from other sources [4].However, it can be seen that scientific experiment is not scientifically valid [13] because there are some limitations, such as time, scope of experiment or scientific inquiry and teaching and learning management using scientific evidence. It practices students to use scientific processes whether it is observation, summary, data prediction leading to the knowledge of science from the facts, principles, assumptions and rules [9].Considering as the misunderstanding, the most important thing is to teach the teachers to have a variety of teaching strategies that encourage students to participate in group discussions with context and scientific methods. It will help students understand the issues [2] In the third aspect of science, it is found that most intern general science pre-service teachers with a view that corresponds to the theory of knowledge creation such as the issue of the relationship of science, technology and society, and the issue of the collaboration of scientists. Additionally, there are issues that are consistent with the original concept of influence on records of scientists; most of them found that scientists report data to match experimental results without distortion. The original knowledge does not affect scientists' recordings. The results of this research are essential in the production of science teachers. The intern pre-service student should gain enough experience on teaching nature of science supported by the recommendation of [1] that the teaching of nature to science is to add more experiences by teaching through science discovery.

8. Conclusions

The nature of science from both perspectives is different: the traditional concept, role of the teacher is the manager and transfer the truth of content, the scientific knowledge available to the students, contrary to the theory of knowledge creation; teachers play an important role in promoting students learn to understand science by focusing on the process of finding knowledge rather than on scientific truth. The students must manage information or experience and to create meaning of itself by providing students in the real context. In teaching and learning management, teachers must strive to create a moral and ethical environment. The students play a full role in learning by taking self and control for self-learning. The roles of the teacher are to cooperate, to facilitate and to assist students in learning. The division taking responsibility for teacher production is required to prepare and understand the nature of science for students before interning.

9. Acknowledgements

The researchers would like to express the deepest appreciation to the school administrators and mentors as well as those who are involved in all research in providing advice and guidance.

References

[1] Abd-El-Khalick F and Lederman N G 2000 Improving science teachers' conceptions of the nature of science: A critical review of the literature. **International Journal of Science Education 22** pp 665–701

[2] Akarsu B 2010 Science education research vs. physics education research: A structural comparison **European Journal of Physics Education 1** pp 13–19

[3] American Association for the Advancement of Science (AAAS) 1993 Science for all Americans (New York: Oxford University Press)

[4] Buaraphan K 2009 Pre-service and in-service science teachers' responses and reasoning about the nature of science **Educational Research and Review 4** pp 561–581

[5] Buaraphan K and Sung-Ong S 2009 Thai pre-service science teachers' conceptions of the Nature of Science **Asia-Pacific Forum on Science Learning and Teaching 10**

[6] Chamrat S and Yutakom N 2008 Chemistry teachers' understanding and practices of the nature of science when teaching atomic structure concepts **Kasetsart Journal: Social Sciences, 29** pp 228–239

[7] Faikhamta, C. 2013. The development of in-service science teachers' understandings of and orientations to teaching the nature of science within a PCK-based NOS course. *Research in Science Education. 43*(2), 847 – 869

[8] Haidar A H 1999 Emirates pre-service and in-service teachers' views about the nature of science **International Journal of Science Education 21** pp 807–822

[9] Khishfe R Relationship between Nature of Science Understandings and Argumentation Skills: A Role for Counterargument and Contextual Factors **Journal of Research in Science Teaching 2012 49** pp 489–514

[10] Lalitanurak P and Faikhamta C 2011 Views on the nature of science of student teachers in the Project for the Promotion of Science and Mathematics Talented Teachers (PSMT)

[11] Lederman N G 2004 Syntax of nature of science within inquiry and science instruction. In L. B. Flick & N. G. Lederman (Eds.) **Science & technology education library 25**

[12] Lederman N G 2007 Nature of science: Past, present, and future **Handbook of research in science education** pp 831–879 (Mahwah, New Jersey: Lawrence Erlbaum Publishers)

[13] Low 2000 Advertising vs. sales promotions: a brand management perspective **Journal of Product and Brand Management 9** 389–414

[14] McComas W F 1996 Ten myths of science: Reexamining what we think we know about the nature of science **School Science and Mathematics 96** pp 10–16

[15] McComas W F, Clough M P and Almazroa H 1998 The role and character of the nature of science in science education. **Science & Education 7** pp 511–532

[16] McComas W F, Clough M P and Almazroa H 1998 The Role and Character of the Nature of Science in Science Education In W. F. McComas (Ed.), The nature of science in science education: Rationales and strategies (Netherlands: Kluwer Academic Publishers)

[17] Mellado V 1997 Preservice Teachers' Classroom Practice and their Conceptions of the Nature of Science **Science & Education 6** pp 331–354

[18] Ministry of Education 2008 Basic education curriculum B.E. 2551 (A.D.2008) (Bangkok: Kurusapa Ladprao Publishing)

[19] Ministry of Education 2017 Basic education curriculum B.E. 2560 (A.D.2017) (Bangkok: Kurusapa Ladprao Publishing)

[20] National Research Council 1996 Lost crops of Africa: volume I: grains (Washington, DC: The National Academic Press)

[21] National Science Teachers Association (NSTA) 2005 Developing a world view for science education: In North America and across the globe; Final Report of the International Task Force

[22] Tairab Hassan H "Pre-service Teachers' Views of the Nature of Science and Technology before and after a Science Teaching Methods Course" **Research in Education 65** pp 81-88

IOP Conf. Series: Journal of Physics: Conf. Series **1340** (2019) 012085 doi:10.1088/1742-6596/1340/1/012085

Teachers as Makers: The Key Provision of Teacher Preparations for STEM Education

Suthida Chamrat

Department of Curriculum, Instruction and Learning, Faculty of Education, Chiang Mai University

Email: suthida.c@cmu.ac.th

Abstract. This study explores how learning activities based on maker concepts prepared pre-service teachers entering the STEM Education pipeline. 25 pre-service science teachers from the faculty of education at Chiang Mai University enrolled in an Independent Study in Science Teaching course. The course was called "Integrating maker concepts and STEM Education." The course sought to integrate existing science objectives with maker concepts and activities. Pre-service teachers were divided into 8 groups. Each group had to write and present a proposal, develop their instruments/Makings, try it out in schools, submit full reports, present and review peers' final reports, attend a regional STEM festival where they participated in poster presentations. Using the collected data from a shared Google Drive, Video recordings, interviews, and extended activities, the findings indicated that STEM concepts and practices could be conveyed to pre-service teachers by instilling the maker mindset. From the 8 groups, 3 themes emerged: research in science education, science project, and STEM Education which consisted of 1, 1 and 6 projects, respectively. The creation of artefacts or Makings required scientific and mathematical concepts, empirical testing, and technological design. These enabled pre-service teachers to understand and practice teaching and learning in the STEM field. Event case study was employed as research methodology. Details of 8 pre-service teacher projects and 5 assertions of research findings are described.

1. Introduction

STEM and STEM education has been recognized as a crucial way of creating in Thailand an innovative economy. The Thailand government declaration in 2016 of a new economic and a social model, a policy known as 'Thailand 4.0', makes STEM and STEM education even more important [10]. According to the Thailand Ministry of Education, STEM education will be implemented at every level, from basic education to higher education. The Ministry aims to promote a STEM workforce, as well as develop STEM literacy for students [3]. By using STEM to reform learning in schools, it is expected that more students will enroll in STEM pipelines. Consequently, concerns about the preparedness of teachers, along with the concomitant concerns for teacher education are growing, for teachers are clearly the most critical factor in the future of STEM education. Research shows that teachers influence the educational pathways of students in the STEM fields [14]. However, teacher education still lacks sufficient STEM preparation for pre-service teachers; even for in-service teachers, STEM concepts are frequently left ambiguous [9].

Among the feasible ways to prepare teachers for the STEM education pipeline is to integrate maker concepts into current teacher education, as proposed by such researchers as [4,5] and Martin [11]. The

Maker Movement was developed by researchers who advocated its usefulness to support STEM education [8]. Dougherty [4] and Martin [11] described Maker Movement or Maker Concepts as a conceptual framework which consists of "a class of activities focused on designing, building, modifying, and/or repurposing material objects, for playful or useful ends, oriented toward making a ''product'' of some sort that can be used, interacted with, or demonstrated. Making often involves traditional craft and hobby techniques and it often involves the use of digital technologies" [11]. The Maker Movement has been considered as a transformative tool which alters how and what people learn in [13]. Bullock [2] integrated Maker Movement into learning and teaching in the classroom. However, he used the term "Maker Pedagogy" and explained it as an approach that utilizes the principles of ethical hacking (i.e., deconstructing existing technology for the purpose of creating knowledge), adapting (i.e., the freedom to use a technology for new purposes), designing (i.e., selecting components and ideas to solve problems), and creating (i.e., archiving contextual knowledge obtained through engaging in the process of making, as well as the actual tangible products) as part of an overall way of working with those interested in learning about science and technology.

By installing maker Mindsets, pre-service teachers engage in topics they are interested in, which encourages tinkering, playing or exploring stuff that concerns them. This study adopted Maker Concepts and applied them to existing course objectives for pre-service teachers. The results showed how pre-service teachers understand and practice STEM education. This research was guided by the question, "how can we best prepare pre-service teachers to be future teachers of STEM Education?" The research method is described below.

2. Method

2.1. Context of the Study

The context of the study was an existing course for science teacher education called "Independent Study in Science Education". Pre-service teachers enrolling in this course were biology and chemistry majors. Of the 25 pre-service teachers, 15 were female. They participated in this research by purposive sampling. The course ran from January - April, 2016. The extended activities finished in July when pre-service teachers participated in poster presentations at the annual STEM festival in Chiang Mai, Thailand. In total, the study took 7 months for research, implementation and data collection.

2.2. Data collection

Google Drive was used for data collection and was shared between teachers and pre-service teachers. The data was compartmentalized and uploaded into folders such as Student hand-outs, class attendance, PowerPoints, Google Slides, Google Forms, pre-service teachers' initiatives, independent study proposals, final reports, posters for presentations, activity photos and videos. Video recordings and group interviews were also used. Overall, the study was designed to examine pre-service teachers' understandings and practices of STEM education based-on maker concepts.

2.3. Data analysis

Event case study [1] was used as a theoretical framework of research design in this study. Data from the shared Google Drive (teacher documents, pre-service teachers' hands out, pre-service teachers' presentation, artifacts, etc.), Video recordings and group interviews were qualitatively analyzed using thematic analysis as described by Fereday and Muir-Cochrane [7].

3. Findings

After discussing a variety of trends in science education in Thailand and international, pre-service teachers focused their interests on 3 main categories: science projects, scientific concepts and STEM education. Each category was identified by student themselves. During the IS (independent study) project proposal presentation, which took place before midterm examination, pre-service teachers were asked to explain their ideas on their IS projects, and how these projects related to current trends in science education. They were also asked to give the reasons why their work should be categorized into one theme or another. The projects were divided, based on the decisions of the pre-service teachers, into 3 themes: Science Projects, Research in Science Education and STEM education. IS Projects that were categorized in the theme "research in Science Education" focused on exploring and developing students' understandings of the mechanisms of sea breezes and land breezes. Pre-service teachers developed conceptual tests that they later administered to students in schools. However, their work demonstrates some degree of maker concepts as shown in Figure 2. The students studying the concepts of sea breeze and land breeze, designated Group 1 (G1), decided to make learning materials for developing concepts of sea breeze and land breeze. Only one of eight groups chose the theme of Science Project. This group was eager to study the effects of lights' wavelength on plant growth. The other 6 groups, developed new material that could benefit teaching and learning science. Makings developed by Groups 1- 8 (G1-G8) all required scientific and mathematical concepts and practices; however, only groups G3-G8 clearly stated that they had integrated engineering and technological designs into their Makings. Hence, G3-G8 explicitly addressed the stated aim of Stem education for integrating technology, engineering and scientific concepts. The pre-service teachers' projects are summarized in Table 1. Examples of pre-service teachers' makings are presented in Figure 1.

Table 1. Summary of Preservice teachers' IS projects

IS project Themes	IS Project Title	Focusing Concept	Makings	Instruments
Research in Science Education (n=1)	G1. Study and development of Sea breeze and Land Breeze Concepts	explanation of sea breeze and land breeze phenomena	Interactive paper board to explain the causes of sea breeze and land breeze phenomena	Concept Test
Science Project (n=1)	G2 Plant growth and Light color	Effects of wavelength on plant growth	Design of plant cover filter	-
STEM Education (n=6)	G3. The Easy Consistometer	Viscosity and measurement by Consistometer	DIY Consistometer from Acrylic	-
	G4 Mobile-Len for Biology Class	Study of Cell using microscope	Adjustable focus ranges mobile phone Len	Interview protocol
	G5. Holographic Display of Molecular Structure Learning	Explanation of molecular structure using VSEPR theory	Holographic pyramid and 4 planar molecular VDOs	Interview Protocol Video Recording
	G6. Behind the Scene of Rosalind Franklin	The discovery of Double helix structure of DNA	Projector from shoe box that can project spiral wire shadow on screen.	Open-ended questionnaire
	G7. Electron Configuration Model	Electron configuration in atoms using Aufbau principle	Rotatable orbit-electron model made from Acrylic plastic	- Questionnaire - Interview Protocol
	G8. Functional Heart Model	Anatomy and Function of Human Hearth	DIY functional heart model	Concept test

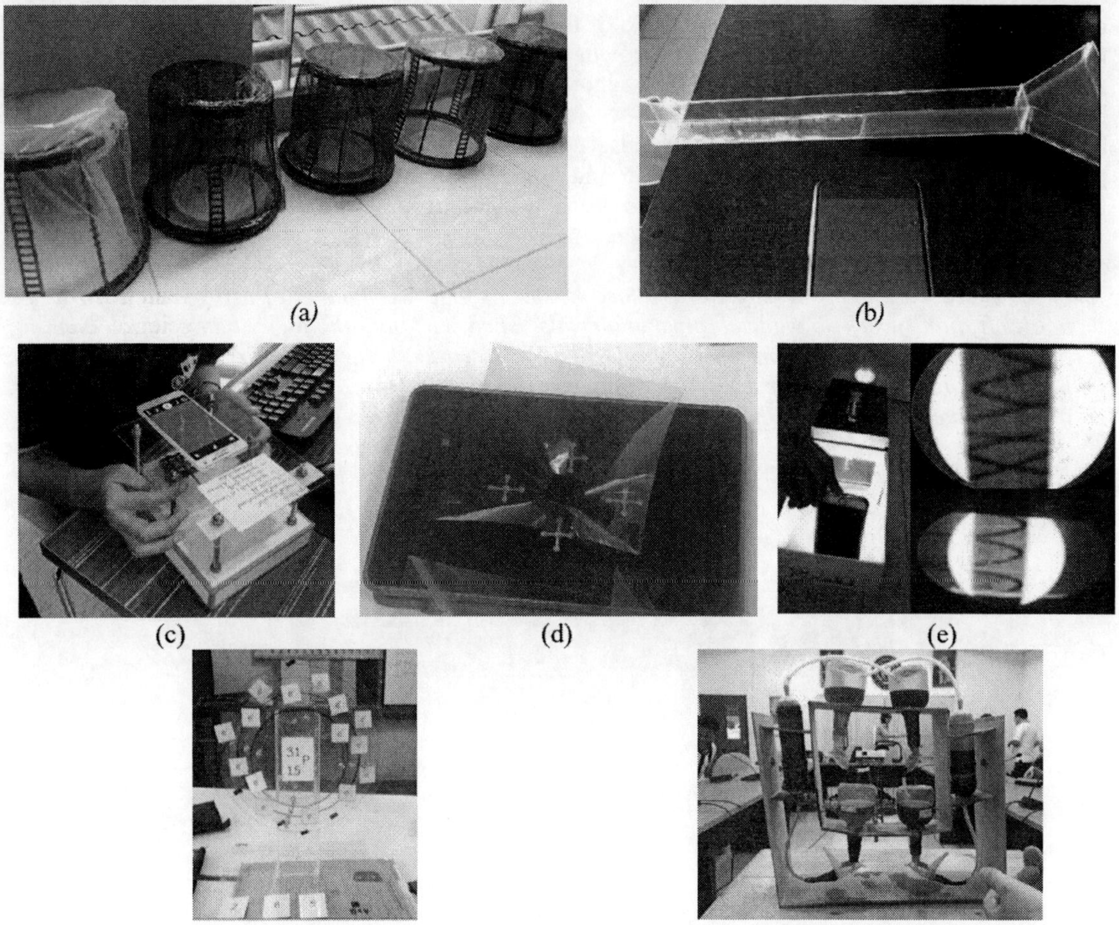

Figure 1. Makings in science project *(a)* G2 and STEM education themes *(b)* G3 *(c)* G4 *(d)* G5 *(e)* G6 *(f)* G7 *(g)* G8

Furthermore, there are 5 issues that emerged from the data analysis. These are follows.

3.1. Giving pre- service teachers a choice based on their interests will enable a higher success rate and higher quality of project, one which better reflects their understanding and practice of STEM education.

Interestingly, given that the pre-service teachers had been allowed to choose their own topics, only one group (G5) changed their title and concept of their project at a very beginning The data from the Google Form asked the pre-service teachers 3 times if they were content with the topics they had choosen : at the beginning, during the process and near the end of the project, prior to submitting their final report. Project titles focus concepts and statement of problems were self- chosen. Hence, most students stated clear goals and obvious processes to accomplish their projects.

3.2. Instilling a Maker mindset in pre- service teachers is excellent preparation for the STEM education pipeline.

During the study, data gathered from personal and group interviews indicated that the highest quality projects, as voted for by their peers, reflected important aspects of Maker Mindsets. These

IOP Conf. Series: Journal of Physics: Conf. Series **1340** (2019) 012085 doi:10.1088/1742-6596/1340/1/012085

aspects can be grouped into the following 5 categories: 1) these groups worked collaboratively to create Maker Spaces; 2) they were not afraid of failure; 3) they loved tinkering; 4) they choose topics they loved to do; 5) and they embraced a Growth Mindset as described by O'Brien and Harlow (2016). During classroom presentations of the final reports, pre-service teachers were asked to tell the stories behind their works. The findings from pre-service teachers when asked what they thought were the best projects proved consistent with the mindsets of makers – i.e., numbers 1 to 5 above. The top three projects as voted for by peers were G4, G5 and G8 (Figure 2). These three groups all had Maker Mindset in common. One of the members from G4, the group which successfully constructed a functional heart model, told her story to the class:

> *"There were many versions of a heart model. At first, we thought of using a squeezing bottle. It didn't work so we tried other materials. Then we found the foggy bottle which can pump the colored water from one to another bottle. It worked! But it needs to be sealed by a hot glue gun. We also constructed the structure using plywood that we bought from a building and construction materials shop."* (the pre-service teacher from G8 group)

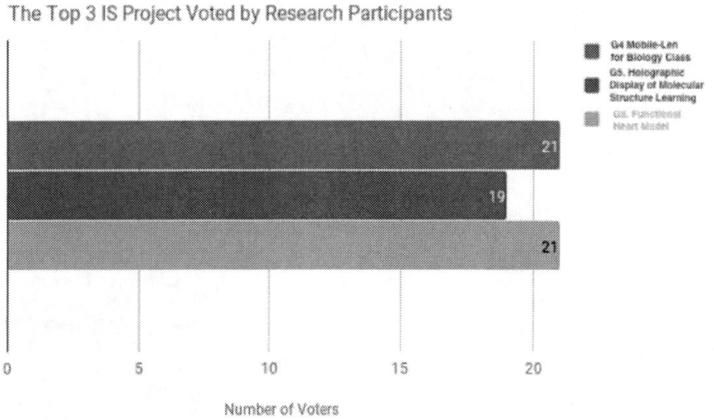

Figure 2. Number of pre-service teachers that voted for the top 3 IS project

3.3. *The complete cycle of a maker pathway is necessary in order for pre-service teachers to experienceSTEM education in areal context.*

The preservice teachers participated in a full cycle of maker pathway. They wrote and presented a proposal; developed their instruments; tried out their learning innovation in schools; submitted a full report; presented their final report to their, as well as reviewing their peers' final reports; finally, they attended a regional annual STEM festival and participated in poster presentations. These processes are portrayed by the diagram in Figure 3. The entire study lasted 7 months: 4 months of the semester, in addition to 3 months of extended activities outside the university, including time spent in classrooms, as well as at the annual STEM festival. There was no single process. Rather, students were exposed to authentic classroom environments, for it is here where they must develop their competencies. Additionally, it is as in-service in the classroom teachers where they will continue to develop the design and implementation for STEM Education

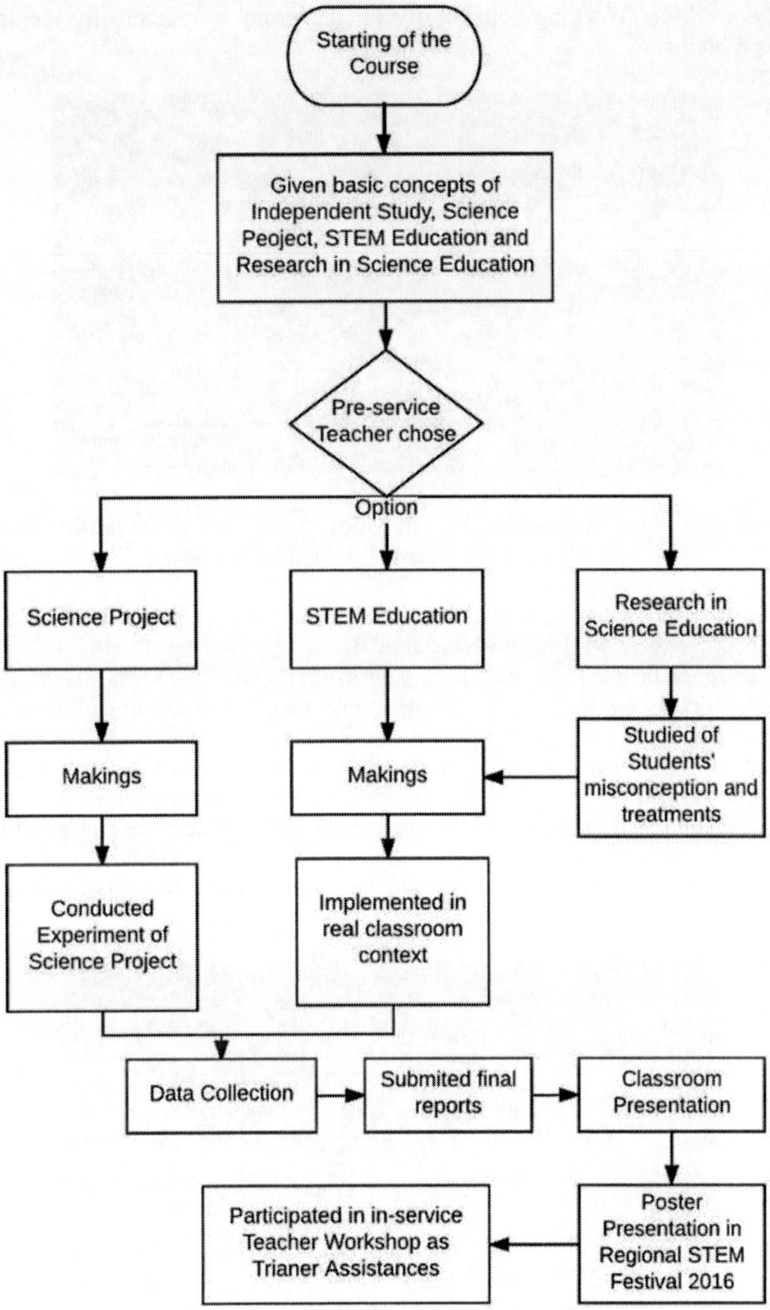

Figure 3. The complete cycle of pre-services' Teachers as Makers pathway in this study

3.4. Maker Concept is a continuous spectrum of progression

The groups showed progressive use of maker concepts. For example, when G8 constructed their Functional Heart Model, they had to construct their design according to a variety of factors, and were challenged to turn a paper design into a physical model. Figure 4 illustrates the original version of their Functional Heart Model. But in the original model, the water pressure caused leakage, as can be clearly seen. Similarly, the color indicator representing low and high oxygen levels in the blood, seemed not to be working. Working collaboratively, the group systematically identified, diagnosed and solved both problems.

Figure 4. The original version of Functional Heart Model (G8), later was modified to a better working version which addressed the deficiencies.

From a conversation with the researcher, the pre-service teacher showed progress in her way of thinking as well as progress in working with her model.

"I tried to solve these problem for 4-5 days. There were many problems during the process. The indicators now are working but the problem of water pressure is still unsolved. The first thing to fix is a valve. The second thing is the flowing system that I cannot control. When it circulates, the water pressure creates leakages at many points. I must address two issues in order to repair this: first, I must find more effective valves, and then I must separate the water system so that I can control the water direction" (Conversation between the research and pre-service teacher from G8 group, Facebook Messenger, 9/5/2016 20:43)

After re-engineering the model, Group 8 brought the model into the classroom for teaching and learning about blood circulation (Figure 5).

Figure 5. Pre-service teachers prepared their final, re-engineered version of a Functional Heart Model

Even after the final version of model had been used, the need for reconstruction remained. On July 27-29, 2016, all pre-service teachers attended the annual STEM Festival 2016. Many of their artifacts were still works in progress. The Maker concepts focused on how the projects gradually developed and progressed during the process. This represents one of the characteristics of Maker which also rely on the conceptual framework of Growth Mindsets.

3.5. Maker concepts does not depend on gender

The data showed no difference in term of quality of makings between males and females. The high-quality groups, G4, G5 and G8 (presented in Figure 2) were composed of 8 females and only one male. There was all female (G4 and G8, 6 females in total) in the best group voted by their peers. There were 2 females and 1 male in G8. Data gathering from the interviews during final report presentations, all members shared certain characteristics in common. All have strong academic backgrounds in biology for G4 (Mobile-Lens) and G8 (the Functional Heart Model). Members of G5 (Holographic Display of Molecular Structure Learning) had a strong chemistry background. However, a strong academic background is only one factor for success. Much depends on how pre-service teachers develop a Maker Mindset, especially, Growth Mindsets.

4. Discussion

The findings showed that pre-service teachers from group 6 to 8 could adopt maker Concepts. The rest were in the beginning steps of Maker Concepts which posed some Maker Mindsets. This confirmed the existing research which shows much promise for Maker movements arising from STEM education. Maker Mindsets appeared to be the most important factor for pre-service teachers developing their understandings and practices of STEM education. Starting from selecting their own topics through to presenting s their Makings to the public at the STEM festival, the pre-service teachers who reflected STEM concepts and practices also reflected Maker Mindsets. They had clears topics and concepts at the very beginning of the course. Further, they loved their topic (G4.-G8). They developed multiple versions of their Makings (G4, G6, G8). Learning progression [6] was also crucial for developing Maker concepts and STEM Education. With integrating Maker Concepts into course activities, gender showed no effects on the quality of work. Female pre-service teachers were equal to males (G3 G5 G6), or even better if judged by the tenets of Maker Mindsets (G4 G8).

5. Conclusion

Maker concepts that use science, technology and mathematics, together with the design process, further the understanding of core STEM subjects. The results of this study support maker concepts as an important role in STEM education as described by Peppler and Bender [13] and Martin [11]. Key components of the maker mindset need to be addressed and integrated into learning activities of science education, in particular in preparing science teachers. Pre-service teachers must be able to choose their own topics of study. They need to have opportunities to explore the real context in classrooms. Further research topics include how to effectively encourage Maker Mindsets in pre-service science education programs, and the long term implications of the Maker Movement or Maker concept for STEM teachers and the advancement of scientific understandings in students. For recommendation, STEM teacher preparation can be readily implemented within the existing course structures of science education for pre-service teachers by, for example, the implementation of Independent Study courses. It is strongly urged that teacher preparation programs give consideration to the many advantages of the maker movement for science education.

IOP Conf. Series: Journal of Physics: Conf. Series **1340** (2019) 012085 doi:10.1088/1742-6596/1340/1/012085

References

[1] Bromley D B 1990 Academic contributions to psychological counselling: I A philosophy of science for the study of individual cases **Counselling Psychology Quarterly 3** pp 299–307

[2] Bullock S M and Sator A J 2015 Maker pedagogy and science teacher education Journal of the Canadian Association for Curriculum Studies 13 pp 60-87

[3] Committee on Mass Communication, Science, Technology and Information 2015 Report of STEM Education Proposed Policy: Active Policy for Youth and Workforce Development in Science Technology Engineering and Mathematics. [Online]. http://library.senate.go.th/document/Ext11101/11101417_0003.PDF (September 6, 2016)

[4] Dougherty D 2012 The maker movement **innovations 7** pp 11–14

[5] Dougherty D 2013 The maker mindset. Design, make, play: Growing the next generation of STEM **Innovators** pp 7–11

[6] Duschl R A and Bismack A S 2016 Reconceptualizing STEM Education: The Central Role of Practices (Routledge)

[7] Fereday J and Muir-Cochrane E 2006 Demonstrating rigor using thematic analysis: A hybrid approach of inductive and deductive coding and theme development **International journal of qualitative methods 5** pp 80–92

[8] Honey M and Kanter D E 2013 Design, make, play: Growing the next generation of STEM innovators (Routledge)

[9] Lederman N G and Lederman J S 2013 Is it STEM or 'S & M"that We Truly Love? **Journal of Science Teacher Education 24** p 1237

[10] Maesincee S 2016 Thailand 4.0 Thriving in the 21st Century through Security, Prosperity & Sustainability. [Online]. http://www.ait.ac.th/news-and-events/2016/news/1thailand-4.0-english-dr.-suvit.pdf (November 20, 2015)

[11] Martin L 2015 The promise of the Maker Movement for education **Journal of Pre-College Engineering Education Research (J-PEER) 5** pp 30–39

[12] O'Brien S, Hansen A K and Harlow D B 2016 Educating Teachers for the Maker Movement: Pre-service Teachers' Experiences Facilitating Maker Activities In Proceedings of the 6th Annual Conference on Creativity and Fabrication in Education ACM, October, pp 99–102

[13] Peppler K and Bender S 2013 Maker movement spreads innovation one project at a time **Phi Delta Kappan 95** 22–25

[14] Wang M T and Degol J 2013 Motivational pathways to STEM career choices: Using expectancy–value perspective to understand individual and gender differences in STEM fields **Developmental Review 33** pp 304–340

IOP Conf. Series: Journal of Physics: Conf. Series **1340** (2019) 012086 doi:10.1088/1742-6596/1340/1/012086

Grade 11 student's mental model of the Nature of Light

Tevarit Chanserm [1], Jiraporn Tupsai[2] and Chokchai Yuenyong[3]

[1] Ban Tha Muang School, Buriram, Thailand, Email: ong_phy@hotmail.com

[2] Khon Kaen University (KKU) Demonstration School (Suksasart), Khon Kaen, Thailand, Email: tjirap@kku.ac.th

[3] Science Education Program, Faculty of Education, Khon Kaen University, Thailand, Email: ychok@kku.ac.th

*Corresponding author's e-mail address: tjirap@kku.ac.th

Abstract. This study aims to explore Grade 11 Student's Mental model of the Nature of Light with the integration of learning about light. The participants were 50 students in Ban Tha Muang School, Buriram, Thailand, Province Thailand. The methodology is in the respect of interpretive paradigm by opened question about Nature of light and interview. The results were student could to explain primary light sources more than secondary light sources because they could students easily find primary light sources in everyday life. Student could to model of understanding mental model about moving light. They could to explain light travels straight out of the light source and travel because its intermediaries because light was electric wave including understanding Mental Model. They could to draw and explain light travels straight from the sun through the vacuum and the atmosphere of the earth and there are some students who paint a wave of light by drawing a wave like a wave in a rope. They could to explain mental model of visibility as see with our eyes, however, there are only some students who associate the impact and reflection to explain the vision. This paper also discusses teaching science about the student's mental model of Nature of Light in physics as a guideline for those who are involved in the development of learning, focusing on learners who learn to light.

1. Introduction

Today, national is expected to be a scientifically literate person by means of being able to understand real life events, having the happiness of knowledge about the natural world, applying scientific processes in personal life, asking and suggesting reasonable solutions for questions, make decisions, able to use technology, and be good at social communication. Physics is such a branch that any event in real life may be explained with the help of real life, which can also encourage student curiosity and make sense why they learn about chemistry. Students encounter many real-life events from an early age and they relate to what they already know and construct coherent understanding of these events. During in physics learning process, students often experience difficulties and lack conceptual understanding of basic concepts [1], [5], [6], [9]. Students usually tend to learn factual knowledge presented and tries to grasp the knowledge that they just need for exams, without making any meaningful connections to what they already know or constructing an understanding of the underlying concepts, they only hold their own coherent understanding based on their prior knowledge and

IOP Conf. Series: Journal of Physics: Conf. Series **1340** (2019) 012086 doi:10.1088/1742-6596/1340/1/012086

experience [6], [10]. In the case of Thailand high school students, students are mostly focused on the university entrance examination and performance rather than learning, the important points for them to take high grades in exams, make good memorization to answer the questions rather than inquiring for scientific facts.

In the current study, Mental model of the Nature of Light were investigated, which is one of the fundamental topics in physics and easily influenced by daily life and experiences [3]. Students' understanding of learning usually is assessed by traditional assessment techniques such as multiple-choice questions, which are often more preferable in science classes since they are easy to apply. Even students are successful in conventional tests, this success does not reveal that the students' understandings of given concepts. A student could get a high score by chance and luck factor; which does not reveal meaning learning. In order to assess students' ideas on nature of light, one of alternative assessment techniques, was used, which included daily life events that students might find interesting to learn.

Norman also speaks of conceptual models. In the article "Some Observations on Mental Models Norman [7] distinguishes between mental models and conceptual models: Conceptual models are devised as tools for the understanding or teaching of physical systems. Mental models are what people really have in their heads and what guides their use of things." In other words, the designer designs a conceptual model into the system in order for it to appear graspable and coherent to the user. If she manages to get the conceptual model right, the correct mental model (in the mind of the user) will follow. At least in theory. Norman's account is of course an over-simplification, but it captures some central dilemmas of interaction design by using mental models as a vehicle for explanation and illustration. The study ideas of existing student's exploration, which is between *10-7* years old, about the light source. Light and vision Found that small students [8]. They can tell their original ideas about many primary sources such as the sun, the light, the flashlight, the light, the TV, but the secondary light source. And may be told secondary light sources in the manner of windows, ceiling, doors, sky, moon, students can relate to the origin. Primary and secondary lighting is very rare. The idea of light travel is related to the idea of a light source if we see light. This idea is hard to understand because it is not visible. The visionary has few students who can understand the scientific explanation of the sight that light must leave the object in our eyes, that could see the object. The reports on the third year of a three-year longitudinal investigation into six secondary students' understanding of optics at a secondary school level [4]. In the third year of this investigation the students, who by now were in Year 12, underwent a teaching sequence that centered on the teaching and learning of physical optics and quantum ideas. The students' mental models of the nature of light were explored prior to, and following this teaching sequence. The researcher took on the dual roles of teacher and researcher. This paper will outline the findings of the third year of this study and the implications they have for the teaching and learning of optics at secondary school level. Concepts related to events students directly or indirectly experience vary according to their structures and the way they occur. In this process, while it is always possible for some concepts to be experienced in daily life, for some it is impossible to understand without having prior knowledge on the subject. This paper also discusses teaching science about the student's mental model of Nature of Light in physics as a guideline for those who are involved in the development of learning, focusing on learners who learn to light in Thailand [12], [13].

2. Methodology

The research regarded interpretive paradigm. Grade 11 student's mental model of the Nature of Light which has been taught by the first author for 16 periods in a month. The participant questionnaire observation and interview, students' discussion and questions, worksheets, and unstructured interview were interpreted to examine mental model of the Nature of Light [3], [12], [13].

2.1 Target Group
The participants were 50 Grade 11 students in in Ban Tha Muang School, Buriram, Thailand

2.2 Research Focus
To this point, studies on light have been conducted on students' mental models and conceptual difficulties concerning such phenomena. The motivation behind this study is to disperse the lack of studies on different physical phenomena related to light. The other phenomena mentioned above are explained by the particle model of light. Enchanting its cue from these ideas, the purpose of this study is to understand primarily the students' mental models of light to explain generation light source, motion of light and visibility of light. Also, the aim was also to show how students use light-related models in explaining physical phenomena. To this end, questions used as data gathering tools were prepared for each context in accordance with the nature of the study. As such, this study puts forth in detail the different forms of models used by students related to physical phenomena. Thus, this study attempted to answer resulting questions: What mental models do students have about the nature of light to explain various phenomena?

2.3 Data Analysis
The data obtained from the question were analyzed by questionnaire to quantitatively by the researcher. As the aim in this study was to determine Student's mental model of the Nature of Light, during the analysis process different contexts were analyzed among themselves, and no generalization was made in students' reasoning with other con- texts. Therefore, students' mental models for each context were determined and as such whether models are context-dependent. First of all, frequently used drawings and descriptions were gathered under three different mental models to explain generation light source, motion of light and visibility of light.

3. Findings

3.1 *Students' mental models about the concept of generation light source*
As a result of the questionnaire, most students have a Mental Model about the light source, rather than drawing or drawing only. And most students know the source of light in daily life. They could tell more than seven light sources. They would to tell of primary sources light and able to tell the Mental Model about a large number of primary sources such as sun, light, flashlight, candle light, etc. However, only a small number of students could describe secondary light sources. Which students usually tell the secondary light source in the manner of the earth, moon, window, etc., as shown in Figure 1.

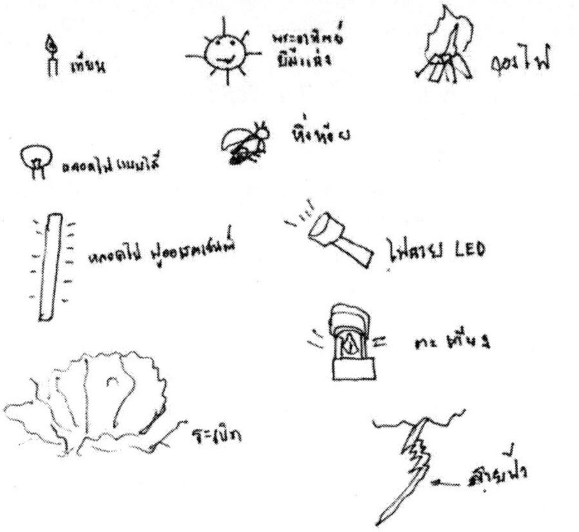

Figure 1. Mental model light source

Student could explain the primary source of light rather than the secondary source of light. They observe the light directly from the source and it can be seen from the drawing or drawing a lecture. They could draw lines out of the light source for a short time. Especially the sun and light bulb, etc. When asked the reason for the interview, students will say "It was painted this way". Students may have seen examples from comic books or from the imagination of students who think that if the bulb goes out, there will be no line from the lamp to differentiate from the light bulb.

They could tell about source of the light emitted by the heat. Incandescence was more than the light source emitted by fluorescence. And why students could tell the light source caused because these light sources can be easily found in everyday life. They were knowing familiar with this source of light ensample in Figure 2. They could to tell the sun, candles, fire, matches, etc., most students can lift up to five examples of this kind of light source, especially the sun. All students can say that the light source caused by the light emitting heat. However, there are some people who can give examples of light sources emitted by glow such as fireflies, fluorescent lights, etc.

Figure 2. Mental model of student of sources light

Most students are unable to link the relationship between primary light sources and secondary light sources. Teachers prepare to provide the correct main ideas about the different types of light sources about primary light sources and secondary light sources. This provided light by reflecting light from one primary light source for students to take advantage of the study of physics in the geometric light in the next order.

3.2 *Students' mental models about the concept of motion of light*

With the first question given below, students' mental models of light were determined for Question. First question: How were you having an idea about the journey of light? Student thought are related to the idea of light source, if They would see the light. This idea is difficult for students to understand because they do not see it. The answer to that question. Second question: How does sunlight come to the world? Some students answer the question by drawing a drawing or drawing a lecture. The answer of the students has the correct understanding and know that light is electromagnetic, so light does not need to rely on moving medium. Some students respond that the sun shines through the atmosphere through the atmosphere. The light moves from the sun to the Earth as shown in Figure 3 (a). There are some students who describe the earth's light through the world without moving media, linked to reflection and refraction. Students explain that the sunlight is pushing the air down. If the object or object is subjected to collision, refraction and refraction will take place.

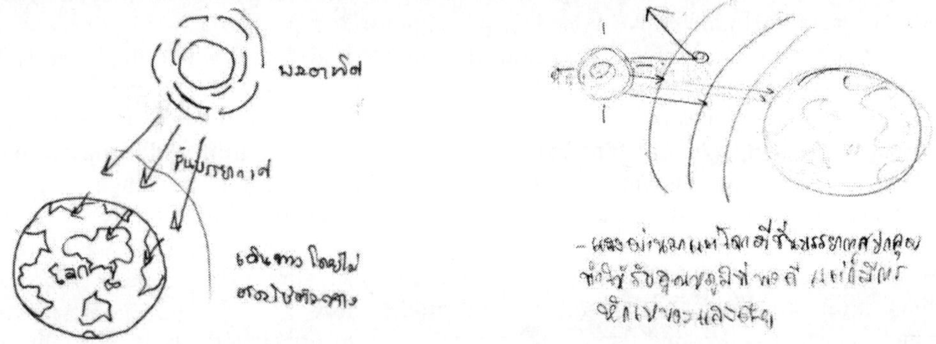

(a) The sun will shine as go down to the Earth. (b) Light form the sun pass environment

Figure 3. Mental model of light way

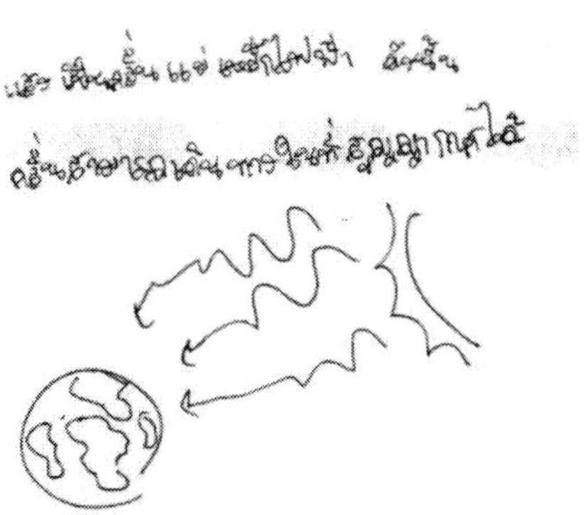

Figure 4. Light was wave

(a) Student writing arrowheads instead of light. directions.

(b) Student writhed without arrowheads. instead of light.

Figure 5. Light travels in a straight line.

In addition, some students use their own understanding patterns to convey the wave of light by drawing. As the wave in the rope, shown in figure 4. Student's caption illustrates did the direction of the arrow. The light of the sun could write lines that were not straight lines instead of light waves. It was important to know that the student knows that light travels straight but arrows. But it was not straight lines, because the light is wave. Also known as light waves.

Interesting points from this study, some students will not understand where the beam is coming from. They did not understand the origin of the sound but students say that the beam is from the sun. According to scientists, the beam was sent down [8], [10], [11] describes the issue. In this way, perhaps the scientist might have been thinking about the journey of light from childhood as well.

From Figure 5 (a), most students visualized the passage of light in a straight line, and the arrowhead indicates the direction of light. To indicate that the directional light ejected from the light source. However, only a small percentage of students draw the path of light in a straight line without the direction of the arrow. The direction of light was shown in Figure 5 (b). The idea of the scientist was that light travels straight out of the light source in all directions.

Some students could understand the scientific explanation of light. Students see the light must be reflection from the object and it could saw the object. In research the questionnaire: How do students see the book? And how did students see candles? They were describing, write, or draw, their own free speech. Based on data analysis. The results show that students had shown model of vision as that can be divided into 2 subjects, which corresponds to research [8] Details were as follows:

: Students present ideas as images but didn't explain the meaning of the image.

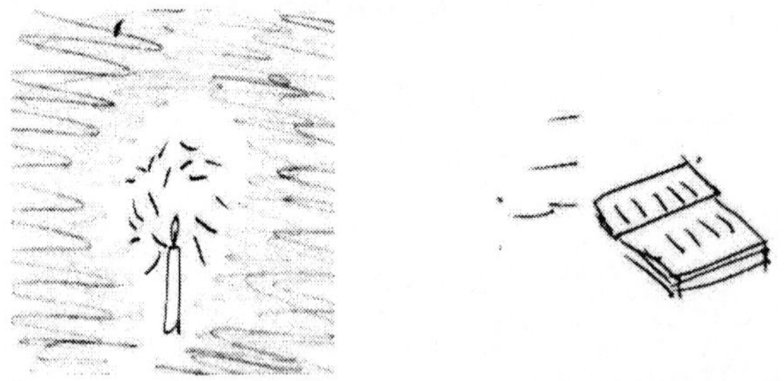

Figure 6. Mental model of vision, which wasn't explain image.

Analyzing data from questionnaire that the visualization of secondary light source was related to the reflection of light. It was difficult for some students because they couldn't explain objects that need to reflect light from the light source. Another eye to saw the object for some students who cannot explain, write, lecture or draw a picture. They would simply explain that when we look at any object. They explain saw that the object was unconnected, that the reflection of light of the visible part. Then draw a picture showing how the student sees the book. What does the candle look like? No explanation or rationale. Details were as follows:

: Students present ideas as images but didn't unconnected the meaning of the image.

Figure 7. Mental model to see the object was unconnected.

Conduction form figure would represent knowledge about light of some student. They represent model of light as necessary went they saw object. They would to draw object or sources of light as show concept of student represent as light help to saw object.

3.3 *Students' mental models about the concept of visibility of light.*

The finding of research has student' mental models about the concept of visibility of light. First, several students could show mental model as who draw line connect between eyes and object to one step. But model wasn't complete because who didn't draw arrow of line of light. Also, example in figure 8(a) and some student could to connect line of light between object and eyes show in figure 8(b). The finding of research, several students would to draw model of light as line direct and had some student draw model of light as point.

(a) **(b)**

Figure 8. Mental model's student about seeing the objects as connected.

Second, several students could show mental model as who draw line connect between eyes and object to two steps. Show in figure 9(b) were concept about in line of light help to saw object because the light reflects form object and light move to eyes. And some time, we saw the misconception of student for example show in figure 9(a) by mean of student draw direction of light go away form eye onto sources of light. They gave respond as eyes will be staring at an object. Actually, mentioning the light, it must leave the light source. However, we found that most students have the ability to write a light path in a straight line. There was a direction from the object to the eye which was described by correct concept.

IOP Conf. Series: Journal of Physics: Conf. Series **1340** (2019) 012086 doi:10.1088/1742-6596/1340/1/012086

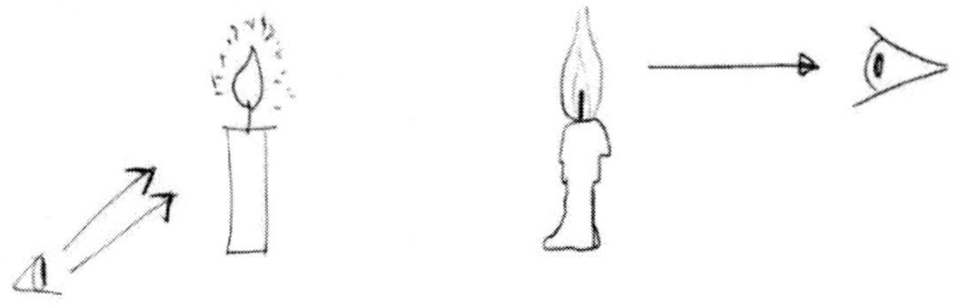

9 (a) 9 (b)

Figure 9. Mental model direct of light.

Third, the finding of research explained to connect way of light 2 step. All most student could to draw line of light connect between object source of light and eyes. They claim understanding as "Light must use help to see" and "We need to use eyes to see." The sentence interpretive understand of student to connect in several as mental models involve light. Student explained understanding about to see of object were to without the direction of light from the light source but can tell that light was needed for vision. For example, in figure 10(a)

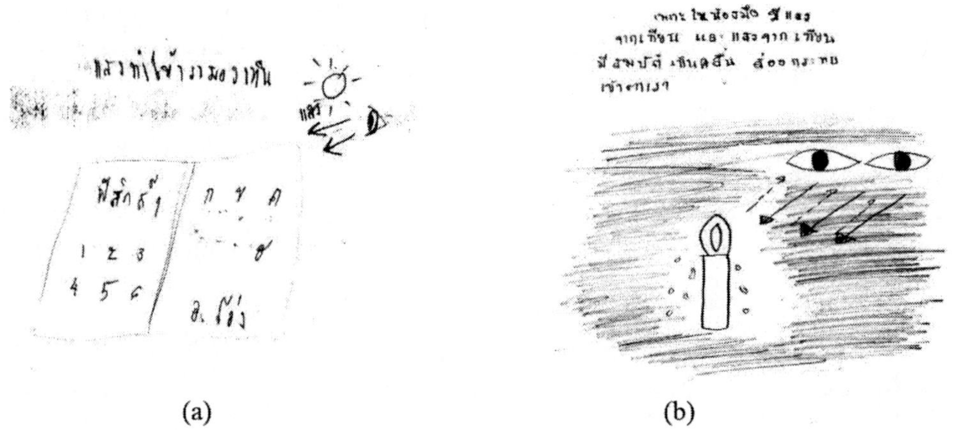

(a) (b)

Figure 10. Mental model of student about to see object as show connect two steps.

Form figure 10(b), represent mental model of student connection thinking and knowledge explain light moving to object and light leave go to eyes by draw line of light. We had understanding some student form mental model about view same scientist as property of light. We can see it because the light struck the object and reflected go to our eyes for example in figure 11.

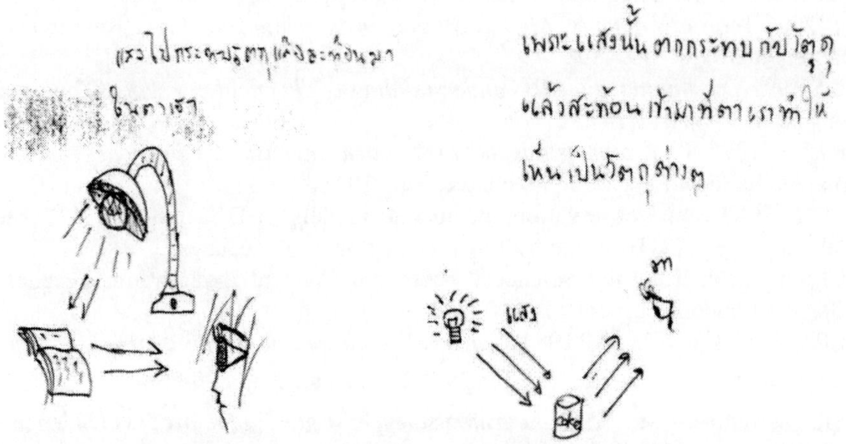

Figure 11 Mental model of line of light.

The results of the student's explanation of the vision follow were most of the students explained the sight of the book different from seeing the flame from the candle. They thought that the candles sent light directly into the eyes in fact, the book couldn't have sent light to eyes. They used light distribution features to explain their vision which makes the description of the vision deep to the point of seeing of various objects. For example, light white in the nature take on object appear reflection of light, absorption of light, and transmission of light. If object could reflection of light all wavelength in equal then it's would to white light. The object absorbs all the light, we would see that the object is black because there was no light from that object reflect out to our eyes. If the object didn't absorb the not equal light in each wavelength when the white light hits, the object would appear as that color. For example; Apple had rad color if when light impact to it could to absorb blue color and green color, then it could to reflection red color out to our eyes.

4. Concluding and Discussion

The result of questionnaire and interview regarding nature of light were study to grade 11 student's mental model of the nature of light to detail follow in below. Frist, student could to explain five primary sources of light and secondary sources of light. Students' mental models about the concept of generation light source. Students know the source of light in daily life involve primary source of light and secondary source of light. They allude to incandescence, luminescence and incandescence. Second, mental models of student about the concept of motion of light who describe the earth light through without moving media, linked to reflection and refraction. They would to tell way of light source to direct to electromagnetic wave it can pass atmosphere of world by draw picture and write to explain understanding. And third, mental model of the vision follow was light help to see object. They thought that the candles sent light directly into the eyes in fact, the book couldn't have sent light to eyes. They used light distribution features to explain their vision which makes the description of the vision deep to the point of seeing of various objects. They could to explain light moving to object and reflect to our eyes by draw picture to understanding.

IOP Conf. Series: Journal of Physics: Conf. Series **1340** (2019) 012086 doi:10.1088/1742-6596/1340/1/012086

References

[1] Ben-Zvi R Eylon B S & Silberstein J 1986 *Is an atom of copper malleable?* (Journal of chemical education vol 631) pp 64

[2] Borges A T & Gilbert J K 1999 *Mental models of electricity International* (Journal Science Education vol 21) chapter 1 pp 95-117

[3] Chantaranima T & Yuenyong C 2014 *Grade 11 Students' Existing Ideas about Natures of Light* (Procedia-Social and Behavioral Sciences vol 143) pp 693-697

[4] Hubber P 2006 *Year 12 Students' Mental Models of the Nature of Light* Research in (Science Education vol 36) pp 419-439

[5] Krajcik J S 1991 *Developing students' understanding of chemical concepts* (The psychology of learning science) pp 117-147

[6] Nakhleh M B 1992 *Why some students don't learn chemistry: Chemical misconceptions* (Journal of chemical Education 69) chapter 3 pp 191

[7] Norman D N 1983 Some observations on mental models In D Gentner & A L Stevens Eds Mental model pp 7-14 Hillsdale NJ Lawrence Erlbaum Associates

[8] Nuffield Primary Science 1990 Science Process and Concept Exploration; Teacher's Guide Key Stage 2 London: Harper Collins

[9] Osborne R J & Cosgrove M M 1983 *Children's conceptions of the changes of state of water* (Journal of research in Science Teaching vol 20) chapter 9 pp 825-838

[10] Osborne R J & Wittrock M C 1983 *Learning science: A generative process* (Science education, vol 67) chapter 4 pp 489-508

[11] Richard KC 2001 Learners' *Mental Model of Chemical Bonding* Research in (Science Education vol 31) pp 357-382

[12] Yuenyong C Jones A & Yutakom N 2008 *A comparison of Thailand and New Zealand students' ideas about energy related to technological and societal issues. International* (Journal of Science and Mathematics Education vol 6) chapter 2 pp 293-311

[13] Yuenyong C & Yuenyong J 2007 *Grade 1 to 6 Thai Students' Existing Ideas about Energy* (Science Education International vol 184) pp 289-298

IOP Conf. Series: Journal of Physics: Conf. Series **1340** (2019) 012087 doi:10.1088/1742-6596/1340/1/012087

Levels of Discourse in Mathematics Classroom using Lesson Study and Open Approach

Korkiat Raethong [1,2], **S Thinwiangthong**[1,2]

[1] Mathematics Education, Faculty of Education, Khon Kaen University, Thailand

[2] Centre of Excellence in Mathematics, CHE, Bangkok 10400, Thailand

Corresponding author's e-mail address: gorjung.rw99@gmail.com, sampan@kku.ac.th

Abstract. Start your abstract here... This research aimed to analyze levels of discourse in mathematics classroom using lesson study and open approach. It use of qualitative research methodology by focusing on protocol analysis and analytical description. Target group was a mathematics teacher and 30 seventh grade students in 2nd semester of school year 2017 at Kaeng-Khro Wittaya School, Chaiyaphum. The research instruments were consisted of lesson plan, field note, students' writing work, video recorder and camera recorder. The collected data was analyzed by 4 steps of open approach [7]. and levels of discourse [19]. Research result found that levels of discourse in mathematics classroom represented as following details: 1) Posing open-ended problem, it was found 4 levels of discourse. In early class of learning unit, it was found the level of 0-1 which the teacher had posed open-ended problem before explained the problem with asking some short question to the students for checking their understanding whereas in latter class of learning unit, it was found the level of 2-3 which the teacher had posed open-ended problem then the student questioned and discussed the problem together; 2) Students' self learning, it was discovered 2 levels of discourse which were the level of 2-3. In this discourse, students had questioned and explained about problem solving methods and the teacher had asked the students to clearly describe their problem solving methods together; 3) Whole class discussion and comparison, it was found 2 levels of discourse which were the levels of 2-3. In this discourse, students demonstrated and expressed their own problem solving ideas and other students questioned about their own doubt about each ideas in classroom. The teacher ordered the students' ideas presentation; 4) Summarization through connecting students' mathematical ideas emerged in the classroom; it was showed 2 levels of discourse which this learning unit was found the levels of 2-3. In this discourse, teacher encouraged the students to confirm the ideas obtained from listening to reasons of each idea together for discussion and summarization in classroom.

1. Introduction

In results of Programme for International Student Assessment (PISA) test, it indicates that mean scores of Thai's students in all skill such as reading, mathematical and science skill are quite smaller than OECD mean score [7]. One of these causes that affect the obtaining low score in PISA of the students is teaching quality which teachers behave as a lecturer, but not indeed promote students' knowledge [15]. This reflects that tools used for learning management are defective, especially, teaching method (Inprasitha, 2014). The National Council of Teachers of Mathematics (NCTM) suggested that learning management in mathematics classroom should supply a student to student conversation activity to develop their understanding in mathematics [13].

Lesson Study is a teacher development innovation which can solve abovementioned problems because the Lesson Study is the innovation that focuses on learners importantly and aims to develop teachers gradually and continuously [6]. In Thailand, it was applied in a learning management innovation called open approach which is teaching approach [7]. Features of open approach classroom are to debate various students' ideas and to develop their attitudes by facing experiences with themselves and to provide reasonable suggestion by teacher which aims at discussion and mathematical idea exchange [14]. This idea exchange is more important in learning mathematics [13], which most teachers have not experienced yet. And the most difficulty of learning-teaching mathematics management is to transform teacher-centered learning discourse to basic concept and student-centered learning discourse.

The discourse in mathematics classroom is a part of learning mathematics. The discourse could help students to explain, express their reasons and discuss each mathematical solving technique and support the students to understand the ideas [20]. Format of discourse can be found from speaking words in mathematics classroom [4]. The discourse in mathematics classroom detected by speaking words can be determined as level of discourse by considering character of participation between teacher and students via questioning in classroom. This will encourage student to explain their mathematical ideas, origin of mathematical ideas and debates which can be classified into 4 levels of discourse. The abovementioned patterns of discourse are advantage to set up mathematical criterion for mathematical conversation and working conversation in mathematics classroom [5].

As above-mentioned, the mathematics classroom problem is the former learning-teaching management which teachers transfer mathematical knowledge to students via explanation or expressing mathematics contents in classroom. This showed the discourse of teacher-centered learning. To enhance the quality in classroom to be a focusing on solving problem and self-studying, especially, the discourse of student-centered learning, it requires estimating the level of discourse in the classroom to be considered and taken to improve its mathematics classroom quality. Then, this study will analyze the level of discourse in mathematics classroom using lesson study and open approach.

2. Research Objectives

To analyze the level of discourse in mathematics classroom using lesson study and open approach.

3. Research Methodology

3.1. Target Group

3.1.1. 30 students in grade 7 of Kaeng-Khro Wittaya School, Kaeng-Khro district, Chaiyaphom, which are in mathematics classroom using lesson study and open approach.

3.1.2. 1 mathematics teacher of Kaeng-Khro Wittaya School, Kaeng-Khro district, Chaiyaphom.

3.2. Research Instruments

3.2.1. Mathematics learning activity plans for grade 7: Unit 1: Positive and Negative number.

3.2.2. Video (VDO) recorder used for recording the activity in mathematics classroom and being raw data taken to analyze.

3.2.3. Camera used for picturing students' work and activity.

3.2.4. Sound recorder used for recording the conversation between teacher and students during mathematics class.

3.2.5. Field Notes used for noting students' behavior during solving problem and ideas generated by students.

4. Data Collection

4.1. Researcher, Researcher assistant and teacher co-work
In education process in classroom for design learning plan consisting of 3 steps; 1) Learning plan co-designing which researcher, researcher assistant and teacher design learning plan by following a books published by Keirinkan and Japanese mathematics book published by Gakkotosho. 2) Co-observation which Researcher, researcher assistant and teacher collect data and teacher teaches students via open approach. 3) Classroom reflection which researcher, researcher assistant and teacher reflect their own opinion after classroom.

4.2. Researcher and researcher assistant
Attended a meeting to understand analysis framework and assign works to researcher and researcher assistant.

5. Data Analysis
This analysis is a qualitative analysis. Researcher will use data in protocol of solving problem of students in mathematics classroom using lesson study and open approach collected by VDO & sound recorder, students' works, protocol interview, field note and protocol reflection of researcher for analysis levels of discourse in mathematics classroom using lesson study and open approach following the idea of Stein (2007) and Inprasitha [7].

6. Results
From the study of level of discourse in mathematics classroom using lesson study and open approach. in the content about positive and negative number, it was found 4 levels of discourse in mathematics classroom using lesson study and open approach as follows,

- Level 0; in this level, the teacher asked a question and confirmed answer correction, while students answered the question shortly. Then teacher suggested and explained about mathematical idea to the students.
- Level 1; in this level, the teacher asked a question frankly about a student's idea, while other students listened to the teacher explaining about the student's idea and students noted data before next student's idea explanation began. Teacher might ask some students to help the student solving the problem.
- Level 2; in this level, the teacher asked open-ended question to encourage the students' thinking and allow them comment about another student's work. Students commented the ideas willingly. The teacher confirmed and determined the direction of learning based on the students' explanation.
- Level 3; in this level, the teacher served the students to discuss by supporting them to ask some question to each other and another student explained his her own idea. The idea of classroom was created from having the student expressed and explained the origin of their ideas intensively and then other students listened to his/her explanation. Teacher asked about solving method and idea for checking their understanding.

From the study of level of discourse in mathematics classroom using lesson study and open approach, it was discovered 4 levels of discourse in each open approach steps as follows,

Posing open-ended problem; in this step, it was found 4 levels of discourse. Level 0; the teacher asked a question to the students for only confirming answer correction. The students gave the short answers about problem situation. Level 1; the teacher asked about problem situation directly and methods that the students used to solve their own problem, while other students listened to the presenting student. The teacher explains strategy of presenting student to other students to check their own ideas. Level 2; the teacher gave a problem situation and asked open-ended question to students and they explained their own solving problem ideas. And level 3; the teacher promoted students to ask the presenting student and he/she explained and understand his/her own problem situation.

Students' self learning; in this step, it was found 2 levels of discourse, namely, level 2; students gave their reason based on their ideas generated from open-ended question from teacher and classmate in classroom. Students commented "agree/disagree" to other students' ideas which they have same/different opinion to, respectively, and level 3; the students asked a presenting student some questions to explain his/her idea to other. The teacher served the presenting student describes his/her own idea for clearly understanding.

Whole class discussion and comparison, it was found 2 levels of discourse, that is, level 2; the teacher ordered students' presentation to activate students' idea presentation. Then other students commented and compared the others' ideas, and level 3; the teacher promoted students to ask the presenting student and the presenting student expressed his/her idea clearly.

Summarization through connecting students' mathematical ideas emerged in the classroom; it was found 2 levels of discourse, namely, level 2; from the idea presentation which support the students commented their own opinion on others' solving problem idea, teacher and students confirmed the idea by giving opinion together. And level 3; from students' idea presentation in this step, student made other students accept his/her idea by describing his/her reason. Teacher supported students to confirm classroom's idea obtained from listening to and sharing ideas together.

7. Conclusion

From the study of level of discourse in mathematics classroom using lesson study and open approach, it can be discussed as step of open approach idea of Inprasitha [7] as follows; Posing open-ended problem; in this step, teacher take students to open-ended problem situation and stimulate students' idea via questioning. Explanation can lead students to correct way. In addition, teacher give a chance to students for asking about idea or possible outcome which has many ideas, then it is found 4 levels of mathematics classroom discourse. However, at the beginning of learning unit, it is found the levels of 0-1. The teacher had posed open-ended problem before explained the problem with asking some short question to the students for checking their understanding and in latter class of learning unit, it was found the level of 2-3 which the teacher had posed open-ended problem and then the student questioned and discussed the problem together. Students' self learning presented 2 levels of discourse which are level of 2 and 3. Teacher allows students to learn by themselves from solving problem together. But sometimes teacher ask open-ended question to students for stimulating students' opinions or asking students' idea during solving problem to demonstrate to others understand together. Moreover, for solving problem in each group, students discuss about strategy to solving problem which emphasize various ideas during solving problem via questioning. Whole class discussion and comparison; it was found 2 levels of discourse which are level 2 and 3. Students posed the idea that they have solved together. Teacher support students to question the presenting student and presenting student describe the idea. Then many ideas generated from self-learning are presented in classroom and students have learned and understand many ideas. Summarization through connecting students' mathematical ideas emerged in the classroom; it was found 2 levels of discourse, namely, level 2 and 3. Teacher encourages students to comment about their own and others' ideas and demonstrates the origin of idea via questioning and find the conclusion idea together.

The learning activity using open approach following idea of Inprasitha [7] supports students using discourse for solving problem in classroom by open-ended question asked by teacher. This shows 4 levels of discourse. Mostly, it was found the level 2 which teacher asked open-ended question to encourage the students' thinking and allow them comment about another student's work. Students commented the ideas willingly. The teacher confirmed and determined the direction of learning based on the students' explanation. And level 3; the students asked a presenting student some questions to explain his/her idea to other. The teacher served the presenting student describes his/her own idea for clearly understanding. These are corresponded to Nampreeda [12] which told that open-ended problem situation can support students to use mathematical discourse in classroom following theory of Lakatos (1976). Students apply conversation and presentation process to express their own idea. And giving a reason is used in group discussion which is corresponded to Phuengphon and Inprasitha [17] said that in mathematics classroom using lesson study and open approach, there is mathematics discourse appearing and it is used all the time by beginning with teacher who starts to use discourse.

8. Recommendation

8.1. Recommendation for Implementation of Research Findings
This research can be used in different content via open approach learning using open-ended problem situation to allow students express, explain their ideas and give a reason for solving problem method. This can promote the level of discourse in classroom.

8.2. Recommendation for Further Research
It should be study on how levels of discourse between teacher and students, or students and students, do support the students understanding, in order to use the result in planning and implement the lesson to promote the students' understanding in mathematics learning meaningfully.

9. Acknowledgments
This research is (partially) supported by The Centre of Excellence in Mathematics, The Commission on Higher Education, Thailand and Centre for Research in Mathematics Education, Faculty of Education, Khon Kaen University and The Project of Development of student's Mathematics Higher-Order Thinking at North East of Thailand, Faculty of Education, Khon Kaen University.

References
[1] Becker J P and Shimada S 1997 *The open-ended approach: A new proposal for teaching mathematics* (Reston, Virginia: National Council Mathematic)
[2] Catherine C Stein 2007 *Let's Talk – Promoting Mathematical Discourse in the Classroom* (Mathematics Teacher November 2007 vol 101) Issue 4 pp 285-289
[3] Charoensin-o-larn C 2006 *Development discourse: power, knowledge, truth, identity and otherness* (Bangkok: Vibhasa, Inc)
[4] Gee J P 1996 *Social linguistics and literacies: Ideology in discourses* (London: Falmer Press)
[5] Hufferd-Ackles K, Fuson K C, and Sherin M G 2004 *Describing levels and components of a math-talk learning community* (Journal for Research in Mathematics Education vol 35) pp 81–116.
[6] Inprasitha M 2010 *One Feature of Adaptive Lesson Study in Thailand Designing Learning Unit* (Proceedings of the 45th National Meeting of Mathematics Education) pp. 193-206. Gyeongju: Korea.
[7] Inprasitha M 2014 *Processes of Problem Solving in School Mathematics* (Khon Kaen: Penprinting, Inc)
[8] Isoda M 2010 *Lesson Study: Problems Solving Approaches in Mathematics Education as a Japanese Experience* (Procedia Social and Behavioral Sciences vol 8) 17-27
[9] Lampert M 1990 "*When the Problem Is Not the Question and the Solution Is Not the Answer: Mathematical Knowing and Teaching*" (American Education Research Journal 27 (Spring 1990)) pp 29–63

IOP Conf. Series: Journal of Physics: Conf. Series **1340** (2019) 012087 doi:10.1088/1742-6596/1340/1/012087

[10] Lewis C 2002 *Lesson Study: A handbook of teacher – led instructional change* (Philadelphia, PA: Research for better School)

[11] Loipha S and Inprasitha M 2004 *Teacher Professional Development to Promote Mathematics Learning* (KKU Journal of Mathematics Education vol 1 January-. June) pp 18-28

[12] Nampreeda C 2008 *An Analysis of Students' Mathematical Discourses in Open-End Problem Solving Situations* Master of Mathematics Education Thesis in Education, Graduate school, Khon Kaen University.

[13] National Council of Teachers of Mathematics [NCTM] 2000 *Principles and Standards for School Mathematics* Reston, VA: NCTM.

[14] Nohda N 2000 *A Study of "Open-approach" Method in School Mathematics: focusing on mathematical Problem Solving Activities* Paper present at the ICME, Tokyo/Makuhari, Japan.

[15] Office of the Education Council [ONEC] 2556 *Measure of Achievement* (Bangkok: Prigwhan Graphic, Inc)

[16] Pasjuso S, Thinwiangthong S and Kongthip Y 2010 *Comparative study of gesture in mathematical communication in Thai traditional and innovation classroom* In Y. Shimizu, Y. Sekiguchi, & K. Hino (Eds) (Proceeding of the 5th East Asia Regional Conference on Mathematics Education vol 1) pp 230

[17] Phuengphon J and Inprasitha N 2015 *Lesson study and open approach: a way to promote mathematics classroom discourse* (Valaya Alongkorn Review Vol 5) No 2

[18] Saengpun J 2012 *Development of Students' Semiotic Activity through Collective Discourse in Mathematics Classroom Innovated by Lesson Study and Open Approach* Doctor of Philosophy Thesis in Mathematics Education. Graduate School, Khon Kaen University.

[19] Stigler & Hiebert 1999 *The Teaching Gap.* (New York: The Free Press)

[20] Stiles J 2015 *Supporting Mathematical Discourse in the Early Grades Interactive Technologies in STEM Teaching and Learning* (Education Development Center, Inc. Copyright 2016)

[21] Sumailah A Sabbagh 2014 *The Discourse Levels Indicators Used by Mathematics Teachers to Promote Students' Understanding and Thinking for the High School in Jordan* (International Proceedings of Economics Development and Research vol 78) pp 74-79 Singapore: Singapore IACSIT Press.

[22] Thinwiangthong S 2012 *Small-group Mathematical Communication in Classroom in Lesson Study and Open Approach Context* Doctor of Philosophy Thesis in Mathematics Education. Graduate School, Khon Kaen University.

[23] Walshaw M and Anthony G 2008 *The teacher's role in classroom discourse: A review of recent research into mathematics classrooms* (Review of Educational Research vol 78) chapter 3 pp 516-551

IOP Conf. Series: Journal of Physics: Conf. Series **1340** (2019) 012088 doi:10.1088/1742-6596/1340/1/012088

Mathematical Conceptual Understanding about Geometry of 8th Grade Students in Classroom Using Lesson Study and Open Approach with The Geometer's Sketchpad

N Samphantakul[1,2], S Thinwiangthong[1,2]

[1] Mathematics Education, Faculty of Education, Khon Kaen University, Thailand
[2] Centre of Excellence in Mathematics, CHE, Bangkok 10400, Thailand

Author's e-mail address: nattisinee_samphantakul@kkumail.com, sampan@kku.ac.th

Abstract. This research aims to analyze the Mathematical Conceptual Understanding (MCU) about Geometry in classroom using Lesson Study and Open Approach [11] with the Geometer's Sketchpad (GSP). It used qualitative research methodology which focuses on protocol analysis through analytic description. The target group was 9 (divided into 3 groups) of 8th grade students from Kaeng Khro Wittaya School, Kaeng Khro District, Chaiyaphum Province, 2016 Academic Year. The research tools were lesson plan of Figures and Geometry Shapes with GSP, Video recorder and field note. The data included students' problem solving protocols, students' interviewing protocols, collaboratively lesson plan and reflection protocols, and the students' written works. Data were analyzed by APOS framework [7]. The results were found that: 1) Posing Open-ended problems, GSP program was used for surveying details and conditions of problems. Students have the MCU in the action level, the students were able to follow the instructions, 2) Students' self-learning, students applied the surveying details and conditions from GSP program in the 1st step to solve their mathematical problems and could have various ideas. Students have the MCU in the process level, they can draw the connection in the problematic problems as well as describe that idea in descriptive manner. They can also present and elaborate their ideas, 3) Whole class discussion and comparison, GSP program was used for expanding and comparing the students' ideas. Students have the MCU in the object level, The students can make conclusion as well as give definitions, principles qualities to make the connection or solve the problems, and 4) Summarization through connecting students' mathematical ideas emerged in the classroom, GSP Program was used to connect and summarize the students' mathematical idea. Students have the MCU in the object and schema levels, the students are able to make connection of their current lesson with the other lesson such as geometry and so on.

1. Introduction

In present days, the classroom in Thailand faced several issues such as the learning methods that focus only on academic achievement. The students are evaluated only by their examination marks, and the process of learning is overlooked completely. It is apparent that in most classrooms, a teacher plays a role as a key person. Learning methods that those teachers has used depend on spoon-feeding, or the technique that teachers provide knowledge to students such as lecture, demonstration from teachers with the use of tools or description [12]. This is similar to Maxwell, Daniel and Peerasit [3]. stated that the main problem of Thai education system is that teachers depend too much on rote-learning method and evaluate students by marking how much content they can memorize without training them to have critical thinking, analyze or conduct research in scientific manners. The consequences of this are included learner's lack of understanding and poor development of emotion quotient (E.Q.) and social quotient (S.Q.). As a result, students tend to be passive. They learn quietly and obey teacher's instruction easily without asking, arguing, giving their opinions or using their creativity. This is much different from the classroom taught in Japan as it encourages students to use their creativity, be themselves as much as possible and dare to think differently. The issues found in Thai classroom seem to be the result of spoon-feeding and rote-learning method without teaching learners to think, analyze or study with a sense of self-reliance [17]. Programme for International Student Assessment (PISA), an international academic evaluation, included open-ended questions that required students to reflect their thought after reading as answers. The marks are determined by the reason given for answers. If this type is used to evaluate the students in Thailand, it is very likely that it would indicate that Thai education is in the serious crisis. Therefore, it can be inferred that the tools we used in classroom, such as teaching methods, are problematic [13].

One of the solutions to these problems is to change the pattern of activities in classroom. For example, a teacher can encourage students to using their ideas to solve problems by themselves, and they should pay attention to study their behaviours while the students are solving problems (Steen, 1990 cited in [12, 13]), therefore, applied and adapted Lesson Study with the co-operation of the teachers. The outstanding aspects of Lesson Study are included gradual and consistent development as well as the alteration of classroom by integrating Open Approach, or the use of open-ended problems in the mathematical activities. Inprasitha [10] adapted some processes of Lesson Study, originated in Japan, and concluded it into 3 steps which are 1) weekly collaborative planning 2) weekly collaborative observing 3) weekly collaborative reflecting. The methods in integrating Open Approach to the mathematics classroom can be divided into 4 steps: posing open-ended problems, students' self-learning, whole class discussion and comparison, summarization through connecting students' mathematical ideas emerged in the classroom [10]. The Open Approach-centered learning patterns aim for preparing learners with variety of open-ended questions to respond to different requirements, interests or mathematical thinking ability of each individual [21].

Lesson Study is applied to study the motivation behind the use of new innovation of The Geometer's Sketchpad Program. It is found that Geometer's Sketchpad used in the group of mathematics teachers through the Lesson Study and co-operation. The data accumulated from teaching plan arrangement and the interview with each participant reveals that classroom study help teachers to learn new skills efficiently, Students also positively responded to the study environment of The Geometer's Sketchpad Program. The outcome shows that the studying in classroom by using The Geometer's Sketchpad Program in mathematics classroom tend to be successful and efficient [17]. By using The Geometer's Sketchpad Program to design learning activities used in mathematics classroom, the changes are found in the activities that used The Geometer's Sketchpad in classroom study group. It is found that the gradual improvement in skills and credibility of the participants in the classroom.By applying Open Approach with The Geometer's Sketchpad-centered teaching plan in classroom, it reflected the variety of mathematical thinking such as surveying, reasoning, investigating and diversity in solving problems [15]. The skill that citizens in the 21st century should develop is learning and invention skills such as creative thinking, inventing, critics thinking, problem solving, communication, co-operation and technological skills [3].

As Thai education is under the consistent development, the use of technology in learning will become an important factor and a significant alternative choice that will be beneficial to classroom arrangement. It is expected to improve education as well as teaching lesson arrangement to be more efficient, and this will give direct impact to children and alter Thailand to be a society of learning [23]. Furthermore, technology is very important to mathematics study, and it affects the advance teaching and learning [20]. The impact of using technology in mathematics study is technology's efficiency in emphasizing on practicing to improve making conclusion, structure connecting and problem solving skills [3]. It can be interpreted that technology can be used for assisting in teaching and enhancing student's ability in learning. With the appropriate use of technology, students are enabled to improve their knowledge and understanding in mathematics as technology can be used to support in investigation of students in any aspects of mathematics studies and enhance learners to see the importance of actions, consequences, reasons and solving existing problems [20]. However, Technology is no genie, and it cannot replace a teacher. A teacher is the one who choose to use or design problematic situations or learning activities so that it would be possible to use technology to its fullest potential [20]. Therefore, the two things that teachers might need to make decisions on the use of technology in the classroom are to select the programs according to the range of importance according to Laborde [19] and secondly, to design problematic situations [8].

The Geometer's Sketchpad (GSP) is the software program for creating and investigating dynamics models of mathematical objects, diagrams and graphs. GSP is an efficient tool for creation, demonstration and investigation, especially for geometry . For example, The Geometer's Sketchpad can be used to investigate geometric shapes. The Institute for the Promotion of Teaching Science and Technology (IPST) has been encouraging the use of this program as a new tool in Thai mathematics study to improve the level of mathematics study. Teachers can use The Geometer's Sketchpad to assist the teaching activities because this program has moving pictures that make it easy to explain complicated content. Also, this program encourages self-learning as a student can practice this program by himself or herself. Last but not least, The Geometer's Sketchpad is designed for the use of creativity, survey and mathematical content analysis [25].

Geometry is the very useful content for mathematical improvement such as inductive reasoning, reductive reasoning, fact proof, estimation as well as categorization and definition of geometric shapes [20]. The understanding of geometry is one of the essential skills in studying mathematics. However, most of students usually reflected their understandings toward geometry with negative attitude such as 'I can't do it', 'it isn't important', 'it's tedious', or even 'it's too challenging' [22]. The best way to study mathematics is to understand it. Therefore, the main objective is in advance mathematics learning, students should deal with it by themselves [18]. Understanding is the fundamental aspect of learning, and it is a pattern of learning that should be involved with problem solving [9]. Hence, the object of many mathematical studies in hope for academic achievement is to encourage students to study mathematics with understandings [9]. In studying mathematics, understanding is the most essential and significant aspect that motivates students to study with their fullest potential. Teaching with Open Approach could provide teachers solutions and 'How to' directions to understand student's mathematical concept more closely, and this is similar to the concept developed by Inprasitha [13]. In studying mathematics understanding after the use of The Geometer's Sketchpad as an assisting tool, it is found that the program can provide opportunities in thinking and solving problems more efficiently, and this would prepare them for the advance study [1].

With these reasons, it is readily apparent that students usually have problems with learning geometry because they find it difficult to understand. Hence, technology can be used to help them understand the lesson better. The researchers aim for using The Geometer's Sketchpad to create problematic situations or teaching activities. However, the limitation of the technology is seen, so the researchers apply the innovation of Lesson Study and Open Approach as well [12]. They also analyze students' by based on APOS theoretical framework [7].

2. Research Objectives

This research aimed to analyze the level of student's understanding in geometry taught in classroom, being applied Lesson Study, Open Approach and The Geometer's Sketchpad.

3. Definition of Technical Terms

3.1. Mathematical Conceptual Understanding (MCU):

This term means an insight or action in each content, and students would show their understanding through responding, the action of problem solving, expressing their ideas, role-play, reasoning, expressing through words, gestures, photos, or symbols as well as connecting and concluding ideas or give the mathematical definitions that are meaningful to oneself .According to Dubinsky & McDonald's APOS theatrical framework, mathematical conceptual understanding can be categorized into 4 types:

1. Action Level: this means student's ability to respond to external stimulation. The students in each are required to adhere to conditions or instructions in order. Each step needs to be done before they can do the next one. Therefore, students who have mathematical conceptual understanding in action level would be able to arrange their actions with both physical and conceptual objects in given order. For example, the students can answer the questions given by teachers, or they can perform orders in problematic situations.
2. Process Level: this is defined as student's ability in elaborating, reflecting, or thinking about the processes retrospectively. They do not need to elaborate the actions taken in each step, but this occurs from the understanding their actions gained from doing repetitively. The students can have pictures in mind to predict the outcome or make conclusion. For instance, the students can draw relationship occurred in problem solving as well as descriptive narration of different relations. The students can explain idea of their own group without showing repetitive actions.
3. Object Level: this is defined as an ability to understand the whole picture level as though everything is the same object. The process arrangement could alter the outcome of the process, and it provides the new advance level concept. The students improve their mathematical conceptual understanding in process level so that they can compare their group ideas with other ideas emerged in the classroom. They can find similarities, differences, explain formula or make conclusion of the ideas they have found. They can also give examples to make it easier to understand and connect the different ideas such as ideas from their own group and other groups. For example, the students can make conclusion, give definition or qualities to explain or solve problems.
4. Schema Level: this is an ability to connect between experiences or new processes with the existing cognitive structure, and it might be connected with the previous separated process to be a part of the cognitive structure. The students would be able to realize their concepts and see the connections of different concepts occurred in the classroom by comparing the pair of concepts or connect the different ideas such as ideas from their own group and other groups to make conclusion of the ideas that meaningful to oneself. For example, the students can connect the content emerged from the horizontal relationship, regarding geometric shapes and other topics.

3.2. Classroom using Lesson Study and Open Approach:

This is the classroom that integrated Lesson Study and Open Approach to create a classroom that concentrate on soling problem with a sense of self-reliance .The classroom is applied with Lesson Study and Open Approach according to Inprasitha [11-13].

3.3. *Lesson Study*

It can be defined as an approach in working together between teachers, researchers and researching assistants .Three of them worked together to design lesson plan and apply it to the classroom .The classroom would be observed and reflected by the team .According to Inprasitha [11], the process of this method can be seen as following details:

1. Collaborative Plan: in this step, the lesson study team analyzes and designs open-ended problems by using The Geometer's Sketchpad and the content in KEIRINKAN's mathematics textbook, the sixth chapter: geometric figures and shapes. They also predict concept and design teaching media by using The Geometer's Sketchpad.
2. Collaborative Observing: in this step, classroom study team implemented their lesson plan in the classroom. Researchers and researching assistants observed the classrooms by concentrating student's ideas in solving problems to analyze their mathematical conceptual understanding.
3. Collaborative Reflecting: this step is to reflect after using the lesson plan in the classroom, the team worked together to reflect the appropriation of the use of The Geometer's Sketchpad in the classroom, student's ideas and their mathematic understanding emerged in the classroom. They also worked to reflect the instructional problems and the way to improve the lesson plan or lesson in the future.

3.4. Open Approach:
This means the method in providing classes that focuses on solving mathematical problems by using open-ended problem, provides opportunities for students to solve problems by themselves and encourages mathematical understandings. According to Inprasitha [11], there are 4 steps as following,

1. Posing open-ended problem: the teacher poses open-ended problem by using The Geometer's Sketchpad to the students in the classroom.
2. Students' self-learning: the students solve the problems by themselves or solve them with the colleagues in the same group. In each group, there are three members. While students were solving problems, there were no interventions from the teachers.
3. Whole class discussion and comparison: students in each group present their group's ideas to the class and discuss them with the classmates. The teachers attempt to elaborate the students' ideas in each group as well as the ideas in The Geometer's Sketchpad so that students can understand the ideas from the colleagues in the class. The teachers also provided other students the chances to ask questions.
4. Summarizing by connecting students' mathematical ideas emerged in the classroom: in this step, the teachers used The Geometer's Sketchpad to make the conclusion of the students' ideas. Both teachers and students worked together in making conclusion of concepts and important topics and connected them together as a conclusion of the lesson.

3.5. The Geometer's Sketchpad program :
This software is created to use for investigation and analysis of various aspects of mathematics content. We can use dynamics geometry to create mathematical models that can react in different ways, ranging from basic research of shapes and figures as well as complicated, dynamic and advance drawing. This can be used to enhance the understandings towards geometry, and it also can be used as a learning tool as it can give figurative information about how to create geometric shapes. The outstanding aspect of this program is dynamics. A user can create figure or geometric shapes with the tools in the program as well as adapting information or pictures to see the changes of geometric shapes, make prediction for learning and having concluding ideas.

3.6. Geometry:
The basic knowledge about dot, line, plane and types of 2-D and 3-D shapes, experiences in seeing and drawing line, angle, triangle and other shapes. The students can imagine the creation of shapes and figures by using what they have seen in daily life. For lower secondary level, geometry depends on the introduction of basic information of geometry. The students can investigate its relationship by

measuring, observing, comparing the geometric alterations and the categorization of geometric objects.

4. Research Methodology

4.1. The Context of Classroom in the Study

This research relies on the context of The Project of Development of Student's Mathematics Higher-Order Thinking at North East of Thailand. This project is developed by Center for Research in Mathenmatics Education, Khon Kaen University. The data is collected from Kang Kraw Wittaya School. This school has been adhering to the project by using Lesson Study and Open Approach. The directors participated in the meeting of the network to commit the co-operation MOU of the project on July 3-4, 2016 at Education Faculty, Khon Kaen University. The project is launched on July 4, 2016. This project had been prepared in advance since June 22, 2016 by the project of Development of Teachers' Formative Assessment by Using Lesson Study and Open Approach which is supported by Centre of Exellence in Mathematics (CEM) in part of mathematics education, Khon Kaen University and University of North Texas America.

4.2. Target Group

The target group of this reseach is the students in Kang Kraw Wittaya School, Chaiyapoom, from class M 2/1, by focusing on their second semester 2016. These students have been taught with Open Approach since they were in Matthayom 1 (or grade 7). The researchers start to observe the class in the first semester, 2016 by focusing on the students' behaviors in solving problems and ideas. They also make students get used to having them in class. The selection is drawn by choosing the students who can work together in sharing opinions while solving open-ended problem. The researchers chose 9 students and divided them into 3 groups, and they remained in the same group for the rest of research process.

4.3. Research Instruments

1) Collecting data tools are included the lesson plans by using The Geometer's Sketchpad, fieldnote, worksheets, voice recorder, video recorder and camera.
2) Analyzing data tool are included Dubinsky & McDonald [7]'s APOS theoretical framework, protocol recorded from voice recorder and video recorder, protocol recorded form interviewing and student's written works.

5. Data Collection

The researchers and lesson study team studied the general context of Kang Kraw Wittaya School, especially in the class that they want to study. They worked together to make lesson plan, which is applied Lesson Study, Open Approach and The Geometer's Sketchpad. Also, their study included observing the behaviors of the target group and attempting to make students accustomed to problem solving. While collecting data, researchers planned together with the lesson study team, the reaearching assistants recorded audio, video and photos. The researchers and their assistants also obseve mathematical ideas of students, problem solving through observation, the chance given for discussion between teachers and students. While studying, the researchers and the assistants could not intervene student's way of thinking.

6. Data Analysis

Qualitative manner is applied to this research. The protocol is made by transferring audio records into texts. It is analyzed together with target students' written works, fieldnote and interviewing records. The data is analyzed by the framework of APOS Theory [7]. The mathematical conceptual understanding can be divided into 4 levels which are 1) Action 2) Process 3) Object and 4) Schema. The example of data analysis can be seen as the following:

6.1. :Step 1Posing Open-Ended Problems.

In this step, GSP program is used to pose the problems and explore problem's details and conditions. It can be inferred that students have mathematical conceptual understanding in the action level.

Figure 1. the steps of posing open-ended problem, regarding lesson unit 3: shapes rotation

6.1.1 Action Level: mathematical conceptual understandings in action level can be seen when the students can solve problem situation by adhering to conditions or instructions given systematically, and this can be seen in the situation below.

The teacher posed open-ended problem by using The Geometer's Sketchpad, like "when I rotate this figure around l-line, what do you think it would happen". The students were expected to gather and observe it together. They were initially instructed so that they can experiment what solutions can be used to solve the given situation. This can be as in the following protocol:

Table 1: An example protocol of students' mathematical conceptual understanding in action level

Item 2	Teacher	When I rotate this rectangle around l-line, what do you think it would happen?
Item 3	Nice	Cylinder.
Item 11	Teacher	And, if I rotate this right triangle?
Item 12	Gear	Cone.
Item 15	Teacher	What happens if I rotate this half circle around l-line?
Item 16	Gear	Sphere.
Item 17	Teacher	Someone said that it is a sphere. Are you sure? Well, let's think that it rotates around l-line.
Item 18	Ryo	A ball.
Item 19	Teacher	A ball? How do we call this ball?
Item 20	Smile	Sphere.

According to Item 2-16, while posing opening-ended problems, the teacher used Geometer's Sketchpad. The students in group 1 can responded the problems, like if plane figures (2D) are rotated, what solid shapes (3D), they would turn into. This can be inferred that students have mathematical conceptual understanding in **action level**.

6.2. Step 2 :Students 'Self-Learning :

In this step, it can be seen that students can use what they have observed from using GSP program in the first step to solve mathematical problems, and this brought them several ideas .Thus, this illustrates that the students have mathematical conceptual understanding in process level.

Figure 2. the steps of student's self-learning, regarding lesson unit 3: shapes rotation

6.2.1 Process Level: The students can develop their understandings from action level to process level .They can draw relationship in the problem situations by drawing picture or writing messages and geometric symbols. This can be seen in the step of student's self-learning, and it can be seen in the following details.

In this step, the students solve problems by themselves. They gathered and discussed about the given problem situations, and they drew solid figures by rotating figures around l-line, as it can be seen in the following protocol:

Table 2. An example protocol of students' mathematical conceptual understanding in process level

Item 31	Smile	Rotate it around l-line, right?
Item 32	Nice	I get it now. From this point, if we rotate it, it would turn into a sphere because it has constant radius.
Item 33	Smile	Rotate around l-line, and it becomes a sphere around l-line.
Item 34	Gear	Let's identify each side.
Item 35	Nice	The first figure is a square. We will name each side a, b, c, and d. For b-side, it is next to l-line. When rotating a-side and c-side around l-line, it will turn into the cross-section of a circle.
Item 36	Smile	What's about D-side? D-side become the lateral surface.
Item 37	Nice	Well, it is an area of lateral surface of a cylinder.
Item 38	Smile	For the other figures, we can write them similar to this.

Figure 3. demonstrates of problem solving and writing of students the steps of students' self-learning, regarding lesson unit 3: shapes rotation

According to Item 31-37 (the protocol of Group 2), it is inferred that the students can explain the figures that can be when rotating the figures around l-line. For example, they can explain that a figure can turn into a circle because it has constant radius. They can also identify each side and give conditions for each side of the given figure so that they can explain the compositions that can make a plane figure become solid figures. The above protocol shows that the students have mathematical conceptual understanding in the process level.

6.3. Step 3 : Whole Class Discussion and Comparison :

In this step, GSP program is used to extend and compare student's mathematical ideas occurred in the classroom. This illustrates that students have mathematical conceptual understanding in the object level.

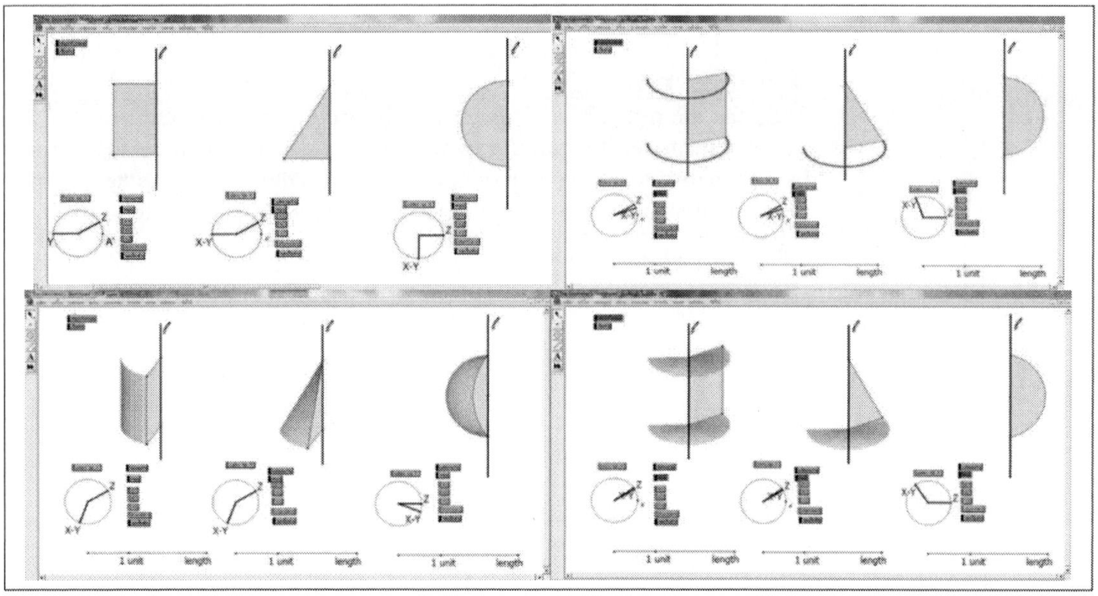

Figure 4. The use of The Geometer's sketchpad for the lesson unit 3: problematic plane in the step of class discussion and comparison

IOP Conf. Series: Journal of Physics: Conf. Series **1340** (2019) 012088 doi:10.1088/1742-6596/1340/1/012088

6.3.1 Object Level: In this level, the students can tell the compositions of all solid shapes that created by rotating plane around l-line once, they can give definitions and principles in creating solid shapes .This happens in the learning activities, in the process of connecting students 'mathematical ideas.

In the step of whole class discussion and comparison, the students in the Group 2 shared ideas after solving the problems by themselves. They tried to find a way to explain this in the different way from other colleagues. They explained ideas and gave examples as the following protocols:

Table 3: An example protocol of students' mathematical conceptual understanding in object level

Item 65	Smile	When a rectangle rotates around l-line, it will become a cylinder. We know this by seeing the distance here. For this figure, if it is rotated, it will become a circle. This is because of a distance here. The length of each rotation is equal, so it becomes a circle.
Item 66	Nice	We use l-line as a center of rotation. For c-side and a-side, when we rotate them by using l-line as a center, we'll get a circle that have c-side and a-side as radius of the circle. It became the cross-section of a cylinder D. For this side, if we rotate it, it will become the lateral surface of cylinder. The height of cylinder is determined by the height of the length side that rotates around l-line and becomes a cylinder.
Item 67	Smile	If this triangle rotates around l-line, it will become a cone. We know it by seeing the base and the length of the base. When it rotates, we'll get a cross-section of a cone in a shape of a circle. The length of the base is the radius of a circle or a cross-section of a cone. The reason why at the end of cone is sharp is that there is no distance from l-line. This is different from the first figure as that the lengths of that figure are equal, so there is no angle.
Item 68	Teacher	Can't you explain more about no angle?
Item 69	Gear	If there is a length from l-line, it'll be a circle, but if it isn't, it'll be a sharp-ended angle.
Item 70	Smile	Let's see the next figure. When the half-circle rotates around l-line, it'll become a sphere. It's because there is no distance between the top or the base of half-circle from l-line. In other words, there isn't an angle. But, there are distance between the other points and l-line, so it'll become a sphere.

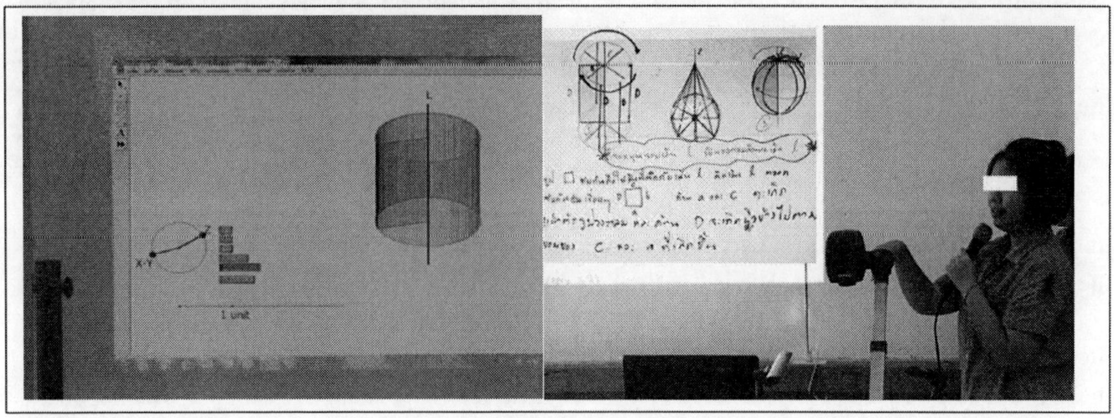

Figure 5. the steps of whole class discussion and comparison, regarding lesson unit 3: shapes rotation

According to protocol Item 65-70, the students presented the compositions that they thought would create solid shapes. They said that l-line is the center of the rotation and determined the conditions so that they could tell what parts in the solid shapes those lines are. For example, a-line and c-line are the radius of a circle and they make a cross-section of cylinder. D-side or the edge of the given figure is the side surface of solid shapes. Therefore, it can be interpreted that those students have mathematical conceptual understanding in **object level.**

6.4. Step 4 :Summarizing by Connecting Students 'Mathematical Ideas Emerged in the Classroom
In the previous step, GSP program is used to extend and compare mathematical ideas that emerged in the classroom. It can be seen that the students have mathematical conceptual understanding in Schema level.

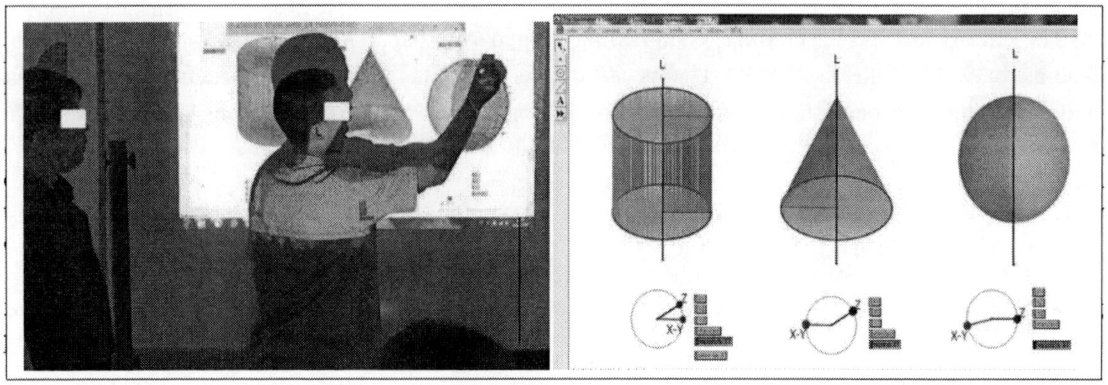

Figure 6. Step 4 :Summarizing by Connecting Students 'Mathematical Ideas Emerged in the Classroom

Table 4. An example protocol of students' mathematical conceptual understanding in schema level

Item 122	Teacher	Everyone can explain it very clearly. However, I'd like each of you try to conclude what you understand differently. Please try to see the connection between things. For example, a part of triangle does what function in transforming it into solid shapes.
Item 132	Gear	After rotating, the area of this figure will create an area of a shape.
Item 133	Teacher	Wow. That's incredible. No one has mentioned that. If we rotating the area of a figure, what would happen?
Item 134	Nice	Volume.
Item 135	Teacher	Kim said that it's an area of a shape, but Nice called it volume. So, what should we call it?
Item 136	Gear	If we rotate an area of a figure, it'll become an area of a shape, or what we call volume.

According to Item 136, "If we rotate an area of a figure, it'll become an area of a shape, or what we call volume", this illustrates that the students in Group 3 can connect the knowledge of area of a figure and volume of a solid shapes. This can be inferred that these students have mathematical understanding in the **level of Schema**.

7. Results of Study

According to the analysis of student's mathematical conceptual understanding about geometry, it is found that all of three groups have mathematical conceptual understanding according to APOS Theory (Dubinsky & McDonald, 2001). It can be seen that after being taught in the class that applied Lesson Study, Open Approach and The Geometer's Sketchpad, students have mathematical conceptual understanding in four levels.

The data is analyzed according to Inprasitha's Open Approach (2011), and it is also analyzed by Dubinsky & McDonald's APOS Theory (2001). As a result, student's mathematical conceptual understanding after being taught in the classroom that used Lesson Study, Open Approach and The Geometer's Sketchpad can be summarized as shown the following table:

8. Results of Study

According to the analysis of student's mathematical conceptual understanding about geometry, it is found that all of three groups have mathematical conceptual understanding according to APOS Theory (Dubinsky & McDonald, 2001). It can be seen that after being taught in the class that applied Lesson Study, Open Approach and The Geometer's Sketchpad, students have mathematical conceptual understanding in four levels.

The data is analyzed according to Inprasitha's Open Approach (2011), and it is also analyzed by Dubinsky & McDonald's APOS Theory (2001). As a result, student's mathematical conceptual understanding after being taught in the classroom that used Lesson Study, Open Approach and The Geometer's Sketchpad can be summarized as shown the the tabl 4:

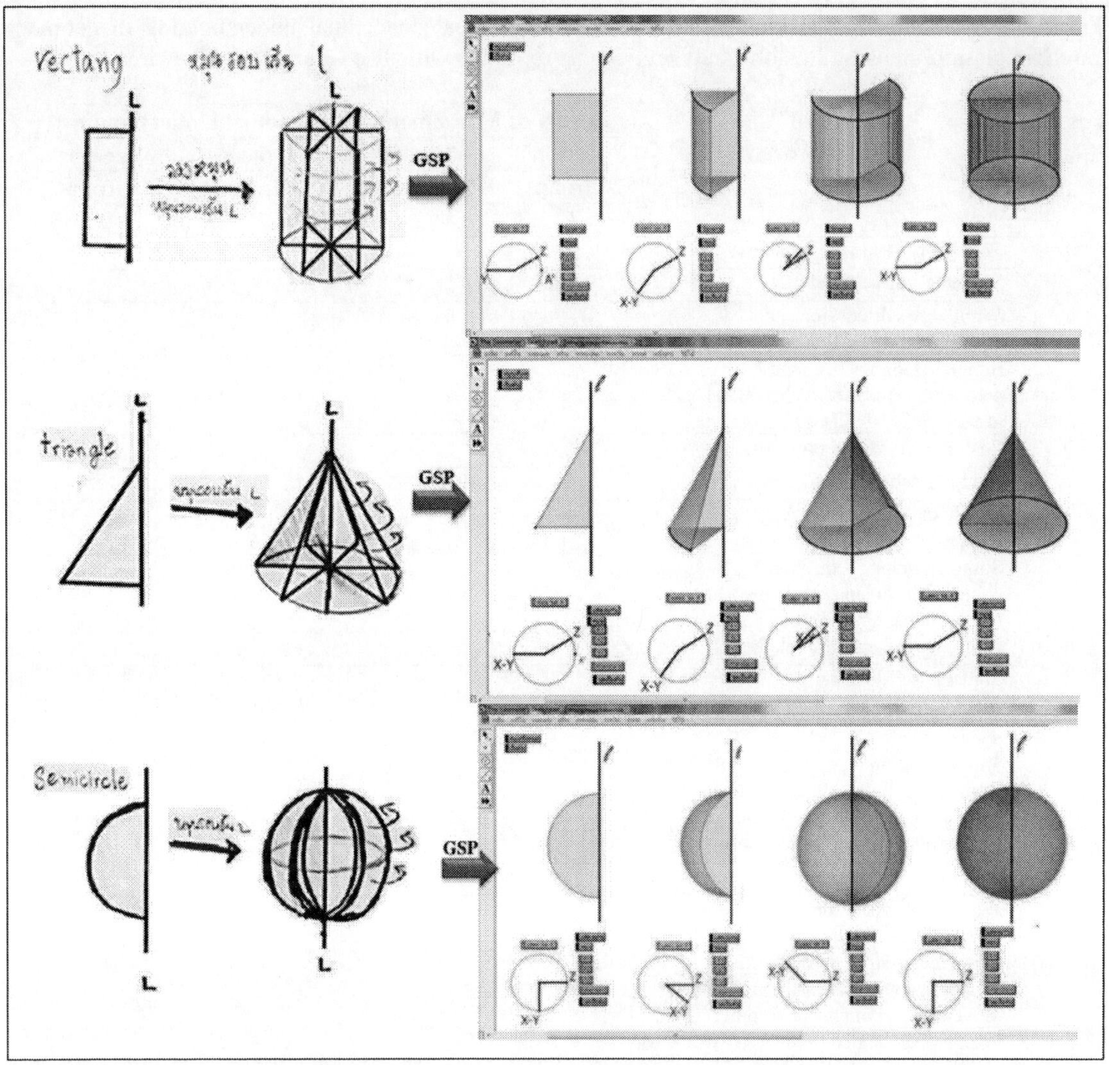

Figure 7. Demonstrates student's writing and using GSP in the steps of summarizing by connecting students' mathematical thinking emerged in the classroom, regarding lesson unit 3: shapes rotation

Table 5: The summary of levels of students' mathematical conceptual understanding in geometry taught in classroom using Lesson Study and Open Approach with The Geometer's Sketchpad

Lesson Plans	Steps of Open Approach	Levels of Mathematical Conceptual Understanding											
		Action			Process			Object			Schema		
		Group			Group			Group			Group		
		1	2	3	1	2	3	1	2	3	1	2	3
1	Posing open- ended problems	✓	✓	✓									
	Students'self-learning				✓	✓	✓						
	Whole class discussion and comparison							✓	✓	✓			
	Summarization through connecting students'mathematical ideas emerged in the classroom										✓	✓	✓
2	Posing open- ended problems	✓	✓	✓									
	Students'self-learning				✓	✓	✓						
	Whole class discussion and comparison							✓	✓	✓			
	Summarization through connecting students'mathematical ideas emerged in the classroom												✓
3	Posing open- ended problems	✓	✓	✓									
	Students'self-learning				✓	✓	✓						
	Whole class discussion and comparison							✓	✓	✓			
	Summarization through connecting students'mathematical ideas emerged in the classroom										✓	✓	✓
4	Posing open- ended problems	✓	✓	✓									
	Students'self-learning				✓	✓	✓						
	Whole class discussion and comparison							✓	✓	✓			
	Summarization through connecting students'mathematical ideas emerged in the classroom											✓	✓
5	Posing open- ended problems	✓	✓	✓									
	Students'self-learning				✓	✓	✓						
	Whole class discussion and comparison							✓	✓	✓			
	Summarization through connecting students'mathematical ideas emerged in the classroom												✓

According to table 4, it is readily apparent that in the step of posing open-ended problems, all of three target groups demonstrated their mathematical conceptual understanding in the action level. In the step of students' self-learning, it is found that the students have mathematical conceptual understanding in the process level. In the step of whole class discussion and comparison, it is found that the students have mathematical conceptual understanding in the object level. In the step of summarizing by connecting students' mathematical ideas emerged in the classroom, mathematical conceptual understanding in the schema level can be found only in some groups of the students.

9. Discussion

According to the research result, it can be seen that the classroom that has been applying Lesson Study, Open Approach and The Geometer's Sketchpad (GSP), the students were enabled to have mathematical conceptual understanding (MCU) in 4 levels: (1) MCU in Action Level can be seen when problem were posed. GSP program is used to pose problems and students investigate the details and conditions of problem situation. The students were able to follow the instructions (2) MCU in process level can be seen in the step of student's self-learning. The students can apply what they had observed form using GSP in the first step to solve problems by themselves, and it also brought them have multiple ideas. They can draw the connection of actions in problem solving and describe their ideas in descriptive manner. They can also present and elaborate their ideas. (3) MCU in object level can be seen in the step of discussion and comparison in class. GSP was used to extend and compare mathematical ides of each students in the classroom. The students can make conclusion as well as give definitions, principles from solving the problems. (4) MCU in schema level can be witnessed during the step of summarizing by connecting students' mathematical ideas emerged in the classroom. By using GSP, the students can make connection of their new knowledge with the previous knowledge and also the other knowledge of geometry.

10. Recommendation

10.1. Recommendation for Implementation of Research Findings

1. Teachers can use the research findings as the guideline for planning the lessons with the use of Lesson Study, Open Approach and The Geometer's Sketchpad so that they can encourage students to have mathematical conceptual understanding.
2. Teachers can use the research findings as guidance of technology usage in the problems solving classroom, and they can also use technology as cognitive technology.

10.2. Recommendation for Further Research

1. to extend the study field by using other program such as GeoGebra, Cabri 3D and so on, and to apply them in the classroom together with Lesson Study and Open Approach by choosing the program that is appropriate with the contents.
2. the learners should use the program to solve the problems by themselves. This will improve their ability in using technology. They would be more skillful in choosing, using and processing technology which are the skills expected according to the 5th item in Fundamental National Curriculum (2008).

11. Acknowledgments

This research is (partially) supported by The Centre of Excellence in Mathematics, The Commission on Higher Education, Thailand and Center for Research in Mathematics Education, Faculty of Education, Khon Kaen University and The Project of Development of Student's Mathematics Higher-Order Thinking at North East of Thailand, Faculty of Education, Khon Kaen University.

References

[1] Bootsikeaw S 2010 *The Study of Matthayom 6 Students 'Understanding in Linear Taught by Using the Geometer's Sketchpad as a Learning Tool* (Thesis .Master of Education)Mathematics(, Graduate School Khon Kaen University (in Thai)

[2] C C Meng and L C Sam 2011 *Encouraging the Innovative Use of Geometer's Sketchpad through Lesson Study* (Creative Education vol 2) pp236

[3] Chiangkool W 2016 *The Situation of Thai Edcation in 2014/2015 "How to Reform Thai Education to Catch Up with the 21st Century World"* (1st ed). Bangkok :Pim Dee Karn Pim. (in Thai)

[4] Corbitt M K) Ed (1985 *Impact of computing technology on school mathematics :Report of an NCTM conference* (Arithmetic Teacher vol 32) chapter 8 pp 14-18 (ERIC Doc. No. EJ318967)
[5] Dencha T 2013 *The Survey of Mathayom 4 students 'Understanding of Exponentiation taught by Using Open Approach* (Thesis .Master of Science and Technology, Graduate School Khon Kaen University (in Thai)
[6] Dimakos G, Nikoloudakis E, Ferentinos S and Choustoulakis E 2007 *Developing a Proof - Writing Tool for novice Lyceum geometry students* (The Teaching of Mathematics) chapter 2 pp 87-106
[7] Dubinsky E and McDonald M A 2001 *APOS :A constructivist theory of learning in undergraduate mathematics education research* In Holton, D) Ed(The Teaching and Learning of Mathematics at University Level :an ICMT Study pp 273 -280 (Dordrechet :Kluwer Academic Publishers)
[8] Habre S and Grundmeier T A 2007 *Prospective mathematics teachers 'views on the role of technology in mathematics education* (IUMPST :The Journal vol 3))Technology .(Retrieved from http//:www.k-12prep.math.ttu.edu/journal/technology/habre01/article.pdf
[9] Hiebert J and Carpenter T P 1992 *Learning and teaching with understanding* In D A Grouws (Ed) Handbook of research on mathematics teaching and learning pp 65 -97 (New York : Macmillan Publishing Company)
[10] Inprasitha M 2010 *Adapting Lesson Study in APEC Member Economies* Proceeding of APEC Conference on Replicating Exemplary Practices in Mathematics Education pp.127-132 (Surat Thani :Human Resources Development Working Group)
[11] Inprasitha M 2011 *One Feature of Adaptive Lesson Study in Thailand Designing Learning Unit* (Journal of Science and Mathematics Education in Southeast Asia vol 34) chapter 1 47-66
[12] Inthaprsitha M 2003 *The Reformation of Mathematic Studies in School by Focusing on Mathematical Approach* (Khon Kaen :Khon Kaen Printin (in Thai)
[13] Inprasitha M 2014 *Processes of Problem Solving in School Mathematics* (Khon Kaen: Penprinting, Inc) (in Thai).
[14] Isoda M 2010 *Lesson Study :Problems Solving Approaches in Mathematics Education as a Japanese Experience* (Procedia Social and Behavioral Sciences vol 8) pp 17-27
[15] Jaion K 2007 *The Development of Learner's Mathematical Thinking by Using Learning Approach that Focuses on Applying the Geometer's Sketchpad* (Thesis, Master of Education)Mathematics(, Graduate School Khon Kaen University) (in Thai)
[16] Kaewdaeng R 1997 *The Reformation of Thai Education* (Bangkok :Phrae Wittaya)) in Thai.(
[17] Kor L K and Lim C S 2009 *Lesson Study :A Potential Driving Force behind the Innovative Use of Geometer's Sketchpad* (Journal of Mathematics Education vol 2) chapter 1 pp 69-82
[18] Kümmerer B 2001 *Trying the Impossible -Teaching Mathematic to Physicists and Engineers . In D .Holton)Ed(* (The teaching and learning of mathematics at university level :An ICMI study) pp 321–334 Dordrecht, The Netherlands :Kluwer Academic Publishers
[19] Laborde C 2001 *Integration of technology in the design of geometry tasks with Cabri-Geometry* (International Journal for Computers in Mathematical Learning vol 6) chapter 3 pp 283 -317
[20] National Council of Teachers of Mathematics)Ed(2000*Principles and Standards for School Mathematics* (Reston, Va :NCTM)
[21] Nohda N 2000 *A Study of "Open-Approach "Method in School Mathematics :Focusing on Mathematical Problem Solving Activities* (Paper presented at ICMI 9, August 2000, Makuhari, Japan)
[22] Pegg J 1995 *Learning and teaching geometry* In L Grimison) (Ed.) (Teaching secondary school mathematics :Theory into practice) pp 36-74 London :Harcourt Brace
[23] Sawat Phuthong 2014 *Institute of Technical Education :Options to improve education and quality of education* (Thailand Education Journal vol 11) chapter 117 pp 57-58 (in Thai)

IOP Conf. Series: Journal of Physics: Conf. Series **1340** (2019) 012088 doi:10.1088/1742-6596/1340/1/012088

[24] Thawatchai Dencha 2015 *Investigation of 10th Grade Students 'Mathematical Understand Levels on Exponent in Classroom Using Taught by Open Approach* Master of Education Thesis in Science and Technology Education, Graduate School, Khon Kaen University .

[25] The Institute for the Promotion of Teaching Science and Technology 2005 *The Manual of The Geometer's Sketchpad* (Bangkok :IPST)

IOP Conf. Series: Journal of Physics: Conf. Series **1340** (2019) 012089 doi:10.1088/1742-6596/1340/1/012089

Student's Concepts of Subtraction in Mathematics Classroom Using Lesson Study and Open Approach

S Phukongchana[1,2]**, S Thinwiangthong**[1,2]

[1] Mathematics Education, Faculty of Education, Khon Kaen University, Thailand
[2] Centre of Excellence in Mathematics, CHE, Bangkok 10400, Thailand

Author's e-mail address: sawitree_pokky@hotmail.co.th , sampan@kku.ac.th

Abstract. The purpose of this research was to analyze mathematical concept of students through mathematical representation in the classroom of using Lesson Study and Open Approach. The target group of this research was 25 Grade 1 students at Bansomkob (Ratradbamrung) School, Kaset Sombun District, Chaiyaphum Province in the academic year 2016. The school has been participating in the Project of Development of student's Mathematics Higher-Order Thinking at North East of Thailand, and also the professional development process through Lesson Study and Open Approach of Inprasitha [6] 's framework. The collected data included: 1) protocol data transcribed from voice and video recorders during classroom instruction 2) students' writings and 3) protocol from interviews with students. The data were analyzed by using Vergnaud [15]'s conceptual framework. The results of this research indicated that student's concepts of subtraction in mathematics classroom using Lesson Study and Open Approach consisted of 1) set of situations: students were involved to the real world situations which was meaningful to them so they could describe the situations while 'take away' and the number of objects in the situations decreased and 'Take different' they could describe situations that there was one more thing than one. 2) set of operational invariant: students' concept of subtraction 'take away' students presented the problem solving by removing objects from the situations so the number was decreased, while the similar or analogous operation occurred in each situation repeatedly in relation to other situations until become operational invariant and students' concept of subtraction 'take different' students presented the problem solving by operation of matching for comparison was also found before figuring out the emergent number of the concept on subtraction of taking different. and 3) set of symbolic representations: students used symbols as sentences or words through the forms of natural language such as being 'expelled, 'scooped out', 'disappeared', 'none', 'given', in concept of 'take away' and 'over' in concept of 'take different' as well as the sentences or words in mathematical language and other symbols which included number sentence, mathematical sign drawing a line, circle, and arrow for representing the operation of being 'took out', 'eaten', 'gone, and 'took away' whereas concept of 'take different' student drew line to matching one to one thing and drawing a line, circle, and arrow for representing the operation of 'over'

1. Introduction

The method of mathematics instruction that involves the component of only teacher, students, and content, as teacher only plays a role of transferring the content to students by teaching them to remember and focusing on informing, explaining, and learning achievement through the test rather than emphasizing on learning process, does not promote students to think reasonably, and there is no mathematical learning process among them. As a result, there is no context to promote mathematical learning process for students. In addition, student's thinking is obstructed. So traditional classroom is required to change as a new classroom, which emphasizes on both process and result [4] Mathematics classroom that employs learning management using Lesson Study and Open Approach innovations is considered as a problem solving classroom through the Open Approach as a teaching approach [5]. These innovations are adapted to use in Thailand by the Center for Research in Mathematics Education, Khon Kaen University. Takahashi [12] stated that the Lesson Study is a process that helps improving mathematics instruction and significantly affects learning process of both teacher and student: it helps promoting the ability of adapting knowledge and understanding gaining from classroom. Inprasitha [7] indicated the clear characteristics of Lesson Study system in which teachers shall jointly discuss and develop learning management tools and plans, and mainly focuses on the use of open-ended questions. This will be based on the contexts of student's learning and thinking in association with Open Approach. The aim of teaching through Open Approach is to emphasize on encouraging all students to be able to think and solve mathematical problems by self-learning, and expand or increase the quality of mathematical process [9].

One of the significant mathematical processes, according to the National Council of Teachers of Mathematics [11], is mathematical representation which is needed for learning mathematics through problem solving. This is consistent with Despina [2] who mentioned that representation is considered as a key of solving mathematical problems. The ability of dealing with mathematical representation is a significant component of mathematical writing and reading abilities in which students must be promoted to use representation in order for figuring out the ways of problem solving [17].

In terms of classroom contexts emphasizing on problem solving, Isoda and Nakamura [8] indicated the significant representations on expressing the ideas in the classroom of problem solving, such as sentence, diagram, arrow. These representations are categorized in the symbol representation, which can be the facilitator for communicating the linkage and relation of operational invariant through the concept [14]. Vergnaud [15] and Inprasitha [6] also stated that the integration of this concept can be initiated in the Open Approach classroom while students are trying to solve the problems throughout searching, collecting, designing and presenting various approaches.

As a result, the mentioned discussion represents a significance of representation which affects the communication of student's idea and problem solving. The representation is related to operational invariant resulting into the concept. This can be happened in the classroom that emphasizes on solving problem using Open Approach for instructional management as it widely opens for problem solving and promotes students towards various thinking. Student also deals with problem solving, searching, collecting, and designing by oneself. Thus, the researcher is interested in analysing student's concepts on emergent representations towards mathematical problem solving on subtraction in the classroom using Lesson Study and Open Approach.

2. Research Objectives

This research aimed to analyze students' concepts on subtraction through mathematical representations in the classroom using Lesson Study and Open Approach.

IOP Conf. Series: Journal of Physics: Conf. Series **1340** (2019) 012089 doi:10.1088/1742-6596/1340/1/012089

3. Research Methodology

3.1 . Target Group

The target group of this research was 25 Grade 1 students at Bansomkob (Ratradbamrung), Sraponthong Sub-district, Kaset Sombun District, Chaiyaphum Province during the first semester of academic year 2016.

3.2 Research Instruments

- Mathematics textbook for Grade 1 students translated from Japanese to Thai by Center for Research in Mathematics Education.
- The 7 lesson plans for unit 5 of mathematics learning management for Grade 1 students entitled Subtraction)1(;
- Video recorder for recording classroom instructional activities, which will be used as a key data for analysis;
- Audio recorder for recording voices of teacher and students during instructional activities, which will be used as a key data for analysis. Voices were recorded by researcher before starting classroom;
- Camera for capturing students 'pieces of work and classroom instructional activities, which will be used as a part of data analysis;

4. Data Collection

4.1 Researcher spent a month to study general background of the school and target classroom of data collection. Learning behaviors of students were observed. Mathematics teacher and homeroom teacher were asked for additional information. School director was also asked for the use of Lesson Study and Open Approach innovations of the classroom.

4.2 Researcher employed the Lesson Study to create the lesson plan of Open Approach. Each class involved 5 persons to jointly create the plan, including the teacher of Grade 1 student, the school director who had jointly observed Grade 1 classroom throughout academic year, and 3 master degree students majoring in Mathematics Education from the Faculty of Education, Khon Kaen University.

4.3 The meeting between researcher and research assistants was held for understanding the framework of data analysis, as well as clarifying the roles of researcher and research assistants on recording voices, still photographs, and videos, as a preparation before conducting data analysis.

After that, the created lesson plans were implemented to instructional management at the target classroom. The classroom was collaboratively observed. There were also comprehensive data record and collection according to the plans mentioned above. The reflection was then conducted by considering students' ideas during classroom activities and their pieces of work, and the issues related to representations raised by students in the class. During the lesson plan that was being created, there was a discussion about students' ideas that came out during instructional activities.

5. Data Analysis

This research employed qualitative research methodology. Data was the behaviors during teaching and learning mathematics on subtraction that were recorded by video recorder and audio recorder. The wordings were transcribed as the item of wordings or dialogue. Students' behaviors were also described as shown in the brackets, which called protocol transcription. In addition, pieces of writing of target students written during classroom activities were brought to analyze towards mathematical concepts of students on subtraction based on conceptual framework of Vergnaud [15] and Inprasitha [6].

6. Results

Components of students' mathematical concepts were as follows.

6.1. Set of Situations:

Problem situations meant to students, which related to students' real world. Students could explain those situations with the concept of subtraction on taking away as it represents the existence and disappearance, or number of the objects in that situation decreased (See Figure 1). And the concept of subtraction on taking different as the problem situations was related to real world and meant to students towards overviewing number of a thing that was much more than the other (See Figure 2).

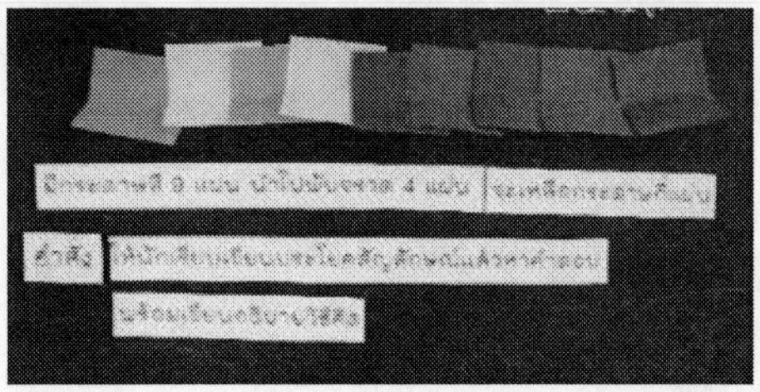

There are 9 sheets of color paper. 4 were brought to make rockets. How many sheets are there left?

Figure 1. An example of situation in concept 'take away' that means to students

How many dogs are more than cats?

Figure 2. An example of situation in concept 'take different' that means to students

6.2. Operational Invariant:

Students presented the problem solving by removing the objects from the situation so the number decreased (See Figure 3 and table 1). The operation of each situation occurred similarly or analogously, as well as occurred repeatedly with other operation until become invariant. The operation of matching for comparison was also found before figuring out the emergent number of the concept on subtraction of taking different (See Figure 4).

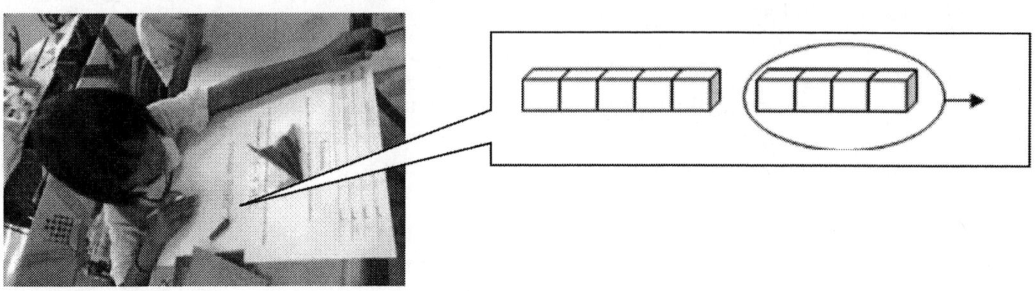

Figure 3: An example of students' operations in concept 'Take away'

Table 1. An example of students' operational protocol

Item 164	Teacher	How many sheets of paper did you fold?
Item 165	Meen	Four sheets.
Item 166	Meen	)draw the pictures of nine blocks on the paper sheet(
Item 167	Teacher	How many folded sheets? How many in total?
Item 168	Meen	Nine.
Item 169	Teacher	So there are nine in total.
Item 170	Meen	Take away) .draw the circle covering four blocks then draw the arrow pointing out(
Item 171	Meen	Four, Five, Six, Seven.

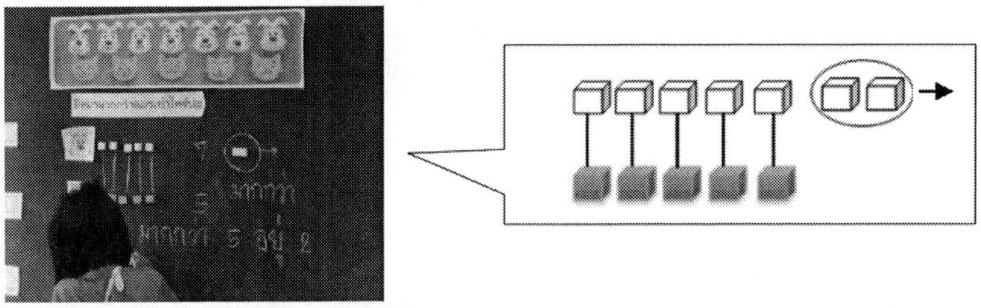

Figure 4. An example of students' operations in concept 'Take different'

Table 2. An example of students' operational protocol

Item 208	Teacher	How to prove that?
Item 209	Pam	Matching
Item 210	Teacher	Ah matching and then, which's it over.
Item 211	Fam	second (He pointed to two blocks in blackboard.)
Item 212	Teacher	Was two pieces that you took away.
Item 213	Fam	Yes.
Item 214	Teacher	What is that arrow mean?
Item 215	Fam	Take away
Item 216	Teacher	Take away. So, are their equal?
Item 217	Student	equal
Item 218	Teacher	Why you take two this block away?
Item 219	Student	It's over

6.2. Set of Symbolic Representations:

The representations were made for explaining and communicating the ideas of solving operational problem in that problem situation through symbolic representations of operational invariant. This was conducted for solving the set of open-ended problem situation towards subtraction, which included 1) sentences of students' natural language: 'expel', 'take out', 'scoop out', 'fly away' for concept of 'take away' (See Figure 5 and table 3) and 'over' for concept of 'take different'(See Figure 8 and table 6) 2) numeral symbol and mathematical symbol: subtraction (See Figure 4 and table 7 for 'take away' and Figure 10 and table 8 for 'take different') and 3) diagram: 'drawing a circle' 'arrow' as a part of operation in order for representing the meaning of that operation for concept of 'take away' (See Figure 6 and table 4) and 'drawing line matching' for comparison was also found before figuring out the emergent number of the concept on subtraction of taking different and 'drawing a circle' 'arrow' for representing. (See Figure 9 and table 7)

IOP Conf. Series: Journal of Physics: Conf. Series **1340** (2019) 012089 doi:10.1088/1742-6596/1340/1/012089

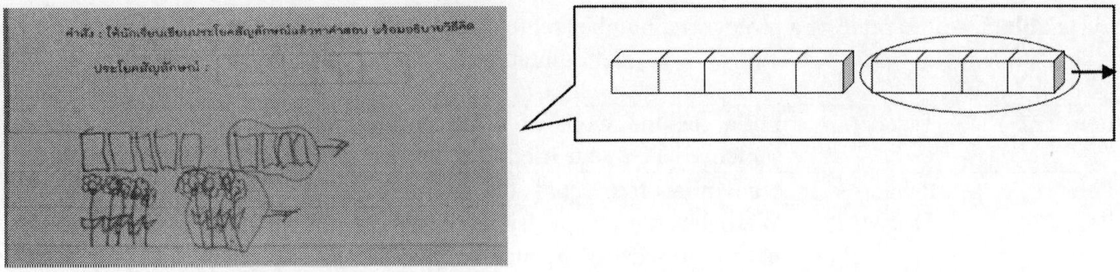

Figure 5. An example of a diagram as a symbolic representation in concept 'take away'

Table 3. An example of a diagram protocol as a symbolic representation

Item 166	Meen	)draw the pictures of nine blocks on the paper sheet(
Item 167	Teacher	How many sheets that were folded? How many in total?
Item 168	Meen	Nine
Item 169	Cheppe	So there are nine in total
Item 170	Meen	Take away) .draw the circle covering four blocks then draw the arrow pointing out(

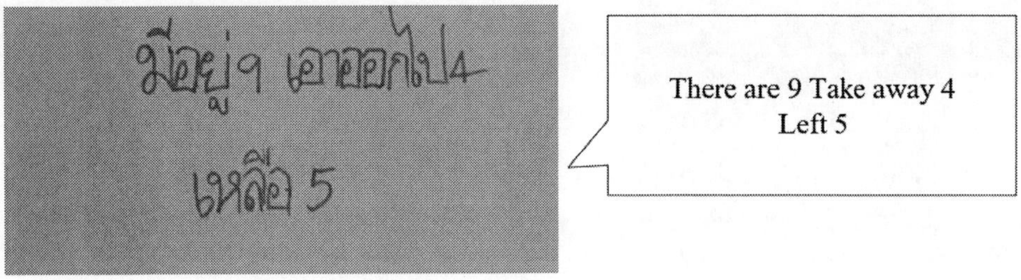

Figure 6. An example of a natural language as a symbolic representation in concept 'take away'

Table 4. An example of a natural language protocol as a symbolic representation

Item 166	Meen	)draw the pictures of nine blocks on the paper sheet(
Item 167	Teacher	How many sheets that were folded? How many in total?
Item 168	Mean	Nine.
Item 169	Cheppe	So there are nine in total.
Item 170	Meen	Take away) .draw the circle covering four blocks then draw the arrow pointing out(
Item 171	Teacher	Look at the circle drawn by your friend .What does it mean?
Item 172	Meen	<u>Take away.</u>
Item 173	Chon	<u>Leave out.</u>

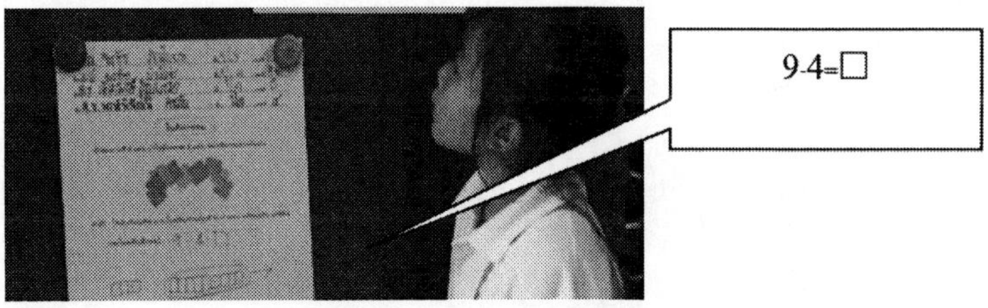

Figure 7. An example of number sentence and mathematics symbol as a symbolic representation

Table 5. An example of a protocol of number sentence and mathematics symbol as a symbolic representation

Item 222	Teacher	Let's see the work of other group .Pam, who wrote this number sentence? Tell your friends about it.
Item 223	Pam	Nine minus four equals the squares
Item 224	Teacher	Well, listen to your friend .Look at the number sentence of your friend .So what is the number sentence of this group? Pam, try to say it again.
Item 225	Pam	Nine minus four equals the squares
Item 226	Teacher	Well, nine minus four equals the squares .So, what is this? right here? What is this? Let's look at your friend's picture .Pam, you drew this one for me .What does it mean? What does it represent?
Item 227	Students	Represent the block.
Item 228	Teacher	Represent the block .Let's see Pam's picture .What is it?
Item 229	Pam	It's block.
Item 230	Teacher	So, what does block represent?
Item 231	Pam	Paper.

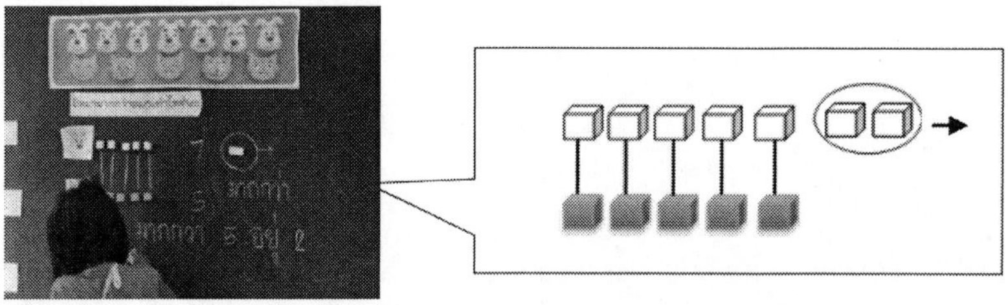

Figure 8: An example of a diagram as a symbolic representation in concept 'Take different'

Table 6. An example of students' operational protocol

Item 208	Teacher	How to prove that?
Item 209	Pam	matching
Item 210	Teacher	Ah matching and then, which's it over.
Item 211	Fam	second (He pointed to two blocks in blackboard.)
Item 212	Teacher	Was two pieces that you took away.
Item 213	Fam	Yes.
Item 214	Teacher	What is that arrow mean?
Item 215	Fam	Take away
Item 216	Teacher	Take away. So, are their equal?
Item 217	Student	equal
Item 218	Teacher	Why do you take two this block away?
Item 219	Student	It's over

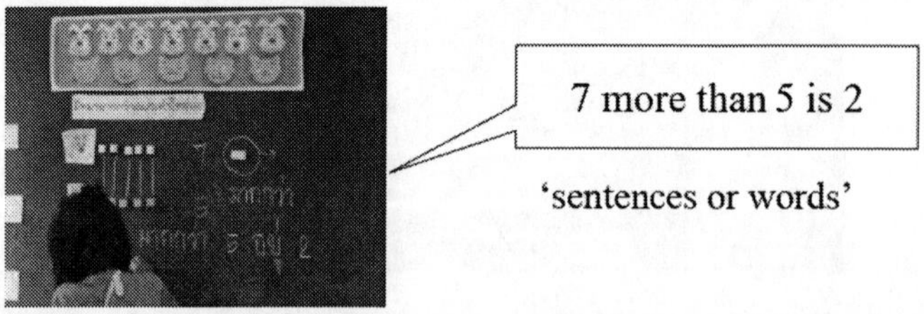

Figure 9. An example of a natural language as a symbolic representation in concept 'Take different'

Table 7. An example of students' operational protocol

Item 218	Teacher	Why do you take two this block away?
Item 219	Student	It's over
Item 220	Teacher	Two blocks are over. What does it mean?
Item 221	Pam	It's over, more than 5
Item 222	Teacher	Can you write?
Item 223	Pam	(She wrote 7 more than 5 is 2)

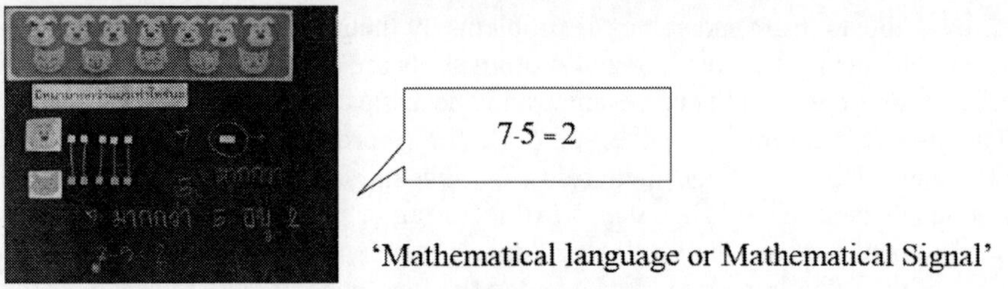

Figure 10: An example of number sentence and mathematics symbol as a symbolic representation

Item 224	Teacher	Seven more than five is two. Can we write in mathematical signal form? What is mathematical signal.
Item 225	Phai	Seven minus two
Item 226	Teacher	Why can you describe
Item 227	Pam	No no it's seven minus five
Item 228	Teacher	Can you write in blackboard.
Item 229	Pam	)wrote 7-5 = 2)
Item 230	Teacher	What is result.
Item 231	Meen	Two

Table 8: An example of students' operational protocol

7. Conclusion

According to the findings of this research, the mathematical concepts were found in the classroom of using Lesson Study and Open Approach, which consisted of 1) Set of Situations: problem situations that were meaningful to students. Students were involved to the real world situations which was meaningful to them so they could describe the situations while 'take away' and the number of objects in the situations decreased and 'Take different' they could describe situations that there was one more thing than the one. 2) Operational Invariant: students' concept of subtraction 'take away' students presented the problem solving by removing objects from the situations so the number was decreased, while the similar or analogous operation occurred in each situation repeatedly in relation to other situations until become operational invariant and students' concept of subtraction 'take different' students presented the problem solving by operation of matching for comparison was also found before figuring out the emergent number of the concept on subtraction of taking different and 3) set of symbolic representations: set of symbolic representations: students used symbols as sentences or words through the forms of natural language such as being 'expelled, 'eaten', 'scooped out', 'disappeared', 'none', 'given', in concept of 'take away' and 'over' in concept of 'take different' as well as the sentences or words in mathematical language and other symbols which included number sentence, mathematical sign drawing a line, circle, and arrow for representing the operation of being 'took out', 'eaten', 'gone, and 'took away' whereas concept of 'take different' student drew line to matching one to one thing and drawing a line, circle, and arrow for representing the operation of 'over'. These represented the connected relation and were used to communicate about it in order for representing the meaning of operation at that time. In the contexts of Lesson Study and Open Approach as a teaching approach, teachers collaboratively develop the teaching as a team according to the steps of Lesson Study and Open Approach. The team collaboratively create lesson plan, observe the classroom activities, and reflect the teaching after instruction These cooperative works facilitate students in the classroom to learn through expressing ideas, self-learning, discussing the results and summarizing them together. So these students could construct mathematical concepts.

Isoda [8] stated that the basic principle of mathematics instruction emphasizing on problem solving is teaching students to learn mathematics by oneself. This means the teacher must develop students to be able to think and solve the problems by themselves. And the representation is necessary for Grade 1 students as their reading and writing skills are at basic level, which is difficult for them to communicate their ideas.So the representation is necessary as a facilitator for communicating ideas of kids. Inprasitha [6] claimed that the use of Open Approach as a guideline for teaching can have teachers to considerably change their roles. During the second stage of teaching through the Open Approach, teachers have to change the role of lecturing or giving knowledge to be an observer in order to make a note of students' ideas during solving problems by themselves. This research finding is consistent with [7] who introduced and adapted the Lesson Study and the model of problem solving to use in Thailand. The findings indicated the meaningful understanding in the classroom of Lesson Study and Open Approach towards integral number. It was also observed that each student had own style to use diagram and arrow. These represent the development of mathematics classroom and the change in students' active learning. In other words, teaching mathematics in the classroom of Open Approach aims at preparing students to reach the challenge situations by using potential open-ended problems, which will respond to individual difference both in terms of abilities and understanding, and in order for supporting the development of the way of students' mathematics thinking [6]

8. Recommendation

8.1. Recommendation for Implementation of Research Findings

8.1.1. This research can be a guideline for teachers to understand the students' concepts in problem solving. And it can be used to predict students' ideas and concepts in order for planning next classes' learning management.

8.1.2. This information can facilitate teachers to use the meaningful situation, invariant feature, and the meaningful symbolic representations and to promote students to understand more and be able to better solve the problems.

8.2. Recommendation for Further Research

8.2.1. There should be other studies on students' mathematical concepts through representations of mathematical operation apart from subtraction. As this research employed the mathematics textbook translated from Japanese textbook, it helped promoting various representations among students. Further research should study other concepts of mathematical skills apart from representations, e.g., communication.

8.2.2. There should be a study on the use of representations for promoting the concept construction among students with special need in order for help promoting mathematical thinking for students with special need.

9. Acknowledgments
This research is (partially) supported by The Centre of Excellence in Mathematics, The Commission on Higher Education, Thailand and Centre for Research in Mathematics Education, Faculty of Education, Khon Kaen University and The Project of Development of student's Mathematics Higher-Order Thinking at North East of Thailand, Faculty of Education, Khon Kaen University

References
[1] Changsri N 2015 *First Grade Students' Mathematical Ideas of Addition in The Context of Lesson Study and Open Approach* (Conference Proceedings of ICMI Study 23. June) pp 3-7 2015 Macau, China.
[2] Despina A 2010 *An examination of middle school students' representation practices in mathematical problem solving through the lens of expert work: towards an organizing scheme* (Springer Science Business Media B.V. 2010)
[3] Hitotsumatsu S 2010 *Mathematics Textbook for Grade 1 Students* (Translated by Maitree Inprasitha) Khon Kaen: Klang Nana Wittaya.
[4] Inprasitha M 2003 *A Reform of Learning Process of Mathematics Subject in Schools Emphasizing on Mathematical Process* (Khon Kaen: Khon Kaen Printing)
[5] Inprasitha M 2011 *One Feature of Adaptive Lesson Study in Thailand: Designing a Learning Unit* (Journal of Science and Mathematics Education in Southeast Asia 2011 vol 34) No 1 pp 47 – 66
[6] Inprasitha M 2015 *An Open Approach Incorporating Lesson Study: An Innovation for Teaching Whole Number Arithmetic* (Conference Proceedings of ICMI Study 23 June) pp 3-7 2015. Macau, China pp 315-322
[7] Inprasitha M 2015 *Preparing Ground for the Introduction of Lesson Study in Thailand* (Lesson Study Challenges in Mathematics Education) pp 109-122 Singapore. World Scientific Publishing Co. Pte.
[8] Isoda and Katagiri 2012 *Mathematical Thinking, How to develop it in the classroom* (Singapore: World Scientific Publishing Co Pte Ltd)
[9] Loifa S and Inorasitha M 2004 *Development of New Teaching Profession Promoting Mathematics Learning* (KKU Journal of Mathematics Education vol 1) chapter 1 pp 18-28
[10] Ministry of Education 2009 *Basic Education Core Curriculum B.E. 2008* (Bangkok: The Agricultural Federative Co-operation of Thailand)
[11] National Council of Teacher of Mathematics 2000 *Principles and standards for school mathematics* (Reston: The National Council of Teachers of Mathematics, Inc)
[12] Takahashi A 2015 *Lesson Study: An Essential Process for Improving Mathematics Teaching and Learning* (Lesson Study Challenges in Mathematics Education) pp 51-58 Singapore. World Scientific Publishing Co. Pte.
[13] Vergnaud G 1982 *Cognitive and Developmental Psychology and Research in Mathematics Education: some theoretical and methodological issues* (For the Learning Mathematics vol 3) chapter 2 (November 1982)
[14] Vergnaud 1996 *The Theory of Conceptual Fields Theories of Mathematical Learning* (New Jersey: Lawrence eribaum associates Inc Publisher, 1996)
[15] Vergnaud 1997 *The Nature of Mathematical Concepts. Learning and teaching mathematics* (England: Psychology Press/Erlbaum (UK) Taylor & Francis)
[16] Vergnaud 2013 *Conceptual Development and Learning* (Revista Qurriculum vol 26) marzo 2013, pp 39-59
[17] Zhe L 2012 *Survey of Primary Students' Mathematical Representation Status and Study on the Teaching Model of Mathematical Representation* (Journal of Mathematics Education, August 2012, Vol 5) No 1 pp 63-76

Students' Words in Mathematical Communication in Classroom Using Lesson Study and Open Approach

T Wiriyaudomsatean [1,2], S Thinwiangthong[1,2]

[1] Mathematics Education, Faculty of Education, Khon Kaen University, Thailand
[2] Centre of Excellence in Mathematics, CHE, Bangkok 10400, Thailand

Author's e-mail address: thanabat_wir@hotmail.com , sampan@kku.ac.th

Abstract. This study aimed to investigate student's words in mathematical communication in classroom using Lesson Study and Open Approach. The target group was 8 eleventh grade students (separated 2 groups and 4 people per group) in which studying in the Intensive Science and Mathematics English Program, the second semester of 2016 school years at Boonwattana School. The data collection tools were 5 lesson plans of probability learning unit, filed note, video, voice recorder and interview note. For the data analysis, Emori [2]'s framework was used to interpret students' mathematical communication followed by Cathcart, Samovar and Henman [1]'s framework in order to identify the type of students' words using. The study results revealed that in the classroom using Lesson Study and Open Approach, the students express 3 characteristics and use words as detail: 1) Rigorousness, while the target group attempt to grasp problem situations and solving problem. The words were used as an express their ideas, the students communicate by using concrete words and descriptive words, after that their friends able to perceive it. 2) Economy, it was noticeable in students' communication which was precisely explained and exactly perceived by their friends, the students communicate by using abstract words, concrete words and judgmental words after that their friends able to show evaluation and personal react which were considered. 3) Freedom, the students have independent and different thinking. The students communicate by using abstract words, concrete words and descriptive words in order to explain their ideas independently. In all of 3 characteristics, it did not found polarised words, the students did not use words to evaluate and argue their friends' ideas as completely difference.

1. Introduction

Learning in the modern world calls for students to interact with others more. Including in the world of work and coordinating with others is very important. Communication skills are a tool to prepare the students for the real world [8]. But most of the teaching and learning management is focused on the interaction between the instructor and the learner, or students with subject matter, but ignore the relationship between the learner and the learner [7]. Most of the teaching and learning management in Thai schools was taught in the school which focuses on learning achievement do not care about the learning process of the students. It is obvious that most teachers focus on teaching by paying attention to teachers as the main way to transfer knowledge from teacher to learner by various forms such as lecture demonstration of teachers with various devices or explanation. The model of the educational activities should be emphasized so that students can learn and apply in everyday life to succeed in the future. The learning activities for students in mathematics are the same. We must strive to provide instructional activities that students can link to solve the problem by themselves. This is an opportunity for students to learn by their potential [4].

Open Approach is a teaching approach that focuses on problem solving. Students can use their existing knowledge to solve problems that have been created until they have developed their way of thinking. Teachers must try to understand the students' ideas that arise during problem solving in order to motivate and enhance students' ideas. Students are encouraged to share their ideas. There are 4 steps of open approach; 1) Posing open-ended problems, 2) Students' self-learning, 3) Whole class discussion and comparison and 4) Summing-up by connecting students' emergent mathematical ideas. There are 3 types of open-ended problem; 1) Processes are open, refers to the problem that have several solving methods, 2) Answers are open, the problem with a wide range of answers or multiple answers, and 3) Problem to problem, it means that students can develop new problem from an original problem [4].The problem situation in the open approach is the openness in terms of having a wide range of solutions or multiple answers. This problem will allow students with different abilities to use their knowledge to solve problems manually. Once students have solved their problems, they will be able to draw their own mathematical ideas. In Thailand, open approach was integrated with lesson study by Assoc. Prof. Maitree Inprasitha since 2002 [3].

Stigler and Hiebert [9] said that lesson study was teacher development system is used in schools. Teachers develop lesson plans before teaching, bring developed plan in the real classroom and observation, reflection of the lesson and develop the lesson plans. The obvious thing is that the teacher has improved the lesson together. The important thing in the classroom is to plan for the development of the learner in the classroom and sharing knowledge with other teachers. Inprasitha, Inprasitha and Pattanajak [4] and Inprasitha, Loipha and Silanoi [5] said that lesson study was challenges in the classroom. It was the development of learning in the classroom that needs to change and sharing knowledge and problems in class with other teachers, Including the teacher's perception of the goal of teaching together. Inprasitha, Loipha and Silanoi [5] stated that lesson study has 3 phases as follows; 1) Collaboratively design research lessons (Plan), 2) Collaboratively observing the research lessons (Do) and 3) Collaboratively reflection or post-discussion (See). Based on the concept of lesson study above, it was a way to develop the teaching of a team of teachers who exchange knowledge for use in developing lesson plans that bring open-ended problems together. Then put the lesson plan into use in the classroom by a teacher and classroom observation by observing teachers, and the teachers together reflect the results of the teaching to develop the next lesson together.

Mathematical communication was integrating activities with problem solving; include discussion, talk, interaction, and reasoning related to solving mathematics problems. In addition, communication in the learning process and teaching mathematics is also an important part and need for mathematics. Because communication is a way of exchanging ideas and creating a clear understanding of mathematical problems when considering the characteristics of communication. Then consider the word "Communication" focus on the mathematical thinking process. Mathematical communication is the mathematical relationship of the participants in the communication process. Mathematical communication is also an important key to developing mathematical problems. Most teachers are interested in the amount of speech a student has, but the effectiveness of such mathematical communication has never been determined through the expressions of the words communicated, and the memory of man is to determine things through words. That is, we will know, we have to create in the form of terminology used in mathematical communication or doing something in the mathematical structure [2].

The main communication problem is caused by the use of specific words. This is because the content of the audience is not consistent with the meaning of the listener. These problems can be disappeared by explaining the idea with the use of words in the reasoning [11]. In addition, the terms used in the classroom in mathematics are based on the ability of each student to express their own words. The development of the appropriate form of the word depends on the student's choice of words to suit the context. Of course, there is a lot of information about the nature of the message itself.

IOP Conf. Series: Journal of Physics: Conf. Series **1340** (2019) 012090 doi:10.1088/1742-6596/1340/1/012090

Cathcart, Samovar, Henman [1] said that the choice of words and the meanings of words can express the self-concept. The concept of the messenger is used to confirm and adjust the language to suit the communication and reduce the distortion of the substance.

From the above that the mathematical communication had important skill and ability. The mathematical communication is the mathematical thought process of participants through communication. Keyword selection and word meanings can express the concept of the dispatcher and the message. Mathematical communication is also important in developing mathematical problems. In teaching and learning at the high school, students do not focus on communication, but mathematical learning management uses open approach innovation. The open approach determines the mathematical activity. Students are encouraged to express their ideas in problem solving. The mathematical instruction is consistent with the concept of mathematical learning. This is a guideline for development of mathematical learning activities that promote meaningful mathematical learning. The study of the use of words in mathematical communication of students in grade 11 using open approach is important issue to study.

2. Research Objectives

The research objective is to survey the 11-grade students' words in mathematical communication in classroom using lesson study and open approach.

3. Research Methodology

3.1. Target Group

Target group were 8 students in grade 11, which studying in the Intensive Science and Mathematics English Program, the second semester of 2016 school years at Boonwattana School.

3.2. Research Instruments

- Lesson plans of 5 mathematics lessons on probability for 11-grade students.
- Field note for recording student's words in mathematical communication during classroom instruction.
- Interview note for taking note of students' opinion on their mathematical communication and words using.

4. Data Collection

The researcher and research assistant were planed data collection collaboratively. Data were collected by fieldnote recording through classroom observing, video recordings during classroom instruction, and audio recording during interviews after classroom activities. Video and audio were transcript to protocols of classroom instruction and interviews.

5. Data Analysis

Data were analyzed by analytic description according to the method of protocol analysis. Data were analyzed by using frameworks of mathematical communication [2] including of 1) rigorousness 2) economy and 3) freedom, and words using [1] including of 1) abstract words, 2) concrete words, 3) descriptive words, 4) judgmental words and 5) polarized words. In analytic description, evidences of students' voices, written works, movements, and facial expression were compound in analysis to indicate students' words in mathematical communication. The exemplary detail of analysis could show as follows.

IOP Conf. Series: Journal of Physics: Conf. Series **1340** (2019) 012090 doi:10.1088/1742-6596/1340/1/012090

5.1. The students' words in rigorousness

Table 1. shows a discussion to find the method of rolling 2 dices providing a sum mark lower than 5 as following protocol.

Item 75	Salin	Probability that is provided from rolling 2 dices has 36 methods and then we write ... what was that? Probability of a sum mark lower than 5, right?
Item 76	Pha	Probability of the mark ...
Item 77	Orn	Probability of a sum mark of 2 dices providing lower than 5 must also be explained that has 10 methods.
Item 78	Su	So how to know it? Distribute it all?
Item 79	Orn	Yes, let's write it down all. There're 10 methods.
Item 85	Su	Why does it have 10 methods?
Item 86	Orn	It is lower than 5 marks, right? For the sum mark lower than 5 methods, first one is 1 1,1 2.
Item 87	Orn	It must have 1 1, 1 2.
Item 88	Salin, Orn, Su	1 2, 1 3, 1 4
Item 89	Orn	This has 4 numbers.
Item 90	Salin	Ahh
Item 91	Orn	The second one has 2 1
Item 92	Salin, Orn	2 2
Item 93	Orn, Su	2 3
Item 94	Orn	The third one is ..
Item 95	Orn, Su	3 1, 3 2
Item 96	Su	The forth one has only 2,1.
Item 97	Pha	Umm
Item 98	Orn	Then it decreases as 4,3,2,1. And its sum is 10. 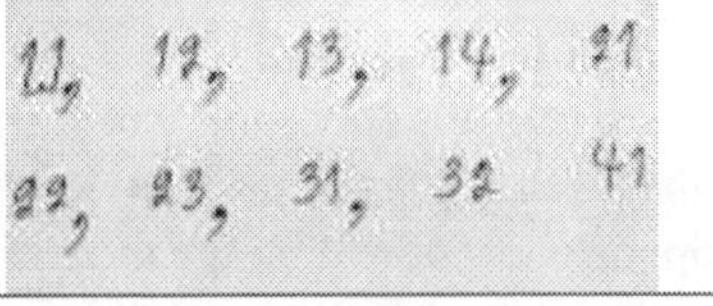

An analysis of rigorousness of mathematical communication is seen from giving mathematical opinions during students' conversation and answering question, which students give opinions perceptively and correctly as mathematical accuracy. From the mathematical commination protocol, students searched the methods to roll 2 dices providing the sum mark lower than 5 as shown in Item 77. Orn said "Probability of a sum mark of 2 dices providing lower than 5 must also be explained that has 10 methods". Orn had given an opinion that there were 10 methods for rolling the dices and expressed this method by saying " 1,1" instead of rolling first dice getting 1 mark and another dice getting 1 mark. In Item 86-Item 96, all student cooperated to express the methods of rolling the dices getting lower than 5 marks as well as Orn had summarized that "it then decreases as 4,3,2,1. And its sum is 10" as seen in Item 98, which indicated that rolling the dices gaining a sum mark of lower than 5. With rolling first dice presenting 1 mark, there are 4 methods. With rolling first dice presenting 2 marks, there are 3 methods. With rolling first dice presenting 3 marks, there are 2 methods. With rolling first dice presenting 4 marks, there is 1 method. The students' idea expression about the number of methods of rolling the dices proving lower than 5 of sum marks was mathematically correct which showed that this characteristic of communication is rigorousness.

In the rigorousness of mathematical communication, the students had said the words as following.

5.1.1. Concrete words: According to Episode 1, as analyzing the students' concrete word using from their conversation that they used specific saying, obvious meaning and understandable explanation. These words could reduce mistake of meaning understanding. According to the protocol, in Item 77, it was seen Orn said " Probability of a sum mark of 2 dices providing lower than 5 must also be explained that has 10 methods". It was seen in Item 77 that Orn presented her idea to roll the dices obtaining lower than 5 marks and further expressed that it would have 10 methods. This demonstrated Orn's saying was specific into the methods to rolling the dices presenting lower than 5 marks and she said further obviously that there were 10 methods to rolling the dices presenting lower than 5 marks. This presented the students used the concrete words.

5.1.2. Descriptive words: According to Episode 1, as analyzing the students' descriptive words from the conversation that the students explained the fact details. The objective is to make the receivers understand more than that of speaker. These words are neutral, not judgmental. According to the protocol, in Item 86, it was seen a phrase that "it is lower than 5 marks, right? For the sum mark lower than 5 methods, first one is 1 1,1 2". This was a description to express the fact by giving an example to others to understand what there are methods of rolling the dices providing lower than 5 marks. This was the communication which purposed the receivers understand better. It indicated that the students had used descriptive words.

5.2. Students' words in economy

Table 2. shows a discussion to find the method of rolling 2 dices which provides a sum mark higher than 5 as following protocol.

Item 69	Orn	Let us find the method of a sum mark lower than 5, first?
Item 70	Salin	Do we use a formula?
Item 71	Orn	Not sure. It's just to find the methods of a sum mark lower than 5 first, and then remove them from all.
Item 72	Pha	(Nodding off) Well, it all has 36 methods, right?
Item 73	Orn	Write it down, all opportunity of rolling the dices is 36 methods.
Item 74	Su	So what? Find the sum mark lower than 5, right?
Item 75	Salin	All opportunity of rolling the dices is 36 methods, and so what? The opportunities of the mark lower than 5?

An analysis of economy of a mathematical communication character is presented from the giving mathematical opinion during students' conversation and answering question which the students' opinions are short communication but contain a mathematical reason. From the students' communication about finding a number of methods that provides lower than 5 marks, in the beginning, the students found all result of the sum mark higher than 5, as seen in Item 69, which said " let us find the method of a sum mark lower than 5, first?". Orn purposed to her group mates to find all method of rolling 2 dices first, and then find the methods of rolling 2 dices providing the marks lower than 5, after that, remove them from all method. Finding all method of rolling the dices providing the marks lower than 5 and then minus by the unconcerned methods is a mathematically reasonable idea expression. This indicated that the students had used word's economy.

In the economy of mathematical communication, the students had said the words as following.

5.2.1. Abstract words: According to Episode 2, as analyzing students' abstract word using from their conversation using overview, unspecific and wide meaning saying. These words are difficult to understand. From the protocol, it was seen in Item 69 that Orn said "let us find the method of a sum mark lower than 5, first?". Orn described her own idea about finding a number of method of rolling 2 dices getting the mark higher than 5 by finding the methods of rolling the dices providing mark lower than 5, before, and removing them from all method of rolling 2 dices. But as seen in Item 69, Orn said " let us find the method of a sum mark lower than 5, first?". It was speaking of only a part of all solution. This made her group mates confused that what it really means by Orn's saying. It then was seen Salin's question in Item 70 that is "Do we use a formula?". Orn saying in Item 69 was indicated that the students had used the abstract word.

5.2.2. Concrete words: According to Episode 2, as analyzing the students' concrete word using from their conversation that they used specific saying, obvious meaning and understandable explanation. These words could reduce mistake of meaning understanding. From the protocol, in Item 71, Orn said "it's just to find the methods of a sum mark lower than 5 first, and we then remove them from all". Orn had explained the solution to find the methods of rolling the dices proving the marks more than 5, which specified to finding the all method of rolling 2 dices first and then minus with the methods of rolling the dices proving the marks lower than 5. This communication obtained the meaning of obvious solution and showed that the students had used the concrete words.

5.2.3. Judgmental words: According to Episode 2, as analyzing the students' judgmental word using from their conversation using to evaluate and individual reaction, these words will present positive or negative reaction of messenger to receiver know the situation. From the protocol it was seen in Item 72, Pha said "(Nodding off) Well, it all has 36 methods, right?" Pha reacted by nodding her head off and said "well", simultaneously. This evaluation meant that she agreed with Orn's idea as in Item 71 that Orn said "it's just to find the methods of a sum mark lower than 5 first, and we then remove them from all". It meant Pha showed positive reaction to Salin's message which indicated that the students had used the judgmental words.

5.3. Students' words in freedom

Table 3. shows a discussion after finding the method of rolling 2 dices which provides a sum mark higher than 5 using distribution method as following protocol.

Item 98	Orn	Then it decreases as 4,3,2,1. And its sum is 10. 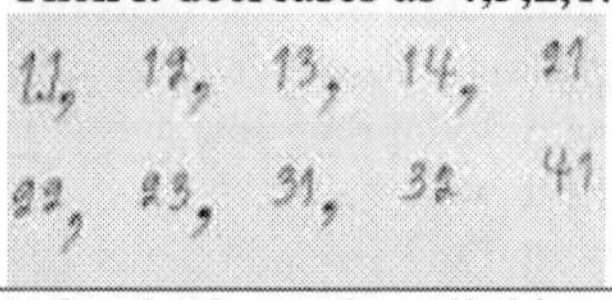
Item 99	Salin	Why don't we draw linking map?
Item 100	Su	Link the mark of first dice to that of second dice, but sum of both is lower than 5. Like this?
Item 100	Orn	Mapping? Or what? Might as well do table.
Item 101	Salin	Mapping is easier. It's just link the first dice to the second dice, provide the first dice as 1 mark then link it to the marks of the second dice as 1 2 3 4 as written. Then provide the first dice as 2 marks and the second dice is 1, 2 and so on as you wrote.
	Pha	It's called Tree map, right? Or Tree? I've seen before.
Item 102	Su	Like this? (Su writes the arrows to link 1 to 1,2,3,4) 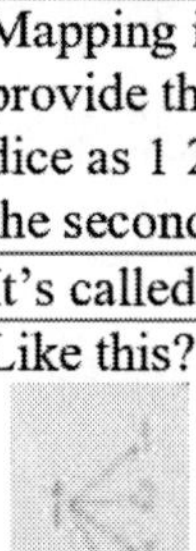
Item 103	Orn	It's probably called tree map.

An analysis of freedom of a mathematical communication character is presented from the giving mathematical opinion during students' conversation and answering question which their giving opinions have expressed in messages or received materials or developed received materials. According to the conversation protocol in Item 98, Orn said "Then it decreases as 4,3,2,1. And its sum is 10". It meant Orn had proposed her idea to her group mates then got result distribution approach of method of rolling the dices providing the sum mark lower than 5 that is 11,12,13,14,21,22,23,31,32,41. After they'd got the conclusion from Orn's idea, Salin proposed her own solution as seen in Item 99 "Why don't we draw linking map?" and proposed further idea by Orn as seen in Item 100, Orn said " Mapping? Or what? Might as well do table". It was seen that for solution of rolling 2 dices getting the sum mark lower than 5, the students showed various ideas to solve the problem such as distribution method, tree mapping and table. This showed that the students' mathematical communication character was freedom

In the economically mathematical communication, the students had said the words as following.

5.3.1. Abstract words: According to Episode 3, as analyzing students' abstract word using from their conversation using overview, unspecific and wide meaning saying. These words are difficult to understand. From the protocol, it was seen in Item 99 that Salin said "… linking map". Salin had proposed her own idea to find the number of methods of rolling the dices proving the mark lower than 5 using drawing tree maps as said "linking map". This described overview feature of tree map which confusing their group mates. It showed that the students had used abstract word.

5.3.2. Concrete words: According to Episode 3, as analyzing the students' concrete word using from their conversation that they used specific saying, obvious meaning and understandable explanation. These words could reduce mistake of meaning understanding. From above protocol, in Item 100, Su said "Link the mark of first dice to that of second dice, but sum of both is lower than 5. Like this?". Su had described the feature of "linking map" as Salin said in Item 99. Su's description approach was to say what she understands with saying the detail of the method of finding marks of 2 dices lower than 5 using linking a couple of the 1st and 2nd dice providing the mark lower than 5. This communication was clearly in solution. It presented that the students had used concrete word.

5.3.3. Descriptive words: According to Episode 3, as analyzing the students' descriptive words from the conversation that the students explained the fact details. The objective is to make the receivers understand more than the understanding of speaker. These words are neutral, not judgmental. According to above protocol, in Item 101, Salin said "Mapping is easier. It's just linked the first dice to the second dice, provide the first dice as 1 mark then link it to the marks of the second dice as 1 2 3 4 as written. Then provide the first dice as 2 marks and the second dice is 1, 2 and so on as you wrote." Salin had explained to see the fact by giving some examples to their group mates to understand her idea of drawing map to express the methods of rolling the dices getting the mark lower than 5 which this communication want the receivers understand better. This indicated that the students had used the descriptive word.

6. Results

The research result reveals the students' words in mathematical communication. There are 3 characteristics of mathematical communication; 1) Rigorousness 2) Economy and 3) Freedom. In mathematical communication, there are 4 types of words as follows: 1) Abstract words: It was a type that students talk about or describe using broad words. There are no specific types of explanations. This type of speech is often unclear. It is difficult to understand. This ambiguous language can confuse others. 2) Concrete words: It is the type that students describe as specific and can explain what is recognized. These words make the meaning clearer. It also reduces the misunderstanding of the meaning of words. 3) Descriptive words: It is the type of student that uses material words that interest recipients and can understand the substance. This type of statement is not a verdict. It is a neutral term used to describe the truth that the messenger knows. And 4) Judgmental words: It is the type that students have to say to express their values and reactions. This type of word expresses the positive or negative reaction of the dispatcher to the receiver, rather than describing the situation. The results are shown in Table 1.

Table 4. The appearance of students' words in mathematical communication in classroom using lesson study and open approach

mathematical communication	Types of Words Abstract words	Concrete words	Descriptive words	Judgmental words	Polarized words
Rigorousness	-	✓	✓	-	-
Economy	✓	✓	-	✓	-
Freedom	✓	✓	✓	-	-

According to table 3., the study results revealed that in the classroom using lesson study and open approach, there are students' mathematical communication includes 3 characteristics and the students' words using in each characteristic as detail: 1) Rigorousness, while the target group attempt to grasp problem situations and solving problem. The words were used as an express their thinking, the students communicate by using concrete words and descriptive words, after that their friends able to perceive and understand it. 2) Economy, it was noticeable in students' mathematical communication which was precisely explained and exactly perceived by their friends, the students communicate by using abstract words, concrete words and judgmental words, after that their friends able to show evaluation and personal react which were considered. 3) Freedom, the students have independent and different thinking. The students communicate by using abstract words, concrete words and descriptive words in order to explain their ideas independently. In all of 3 characteristics, it was did not found polarised words, the students did not use this type of words to evaluate and argue their friends' ideas as completely difference.

7. Conclusion and discussion

The study results revealed that in the classroom using lesson study and open approach, the students expressed 3 characteristics of mathematical communication and used types of words in each characteristic. For the rigorousness, the students used concrete and descriptive words to make sense the problem situations and to express their ideas in solving mathematics problem. They can understand the problem situation and the ideas for solving problem. For the economy, the students used concrete and descriptive and judgmental words to precisely communicate their ideas. Their friends can understand the words exactly according to student' s intention. For the freedom, the students used abstract, concrete and descriptive words to communicate their different ideas. They have independent to think by themselves and explain their ideas independently. In all of 3 characteristics, it was did not found polarised words, the students did not use words to evaluate and argue their friends' ideas as completely difference. Because in classroom using lesson study and open approach, mathematical communication is a tool for potential compiling ideas by speaking, acting and writing to express and explain the ideas. It provides the students to explain, collect, expand and share ideas with the others. Therefore, they can use the words in mathematical communication.

The research results consistent with Inpanya [3] who studied of mathematical communication between teacher and students in open approach classroom. The study results revealed that mathematical communication among teacher and students through open approach could generate three chains of mathematic communication, namely coordinate chains of mathematical communication, resonant chains of mathematical communication, and emergent chains of mathematical communication. Moreover, it consistent with Inprasitha, Pattanajak and Inprasitha [5] that studied of student' s mathematical communication in teacher professional development. The research used qualitative research methodology consisting of ethnographic approach; teaching experiment; protocol analysis and analytic description. The study results revealed that in mathematics classrooms taught by open approach, it provides an opportunity for the students to communicate mathematics by the verbal and non-verbal language. The students have mathematical communication with 3 communication features include rigorousness, economy, freedom of thinking while they participated in communication. Additionally, the research result according with Thinwiangthong [10] found that the students have mathematical communication, includes rigorousness, economy and freedom characteristics, in small group within the context of lesson study and open approach classroom.

8. Recommendation

8.1. Recommendation for Implementation of Research Findings
This study found the students' words in mathematical communication in classroom using lesson study and open approach. In lesson planning, the students' words can be used as a guideline for mathematical ideas in problem solving through open approach. In lesson implementing, teachers can use words of students to understand and utilize by highlighting, expand, connect, and summarize the student's mathematical ideas. In lesson reflection, teachers can use the students' words in mathematical communication to adjust their instruction.

8.2. Recommendation for Further Research
The further research should be the study of how teachers use the students' words in mathematical communication to improve the lessons in context of lesson study and open approach classroom. Another one is the study of complexity and relation between words using and mathematics contents for high school students, because the high school mathematics contents is quite complex and difficult for teaching and learning by using lesson study and open approach.

Acknowledgments
This research is (partially) supported by The Centre of Excellence in Mathematics, The Commission on Higher Education, Thailand and Centre for Research in Mathematics Education, Faculty of Education, Khon Kaen University and The Project of Development of student's Mathematics Higher-Order Thinking at North East of Thailand, Faculty of Education, Khon Kaen University.

References
[1] Cathcart R S, Samovar L A and Henman L D 1996 *Small Group Communication :Theory & Practice* (United States :Times Mirror Higher Education Group)
[2] Emori H 2005 *Building up the foundation as a researcher by learning communication study as a ground theory on mathematics education* (The Workshop and Intensive Lecture for Young Mathematics Educations in Thailand 2005) pp 11 Khon Kean :Khon Kean University
[3] Inpanya T 2011 *Mathematical communication between teachers and students in an open-class classroom* (Master of Education Thesis, Mathematics Education, Faculty of Education, Khon Kaen University)
[4] Inprasitha M, Inprasitha N and Pattanajak A 2004 *Teaching by open-Approach Method in Japanese Mathematics Classroom* (KKU Journal of Mathematics Education vol 1) chapter 1 pp 5-6
[5] Inprasitha M, Inprasitha N and Pattanajak A 2012 *A Study of Student's Mathematical Communication in Teacher Professional Development* (Journal of Modern Education Review vol 2) chapter 1 pp 38–46.
[6] Johnson D W and Johnson E P 1994 *Joining Together: Group Theory and Group Skills* (Fifth Additionally and bacon)

[7] Khammanee T 2007 *Teaching: Knowledge for effective learning process (6th ed)* (Bangkok Chulalongkorn University publisher)
[8] Kotrakoon S 2010 *Educational Psychology* (Bangkok :Chulalongkorn University publisher)
[9] Stigler J and Hiebert J 1999 *The teaching gap: Best ideas from the world's teachers for improving education in the classroom* (New York: The Free Press)
[10] Thinwiangthong S 2012. *Small-group Mathematical Communication in Classroom in Lesson Study and Open Approach Context* Doctor of Philosophy Thesis in Mathematics Education, Graduate School, Khon Kaen University. (in Thai).
[11] Weaver W W 1949. *The Mathematical Theory of Communication* (Urbana, Ill: University of Illinois Press) pp 95

IOP Conf. Series: Journal of Physics: Conf. Series **1340** (2019) 012091 doi:10.1088/1742-6596/1340/1/012091

Local Wisdom Related to STEM Education

Virapong Saeng-Xuto

Faculty of Education, Chiangmai University, Khon Kaen, Thailand

*Corresponding author's e-mail address: virapongxuto@hotmail.co.th

Abstract. This study aimed to clarify plausibility of developing STEM education learning activities based on Thai local wisdom. Research method regarded document analysis. The author previous research about toys and local wisdom were analysed for STEM education. The local toys in northern part of Thailand were selected to represent how they could be used in classroom for STEM education. Two kinds of toys were explained. These included (1) Bunbeu and (2) E-plor. The paper clarified the components of toys, how to play the toys, and STEM knowledge behind playing the toys. This study may have implications for STEM education learning activities.

1. Introduction

STEM education is the goal of science and technology teaching and learning for providing in 21st century. In order to enhance students to apply knowledge for real life problem solving, STEM education becomes the vision of education. STEM education provided the teaching integrated among Science, Technology, Engineering and Mathematics. Explore contents or theory in science, select the appropriate technology, design with engineering and use mathematic to calculate [2].

Current understandings of science, technology, engineering and mathematics (STEM) learning among children and adults, in a variety of settings, including schools, but also contexts such as the home, community-based activities, after-school programs, the workplace, museums, science centers, zoos and aquariums, as well as across a wide range of digital media (Dierking and Falk, 2016)

The learning activities can create with local learning resources in the school and around the community also local wisdom. Students will learn about 4 topics of STEM and at the same time they will learn about customs traditionalism and folklore.

Science is the contents to explain the phenomena of nature of the world also create knowledge and technology to improve the way of life.

By the way local wisdom is the knowledge in the community transfer from generation to generation and integrates knowledge to improve appropriate way of life too [3].

2. Content

Local wisdom was the knowledge of predecessor to improve the way of life and it whole of interdisciplinary such as art, culture, mathematics, science etc. and separate to 7 groups: food, residence, cloths, medicine, career, recreation and transportation as diagram 1 [3].

IOP Conf. Series: Journal of Physics: Conf. Series **1340** (2019) 012091 doi:10.1088/1742-6596/1340/1/012091

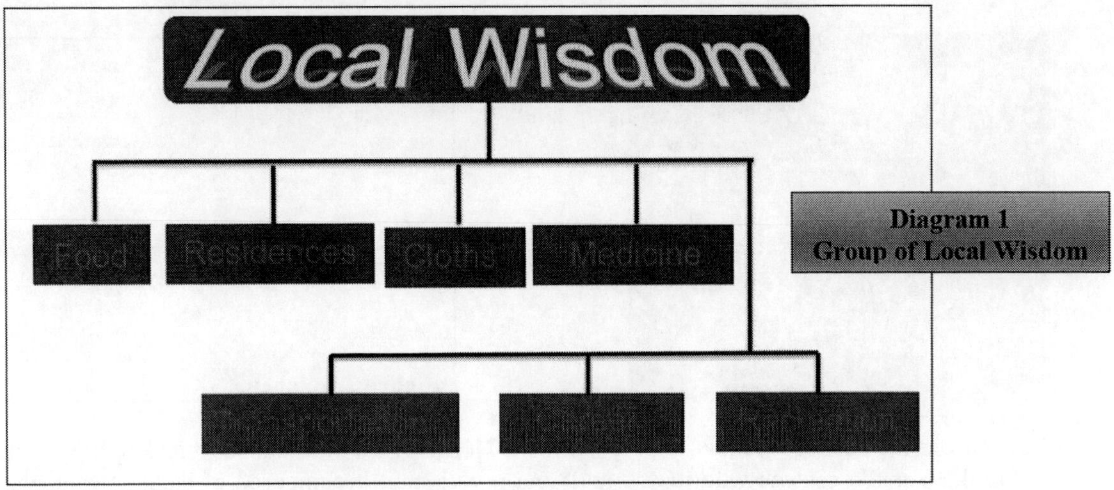

Figure 1. Diagram group of Local Wisdom.

Table 1. The example of local wisdom in each groups.

Local wisdom Group	Physics	Chemistry	Biology
1.Food Trapper trap hook	Lever, Tension, elasticity, Gravity Potensial energy, Moment, Circular motion	Rasin	Living Source of animal
2. Residences Kalae House	Equilibrium, Machine, transfer weight, air flow, roof slope for pour rain	Glue, Rubber	Kind of trees for instruction and plant
3. Cloths Fiber, Cotton, Silk	Toughness, Flexibility, Thermal insulation	Composition of fiber, Strength of acid and base, Adsorption	Kind of cotton and silk
4. Medicine Traditional massage	Kind of force of finger hand and foot of massager	Thai herbal medicine	Tendon of human body Blood circulation system Muscle system, Psychology
5. Transportation Wagon	Wheel and axis, lever, Wedge, Force, Power, Velocity, acceleration, Inertia, Expantion and contraction of matters	Propperties of metal	Kind of trees to make wagons
6. Career Wickerwork	Stress and strain, Equilibrium	Insecticide	Kind of bamboo
7. Recreation Local toys	Newton's law, Circular motion, air pressure, Momentum	Glue, Rubber	Kind of bamboo to make local toys

Except the science knowledge that found in local wisdom, local technology could find too and satisfy with STEM education

This example of local wisdom will show you how to use for STEM education.

- Bunbeu

Figure 2. Bunbeu.

Bunbeu was the kind of local toys that found in the North region of Thailand. It looked like a windmill put in the bamboo canteen with rope to pull and it will spin. It composed of bamboo canteen, stick, propeller, and rope [4].

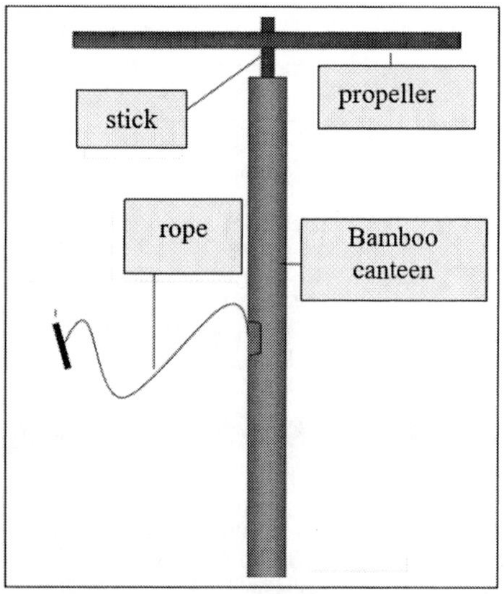

Figure 3. Compose of Bunbeu.

1. Bamboo canteen
2. Stick
3. Propeller
4. Rope

How to play

Round the rope on the stick then pull it the propeller will spin and made sound "breu breu". It made fun for children.

Science knowledge found in Bunbeu

If you see Bunbeu in the top view it look like a wheel and axle [6].

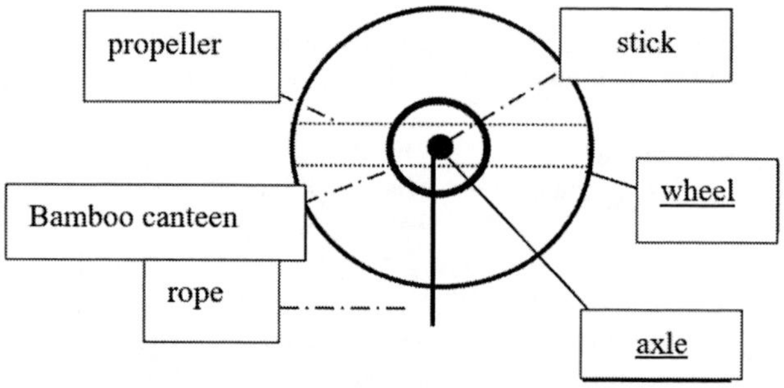

Figure 4. Bunbeu on the top view.

Bunbeu turned round and round because it had inertia and sound "breu breu" was the propeller pass through air. The circular motion of it was the knowledge that can teach for children.

2.1. How to use Bunbeu in the STEM education

The science knowledge that analyze from Bunbeu can develop the activities in STEM education as follows

Wheel and axle student can design size of propeller and stick to decrees force to turn round the stick but the propeller will turn round faster (Engineer) and calculate the length of the stick relate to propeller (Mathematic). How to sharpen the stick and propeller to best turn round by ask from the Bunbeu maker (Technology).

Inertia depend on the weight, student can design size and thick of propeller or take something put on the top of propeller to increase weight (Engineer). Calculate size and wight of propeller and length of rope to pull stick (Mathematic). Ask the Bunbeu maker or search data how to sharpen angle of propeller to catch the wind (Technology).

Sound student can design thick of propeller (Engineer) or design several geometry shape of propeller (Mathematic) to make several of sound. Select several materials to make propeller by enquire about this from Bunbeu maker or search data from internet to make pitch or loud of sound (Technology).

Circular motion Student can design several length of propeller to make several radius for linear velocity and angular velocity (Engineer). Calculate radius from length of propeller was diameter (Mathematic) and will know velocity (Technology)

The other local toy that express is call "E-plor"

E-plor is the local toy that to shoot like a gun it made from bamboo canteen with stick to push bullet to shoot. It composed of

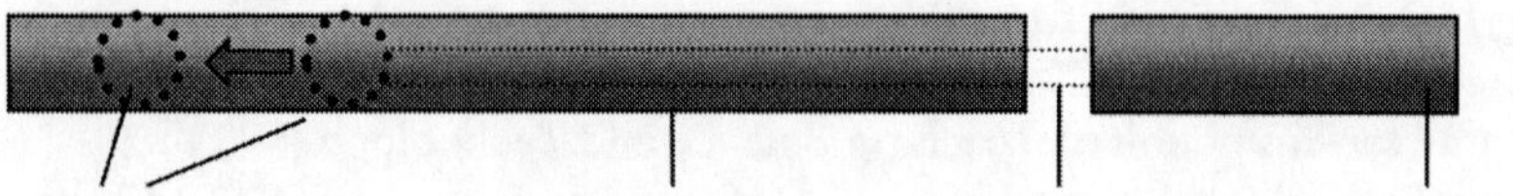

Figure 5. Compose of E-plor.

1. bamboo canteen
2. bamboo stick
3. bullet made from kaffir lime fruit or jute fruit or grass leave or wet paper

How to play

Put the bullets in edge of bamboo canteen then push stick in the hole of bamboo canteen until bullet shoot with sound.

- Science knowledge found in E-plor

To shoot the bullet of E-plor used <u>air pressure</u> that show in the picture 4 [5].

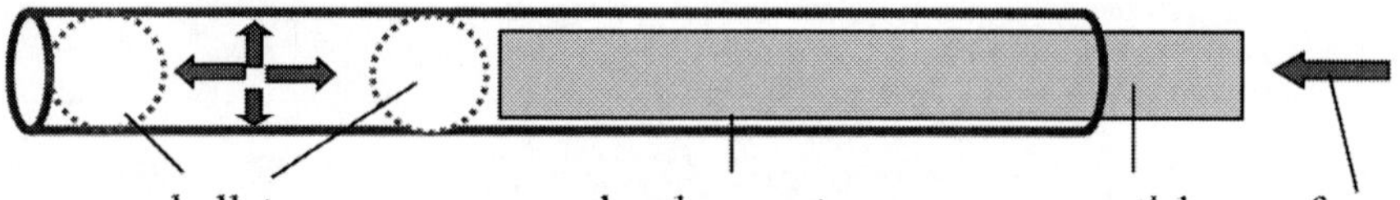

Figure 6. air pressure to shoot the bullet.

The other science knowledge that found in E-plor was <u>friction</u> about bullet and bamboo canteen, velocity of bullet and projectile motion

2.2. *How to use E-plor in the STEM education*

The science knowledge that analyze from E-plor can develop the activities in STEM education as follows

Air pressure student can explore size and lengths of bamboo canteen relate air pressure (Engineer). Calculate size relate length of bamboo canteen to impact air pressure (Mathematic). Use several material to make canteen by ask from E-plor maker or search data from documents (Technology).

Friction student can try to make bullet from several materials to shoot (Engineer) by calculate distance of bullet fall (Mathematic) and shape of bullet to shoot (Technology)

Velocity student can design size and lengths of bamboo canteen relate velocity (Engineer) and calculate velocity of bullet that shoot from bamboo canteen (Mathematic). Adjust bamboo canteen to improve velocity of bullet (Technology).

Projectile motion student can design activity to test E-plor how far of projectile motion to shoot bullet (Engineer), calculate distance of bullet that shoot fall (Mathematic)

And design E-plor how to shoot bullet far (Technology).

3. Conclusion

Local toys are the example of local wisdom that can create several activities for STEM education. Teacher can apply local wisdom to teach in class suitability but should survey and analyze local wisdom obviously before to use.

References

[1] Dierking L D and Falk J H 2016 2020 Vision: Envisioning a new generation of STEM learning research *Cult Stud of Sci Educ* **11** pp 1–10
[2] The Institute for Promotion of Teaching Science and Technology (IPST) 2014 STEM Education Science Technology Engineering and Mathematics Education (STEM Education) (Bangkok: Chulalongkorn University Press)
[3] Saeng-Xuto V 2009 Guidline to make Science Activity with Local Wisdom (Chiang Mai: Chotana Print)
[4] Krutmuang S 1992 Encyclopedia of Former Local Toys (Bangkok: Ton-or Press)
[5] Teerathanakul S and Peuorn K 2011 Physics 1: Machanic for Graduate (Bangkok: Soonseusermkrungthep)
[6] Supap S 2010 Physics 1 for Graduate Book 1 (Bangkok: Sine Publishing)

Ice Cream STEM Education Learning Activity: Inquiry from the Context

Eileen J. Villaruz[1*], Maria Cindy F. Cardona[1], Amelia T. Buan[1], Manuel B. Barquilla[1], and Chokchai Yuenyong[2]
[1] College of Education, MSU – Iligan Institute of Technology, Iligan City, 9200 Philippines
[2] Faculty of Education, Khon Khaen University, Thailand
*Corresponding author's e-mail address: etjvillaruz@gmail.com

Abstract. The paper will clarify the ice cream making STEM education learning activities which provide through Sutaphan nad Yuenyong [11] context based STEM education learning approach. This approach consists of 7 stages including (1) Identification of social issues, (2) Identification of potential solution, (3) Need for knowledge, (4) Decision-making, (5) Development of prototype or product, (6) Test and evaluation of the solution, and (7) Socialization and completion decision stage. Icecream STEM Education Learning Activity will bring students to practice integrated knowledge among chemistry, mathematics, art, and economics in order to design their products of icecream. This may has implication for organizing STEM education in school setting.

1. Introduction

In the 21st century, scientific and technological innovations have become increasingly important as we face the benefits and challenges of both globalization and a knowledge-based economy. To succeed in this new information-based and highly technological society, students need to develop their capabilities in STEM to levels much beyond what was considered acceptable in the past." [1]

The Philippines had just implemented the K to 12 Basic Education Program. It aims to create a functional basic education system that will produce productive and responsible citizens equipped with the essential competencies and skills for both life-long learning and employment. Today, the education system tries to impose on students the attitude of lifelong learning. Life-long learning (LLL) as viewed involves all strategies that are put in place to create opportunities for people to learn throughout life by providing both the individual needs and that of the relevant community [2]. Lifelong learning can be associated to the following notions: (1) individuals take ownership of the need for and the content of learning, (2) learning focuses on how students think rather than on what to think about, (3) teachers become role models and mentors for lifelong learning rather than distributors of information, (4) evaluation encourages students for self-development and becomes a facilitator in the process rather than classifying them according to a norm, and (5) learning is considered as an enjoyable and essential part of the individual [3]. Methods such as problem-based learning, conscious learning, mutual teaching and cognitive apprenticeship, and such naturally motivating activities and similar teaching strategies as cooperation, reflection and student autonomy are employed by individuals in order to remain updated in today' world [4]. It is also important to note that problem solving, critical thinking, and reasoning skills are essential to lifelong learning for these skills allow learners to acquire the knowledge, skills and attitudes necessary not only for academic success but also for their overall well-being [5]. In order to help students become lifelong learners, the schools should be organized in such a way that the students will have the opportunities to develop these skills.

Science education in the Philippines cannot be considered as a strength. Based on results from 2014, the passing rate for the national achievement test (NAT) for grade six pupils is only 69.21%. The passing rate for high school seems far worse, with a passing rate of only 46.38% from 2010 statistics. Aside from the aforementioned national examination, the Philippines has exhibited a consistently abysmal performance in international surveys such as Trends in International Mathematics and Science Study (TIMSS). In 2003, for high school, the country ranked only 34th out of 38 countries in HS II Math, and 43rd out of the 46 participating countries in HS II Science. For elementary, fourth grade participants ranked 23rd out of the 25 countries in both math and science. Our country stopped participating in the survey in 2008, perhaps after getting such lackluster scores [6].

The current situation of STEM education in the Philippines proves that we have a long road ahead of us. Nevertheless, this should not stop us from pausing and admitting that yes, there is a problem that we should work upon. We can still encourage more people to become STEM educators. We can still improve our NAT scores, and confidently participate in international surveys once more. Improving science education will not only allow our country to have citizens who are thinkers first and foremost, but also increase the number of STEM graduates to address the glaring industrial and agricultural needs that will boost the economic growth of our country [7].

The main goal of this context based STEM Education Learning activity is to investigate the colligative properties of solution specifically the freezing point depression by producing a locally-made nutritious ice cream. There are various considerations of coming up with the product such as its costs, nutritional benefits, physical appearance as well as its packaging. Each of these will be addressed by providing appropriate activities to scaffold the attainment of its objectives and goals.

2. Developing STEM Education Learning Activities

Icecream STEM Education Learning Activity will bring students to practice integrated knowledge among chemistry, mathematics, art, and economics in order to design their products of icecream. Chemistry concept may include colloidal emulsion, freezing point depression, enthalpy change, and basic measurement skills. Mathematics may include ratio of the ingredients, measurement skills, and conversion of units. Arts include creative packaging and presentation of their ice cream. And, economics include profitability and low cost.

The lesson plan of icecream STEM Education learning activity was developed on concept of Context-based STEM education learning approach which the STEM education learning activities should provide not only the ways of investigation and solving problem but also a real world problem solving. Regarding on Sutaphan and Yuenyong [8], the context-based STEM education learning approach included (1) Identification of social issues, (2) Identification of potential solution, (3) Need for knowledge, (4) Decision-making, (5) Development of prototype or product, (6) Test and evaluation of the solution, and (7) Socialization and completion decision stage. The lesson plan could be highlighted as the table 1.

IOP Conf. Series: Journal of Physics: Conf. Series **1340** (2019) 012092 doi:10.1088/1742-6596/1340/1/012092

Table 1: Lesson plan on ice cream making on context based STEM education learning approach (Sutaphan and Yuenyong [8]).

STAGE	ACTIVITY
1. Identification of social issues	1. Social Issue: Filipinos love to eat ice cream especially during hot season. Discuss with the students about their favorite ice cream, why they like it and their opinion about health issue of the ice cream. 2. Teachers raise the issue "How can you make a local ice cream that has nutritional value?" 3. PRODUCT: ICE CREAM with a healthy twist Students may design a label or branding/packaging in their ice cream.
2. Identification of potential solution	1. Students and teachers share the cost analysis in the design of their ice cream product. 2. Students may discuss their possible design on the ice cream regarding on the five capitals: physical, financial, social/technology, human and natural capitals. Physical – The aesthetic value of their ice cream, interesting packaging Financial – Using good but affordable local ingredients Social/Technology – Material selection using appropriate materials, find ways to solve the problem Human – The safety of food and its nutritional value Natural – The use of organic ingredients
3. Need for knowledge	1. Project Based Learning (PBL). Each group will have 4 to 5 members. 2. Students need to gather related information in order to create a good product. 3. Students may tap or ask help from local ice cream manufacturers on how to make an ice cream. They could also interview food technologists on what ingredients to be used in order to produce a healthy ice cream. 4. Students may research more information in the internet. Based on the information gathered, the students now have many considerations to build their idea of making a healthy ice cream. CHEMISTRY – colloidal emulsion, freezing point depression, enthalpy change, basic measurement skills, conversion of units MATHEMATICS – ratio of the ingredients, measurement skills, conversion of units ARTS – creative packaging and presentation of their ice cream ECONOMICS – profitability and low cost Basically, the students develop Scientific and Technological Literacy which is the aim of K to 12 Education.

IOP Conf. Series: Journal of Physics: Conf. Series **1340** (2019) 012092 doi:10.1088/1742-6596/1340/1/012092

Table 1 (Cont')

4. Decision-making	1. Students as a group will present their idea in front of the whole class. 2. Non-presenters will give comments, ask questions to further refine the idea. 3. Each group will be asked to refine and finalize their experiment design before starting to develop their product.
5. Development of prototype or product	Students create the ice cream at school during their laboratory activity schedule. Documentation (still pictures and videos) are required in every step of their experiment. Students will be guided with the following questions during their activity: 1. What are the ingredients you are using in your ice cream? 2. Cite the specific ratio and amounts of ingredients? 3. What are the health benefits involve in your ice cream? 4. In what step in your procedure where you can see the colloidal emulsion? Freezing point depression? Enthalpy change? 5. What is the story behind your packaging/branding? 6. Did you enjoy the activity? Why or why not? 7. What values you have learned from the activity?
6. Test and Evaluation of the solution	An ICE CREAM CHALLENGE will commence and there will be honourable panel of judges to be invited. Their ice cream will be evaluated through the aesthetic value/packaging/branding, texture, taste, cost and nutritional value. An open forum will commence so that the students will use these information to better improve their product. Teacher may ask this question: "Aside from ice cream making activity, cite other examples where the concepts of colloidal emulsion, freezing point depression and enthalpy change can be applied?"
7. Socialization and completion decision stage	Each group will make a vlog on their final product and publish it in social media. They can also display their output in the school and sell it. In this way, they will also learn the value of entrepreneurship.

IOP Conf. Series: Journal of Physics: Conf. Series **1340** (2019) 012092 doi:10.1088/1742-6596/1340/1/012092

3. Conclusion

This paper shared the ideas of developing STEM learning activities through ice cream STEM education based on context based approach. The the 7 stages of Sutaphan and Yuenyong [8] context based STEM education teaching approach could guide ideas of developing learning activities from social issues to develop some products. The ice cream STEM Education Learning Activity will allow students using applying scientific and other knowledge for designing the solutions, and provided the context of instruction requires solving a real-world problem or task through teamwork. As students'products of icecream, students will have also chance to apply their scientific and other knowledge for problem solving in context of engineers, technology, or entrepreneurship [8].

References

[1] National Science Foundation (2016). *Science and Engineering Indicator*. Retrieved from https://www.nsf.gov/nsb/publications/2016/nsb20161.pdf

[2] Abukari, A. (2005). Conceptualising Lifelong Learning: A Reflection on Lifelong Learning at Lund University (Sweden) and Middlesex University (UK). *European Journal of Education, 40*(2), 143-154. Retrieved from http://www.jstor.org/stable/3496962

[3] Bryce, J., & Withers, G. (2003). Engaging secondary school students in lifelong learning. *Indigenous Education*, 14.

[4] Demirel, M. (2009). Lifelong learning and schools in the twenty-first century. *Procedia-Social and Behavioral Sciences*, *1*(1), 1709-1716.

[5] Hager, P., & Holland, S. (Eds.). (2007). *Graduate attributes, learning and employability* (Vol. 6). Springer Science & Business Media.

[6] Ambag (2018). *Regional Growth Forecast*. Retrieved from: https://ambag.org/programs-services/planning/regional-growth-forecast

[7] FlipScience, (2018). *2018 In Review: The Biggest Science Stories Of The Year*. Retrieved from https://www.flipscience.ph/news/features-news/2018-review-biggest-science-stories/

[8] Sutaphan, S. and Yuenyong, C. (2019, in press). STEM Education Teaching approach: Inquiry from the Context Based. Journal of Physics: Conference Series.

AUTHOR INDEX